EPIGENETICS

EPIGENETICS

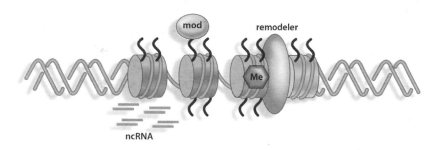

EDITED BY

C. David Allis
The Rockefeller University, New York

Thomas Jenuwein
Research Institute of Molecular Pathology (IMP), Vienna

Danny Reinberg
HHMI/Robert Wood Johnson Medical School
University of Medicine and Dentistry of New Jersey

Marie-Laure Caparros
Associate Editor, London

COLD SPRING HARBOR LABORATORY PRESS
Cold Spring Harbor, New York

EPIGENETICS

Publisher	John Inglis
Acquisition Editors	Alexander Gann and David Crotty
Development Director	Jan Argentine
Project Coordinator	Inez Sialiano
Permissions Coordinator	Carol Brown
Production Editor	Pat Barker
Desktop Editor	Lauren Heller
Production Manager	Denise Weiss
Cover Designer	Lewis Agrell

Front cover artwork: Depicted is a schematic representation of the chromatin template. Epigenetic regulation affects and modulates this template through noncoding RNAs (ncRNA) that associate with it, covalent modification of histone tails (mod), methylation of DNA (Me), remodeling factors (blue oval), and nucleosomes that contain standard as well as variant histone proteins (the yellow nucleosome). In the background is a representation of several model organisms in which epigenetic control has been studied. From top left: Pair of mouse chromosomes that may differ in their genomic imprint; a *S. cerevisiae* colony, showing epigenetically inherited variegation of gene expression; anatomy of *C. elegans*; illustration of *T. thermophila*, showing the large "active" macronucleus and the smaller "silent" micronucleus; *D. melanogaster*; maize section with kernel color variegation; *Arabidopsis* flower.

Library of Congress Cataloging-in-Publication Data

Epigenetics / edited by C. David Allis, Thomas Jenuwein, Danny Reinberg ;
Marie-Laure Caparros, associate editor.
    p. cm.
 Includes bibliographical references and index.
 ISBN-13: 978-0-87969-724-2 (hardcover : alk. paper)
 1. Genetic regulation. I. Allis, C. David. II. Jenuwein, Thomas. III.
Reinberg, Danny.
 [DNLM: 1. Epigenesis, Genetic. 2. Gene Expression Regulation. QU 475
E64 2006] I. Title.

QH450.E655 2006
572.8'65--dc22

i 0050 1 1 838 ‹

2006028894

10 9 8 7 6 5 4 3 2 1

All Cold Spring Harbor Laboratory Press publications may be ordered directly from Cold Spring Harbor Laboratory Press, 500 Sunnyside Blvd., Woodbury, New York 11797-2924. Phone: 1-800-843-4388 in Continental U.S. and Canada. All other locations: (516) 422-4100. FAX: (516) 422-4097. E-mail: cshpress@cshl.edu. For a complete catalog of all Cold Spring Harbor Laboratory Press publications, visit our World Wide Web Site http://www.cshlpress.com./

Long before epigenetics changed from little more than a diverse collection of bizarre phenomena to a well-respected field covered by its own textbook, a talented group of foresighted molecular biologists laid a rich foundation upon which the modern era of chromatin biology and epigenetics is based. This group includes Vince Allfrey, Wolfram Hörz, Hal Weintraub, Alan Wolffe, and Abe Worcel. This book is dedicated to their collective memory. Their passion and commitment to the study of chromatin biology inspired all of us who followed their work, and we now profit from their many insights.

Contents

Preface, ix

1 Epigenetics: From Phenomenon to Field, 1
   Daniel E. Gottschling

2 A Brief History of Epigenetics, 15
   Gary Felsenfeld

3 Overview and Concepts, 23
   C. David Allis, Thomas Jenuwein, and Danny Reinberg

4 Epigenetics in *Saccharomyces cerevisiae*, 63
   Michael Grunstein and Susan M. Gasser

5 Position-Effect Variegation, Heterochromatin Formation, and Gene Silencing in *Drosophila*, 81
   Sarah C.R. Elgin and Gunter Reuter

6 Fungal Models for Epigenetic Research: *Schizosaccharomyces pombe* and *Neurospora crassa*, 101
   Robin C. Allshire and Eric U. Selker

7 Epigenetics of Ciliates, 127
   Eric Meyer and Douglas L. Chalker

8 RNAi and Heterochromatin Assembly, 151
   Robert Martienssen and Danesh Moazed

9 Epigenetic Regulation in Plants, 167
   Marjori Matzke and Ortrun Mittelsten Scheid

10 Chromatin Modifications and Their Mechanism of Action, 191
    Tony Kouzarides and Shelley L. Berger

11 Transcriptional Silencing by Polycomb Group Proteins, 211
    Ueli Grossniklaus and Renato Paro

12 Transcriptional Regulation by Trithorax Group Proteins, 231
    Robert E. Kingston and John W. Tamkun

13 Histone Variants and Epigenetics, 249
    Steven Henikoff and M. Mitchell Smith

14 Epigenetic Regulation of Chromosome Inheritance, 265
    Gary H. Karpen and R. Scott Hawley

15 Epigenetic Regulation of the X Chromosomes in *C. elegans*, 291
    Susan Strome and William G. Kelly

16 Dosage Compensation in *Drosophila*, 307
    John C. Lucchesi and Mitzi I. Kuroda

17 Dosage Compensation in Mammals, 321
    Neil Brockdorff and Bryan M. Turner

18 DNA Methylation in Mammals, 341
    En Li and Adrian Bird

19 Genomic Imprinting in Mammals, 357
    Denise P. Barlow and Marisa S. Bartolomei

20 Germ Line and Pluripotent Stem Cells, 377
    M. Azim Surani and Wolf Reik

21 Epigenetic Control of Lymphopoiesis, 397
    Meinrad Busslinger and Alexander Tarakhovsky

22 Nuclear Transplantation and the Reprogramming of the Genome, 415
    Rudolf Jaenisch and John Gurdon

23 Epigenetics and Human Disease, 435
    Huda Y. Zoghbi and Arthur L. Beaudet

24 Epigenetic Determinants of Cancer, 457
    Stephen B. Baylin and Peter A. Jones

Appendices

1 WWW Resources, 477

2 Histone Modifications and References, 479

Index, 491

Preface

This advanced textbook on "Epigenetics" is truly a reflection of many talented colleagues and individuals, all of whom made this book possible and a rewarding experience. However, without hesitation, the editors want to thank Marie-Laure Caparros (London), without whom this project would have never materialized. Early in the process, it became evident that the editorial team needed help in coordinating such a large project, particularly for keeping the dialogue and editorial feedback with the >40 colleagues who agreed to provide outstanding chapter contributions, only to realize that we wanted more than their expert reviews and attention to detail. Marie-Laure has been instrumental in keeping the momentum moving forward, has bravely exchanged critical comments when needed, has informed all of us on the many deadlines, and has provided necessary coherence to make embryonic chapters come to life. Without her, this book would not have been possible. We are also grateful to our individual assistants, who forever kept us on our toes: Elizabeth Conley (David Allis), Christopher Robinson (Thomas Jenuwein), and Shelli Altman (Danny Reinberg). All of them are the unsung heroes of this book. We thank all of them for their innumerable contributions, large and small, and their unending patience with each of us and our quirky styles and shortcomings as editors.

Discussions for such a book took initial form on the coattails of the outstanding 69th Cold Spring Harbor Symposium on *Epigenetics* in the summer of 2004, but were seeded in early 2003 and formally commissioned by CSHL Press through Alex Gann and other colleagues. This was followed by formulating an editorial team between David Allis, Thomas Jenuwein, and Danny Reinberg. The first concrete outline for this project, including the brainstorming of various chapters and potential contributing authors, was done on the picnic bench at the FASEB meeting on *Chromatin and Transcription* in Snowmass, Colorado, July 2003. We were then very fortunate to confirm the lineup of contributing colleagues who are the leaders in their field.

In the early planning stages, a vision crystallized for a different concept. Ideally, we sought to ask not for a compilation of expert reviews which might soon be outdated. Rather, we wanted to compile a set of conceptual chapters, from pairs of experts, that highlight important discoveries for students in chromatin biology and for colleagues outside the epigenetics field. In keeping to a conceptual outline, we hoped to have a more long-lasting impact. Also, by including many diagrams and illuminating figures, and appendices, we hoped to list most of the systems and epigenetic marks currently known. The General Summaries were aimed as a stand-alone précis of the topics covered in each chapter, preceded by "teaser" images to entice the reader to investigate.

The figures have been another important hallmark for this book; particularly, the examples provided in the Overview and Concepts chapter. Here, Stefan Kubicek, a Ph.D. student from the Jenuwein lab at the IMP (Vienna), and Marie-Laure Caparros have been the masters of the diagrams. They honed draft upon draft of figures (sometimes only from sketches) for the chapters, such that we could gain a more coherent presentation. Several postdocs and Ph.D. students (Gabriella Farkas, Fatima Santos, Heike Wöhrmann, and others) in the labs of several authors also kindly contributed to the excellent illustrations in this book. However, we were unable to convert all of the contributions, and some figures have remained as submitted. We are also particularly grateful to Monika Lachner, Mario Richter, Roopsha Sengupta, Patrick Trojer, and other Ph.D. students and Postdocs in the Allis, Jenuwein, and Reinberg laboratories for amending, proofreading, and finalizing the tables and summaries that are displayed in the appendices. Here, Dr. Steven Gray (St. James Hospital, Dublin) has been particularly instrumental in validating and providing additional information for the table that lists all the currently known histone modifications.

Where appropriate, submitted chapters were sent out for comments from other colleagues who provided important input for streamlining and clarifying some of the complex concepts. Not all of this input could be converted into the revised and final versions, but the comments and suggestions helped to shape many of the chapters and the overall framework of the book. Here, we are indebted to G. Almouzni, P. Becker, H. Cedar, V. Chandler, W. Dean, R. Feil, A. Ferguson-Smith, M. Gartenberg, S. Grewal, M. Hampsey, E. Heard, R. Metzenberg, V. Pirrotta, F. Santos, T. Schedl, D. Solter, R. Sternglanz, S. Tilghman, and others.

Finally, we acknowledge the intellectual and, in some cases, emotional contributions made by all of our colleagues in the field who provided the chapters to make this book what it is. Their contributions, by way of written chapters and drawings, stand by themselves. [...] may not be obvious is the feedback and cross-fertil[...] that all of them had with the editorial team to help sh[...] and guide the book as it took form. The Overview an[...] Concepts chapter itself reflects their feedback, as in early drafts, we put too much of our own colors and bias into the sentences. For their wisdom and for bringing us a deeper perspective and balance, we thank them, and we admit that any deficiencies and mistakes there are ours.

Financial support for this book has come from CSHL Press (New York), the *Epigenome* FP6 NoE (European Union), IMP (Vienna), the Rockefeller University (New York), and the Howard Hughes Medical School-Robert Wood Johnson Medical School (Piscataway, New Jersey). Critical contributions were also made by Upstate Serologicals (Lake Placid, New York) and AbCam (Cambridge, UK), leading suppliers of epigenetic-based reagents and tools.

CDA, TJ, DR

Epigenetics: From Phenomenon to Field

Daniel E. Gottschling

Fred Hutchinson Cancer Research Center, Seattle, Washington 98109

CONTENTS

1. Introduction, 2
2. A History of Epigenetics at Cold Spring Harbor Symposia, 2
3. The 69th Symposium, 8
   3.1 The Histone Code Hypothesis, 8
   3.2 Dynamic Silent Chromatin, 8
   3.3 Nuclear Organization, 9
   3.4 Prions, 9
   3.5 New Phenomenon, 10
4. Closing Thoughts, 10

Acknowledgments, 11

References, 11

1 Introduction

In the summer of 2004, the 69th Cold Spring Harbor Symposium on Quantitative Biology covered the topic of "Epigenetics," and many of the authors of this book were in attendance. As an observer at this Symposium, I knew this was going to be an interesting meeting. It started simply enough by trying to define epigenetics. After a week of querying participants about this, it became clear that such a request was akin to asking someone to define "family values"—everyone knew what it meant, but it had a different meaning for each person. Part of the reason for the range of opinions may be understood from the etymology of "epigenetics" as explained by David Haig: The word had two distinct origins in the biological literature in the past century, and the meaning has continued to evolve. Waddington first coined the term for the study of "causal mechanisms" by which "the genes of the genotype bring about phenotypic effects" (see Haig 2004). Later, Nanney used it to explain his realization that cells with the same genotype could have different phenotypes that persisted for many generations.

I define an epigenetic phenomenon as a change in phenotype that is heritable but does not involve DNA mutation. Furthermore, the change in phenotype must be switch-like, "ON" or "OFF," rather than a graded response, and it must be heritable even if the initial conditions that caused the switch disappear. Thus, I consider epigenetic phenomena to include the lambda bacteriophage switch between lysis and lysogeny (Ptashne 2004), pili switching in uropathogenic *Escherichia coli* (Hernday et al. 2003), position-effect variegation in *Drosophila* (Henikoff 1990), heritable changes in cortical patterning of *Tetrahymena* (Frankel 1990), prion diseases (Wickner et al. 2004a), and X-chromosome inactivation (Lyon 1993).

The 69th Symposium came on the 100th anniversary of genetics as a field of study at Cold Spring Harbor Laboratory, making it very timely to consider epigenetics. Given this historical context, I thought it appropriate to provide an examination of epigenetics through the portal of previous Cold Spring Harbor Symposia. Although the 69th Symposium was the first dedicated to the topic, epigenetic phenomena and their study have been presented throughout the history of this distinguished series. The history I present is narrowed further by my limitations and likings. For a more complete and scholarly portrayal, I can recommend the more than 1000 reviews on epigenetics that have been written in the past five years.

In presenting this chronological account, I hope to convey a sense of how a collection of apparently disparate phenomena coalesced into a field of study that affects all areas of biology, and that the study of epigenetics is founded upon trying to explain the unexpected—perhaps more than any other field of biological research.

2 A History of Epigenetics at Cold Spring Harbor Symposia

In 1941 during the 9th Symposium, the great *Drosophila* geneticist H.J. Muller described developments on his original "eversporting displacement," in which gross chromosomal rearrangements resulted in the mutant mosaic expression of genes near the breakpoint (Muller 1941). By the time of this meeting, he referred to it as "position effect variegation." It was well established that the affected genes had been transferred "into the neighborhood of a heterochromatic region," that the transferred euchromatic regions had been "partly, but variably, transformed into a heterochromatic condition—'heterochromatized,'" and that *addition* of extra copies of heterochromatic chromosomes "allowed the affected gene to become more normal in its functioning." This latter observation was an unexpected quandary at the time, which we now know to be the result of a titration of limiting heterochromatin components.

At the 16th Symposium (1951), a detailed understanding of the gene was of high priority. This may explain why little progress had been made on understanding position-effect variegation (PEV), although more examples were being discovered. However, the opening speaker noted that PEV would be an exciting area for future research (Goldschmidt 1951). Barbara McClintock noted that chromosomal position effects were the basis of differences in "mutable loci" of maize, and she speculated that the variation of mutability she observed likely had its roots in the same mechanisms underlying PEV in *Drosophila* (McClintock 1951).

By the time of the 21st Symposium, McClintock's ideas about "controlling elements" had developed (McClintock 1956). Two were particularly poignant with regard to epigenetics. In the *Spm* controlling element system, she had uncovered variants that allowed her to distinguish between *trans*-acting factors that could "suppress" a gene (reduce or eliminate its phenotypic expression) rather than mutate it. She also noted that some controlling elements could suppress gene action not only at the locus where it had inserted, but also at loci that were located some distance on either side of it. Others were discovering this "spreading effect" as well. J. Schultz presented a biochemical and physical characterization of whole *Drosophila* that contained different amounts of heterochromatin (Schultz 1956). Although the work was quite primitive and the conclusions drawn were limited, the work represented early attempts to dissect the structure of heterochromatin and demonstrated just how difficult the problem would be.

Two talks at the 23rd Symposium were landmarks with respect to our present-day Symposium. First, R.A. Brink described his stunning observations of "paramutation" at the R locus in maize. If two alleles (R^{st} and R^{r}) with distinct phenotypes as homozygotes are combined to form a heterozygote, and this R^{st}/R^{r} plant is in turn crossed again, the resulting progeny that contain the Rr allele will *always* have an R^{st} phenotype, even though the R^{st} is no longer present (Brink 1958). However, this phenotype is metastable—in subsequent crosses the phenotype reverts to the normal R^{r} phenotype. He meant for the word paramutation "to be applied in this context in its literal sense, as referring to a phenomenon distinct from, but not wholly unlike, mutation." Second, D.L. Nanney went to great lengths to articulate "conceptual and operational distinctions between genetic and epigenetic systems" (Nanney 1958). In essence, he defined epigenetics differently from how it had been originally intended by Waddington (for details, see Haig 2004). He found it necessary to do so in order to describe phenomena he observed in *Tetrahymena*. He found evidence that the cytoplasmic history of conjugating parental cells influenced the mating-type determination of resulting progeny. His definition encompassed observations made by others as well, including Brink's work on the R locus and McClintock's work noted in the 21st Symposium.

Mary Lyon's recently proposed hypothesis of X-chromosome inactivation in female mammals (Lyon 1961) was of considerable interest at the 29th Symposium. S. Gartler, E. Beutler, and W.E. Nance presented further experimental evidence in support of it (Beutler 1964; Gartler and Linder 1964; Nance 1964). Beutler reviewed multiple examples of mosaic expression of X-linked genes in women, supporting the random nature of X inactivation. From careful quantitative analysis of an X-linked gene product, Nance deduced that X inactivation occurred before the 32-cell stage of the embryo.

The 38th Symposium on "Chromosome Structure and Function" represented a return to examining eukaryotic chromosomes—significant progress had been made studying prokaryotic and phage systems, and consequently, bacterial gene expression had dominated much of the thinking in the burgeoning field of molecular biology. However, an appreciation for chromatin (DNA with histones and nonhistone proteins) in eukaryotes was building, but it was unclear whether it played a role in chromosome structure

or function, or both (Swift 1974). Nevertheless, several groups began to speculate that posttranslational modification of chromatin proteins, including histones, was associated with gene transcription or overall chromosome structure (Allfrey et al. 1974; Louie et al. 1974; Weintraub 1974). There was only a hint of epigenetic phenomena in the air. It had been hypothesized that repetitive DNA regulated most genes in eukaryotes, partly based on the fact that McClintock's controlling elements were repeated in the genome. However, it was reported that most repeated DNA sequences were unlinked to genes (Peacock et al. 1974; Rudkin and Tartof 1974). From these observations, the idea that repeated elements regulated gene expression lost significant support from those in attendance. More importantly, however, these same studies discovered that most of the repetitive DNA was located in heterochromatin.

The 42nd Symposium demonstrated that in four years, an amazing number of technical and intellectual advances had transformed the study of eukaryotic chromosomes (Chambon 1978). This included the use of DNA restriction enzymes, development of recombinant DNA technology, routine separation of proteins and nucleic acids, the ability to perform Southern and northern analysis, rapid DNA and RNA sequencing, and immunofluorescence on chromosomes. The nucleosome hypothesis had been introduced, and mRNA splicing had been discovered. Biochemical and cytological differences in chromatin structure, especially between actively transcribed and inactive genes, comprised the primary interest at this meeting. However, most relevant to epigenetics, Hal Weintraub and colleagues presented ideas about how chromatin could impart variegated gene expression in an organism (Weintraub et al. 1978).

The 45th Symposium was a celebration of Barbara McClintock's discoveries—movable genetic elements (Yarmolinsky 1981). Mechanistic studies of bacterial transposition had made enormous progress and justifiably represented about half the presentations, whereas others presented evidence that transposition and regulated genomic reorganization occurred not only in maize, but also in other eukaryotes—including flies, snapdragons, *Trypanosomes*, *Ascobolus*, and budding yeast. In the context of this meeting, all observed variegated expression events were ascribed to transposition. Moreover, there was a reticence to seriously consider that controlling elements were responsible for most gene regulation (Campbell 1981), which led some to suggest that "the sole function of these elements is to promote genetic variability." In essence, the idea that heterochromatin was responsible for the regulated expression in position-effect variegation was called into question. With respect to future epigenetic studies, perhaps the most noteworthy discussion was the firm establishment of "silent mating cassettes" in *Saccharomyces cerevisiae* (Haber et al. 1981; Klar et al. 1981; Nasmyth et al. 1981; Rine et al. 1981).

Leading up to the 47th Symposium, a general correlation had been established in vertebrate systems that the overall level of cytosine methylation in CpG DNA sequences was lower for genes that were transcribed than for those that were not. However, there were exceptions to this generalization, and more detailed analysis was presented that methylation of a specific area of a gene's promoter was most important (Cedar et al. 1983; Doerfler et al. 1983; La Volpe et al. 1983). On the basis of restriction/modification systems of bacteria, it was thought that DNA methylation prevented binding of key regulatory proteins. Furthermore, it had been shown that DNA methylation patterns could be mitotically inherited in vertebrates, which led to the hypothesis that DNA methylation could serve as a means of transcriptional "memory" as cells divided through development (Shapiro and Mohandas 1983). Another major epigenetic-related finding was the identification of DNA sequences on either side of the "silent mating cassettes" in budding yeast that were responsible for transcriptional repression of genes within the cassettes—these defined the first DNA sequences required for chromosomal position effects (Abraham et al. 1983).

"The Molecular Biology of Development" was the topic for the 50th Symposium, and it too encompassed a number of important advances. Perhaps one of the most exciting developments was the overall awareness that fundamental molecular properties were conserved throughout evolution—e.g., human RAS functioned in budding yeast, homeo box proteins were conserved between flies and humans (Rubin 1985). New efforts to understand chromosome imprinting began with the development of nuclear transfer in mice (Solter et al. 1985). These studies revealed that parent-of-origin information was stored within the paternal and maternal genomes of a new zygote; it was not just the DNA that was important, but the chromosomes contained additional information about which parent they had passed through, and the information was required for successful development of an embryo. Part of the answer was thought to lie in the fact that differential gene expression was dependent on the parental origin of a chromosome (Cattanach and Kirk 1985).

There were a number of studies aimed at understanding the complex regulation of the bithorax complex, but

notably, E.B. Lewis made special mention of the curious nature of known *trans* regulators of the locus; nearly all were repressors of the locus (Lewis 1985). Thus, maintaining a gene in a silenced state for many cell doublings was imperative for normal development. This contrasted with much of the thinking at the time—that gene activation/induction was where the critical regulatory decisions of development would be.

DNA transformation and insertional mutagenesis techniques had recently been achieved for a number of organisms. One particularly creative and epigenetic-related use of this technology came in *Drosophila*. A P-element transposon with the *white* eye-color gene on it was created and "hopped" throughout the genome (Rubin et al. 1985). This provided a means to map sites throughout the *Drosophila* genome where PEV could occur.

This meeting also highlighted the first genetic approaches to dissecting sex determination and sex chromosome dosage compensation in *Drosophila* (Belote et al. 1985; Maine et al. 1985) and *Caenorhabditis elegans* (Hodgkin et al. 1985; Wood et al. 1985).

The 58th Symposium highlighted the celebration of the 40th anniversary of Watson and Crick's discovery. Part of the celebration was a coming-out party for epigenetic phenomena: There was identification of new phenomena, beginnings of molecular analysis of other phenomena, and sufficient progress had been made in a number of systems to propose hypotheses and to test them.

In trypanosomes, the family of *Variable Surface antigen Genes (VSG)* located near telomeres are largely silenced, with only one *VSG* expressed at a time. Although this organism does not appear to contain methylated DNA, it was reported that the silenced *VSG* genes contained a novel minor base: β-D-glucosylhydroxymethyluracil (Borst et al. 1993). This base appeared to be in place of thymidine in the DNA. Parallels between this base and cytosine methylation in other organisms were easy to draw—the modifications were important for maintaining a silenced gene. But how the base was introduced into the DNA, or how it imparted such a function, was unclear.

Progress had also been made in vertebrate epigenetic phenomena, including chromosomal imprinting and X inactivation (Ariel et al. 1993; Li et al. 1993; Tilghman et al. 1993; Willard et al. 1993). It had become clear by this time that numerous loci were subject to imprinting in mammals; only one allele was expressed in diploid cells, and expression was dependent on parental origin. The *Igf2-H19* locus was of particular interest, primarily because it contained two nearby genes that were regulated in opposing fashion. *Igf2* is expressed from the paternal chromosome while the maternal copy is repressed, whereas the paternal allele of *H19* is repressed and its maternal allele is expressed. Interestingly, methylated CpG was observed just upstream of both genes on the paternal chromosome. It was proposed that the differential methylation regulated access of the two genes to a nearby enhancer element—the enhancer was closer to, and just downstream of, *H19* (Tilghman et al. 1993). A mutually exclusive competition between the two genes for the enhancer was envisioned; when the *H19* gene was methylated, the enhancer was free to activate the more distant *Igf2* gene. Support for the idea that DNA methylation played a regulatory role in this process came from mouse studies. Mutation of the first vertebrate gene encoding a 5-methyl-cytosine DNA methyltransferase in ES cells showed that as embryos developed, the paternal copy of *H19* became hypomethylated and the gene became transcriptionally active (Li et al. 1993).

An important step in the way in which ^{5Me}CpG mediated its effects came from the purification of the first ^{5Me}CpG DNA-binding complex (MeCP1) (Bird 1993). Not only did it bind DNA, but when tethered upstream of a reporter gene, MeCP1 caused the gene to be repressed. Although this did not explain regulation at the *Igf2-H19* locus, it did provide a potential mechanism to explain the general correlation between DNA methylation and gene repression.

Genetic mapping over a number of years had identified a portion of the human X chromosome as being critical for imparting X inactivation. Molecular cloning studies of this X-inactivation center led to the discovery of the *Xist* gene (Willard et al. 1993), an ~17-kb noncoding RNA that was expressed only on the inactive X chromosome. The mouse version of *Xist* was surprisingly homologous in structure and sequence and held the promise of being an excellent model system to dissect the way in which this RNA functioned to repress most of the X chromosome.

Two notable findings were described in *Neurospora* (Selker et al. 1993). First, it was shown that cytosine DNA methylation was not limited to CpG dinucleotides but could occur in seemingly any DNA context. Second was the amazing description of the phenomenon of repeat-induced point mutation (RIP). Sequences become "RIP'd" when there is a sequence duplication (linked or unlinked) in a haploid genome and the genome is put through the sexual cycle via conjugation. Two events occur: Both copies of the duplicated DNA pick up G:C → A:T mutations, and DNA within a few hundred base pairs of the RIP'd sequences becomes methylated. This double attack on the genome is quite efficient—50% of unlinked loci succumb to RIP, whereas tightly linked loci approach 100%—and readily abolishes gene function.

The *brown* gene in *Drosophila*, when translocated near heterochromatin, displays dominant PEV; the translocated copy can cause repression of the wild-type copy. In searching for enhancers and suppressors of this *trans*-inactivation phenomenon, Henikoff discovered that duplication of the gene located near heterochromatin *increased* the level of repression on the normal copy (Martin-Morris et al. 1993). Although the mechanism underlying this event remained mysterious, it was postulated that the phenomenon might be similar to RIP in *Neurospora*, although it had to occur in the absence of DNA methylation, which does not occur in *Drosophila*.

Paul Schedl elucidated the concept of chromosomal "boundary elements" (Vazquez et al. 1993). The first were located on either side of the "puff" region at a heat shock locus in *Drosophila* and were defined by their unusual chromatin structure—an ~300-bp nuclease-resistant core bordered by nuclease hypersensitive sites. It was postulated that such elements separated chromatin domains along the chromosome. Two in vivo assays supported this hypothesis: (1) When bordering either side of a reported gene, boundary elements effectively eliminated chromosomal position effects when the construct was inserted randomly throughout the genome. (2) The boundary element was also defined by its ability to block enhancer function. When inserted between a gene promoter and its enhancer, the boundary element blocked the gene's expression. Although not as well defined, the concept of boundary elements was also developing in other organisms, especially at the globin locus in mammals (Clark et al. 1993).

Budding yeast shone the light on a mechanistic inroad to chromatin-related epigenetic phenomena. It had already been established that the silencers at the silent mating-type loci were sites for several DNA-binding proteins. Their binding appeared to be context-dependent, as exemplified by the Rap1 protein, which not only was important in silencing, but also bound upstream of a number of genes to activate transcription (for review, see Laurenson and Rine 1992).

Over the years, numerous links had been made between DNA replication and transcriptionally quiescent regions of the genome. The inactive X chromosome, heterochromatin, and silenced imprinted loci had all been reported to replicate late in S phase relative to transcriptionally active regions of the genome. In addition, it had been shown that the establishment of silencing at the silent mating-type loci required passage through S phase, suggesting that silent chromatin had to be built on newly replicated DNA. Thus, there was great interest when one of the silencers was found to be an origin of DNA replication, and its origin activity could not be separated from silencing function (Fox et al. 1993). Furthermore, mutants in the recently identified origin recognition complex (ORC) were found to cripple silencing (Bell et al. 1993; Fox et al. 1993).

The discovery that telomeres in *Saccharomyces cerevisiae* exerted PEV, just like that seen in *Drosophila*, brought another entrée into dissecting heterochromatic structure and its influence on gene expression. Reporter genes inserted near telomeres give variegated expression in a colony. The repressed state is dependent on many of the same genes (*SIR2*, *SIR3*, *SIR4*) as those required for silencing at the silent mating-type loci. Several key aspects about the silent chromatin structure and the regulation of the variegated expression were described. It is worth noting that heterochromatin is defined cytologically as condensed chromatin, but silent chromatin in *S. cerevisiae* has never been visualized in this way. Nevertheless, because of similarities to PEV in *Drosophila*, there was enthusiasm to consider silent chromatin in yeast to be a functional equivalent of heterochromatin (described in Weintraub 1993).

From the yeast studies, a number of fundamental concepts began to come to light. First, the importance of histone H3 and H4 became evident. In particular, the amino-terminal tail of histones H3 and H4 appeared to be directly involved in the formation of silent heterochromatin (Thompson et al. 1993). Specific mutants in the tails of these histones alleviated or crippled silencing and led to the notion that both the net charge of the residues on the tails and specific residues within the tails contributed to silencing. In addition, these early days of chromatin immunoprecipitation (ChIP) demonstrated that the lysines in the amino-terminal tail of histone H4 were hypoacetylated in regions of silent chromatin relative to the rest of the genome. Moreover, one of the histone mutants identified histone H4 K16, which could be acetylated, as critical for forming silent chromatin.

Telomeres appeared to provide the simplest system in which to develop an understanding of how Sir proteins mediated silencing. The concept of recruiting silencing proteins was being developed. Briefly, the telomeric DNA-binding protein, Rap1, was found to interact with Sir3 and Sir4 by two-hybrid methods (described in Palladino et al. 1993). Thus, Rap1 could "recruit" these Sir proteins to the telomeric region of the genome. There was evidence that Sir3 and Sir4 could bind to one another, and most importantly, Sir3 and perhaps Sir4 interacted with the tails of histones H3 and H4 (Thompson et al. 1993).

Furthermore, overexpression for Sir3 caused it to "spread" inward along the chromatin fiber from the telomere, suggesting that it was a limiting component of silent chromatin and could "polymerize" along the chromatin (Renauld et al. 1993). Taken together, there appeared to be a large interaction network important for silencing—the Sir proteins initiated assembly at telomeric DNA, due to their interaction with Rap1, and then polymerized from the telomere along the chromatin fiber, presumably by binding to the tails of histones H3 and H4.

Switching between transcriptional states in variegated telomeric expression appeared to be the result of a competition between silent and active gene expression (Aparicio and Gottschling 1994; described in Weintraub 1993). If the transcriptional activator for a telomeric gene was deleted, the gene's basal transcriptional machinery was insufficient for expression and the gene was constitutively silenced. Conversely, overexpression of the activator caused the telomeric gene to be expressed continuously—the gene was never silenced. In the absence of *SIR3* (or *SIR2* or *SIR4*), basal gene expression was sufficient, whereas increased dosage of *SIR3* increased the fraction of cells that were silenced. Although a transcriptional activator could overcome silencing throughout the cell cycle, it was most effective when cells were arrested in S phase, presumably when chromatin was being replicated and, hence, most susceptible to competition. Somewhat surprisingly, cells arrested in G_2/M also could be easily switched, suggesting that silent chromatin had not yet been fully assembled by this time.

Silent chromatin in yeast was shown to be recalcitrant to nucleases and DNA modification enzymes, suggesting that the underlying DNA was much less accessible relative to most of the genome (described in Thompson et al. 1993).

It also appeared that there was a hierarchy of silencing within the yeast genome: The telomeres were the most sensitive to perturbation, *HML* was next, and *HMR* was the least sensitive. In fact, when the *SIR1* gene was mutated, the normally completely silenced *HM* loci displayed variegated expression (Pillus and Rine 1989).

Finally, Sir3 and Sir4 were localized to the nuclear periphery, as were the telomeres. It was proposed that the nucleus was organized such that the nuclear envelope provided a special environment for silencing (Palladino et al. 1993).

Schizosaccharomyces pombe also has silent mating cassettes that were suspected to behave similarly to those in *S. cerevisiae*. However, in *S. pombe*, there was an added twist to the story of mating-type switching. In an elegant set of experiments, Amar Klar proposed how a "mark" is imprinted on one strand of DNA in a cell (Klar and Bonaduce 1993). The mark is manifested, after two cell divisions in one of the four granddaughter cells, as a double-stranded break that facilitates mating-type switching. This yeast does not have any known DNA modifications (methylation, etc.), hence, a different type of mark was postulated to be left on the DNA strand.

The topic of the 59th Symposium was "The Molecular Genetics of Cancer." The concept of epigenetic regulation in oncogenesis had begun to develop after the idea of tumor suppressor genes became established. There had been a couple of studies supporting such a notion, but an interesting twist to the story came in studies of Beckwith-Wiedemann syndrome and Wilms' tumor patients. Mutations in both types of patients had been mapped to a locus that included the imprinted *H19-IGF2* genes. Feinberg et al. (1994) discovered "loss of imprinting" (LOI) for these genes in affected patients—the maternal locus lost its imprint, *H19* was repressed, and *IGF2* was expressed. Thus LOI, which in principle could occur elsewhere in the genome, could cause either biallelic expression and/or extinction of genes critical in oncogenesis.

In the couple of years leading up to the 63rd Symposium on "Mechanisms of Transcription," several important developments occurred that would affect the molecular understanding of several epigenetic phenomena. Histone-modifying enzymes were identified—specifically, histone acetylases and deacetylases. Some of these enzymes played critical roles in regulating gene expression and provided an entry into gene products that directly affected PEV and silencing. The tip of this iceberg was presented at the Symposium (see Losick 1998). Molecular dissection of the Sir3 and Sir4 silencing proteins in yeast revealed the polyvalent nature of their interactions and revealed how the network of interactions between all the Sir proteins, the histones, and various DNA-binding factors set up silent chromatin. In addition, the molecular details of how various loci (telomeres, the rDNA, *HM* loci, and double-stranded breaks) could compete for the limited supply of Sir proteins were shown. By crippling the ability of a specific locus to recruit silencing factors, Sir protein levels were increased at the other loci (Cockell et al. 1998). This provided direct evidence that principles of mass action were at work and that silencing at one locus could affect the epigenetic silencing at other loci—an idea originally put forth in studies on PEV in *Drosophila*, but not yet tested (Locke et al. 1988).

Another finding explained how DNA methylation could regulate gene expression through chromatin. This came with the identification of protein complexes composed of MeCP2, which bind both methylated DNA and histone deacetylases (Wade et al. 1998). Methylated DNA could serve as a point of recruiting deacetylases to a locus and thus facilitate silencing of nearby genes.

The concept of boundary elements was extended from *Drosophila* to mammals, with clear evidence provided at the β-globin locus, thus indicating that chromatin boundaries were indeed likely conserved in metazoans and perhaps all eukaryotes (Bell et al. 1998).

The 64th Symposium on "Signaling and Gene Expression in the Immune System" provided evidence about how monoallelic expression arose, and that it might be more widespread than previously thought. Monoallelic expression at the immunoglobulin loci had been obvious in lymphocytes for some time—it guaranteed the production of a single receptor type per lymphoid cell (Mostoslavsky et al. 1999). The allele to be expressed was chosen early in development, apparently at random: Both alleles began in a repressed state, but over time one became demethylated. It was unclear how a single allele was chosen, but the phenomenon appeared at other loci, too, where the necessity of monoallelism was not obvious. For instance, only one allele of genes encoding the cytokines IL-2 and IL-4 was expressed (Pannetier et al. 1999).

The most significant epigenetics-related talk at the 65th Symposium concerned the discovery that the Sir2 protein was a histone deacetylase (Imai et al. 2000). This was the only Sir protein that had clear homologs in all other eukaryotes and that regulated PEV. It seemed to be the enzyme primarily responsible for removing acetyl moieties from histones in silent chromatin. Furthermore, because it was an NAD-dependent enzyme, it linked the regulation of silencing (heterochromatin) to cellular physiology.

The 68th Symposium on "The Genome of *Homo sapiens*" was an important landmark in genetics, and although there is still much genetic work to be done, the complete sequencing of this and other genomes signified that it was time to move "above genetics"—a literal meaning of epigenetics.

This historical account highlights several themes shared with many other areas of research. First, it demonstrates the episodic nature of advances in epigenetics. Second, as molecular mechanisms underlying epigenetic phenomena began to be understood, it made it easier to connect epigenetics to biological regulation in general. Third, it showed that people whom we now consider to be scientific luminaries had made these connections early on—it just took a while for most others to "see" the obvious.

3 The 69th Symposium

A few general principles have been identified over the years that are common to all epigenetic phenomena, and they serve to guide experimental approaches in the search for a detailed understanding. First, the differences between the two phenotypic states ("OFF" and "ON") always have a corresponding difference in structure at a key regulatory point—form translates into function. Hence, identifying the two distinct structures, the components that compose them, and the compositional differences between them have been the primary tasks. Second, the distinct structures must have the ability to be maintained and perpetuated in a milieu of competing factors and entropic forces. Thus, each structure requires self-reinforcement or positive feedback loops which ensure that it is maintained and propagated over many cellular divisions; in some cases, such as X-chromosome inactivation, this appears to be on the order of a lifetime.

Many of the mechanistic principles defined in the earlier symposia continued to be refined in the 69th Symposium, but there were also new developments. To put these new developments in context, it is important to note that two other discoveries had a major impact on epigenetics. One was the discovery of RNA interference and related RNA-based mechanisms of regulation. The other was the discovery of mechanisms underlying the prion hypothesis. Both of these fields have advanced rapidly in the past decade, with some of the studies contributing to knowledge about chromatin-based epigenetics and others providing new perspectives about heritable transmission of phenotypes.

Many of the accomplishments reported at the Symposium are detailed in the chapters of this book, so I eschew discussing these topics here. However, I will touch upon a few advances that caught my fancy and are not covered within these pages. At the end, I will try to distill the most important concepts I took away from the meeting.

3.1 The Histone Code Hypothesis

In considering histone modifications and their potential information content, there were many discussions about the "histone code hypothesis" (Jenuwein and Allis 2001). Most of those I participated in, or overheard, were informal and rather lively. The proponents of the "code" cite examples such as tri-methylation of histone H3 at K9 and its greater affinity for the HP1 class of heterochromatin proteins (Jenuwein and Allis 2001). Those on the other side cite biochemical and genetic evidence that the net charge on the amino-terminal tail of histone H4, irrespective of which position the charge is at, has dramatic effects on DNA binding or phenotype (Megee et al. 1995; Zheng and Hayes 2003).

Grunstein presented data that included genome-wide analysis of histone acetylation modifications and chromatin-associated proteins using specific antibodies and ChIP-Chip in *S. cerevisiae* (Millar et al. 2004). His focus was on the epigenetic switch associated with H4K16 acetylation for binding, or not binding, particular chromatin proteins—thus supporting the histone code hypothesis. Although not discussed, some of his data appeared to support reports from others that for much of the genome, there is no correlation between specific histone modifications and gene expression (i.e., all active genes have the same marks, and these marks are not present on inactive genes) (Schubeler et al. 2004; Dion et al. 2005). Taking all the results together, I suspect that both specific modifications *and* general net charge effects will be used as mechanisms for regulating chromatin structure and gene expression.

3.2 Dynamic Silent Chromatin

I must confess that, on the basis of static images of heterochromatin and the refractory nature of silent chromatin, I was convinced that once established, a heterochromatic state was as solid as granite. Only when it was time for DNA replication would the impervious structure become relaxed. In thinking this way, I foolishly ignored principles of equilibrium dynamics I had learned in undergraduate chemistry. However, these lessons were brought home again by studies of silent chromatin and heterochromatin, where it was shown that silencing proteins of yeast (Sir3), and heterochromatin proteins in mammalian cells (HP1), were in a dynamic equilibrium—proteins were rapidly exchanged between heterochromatin and the soluble compartment—even when the chromatin was in its most impervious state (Cheng and Gartenberg 2000; Cheutin et al. 2003). The realization of its dynamic qualities forced a different view of how an epigenetic chromatin state is maintained and propagated. It suggests that in some systems the epigenetic state can be reversed at any time, not just during DNA replication. Hence, we can infer that mechanisms of reinforcement and propagation for silenced chromatin must function constantly.

Methylation of histones was widely held to be the modification that would indeed impart a "permanent" mark on the chromatin (for review, see Kubicek and Jenuwein 2004). In contrast to all other histone modification (e.g., phosphorylation, acetylation, ubiquitination), there were no enzymes known that could reversibly remove a methyl group from the amine of lysine or arginine. Furthermore, removing the methyl group under physiological conditions by simple hydrolysis was considered thermodynamically disfavored and thus unlikely to occur spontaneously.

Those thinking that methylation marks were permanent had their belief system shaken a bit by several reports. First, it was shown that a nuclear peptidylarginine deiminase (PAD4) could eliminate monomethylation from histone H3 at arginine (R) residues (Cuthbert et al. 2004; Wang et al. 2004). Although this methyl removal process results in the arginine residue being converted into citrulline, and hence is not a true reversal of the modification, it nevertheless provided a mechanism for eliminating a permanent methyl mark.

Robin Allshire provided a tantalizing genetic argument that the *tis2* gene from *S. pombe* reversed dimethylation on histone H3 at K9 (R. Allshire, pers. comm.). He may have been on the right track, because a few months after the meeting, the unrelated LSD1 enzyme from mammals was shown to specifically demethylate di- and monomethyl on histone H3 at K4 (Shi et al. 2004), reversing an "active" chromatin mark. Quite interestingly, LSD1 did not work on trimethylated H3K4—thus, methylation could be reversed during the marking process, but reversal was not possible once the mark was fully matured.

However, Steve Henikoff presented a way by which a permanent trimethyl lysine mark could be eliminated. He showed that the variant histone H3.3 could replace canonical histone H3 in a replication-independent transcription-coupled manner (Henikoff et al. 2004). In essence, a histone that contained methyl marks for silencing could be removed and replaced with one that was more conducive to transcription. When total chromatin was isolated, histone H3.3 had many more active chromatin methylation marks (e.g., K79me) on it than canonical histone H3 did.

In considering all these results, it seems that there may not be a simple molecular modification within histones that serves as a memory mark for propagating the silent chromatin state through cell division. Rather, there must be a more tenuous set of interactions that increase the probability that a silent state will be maintained, although they do not guarantee it.

3.3 Nuclear Organization

Correlations between nuclear location and gene expression have been made for many years (Mirkovitch et al. 1987). These observations began to drive the notion that there were special compartments within the cell where gene expression or silencing was restricted. It was argued that this organization was necessary to keep the complexity of the genome and its regulation in a workable order. This idea was supported by studies in *S. cerevisiae*, where telomeres are preferentially located at the nuclear periphery, as are key components of the silencing complex, such as Sir4 (Palladino et al. 1993). Mutations that released the telomeres, or Sir4, from the nuclear periphery resulted in a loss of telomeric silencing (Laroche et al. 1998; Andrulis et al. 2002). Furthermore, artificially tethering a partially silenced gene to the periphery caused it to become fully silenced (Andrulis et al. 1998).

In an insightful experiment, Gasser showed that if the telomeres and the silencing complex were both released from the periphery, and free to move throughout the nucleus, telomeric silencing was readily established (Gasser et al. 2004). Thus, there does not appear to be a special need for localizing loci to a compartment. This is more consistent with the findings that rapid movement of chromatin proteins on and off chromosomes can still mediate effective regulation such as silencing. Perhaps some of the localization is necessary to keep high local concentrations of relevant factors under special (stressful?) conditions. Alternatively, this may represent a combination of domains put together through evolution that worked long ago, but had no ultimate purpose.

3.4 Prions

Wickner provided an overview and criteria for defining prions, and from his description it is clear that they are part of the epigenetic landscape (Wickner et al. 2004a,b). In the simplest molecular sense, prions are proteins that can cause heritable phenotypic changes, by acting upon and altering their cognate gene product. No DNA sequence changes occur; rather, the prion typically confers a structural change in its substrate. The best-studied and understood class of prions causes soluble forms of a protein to change into amyloid fibers. In many cases, the amyloid form reduces or abolishes normal activity of the protein, thus producing a change in phenotype. Wickner defined another class of prions that do not form amyloid filaments. These are enzymes that require activation by their own enzymatic activity. If a cell should have only inactive forms of the enzyme, then an external source of the active enzyme is required to start what would then become a self-propagating trait, as long as at least one active molecule was passed on to each cell. He provided two examples and the expectation that this class of proteins will define a new set of epigenetic mechanisms to pursue.

Si presented preliminary evidence that a prion model may explain learned memory in *Aplysia* (Si et al. 2004). Protein translation of a number of stored mRNAs in neuronal cells is important for the maintenance of short-term memory in this snail. He found that a regulator of protein translation, CPEB, can exist in two forms, and that the activated form of CPEB acts dominantly to perpetuate itself. Testing of this idea is still in its early days, but it offers an exciting new way of considering the issue of how we remember.

3.5 New Phenomenon

The description of a new and unexpected phenomenon always holds our imagination. One presentation in particular held my thoughts for weeks after the Symposium. Standard genetic analysis of mutant alleles of the *HOT-HEAD* gene, which regulates organ fusion in *Arabidopsis*, revealed that normal rules of Mendelian genetics were not being followed (Lolle et al. 2005). It was discovered that if heterozygous *HOTHEAD/hothead* plants self-fertilized and produced a homozygous *hothead/hothead* plant, and then this homozygous *hothead/hothead* plant was allowed to self-fertilize, the progeny from this homozygous parent reverted to a *HOTHEAD/hothead* genotype at a frequency of up to 15%. This stunning level of wild-type reversion produced an exact duplicate, at the nucleotide level, of the wild-type gene seen in the earlier generations. This reversion was not limited to the *HOTHEAD* locus—several other loci had similar frequencies of reversion to wild-type alleles. However, all the reversions required that the parent be homozygous *hothead/hothead*. The gene product of *HOTHEAD* did not offer an obvious explanation as to how this could occur, but discussions certainly suggested that an archival copy of the wild-type gene was transmitted, perhaps via RNA, through successive generations. Although it could be argued that this phenomenon is outside the purview of "epigenetics"—due to the change in DNA sequence—the heritable transmission of the putative archived copy does not follow normal genetic rules. Nevertheless, this phenomenon has enormous implications for genetics, especially in evolutionary thinking.

4 Closing Thoughts

So, what more needs to be done to understand epigenetic mechanisms? For the most part, we are still collecting (discovering) the components. Just as the full sequence of a genome has greatly facilitated progress in genetics, a clearer understanding for epigenetics will likely come when all the parts are known. It is encouraging to see the great strides that have been made in the last decade.

I confess that I cannot discern whether we are close to, or far away from, having an accurate mechanistic understanding about how epigenetic states are maintained and propagated. The prion-based phenomenon may be the first to be understood, but those that are chromatin-based seem the farthest off. The polyvalent nature of interactions that seem to be required to establish a silenced state on a chromosome increases the complexity of the problem. This is further compounded by the dynamic nature of silent chromatin. The ability to know more about movement of components in and out of chromatin structures requires application of enhanced or new methods for an eventual understanding. Whereas chromatin immunoprecipitation has been important in establishing which components reside in a structure, it has temporarily blinded us to the dynamics.

I suspect that, given the complexity, simply measuring binding and equilibrium constants between all the components and trying to derive a set of differential equations to simulate epigenetic switches may not be an effective use of resources, nor will it necessarily result in better comprehension. Rather, I speculate that a new type of mathematical approach will need to be developed and combined with new experimental measuring methods, in order to eventually understand epigenetic events. Part of this may require development of in vitro systems that faithfully recapitulate an epigenetic switch between states.

The idea of competition between two states in most epigenetic phenomena likely reflects an "arms race" that is happening at many levels in the cell, followed by attempts to rectify "collateral damage." For instance, silencing proteins may have evolved to protect the genome from transposons. However, because silencing proteins work through the ubiquitous nucleosomes, some critical genes become repressed. To overcome this, histone modifications (e.g., methylation of H3K4 and H3K79) and variant replacement histones (H2A.Z) evolved to prevent silencing proteins from binding to critical genes. Depending on subsequent events, these changes may be co-opted for other processes—e.g., repression of some of the genes by the silencing proteins may have become useful (silent mating loci). The silencing mechanisms may have been co-opted for other functions as well, such as promoting chromosome segregation. And so it goes...

I look forward to having the genomes of more organisms sequenced, because this might lead us to understand an order of events through evolution that set up the epigenetic processes we see today. For instance, *S. cerevisiae* does not have RNAi machinery, but many other fungi do. By filling in some of the phylogenetic gaps between species, we may discover what events led to *S. cerevisiae* no longer "needing" this system.

Perhaps more than any other field of biological research, the study of epigenetics is founded on trying to understand unexpected observations, ranging from H.J. Muller's position-effect variegation, to polar overdominance in the *callipyge* phenotype (Georges et al. 2004). The hope of understanding something unusual serves as the bait to draw us in, but we soon become entranced by the cleverness of the mechanisms employed. This may explain why this field has drawn more than its share of light-hearted and clever minds. I suspect it will continue to do so, as we develop a deeper understanding of the cleverness, and as new and unexpected epigenetic phenomena are discovered.

Acknowledgments

I thank my colleagues at the University of Chicago and the Fred Hutchinson Cancer Research Center for making my own studies on epigenetics so enjoyable, and I thank the National Institutes of Health for financial support.

References

Abraham J., Feldman J., Nasmyth K.A., Strathern J.N., Klar A.J., Broach J.R., and Hicks J.B. 1983. Sites required for position-effect regulation of mating-type information in yeast. *Cold Spring Harbor Symp. Quant. Biol.* **47:** 989–998.

Allfrey V.G., Inoue A., Karn J., Johnson E.M., and Vidali G. 1974. Phosphorylation of DNA-binding nuclear acidic proteins and gene activation in the HeLa cell cycle. *Cold Spring Harbor Symp. Quant. Biol.* **38:** 785–801.

Andrulis E.D., Neiman A.M., Zappulla D.C., and Sternglanz R. 1998. Perinuclear localization of chromatin facilitates transcriptional silencing. *Nature* **394:** 592–595.

Andrulis E.D., Zappulla D.C., Ansari A., Perrod S., Laiosa C.V., Gartenberg M.R., and Sternglanz R. 2002. Esc1, a nuclear periphery protein required for Sir4-based plasmid anchoring and partitioning. *Mol. Cell. Biol.* **22:** 8292–8301.

Aparicio O.M. and Gottschling D.E. 1994. Overcoming telomeric silencing: A *trans*-activator competes to establish gene expression in a cell cycle-dependent way. *Genes Dev.* **8:** 1133–1146.

Ariel M., Selig S., Brandeis M., Kitsberg D., Kafri T., Weiss A., Keshet I., Razin A., and Cedar H. 1993. Allele-specific structures in the mouse Igf2-H19 domain. *Cold Spring Harbor Symp. Quant. Biol.* **58:** 307–313.

Bell A., Boyes J., Chung J., Pikaart M., Prioleau M.N., Recillas F., Saitoh N., and Felsenfeld G. 1998. The establishment of active chromatin domains. *Cold Spring Harbor Symp. Quant. Biol.* **63:** 509–514.

Bell S.P., Marahrens Y., Rao H., and Stillman B. 1993. The replicon

model and eukaryotic chromosomes. *Cold Spring Harbor Symp. Quant. Biol.* **58:** 435–442.

Belote J.M., McKeown M.B., Andrew D.J., Scott T.N., Wolfner M.F., and Baker B.S. 1985. Control of sexual differentiation in *Drosophila* melanogaster. *Cold Spring Harbor Symp. Quant. Biol.* **50:** 605–614.

Beutler E. 1964. Gene inactivation: The distribution of gene products among populations of cells in heterozygous humans. *Cold Spring Harbor Symp. Quant. Biol.* **29:** 261–271.

Bird A.P. 1993. Functions for DNA methylation in vertebrates. *Cold Spring Harbor Symp. Quant. Biol.* **58:** 281–285.

Borst P., Gommers-Ampt J.H., Ligtenberg M.J., Rudenko G., Kieft R., Taylor M.C., Blundell P.A., and van Leeuwen F. 1993. Control of antigenic variation in African trypanosomes. *Cold Spring Harbor Symp. Quant. Biol.* **58:** 105–114.

Brink R.A. 1958. Paramutation at the R locus in maize. *Cold Spring Harbor Symp. Quant. Biol.* **23:** 379–391.

Campbell A. 1981. Some general questions about movable elements and their implications. *Cold Spring Harbor Symp. Quant. Biol.* **45:** 1–9.

Cattanach B.M. and Kirk M. 1985. Differential activity of maternally and paternally derived chromosome regions in mice. *Nature* **315:** 496–498.

Cedar H., Stein R., Gruenbaum Y., Naveh-Many T., Sciaky-Gallili N., and Razin A. 1983. Effect of DNA methylation on gene expression. *Cold Spring Harbor Symp. Quant. Biol.* **47:** 605–609.

Chambon P. 1978. Summary: The molecular biology of the eukaryotic genome is coming of age. *Cold Spring Harbor Symp. Quant. Biol.* **42:** 1209–1234.

Cheng T.H. and Gartenberg M.R. 2000. Yeast heterochromatin is a dynamic structure that requires silencers continuously. *Genes Dev.* **14:** 452–463.

Cheutin T., McNairn A.J., Jenuwein T., Gilbert D.M., Singh P.B., and Misteli T. 2003. Maintenance of stable heterochromatin domains by dynamic HP1 binding. *Science* **299:** 721–725.

Clark D., Reitman M., Studitsky V., Chung J., Westphal H., Lee E., and Felsenfeld G. 1993. Chromatin structure of transcriptionally active genes. *Cold Spring Harbor Symp. Quant. Biol.* **58:** 1–6.

Cockell M., Gotta M., Palladino F., Martin S.G., and Gasser S.M. 1998. Targeting Sir proteins to sites of action: A general mechanism for regulated repression. *Cold Spring Harbor Symp. Quant. Biol.* **63:** 401–412.

Cuthbert G.L., Daujat S., Snowden A.W., Erdjument-Bromage H., Hagiwara T., Yamada M., Schneider R., Gregory P.D., Tempst P., Bannister A.J., and Kouzarides T. 2004. Histone deimination antagonizes arginine methylation. *Cell* **118:** 545–553.

Dion M.F., Altschuler S.J., Wu L.F., and Rando O.J. 2005. Genomic characterization reveals a simple histone H4 acetylation code. *Proc. Natl. Acad. Sci.* **102:** 5308–5309.

Doerfler W., Kruczek I., Eick D., Vardimon L., and Kron B. 1983. DNA methylation and gene activity: The adenovirus system as a model. *Cold Spring Harbor Symp. Quant. Biol.* **47:** 593–603.

Feinberg A.P., Kalikin L.M., Johnson L.A., and Thompson J.S. 1994. Loss of imprinting in human cancer. *Cold Spring Harbor Symp. Quant. Biol.* **59:** 357–364.

Fox C.A., Loo S., Rivier D.H., Foss M.A., and Rine J. 1993. A transcriptional silencer as a specialized origin of replication that establishes functional domains of chromatin. *Cold Spring Harbor Symp. Quant. Biol.* **58:** 443–455.

Frankel J. 1990. Positional order and cellular handedness. *J. Cell Sci.* **97:** 205–211.

Gartler S.M., and Linder D. 1964. Selection In mammalian mosaic cell populations. *Cold Spring Harbor Symp. Quant. Biol.* **29:** 253–260.

Gasser S.M., Hediger F., Taddei A., Neumann F.R., and Gartenberg M.R. 2004. The function of telomere clustering in yeast: The circe effect. *Cold Spring Harbor Symp. Quant. Biol.* **69:** 327–337.

Georges M., Charlier C., Smit M., Davis E., Shay T., Tordoir X., Takeda H., Caiment F., and Cockett N. 2004. Toward molecular understanding of polar overdominance at the ovine callipyge locus. *Cold Spring Harbor Symp. Quant. Biol.* **69:** 477–483.

Goldschmidt R.B. 1951. The theory of the gene: Chromosomes and genes. *Cold Spring Harbor Symp. Quant. Biol.* **16:** 1–11.

Haber J.E., Weiffenbach B., Rogers D.T., McCusker J., and Rowe L.B. 1981. Chromosomal rearrangements accompanying yeast mating-type switching: Evidence for a gene-conversion model. *Cold Spring Harbor Symp. Quant. Biol.* **45:** 991–1002.

Haig D. 2004. The (dual) origin of epigenetics. *Cold Spring Harbor Symp. Quant. Biol.* **69:** 67.

Henikoff S. 1990. Position-effect variegation after 60 years. *Trends Genet.* **6:** 422–426.

Henikoff S., McKittrick E., and Ahmad K. 2004. Epigenetics, histone H3 variants, and the inheritance of chromatin states. *Cold Spring Harbor Symp. Quant. Biol.* **69:** 235–243.

Hernday A.D., Braaten B.A., and Low D.A. 2003. The mechanism by which DNA adenine methylase and PapI activate the pap epigenetic switch. *Mol. Cell* **12:** 947–957.

Hodgkin J., Doniach T., and Shen M. 1985. The sex determination pathway in the nematode *Caenorhabditis elegans:* Variations on a theme. *Cold Spring Harbor Symp. Quant. Biol.* **50:** 585–593.

Imai S., Johnson F.B., Marciniak R.A., McVey M., Park P.U., and Guarente L. 2000. Sir2: An NAD-dependent histone deacetylase that connects chromatin silencing, metabolism, and aging. *Cold Spring Harbor Symp. Quant. Biol.* **65:** 297–302.

Jenuwein T. and Allis C.D. 2001. Translating the histone code. *Science* **293:** 1074–1080.

Klar A.J., and Bonaduce M.J. 1993. The mechanism of fission yeast mating-type interconversion: Evidence for two types of epigenetically inherited chromosomal imprinted events. *Cold Spring Harbor Symp. Quant. Biol.* **58:** 457–465.

Klar A.J., Hicks J.B., and Strathern J.N. 1981. Irregular transpositions of mating-type genes in yeast. *Cold Spring Harbor Symp. Quant. Biol.* **45:** 983–990.

Kubicek S. and Jenuwein T. 2004. A crack in histone lysine methylation. *Cell* **119:** 903–906.

La Volpe A., Taggart M., Macleod D., and Bird A. 1983. Coupled demethylation of sites in a conserved sequence of *Xenopus* ribosomal DNA. *Cold Spring Harbor Symp. Quant. Biol.* **47:** 585–592.

Laroche T., Martin S.G., Gotta M., Gorham H.C., Pryde F.E., Louis E.J., and Gasser S.M. 1998. Mutation of yeast Ku genes disrupts the subnuclear organization of telomeres. *Curr. Biol.* **8:** 653–656.

Laurenson P. and Rine J. 1992. Silencers, silencing, and heritable transcriptional states. *Microbiol. Rev.* **56:** 543–560.

Lewis E.B. 1985. Regulation of the genes of the bithorax complex in *Drosophila. Cold Spring Harbor Symp. Quant. Biol.* **50:** 155–164.

Li E., Beard C., Forster A.C., Bestor T.H., and Jaenisch R. 1993. DNA methylation, genomic imprinting, and mammalian development. *Cold Spring Harbor Symp. Quant. Biol.* **58:** 297–305.

Locke J., Kotarski M.A., and Tartof K.D. 1988. Dosage-dependent modifiers of position effect variegation in *Drosophila* and a mass action model that explains their effect. *Genetics* **120:** 181–198.

Lolle S.J., Victor J.L., Young J.M., and Pruitt R.E. 2005. Genome-wide non-mendelian inheritance of extra-genomic information in *Arabidopsis. Nature* **434:** 505–509.

Losick R. 1998. Summary: Three decades after sigma. *Cold Spring Har-* bor Symp. Quant. Biol.* **63:** 653–666.

Louie A.J., Candido E.P., and Dixon G.H. 1974. Enzymatic modifications and their possible roles in regulating the binding of basic proteins to DNA and in controlling chromosomal structure. *Cold Spring Harbor Symp. Quant. Biol.* **38:** 803–819.

Lyon M.F. 1961. Gene action in the X-chromosome of the mouse (*Mus musculus* L.). *Nature* **190:** 372–373.

———. 1993. Epigenetic inheritance in mammals. *Trends Genet.* **9:** 123–128.

Maine E.M., Salz H.K., Schedl P., and Cline T.W. 1985. Sex-lethal, a link between sex determination and sexual differentiation in *Drosophila melanogaster. Cold Spring Harbor Symp. Quant. Biol.* **50:** 595–604.

Martin-Morris L.E., Loughney K., Kershisnik E.O., Poortinga G., and Henikoff S. 1993. Characterization of sequences responsible for *trans*-inactivation of the *Drosophila brown* gene. *Cold Spring Harbor Symp. Quant. Biol.* **58:** 577–584.

McClintock B. 1951. Chromosome organization and genic expression. *Cold Spring Harbor Symp. Quant. Biol.* **16:** 13–47.

———. 1956. Controlling elements and the gene. *Cold Spring Harbor Symp. Quant. Biol.* **21:** 197–216.

Megee P.C., Morgan B.A., and Smith M.M. 1995. Histone H4 and the maintenance of genome integrity. *Genes Dev.* **9:** 1716–1727.

Millar C.B., Kurdistani S.K., and Grunstein M. 2004. Acetylation of yeast histone H4 lysine 16: A switch for protein interactions in heterochromatin and euchromatin. *Cold Spring Harbor Symp. Quant. Biol.* **69:** 193–200.

Mirkovitch J., Gasser S.M., and Laemmli U.K. 1987. Relation of chromosome structure and gene expression. *Philos. Trans. R. Soc. Lond. B Biol. Sci.* **317:** 563–574.

Mostoslavsky R., Kirillov A., Ji Y.H., Goldmit M., Holzmann M., Wirth T., Cedar H., and Bergman Y. 1999. Demethylation and the establishment of κ allelic exclusion. *Cold Spring Harbor Symp. Quant. Biol.* **64:** 197–206.

Muller H.J. 1941. Induced mutations in *Drosophila. Cold Spring Harbor Symp. Quant. Biol.* **9:** 151–167.

Nance W.E. 1964. Genetic tests with a sex-linked marker: Glucose-6-phosphate dehydrogenase. *Cold Spring Harbor Symp. Quant. Biol.* **29:** 415–425.

Nanney D.L. 1958. Epigenetic factors affecting mating type expression in certain ciliates. *Cold Spring Harbor Symp. Quant. Biol.* **23:** 327–335.

Nasmyth K.A., Tatchell K., Hall B.D., Astell C., and Smith M. 1981. Physical analysis of mating-type loci in *Saccharomyces cerevisiae. Cold Spring Harbor Symp. Quant. Biol.* **45:** 961–981.

Palladino F., Laroche T., Gilson E., Pillus L., and Gasser S.M. 1993. The positioning of yeast telomeres depends on SIR3, SIR4, and the integrity of the nuclear membrane. *Cold Spring Harbor Symp. Quant. Biol.* **58:** 733–746.

Pannetier C., Hu–Li J., and Paul W.E. 1999. Bias in the expression of IL-4 alleles: The use of T cells from a GFP knock-in mouse. *Cold Spring Harbor Symp. Quant. Biol.* **64:** 599–602.

Peacock W.J., Brutlag D., Goldring E., Appels R., Hinton C.W., and Lindsley D.L. 1974. The organization of highly repeated DNA sequences in *Drosophila melanogaster* chromosomes. *Cold Spring Harbor Symp. Quant. Biol.* **38:** 405–416.

Pillus L. and Rine J. 1989. Epigenetic inheritance of transcriptional states in *S. cerevisiae. Cell* **59:** 637–647.

Ptashne M. 2004. *A genetic switch: Phage lambda revisited*, 3rd edition. Cold Spring Harbor Laboratory Press, Cold Spring Harbor, New York.

Renauld H., Aparicio O.M., Zierath P.D., Billington B.L., Chhablani

S.K., and Gottschling D.E. 1993. Silent domains are assembled continuously from the telomere and are defined by promoter distance and strength, and by SIR3 dosage. *Genes Dev.* **7:** 1133–1145.

Rine J., Jensen R., Hagen D., Blair L., and Herskowitz I. 1981. Pattern of switching and fate of the replaced cassette in yeast mating-type interconversion. *Cold Spring Harbor Symp. Quant. Biol.* **45:** 951–960.

Rubin G.M. 1985. Summary. *Cold Spring Harbor Symp. Quant. Biol.* **50:** 905–908.

Rubin G.M., Hazelrigg T., Karess R.E., Laski F.A., Laverty T., Levis R., Rio D.C., Spencer F.A., and Zuker C.S. 1985. Germ line specificity of P-element transposition and some novel patterns of expression of transduced copies of the white gene. *Cold Spring Harbor Symp. Quant. Biol.* **50:** 329–335.

Rudkin G.T. and Tartof K.D. 1974. Repetitive DNA in polytene chromosomes of *Drosophila melanogaster*. *Cold Spring Harbor Symp. Quant. Biol.* **38:** 397–403.

Schubeler D., MacAlpine D.M., Scalzo D., Wirbelauer C., Kooperberg C., van Leeuwen F., Gottschling D.E., O'Neill L.P., Turner B.M., Delrow J., et al. 2004. The histone modification pattern of active genes revealed through genome-wide chromatin analysis of a higher eukaryote. *Genes Dev.* **18:** 1263–1271.

Schultz J. 1956. The relation of the heterochromatic chromosome regions to the nucleic acids of the cell. *Cold Spring Harbor Symp. Quant. Biol.* **21:** 307–328.

Selker E.U., Richardson G.A., Garrett-Engele P.W., Singer M.J., and Miao V. 1993. Dissection of the signal for DNA methylation in the ζ–η region of *Neurospora*. *Cold Spring Harbor Symp. Quant. Biol.* **58:** 323–329.

Shapiro L.J. and Mohandas T. 1983. DNA methylation and the control of gene expression on the human X chromosome. *Cold Spring Harbor Symp. Quant. Biol.* **47:** 631–637.

Shi Y., Lan F., Matson C., Mulligan P., Whetstine J.R., Cole P.A., and Casero R.A. 2004. Histone demethylation mediated by the nuclear amine oxidase homolog LSD1. *Cell* **119:** 941–953.

Si K., Lindquist S., and Kandel E. 2004. A possible epigenetic mechanism for the persistence of memory. *Cold Spring Harbor Symp. Quant. Biol.* **69:** 497–498.

Solter D., Aronson J., Gilbert S.F., and McGrath J. 1985. Nuclear transfer in mouse embryos: Activation of the embryonic genome. *Cold Spring Harbor Symp. Quant. Biol.* **50:** 45–50.

Swift H. 1974. The organization of genetic material in eukaryotes: Progress and prospects. *Cold Spring Harbor Symp. Quant. Biol.* **38:** 963–979.

Thompson J.S., Hecht A., and Grunstein M. 1993. Histones and the regulation of heterochromatin in yeast. *Cold Spring Harbor Symp. Quant. Biol.* **58:** 247–256.

Tilghman S.M., Bartolomei M.S., Webber A.L., Brunkow M.E., Saam J., Leighton P.A., Pfeifer K., and Zemel S. 1993. Parental imprinting of the H19 and Igf2 genes in the mouse. *Cold Spring Harbor Symp. Quant. Biol.* **58:** 287–295.

Vazquez J., Farkas G., Gaszner M., Udvardy A., Muller M., Hagstrom K., Gyurkovics H., Sipos L., Gausz J., Galloni M., et al. 1993. Genetic and molecular analysis of chromatin domains. *Cold Spring Harbor Symp. Quant. Biol.* **58:** 45–54.

Wade P.A., Jones P.L., Vermaak D., Veenstra G.J., Imhof A., Sera T., Tse C., Ge H., Shi Y.B., Hansen J.C., and Wolffe A.P. 1998. Histone deacetylase directs the dominant silencing of transcription in chromatin: Association with MeCP2 and the Mi-2 chromodomain SWI/SNF ATPase. *Cold Spring Harbor Symp. Quant. Biol.* **63:** 435–445.

Wang Y., Wysocka J., Perlin J.R., Leonelli L., Allis C.D., and Coonrod S.A. 2004. Linking covalent histone modifications to epigenetics: The rigidity and plasticity of the marks. *Cold Spring Harbor Symp. Quant. Biol.* **69:** 161–169.

Weintraub H. 1974. The assembly of newly replicated DNA into chromatin. *Cold Spring Harbor Symp. Quant. Biol.* **38:** 247–256.

———. 1993. Summary: Genetic tinkering local problems, local solutions. *Cold Spring Harbor Symp. Quant. Biol.* **58:** 819–836.

Weintraub H., Flint S.J., Leffak I.M., Groudine M., and Grainger R.M. 1978. The generation and propagation of variegated chromosome structures. *Cold Spring Harbor Symp. Quant. Biol.* **42:** 401–407.

Wickner R.B., Edskes H.K., Ross E.D., Pierce M.M., Baxa U., Brachmann A., and Shewmaker F. 2004a. Prion genetics: New rules for a new kind of gene. *Annu. Rev. Genet.* **38:** 681–707.

Wickner R.B., Edskes H.K., Ross E.D., Pierce M.M., Shewmaker F., Baxa U., and Brachmann A. 2004b. Prions of yeast are genes made of protein: Amyloids and enzymes. *Cold Spring Harbor Symp. Quant. Biol.* **69:** 489–496.

Willard H.F., Brown C.J., Carrel L., Hendrich B., and Miller A.P. 1993. Epigenetic and chromosomal control of gene expression: Molecular and genetic analysis of X chromosome inactivation. *Cold Spring Harbor Symp. Quant. Biol.* **58:** 315–322.

Wood W.B., Meneely P., Schedin P., and Donahue L. 1985. Aspects of dosage compensation and sex determination in *Caenorhabditis elegans*. *Cold Spring Harbor Symp. Quant. Biol.* **50:** 575–583.

Yarmolinsky M.B. 1981. Summary. *Cold Spring Harbor Symp. Quant. Biol.* **45:** 1009–1015.

Zheng C., and Hayes J.J. 2003. Structures and interactions of the core histone tail domains. *Biopolymers* **68:** 539–546.

A Brief History of Epigenetics

Gary Felsenfeld

National Institute of Diabetes and Digestive and Kidney Diseases, National Institutes of Health,
Bethesda, Maryland 20892–0540

CONTENTS

1. Introduction, 16

2. Clues from Genetics and Development, 16

3. DNA Is the Same in All Somatic Cells of an Organism, 17

4. The Role of DNA Methylation, 17

5. The Role of Chromatin, 18

6. All Mechanisms Are Interrelated, 19

References, 21

1 Introduction

The history of epigenetics is linked with the study of evolution and development. But during the past 50 years, the meaning of the term "epigenetics" has itself undergone an evolution that parallels our dramatically increased understanding of the molecular mechanisms underlying regulation of gene expression in eukaryotes. Our present working definition is "the study of mitotically and/or meiotically heritable changes in gene function that cannot be explained by changes in DNA sequence" (Riggs et al. 1996). Until the 1950s, however, the word epigenetics was used in an entirely different way to categorize all of the developmental events leading from the fertilized zygote to the mature organism—that is, all of the regulated processes that, beginning with the genetic material, shape the final product (Waddington 1953). This concept had its origins in the much earlier studies in cell biology and embryology, beginning in the late 19th century, that laid the groundwork for our present understanding of the relationship between genes and development. There was a long debate among embryologists about the nature and location of the components responsible for carrying out the developmental plan of the organism. In trying to make sense of a large number of ingenious but ultimately confusing experiments involving the manipulation of cells and embryos, embryologists divided into two schools: those who thought that each cell contained preformed elements that enlarged during development, and those who thought the process involved chemical reactions among soluble components that executed a complex developmental plan. These views focused on the relative importance of the nucleus and cytoplasm in the developmental process. Following Flemming's discovery of the existence of chromosomes in 1879, experiments by many investigators, including Wilson and Boveri, provided strong evidence that the developmental program resided in the chromosomes. Thomas Hunt Morgan (1911) ultimately provided the most persuasive proof of this idea through his demonstration of the genetic linkage of several *Drosophila* genes to the X chromosome.

From that point onward, rapid progress was made in creating linear chromosome maps in which individual genes were assigned to specific sites on the *Drosophila* chromosomes (Sturtevant 1913). Of course, the questions of classic "epigenesis" remained: What molecules within the chromosomes carried the genetic information, how did they direct the developmental program, and how was the information transmitted during cell division? It was understood that both nucleic acid and proteins were present in chromosomes, but their relative contributions were not obvious; certainly, no one believed that the nucleic acid alone could carry all of the developmental information. Furthermore, earlier questions persisted about the possible contribution of the cytoplasm to developmental events. Evidence from *Drosophila* genetics (see below) suggested that heritable changes in phenotype could occur without corresponding changes in the "genes." This debate was dramatically altered by the identification of DNA as the primary carrier of genetic information. Ultimately, it became useful to redefine epigenetics so as to distinguish heritable changes that arise from sequence changes in DNA from those that do not.

2 Clues from Genetics and Development

Whatever the vagaries of the definition, the ideas and scientific data that underlie the present concept of epigenetics had been accumulating steadily since the early part of the 20th century. In 1930, H.J. Muller (Muller 1930) described a class of *Drosophila* mutations he called "eversporting displacements" ("eversporting" denoting the high rate of phenotypic change). These mutants involved chromosome translocations (displacements), but "even when all parts of the chromatin appeared to be represented in the right dosage—though abnormally arranged—the phenotypic result was not always normal." In some of these cases, Muller observed flies that had mottled eyes. He thought that this was probably due to a "genetic diversity of the different eye-forming cells," but further genetic analysis led him to connect the unusual properties with chromosomal rearrangement, and to conclude that "chromosome regions, affecting various characters at once, are somehow concerned, rather than individual genes or suppositious 'gene elements.'" Over the next 10 to 20 years, strong evidence provided by many laboratories (see Hannah 1951) confirmed that this variegation arose when rearrangements juxtaposed the white gene with heterochromatic regions.

During that period, chromosomal rearrangements of all kinds were the object of a great deal of attention. It was apparent that genes were not completely independent entities; their function could be affected by their location within the genome—as amply demonstrated by the many *Drosophila* mutants that led to variegation, as well as by other mutants involving translocation to euchromatic regions, in which more general (non-variegating) position effects could be observed. The role of transposable elements in plant genetics also became clear, largely through the work of McClintock (1965).

A second line of reasoning came from the study of developmental processes. It was evident that during development there was a divergence of phenotypes among differentiating cells and tissues, and it appeared that such distinguishing features, once established, could be clonally inherited by the dividing cells. Although it was understood at this point that cell-specific programming existed, and that it could be transmitted to daughter cells, how this was done was less clear.

A number of mechanisms could be imagined, and were considered. Particularly for those with a biochemical point of view, a cell was defined by the multiple interdependent biochemical reactions that maintained its identity. For example, it was suggested in 1949 by Delbruck (quoted in Jablonka and Lamb 1995) that a simple pair of biochemical pathways, each of which produced as an intermediate an inhibitor of the other pathway, could establish a system that could switch between one of two stable states. Actual examples of such systems were found somewhat later in the lac operon of *Escherichia coli* (Novick and Weiner 1957) and in the phage switch between lysogenic and lytic states (Ptashne 1992). Functionally equivalent models could be envisioned in eukaryotes. The extent to which nucleus and cytoplasm each contributed to the transmission of a differentiated state in the developing embryo was of course a matter of intense interest and debate; a self-stabilizing biochemical pathway would presumably have to be maintained through cell division. A second kind of epigenetic transmission was clearly demonstrated in *Paramecia* and other ciliates, in which the ciliary patterns may vary among individuals and are inherited clonally (Beisson and Sonneborn 1965). Altering the cortical pattern by microsurgery results in transmission of a new pattern to succeeding generations. It has been argued that related mechanisms are at work in metazoans, in which the organization of cellular components is influenced by localized cytoplasmic determinants in a way that can be transmitted during cell division (Grimes and Aufderheide 1991).

3 DNA Is the Same in All Somatic Cells of an Organism

Although chromosome morphology indicated that all somatic cells possessed all of the chromosomes, it could not have been obvious that all somatic cells retained the full complement of DNA present in the fertilized egg. Nor until the work of Avery, MacLeod, and McCarty in 1944, and that of Hershey and Chase (1952), was it even clear that a protein-free DNA molecule could carry genetic information, a conclusion strongly reinforced by Watson and Crick's solution of the structure of DNA in 1953. Work by Briggs and King (1952) in *Rana pipiens* and by Laskey and Gurdon (1970) in *Xenopus* had demonstrated that introduction of a nucleus from early embryonic cells into enucleated oocytes could result in development of an embryo. But as late as 1970, Laskey and Gurdon could state that "It has yet to be proved that somatic cells of an adult animal possess genes other than those necessary for their own growth and differentiation." In the paper containing this statement, they went on to show that to a first approximation, the DNA of a somatic cell nucleus was competent to direct embryogenesis when introduced into an enucleated egg. It was now clear that the program of development, and the specialization of the repertoire of expression seen in somatic cells, must involve signals that are not the result of some deletion or mutation in the germ-line DNA sequence when it is transmitted to somatic cells.

Of course, there are ways in which the DNA of somatic cells can come to differ from that of the germ line, with consequences for the cellular phenotype: For example, transposable elements can alter the pattern of expression in somatic cells, as demonstrated by the work of Barbara McClintock and other plant geneticists. Similarly, the generation of antibody diversity involves DNA rearrangement in a somatic cell lineage. This rearrangement (or more precisely its consequences) can be considered a kind of epigenetic event, consistent with the early observations of position-effect variegation described by Muller. However, much of the work on epigenetics in recent years has focused on systems in which no DNA rearrangements have occurred, and the emphasis has therefore been on modifications to the bases, and to the proteins that are complexed with DNA within the nucleus.

4 The Role of DNA Methylation

X-chromosome inactivation provided an early model of this kind of epigenetic mechanism (Ohno et al. 1959; Lyon 1961); the silenced X chromosome was clearly chosen at random in somatic cells, and there was no evidence of changes in the DNA sequence itself. In part to account for this kind of inactivation, Riggs (1975) and Holliday and Pugh (1975) proposed that DNA methylation could act as an epigenetic mark. The key elements in this model were the ideas that sites of methylation were palindromic, and that distinct enzymes were responsible for methylation of unmodified DNA and DNA already methylated on one strand. It was postulated that the first methylation event would be much more difficult than the second; once the first strand was modified, however, the complementary strand would quickly be modified at the same palindromic site. A methylation mark present on a parental strand would be copied on the daughter strand following replication, resulting in faithful transmission of the methylated state to the next generation. Shortly thereafter, Bird took advantage of the fact that the principal target of methylation in animals is the sequence CpG (Doskocil and Sorm 1962) to introduce the use of methylation-sensitive restriction enzymes as a way of detecting the methylation state. Subsequent studies (Bird 1978; Bird and Southern 1978) then showed that endogenous CpG sites were either completely unmethylated or completely methylated. The predictions of the model were thus confirmed, establishing a mechanism for epigenetic transmission of the methylation mark through semiconservative propagation of the methylation pattern.

In the years following these discoveries, a great deal of attention has been focused on endogenous patterns of DNA methylation, on the possible transmission of these patterns through the germ line, on the role of DNA methylation in silencing gene expression, on possible mechanisms for initiation or inhibition of methylation at a fully unmethylated site, and on the identification of the enzymes responsible for de novo methylation and for maintenance of methylation on already methylated sites. Although much of the DNA methylation seen in vertebrates is associated with repetitive and retroviral sequences and may serve to maintain these sequences in a permanently silent state, there can be no question that in many cases this modification provides the basis for epigenetic transmission of the state of gene activity. This is most clearly demonstrated at imprinted loci (Cattanach and Kirk 1985) such as the mouse or human *Igf2/H19* locus, where one allele is marked by DNA methylation, which in turn controls expression from both genes (Bell and Felsenfeld 2000; Hark et al. 2000). At the same time, it was clear that this could not be the only mechanism for epigenetic transmission of information. For example, as noted above, position-effect variegation had been observed many years earlier in *Drosophila*, an organism that has extremely low levels of DNA methylation. Furthermore, in subsequent years, *Drosophila* geneticists had identified the *Polycomb* and *Trithorax* groups of genes, which appeared to be involved in permanently "locking in" the state of activity, either off or on, respectively, of clusters of genes during development. The fact that these states were stably transmitted during cell division suggested an underlying epigenetic mechanism.

5 The Role of Chromatin

It had been recognized for many years that the proteins bound to DNA in the eukaryotic nucleus, especially the histones, might be involved in modifying the properties of DNA. Well before most of the work on DNA methylation began, Stedman and Stedman (1950) proposed that the histones could act as general repressors of gene expression. They argued that since all somatic cells of an organism had the same number of chromosomes, they had the same genetic complement (although this was not demonstrated until some years later, as noted above). Understanding the subtlety of histone modifications was far in the future, so the Stedmans operated on the assumption that different kinds of cells in an organism must have different kinds of histones in order to generate the observed differences in phenotype. Histones can indeed reduce levels of transcript far below those commonly observed for inactive genes in prokaryotes. Subsequent work addressed the capacity of chromatin to serve as a template for transcription, and asked whether that capacity was restricted in a cell-type-specific manner. In a 1963 paper, Bonner (Bonner et al. 1963) prepared chromatin from a globulin-producing tissue of the pea plant, and showed that when *E. coli* RNA polymerase was added, and the resulting transcript translated in an in vitro system, globulin could be detected. The result was specific to this tissue. With the advent of hybridization methods, the transcript populations from such in vitro experiments could be examined (Paul and Gilmour 1968) and shown to be specific for the particular tissue from which the chromatin was derived. Other results suggested that this specificity reflected a restriction in access to transcription initiation sites (Cedar and Felsenfeld 1973). Nonetheless, there was a period in which it was commonly believed that the histones were suppressor proteins that passively silenced gene expression. In this view, activating a gene simply meant stripping off the histones; once that was done, it was thought, transcription would proceed pretty much as it did in prokaryotes. There was, however, some evidence that extended regions of open DNA did not exist in eukaryotic cells (Clark and Felsenfeld 1971). Furthermore, even if the naked DNA model was correct, it was not clear how the decision would be made as to which histone-covered regions should be cleared.

The resolution of this problem began as early as 1964, when Allfrey (Allfrey et al. 1964) had speculated that histone acetylation might be correlated with gene activation, and that "active" chromatin might not necessarily be stripped of histones. In the ensuing decade, there was great interest in examining the relationship between histone modifications and gene expression. Modifications other than acetylation (methylation and phosphorylation) were identified, but their functional significance was unclear. It became much easier to address this problem after the discovery by Kornberg and Thomas (1974) of the structure of the nucleosome, the fundamental chromatin subunit. The determination of the crystal structure of the nucleosome, first at 7 Å and then at 2.8 Å resolution, also provided important structural information, particularly evidence for the extension of the histone amino-terminal tails beyond the DNA–protein octamer core, making evident their accessibility to modification (Richmond et al. 1984; Luger et al. 1997). Beginning in 1980 and extending over some years, Grunstein and his collaborators (Wallis et al. 1980; Durrin et al. 1991), applying yeast genetic analysis, were able to show that the histone amino-terminal tails were essential for regulation of gene expression, and for the establishment of silent chromatin domains.

The ultimate connection to detailed mechanisms began with the critical demonstration by Allis (Brownell et al. 1996) that a histone acetyltransferase from *Tetrahymena* was homologous to yeast transcriptional regulatory protein Gcn5, providing direct evidence that histone acetylation was connected to control of gene expression. Since then, of course, there has been an explosion of discovery of histone modifications, as well as a reevaluation of the roles of those that were known previously.

This still did not answer the question of how the sites for modification were chosen in vivo. It had been shown, for example (Pazin et al. 1994), that Gal4-VP16 could activate transcription from a reconstituted chromatin template in an ATP-dependent manner. Activation was accompanied by repositioning of nucleosomes, and it was suggested that this was the critical event in making the promoter accessible. A fuller understanding of the significance of these findings required the identification of ATP-dependent nucleosome remodeling complexes such as SWI/SNF and NURF (Peterson and Herskowitz 1992; Tsukiyama and Wu 1995), and the realization that both histone modification and nucleosome remodeling were involved in preparing the chromatin template for transcription.

It was not clear how information about the state of activity could, employing these mechanisms, be transmitted through cell division; their role in epigenetic transmission of information was thus unclear. The next important step came from the realization that modified histones recruited, in a modification-specific way, proteins that could affect the local structural and functional

states of chromatin. It was found, for example, that methylation of histone H3 lysine 9 resulted in the recruitment of the heterochromatin protein HP1 (Bannister et al. 2001; Lachner et al. 2001; Nakayama et al. 2001). Furthermore, HP1 could recruit the enzyme (Suv39 h1) that is responsible for that methylation. This led to a model for propagation of the silenced chromatin state along the region through a processive mechanism (Fig. 1a). Equally important, it provided a reasonable explanation of how that state could be transmitted and survive through the replication cycle (Fig. 1b). Analogous mechanisms for propagation of an active state have been proposed that involve methylation of histone H3 lysine 4 and the recruitment of Trithorax group proteins (Wysocka et al. 2005).

Different kinds of propagation mechanisms have been suggested that depend on variant histones rather than modified histones (Ahmad and Henikoff 2002; McKittrick et al. 2004). Histone H3 is incorporated into chromatin only during DNA replication. In contrast, the histone variant H3.3, which differs from H3 by four amino acids, is incorporated into nucleosomes in a replication-independent manner, and it tends to accumulate in active chromatin, where it is enriched in the "active" histone modifications (McKittrick et al. 2004). It has been proposed that the presence of H3.3 is sufficient to maintain the active state, and that after replication, although it would be diluted twofold, enough H3.3 would remain to maintain the active state. The consequent transcription would result in replacement of H3 containing nucleosomes with H3.3, thus perpetuating the active state in the next generation.

6 All Mechanisms Are Interrelated

These models finally begin to complete the connection between modified or variant histones, specific gene activation, and epigenetics, although of course there is much more to be done. Whereas these mechanisms give us some ideas about how the heterochromatic state may be maintained, they do not explain how silencing chromatin structures are first established. It has only recently become clear that this involves the production of RNA transcripts, particularly from repeated sequences, which are processed into small RNAs through the action of proteins such as Dicer, Argonaute, and RNA-dependent RNA polymerase. These RNAs are subsequently recruited to the homologous DNA sites as part of complexes that include components of the Polycomb group of proteins, thus initiating the formation of heterochromatin. There is now also evi-

a

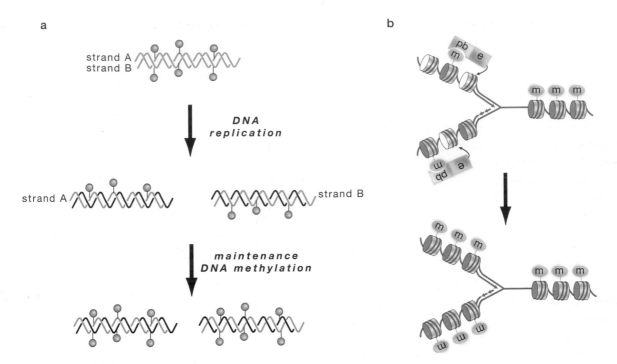

b

Figure 1. Mechanisms for Maintaining a Pattern of DNA Methylation and a Histone Modification during DNA Replication

(*a*) A mechanism for maintaining a pattern of DNA methylation during DNA replication. During replication, the individual DNA strands, with a specific methylation pattern at CpG or CpXpG residues, become paired with a strand of newly synthesized, unmethylated DNA. CpG on one strand has a corresponding CpG on the other. The maintenance DNA methyltransferase recognizes a hemimethylated site, and methylates the cytosine on the new strand, so that the pattern of methylation is undisturbed. (*b*) A general mechanism for maintaining a histone modification during replication. The modified histone tail (m) interacts with a protein binder (pb) that has a binding site specific for that modification. pb, in turn, has a specific site for the enzyme (e) which carries out that histone modification. e, in turn, can then modify an adjacent nucleosome. During replication, the newly deposited histones which are interspersed with parental histones can thus acquire the parental modification. A similar mechanism would allow propagation of histone modifications from a modified region into an unmodified one at any stage of the cell cycle.

dence that the same mechanisms are required for maintenance of at least some heterochromatic regions. In a way, these stable cyclic reaction pathways are reminiscent of Delbruck's 50-year-old model, of a stable biochemical cycle that maintains the state of the organism.

We now know of countless examples of epigenetic mechanisms at work in the organism. In addition to imprinting at many loci, and the allele-specific and random X-chromosome inactivation described above, there are epigenetic phenomena involved in antibody expression, where the rearrangement of the immunoglobulin genes on one chromosome is selectively inhibited, and in the selection for expression of single odorant receptor genes in olfactory neurons (Chess et al. 1994; Shykind et al. 2004). In *Drosophila*, the Polycomb group genes are responsible for establishing a silenced chromatin domain that is maintained through all subsequent cell divisions.

Epigenetic changes are also responsible for paramutation in plants, in which one allele can cause a heritable change in expression of the homologous allele (Stam et al. 2002). This is an example of an epigenetic state that is inherited meiotically as well as mitotically, a phenomenon documented in plants but only rarely in animals (Jorgensen 1993). Much of the evidence for the mechanisms described above has come from work on the silencing of mating-type locus and centromeric sequences in *Schizosaccharomyces pombe* (Hall et al. 2002). In addition, the condensed chromatin structure characteristic of centromeres in organisms as diverse as flies and humans has been shown to be transmissible through centromere-associated proteins rather than DNA sequence. In all of these cases, the DNA sequence remains intact, but its capacity for expression is suppressed. This is likely in all cases to be mediated by DNA methylation, histone modification, or both; in some cases, we already know that to be true. Finally, the epigenetic transmission of "patterns," described above for *Paramecia*, now extends to the prion proteins, which maintain and propagate their alternatively folded state to daughter cells.

Although this has been presented as a sequential story, it should more properly be viewed as a series of parallel and overlapping attempts to define and explain epigenetic phenomena. The definition of the term epigenetics has changed, but the questions about mechanisms of development raised by earlier generations of scientists have not. Contemporary epigenetics still addresses those central questions. Seventy years have passed since Muller described what is now called position-effect variegation. It is gratifying to trace the slow progress from observation of phenotypes, through elegant genetic studies, to the recent analysis and resolution at the molecular level. With this knowledge has come the understanding that epigenetic mechanisms may in fact be responsible for a considerable part of the phenotype of complex organisms. As is often the case, an observation that at first seemed interesting but perhaps marginal to the main issues turns out to be central, although it may take a long time to come to that realization.

References

Ahmad K. and Henikoff S. 2002. The histone variant H3.3 marks active chromatin by replication-independent nucleosome assembly. *Mol. Cell* **9:** 1191–1200.

Allfrey V.G., Faulkner R., and Mirsky A.E. 1964. Acetylation and methylation of histones and their possible role in the regulation of RNA synthesis. *Proc. Natl. Acad. Sci.* **51:** 786–794.

Avery O.T., MacLeod C.M., and McCarty M. 1944. Studies on the chemical nature of the substance inducing transformation of pneumococcal types. *J. Exp. Med.* **79:** 137–158.

Bannister A., Zegerman P., Partridge J., Miska E., Thomas J., Allshire R., and Kouzarides T. 2001. Selective recognition of methylated lysine 9 on histone H3 by the HP1 chromo domain. *Nature* **410:** 120–124.

Beisson J. and Sonneborn T.M. 1965. Cytoplasmic inheritance of the organization of the cell cortex in *Paramecium aurelia*. *Proc. Natl. Acad. Sci.* **53:** 275–282.

Bell A.C. and Felsenfeld G. 2000. Methylation of a CTCF-dependent boundary controls imprinted expression of the Igf2 gene. *Nature* **405:** 482–485.

Bird A.P. 1978. Use of restriction enzymes to study eukaryotic DNA methylation. II. The symmetry of methylated sites supports semiconservative copying of the methylation pattern. *J. Mol. Biol.* **118:** 49–60.

Bird A.P. and Southern E.M. 1978. Use of restriction enzymes to study eukaryotic DNA methylation. I. The methylation pattern in ribosomal DNA from *Xenopus laevis*. *J. Mol. Biol.* **118:** 27–47.

Bonner J., Huang R.C., and Gilden R.V. 1963. Chromosomally directed protein synthesis. *Proc. Natl. Acad. Sci.* **50:** 893–900.

Briggs R. and King T.J. 1952. Transplantation of living nuclei from blastula cells into enucleated frogs' eggs. *Proc. Natl. Acad. Sci.* **38:** 455–463.

Brownell J.E., Zhou J., Ranalli T., Kobayashi R., Edmondson D.G., Roth S.Y., and Allis C.D. 1996. Tetrahymena histone acetyltransferase A: A homolog to yeast Gcn5p linking histone acetylation to gene activation. *Cell* **84:** 843–851.

Cattanach B.M. and Kirk M. 1985. Differential activity of maternally and paternally derived chromosome regions in mice. *Nature* **315:** 496–498.

Cedar H. and Felsenfeld G. 1973. Transcription of chromatin in vitro. *J. Mol. Biol.* **77:** 237–254.

Chess A., Simon I., Cedar H., and Axel R. 1994. Allelic inactivation regulates olfactory receptor gene expression. *Cell* **78:** 823–834.

Clark R.J. and Felsenfeld G. 1971. Structure of chromatin. *Nat. New Biol.* **229:** 101–106.

Doskocil J. and Sorm F. 1962. Distribution of 5-methylcytosine in pyrimidine sequences of deoxyribonucleic acids. *Biochim. Biophys. Acta* **55:** 953–959.

Durrin L.K., Mann R.K., Kayne P.S., and Grunstein M. 1991. Yeast histone H4 N-terminal sequence is required for promoter activation in vivo. *Cell* **65:** 1023–1031.

Grimes G.W. and Aufderheide K.J. 1991. Cellular aspects of pattern formation: The problem of assembly. *Monogr. Dev. Biol.* **22:** 1–94.

Hall I.M., Shankaranarayana G.D., Noma K., Ayoub N., Cohen A., and Grewal S.I. 2002. Establishment and maintenance of a heterochromatin domain. *Science* **297:** 2215–2218.

Hannah A. 1951. Localization and function of heterochromatin in *Drosophila melanogaster*. *Adv. Genet.* **4:** 87–125.

Hark A.T., Schoenherr C.J., Katz D.J., Ingram R.S., Levorse J.M., and Tilghman S.M. 2000. CTCF mediates methylation-sensitive enhancer-blocking activity at the H19/Igf2 locus. *Nature* **405:** 486–489.

Hershey A.D. and Chase M. 1952. Independent functions of viral protein and nucleic acid in growth of bacteriophage. *J. Gen. Physiol.* **36:** 39–56.

Holliday R. and Pugh J.E. 1975. DNA modification mechanisms and gene activity during development. *Science* **187:** 226–232.

Jablonka E. and Lamb M.J. 1995. *Epigenetic inheritance and evolution: The Lamarckian dimension*. Oxford University Press, New York, p. 82.

Jorgensen R. 1993. The germinal inheritance of epigenetic information in plants. *Philos. Trans. R. Soc. Lond. B Biol. Sci.* **339:** 173–181.

Kornberg R.D. and Thomas J.O. 1974. Chromatin structure; oligomers of the histones. *Science* **184:** 865–868.

Lachner M., O'Carroll D., Rea S., Mechtler K., and Jenuwein T. 2001. Methylation of histone H3 lysine 9 creates a binding site for HP1 proteins. *Nature* **410:** 116–120.

Laskey R.A. and Gurdon J.B. 1970. Genetic content of adult somatic cells tested by nuclear transplantation from cultured cells. *Nature* **228:** 1332–1334.

Luger K., Mader A.W., Richmond R.K., Sargent D.F., and Richmond T.J. 1997. Crystal structure of the nucleosome core particle at 2.8 Å resolution. *Nature* **389:** 251–260.

Lyon M.F. 1961. Gene action in the X-chromosome of the mouse. *Nature* **190:** 372–373.

McClintock B. 1965. The control of gene action in maize. *Brookhaven Symp. Biol.* **18:** 162–184.

McKittrick E., Gafken P.R., Ahmad K., and Henikoff S. 2004. Histone H3.3 is enriched in covalent modifications associated with active chromatin. *Proc. Natl. Acad. Sci.* **101:** 1525–1530.

Morgan T. 1911. An attempt to analyze the constitution of the chromosomes on the basis of sex-linked inheritance in *Drosophila. J. Exp. Zool.* **11:** 365–414.

Muller H.J. 1930. Types of visible variations induced by X-rays in *Drosophila. J. Genet.* **22:** 299–334.

Nakayama J., Rice J.C., Strahl B.D., Allis C.D., and Grewal S.I. 2001. Role of histone H3 lysine 9 methylation in epigenetic control of heterochromatin assembly. *Science* **292:** 110–113.

Novick A. and Weiner M. 1957. Enzyme induction as an all-or-none phenomenon. *Proc. Natl. Acad. Sci.* **43:** 553–566.

Ohno S., Kaplan W.D., and Kinosita R. 1959. Formation of the sex chromatin by a single X-chromosome in liver cells of *Rattus norvegicus. Exp. Cell Res.* **18:** 415–418.

Paul J. and Gilmour R.S. 1968. Organ-specific restriction of transcription in mammalian chromatin. *J. Mol. Biol.* **34:** 305–316.

Pazin M.J., Kamakaka R.T., and Kadonaga J.T. 1994. ATP-dependent nucleosome reconfiguration and transcriptional activation from preassembled chromatin templates. *Science* **266:** 2007–2011.

Peterson C.L. and Herskowitz I. 1992. Characterization of the yeast SWI1, SWI2, and SWI3 genes, which encode a global activator of transcription. *Cell* **68:** 573–583.

Ptashne M. 1992. *A genetic switch: Phage λ and higher organisms*, 2nd edition. Blackwell Science, Malden, Massachusetts and Cell Press, Cambridge, Massachusetts.

Richmond T.J., Finch J.T., Rushton B., Rhodes D., and Klug A. 1984. Structure of the nucleosome core particle at 7 Å resolution. *Nature* **311:** 532–537.

Riggs A.D. 1975. X inactivation, differentiation, and DNA methylation. *Cytogenet. Cell Genet.* **14:** 9–25.

Riggs A.D. and Porter T.N. 1996. Overview of epigenetic mechanisms. In *Epigenetic mechanisms of gene regulation* (ed. V.E.A. Russo et al.), pp. 29–45. Cold Spring Harbor Laboratory Press, Cold Spring Harbor, New York.

Riggs A.D., Martienssen R.A., and Russo V.E.A. 1996. Introduction. In *Epigenetic mechanisms of gene regulation* (ed. V.E.A. Russo et al.), pp. 1–4. Cold Spring Harbor Laboratory Press, Cold Spring Harbor, New York.

Shykind B.M., Rohani S.C., O'Donnell S., Nemes A., Mendelsohn M., Sun Y., Axel R., and Barnea G. 2004. Gene switching and the stability of odorant receptor gene choice. *Cell* **117:** 801–815.

Stam M., Belele C., Dorweiler J., and Chandler V. 2002. Differential chromatin structure with a tandem array 100 kb upstream of the maize b1 locus is associated with paramutation. *Genes Dev.* **16:** 1906–1918.

Stedman E. and Stedman E. 1950. Cell specificity of histones. *Nature* **166:** 780–781.

Sturtevant A. 1913. The linear arrangement of six sex-linked factors in *Drosophila*, as shown by their mode of association. *J. Exp. Zool.* **14:** 43–59.

Tsukiyama T. and Wu C. 1995. Purification and properties of an ATP-dependent nucleosome remodeling factor. *Cell* **83:** 1011–1020.

Waddington C.H. 1953. Epigenetics and evolution. *Symp. Soc. Exp. Biol.* **7:** 186–199.

Wallis J.W., Hereford L., and Grunstein M. 1980. Histone H2B genes of yeast encode two different proteins. *Cell* **22:** 799–805.

Wysocka J., Swigut T., Milne T. Dou Y., Zhang X., Burlingame A., Roeder R., Brivanlou A., and Allis C.D. 2005. WDR5 associates with histone H3 methylated at K4 and is essential for H3 K4 methylation and vertebrate development. *Cell* **121:** 859–872.

Overview and Concepts

C. David Allis,[1] Thomas Jenuwein,[2] and Danny Reinberg[3]

[1]The Rockefeller University, New York, New York; [2]Research Institute of Molecular Pathology, Vienna, Austria; [3]UMDNJ-Robert Wood Johnson Medical School, Piscataway, New Jersey

CONTENTS

1. Genetics Versus Epigenetics, 25

2. Model Systems for the Study of Epigenetics, 26

3. Defining Epigenetics, 28

4. The Chromatin Template, 29

5. Higher-Order Chromatin Organization, 31

6. The Distinction between Euchromatin and Heterochromatin, 34

7. Histone Modifications and the Histone Code, 36

8. Chromatin-remodeling Complexes and Histone Variants, 39

9. DNA Methylation, 41

10. RNAi and RNA-directed Gene Silencing, 42

11. From Unicellular to Multicellular Systems, 44

12. Polycomb and Trithorax, 45

13. X Inactivation and Facultative Heterochromatin, 47

14. Reprogramming of Cell Fates, 49

15. Cancer, 50

16. What Does Epigenetic Control Actually Do? , 52

17. Big Questions in Epigenetic Research, 55

References, 56

GENERAL SUMMARY

The DNA sequencing of the human genome and the genomes of many model organisms has generated considerable excitement within the biomedical community and the general public over the past several years. These genetic "blueprints" that exhibit the well-accepted rules of Mendelian inheritance are now readily available for close inspection, opening the door to improved understanding of human biology and disease. This knowledge is also generating renewed hope for novel therapeutic strategies and treatments. Many fundamental questions nonetheless remain. For example, how does normal development proceed, given that every cell has the same genetic information, yet follows a different developmental pathway, realized with exact temporal and spatial precision? How does a cell decide when to divide and differentiate, or when to retain an unchanged cellular identity, responding and expressing according to its normal developmental program? Mistakes made in the above processes can lead to the generation of disease states such as cancer. Are these mistakes encoded in faulty genetic blueprints that we inherited from one or both of our parents, or are there other layers of regulatory information that are not being properly read and decoded?

In humans, the genetic information (DNA) is organized into 23 chromosome pairs consisting of approximately 25,000 genes. These chromosomes can be compared to libraries with different sets of books that together instruct the development of a complete human being. The DNA sequence of our genome is composed of about 3×10^9 bases, abbreviated by the four letters (or bases) A, C, G, and T within its sequence, giving rise to well-defined words (genes), sentences, chapters, and books. However, what dictates when the different books are read, and in what order, remains far from clear. Meeting this extraordinary challenge is likely to reveal insights into how cellular events are coordinated during normal and abnormal development.

When summed across all chromosomes, the DNA molecule in higher eukaryotes is about 2 meters long and therefore needs to be maximally condensed about 10,000-fold to fit into a cell's nucleus, the compartment of a cell that stores our genetic material. The wrapping of DNA around "spools" of proteins, so-called histone proteins, provides an elegant solution to this packaging problem, giving rise to a repeating protein:DNA polymer known as chromatin. However, in packaging DNA to better fit into a confined space, a problem develops, much as

when one packs too many books onto library shelves: It becomes harder to find and read the book of choice, and thus, an indexing system is needed. Chromatin, as a genome-organizing platform, provides this indexing. Chromatin is not uniform in structure; it comes in different packaging designs from a highly condensed chromatin fiber (known as heterochromatin) to a less compacted type where genes are typically expressed (known as euchromatin). Variation can enter into the basic chromatin polymer through the introduction of unusual histone proteins (known as histone variants), altered chromatin structures (known as chromatin remodeling), and the addition of chemical flags to the histone proteins themselves (known as covalent modifications). Moreover, addition of a methyl group directly to a cytosine (C) base in the DNA template (known as DNA methylation) can provide docking sites for proteins to alter the chromatin state or affect the covalent modification of resident histones. Recent evidence suggests that noncoding RNAs can "guide" specialized regions of the genome into more compacted chromatin states. Thus, chromatin should be viewed as a dynamic polymer that can index the genome and potentiate signals from the environment, ultimately determining which genes are expressed and which are not.

Together, these regulatory options provide chromatin with an organizing principle for genomes known as "epigenetics," the subject of this book. In some cases, epigenetic indexing patterns appear to be inherited through cell divisions, providing cellular "memory" that may extend the heritable information potential of the genetic (DNA) code. Epigenetics can thus be narrowly defined as changes in gene transcription through modulation of chromatin, which is not brought about by changes in the DNA sequence.

In this overview, we explain the basic concepts of chromatin and epigenetics, and we discuss how epigenetic control may give us the clues to solve some long-standing mysteries, such as cellular identity, tumorigenesis, stem cell plasticity, regeneration, and aging. As readers comb through the chapters that follow, we encourage them to note the wide range of biological phenomena uncovered in a diverse range of experimental models that seem to have an epigenetic (non-DNA) basis. Understanding how epigenetics operates in mechanistic terms will likely have important and far-reaching implications for human biology and human disease in this "post-genomic" era.

1 Genetics Versus Epigenetics

Determining the structural details of the DNA double helix stands as one of the landmark discoveries in all of biology. DNA is the prime macromolecule that stores genetic information (Avery et al. 1944), and it propagates this stored information to the next generation through the germ line. From this and other findings, the "central dogma" of modern biology emerged. This dogma encapsulates the processes involved in maintaining and translating the genetic template required for life. The essential stages are (1) the self-propagation of DNA by semiconservative replication; (2) transcription in a unidirectional 5′ to 3′ direction, templated by the genetic code (DNA), generation of an intermediary messenger RNA (mRNA); (3) translation of mRNA to produce polypeptides consisting of linear amino to carboxyl strings of amino acids that are colinear with the 5′ to 3′ order of DNA. In simple terms: $DNA \leftrightarrow RNA \rightarrow protein$. The central dogma accommodates feedback from RNA to DNA by the process of reverse transcription, followed by integration into existing DNA (as demonstrated by retroviruses and retrotransposons). However, this dogma disavows feedback from protein to DNA, although a new twist to the genetic dogma is that rare proteins, known as prions, can be inherited in the absence of a DNA or RNA template. Thus, these specialized self-aggregating proteins have properties that resemble some properties of DNA itself, including a mechanism for replication and information storage (Cohen and Prusiner 1998; Shorter and Lindquist 2005). Additionally, emerging evidence suggests that a remarkably large fraction of our genome is transcribed into "noncoding" RNAs. The function of these noncoding RNAs (i.e., non-protein-encoding except tRNAs, rRNAs, snoRNAs) is under active investigation and is only beginning to become clear in a limited number of cases.

The origin of epigenetics stems from long-standing studies of seemingly anomalous (i.e., non-Mendelian) and disparate patterns of inheritance in many organisms (see Chapters 1 and 2 for a historical overview). Classic Mendelian inheritance of phenotypic traits (e.g., pea color, number of digits, or hemoglobin insufficiency) results from allelic differences caused by mutations of the DNA sequence. Collectively, mutations underlie the definition of phenotypic traits, which contributes to the determination of species boundaries. These boundaries are then shaped by the pressures of natural selection, as explained by Darwin's theory of evolution. Such concepts place mutations at the heart of classic genetics. In contrast, non-Mendelian inheritance (e.g., variation of embryonic growth, mosaic skin coloring, random X inactivation, plant paramutation) (Fig. 1) can manifest, to take one example, from the expression of only one (of two) alleles within the same nuclear environment. Importantly, in these circumstances, the DNA sequence is not altered. This is distinct from another commonly referred to non-Mendelian inheritance pattern that arises from the maternal inheritance of mitochondria (Birky 2001).

The challenge for epigenetic research is captured by the selective regulation of one allele within a nucleus. What distinguishes two identical alleles, and how is this distinction mechanistically established and maintained through successive cell generations? What underlies differences observed in monozygotic ("identical") twins that make them not totally identical? Epigenetics is sometimes cited as one explanation for the differences in outward traits, by translating the influence of the environment, diet, and potentially other external sources to the expression of the genome (Klar 2004; see Chapters 23 and 24). Determining what components are affected at a molecular level, and how alterations in these components affect human biology and human disease, is a major challenge for future studies.

Another key question in the field is, How important is the contribution of epigenetic information for normal development? How do normal pathways become dysfunctional, leading to abnormal development and neoplastic transformation (i.e., cancer)? As mentioned above, "identical" twins share the same DNA sequence, and as such, their phenotypic identity is often used to underscore the defining power of genetics. However, even twins such as these can exhibit outward phenotypic differences, likely imparted by epigenetic modifications that occur over the lifetime of the individuals (Fraga et al. 2005). Thus, the extent to which epigenetics is important in defining cell fate, identity, and phenotype remains to be fully understood. In the case of tissue regeneration and aging, it remains unclear whether these processes are dictated by alterations in the genetic program of cells or by epigenetic modifications. The intensity of research on a global scale testifies to the recognition that the field of epigenetics is a critical new frontier in this post-genomic era.

In the words of others, "We are more than the sum of our genes" (Klar 1998), or "You can inherit something beyond the DNA sequence. That's where the real excitement in genetics is now" (Watson 2003). The overriding motivation for deciding to edit this book was the general belief that we and all the contributors to this volume could transmit this excitement to future generations of students, scientists, and physicians, most of whom were taught genetic, but not epigenetic, principles governing inheritance and chromosome segregation.

twins

Barr body

polytene
chromosomes

cloned cat

epigenetic
biology

yeast mating types

mutant plant

tumor tissue

blood smear

Figure 1. Biological Examples of Epigenetic Phenotypes

Epigenetic phenotypes in a range of organisms and cell types, all attributable to *non-genetic* differences. *Twins:* Slight variations partially attributable to epigenetics (©Randy Harris, New York). *Barr body:* The epigenetically silenced X chromosome in female mammalian cells, visible cytologically as condensed heterochromatin. *Polytene chromosomes:* Giant chromosomes in *Drosophila* salivary glands, ideally suited for correlating genes with epigenetic marks (reprinted from Schotta et al. 2003 [©Springer]). *Yeast mating type:* Sex is determined by the active MAT locus, while copies of both mating-type genes are epigenetically silenced (©Alan Wheals, University of Bath). *Blood smear:* Heterogeneous cells of the same genotype, but epigenetically determined to serve different functions (courtesy Prof. Christian Sillaber). *Tumor tissue:* Metastatic cells (*left*) showing elevated levels of epigenetic marks in the tissue section (reprinted, with permission, from Seligson et al. 2005 [©Macmillan]). *Mutant plant: Arabidopsis* flower epiphenotypes, genetically identical, with epigenetically caused mutations (reprinted, with permission, from Jackson et al. 2002 [©Macmillan]). *Cloned cat:* Genetically identical, but with varying coat-color phenotype (reprinted, with permission, from Shin et al. 2002 [©Macmillan]).

2 Model Systems for the Study of Epigenetics

The study of epigenetics necessarily requires good experimental models, and as often is the case, these models seem at first sight far removed from studies using human (or mammalian) cells. Collectively, however, results from many systems have yielded a wealth of knowledge. The historical overviews (Chapters 1 and 2) make reference to several important landmark discoveries that have emerged from early cytology, the growth of genetics, the birth of molecular biology, and relatively new advances in chromatin-mediated gene regulation. Different model organisms (Fig. 2) have been pivotal in addressing and solving the various questions raised by epigenetic research. Indeed, seemingly disparate epigenetic discoveries made in various model organisms have served to unite the research community. The purpose of this section is to highlight some of these major findings, which are discussed in more detail in the following chapters of this book. As readers note these discoveries, they should focus on the fundamental principles that investigations using these model systems have exposed; their collective contributions point more often to common concepts than to diverging details.

Unicellular and "lower" eukaryotic organisms—*Saccharomyces cerevisiae, Schizosaccharomyces pombe,* and *Neurospora crassa*—permit powerful genetic analyses, in part facilitated by a short life cycle. Mating-type (MAT) switching that occurs in *S. cerevisiae* (Chapter 3) and *S. pombe* (Chapter 6) has provided remarkably instructive examples, demonstrating the importance of chromatin-mediated gene control. In the budding yeast *S. cerevisiae,* the unique silent information regulator (SIR) proteins were shown to engage specific modified histones. This was preceded by elegant experiments using genetics to document the active participation of histone proteins in gene regulation (Clark-Adams et al. 1988; Kayne et al. 1988). In the fission yeast *S. pombe,* the patterns of histone modification operating as activating and repressing signals are remarkably similar to those in metazoan organisms. This has opened the door for powerful genetic screens being employed to look for gene products that suppress or enhance the silencing of genes. Most recently, a wealth of mechanistic insights linking the RNA interference (RNAi) machinery to the induction of histone modifications acting to repress gene expression was discovered in fission yeast (Hall et al. 2002; Volpe et al. 2002). Shortly afterward, the RNAi machinery was also implicated in transcriptional gene silencing in the plant *Arabidopsis thaliana,* underscoring the potential importance of this regulation in a wide range of organisms (see Section 10).

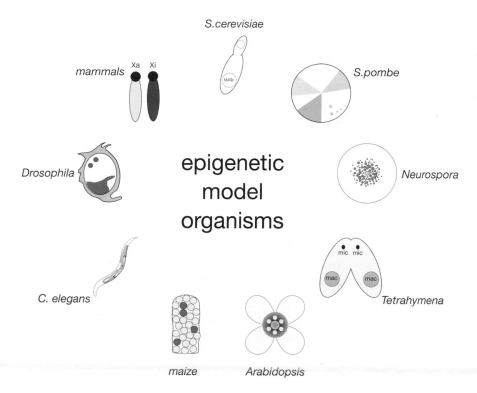

mammals

S.cerevisiae

S.pombe

Drosophila

epigenetic
model
organisms

Neurospora

C. elegans

Tetrahymena

maize   Arabidopsis

Figure 2. Model Organisms Used in Epigenetic Research

Schematic representation of model organisms used in epigenetic research. *S. cerevisiae*: Mating-type switching to study epigenetic chromatin control. *S. pombe:* Variegated gene silencing manifests as colony sectoring. *Neurospora crassa:* Epigenetic genome defense systems include repeat-induced point mutation, quelling, and meiotic silencing of unpaired DNA, revealing an interplay between RNAi pathways, DNA and histone methylation. *Tetrahymena:* Chromatin in somatic and germ-line nuclei are distinguished by epigenetically regulated mechanisms. *Arabidopsis:* Model for repression by DNA, histone, and RNA-guided silencing mechanisms. *Maize:* Model for imprinting, paramutation, and transposon-induced gene silencing. *C. elegans:* Epigenetic regulation in the germ line. *Drosophila:* Position-effect variegation (PEV) manifest by clonal patches of expression and silencing of the white gene in the eye. *Mammals:* X-chromosome inactivation.

Other "off-beat" organisms have also made disproportionate contributions toward unraveling epigenetic pathways that at first seemed peculiar. The fungal species, *N. crassa*, revealed the unusual non-Mendelian phenomenon of repeat induced point mutation (RIP) as a model for studying epigenetic control (Chapter 6). Later, this organism was used to demonstrate the first functional connection between histone modifications and DNA methylation (Tamaru and Selker 2001), a finding later extended to "higher" organisms (Jackson et al. 2002). Ciliated protozoa, such as *Tetrahymena* and *Paramecium*, commonly used in biology laboratories as convenient microscopy specimens, facilitated important epigenetic discoveries because of their unique nuclear dimorphism. Each cell carries two nuclei: a somatic macronucleus that is transcriptionally active, and a germ-line micronucleus that is transcriptionally inactive. Using macronuclei as an enriched starting source of "active" chromatin, the biochemical purification of the first nuclear histone-modifying enzyme—a histone acetyltransferase or HAT—was made (Brownell et al. 1996). Ciliates are also well known for their peculiar phenomenon of programmed DNA elimination during their sexual life cycle, triggered by small noncoding RNAs and histone modifications (Chapter 7).

In multicellular organisms, genome size and organismal complexity generally increase from invertebrate (*Caenorhabditis elegans*, *Drosophila melanogaster*) or plant (*A. thaliana*) species to "higher," and to some, "more relevant," vertebrate organisms (mammals). Plants, however, have been pivotal to the field of epigenetics, providing a particularly rich source of epigenetic discoveries (Chapter 9) ranging from transposable elements and paramutation (McClintock 1951) to the first description of noncoding RNAs involved in transcriptional silencing (Ratcliff et al. 1997). Crucial links between DNA methylation, histone modification, and components of the RNAi machinery came through plant studies. The discovery of plant epialleles, with comic names such as SUPERMAN and KRYPTONITE (e.g., Jackson et al. 2002), and several vernalizing genes (Bastow et al. 2004; Sung and Amasino 2004) have further provided the research field with insights into understanding the developmental role of epigenetics and cellular memory. Plant meristem cells have also offered the opportunity to study crucial questions such as somatic regeneration and stem cell plasticity (see Chapters 9 and 11).

For understanding animal development, *Drosophila* has been an early and continuous genetic powerhouse. Based on the pioneering work of Muller (1930), many developmental mutations were generated, including the homeotic transformations and position-effect variegation (PEV) mutants explained below (also see Chapter 5). The homeotic transformation mutants led to the idea that there could be regulatory mechanisms for establishing and maintaining cellular identity/memory which was later shown to be regulated by the Polycomb and trithorax systems (see Chapters 11 and 12). For PEV, gene activity is dictated by the surrounding chromatin structure and not by primary DNA sequence. This system has been a particularly informative source for dissecting factors involved in epigenetic control (Chapter 5). Over 100 suppressors of variegation [*Su(var)*] genes are believed to encode components of heterochromatin. Without the foundation established by these landmark studies, the discovery of the first histone lysine methyltransferases (HKMTs) (Rea et al. 2000) and the resultant advances in histone lysine methylation would not have been possible. As is often the case in biology, comparable screens have been carried out in fission yeast and in plants, identifying silencing mutants with functional conservation with the *Drosophila Su(var)* genes.

The use of reverse genetics via RNAi libraries in the nematode worm *C. elegans* has contributed to our understanding of epigenetic regulation in metazoan development. There, comprehensive cell-fate tracking studies, detailing all the developmental pathways of each cell, have highlighted the fact that Polycomb and trithorax systems probably arose with the emergence of multicellularity (see Sections 12 and 13). In particular, these mechanisms of epigenetic control are essential for gene regulation in the germ line (see Chapter 15).

The role of epigenetics in mammalian development has mostly been elucidated in the mouse, although a number of studies have been translated to diverse human cell lines and primary cell cultures. The advent of gene "knock-out" and "knock-in" technologies has been instrumental for the functional dissection of key epigenetic regulators. For instance, the Dnmt1 DNA methyltransferase mutant mouse provided functional insight for the role of DNA methylation in mammals (Li et al. 1992). It is embryonic-lethal and shows impaired imprinting (see Chapter 18). Disruption of DNA methylation has also been shown to cause genomic instability and reanimation of transposon activity, particularly in germ cells (Walsh et al. 1998; Bourc'his and Bestor 2004). There are approximately 100 characterized chromatin-regulating factors (i.e., histone and DNA-modifying enzymes, components of nucleosome remodeling complexes and of the RNAi machinery) that have been disrupted in the mouse. The mutant phenotypes affect cell proliferation, lineage commitment, stem cell plasticity, genomic stability, DNA repair, and chromosome segregation processes, in both somatic and germ cell lineages. Not surprisingly, most of these mutants are also involved in disease development and cancer. Thus, many of the key advances in epigenetic

control took advantage of unique biological features exhibited by many, if not all, of the above-mentioned model organisms. Without these biological processes and the functional analyses (genetic and biochemical) that delved into them, many of the recent advances in epigenetic control would have remained elusive.

3 Defining Epigenetics

The above discussion begs the question, What is the common thread that allows diverse eukaryotic organisms to be connected with respect to fundamental epigenetic principles? Different epigenetic phenomena are linked largely by the fact that DNA is not "naked" in all organisms that maintain a true nucleus (eukaryotes). Instead, the DNA exists as an intimate complex with specialized proteins, which together comprise chromatin. In its simplest form, chromatin—i.e., DNA spooled around nucleosomal units consisting of small histone proteins (Kornberg 1974)—was initially regarded as a passive packaging molecule to wrap and organize the DNA. Distinctive forms of chromatin arise, however, through an array of covalent and non-covalent mechanisms that are being uncovered at a rapid pace (see Section 6). This includes a plethora of posttranslational histone modifications, energy-dependent chromatin-remodeling steps that mobilize or alter nucleosome structures, the dynamic shuffling of new histones (variants) in and out of nucleosomes, and the targeting role of small noncoding RNAs. DNA itself can also be modified covalently in many higher eukaryotes, by methylation at the cytosine residue, usually but not always, of CpG dinucleotides. Together, these mechanisms provide a set of interrelated pathways that all create variation in the chromatin polymer (Fig. 3).

Many, but not all, of these modifications and chromatin changes are reversible and, therefore, are unlikely to be propagated through the germ line. Transitory marks are attractive because they impose changes to the chromatin template in response to intrinsic and external stimuli (Jaenisch and Bird 2003), and in so doing, regulate the access and/or processivity of the transcriptional machinery, needed to "read" the underlying DNA template (Sims et al. 2004; Chapter 10). Some histone modifications (like lysine methylation), methylated DNA regions, and altered nucleosome structures can, however, be stable through several cell divisions. This establishes "epigenetic states" or means of achieving cellular memory, which remain poorly appreciated or understood. From this perspective, chromatin "signatures" can be viewed as a highly organized system of information storage that can index distinct regions

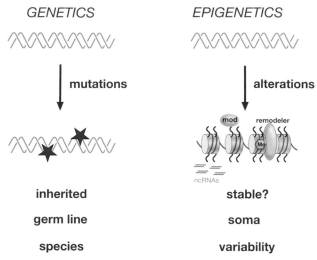

Figure 3. Genetics Versus Epigenetics

GENETICS: Mutations (*red stars*) of the DNA template (*green helix*) are heritable somatically and through the germ line. *EPIGENETICS:* Variations in chromatin structure modulate the use of the genome by (1) histone modifications (mod), (2) chromatin remodeling (*remodeler*), (3) histone variant composition (*yellow nucleosome*), (4) DNA methylation (*Me*), and (5) noncoding RNAs. Marks on the chromatin template may be heritable through cell division and collectively contribute to determining cellular phenotype.

of the genome and accommodate a response to environmental signals that dictate gene expression programs.

The significance of having a chromatin template that can potentiate the genetic information is that it provides multidimensional layers to the readout of DNA. This is perhaps a necessity, given the vast size and complexity of the eukaryotic genome, particularly for multicellular organisms (see Section 11 for further details). In such organisms, a fertilized egg progresses through development, starting with a single genome that becomes epigenetically programmed to generate a multitude of distinct "epigenomes" in more than 200 different types of cells (Fig. 4). This programmed variation has been proposed to constitute an "epigenetic code" that significantly extends the information potential of the genetic code (Strahl and Allis 2000; Turner 2000; Jenuwein and Allis 2001). Although this is an attractive hypothesis, we stress that more work is needed to test this and related provocative theories. Other alternative viewpoints are being advanced which argue that clear combinatorial "codes," like the triplet genetic code, are not likely in histones or are far from established (Schreiber and Bernstein 2002; Henikoff 2005). Despite these uncertainties, we favor the general view that a combination of covalent and non-covalent mechanisms will act to create chromatin states that can be templated through cell division and development by mechanisms that are just beginning to be defined. Exactly how these altered chromatin states are faithfully propagated during DNA replication and mitosis remains one of the fundamental challenges of future studies.

The phenotypic alterations that occur from cell to cell during the course of development in a multicellular organism were described by Waddington as the "epigenetic landscape" (Waddington 1957). Yet the spectrum of cells, from stem cells to fully differentiated cells, all share identical DNA sequences but differ remarkably in the profile of genes that they actually express. With this knowledge, epigenetics later came to be defined as the "Nuclear inheritance which is not based on differences in DNA sequence" (Holliday 1994).

Since the discovery of the DNA double helix and the early explanations of epigenetics, our understanding of epigenetic control and its underlying mechanisms has greatly increased, causing some to describe it in more lofty terms as a "field" rather than just "phenomena" (see Wolffe and Matzke 1999; Roloff and Nuber 2005; Chapter 1). In the past decade, considerable progress has been gained regarding the many enzyme families that actively modify chromatin (see below). Thus, in today's modern terms, epigenetics can be molecularly (mechanistically) defined as "The sum of the alterations to the chromatin template that collectively establish and propagate different patterns of gene expression (transcription) and silencing from the same genome."

4 The Chromatin Template

The nucleosome is the fundamental repeating unit of chromatin (Kornberg 1974). On the one hand, the basic chromatin unit consists of a protein octamer containing two molecules of each canonical (or core) histone (H2A, H2B, H3, and H4), around which is wrapped 147 bp of DNA. Detailed intermolecular interactions between the core histones and the DNA were determined from landmark studies leading to an atomic (2.8 Å) resolution X-ray picture of the nucleosome assembled from recombinant parts (Fig. 5) (Luger et al. 1997). Higher-resolution images of mononucleosomes, as well as emerging higher-order structures (tetranucleosomes) (Schalch et al. 2005), continue to capture our attention, promising to better explain the physiologically relevant substrate upon which most, if not all, of the chromatin remodeling and transcriptional machinery operates.

The core histone proteins that make up the nucleosome are small and highly basic. They are composed of a

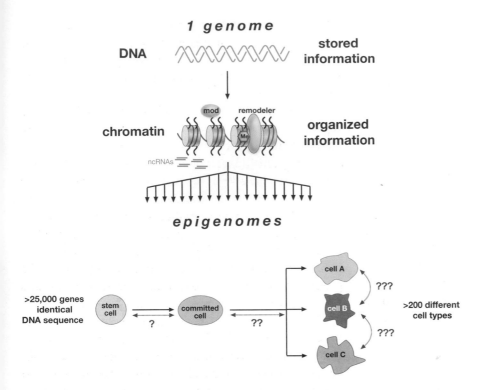

1 genome

DNA — stored information

chromatin — organized information

epigenomes

>25,000 genes identical DNA sequence

stem cell ? committed cell ?? cell A / cell B / cell C ??? >200 different cell types

Figure 4. DNA Versus Chromatin

The genome: Invariant DNA sequence (*green double helix*) of an individual. The epigenome: The overall chromatin composition, which indexes the entire genome in any given cell. It varies according to cell type, and response to internal and external signals it receives. (*Lower panel*) Epigenome diversification occurs during development in multicellular organisms as differentiation proceeds from a single stem cell (the fertilized embryo) to more committed cells. Reversal of differentiation or transdifferentiation (*blue lines*) requires the reprogramming of the cell's epigenome.

globular domain and flexible (relatively unstructured) "histone tails," which protrude from the surface of the nucleosome (Fig. 5). Based on amino acid sequence, histone proteins are highly conserved from yeast to humans. Such a high degree of conservation lends support to the general view that these proteins, even the unstructured tail domains, are likely to serve critical functions. The tails, particularly of histones H3 and H4, in fact hold important clues to nucleosomal variability (and hence chromatin), as many of the residues are subject to extensive posttranslational modifications (see back end paper for standard nomenclature used in this textbook and Appendix 2 for a listing of known histone modifications).

Acetylation and methylation of core histones, notably H3 and H4, were among the first covalent modifications to be described, and were long proposed to correlate with positive and negative changes in transcriptional activity. Since the pioneering studies of Allfrey and coworkers (Allfrey et al. 1964), many types of covalent histone modifications have been identified and characterized; these include histone phosphorylation, ubiquitination, sumoylation, ADP-ribosylation, biotinylation, proline isomerization, and likely others that await description (Vaquero et al. 2003). These modifications occur at specific sites and residues, some of which are illustrated in Figure 6 and listed in Appendix 2. Specific enzymes and enzymatic complexes, some of which are highlighted in the follow-

ing overview and individual chapters, catalyze these covalent markings. Because these lists will continue to grow in years to come, our intent was to mention only individual marks and enzymes that can illustrate what we feel are important general concepts and principles.

In certain chromatin regions, nucleosomes may contain histone variant proteins in place of a core (canonical) histone. Ongoing research is showing that this compositional difference contributes to marking regions of the

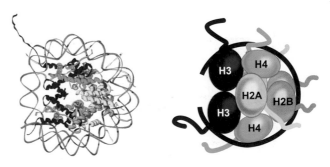

Figure 5. Nucleosome Structure

(*Left*) A 2.8 Å model of a nucleosome. (*Right*) A schematic representation of histone organization within the octamer core around which the DNA (*black line*) is wrapped. Nucleosome formation occurs first through the deposition of an H3/H4 tetramer on the DNA, followed by two sets of H2A/H2B dimers. Unstructured aminoterminal histone tails extrude from the nucleosome core, which consists of structured globular domains of the eight histone proteins.

chromosomes for specialized functions. Variant proteins for core histones H2A and H3 are currently known, but none exists for histones H2B and H4. We suspect that histone variants, although often minor in terms of amount and accordingly more difficult to study, are bountiful in the information they contain and essential to contributing to epigenetic regulation (for more detail, see Section 8 and Chapter 13).

5 Higher-Order Chromatin Organization

Chromatin, the DNA-nucleosome polymer, is a dynamic molecule existing in many configurations. Historically, chromatin has been classified as either euchromatic or heterochromatic, stemming from the nuclear staining patterns of dyes used by cytologists to visualize DNA. Euchromatin is decondensed chromatin, although it may be transcriptionally active or inactive. Heterochromatin can broadly be defined as highly compacted and silenced chromatin. It may exist as permanently silent chromatin (constitutive heterochromatin), where genes will rarely be expressed in any cell type of the organism, or repressed (facultative heterochromatin) in some cells during a specific cell cycle or developmental stage. Thus, there is a spectrum of chromatin states and a long-standing literature suggesting that chromatin is a highly dynamic macromolecular structure, prone to remodel-

ing and restructuring as it receives physiologically relevant input from upstream signaling pathways. Only recently, however, has excellent progress been made unraveling molecular mechanisms that govern these remodeling steps.

The textbook, 11-nm "beads on a string" template represents an active and largely "unfolded" interphase configuration wherein DNA is periodically wrapped around repeating units of nucleosomes (Fig. 7). The chromatin fiber, however, is not always made up of regularly spaced nucleosomal arrays. Nucleosomes may be irregularly packed and fold into higher-order structures that are only beginning to be observed at atomic resolution (Khorasanizadeh 2004). Differential and higher-order chromatin conformations occur in diverse regions of the genome during cell-fate specification or in distinct stages of the cell cycle (interphase versus mitotic chromatin).

The arrangement of nucleosomes on the 11-nm template can be altered by *cis*-effects and *trans*-effects of covalently modified histone tails (Fig. 8). *cis*-Effects are brought about by changes in the physical properties of modified histone tails, such as a modulation in the electrostatic charge or tail structure that, in turn, alters internucleosomal contacts. A well-known example, histone acetylation, has long been suspected to neutralize positive charges of highly basic histone tails, generating a localized expansion of the chromatin fiber, thereby enabling better access of

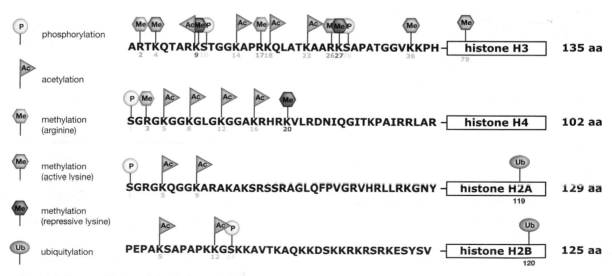

Figure 6. Sites of Histone Tail Modifications

The amino-terminal tails of histones account for a quarter of the nucleosome mass. They host the vast majority of known covalent modification sites as illustrated. Modifications do also occur in the globular domain (*boxed*), some of which are indicated. In general, active marks include acetylation (*turquoise Ac flag*), arginine methylation (*yellow Me hexagon*), and some lysine methylation such as H3K4 and H3K36 (*green Me hexagon*). H3K79 in the globular domain has anti-silencing function. Repressive marks include H3K9, H3K27, and H4K20 (*red Me hexagon*), Green = active mark, red = repressive mark.

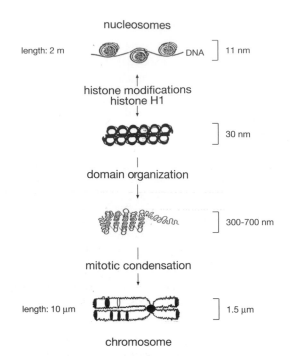

Figure 7. Higher-Order Structuring of Chromatin

The 11-nm fiber represents DNA wrapped around nucleosomes. The 30-nm fiber is further compacted into an as-yet-unconfirmed structure (illustrated as solenoid conformation here), involving linker histone H1. The 300–700-nm fiber represents dynamic higher-order looping that occurs in both interphase and metaphase chromatin. The 1.5-μm condensed chromosome represents the most compacted form of chromatin that occurs only during nuclear division (mitosis or meiosis). It is not yet clear how mitotic chromosome-banding patterns (i.e., G- or R-banding) correlate with particular chromatin structures.

transcription machinery to the DNA double helix. Phosphorylation, through the addition of net negative charge, can generate "charge patches" (Dou and Gorovsky 2000) that are believed to alter nucleosome packaging or to expose histone amino termini by altering the higher-order folded state of the chromatin polymer (Wei et al. 1999; Nowak and Corces 2004). In much the same way, linker histones (H1) are believed to promote the packaging of higher-order fibers by shielding the negative charge of linker DNA between adjacent nucleosomes (Thomas 1999; Khochbin 2001; Harvey and Downs 2004; Kimmins and Sassone-Corsi 2005). The addition of bulky adducts, such as ubiquitin and ADP-ribose, may also induce different arrangements of the histone tails and open up nucleosome arrays. The extent to which histone tails can induce chromatin compaction through modification-dependent and -independent mechanisms is not clear.

Histone modifications may also elicit what we refer to as *trans*-effects by the recruitment of modification-binding partners to the chromatin. This can be viewed as "reading" a particular covalent histone mark in a context-dependent fashion. Certain binding partners have a particular affinity and hence are known to "dock" onto specific histone tails and often do so by serving as the chromatin "Velcro" for one polypeptide within a much larger enzymatic complex that needs to engage the chromatin polymer. For instance, the bromodomain—a motif that recognizes acetylated histone residues—is often, but not always, part of a histone acetyltransferase (HAT) enzyme that exists to acetylate target histones (see Fig. 10 in Section 7) as part of a larger chromatin-remodeling complex (Dhalluin et al. 1999; Jacobson et al. 2000). Similarly, methylated lysine residues embedded in histone tails can be read by chromodomains (Bannister et al. 2001; Lachner et al. 2001; Nakayama et al. 2001) or similar domains (e.g., MBT, tudor) (Maurer-Stroh et al. 2003; Kim et al. 2006) to facilitate downstream chromatin-modulating events. In some cases, for instance, the association of chromodomain proteins precipitates the spreading of heterochromatin by the histone methyltransferase (HKMT)-catalyzed methylation of adjacent histones which can then be read by chromodomain proteins (Chapter 5).

Histone modifications of both the tail regions and the globular core region (Cosgrove et al. 2004) can also target ATP-dependent remodeling complexes to the 11-nm fiber required for the transition from poised euchromatin to a transcriptionally active state. This mobilization of nucleosomes may occur by octamer sliding, alteration of nucleosome structure by DNA looping (for more detail, see Chapter 12) or replacement of specific core histones with histone variants (Chapter 13). ATP-dependent chromatin remodelers (such as SWI/SNF, an historically important example) hydrolyze energy to bring about significant changes in histone:DNA contacts, resulting in looping, twisting, and sliding of nucleosomes. These non-covalent mechanisms have been shown to be critically important for gene regulatory events (Narlikar et al. 2002) as much as those involving covalent histone modifications (see Chapter 10). The finding that specific ATP-dependent remodelers can shuffle histone variants into and out of chromatin provides a means to link *cis*, *trans*, and remodeling mechanisms. Understanding, in turn, how these interconnected mechanisms act in a concerted fashion to vary epigenetic states in chromatin is far from complete.

More compact and repressive higher-order chromatin structures (30-nm) can also be achieved through the recruitment of linker histone H1 and/or modification-dependent or "architectural" chromatin-associated factors

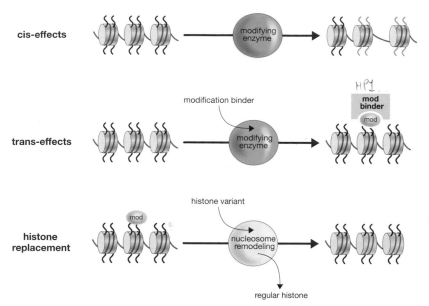

cis-effects

trans-effects

histone replacement

modification binder

histone variant

regular histone

Figure 8. Transitions in the Chromatin Template (*cis/trans*)

cis-effects: A covalent modification of a histone tail residue results in an altered structure or charge that manifests as a change in chromatin organization. *trans-effects:* The enzymatic modification of a histone tail residue (e.g., H3K9 methylation) results in an affinity for chromatin-associated protein (mod binder, e.g., HP1). The association of a mod binder (or associated protein complexes) causes downstream alterations in chromatin structure. *Histone replacement:* A covalent histone modification (or other stimulus) can signal the replacement of a core histone with a histone variant through a nucleosome-remodeling exchanger complex.

such as heterochromatin protein 1 (HP1) or Polycomb (PC). Although it is commonly held that compaction of nucleosomal chromatin (11-nm) into a 30-nm transcriptionally incompetent conformation is accomplished by the incorporation of linker histone H1 during interphase, the functional and structural dissection of this histone has, until recently, been difficult (Fan et al. 2005). One likely problem underlying these studies is the fact that histone H1 occurs as different isoforms (~8 in mammals), making it difficult to do detailed genetic analyses. Thus, there is redundancy between some H1 isoforms whereas others may hold tissue-specific functions (Kimmins and Sassone-Corsi 2005). Interestingly, H1 itself can be covalently modified (phosphorylated, methylated, poly(ADP) ribosylated, etc.), raising the possibility that *cis* and *trans* mechanisms currently being dissected on core histones may well extend to this important class of linker histone, and also to nonhistone proteins (Sterner and Berger 2000).

Considerable debate has taken place over the details of the way in which the 30-nm chromatin fiber is organized. In general, either "solenoid" (one-start helix) models, wherein the nucleosomes are gradually coiled around a central axis (6–8 nucleosomes/turn), or more open "zigzag" models, which adopt higher-order self-assemblies (two-start helix), have been described. New evidence, including that collected from X-ray structure using a model system containing four nucleosomes, suggests a fiber arrangement more consistent with a two-start, zigzag arrangement of linker DNA connecting two stacks of nucleosome particles (Khorasanizadeh 2004; Schalch et al. 2005). Despite this progress, we note that linker his-

tone is not present in the current structures, and even if it were present, the 30-nm chromatin fiber compacts the DNA only approximately 50-fold. Thus, considerably more levels of higher-order chromatin organization exist that have yet to be resolved outside of light- and electron-microscopic examination, whether leading to interphase or mitotic chromatin states. Despite structural uncertainties, recent results in living cells have now established the existence of multiple levels of chromatin folding above the 30-nm fiber within interphase chromosomes. A noteworthy advance was the development of new approaches to label specific DNA sequences in live cells, making it possible to study the dynamics of chromatin opening and closing in vivo in real time. Interestingly, these results reveal a dynamic interplay of positive and negative chromatin-remodeling factors in setting higher-order chromatin structures for states more or less compatible with gene expression (Fisher and Merkenschlager 2002; Felsenfeld and Groudine 2003; Misteli 2004).

Organization into larger looped chromatin domains (300–700 nm) occurs, perhaps through anchoring the chromatin fiber to the nuclear periphery or other nuclear scaffolds via chromatin-associated proteins such as nuclear lamins. The extent to which these associations give rise to meaningful functional "chromosome territories" remains unclear, but numerous reports are showing that this concept deserves serious attention. For instance, clustering of multiple active chromatin sites to RNA polymerase II (RNA pol II) transcription factors has been observed, and similar concepts seem to apply to the clustering around replicating DNA and DNA polymerase. In contrast, clustering of "silent" heterochromatin (particularly pericentromeric foci) and genes localized in *trans* has also been documented (see Chapters 4 and 21). How these associations are controlled and the extent to which nuclear localization of chromatin domains affects genome regulation are not yet clear. There is, nonetheless, an increasing body of evidence showing correlations of an active or silent chromatin configuration with a particular nuclear territory (Cremer and Cremer 2001; Gilbert et al. 2004; Janicki et al. 2004; Chakalova et al. 2005).

The most condensed DNA structure is observed during the metaphase stage of mitosis or meiosis. This permits the faithful segregation of exact copies of our genome (one or two copies of each chromosome, depending on the division at hand), via chromosomes, to each daughter cell. This condensation involves a dramatic restructuring of the DNA from a 2-m molecule when fully extended, into discrete chromosomes measuring on average 1.5 μm in diameter (Fig. 7). This is no less than a 10,000-fold compaction and is achieved by the hyperphosphorylation of linker (H1) and core histone H3, and the ATP-dependent action of the condensin and cohesin complexes, and topoisomerase II. Exactly how non-histone complexes engage mitotic chromatin (or M-phase chromatin modifications), and what rules dictate their association and release from chromatin in a cell-cycle-regulated fashion, remain to be determined (Bernard et al. 2001; Watanabe et al. 2001). Here, the well-known mitotic phosphorylation of histone H3 (i.e., serines 10 and 28) and members of the H1 family may provide important clues, but genetic and biochemical experiments have yet to yield full insights into what the function of these mitotic marks is. Interestingly, a formal theory has been proposed that specific methylation marks, when paired with more dynamic and reversible phosphorylation marks, may act as a "binary switch" in histone proteins, governing the binding and release of downstream effectors that engage the chromatin template (Fischle et al. 2003a). Using HP1 binding to methylated histone H3 on lysine 9 (H3K9me) and mitotic serine 10 phosphorylation (H3S10ph) as a paradigm, evidence in support of a mitotic "methyl/phos switch" has recently been provided (Daujat et al. 2005; Fischle et al. 2005; Hirota et al. 2005).

Specialized chromosomal domains, such as telomeres and centromeres, serve distinct functions dedicated to proper chromosome dynamics. Telomeres act as chromosomal ends, providing protection and unique solutions to how the very ends of DNA molecules are replicated. Centromeres provide an attachment anchor for spindle microtubules during nuclear division. Both of these specialized

domains have a fundamental role in the events that lead to faithful chromosome segregation. Interestingly, both telomeric and centromeric heterochromatin is distinguishable from euchromatin, and even other heterochromatic regions (see below), by the presence of unique chromatin structures that are largely repressive for gene activity and recombination. Moving expressed genes from their normal positions in euchromatin to new positions at or near centromeric and telomeric heterochromatin (see Chapters 4–6) can silence these genes, giving rise to powerful screens described earlier that sought to identify suppressors or enhancers of position-effect variegation (PEV) or telomere-position effects (TPE; Gottschling et al. 1990; Aparicio et al. 1991). Centromeres and telomeres have molecular signatures that include, for example, hypoacetylated histones. Interestingly, centromeres are also "marked" by the presence of the histone variant CENP-A, which plays an active role in chromosome segregation (Chapter 14). Thus, the proper assembly and maintenance of distinct centromeric and pericentromeric heterochromatin is critical for the completion of mitosis or meiosis, and hence, cellular viability. In addition to the well-studied centromeric and pericentromeric forms of constitutive heterochromatin, progress is being made into mechanisms of epigenetic control for centromeric (and telomeric) "identity." Clever experiments have shown that "neocentromeres" can function in place of normal centromeres, demonstrating that DNA sequences do not dictate the identity of centromeres (Chapters 13 and 14). Instead, epigenetic hallmarks, including centromere-specific modification patterns and histone variants, mark this specialized chromosomal domain. Considerable progress is being made into how other coding, noncoding, and repetitive regions of chromatin contribute to these epigenetic signatures. How any of these mechanisms relate, if at all, to chromosomal banding patterns is not known, but remains an intriguing possibility. Achieving an understanding of the epigenetic regulation of these portions of unique chromosomal regions is needed, highlighted by the fact that numerous human cancers are characterized by genomic instability, which is a hallmark of certain disease progression and neoplasia.

6 The Distinction between Euchromatin and Heterochromatin

This overview has been divided into discussions of euchromatin and heterochromatin, although we acknowledge that multiple forms of both classes of chromatins exist. Euchromatin, or "active" chromatin, consists largely of coding sequences, which only account for a small fraction (less than 4%) of the genome in mammals. What molecular signals then mark coding sequences with the potential for productive transcription, and how does chromatin structure contribute to the process? An extensive literature has suggested that euchromatin exists in an "open" (decompacted), more nuclease-sensitive configuration, making it "poised" for gene expression, although not necessarily transcriptionally active. Some of the genes are ubiquitously expressed (housekeeping genes); others are developmentally regulated or stress-induced in response to environmental cues. The cooperation of selected *cis*-acting DNA sequences (promoters, enhancers, and locus control regions), bound by combinations of *trans*-acting factors, triggers gene transcription in concert with RNA polymerase and associated factors (Sims et al. 2004). Together these factors have been highly selected during evolution to orchestrate an elaborate series of biochemical reactions that must occur in the appropriate spatial and temporal setting. Does chromatin provide an "indexing system" which better ensures that the above machinery can access its target sequences in the appropriate cell type?

At the DNA level, the AT-rich vicinity of promoters is often devoid of nucleosomes and may exist in a rigid noncanonical B-form DNA configuration, promoting transcription factor (TF) occupancy (Mito et al. 2005; Sekinger et al. 2005). However, TF occupancy is not enough to ensure transcription. The recruitment of nucleosome-remodeling machines, through the induction of activating histone modifications (e.g., acetylation and H3K4 methylation), facilitates the engagement of the transcription machinery by pathways that are currently being defined (Fig. 9 and Chapter 10). Exchange of displaced histones with histone variants after the transcription machinery has unraveled and transcribed the chromatin fiber ensures integrity of the chromatin template (Ahmad and Henikoff 2002). Achieving fully mature mRNAs, however, also requires posttranscriptional processes involving splicing, polyadenylation, and nuclear export. Thus, the collective term "euchromatin" likely represents a complex chromatin state(s) that encompasses a dynamic and elaborate mixture of dedicated machines that interact together and closely with the chromatin fiber to bring about the transcription of functional RNAs. Learning the "rules" as to how, in the most general sense, the "activating machinery" interacts with the transcription apparatus as well as the chromatin template is an exciting area of current research, although due to its dynamic nature, it may not strictly classify as epigenetics, but more as transcription and chromatin dynamics studies.

What then defines "heterochromatin?" Although it is historically less well studied than euchromatin, new insights suggest that heterochromatin plays a critically important role in the organization and proper functioning of genomes from yeast to humans (although *S. cerevisiae* has a distinct form of heterochromatin). Underscoring its potential importance is the fact that 96% of the mammalian genome consists of noncoding and repetitive sequences. New mechanistic insights, underlying the formation of heterochromatin, have revealed unexpected findings. For example, non-sequence-specific transcription, which produces double-stranded RNA (dsRNA), is subject to silencing by an RNA interference (RNAi)-like mechanism (see Section 10 below). The production of such dsRNAs acts as an "alarm signal" reflecting the fact that the underlying DNA sequence cannot generate a functional product, or has been invaded by RNA transposons or viruses. The dsRNA is then processed by Dicer and targeted to chromatin by complexes dedicated to initiating a cascade of events leading to the formation of heterochromatin. Using a variety of model systems, remarkable progress has been made dissecting what appears to be a highly conserved pathway leading to a heterochromatin "locked-down" state. Although the exact order and details may vary, this general pathway involves histone tail deacetylation, methylation of specific lysine residues (e.g., H3K9), recruitment of heterochromatin-associated proteins (e.g., HP1), and establishment of DNA methylation (Fig. 9). It is likely that sequestering of selective genomic regions to repressive nuclear domains or territories may enhance heterochromatin formation. Interestingly, increasing evidence suggests that heterochromatin may be the "default state," at least in higher organisms, and that the presence of a strong promoter or enhancer, producing a productive transcript, can override heterochromatin. Even in lower eukaryotes, the general concepts of heterochromatin assembly seem to apply. Hallmark features include hypoacetylated histone tails, followed by the binding of acetylation-sensitive heterochromatin proteins (e.g., SIR proteins; for details, see Chapter 4). Depending on the fungal species (e.g., budding vs. fission yeast), varying amounts of histone methylation and HP1-like proteins exist. Even though these genomes are more set to a general default state of being poised for transcription, some heterochromatin-like genomic regions are present (mating loci, telomeres, centromeres, etc.) that are able to suppress gene transcription and genetic recombination when test genes are placed in these new neighborhoods.

transcription unit (gene)

DNA repeats (noncoding)

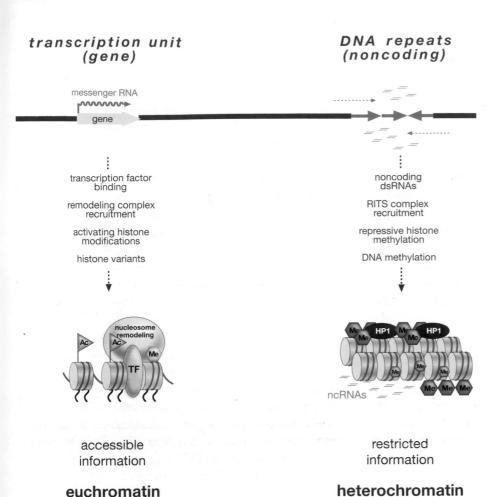

messenger RNA

gene

transcription factor
binding

remodeling complex
recruitment

activating histone
modifications

histone variants

nucleosome
remodeling

Ac

Ac

Me

TF

accessible
information

euchromatin

noncoding
dsRNAs

RITS complex
recruitment

repressive histone
methylation

DNA methylation

Me

HP1

Me

HP1

Me

Me

Me

Me

Me

Me

Me

ncRNAs

restricted
information

heterochromatin

Figure 9. Distinction between Euchromatic and Heterochromatic Domains

Summary of common differences between euchromatin and constitutive heterochromatin. This includes differences in the type of transcripts produced, recruitment of DNA-binding proteins (i.e., transcription factor [TF]), chromatin-associated proteins and complexes, covalent histone modifications, and histone variant composition.

What useful functions might heterochromatin serve? The definition of centromeres, a region of constitutive heterochromatin, correlates well with a heritable epigenetic state and is thought to be evolutionarily driven by the largest clustering of repeats and repetitive elements on a chromosome. This partitioning ensures large and relatively stable heterochromatic domains marked by repressive "epigenetic signatures," facilitating chromosome segregation during mitosis and meiosis (Chapter 14). Here, it is noteworthy that centromeric repeats and the corresponding epigenetic marks that associate with them have been duplicated and moved onto other chromosome arms to create "silencing domains" in organisms such as fission yeast. Constitutive heterochromatin at telomeres (the protective ends of chromosomes) similarly ensures stability of the genome by serving as chromosomal "caps." Last, heterochromatin formation is known to be a defense mechanism against invading DNA. Collectively, these findings underscore a general view that heterochromatin serves important genome maintenance functions which may rival even that of euchromatin itself.

In summary, the broad functional distinction between euchromatin and heterochromatin can thus far be attributed to three known characteristics of chromatin. First is the nature of the DNA sequence—e.g., whether it contains AT-rich "rigid" DNA around promoters, repetitive sequences and/or repressor-binding sequences that signal factor association. Second, the quality of the RNA produced during transcription determines whether it is fully processed into an mRNA that can be translated, or whether the RNA is degraded or earmarked for use by the RNAi machinery to target heterochromatinization. Third, spatial organization within the nucleus can play a significant sequestering role for the maintenance of localized chromatin configurations.

7 Histone Modifications and the Histone Code

We have explored how histone modifications may change the chromatin template by *cis*-effects that alter internucleosomal contacts and spacing, or the *trans*-effects caused by histone and non-histone protein associations with the template. What is the contribution and biological output of histone modifications? Patterns of chromatin structure that correlate with histone tail modifications have emerged from studies using bulk histones, suggesting that epigenetic marks may provide "ON" (i.e., active) or "OFF" (inactive) signatures. This has come through a long history of mostly correlative studies showing that certain histone modifications, notably histone acetylation, are associated with active chromatin domains or regions that are generally permissive for transcription. In contrast, other marks, such as certain phosphorylated histone residues, have long been associated with condensed chromatin that, in general, fails to support transcriptional activity. The histone modifications shown in Appendix 2 summarize the sites of modification that are known at this time. Here, we stress that these reflect modifications and sites that may well not be exhibited by every organism.

How are histone modifications established or removed in the first place? A wealth of work in the chromatin field has suggested that histone tail modifications are established ("written") or removed ("erased") by the catalytic action of chromatin-associated enzymatic systems. However, the identity of these enzymes eluded researchers for years. Over the last decade, a remarkably large number of chromatin-modifying enzymes have been identified from many sources, most of which are compiled in Appendix 2. This has been achieved through numerous biochemical and genetic studies. The enzymes often reside in large multi-subunit complexes that can catalyze the incorporation or removal of covalent modifications from both histone and non-histone targets. Moreover, many of these enzymes catalyze their reactions with remarkable specificity to target residue and cellular context (i.e., dependent on external or intrinsic signals). For clarity, and by way of example, we discuss briefly the four major enzymatic systems that catalyze histone modifications, together with their counterpart enzymatic systems that reverse the modifications (Fig. 10) (Vaquero et al. 2003; Holbert and Marmorstein 2005). Together, these antagonistic activities govern the steady-state balance of each modification in question.

Histone acetylases (HATs) acetylate specific lysine residues in histone substrates (Roth et al. 2001) and are reversed by the action of histone deacetylases (HDACs) (Grozinger and Schreiber 2002). The histone kinase family of enzymes phosphorylate specific serine or threonine residues, and the phosphatases (PPTases) remove phosphorylation marks. Particularly well known are the mitotic kinases, such as cyclin-dependent kinase or aurora kinase, which catalyze the phosphorylation of core (H3) and linker (H1) histones. Less clear in each case are the opposing PPTases that act to reverse these phosphorylations as cells exit mitosis.

Two general classes of methylating enzymes have been described: the PRMTs (protein arginine methyltransferases) whose substrate is arginine (Lee et al. 2005), and the HKMTs (histone lysine methyltransferases) that act on lysine residues (Lachner et al. 2003). Arginine methylation is indirectly reversed by the action of deiminases, which convert methyl-arginine (or arginine) to a citrulline residue (Bannister and Kouzarides 2005). Methylated lysine residues appear to be chemically more stable. Lysine methylation has been shown to be present in mono-, di-, or tri-methylated states. Several tri-methylated residues in the H3 and H4 amino termini appear to have the potential to be stably propagated during cell divisions (Lachner et al. 2004), as well as the H4K20me1 mark in Drosophila imaginal discs (Reinberg et al. 2004). Recently, a lysine-specific "demethylase" (LSD1) was described as an amine oxidase that is able to remove H3K4 methylation (Shi et al. 2004). The enzyme acts by FAD-dependent oxidative destabilization of the aminomethyl bond, resulting in the formation of unmodified lysine and formaldehyde. LSD1 was shown to be selective for the activating H3K4 methylation mark and can only destabilize mono- and di-, but not tri-methylation. This demethylase is part of a large repressive protein complex that also contains HDACs and other enzymes. Other evidence suggests that LSD1 can associate in a complex together with the androgen receptor at target loci and demethylate the H3K9me2 repressive histone mark to contribute to transcriptional activation (Metzger et al. 2005). A different class of histone demethylases has been characterized to work via a more potent mechanism—radical attack—known as hydroxylases or dioxygenases (Tsukada et al. 2006). One of these only destabilizes H3K36me2 (an active mark), but not in the tri-methyl state. This novel jumonji histone demethylase (JHDM1) contains the conserved jumonji domain, of which there are around 30 known in the mammalian genome, suggesting that some of these enzymes may also be able to attack other residues as well as a tri-methyl state (Fodor et al 2006; Whetstine et al 2006).

Considerable progress has been made in dissecting the enzyme systems that govern the steady-state balance of these modifications, and we suspect that much more progress will be made in this exciting area. It remains a challenge to understand how these enzyme complexes are regulated and how their physiologically relevant substrates

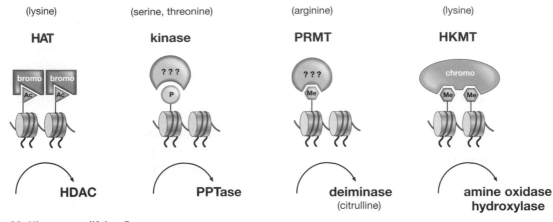

Figure 10. Histone-modifying Enzymes

Covalent histone modifications are transduced by histone-modifying enzymes ("writers") and removed by antagonizing activities. They are classified into families according to the type of enzymatic action (e.g., acetylation or phosphorylation). Protein domains with specific affinity for a histone tail modification are termed "readers." (HAT) Histone acetyltransferase; (PRMT) protein arginine methyltransferase; (HKMT) histone lysine methyltransferase; (HDAC) histone deacetylase; (PPTase) protein phosphatases; (Ac) acetylation; (P) phosphorylation; (Me) methylation.

and sites are targeted. In addition, it remains unclear how covalent mechanisms affect epigenetic phenomena.

Histone modifications do not occur in isolation, but rather in a combinatorial manner as proposed for modification cassettes (i.e., covalent modifications in adjacent residues of a particular histone tail, e.g., H3K9me and H3S10ph or H4S1ph, H4R3me, and H4K4ac) and *trans* histone pathways (covalent modifications between different histone tails or nucleosomes; see Fig. 11). Intriguingly, almost all of the known histone modifications correlate with activating or repressive function, dependent on which amino acid residue(s) in the histone amino termini is modified. Both synergistic and antagonistic pathways have been described (Zhang and Reinberg 2001; Berger 2002; Fischle et al. 2003b) that can progressively induce combinations of active marks, while simultaneously counteracting repressive modifications. It is, however, not known how many distinct combinations of modifications across the various amino-terminal histone positions exist for any given nucleosome, because most of the studies have been carried out on bulk histone preparations. In addition to the amino termini, modifications in the globular histone fold domains have recently been shown to affect chromatin structure and assembly (Cosgrove et al. 2004), thereby influencing gene expression and DNA damage repair (van Attikum and Gasser 2005; Vidanes et al. 2005). It is also worth noting that several of the histone-modifying enzymes also target non-histone substrates (Sterner and Berger 2000; Chuikov et al. 2004). Figure 11 illustrates two examples of established hierarchies of histone modifications that seem to index tran-

scription of active chromatin or, in contrast, pattern heterochromatic domains.

These studies provoke the question of whether there is a "histone code" or even an "epigenetic code." Although this theoretical concept has been highly stimulating, and has been shown to be correct in some of its predictions, the issue as to whether a code actually exists has remained largely open. As a comparison, the genetic code has proven extremely useful, because of its predictability and near universality. It uses for the most part a four-base "alphabet" in the DNA (i.e., nucleotides), forming what is generally an invariant and nearly universal language. In contrast, current evidence suggests that histone-modification patterns are likely to vary considerably from one organism to the next, especially between lower and higher eukaryotes, such as yeast and humans. Thus, even if a histone code exists, it is not likely to be universal. This situation is made considerably more complicated when one considers the dynamic nature of histone modifications, varying in space and in time. Furthermore, the chromatin template engages a staggering array of remodeling factors (Vignali et al. 2000; Narlikar et al. 2002; Langst and Becker 2004; Smith and Peterson 2005). However, chromatin immunoprecipitation assays (ChIP), when examined on genome-wide levels (ChIP on chip), have begun to decipher nonrandom and somewhat predictable patterns in several genomes (e.g., *S. pombe*, *A. thaliana*, mammalian cells), such as strong correlations of H3K4me3 with activated promoter regions (Strahl et al. 1999; Santos-Rosa et al. 2002; Bernstein et al. 2005) and of H3K9 (Hall et al. 2002; Lippman et al. 2004; Martens et al. 2005) and H3K27 (Litt et al. 2001; Ringrose et al. 2004) methylation with silenced heterochromatin. Perhaps the limitation of the histone code is that one modification does not invariantly translate to one biological output. However, modifications combinatorially or cumulatively do appear to define and contribute to biological functions (Henikoff 2005).

8 Chromatin-remodeling Complexes and Histone Variants

Another major mechanism by which transitions in the chromatin template are induced is by signaling the recruitment of chromatin "remodeling" complexes that use energy (ATP-hydrolysis) to change chromatin and nucleosome composition in a non-covalent manner. Nucleosomes, particularly when bound by repressive chromatin-associated factors, often impose an intrinsic inhibition to the transcription machinery. Hence, only some sequence-specific transcription factors and regulators (although not the basal transcription machinery) are able to gain access to their binding site(s). This accessibility problem is solved, in part, by protein complexes that

mobilize nucleosomes and/or alter nucleosomal structure. Chromatin-remodeling activities often work in concert with activating chromatin-modifying enzymes and can generally be categorized into two families: the SNF2H or ISWI, and the Brahma or SWI/SNF family. The SNF2H/ISWI family mobilizes nucleosomes along the DNA (Tsukiyama et al. 1995; Varga-Weisz et al. 1997), whereas Brahma/SWI/SNF transiently alter the structure of the nucleosome, thereby exposing DNA:histone contacts in ways that are currently being unraveled (see Chapter 12).

Additionally, some of the ATP-hydrolyzing activities resemble "exchanger complexes" that are themselves dedicated to the replacement of conventional core histones with specialized histone "variant" proteins. This ATP-costing shuffle may actually be a means by which existing modified histone tails are replaced with a clean slate of variant histones (Schwartz and Ahmad 2005). Alternatively, recruitment of chromatin-remodeling complexes, such as SAGA (Spt-Ada-Gcn5-acetyltransferase) can also be enhanced by preexisting histone modifications to ensure transcriptional competence of targeted promoters (Grant et al. 1997; Hassan et al. 2002).

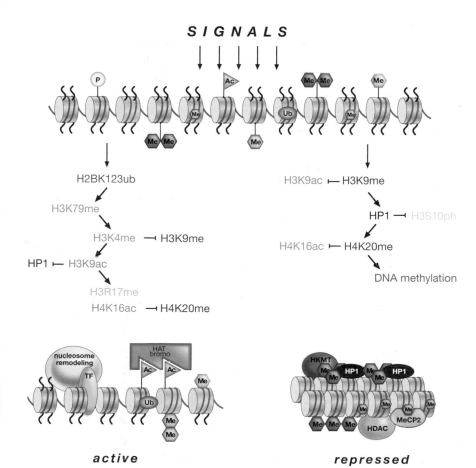

Figure 11. Coordinated Modification of Chromatin

The transition of a naïve chromatin template to active euchromatin (*left*) or the establishment of repressive heterochromatin (*right*), involving a series of coordinated chromatin modifications. In the case of transcriptional activation, this is accompanied by the action of nucleosome-remodeling complexes and the replacement of core histones with histone variants (*yellow*, namely H3.3).

In addition to transcriptional initiation and establishing the primary contact with a promoter region, the passage of RNA pol II (or of RNA pol I) during transcriptional elongation is further obstructed by the presence of nucleosomes. Mechanisms are therefore required to ensure the completion of nascent transcripts (particularly of long genes). In particular, a series of histone modifications and docking effectors act in concert with chromatin-remodeling complexes such as SAGA and FACT (for facilitate chromatin transcription) (Orphanides et al. 1998) to allow RNA pol II passage through nucleosomal arrays. These concerted activities will, for example, induce increased nucleosomal mobility, displace H2A/H2B dimers, and promote the exchange of core histones with histone variants. As such, they provide an excellent example of the close interplay between histone modifications, chromatin remodeling, and histone variant exchange to facilitate transcriptional initiation and elongation (Sims et al. 2004). Other remodeling complexes have also been characterized, such as Mi-2 (Zhang et al. 1998; Wade et al. 1999) and INO-80 (Shen et al. 2000), which are involved in stabilizing repressed rather than active chromatin.

Compositional differences of the chromatin fiber that occur through the presence of histone variants contribute to the indexing of chromosome regions for specialized functions. Each histone variant represents a substitute for a particular core histone (Fig. 12), although histone variants are often a minor proportion of the bulk histone content, and thus more difficult to study than regular histones. An increasing body of literature (for review, see Henikoff and Ahmad 2005; Sarma and Reinberg 2005) documents that histone variants have their own pattern of susceptibility to modifications, likely specified by the small number of amino acid changes that distinguish them from their family members. On the other hand, some histone variants have distinct amino- and carboxy-terminal domains with unique chromatin-regulating activity and different affinities to binding factors. By way of example, transcriptionally active genes have general histone H3 exchanged by the H3.3 variant, in a transcription-coupled mechanism that does not require DNA replication (Ahmad and Henikoff 2002). The replacement of core histone H2A with the H2A.Z variant correlates with transcriptional activity and can index the 5′ end of nucleosome-free promoters. However, H2A.Z has also been associated with repressed chromatin. CENP-A, the centromere-specific H3 variant, is essential for centromeric function and hence chromosome segregation. H2A.X, together with other histone marks, is associated with

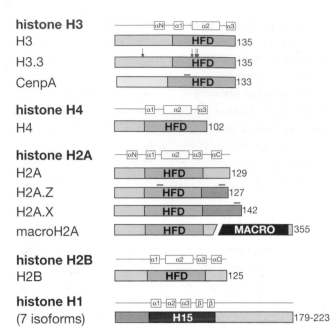

Figure 12. Histone Variants

Protein domain structure for the core histones (H3, H4, H2A, H2B), linker histone H1, and variants of histones H3 and H2A. The histone fold domain (HFD) where histone dimerization occurs, and regions of the protein that differ in histone variants (shown in red) are indicated.

sensing DNA damage and appears to index a DNA lesion for recruitment of DNA repair complexes. MacroH2A is a histone variant that specifically associates with the inactive X chromosome (Xi) in mammals (for more details on histone variants, see Chapter 13).

Importantly, and in contrast to the commonly held textbook notion that histones are synthesized and deposited only during S phase, synthesis and substitution of many of these histone variants occurs independently of DNA replication. Hence, the replacement of core histones by histone variants is not restricted to cell cycle stages (i.e., S phase), but can take immediate effect in response to ongoing mechanisms (e.g., transcriptional activity or kinetochore tension during cell division) or stress signals (e.g., DNA damage or nutrient starvation). Elegant biochemical studies have documented chromatin remodeling or exchanger complexes that are specific for replacement of distinct histone variants, such as H3.3, H2A.Z, or H2A.X (Cairns 2005; Henikoff and Ahmad 2005; Sarma and Reinberg 2005). For instance, replacement of H3 with the H3.3 variant occurs via the action of the HIRA (histone regulator A) exchanger complex (Tagami et al. 2004), and H2A is replaced by H2A.Z through the activity of the SWR1 (Swi2/Snf2-related ATPase 1) exchanger complex

(Mizuguchi et al. 2004). Together, these substitutions allow variant nucleosomes to build particularly active chromatin. For some of the other histone variants, the mechanism of targeting and exchange remains to be determined, whether via an ATP-dependent histone exchanger complex or a histone chaperone protein. It has now even been postulated that exchanger complexes may exist to substitute modified histones with their unmodified counterparts, as a mechanism for erasing more robust epigenetic marks that reside in the histone amino termini.

9 DNA Methylation

DNA methylation is the oldest epigenetic mechanism known to correlate with gene repression (Razin and Riggs 1980). It is present to varying degrees in all eukaryotes except yeast. This modification consists of the addition of a methyl group at cytosine residues of the DNA template. It occurs at CpG dinucleotides in mammals, whereas other symmetric, asymmetric, and non-CpG methylation patterns are known in N. crassa and plants. The distribution of methylated DNA along the genome shows enrichment at noncoding regions (e.g., centromeric heterochromatin) and interspersed repetitive elements (transposons) but not in the CpG islands of active genes (Bird 1986). In fact, the increasing levels of DNA methylation correlate with a relative increase in noncoding and repetitive DNA content in the genomes of higher eukaryotes (see Fig. 15 in Section 11). Experimental evidence indicates that this is because DNA methylation serves mainly as a host defense mechanism to silence much of the genome of foreign origin (i.e., replicated transposable elements, viral sequences, and other repeated sequences).

DNA methyltransferases (DNMTs) are the "effectors" of DNA methylation, and catalyze either de novo (i.e., at novel sites) or maintenance methylation of hemimethylated DNA following DNA replication (see Chapter 18). Loss of the ability to maintain DNA methylation can result in several diseases, such as ICF (Immunodeficiency, Centromeric instability, and Facial abnormalities) (see Chapters 18 and 23). Deregulation in the levels of DNA methylation is also a contributing factor to cancer progression (Chapter 24).

What are the signals that direct DNMTs to methylate certain regions of DNA? Currently it is known that highly repetitive tandem repeat sequences of the genome (e.g., pericentromeric chromatin) rely on the repressive H3K9 methylation marks to direct DNA methylation de novo, as evidenced in N. crassa and plants (see Chapters 6 and 9). Interspersed repeats can also signal de novo DNA

methylation, described in the context of RIP in N. crassa (Tamaru and Selker 2001) and retrotransposon silencing in the male germ line of mammals. A protein responsible for the latter has been identified—Dnmt3L—and may function by scanning the genome to identify high levels of homology–heterology junctions that signal the requirement for DNA methylation (Bourc'his and Bestor 2004). In plants, RNAs provide the signal for de novo DNA methylation, through a unique mechanism termed RNA-dependent DNA methylation (RdDM; see Chapter 9). There is evidence that chromatin remodelers of the SWI/SNF family are in some way necessary for the global patterns of DNA methylation, as demonstrated in plants (Jeddeloh et al. 1999) for the DDM1 protein and mammals by the Lsh1 homolog (Yan et al. 2003). Last, the HKMT-PcG protein Ezh2 may also be involved in directing DNA methylation at certain promoters in mammals (Vire et al. 2005).

Once established, the way in which DNA methylation may function to silence chromatin is not entirely clear, although evidence points to trans-regulation. Binders of methylated cytosines, called methyl-CpG-binding domain proteins (MBD), can be considered the DNA methylation equivalent to binders (or readers) of modified histone motifs (Fig. 13). For example, the methylcytosine-binding protein (MeCP2) binds methylated CpGs and recruits HDACs to mediate repressive histone marks (see Chapter 18). DNA methylation is also known to disturb the recognition sites of transcriptional regulators (e.g., CTCF) that are involved in genomic imprinting (see Chapter 19).

The existence of methylated DNA at imprinted loci that silence either the maternal or paternal allele in plants and placental mammals suggests that in the course of evolution they uniquely harnessed this epigenetic mechanism to stabilize gene repression. Interestingly, in marsupials, there is a lack of DNA methylation at imprinted loci, indicating that its involvement in mammalian imprinting is a relatively recent evolutionary event (see Chapters 17 and 19). Conversely, in dipteran insects such as Drosophila, DNA methylation has largely been lost as a functional epigenetic mechanism (Lyko 2001).

Highly repetitive regions of the mammalian genome that are typically methylated become increasingly mutagenic when unmethylated, to the extent of causing global genomic instability (Chen et al. 1998). Chromosomal abnormalities ensue, which are a major cause of many diseases and cancer progression (see Section 15). This underlines the crucial role that DNA methylation plays in genome integrity. Conversely, individual methylated cyto-

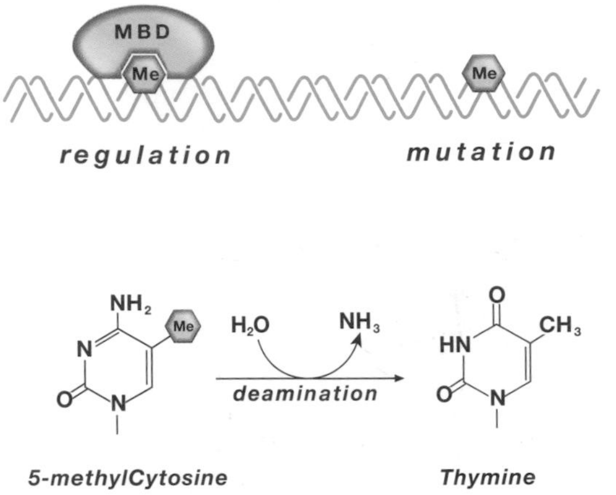

Figure 13. DNA Methylation and Deamination

Cytosine nucleotides that are methylated (Me) may be bound by methyl DNA-binding proteins (MBD). Unbound 5-methylcytosine is prone to spontaneous mutation through deamination (reaction shown in the lower panel), resulting in a 5-methyl CpG to TpA transition in the DNA sequence.

sine bases have a high propensity to spontaneously mutate. Thus, over time, C-T transitions occur through a deamination reaction (Fig. 13), but this characteristic is also thought to be beneficial for protecting the host genome because it permanently deactivates parasitic DNA sequences such as transposons. In a different context, this same chemical reaction is actively catalyzed by the activation-induced deaminase, or AID. Expression of the enzyme in B and T cells causes "somatic hypermutation" at the immunoglobulin (Ig) locus. This is an important mechanism for expanding the repertoire of antigen receptors and hence strengthening the immunity of mammals (Petersen-Mahrt 2005; Chapter 21). AID expression observed in early mammalian development has led to the suggestion that it may provide an alternative route to demethylating DNA, although this would happen at the risk of increased point mutation rates.

DNA methylation and histone methylation are prominent mechanisms for epigenetic regulation of the genome. Noncoding RNAs, as described in the next section, are important primary triggers for inducing silent chromatin. It is also known that RNA molecules can be heavily methylated at the sugar or nucleoside backbone. Moreover, methylation at the 3′ end of small noncoding RNAs has been shown to stabilize these molecules (Yu et al. 2005). Intriguingly, Dnmt2 was recently identified as a tRNA methyltransferase (Goll et al 2006). It is therefore plausible that, similar to DNMTs and HKMTs, RNA methyltransferases may exist as "writers" of epigenetic information, although there is no direct evidence for this. However, RNA methylation appears to be "sensed" by certain Toll-like receptors (transmembrane receptors that recognize common pathogen molecular motifs) to mediate innate immunity (Ishii and Akira 2005), corroborating such a hypothesis. This raises the interesting possibility that RNA methylation may yet prove to be a third form of methylation-based epigenetic modulation.

10 RNAi and RNA-directed Gene Silencing

The knowledge that constitutive heterochromatin at centromeres and telomeres plays an instructive role in genome integrity has contributed to a paradigm shift in the way that repetitive noncoding "junk" DNA is viewed. Is it possible that these repetitive sequences serve a non-wasteful purpose that is only beginning to be elucidated? Is it even possible that such DNA sequences are not completely "silent?"

This possibility has stemmed from a fundamental series of discoveries that linked RNAi to the formation of silent chromatin (heterochromatin). RNAi is a host defense mechanism that breaks down dsRNA species into small RNA molecules (known as short interfering RNA or siRNA). This process ultimately leads to RNA degradation or the use of the small RNAs to inhibit translation, known as posttranscriptional silencing (PTGS). The more recently discovered transcriptional gene silencing (TGS) mechanism, leading to heterochromatin formation, was discovered through the convergence of independent lines of investigation into chromatin and the RNAi machinery. On the one hand, much was known about repressive DNA methylation (in fungi, plants, and mammals), chromatin modifications (e.g., H3K9me3), and chromatin-associated factors (HP1) that are characteristic of heterochromatin domains. On the other hand, researchers were making headway in identifying factors of the RNAi machinery (e.g., Dicer, Argonaute, RNA-dependent RNA polymerase or RdRP). The most convincing progress that tied together these two seemingly divergent fields came from elegant studies in *S. pombe*, where mutations of any component of the RNAi machinery resulted in defects in chromosome segregation (Hall et al. 2002; Reinhart and Bartel 2002; Volpe et al. 2002). This was brought about by the inability to stabilize centromeric heterochromatin and underscored the likely widespread role of RNAi-mediated mechanisms in producing silent heterochromatin domains. It also highlighted the importance of heterochromatin beyond transcriptionally silencing genes, to a role in maintaining genome integrity and hence viability, as shown by the requirement of centromeric heterochromatin for the process of chromosome segregation. Emerging evidence also suggests that siRNAs are required in defining other specialized regions of functional heterochromatin, such as telomeres.

Transcription from both DNA strands of *S. pombe* pericentromeric repeats and the detection of processed siRNA derivatives provided strong evidence that the dsRNA derivative was the critical substrate to target the RITS complex to the centromeres for silencing (Fig. 14) (Verdel et al. 2004). Furthermore, *clr4* mutants (the *S. pombe* ortholog of mammalian *Suv39h* HKMT) failed to process dsRNA into siRNAs, strengthening the case for the interplay between the RNAi machinery and heterochromatin assembly (Motamedi et al. 2004). Exactly how siRNAs, generated by the RNAi machinery (i.e., Dicer, Argonaute, RdRP), initiate heterochromatin assembly or guide it to appropriate genomic loci is still unknown. A model has emerged in which a complex interaction between the RNAi machinery complex RITS and centromeric repeats leads to a self-reinforcing cycle of heterochromatin formation involving Clr4, HDACs, Swi6 (ortholog to mammalian HP1), and cohesin, probably via Ago-directed annealing of RNA:RNA hybrids to the nascent transcript (Fig. 14) (see Chapters 6 and 8).

In *Tetrahymena,* a similar RNA-mediated targeting mechanism has been recognized to direct the unique case of DNA elimination that occurs in the somatic nucleus. In this case, transcription occurs from both strands of the internal eliminated segment (IES) sequences in the "silent," germ-line (micronuclear) genome at the appropriate stage of the sexual pathway (Chalker and Yao 2001; Mochizuki et al. 2002). Along the same lines as the RNAi-dependent TGS model, a scan RNA (scnRNA) model was proposed to explain how DNA sequences in the parental macronucleus can epigenetically control genomic alterations in the new macronucleus, involving small RNAs (for more detail, see Chapter 7). These exciting results provide the first demonstration of an RNAi-like process directly altering a somatic genome. This raises the intriguing possibility that intergenic RNAs produced at the V-DJ locus (Bolland et al. 2004) may potentially direct DNA sequence elimination during V-DJ recombination of the immunoglobulin heavy chain (IgH) locus in B cells and the T-cell receptor (TCR) loci in T cells.

In plants, there are a number of orthologs for many of the RNAi components, resulting in a variety of RNA silencing pathways that can act with greater specificity for particular DNA sequences, although there is some redun-

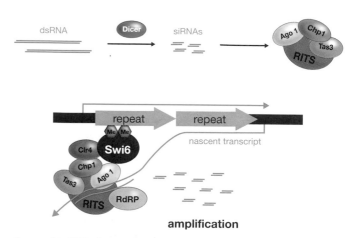

Figure 14. RNA-directed Heterochromatin Formation

Complementary dsRNA transcripts, produced by transcription of both strands or the folding back of inverted repeat transcripts, result in the generation of siRNAs through the action of Dicer (*top*). Incorporation of the siRNAs into the RITS complex through binding with the Argonaute protein (Ago) activates the complex for targeting to complementary DNA or nascent RNA. The complex attracts Clr4, which transduces histone H3K9me2. The modification-specific binder, Swi6, binds to these modified histones, facilitating the spread of a repressive chromatin domain. The action of RdRP amplifies the levels of siRNAs by using existing siRNAs as primers, reinforcing the targeting capacity of the RITS complex to specific regions of DNA.

dancy between factors. Studies of RNAi-mediated TGS in plants have revealed a novel class of RNA polymerases—RNA polymerase IV (or RNA pol IV)—that may transcribe DNA solely at heterochromatic regions (Herr et al. 2005; Pontier et al. 2005). Also unique to plants is the demonstration that RNAi pathways directly affect DNA methylation (Chan et al. 2004) (explained in detail in Chapter 9).

RNAi-like chromatin effects have also been uncovered in *Drosophila* and mammals. For instance, RNase A treatment of permeabilized mammalian cells rapidly removes heterochromatic H3K9me3 marks, suggesting that an RNA moiety may be a structural component of pericentromeric heterochromatin (Maison et al. 2002). Ablation of siRNA processing factors in vertebrates impairs H3K9 methylation and HP1 binding at pericentromeric heterochromatin (Fukagawa et al. 2004). Intriguingly, embryonic stem (ES) cells still proliferate, but fail to differentiate, in dicer-null mutants (Kanellopoulou et al. 2005), suggesting a currently not-understood connection between the RNAi machinery, noncoding RNAs, and mammalian development. In *Drosophila*, silencing of the tandem arrays of the mini-white gene, subject to PEV also appears to be dependent on the RNAi machinery (Pal-Bhadra et al. 2004).

Collectively, these studies indicate a crucial, and probably primary, role for noncoding RNAs in triggering epigenetic transitions and heritably maintaining specific chromatin states of the chromatin template. In fact, these noncoding RNAs have provided the answer for how diverse repetitive sequences in different organisms achieve heterochromatinization through an RNA-targeted mechanism. In an effort to identify more targets of RNAi, the sequencing of small RNAs has revealed that they are largely transcribed from endogenous transposons and other repetitive sequences in plants, *Drosophila*, and mammals, among other organisms (Almeida and Allshire 2005; Bernstein and Allis 2005). Together, these results indicate that RNAi has evolved, in part, to maintain genomic stability by silencing mobile DNA elements and viruses, and is a conserved mechanism across most eukaryotic species. It now appears, however, not only that RNA silencing represses invading sequences, but also that this basic mechanism has been harnessed by the cell for the heterochromatinization of centromeres, thereby ensuring correct chromosome segregation and genome integrity.

Together, the above examples indicate a striking variation to the central dogma of gene control that is beginning to emerge as follows: *DNA→noncoding RNA→chromatin→gene function*. The idea that noncoding RNAs would actively participate in RNAi-like mechanisms which also target locus-specific domains for chromatin remodeling and gene silencing was never anticipated.

11 From Unicellular to Multicellular Systems

The 5,000–6,000 genes contained in the genomes of budding and fission yeasts are sufficient to regulate basic metabolic and cell division processes. There is, however, no requirement for cell differentiation, because these unicellular organisms are essentially clonal and, as such, repetitive "immortal" entities. In contrast, mammals code for ~25,000 genes required in ~200 different cell types. Understanding how multicellular complexity is generated and coordinated from the same genetic template is a key question in epigenetic research.

A comparison of the genome sizes between yeasts, flies, plants, and mammals indicates that genome size significantly expands with the complexity of the respective organism. There is a more than 300-fold difference between the genome sizes of yeast and mammals, but only a modest 4–5-fold increase in overall gene number (Fig. 15). However, the ratio of coding to noncoding and repetitive sequences is indicative of the complexity of the genome: The largely "open" genomes of unicellular fungi have relatively little noncoding DNA compared with the highly heterochromatic genomes of multicellular organisms. In particular, mammals have accumulated considerable repetitive elements and noncoding regions, which account for the majority of its DNA sequence (52% noncoding and 44% repetitive DNA). Only 4% of the mammalian genome thus encodes for protein function (including intronic sequences). This massive expansion of repetitive and noncoding sequences in multicellular organisms is most likely due to the incorporation of invasive elements, such as DNA transposons, retrotransposons, and other repetitive elements. Although these represent a burden for coordinated gene expression programs, they also allow genome evolution and plasticity, and a certain degree of stochastic gene regulation. The expansion of repetitive elements has even infiltrated the transcriptional units of the mammalian genome. This results in transcription units that are frequently much larger (30–200 kb), commonly containing multiple promoters and DNA repeats within untranslated introns. In contrast, plants, with similarly large genomes, generally possess smaller transcription units with smaller introns, because they have evolved defense mechanisms to ensure that transposon insertion within transcription units is not tolerated.

There are important organismal differences that manifest in the types of epigenetic pathways utilized despite high degrees of functional conservation for many mechanisms across species. Differences are, in part, believed to be related to genome size. The vast expansion of the genome with noncoding and repetitive DNA in higher eukaryotes requires more extensive epigenetic silencing mechanisms. This correlates with the fact that mammals and plants employ a full range of repressive histone lysine methylation, DNA methylation, and RNAi silencing mechanisms. Another challenge accompanying multicellularity is how to coordinate and maintain multiple cell types (cellular identity). This is a delicate balance, involving the Polycomb (PcG) and Trithorax (trxG) groups of protein complexes for genome regulation. The PcG proteins, in particular, correlate with the emergence of multicellularity (see Section 12).

Cells within multicellular organisms can be functionally divided into two major compartments: germ cells (totipotent and required for transmission of genetic information to the next generation) and somatic cells (the differentiated "powerhouse" of an organism). There are important questions of how the germ-cell compartment

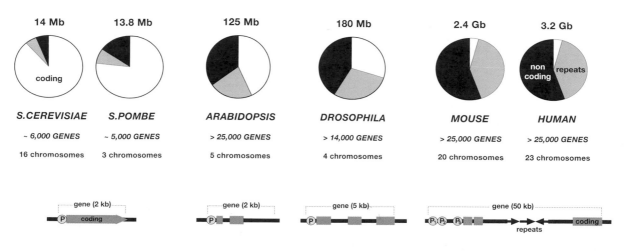

Figure 15. Pie Charts of Organismal Genome Organization

Genome sizes are indicated for the major model organisms used in epigenetic research at the top of each pie chart. The increase in genome size correlates with the vast expansion of noncoding (i.e., intronic, intergenic, and interspersed repeat sequences) and repeat DNA (e.g., satellite, LINE, SINE DNA) sequences in more complex multicellular organisms. This expansion is accompanied by an increase in the number of epigenetic mechanisms (particularly repressive) that regulate the genome. Expansion of the genome also correlates with an increase in size and complexity of transcription units, with the exception of plants; they have evolved mechanisms that are intolerant to insertions or duplications within the transcription unit. P = Promoter DNA element.

maintains totipotency of its epigenome and what mechanisms are involved in erasing, establishing, and maintaining cell fate (cell memory). Because one germ cell can give rise to another germ cell, it essentially has an infinite proliferative potential, as do unicellular "immortal" organisms. However, to fulfill this role, germ cells are for the most part "resting" and unresponsive to external stimuli, so that integrity of their epigenome can be protected. Indeed, mammalian oocytes can be retained in a resting state for more than 40 years. Similarly, adult stem cells (multipotent) are largely a dormant cell population, proliferating (and self-renewing) only when activated by mitogenic stimuli to enter a restricted number of cell divisions. Thus, the makeup of the epigenome is challenged by many intrinsic (e.g., transcription, DNA replication, chromosome segregation) and external (e.g., cytokines, hormones, DNA damage, or general stress responses) signals, particularly if somatic differentiation has forced cells to leave the protective germ-cell and stem-cell environment.

12 Polycomb and Trithorax

Among some of the main effectors that can transduce signals to the chromatin template and participate in maintaining cellular identity (i.e., provide cellular memory) are members of the PcG and trxG groups of genes (Ringrose and Paro 2004). These genes were discovered in *Drosophila* by virtue of their role in the developmental regulation of the *Hox* gene cluster and homeotic gene regulation. PcG and trxG have since been shown to be key regulators for cell proliferation and cellular identity in multicellular eukaryotes. In addition, these groups of genes are involved in several signaling cascades that respond to mitogens and morphogens; regulate stem cell identity and proliferation, vernalization in plants, homeotic transformations and transdetermination, lineage commitment during B- and T-cell differentiation, and many other aspects of metazoan development (see Chapters 11 and 12). We now briefly address what is known about how the PcG and trxG families of genes convert developmental cues into an "epigenetic memory" through chromatin structure.

The PcG and trxG groups of proteins function for the most part antagonistically: The PcG family of proteins establish a silenced chromatin state and the trxG family of proteins in general propagate gene activity. The molecular identification of the *Pc* gene known to stabilize patterns of gene repression over several cell generations provided the first evidence for a molecular mechanism for cellular or epigenetic memory. As well, PC provided an example of a chromodomain-containing protein with a high degree of similarity to the chromodomain of the heterochromatin-associated protein HP1 (Paro and Hogness 1991). As mentioned above, chromodomains are well documented to be specific histone methyl-lysine binding modules (illustrated in Fig. 10).

Approximately 20 PcG genes and at least 15 distinct trxG genes have been identified in *Drosophila*. Functional analyses have shown that these groups of genes constitute a spectrum of diverse proteins yet are highly conserved between eukaryotes. PcG genes encode products that include DNA-binding proteins (e.g., YY1), histone-modifying enzymes (e.g., Ezh2), and other repressive chromatin-associated factors that contain a chromodomain with affinity for H3K27me3 (e.g., PC). trxG genes encode transcription factors (e.g., GAGA or Zeste), ATP-dependent chromatin-remodeling enzymes (e.g., Brahma), and HKMTs such as Ash1 and Trx (or its mammalian homologs MLL, Set1, and the MLL family). In most instances, the trxG and PcG families of proteins function as components of diverse complexes to establish stable chromatin structures that facilitate the expression or silencing of developmentally regulated genes (see Chapters 11 and 12).

Despite recent advances, the mechanism by which PcG- or trxG-containing complexes are targeted to developmentally regulated chromatin regions is not well understood. In *Drosophila*, heritable gene repression requires the recruitment of PcG protein complexes to DNA elements called polycomb response elements (PREs). Equivalent sequences in mammals have remained elusive. It is unclear how PcG protein complexes cause long-range silencing in a PRE-dependent manner, because PREs are usually located kilobases from the transcription start site of target genes. It can be postulated that repulsion or recruitment of PcG complexes may be discriminated by changes in transcriptional activity, or differences in productive versus nonproductive mRNA processing (Pirrotta 1998; Dellino et al. 2004; Schmitt et al. 2005). Current models support PcG binding through interaction with DNA-binding proteins and the affinity of the chromodomain within the PC protein for H3K27me3-modified histones (Cao et al. 2002). However, PcG complexes can also associate in vitro with nucleosomes that lack histone tails (Francis et al. 2004), and furthermore, PRE elements have reduced nucleosome density (Schwartz et al. 2005). The most logical explanation for some of these disparate observations is that PcG binding in vivo would initially require interaction with DNA-bound factors that is then stabilized by association with nucleosomes and modified H3K27me3 in the adjacent chromatin region. Clearly, more research is needed to link existing evidence of how PcG complexes are targeted to regions of chromatin and how they medi-

ate repression. This is likely to be organism-dependent, because there is great heterogeneity in the PcG complexes (see Chapter 11).

Trithorax group proteins maintain in general an active state of gene expression at target genes and overcome (or prevent) PcG-mediated silencing. This transition is even less well understood, but recent evidence suggests that an RNA-based mechanism could provide the trigger for the recruitment of Ash1 to target promoters (Sanchez-Elsner et al. 2006). A number of transient and stable changes in chromatin structure are thought to ensue, perhaps facilitated by intergenic transcription that can establish an open chromatin domain and mediate active histone replacement. Documented chromatin changes include the incorporation of "active" histone-lysine methylation marks by trxG HKMTs such as Trx and Ash1, and the reading of these marks (e.g., the WDR5 recognition of H3K4me; Wysocka et al. 2005). The action of trxG ATP-dependent chromatin-remodeling factors such as Brahma is also required, although how these mechanisms interrelate has yet to be fully determined (for more detail, see Chapter 12).

Many PcG and trxG proteins cooperate to maintain a tightly controlled level of repressed heterochromatin versus active euchromatin in a normal cell. In mammalian somatic interphase nuclei, the nuclear morphology reveals that constitutive domains of pericentromeric heterochromatin are grouped into 15–20 foci (see Fig. 16). Deregulation of cell fate and proliferation control, which leads to developmental abnormalities and cancer, frequently displays abnormal nuclear morphologies. For example, the nuclear organization in PML-leukemia (related to mixed lymphocyte leukemia [MLL]) cells shows an absence of pericentromeric foci (Di Croce 2005). In contrast, senescent (nonproliferating) cells display a nuclear morphology with large ectopic heterochromatin clusters (Narita et al. 2003; Scaffidi et al. 2005). Thus, nuclear morphology appears to be a good marker for distinguishing between normal and aberrant cell states, indicating that nuclear architecture may yet play a regulatory role in maintaining specialized domains of chromatin.

The study of histone modification levels is another indicator of cell normality or abnormality. Many of these changes are attributed to the deregulation of PcG (e.g., Ezh2) or trxG (e.g., MLL) HKMTs, contributing to the progression and even metastatic potential of a tumor (see Section 15). Indeed, the increase in overall levels of either of the above-mentioned proteins is associated with increased risk of prostate cancer, breast cancer, multiple myeloma, or leukemia (Lund and van Lohuizen 2004;

Valk-Lingbeek et al. 2004). In other cases of neoplastic transformation, there is a manifest decrease in repressive histone marks and increase in overall acetylation states (Seligson et al. 2005) causing elevated levels of gene transcription and genomic instability. Clearly, changes in the global control of chromatin, possibly through perturbation of histone-modifying enzymes, affects the functionality of the genome and disrupts the proper gene expression profile of a normal cell.

In the case of cellular senescence, an increase in repressive histone marks is also an indicator of cellular dysfunction. This, concomitant with reduced definition of histone acetylation, can reinforce and even increase the levels of silent chromatin, blocking cellular plasticity and driving cells into an antiproliferative state (Scaffidi et al. 2005). This is largely an age-related effect, although the disease state, progeria, can prematurely advance aging. Conversely, when repressive pericentromeric methyl marks are decreased in mutants lacking the transducing enzyme (Suv39h), cells display increased rates of immortalization, no longer senesce, and show greater rates of genomic instability (Braig et al. 2005). These examples illustrate that chromatin deregulation, demonstrated by the levels of characteristic histone marks, often trans-

duced by PcG and trxG enzymes, and nuclear morphology, is proving to be an important indicator of disease progression.

13 X Inactivation and Facultative Heterochromatin

PcG-mediated gene silencing and X-chromosome inactivation are prime examples for developmentally regulated transitions between active and inactive chromatin states (see Fig. 17), often referred to as facultative heterochromatin. This is in contrast to constitutive heterochromatin (at, e.g., pericentromeric domains), which may by default be induced at noncoding and highly repetitive regions. Facultative heterochromatin occurs at *coding regions* of the genome, where gene silencing is dependent on, and sometimes reversible by, developmental decisions specifying distinct cell fates.

One of the best-studied examples for facultative heterochromatin formation is the inactivation of one of the two X chromosomes in female mammals to equalize the dosage of X-linked gene expression with males that possess only one X (and a heteromorphic Y) chromosome (Chapter 17). Here, chromosome-wide gene silencing of

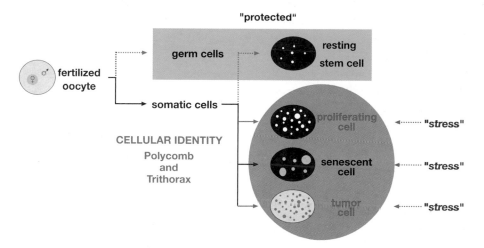

Figure 16. Cellular Identity by PcG and trxG Proteins

Two cell compartments are established during embryogenesis, distinguished by their differentiation potency: They are germ cells (totipotent) and somatic cells (including stem cells) with restricted differentiation potentials. The plasticity of a germ or stem cell's genome expression potential is reflected in reduced levels of repressive histone marks which are no longer visible at pericentromeric foci. Normal proliferating cells typically have a nuclear morphology showing 15–20 heterochromatic foci. Polycomb- and Trithorax-containing complexes operate in specifying the epigenetic and, hence, cellular identity of different lineages. They also function in response to external "stress" stimuli, promoting cellular proliferation and appropriate gene expression. Loss of genome plasticity and proliferation potential occurs in senescent (aging) cells, reflected by abnormally large heterochromatic foci and an overall increased level of repressive histone marks. Highly proliferating tumor cells, however, exhibit changes in the balance of repressive and activating histone marks through the deregulation of PcG and trxG histone-modifying enzymes. This is accompanied by perturbed nuclear morphology.

the inactive X chromosome (Xi) induces a high degree of Xi compaction that is visible as the Barr body, localized in the nuclear periphery of female mammalian cells. How the two alleles of the X chromosomes are counted and how one particular X chromosome is chosen for inactivation are challenging questions in today's epigenetic research.

X inactivation involves a large (~17 kb) noncoding RNA, *Xist*, which appears to act as the primary trigger for chromatin remodeling at the Xi. Although there is the potential to form dsRNA between *Xist* and the antisense transcript *Tsix* (expressed only *before* the onset of X inactivation), no compelling evidence exists for RNAi-dependent mechanisms being involved in the initiation of X inactivation. The X-inactivation center (XIC) and likely DNA "entry" or "docking" sites (postulated to be specialized repetitive DNA elements that are enriched on the X chromosomes) play a role for *Xist* RNA to associate and function as a scaffolding molecule, decorating the Xi in *cis*. *Xist* promotes the recruitment and action of both PRC1 (polycomb respressive complex) and PRC2 complexes, involved in establishing a stable inactive X chromosome. PRC2 components include, for example, the HKMT chromatin-modifying enzyme, EZH2, which catalyzes H3K27me3. PRC1 complex binding may be promoted by both H3K27me3 and histone-modification-independent means, whereas other components of

the complex, such as the Ring1 proteins, ubiquitinate H2A. Such is the heterogeneity of PcG complexes that different components can act independently of other complex components. The chromatin modifications, PcG complex binding, the subsequent incorporation of the histone variant macroH2A along the Xi, and extensive DNA methylation all contribute to generating a facultative heterochromatin structure along the entire Xi chromosome. Once a stable heterochromatic structure is established, *Xist* RNA is no longer required for its maintenance (Avner and Heard 2001; Heard 2005). A similar form of monoallelic silencing is genomic imprinting, which also uses a noncoding or antisense RNA to silence one allelic copy in a parent-of-origin-specific manner (Chapter 19). It is currently not clear whether and how *Dicer*-mutant mouse ES cells would affect the processes of X inactivation or genomic imprinting.

The general paradigm of dosage compensation, a classic epigenetically controlled mechanism, has also been addressed in other model organisms, notably *C. elegans* (Meyer et al. 2004; Chapter 15) and *Drosophila* (Gilfillan et al. 2004; Chapter 16). It is not yet clear whether dosage compensation occurs in birds, despite the fact that they are heterogametic organisms. In *Drosophila*, dosage compensation between the sexes occurs not by X inactivation in the female, but by a twofold up-regulation from the single X chromosome in the male. Intriguingly, two non-

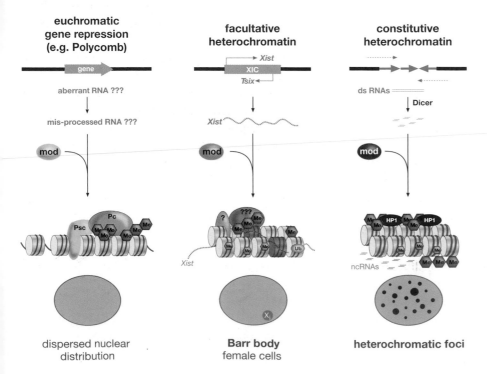

Figure 17. RNA Directed Induction of Repressed Chromatin States

Different forms of silent chromatin have different primary signals, but many are likely to be RNA transcript-related (from aberrant transcripts, to *Xist* RNA, to dsRNAs), depending on the nature of the underlying DNA sequence. This triggers the establishment of a collection of chromatin changes, including a combination of histone modifications (H3K9, H3K27, and H4K20 methylation), the binding of repressive proteins or complexes (e.g., PC or HP1) to the chromatin, DNA methylation, and the presence of histone variants (e.g., macroH2A on the inactive X chromosome). Facultative or constitutive heterochromatin shows visible clustering in the nucleus. Euchromatic repression cannot be determined by nuclear morphology patterns.

coding RNAs, *roX1* and *roX2*, are known to be essential components, and their expression is male-specific. Although similar mechanistic details probably exist between flies and mammals, it is clear that activating chromatin remodeling and histone modifications, notably MOF-dependent H4K16 acetylation on the male X chromosome, plays a key role in *Drosophila* dosage compensation. Exactly how histone-modifying activities, such as the MOF histone acetyltransferase, are targeted to the male X chromosome remains a challenge for future studies. Furthermore, ATP-dependent chromatin-remodeling activities, such as nucleosome-remodeling factor (NURF), are thought to antagonize the activities of the dosage compensation complex (DCC).

Together, this section and Sections 10 and 11 have described mechanisms for RNA-directed chromatin modifications, as they occur for constitutive heterochromatin, the Xi chromosome, and, possibly, also PcG-mediated gene silencing. On the basis of the intriguing parallels, one might postulate that an RNA moiety(s) or unpaired DNA would provide an attractive primary trigger for stabilizing PcG complexes at PREs or compromised promoter function, where they may "sense" the quality of transcriptional processing. Aberrant or stalled elongation and/or splicing errors could spur the interaction between PRE-bound PcG and a promoter, resulting in transcriptional shutdown. Thus, initiation of PcG silencing would be induced by the transition from productive to nonproductive transcription. The extent to which trxG complexes may utilize RNA quality control and/or processing of primary RNA transcripts as part of maintaining transcriptional "ON" states is beginning to be unraveled (Sanchez-Elsner et al. 2006).

14 Reprogramming of Cell Fates

The question of how cell fate can be altered or reversed has long intrigued scientists. The germ cell and early embryonic cells distinguish themselves from other cell compartments as the "ultimate" stem cell by their innate totipotency. Although cell-fate specification in mammals allows for around 200 different cell types, there are, in principle, two major differentiation transitions: from a stem (pluripotent) cell to a fully differentiated cell, and between a resting (quiescent or G_0) and a proliferating cell. These represent the extreme endpoints among many intermediates, consistent with a multitude of different makeups of the epigenome in mammalian development. During embryogenesis, a dynamic increase of epigenetic modifications is detected in the transition from the fertil-

ized oocyte to the blastocyst stage, and then at implantation, gastrulation, organ development, and fetal growth. Most of these modifications or imprints may be erased via transfer of a differentiated cell nucleus to the cytoplasm of an enucleated oocyte. However, some marks may persist, thereby restricting normal development of cloned embryos, and a few could even be inherited as germ-line modifications (g-mod) (see Fig. 18), which, in mammals, are likely to include DNA methylation.

Liver regeneration and muscle cell repair are exceptions of mammalian tissues that can regenerate in response to damage or injury, although most other tissues are unable to be reprogrammed. In other organisms, such as plants and *Axolotl*, certain somatic cells can actually reprogram their epigenome and reenter the cell cycle to regenerate lost or damaged tissue (Tanaka 2003). In general, however, reprogramming of somatic cells is not possible unless they are engineered to recapitulate early development upon nuclear transfer (NT) into an enucleated oocyte. This was first demonstrated in cloned frogs (*Xenopus*), and more recently by the generation of Dolly, the first cloned mammal (Campbell et al. 1996; see Chapter 22).

Three major obstacles to efficient somatic reprogramming in mammals have been identified. First, certain somatic epigenetic marks (e.g., repressive H3K9me3) are stably transmitted through somatic cell divisions and resist reprogramming in the oocyte. Second, a somatic cell nucleus is unable to recapitulate the asymmetry of reprogramming that occurs in the fertilized embryo as a consequence of the differential epigenetic marks inherited by the male and female haploid genomes (see Mayer et al. 2000; van der Heijden et al. 2005; Chapter 20). Third, transmission of imprinted loci that are particularly important in fetal and placental development is not faithfully maintained upon NT (Morgan et al. 2005). Most cloned embryos abort, suggesting that perturbed epigenetic imprints represent a major bottleneck for normal development and could be the cause for the poor efficiencies of assisted reproductive technologies (ART) and the reduced vigor of cloned animals.

The use of embryonic stem cells versus somatic cells shows greatly enhanced reprogramming potential. The demonstration that quiescent cells (a frequent characteristic of stem cells) have a reduction in global H3K9me3 and H4K20me3 states could be a factor indicating enhanced plasticity of the epigenome (Baxter et al. 2004). This is also consistent with the fact that "immortal" unicellular organisms (e.g., yeast) with a largely open and active genome lack several repressive epigenetic mechanisms.

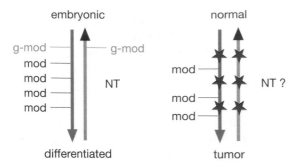

Figure 18. Reprogramming by Nuclear Transfer

During the lifetime of an individual, epigenetic modifications (mod) are acquired in different cell lineages (*left*). Nuclear transfer (NT) of a somatic cell reverses the process of terminal differentiation, eradicating the majority of epigenetic marks (mod); however, some modification that would also be present in the germ line (g-mod) cannot be removed. During neoplastic transformation (from a normal to tumor cell), caused by a series of genetic mutations (*red stars*), epigenetic lesions accumulate. The epigenetic lesions (mod), but not the mutations, can be erased through reprogramming upon NT. This approach evaluates the interplay between genetic and epigenetic contributions to tumorigenesis. (Figure adapted from R. Jaenisch.)

Another feature of normal epigenetic reprogramming in mammals, postfertilization, is its distinct asymmetry. This can first be attributed to different programs of epigenetic specification in the male and female germ cells (Chapters 19 and 20). The sperm genome is largely made up of protamines, although there is a residual but significant level of CENP-A (an H3 histone variant) and other putative epigenetic imprints (Kimmins and Sassone-Corsi 2005), whereas the oocyte is made up of regular nucleosome-containing chromatin. Once fertilized, the sperm and oocyte haploid genomes have another cycle of reprogramming involving DNA demethylation and exchange of histone variants. The modifications can either enhance or balance epigenetic differences of the two parental genomes before nuclear fusion, in the first cell cycle. During differentiation of embryonic (i.e., inner cell mass [ICM]) and extraembryonic (i.e., trophectoderm [TE] and placenta) tissues, different DNA-methylation and histone-modification profiles are established between lineages (Morgan et al. 2005). Somatic cloning cannot faithfully recapitulate these patterns of reprogramming, showing rapid but less extensive demethylation in the first cell cycle, and perturbed DNA methylation and histone lysine methylation between ICM and TE cells.

A closely related concern in somatic cell reprogramming is the fate of imprinted gene loci. For normal embryonic development to proceed, correct allelic expression at imprinted loci is required (Chapter 19).

This was demonstrated by the seminal experiments that generated uniparental embryos (Barton et al. 1984; McGrath and Solter 1984; Surani et al. 1984). Androgenetic embryos (both genomes are of male origin) exhibited retarded embryonic development but hyperproliferation of extraembryonic tissues (e.g., placenta). In gyno- or parthenogenetic embryos (both genomes are of female origin), the placenta is underdeveloped. A parent-specific imprint must therefore be established in the germ cell following erasure of preexisting marks (Chapter 20). It is believed that this occurs for approximately 100 or more imprinted genes, largely involved in systems of resource provision for embryonic and placental development (e.g., Igf2 growth factor). Intriguingly, there is evidence that imprinting may be perturbed during in vitro culture of embryos produced by ART or nuclear transfer (Maher 2005).

15 Cancer

There is a delicate balance between self-renewal and differentiation. Neoplastic transformation (also similarly referred to as tumorigenesis) is regarded as the process whereby cells undergo a change involving uncontrolled cell proliferation, a loss of checkpoint control tolerating the accumulation of chromosomal aberrations and genomic aneuploidies, and mis-regulated differentiation (Lengauer et al. 1998). It is commonly thought to be caused by at least one genetic lesion, such as a point mutation, a deletion, or a translocation, disrupting either a tumor suppressor gene or an oncogene (Hanahan and Weinberg 2000). Tumor suppressor genes become silenced in tumor cells. Oncogenes are activated through dominant mutations or overexpression of a normal gene (proto-oncogene). Importantly, an accumulation of aberrant epigenetic modifications is also associated with tumor cells (see Chapter 24). The epigenetic changes involve altered DNA methylation patterns, histone modifications, and chromatin structure (see Fig. 19). Thus, neoplastic transformation is a complex multistep process involving the random activation of oncogenes and/or the silencing of tumor suppressor genes, through genetic or epigenetic events, and is referred to as the "Knudson two-hit" theory (Feinberg 2004; Feinberg and Tyko 2004). To illustrate, silencing of the retinoblastoma (Rb) gene, a tumor suppressor, causes loss of checkpoint control, which not only provides a proliferative advantage, but also promotes a "second hit" by affecting downstream functions related to chromatin structure which maintain genome integrity (Gonzalo and Blasco 2005). Inappropriate activation of an oncogenic product such as the *myc* gene can have a similar effect (Knoepfler et al 2006).

One question raised by current research is, To what extent do aberrant epigenetic changes contribute to the incidence and overall behavior of a tumor? This was addressed by NT experiments using a melanoma cell nucleus as the donor (Hochedlinger et al. 2004). Any genetic lesions of the donor cell remain; however, NT erases the epigenetic makeup. The tumor incidence of cloned mouse fetuses was then studied, indicating that the spectrum of tumors that arose de novo varied greatly, consistent with different contributions of epigenetic modifications in different tissues that trigger neoplastic progression.

DNA hypomethylation (as opposed to hypermethylation) can occur at discrete loci or over widespread chromosomal regions. DNA *hypo*methylation was, in fact, the first type of epigenetic transition to be associated with cancer (Feinberg and Vogelstein 1983). This has turned out to be a widespread phenotype of cancer cells. At the individual gene level, DNA hypomethylation can be neoplastic due to the activation of proto-oncogenes, the derepression of genes that cause aberrant cell function, or the biallelic expression of imprinted genes (also termed loss of imprinting or LOI) (see Chapters 23 and 24). On a more global genomic scale, broad DNA hypomethylation, particularly at regions of constitutive heterochromatin, predisposes cells to chromosomal translocations and aneuploidies that contribute to cancer progression. This effect is recapitulated in Dnmt1 mutants (Chen et al. 1998). The genomic instability that ensues when there is DNA hypomethylation is due likely to the mutagenic effect of transposon reactivation. With attention turning to the essential role that repressive histone modifications play in maintaining heterochromatin at centromeres and telomeres, evidence has emerged that if these marks are lost, genome instability also results, contributing to cancer progression (Gonzalo and Blasco 2005).

Conversely, DNA *hyper*methylation is concentrated at the promoter regions of CpG islands in many cancers. Silencing of tumor suppressor genes through such aberrant DNA hypermethylation is particularly critical in cancer progression. Recent studies have revealed that there is considerable cross talk between chromatin modifications and DNA methylation, demonstrating that more than one epigenetic mechanism can be involved in the silencing of a tumor suppressor gene. As an illustration, it is known that the tumor suppressor genes, p16 and hMLH1, are silenced by both DNA methylation and repressive histone lysine methylation in cancer (McGarvey et al. 2006).

The deregulation of chromatin modifiers is implicated in many forms of cancer. Certain histone-modifying enzymes become oncogenic, such as the PcG protein EZH2 and the trxG protein MLL, and exert their effect through perturbing a cell's epigenetic identity, which consequently either transcriptionally silences or activates inappropriate genes (Schneider et al. 2002; Valk-Lingbeek et al. 2004). It is clear that the epigenetic identity is crucial to cellular function. In fact, the pattern of global acetyl and methyl histone marks is proving to be a hallmark for the progression of certain cancers, as demonstrated by a study in prostate tumor progression (Seligson et al. 2005).

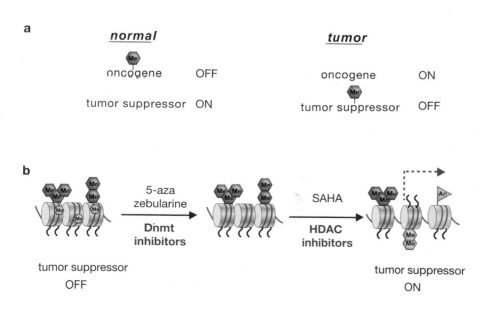

Figure 19. Epigenetic Modifications in Cancer

(*a*) Aberrant epigenetic marks at cancer-causing loci typically involve the derepression of oncogenes or silencing of tumor suppressor genes. Epigenetic marks known to alter a normal cell include DNA methylation, repressive histone methylation, and histone deacetylation. (*b*) The use of epigenetic therapeutic agents for the treatment of cancer has consequences on the chromatin template, illustrated for a tumor suppressor locus. Exposure to Dnmt inhibitors results in a loss of DNA methylation, and exposure to HDAC inhibitors results in the acquisition of histone acetyl marks and subsequent downstream modifications, including active histone methyl marks and the incorporation of histone variants. The cumulative chromatin changes lead to gene re-expression.

The development of drug targets inhibiting the function of the chromatin-modifying effector enzymes has opened up a new horizon for cancer therapeutics (see Fig. 19). The use of DNMT and HDAC inhibitors is in the most advanced stages of clinical trials in this new generation of cancer therapeutics. Zebularine and SAHA are, respectively, two such inhibitors. They are particularly beneficial for cancer cells that have repressed tumor suppressor genes (J.C. Cheng et al. 2004; Garcia-Manero and Issa 2005; Marks and Jiang 2005), because treatment leads to transcriptional stimulation. A major proportion of repressive histone lysine methylation is lost during treatment, most probably due to transcription-coupled histone exchange and nucleosome replacement; however, these inhibitors do not significantly alter H3K9me3 at target promoter regions (McGarvey et al. 2006). It remains to be resolved whether repressive marks that persist could induce subsequent re-silencing of tumor suppressor genes when treatment is paused, thereby counteracting the benefit of "epigenetic therapy." It is possible that a dual epigenetic therapy strategy, using DNMT and HDAC inhibitors, may promise a better prognosis in clinical trials.

Identification of inhibitors to other classes of histone modifiers, namely HKMTs and PRMTs, is currently in the development phase. There are approximately 50 SET domain HKMTs alone in the mammalian genome. Most of the well-characterized enzymes, such as SUV39H, EZH2, MLL, and RIZ, have already been implicated in tumor development (Schneider et al. 2002). Thus, high-throughput screens (HTS) are being employed in efforts to identify small-molecule inhibitors that could be used in exploratory research and, eventually, cancer therapy. All the classes of histone-modifying enzymes are suited for such an approach, as their specific substrate-binding sites (i.e., to histone peptides), in contrast to generic cofactor (e.g., acetyl-CoA and SAM) binding sites, would allow more selective drug development. HTS have been successful for HDACs (Su et al. 2000), PRMTs (D. Cheng et al. 2004), and HKMTs (Greiner et al. 2005).

For the transfer of knowledge to occur from basic to applied research, both hypothesis-driven and empirical approaches are required to ultimately define the efficacy and usefulness of any histone-modifying enzyme inhibitor. For instance, selective HKMT inhibitors against MLL or EZH2 may be valuable therapeutic agents for leukemia or prostate cancer. Alternatively, the use of a SUV39H HKMT inhibitor, which would seem counterintuitive because of the necessity of this enzyme in maintaining constitutive heterochromatin and genome stability, may still preferentially sensitize tumor cells. In addition, analysis of the HDAC inhibitor SAHA has revealed that it may operate through additional pathways that are distinct from transcriptional reactivation (Marks and Jiang 2005). For example, HDAC inhibitors can also sensitize chromatin lesions, inhibiting efficient DNA repair and permitting genomic instabilities that can trigger apoptosis in tumor cells. These observations will have to be monitored when assessing the efficacy of dual combination therapies. Judging from the results to date, however, it is conceivable that combination therapy using HDAC and HKMT inhibitors may be more selective in killing pro-neoplastic cells by driving them into information overflow and chromatin catastrophe. It is hoped that continued research will identify the viable candidates for efficient epigenetic cancer therapy.

16 What Does Epigenetic Control Actually Do?

Approximately 10% of the protein pool encoded by the mammalian genome plays a role in transcription or chromatin regulation (Swiss-Prot database). Given that the mammalian genome consists of 3×10^9 bp, it must accommodate $\sim 1 \times 10^7$ nucleosomes. This gives rise to an overwhelming array of possible regulatory messages, including DNA-binding interactions, histone modifications, histone variants, nucleosome remodeling, DNA methylation, and noncoding RNAs. Yet, the process of transcriptional regulation alone is quite intricate, often requiring the assembly of large multiprotein complexes (>100 proteins) to ensure initiation, elongation, and correct processing of messenger RNA from a single selected promoter. If DNA sequence-specific regulation is so elaborate, one would expect the lower-affinity associations along the dynamic DNA–histone polymer to be even more so. On the basis of these considerations, rarely will there will be one modification that correlates with one epigenetic state. More likely, and as experimental evidence suggests, it is the combination or cumulative effect of several (probably many) signals over an extended chromatin region that stabilizes and propagates epigenetic states (Fischle et al. 2003b; Lachner et al. 2003; Henikoff 2005).

For the most part, transcription factor binding is transient and lost in successive cell divisions. For persistent gene expression patterns, transcription factors are required at each subsequent cell division. As such, epigenetic control can potentiate a primary signal (e.g., promoter stimulation, gene silencing, centromere definition) to successive (but not indefinite) cell generations by the heritable transmission of information through the chromatin template (Fig. 20). Interestingly, in *S. pombe*, Swi6-dependent epigenetic variegation can be suppressed for many cell divisions during both mitosis and meiosis (Grewal and Klar 1996) by histone modifications (most probably H3K9me2). Analogous studies were performed in *Drosophila* using a pulse of an activating transcription factor to transmit cellular memory for *Hox* gene expression during the female germ line (Cavalli and Paro 1999). In both of these examples, epigenetic memory is mediated by chromatin alterations that comprise distinct histone modifications and, most likely, also the incorporation of histone variants.

If histone modifications function together, an imprint may be left on the chromatin template that will help to mark nucleosomes, particularly if a signal is reestablished after DNA replication (Fig. 20). For even more stable inheritance, collaboration between histone modifications, histone variant incorporation, and chromatin remodeling will convert an extended chromatin region into persistent structural alterations that can then be propagated over many cell divisions. Although explained for the inheritance of transcriptional "ON" states, a similar synergy between repressive epigenetic mechanisms will more stably lock silenced chromatin regions, which is further reinforced by additional DNA methylation.

The DNA double helix can be viewed then as a self-organizing polymer which, through its ordering into chromatin, can respond to epigenetic control and amplify a primary signal into a more long-term "memory." In addition, many histone modifications probably evolved in response to intrinsic and external stimuli. In keeping with this, chromatin-modifying enzymes require cofactors, such as ATP (kinases), acetyl-CoA (HATs), and SAM (HKMTs), whose levels are dictated by environmental changes (e.g., diet). Thus, the altered conditions can be translated into a more dynamic or stable DNA–histone polymer. An excellent example is the NAD-dependent HDAC, Sir2, which acts as "sensor" for nutrients and life span/aged cells (Guarente and Picard 2005; Rine 2005). Understanding how these environmental cues are cast into biologically relevant epigenetic signatures, and how they are read, translated, and inherited, lies at the heart of current epigenetic research. It is, however, important to stress that epigenetic control requires an intricate balance between many factors and that functional interaction is not always faithfully reestablished after each cell division. This is a functional contrast with genetics, which involves alteration of the DNA sequence, which is always stably propagated through mitosis and meiosis, if the mutation occurs in the germ line.

An important question arising from the above considerations is how the information contained in the chromatin is maintained from mother to daughter cells. If a cell loses its identity, through disease, misregulation, or reprogramming, is this identity loss accompanied by changes in chromatin structure? Bulk synthesis of most core histones is highly regulated during the cell cycle. Transcription of the core histone genes generally occurs during the S phase, the stage when DNA is replicated (replication coupled). This "coordination" assures that as the amount of DNA is doubled in the cell, there are sufficient core histones to be deposited onto the newly replicated DNA, and thus, the packaging of the DNA occurs simultaneously with DNA replication. As presented above, various regions of chromatin may have distinct differences in histone modifications that program the region to be either transcribed or not. How do domains of the newly synthesized daughter chromatin retain this crucial information for appropriate gene expression? How is the program faithfully templated from one cell generation to the next, or through meiosis and germ-cell formation (sperm and egg)? These central questions await future investigation.

Although initial studies indicated a semiconservative process, wherein a new H3/H4 tetramer is deposited, followed by the incorporation of two new H2A/H2B dimers, recent data have challenged this hypothesis. In this recent model, the "new" H3 and H4 polypeptides, which may already carry several posttranslational modifications, are incorporated as newly synthesized H3/H4 histone dimers together with the "old" H3/H4 dimers segregating between the mother and daughter DNA. If this is the case, then the modified, parental H3/H4 dimers would now also be present with the newly synthesized dimers on the same DNA. Their co-presence may then dictate that appropriate modifications are placed on the newly added dimers (Tagami et al. 2004). This model is attractive and might help explain the inheritance of histone modifications, and thus, the propagation of epigenetic information through DNA replication and cell division. However, more evidence is needed to support the validity of this or other intriguing models to explain the transmission of chromatin marks through cell division.

In closing this chapter, we ask, Does epigenetic control differ in a fundamental way from basic genetic principles? Although we may wish to view Waddington's epigenetic landscape as being demarcated patches of activating versus repressive histone modifications along

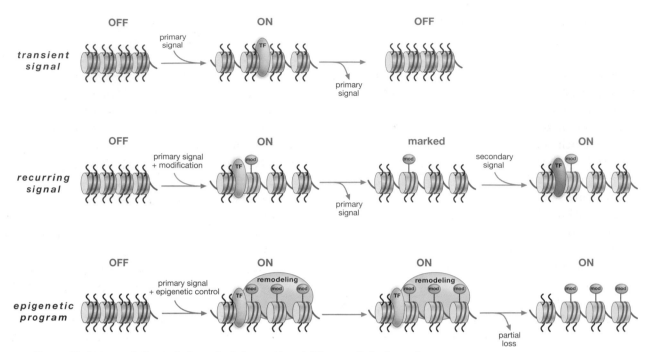

Figure 20. Epigenetic Potentiation of a Primary Signal (Memory/Inheritance)

Classic genetics predicts that gene expression is dependent on the availability and binding of the appropriate panel of transcription factors (TF). Removal of such factors (i.e., a primary signal) results in the loss of gene expression, and thus constitutes a transient activating signal (*top*). Chromatin structure contributes to gene expression, where some conformations are repressive and others active. The activation of a locus may therefore occur through a primary signal and result in the downstream change in chromatin structure, involving active covalent histone marks (mod) and the replacement of core histones with variants (e.g., H3.3). Through cell division, this chromatin structure may only be reestablished in the presence of an activating signal (denoted "recurring signal"). Epigenetic memory results in the maintenance of a chromatin state through cell division, even in the absence of the primary activating signal. Such a memory system is not absolute, but involves multiple levels of epigenetic regulation for remodeling chromatin structure. The dynamic nature of chromatin means that although a chromatin state may be mitotically stable, it is nonetheless prone to change, hence affecting the longevity of epigenetic memory.

the continuum of the chromatin polymer, this notion could easily be overinterpreted. It is only in recent years that we have learned about the major enzymatic systems through which histone modifications might be propagated. This has shaped our current thinking about the stability, and hence the inheritance, of certain histone marks. In addition, it is underscored by the recent discoveries showing that mutations in chromatin-modifying activities, such as nucleosome remodelers (Cho et al. 2004; Mohrmann and Verrijzer 2005), DNMTs (Robertson 2005), HDACs or HMKTs (Schneider et al. 2002), as they are frequently found in abnormal development and neoplasia, are telling examples of the ultimate power of genetic control. As such, tumor incidence in these mutant mice is generally regarded as a *genetic* disease. In contrast, alterations in nucleosome structure, DNA methylation, and histone modification profiles—that are not caused by a mutated gene—would classify as "true"

epigenetic aberrations. Excellent examples of these more plastic systems are stochastic decisions in early embryonic development, reprogramming by nuclear transfer, transcriptional memory, genomic imprinting, mosaic X inactivation, centromere identity, and tumor progression. Genetics and epigenetics are thus closely related phenomena, and inherent to both is their propagation through cell division, which, for genetic control, also comprises the germ line, if mutations occur in germ cells. In the case of other—often too easily categorized—epigenetic modifications, we do not know whether they only reflect a minor and transient response to changes in the external environment or significantly contribute to phenotypic differences that can then be maintained over many, but not indefinite, somatic cell divisions, and occasionally affect the germ line. Even with our greatly improved knowledge of epigenetic mechanisms today, there is little, or no, novel support for Lamarckism.

17 Big Questions in Epigenetic Research

This book discusses the fundamental concepts and general principles that explain how epigenetic phenomena occur, as puzzling as they may seem. Our ultimate goal is to expose the reader to the current understanding of mechanisms that guide and shape these concepts, drawing upon the rich biology from which they emerge. In just a few years, epigenetic research has prompted exciting and remarkable insights and breakthrough discoveries, yet many long-standing questions remain unanswered (see Fig. 21). Although it is tempting to draw broad-brush conclusions and to propound general rules from this progress, we caution against this tendency, suspecting that there will be many exceptions that break the rules. For example, it is clear that striking organismal differences occur. Notably, from unicellular to multicellular organisms, the extent and type of histone modifications, histone variants, DNA methylation, and use of the RNAi machinery does vary.

There are, however, plenty of reasons for renewed energy in research programs designed to gain molecular insights into epigenetic phenomena. Elegant biochemical and genetic studies have already successfully dissected many of the functional aspects of these pathways, in an unprecedented manner. It could therefore be predicted that careful analysis of epigenetic transitions in different cell types (e.g., stem versus differentiated; resting versus proliferating) will uncover hallmarks of pluripotency (Bernstein et al. 2006; Boyer et al. 2006; Lee et al. 2006). This will most likely be valuable in diagnosing which chromatin alter-

ations are significant during normal differentiation as compared with disease states and tumorigenesis. For example, using large-scale mapping approaches with normal, tumor, or ES cells—"epigenetic landscaping" along entire chromosomes (Brachen et al. 2006b; Squazzo et al. 2006; Epigenomics AG, ENCODE, GEN-AU, *EPIGENOME* NoE)—it is anticipated that the knowledge generated could be harnessed for novel therapeutic intervention approaches and work toward promoting a worldwide consortium to map the entire human epigenome (Jones and Martienssen 2005). It is conceivable that differences in the relative abundance between distinct histone modifications, such as the apparent underrepresentation of repressive histone lysine tri-methylation in *S. pombe* and *A. thaliana*, may reflect the greater proliferative and regenerative potential in these organisms as compared to the more restricted developmental programs of metazoan systems. In addition, the functional links between the RNAi machinery, histone lysine methylation, and DNA methylation will continue to provide exciting surprises into the complex mechanisms required for cell-fate determination during development. Similarly, an enhanced understanding of the dynamics and specificity of nucleosome-remodeling machines will contribute to this end. We predict that more "exotic" enzymatic activities will be uncovered, catalyzing epigenetic transitions through modifications of histone and non-histone substrates. It would appear that chromatin alterations, as induced by the above mechanisms, act largely as a response filter to the environment. Thus, it is hoped that this knowledge can ultimately be applied to enhanced therapeutic strategies for resetting

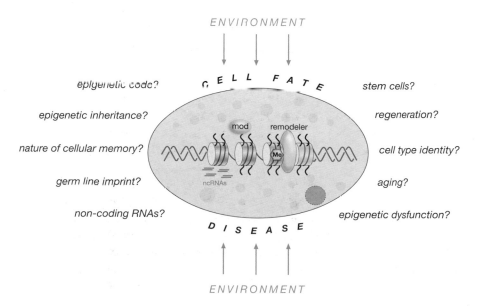

Figure 21. Big Questions in Epigenetic Research

The many experimental systems used in epigenetic research have unveiled numerous pathways and novel insights into the mechanisms of epigenetic control. Many questions, as shown in the figure, still remain and require further elucidation or substantiation in new and existing model systems and methods.

some of an individual's epigenetic response that contribute to aging, disease, and cancer. This includes tissue regeneration, therapeutic cloning (using ES cells and their derivatives), and adult stem cell therapy strategies. It is believed such strategies will extend cellular life span, modulate stress responses to external stimuli, reverse disease progression, and improve assisted reproductive technologies. We predict that understanding the "chromatin basis" of pluripotency and totipotency will lie at the heart of understanding stem cell biology and its potential for therapeutic intervention.

Many fundamental epigenetic questions remain. For example, What distinguishes one chromatin strand from the other allele when both contain the same DNA sequence in the same nuclear environment? What defines the mechanisms conferring inheritance and propagation of epigenetic information? What is the molecular nature of cellular memory? Are there epigenetic imprints in the germ line that serve to keep this genome in a totipotent state? If so, how are these marks erased during development? Alternatively, or in addition, are new imprints added during development that serve to "lock in" differentiated states? We look forward to the next generation of studies (and students) bold enough to tackle these questions with the heart and passion of previous generations of genetic and epigenetic researchers.

In summary, the genetic principles described by Mendel likely govern the vast majority of our development and our outward phenotypes. However, exceptions to the rule can sometimes reveal new principles and new mechanisms leading to inheritance that have been underestimated, and in some cases, poorly understood previously. This book hopes to expose its readers to the newly appreciated basis of phenotypic variation—one that lies outside of DNA alteration. It is our hope that the systems and concepts described in this book will provide a useful foundation for future generations of students and researchers alike who become intrigued by the curiosities of epigenetic phenomena.

References

Ahmad K. and Henikoff S. 2002. The histone variant H3.3 marks active chromatin by replication-independent nucleosome assembly. *Mol. Cell* **9:** 1191–1200.

Allfrey V.G., Faulkner R., and Mirsky A.E. 1964. Acetylation and methylation of histones and their possible role in the regulation of RNA synthesis. *Proc. Natl. Acad. Sci.* **51:** 786–794.

Almeida R. and Allshire R.C. 2005. RNA silencing and genome regulation. *Trends Cell Biol.* **15:** 251–258.

Aparicio O.M., Billington B.L., and Gottschling D.E. 1991. Modifiers of position effect are shared between telomeric and silent mating-type loci in *S. cerevisiae*. *Cell* **66:** 1279–1287.

Avery O.T., Macleod C.M., and McCarty M. 1944. Studies on the chemical nature of the substance inducing transformation of pneumococcal types. Induction of transformation by a desoxyribonucleic acid fraction isolated from pneumococcus Type III. *J. Exp. Med.* **79:** 137–158.

Avner P. and Heard E. 2001. X-chromosome inactivation: Counting, choice and initiation. *Nat. Rev. Genet.* **2:** 59–67.

Bannister A.J. and Kouzarides T. 2005. Reversing histone methylation. *Nature* **436:** 1103–1106.

Bannister A.J., Zegerman P., Partridge J.F., Miska E.A., Thomas J.O., Allshire R.C., and Kouzarides T. 2001. Selective recognition of methylated lysine 9 on histone H3 by the HP1 chromo domain. *Nature* **410:** 120–124.

Barton S.C., Surani M.A., and Norris M.L. 1984. Role of paternal and maternal genomes in mouse development. *Nature* **311:** 374–376.

Bastow R., Mylne J.S., Lister C., Lippman Z., Martienssen R.A., and Dean C. 2004. Vernalization requires epigenetic silencing of FLC by histone methylation. *Nature* **427:** 164–167.

Baxter J., Sauer S., Peters A., John R., Williams R., Caparros M.L., Arney K., Otte A., Jenuwein T., Merkenschlager M., and Fisher A.G. 2004. Histone hypomethylation is an indicator of epigenetic plasticity in quiescent lymphocytes. *EMBO J.* **23:** 4462–4472.

Berger S.L. 2002. Histone modifications in transcriptional regulation. *Curr. Opin. Genet. Dev.* **12:** 142–148.

Bernard P., Maure J.F., Partridge J.F., Genier S., Javerzat J.P., and Allshire R.C. 2001. Requirement of heterochromatin for cohesion at centromeres. *Science* **294:** 2539–2542.

Bernstein B.E., Kamal M., Lindblad-Toh K., Bekiranov S., Bailey D.K., Huebert D.J., McMahon S., Karlsson E.K., Kulbokas E.J., III, Gingeras T.R., et al. 2005. Genomic maps and comparative analysis of histone modifications in human and mouse. *Cell* **120:** 169–181.

Bernstein B.E., Mikkelsen T.S., Xie X., Kamal M., Huebert D.J., Cuff J., Fry B., Meissner A., Wernig M., Plath K., et al. 2006. A bivalent chromatin structure marks key developmental genes in embryonic stem cells. *Cell* **125:** 315–326.

Bernstein E. and Allis C.D. 2005. RNA meets chromatin. *Genes Dev.* **19:** 1635–1655.

Bird A.P. 1986. CpG-rich islands and the function of DNA methylation. *Nature* **321:** 209–213.

Birky C.W., Jr. 2001. The inheritance of genes in mitochondria and chloroplasts: Laws, mechanisms, and models. *Annu. Rev. Genet.* **35:** 125–148.

Bolland D.J., Wood A.L., Johnston C.M., Bunting S.F., Morgan G., Chakalova L., Fraser P.J., and Corcoran A.E. 2004. Antisense intergenic transcription in V(D)J recombination. *Nat. Immunol.* **5:** 630–637.

Bourc'his D. and Bestor T.H. 2004. Meiotic catastrophe and retrotransposon reactivation in male germ cells lacking Dnmt3L. *Nature* **431:** 96–99.

Boyer L.A., Plath K., Zeitlinger J., Brambrink T., Medeiros L.A., Lee T.I., Levine S.S., Wernig M., Tajonar A., Ray M.K., et al. 2006. Polycomb complexes repress developmental regulators in murine embryonic stem cells. *Nature* **441:** 349–353.

Bracken A.P., Dietrich N., Pasini D., Hansen K.H., and Helin K. 2006. Genome-wide mapping of Polycomb target genes unravels their roles in cell fate transitions. *Genes Dev.* **20:** 1123–1136.

Braig M., Lee S., Loddenkemper C., Rudolph C., Peters A.H., Schlegelberger B., Stein H., Dorken B., Jenuwein T., and Schmitt C.A. 2005. Oncogene-induced senescence as an initial barrier in lymphoma development. *Nature* **436:** 660–665.

Brownell J.E., Zhou J., Ranalli T., Kobayashi R., Edmondson D.G.,

Roth S.Y., and Allis C.D. 1996. Tetrahymena histone acetyltransferase A: A homolog to yeast Gcn5p linking histone acetylation to gene activation. *Cell* **84:** 843–851.

Cairns B.R. 2005. Chromatin remodeling complexes: Strength in diversity, precision through specialization. *Curr. Opin. Genet. Dev.* **15:** 185–190.

Campbell K.H., McWhir J., Ritchie W.A., and Wilmut I. 1996. Sheep cloned by nuclear transfer from a cultured cell line. *Nature* **380:** 64–66.

Cao R., Wang L., Wang H., Xia L., Erdjument-Bromage H., Tempst P., Jones R.S., and Zhang Y. 2002. Role of histone H3 lysine 27 methylation in Polycomb-group silencing. *Science* **298:** 1039–1043.

Cavalli G. and Paro R. 1999. Epigenetic inheritance of active chromatin after removal of the main transactivator. *Science* **286:** 955–958.

Chakalova L., Debrand E., Mitchell J.A., Osborne C.S., and Fraser P. 2005. Replication and transcription: Shaping the landscape of the genome. *Nat. Rev. Genet.* **6:** 669–677.

Chalker D.L. and Yao M.C. 2001. Nongenic, bidirectional transcription precedes and may promote developmental DNA deletion in *Tetrahymena thermophila*. *Genes Dev.* **15:** 1287–1298.

Chan S.W., Zilberman D., Xie Z., Johansen L.K., Carrington J.C., and Jacobsen S.E.2004. RNA silencing genes control de novo DNA methylation. *Science* **303:** 1336.

Chen R.Z., Pettersson U., Beard C., Jackson-Grusby L., and Jaenisch R. 1998. DNA hypomethylation leads to elevated mutation rates. *Nature* **395:** 89–93.

Cheng D., Yadav N., King R.W., Swanson M.S., Weinstein E.J., and Bedford M.T. 2004. Small molecule regulators of protein arginine methyltransferases. *J. Biol. Chem.* **279:** 23892–23899.

Cheng J.C., Yoo C.B., Weisenberger D.J., Chuang J., Wozniak C., Liang G., Marquez V.E., Greer S., Orntoft T.F., Thykjaer T., and Jones P.A. 2004. Preferential response of cancer cells to zebularine. *Cancer Cell* **6:** 151–158.

Cho K.S., Elizondo L.I., and Boerkoel C.F. 2004. Advances in chromatin remodeling and human disease. *Curr. Opin. Genet. Dev.* **14:** 308–315.

Chuikov S., Kurash J.K., Wilson J.R., Xiao B., Justin N., Ivanov G.S., McKinney K., Tempst P., Prives C., Gamblin S.J., et al. Regulation of p53 activity through lysine methylation. *Nature* **432:** 353–360.

Clark-Adams C.D., Norris D., Osley M.A., Fassler J.S., and Winston F. 1988. Changes in histone gene dosage alter transcription in yeast. *Genes Dev.* **2:** 150–159.

Cohen F.E. and Prusiner S.B. 1998. Pathologic conformations of prion proteins. *Annu. Rev. Biochem.* **67:** 793–819.

Cosgrove M.S., Boeke J.D., and Wolberger C. 2004. Regulated nucleosome mobility and the histone code. *Nat. Struct. Mol. Biol.* **11:** 1037–1043.

Cremer T. and Cremer C. 2001. Chromosome territories, nuclear architecture and gene regulation in mammalian cells. *Nat. Rev. Genet.* **2:** 292–301.

Daujat S., Zeissler U., Waldmann T., Happel N., and Schneider R. 2005. HP1 binds specifically to Lys26-methylated histone H1.4, whereas simultaneous Ser27 phosphorylation blocks HP1 binding. *J. Biol. Chem.* **280:** 38090–38095.

Dellino G.I., Schwartz Y.B., Farkas G., McCabe D., Elgin S.C., and Pirrotta V. 2004. Polycomb silencing blocks transcription initiation. *Mol. Cell* **13:** 887–893.

Dhalluin C., Carlson J.E., Zeng L., He C., Aggarwal A.K., and Zhou M.M. 1999. Structure and ligand of a histone acetyltransferase bromodomain. *Nature* **399:** 491–496.

Di Croce L. 2005. Chromatin modifying activity of leukaemia associated fusion proteins. *Hum. Mol. Genet.* **14 Spec. No. 1:** R77–R84.

Dou Y. and Gorovsky M.A. 2000. Phosphorylation of linker histone H1 regulates gene expression in vivo by creating a charge patch. *Mol. Cell* **6:** 225–231.

Fan Y., Nikitina T., Zhao J., Fleury T.J., Bhattacharyya R., Bouhassira E.E., Stein A., Woodcock C.L., and Skoultchi A.I. 2005. Histone H1 depletion in mammals alters global chromatin structure but causes specific changes in gene regulation. *Cell* **123:** 1199–1212.

Feinberg A.P. 2004. The epigenetics of cancer etiology. *Semin. Cancer Biol.* **14:** 427–432.

Feinberg A.P. and Tycko B. 2004. The history of cancer epigenetics. *Nat. Rev. Cancer* **4:** 143–153.

Feinberg A.P. and Vogelstein B. 1983. Hypomethylation distinguishes genes of some human cancers from their normal counterparts. *Nature* **301:** 89–92.

Felsenfeld G. and Groudine M. 2003. Controlling the double helix. *Nature* **421:** 448–453.

Fischle W., Wang Y., and Allis C.D. 2003a. Binary switches and modification cassettes in histone biology and beyond. *Nature* **425:** 475–479.

———. 2003b. Histone and chromatin cross-talk. *Curr. Opin. Cell Biol.* **15:** 172–183.

Fischle W., Tseng B.S., Dormann H.L., Ueberheide B.M., Garcia B.A., Shabanowitz J., Hunt D.F., Funabiki H., and Allis C.D. 2005. Regulation of HP1-chromatin binding by histone H3 methylation and phosphorylation. *Nature* **438:** 1116–1122.

Fisher A.G. and Merkenschlager M. 2002. Gene silencing, cell fate and nuclear organisation. *Curr. Opin. Genet. Dev.* **12:** 193–197.

Fodor B.D., Kubicek S., Yonezawa M., O'Sullivan R.J., Sengupta R., Perez-Burgos L., Opravil S., Mechtler K., Schotta G., and Jenuwein T. 2006. Jmjd2b antagonizes H3K9 trimethylation at pericentric heterochromatin in mammalian cells. *Genes Dev.* **20:** 1557–1562.

Fraga M.F., Ballestar E., Paz M.F., Ropero S., Setien F., Ballestar M.L., Heine-Suner D., Cigudosa J.C., Urioste M., Benitez J., et al. 2005. Epigenetic differences arise during the lifetime of monozygotic twins. *Proc. Natl. Acad. Sci.* **102:** 10604–10609.

Francis N.J., Kingston R.E., and Woodcock C.L. 2004. Chromatin compaction by a polycomb group protein complex. *Science* **306:** 1574–1577.

Fukagawa T., Nogami M., Yoshikawa M., Ikeno M., Okazaki T., Takami Y., Nakayama T., and Oshimura M. 2004. Dicer is essential for formation of the heterochromatin structure in vertebrate cells. *Nat. Cell Biol.* **6:** 784–791.

Garcia-Manero G. and Issa J.P. 2005. Histone deacetylase inhibitors: A review of their clinical status as antineoplastic agents. *Cancer Invest.* **23:** 635–642.

Gilbert N., Boyle S., Fiegler H., Woodfine K., Carter N.P., and Bickmore W.A. 2004. Chromatin architecture of the human genome: Gene-rich domains are enriched in open chromatin fibers. *Cell* **118:** 555–566.

Gilfillan G.D., Dahlsveen I.K., and Becker P.B. 2004. Lifting a chromosome: Dosage compensation in *Drosophila melanogaster*. *FEBS Lett.* **567:** 8–14.

Goll M.G., Kirpekar F., Maggert K.A., Yoder J.A., Hsieh C.L., Zhang X., Golic K.G., Jacobsen S.E., and Bestor T.H. 2006. Methylation of tRNAAsp by the DNA methyltransferase homolog Dnmt2. *Science* **311:** 395–398.

Gonzalo S. and Blasco M.A. 2005. Role of Rb family in the epigenetic definition of chromatin. *Cell Cycle* **4:** 752–755.

Gottschling D.E., Aparicio O.M., Billington B.L., and Zakian V.A. 1990. Position effect at *S. cerevisiae* telomeres: Reversible repression of Pol II transcription. *Cell* **63:** 751–762.

Grant P.A., Duggan L., Cote J., Roberts S.M., Brownell J.E., Candau R., Ohba R., Owen-Hughes T., Allis C.D., Winston F., et al. 1997. Yeast Gcn5 functions in two multisubunit complexes to acetylate nucleosomal histones: Characterization of an Ada complex and the SAGA (Spt/Ada) complex. *Genes Dev.* **11:** 1640–1650.

Greiner D., Bonaldi T., Eskeland R., Roemer E., and Imhof A. 2005. Identification of a specific inhibitor of the histone methyltransferase SU(VAR)3-9. *Nat. Chem. Biol.* **1:** 143–145.

Grewal S.I. and Klar A.J. 1996. Chromosomal inheritance of epigenetic states in fission yeast during mitosis and meiosis. *Cell* **86:** 95–101.

Grozinger C.M. and Schreiber S.L. 2002. Deacetylase enzymes: Biological functions and the use of small-molecule inhibitors. *Chem. Biol.* **9:** 3–16.

Guarente L. and Picard F. 2005. Calorie restriction—The SIR2 connection. *Cell* **120:** 473–482.

Hall I.M., Shankaranarayana G.D., Noma K., Ayoub N., Cohen A., and Grewal S.I. 2002. Establishment and maintenance of a heterochromatin domain. *Science* **297:** 2232–2237.

Hanahan D. and Weinberg R.A. 2000. The hallmarks of cancer. *Cell* **100:** 57–70.

Harvey A.C. and Downs J.A. 2004. What functions do linker histones provide? *Mol. Microbiol.* **53:** 771–775.

Hassan A.H., Prochasson P., Neely K.E., Galasinski S.C., Chandy M., Carrozza M.J., and Workman J.L. 2002. Function and selectivity of bromodomains in anchoring chromatin-modifying complexes to promoter nucleosomes. *Cell* **111:** 369–379.

Heard E. 2005. Delving into the diversity of facultative heterochromatin: The epigenetics of the inactive X chromosome. *Curr. Opin. Genet. Dev.* **15:** 482–489.

Henikoff S. 2005. Histone modifications: Combinatorial complexity or cumulative simplicity? *Proc. Natl. Acad. Sci.* **102:** 5308–5309.

Henikoff S. and Ahmad K. 2005. Assembly of variant histones into chromatin. *Annu. Rev. Cell Dev. Biol.* **21:** 133–153.

Herr A.J., Jensen M.B., Dalmay T., and Baulcombe D.C. 2005. RNA polymerase IV directs silencing of endogenous DNA. *Science* **308:** 118–120.

Hirota T., Lipp J.J., Toh B.H., and Peters J.M. 2005. Histone H3 serine 10 phosphorylation by Aurora B causes HP1 dissociation from heterochromatin. *Nature* **438:** 1176–1180.

Hochedlinger K., Blelloch R., Brennan C., Yamada Y., Kim M., Chin L., and Jaenisch R. 2004. Reprogramming of a melanoma genome by nuclear transplantation. *Genes Dev.* **18:** 1875–1885.

Holbert M.A. and Marmorstein R. 2005. Structure and activity of enzymes that remove histone modifications. *Curr. Opin. Struct. Biol.* **15:** 673–680.

Holliday R. 1994. Epigenetics: An overview. *Dev. Genet.* **15:** 453–457.

Ishii K.J. and Akira S. 2005. TLR ignores methylated RNA? *Immunity* **23:** 111–113.

Jackson J.P., Lindroth A.M., Cao X., and Jacobsen S.E. 2002. Control of CpNpG DNA methylation by the KRYPTONITE histone H3 methyltransferase. *Nature* **416:** 556–560.

Jacobson R.H., Ladurner A.G., King D.S. and Tjian R. 2000. Structure and function of a human TAFII250 double bromodomain module. *Science.* **288:** 1422–1425.

Jaenisch R. and Bird A. 2003. Epigenetic regulation of gene expression: How the genome integrates intrinsic and environmental signals. *Nat. Genet.* (suppl.) **33:** 245–254.

Janicki S.M., Tsukamoto T., Salghetti S.E., Tansey W.P., Sachidanandam R., Prasanth K.V., Ried T., Shav-Tal Y., Bertrand E., Singer R.H., and Spector D.L. 2004. From silencing to gene expression: Real-time analysis in single cells. *Cell* **116:** 683–698.

Jeddeloh J.A., Stokes T.L., and Richards E.J. 1999. Maintenance of genomic methylation requires a SWI2/SNF2-like protein. *Nat. Genet.* **22:** 94–97.

Jenuwein T. and Allis C.D. 2001. Translating the histone code. *Science* **293:** 1074–1080.

Jones P.A. and Martienssen R. 2005. A blueprint for a Human Epigenome Project: The AACR Human Epigenome Workshop. *Cancer Res.* **65:** 11241–11246.

Kanellopoulou C., Muljo S.A., Kung A.L., Ganesan S., Drapkin R., Jenuwein T., Livingston D.M., and Rajewsky K. 2005. Dicer-deficient mouse embryonic stem cells are defective in differentiation and centromeric silencing. *Genes Dev.* **19:** 489–501.

Kayne P.S., Kim U.J., Han M., Mullen J.R., Yoshizaki F., and Grunstein M. 1988. Extremely conserved histone H4 N terminus is dispensable for growth but essential for repressing the silent mating loci in yeast. *Cell* **55:** 27–39.

Khochbin S. 2001. Histone H1 diversity: Bridging regulatory signals to linker histone function. *Gene* **271:** 1–12.

Khorasanizadeh S. 2004. The nucleosome: From genomic organization to genomic regulation. *Cell* **116:** 259–272.

Kim J., Daniel J., Espejo A., Lake A., Krishna M., Xia L., Zhang Y., and Bedford M.T. 2006. Tudor, MBT and chromo domains gauge the degree of lysine methylation. *EMBO Rep.* **4:** 397–403.

Kimmins S. and Sassone-Corsi P. 2005. Chromatin remodeling and epigenetic features of germ cells. *Nature* **434:** 583–589.

Klar A.J. 1998. Propagating epigenetic states through meiosis: Where Mendel's gene is more than a DNA moiety. *Trends Genet.* **14:** 299–301.

———. 2004. An epigenetic hypothesis for human brain laterality, handedness, and psychosis development. *Cold Spring Harbor Symp. Quant. Biol.* **69:** 499–506.

Knoepfler P.S., Zhang X.-Y., Cheng P.F., Gafken P.R., McMahon S.B., and Eisenman R.N. 2006. Myc influences global chromatin structure. *EMBO J.* **25:** 2723–2734.

Kornberg R.D. 1974. Chromatin structure: A repeating unit of histones and DNA. *Science* **184:** 868–871.

Lachner M., O'Sullivan R.J., and Jenuwein T. 2003. An epigenetic road map for histone lysine methylation. *J. Cell Sci.* **116:** 2117–2124.

Lachner M., Sengupta R., Schotta G., and Jenuwein T. 2004. Trilogies of histone lysine methylation as epigenetic landmarks of the eukaryotic genome. *Cold Spring Harbor Symp. Quant. Biol.* **69:** 209–218.

Lachner M., O'Carroll D., Rea S., Mechtler K., and Jenuwein T. 2001. Methylation of histone H3 lysine 9 creates a binding site for HP1 proteins. *Nature* **410:** 116–120.

Langst G. and Becker P.B. 2004. Nucleosome remodeling: One mechanism, many phenomena? *Biochim. Biophys. Acta* **1677:** 58–63.

Lee D.Y., Teyssier C., Strahl B.D., and Stallcup M.R. 2005. Role of protein methylation in regulation of transcription. *Endocr. Rev.* **26:** 147–170.

Lee T.I., Jenner R.G., Boyer L.A., Guenther M.G., Levine S.S., Kumar R.M., Chevalier B., Johnstone S.E., Cole M.F., Isono K., et al. 2006. Control of developmental regulators by Polycomb in human embryonic stem cells. *Cell* **125:** 301–313.

Lengauer C., Kinzler K.W., and Vogelstein B. 1998. Genetic instabilities in human cancers. *Nature* **396:** 643–649.

Li E., Bestor T.H., and Jaenisch R. 1992. Targeted mutation of the DNA methyltransferase gene results in embryonic lethality. *Cell* **69:** 915–926.

Lippman Z., Gendrel A.V., Black M., Vaughn M.W., Dedhia N., McCombie W.R., Lavine K., Mittal V., May B., Kasschau K.D., et al. 2004. Role of transposable elements in heterochromatin and epigenetic control. *Nature* **430:** 471–476.

Litt M.D., Simpson M., Gaszner M., Allis C.D., and Felsenfeld G. 2001. Correlation between histone lysine methylation and developmental changes at the chicken beta-globin locus. *Science* **293:** 2453–2455.

Luger K., Mader A.W., Richmond R.K., Sargent D.F., and Richmond T.J. 1997. Crystal structure of the nucleosome core particle at 2.8 Å resolution. *Nature* **389:** 251–260.

Lund A.H. and van Lohuizen M. 2004. Polycomb complexes and silencing mechanisms. *Curr. Opin. Cell Biol.* **16:** 239–246.

Lyko F. 2001. DNA methylation learns to fly. *Trends Genet.* **17:** 169–172.

Maher E.R. 2005. Imprinting and assisted reproductive technology. *Hum. Mol. Genet.* **14 Spec. No. 1:** R133–R138.

Maison C., Bailly D., Peters A.H., Quivy J.P., Roche D., Taddei A., Lachner M., Jenuwein T., and Almouzni G. 2002. Higher-order structure in pericentric heterochromatin involves a distinct pattern of histone modification and an RNA component. *Nat. Genet.* **30:** 329–334.

Marks P.A. and Jiang X. 2005. Histone deacetylase inhibitors in programmed cell death and cancer therapy. *Cell Cycle* **4:** 549–551.

Martens J.H., O'Sullivan R.J., Braunschweig U., Opravil S., Radolf M., Steinlein P., and Jenuwein T. 2005. The profile of repeat-associated histone lysine methylation states in the mouse epigenome. *EMBO J.* **24:** 800–812.

Maurer-Stroh S., Dickens N.J., Hughes-Davies L., Kouzarides T., Eisenhaber F., and Ponting C.P. 2003. The Tudor domain 'Royal Family': Tudor, plant Agenet, Chromo, PWWP and MBT domains. *Trends Biochem. Sci.* **28:** 69–74.

Mayer W., Niveleau A., Walter J., Fundele R., and Haaf T. 2000. Demethylation of the zygotic paternal genome. *Nature* **403:** 501–502.

McClintock B. 1951. Chromosome organization and genic expression. *Cold Spring Harbor Symp. Quant. Biol.* **16:** 13–47.

McGarvey K., Fahrner J., Green E., Martens J., Jenuwein T., and Baylin S.B. 2006. Silenced tumor suppressor genes reactivated by DNA demethylation do not return to a fully euchromatic chromatin state. *Cancer Res.* **66:** 3541–3549.

McGrath J. and Solter D. 1984. Completion of mouse embryogenesis requires both the maternal and paternal genomes. *Cell* **37:** 179–183.

Metzger E., Wissmann M., Yin N., Muller J.M., Schneider R., Peters A.H., Gunther T., Buettner R., and Schule R. 2005. LSD1 demethylates repressive histone marks to promote androgen-receptor-dependent transcription. *Nature* **437:** 436–439.

Meyer B.J., McDonel P., Csankovszki G., and Ralston E. 2004. Sex and X-chromosome-wide repression in *Caenorhabditis elegans*. *Cold Spring Harbor Symp. Quant. Biol.* **69:** 71–79.

Misteli T. 2004. Spatial positioning, a new dimension in genome function. *Cell* **119:** 153–156.

Mito Y., Henikoff J.G., and Henikoff S. 2005. Genome-scale profiling of histone H3.3 replacement patterns. *Nat. Genet.* **37:** 1090–1097.

Mizuguchi G., Shen X., Landry J., Wu W.H., Sen S., and Wu C. 2004. ATP-driven exchange of histone H2AZ variant catalyzed by SWR1 chromatin remodeling complex. *Science* **303:** 343–348.

Mochizuki K., Fine N.A., Fujisawa T., and Gorovsky M.A. 2002. Analysis of a piwi-related gene implicates small RNAs in genome rearrangement in tetrahymena. *Cell* **110:** 689–699.

Mohrmann L. and Verrijzer C.P. 2005. Composition and functional specificity of SWI2/SNF2 class chromatin remodeling complexes. *Biochim. Biophys. Acta* **1681:** 59–73.

Morgan H.D., Santos F., Green K., Dean W., and Reik W. 2005. Epigenetic reprogramming in mammals. *Hum. Mol. Genet.* **14 Spec. No. 1:** R47–R58.

Motamedi M.R., Verdel A., Colmenares S.U., Gerber S.A., Gygi S.P., and Moazed D. 2004. Two RNAi complexes, RITS and RDRC, physically interact and localize to noncoding centromeric RNAs. *Cell* **119:** 789–802.

Muller H.J. 1930. Types of visible variations induced by X-rays in *Drosophila*. *J. Genet.* **22:** 299–334.

Nakayama J., Rice J.C., Strahl B.D., Allis C.D., and Grewal S.I. 2001. Role of histone H3 lysine 9 methylation in epigenetic control of heterochromatin assembly. *Science* **292:** 110–113.

Narita M., Nunez S., Heard E., Narita M., Lin A.W., Hearn S.A., Spector D.L., Hannon G.J., and Lowe S.W. 2003. Rb-mediated heterochromatin formation and silencing of E2F target genes during cellular senescence. *Cell* **113:** 703–716.

Narlikar G.J., Fan H.Y., and Kingston R.E. 2002. Cooperation between complexes that regulate chromatin structure and transcription. *Cell* **108:** 475–487.

Nowak S.J. and Corces V.G. 2004. Phosphorylation of histone H3: A balancing act between chromosome condensation and transcriptional activation. *Trends Genet.* **20:** 214–220.

Orphanides G., LeRoy G., Chang C.H., Luse D.S., and Reinberg D. 1998. FACT, a factor that facilitates transcript elongation through nucleosomes. *Cell* **92:** 105–116.

Pal-Bhadra M., Leibovitch B.A., Gandhi S.G., Rao M., Bhadra U., Birchler J.A., and Elgin S.C. 2004. Heterochromatic silencing and HP1 localization in *Drosophila* are dependent on the RNAi machinery. *Science* **303:** 669–672.

Paro R. and Hogness D.S. 1991. The Polycomb protein shares a homologous domain with a heterochromatin-associated protein of *Drosophila*. *Proc. Natl. Acad. Sci.* **88:** 263–267.

Petersen-Mahrt S. 2005. DNA deamination in immunity. *Immunol. Rev.* **203:** 80–97.

Pirrotta V. 1998. Polycombing the genome: PcG, trxG, and chromatin silencing. *Cell* **93:** 333–336.

Pontier D., Yahubyan G., Vega D., Bulski A., Saez-Vasquez J., Hakimi M.A., Lerbs-Mache S., Colot V., and Lagrange T. 2005. Reinforcement of silencing at transposons and highly repeated sequences requires the concerted action of two distinct RNA polymerases IV in Arabidopsis. *Genes Dev.* **19:** 2030–2040.

Ratcliff F., Harrison B.D., and Baulcombe D.C. 1997. A similarity between viral defense and gene silencing in plants. *Science* **276:** 1558–1560.

Razin A. and Riggs A.D. 1980. DNA methylation and gene function. *Science* **210:** 604–610.

Rea S., Eisenhaber F., O'Carroll D., Strahl B.D., Sun Z.W., Schmid M., Opravil S., Mechtler K., Ponting C.P., Allis C.D., and Jenuwein T. 2000. Regulation of chromatin structure by site-specific histone H3 methyltransferases. *Nature* **406:** 593–599.

Reinberg D., Chuikov S., Farnham P., Karachentsev D., Kirmizis A., Kuzmichev A., Margueron R., Nishioka K., Preissner T.S., Sarma K., et al. 2004. Steps toward understanding the inheritance of repressive methyl-lysine marks in histones. *Cold Spring Harbor Symp. Quant. Biol.* **69:** 171–182.

Reinhart B.J. and Bartel D.P. 2002. Small RNAs correspond to centromere heterochromatic repeats. *Science* **297:** 1831.

Rine J. 2005. Cell biology. Twists in the tale of the aging yeast. *Science* **310:** 1124–1125.

Ringrose L. and Paro R. 2004. Epigenetic regulation of cellular memory by the Polycomb and Trithorax group proteins. *Annu. Rev. Genet.* **38:** 413–443.

Ringrose L., Ehret H., and Paro R. 2004. Distinct contributions of histone H3 lysine 9 and 27 methylation to locus-specific stability of polycomb complexes. *Mol. Cell* **16:** 641–653.

Robertson K.D. 2005. DNA methylation and human disease. *Nat. Rev. Genet.* **6:** 597–610.

Roloff T.C. and Nuber U.A. 2005. Chromatin, epigenetics and stem cells. *Eur. J. Cell Biol.* **84:** 123–135.

Roth S.Y., Denu J.M., and Allis C.D. 2001. Histone acetyltransferases. *Annu. Rev. Biochem.* **70:** 81–120.

Sanchez-Elsner T., Gou D., Kremmer E., and Sauer F. 2006. Noncoding RNAs of trithorax response elements recruit *Drosophila* Ash1 to ultrabithorax. *Science* **311:** 1118–1123.

Santos-Rosa H., Schneider R., Bannister A.J., Sherriff J., Bernstein B.E., Emre N.C., Schreiber S.L., Mellor J., and Kouzarides T. 2002. Active genes are tri-methylated at K4 of histone H3. *Nature* **419:** 407–411.

Sarma K. and Reinberg D. 2005. Histone variants meet their match. *Nat. Rev. Mol. Cell Biol.* **6:** 139–149.

Scaffidi P., Gordon L., and Misteli T. 2005. The cell nucleus and aging: Tantalizing clues and hopeful promises. *PLoS. Biol.* **3:** e395.

Schalch T., Duda S., Sargent D.F., and Richmond T.J. 2005. X-ray structure of a tetranucleosome and its implications for the chromatin fibre. *Nature* **436:** 138–141.

Schmitt S., Prestel M., and Paro R. 2005. Intergenic transcription through a polycomb group response element counteracts silencing. *Genes Dev.* **19:** 697–708.

Schneider R., Bannister A.J., and Kouzarides T. 2002. Unsafe SETs: Histone lysine methyltransferases and cancer. *Trends Biochem. Sci.* **27:** 396–402.

Schreiber S.L. and Bernstein B.E. 2002. Signaling network model of chromatin. *Cell* **111:** 771–778.

Schwartz B.E. and Ahmad K. 2005. Transcriptional activation triggers deposition and removal of the histone variant H3.3. *Genes Dev.* **19:** 804–814.

Schwartz Y.B., Kahn T.G., and Pirrotta V. 2005. Characteristic low density and shear sensitivity of cross-linked chromatin containing polycomb complexes. *Mol. Cell Biol.* **25:** 432–439.

Sekinger E.A., Moqtaderi Z., and Struhl K. 2005. Intrinsic histone-DNA interactions and low nucleosome density are important for preferential accessibility of promoter regions in yeast. *Mol. Cell* **18:** 735–748.

Seligson D.B., Horvath S., Shi T., Yu H., Tze S., Grunstein M., and Kurdistani S.K. 2005. Global histone modification patterns predict risk of prostate cancer recurrence. *Nature* **435:** 1262–1266.

Shen X., Mizuguchi G., Hamiche A., and Wu C. 2000. A chromatin remodeling complex involved in transcription and DNA processing. *Nature* **406:** 541–544.

Shi Y., Lan F., Matson C., Mulligan P., Whetstine J.R., Cole P.A., Casero R.A., and Shi Y. 2004. Histone demethylation mediated by the nuclear amine oxidase homolog LSD1. *Cell* **119:** 941–953.

Shorter J. and Lindquist S. 2005. Prions as adaptive conduits of memory and inheritance. *Nat. Rev. Genet.* **6:** 435–450.

Sims R.J., III, Belotserkovskaya R., and Reinberg D. 2004. Elongation by RNA polymerase II: The short and long of it. *Genes Dev.* **18:** 2437–2468.

Smith C.L. and Peterson C.L. 2005. ATP-dependent chromatin remodeling. *Curr. Top. Dev. Biol.* **65:** 115–148.

Squazzo S.L., O'Geen H., Komashko V.M., Krig S.R., Jin V.X., Jang S.W., Margueron R., Reinberg D., Green R., and Farnham P.J. 2006. Suz12 binds to silenced regions on the genome in a cell-type-specific manner. *Genome Res.* **16:** 890–900.

Sterner D.E. and Berger S.L. 2000. Acetylation of histones and transcription-related factors. *Microbiol. Mol. Biol. Rev.* **64:** 435–459.

Strahl B.D. and Allis C.D. 2000. The language of covalent histone modifications. *Nature* **403:** 41–45.

Strahl B.D., Ohba R., Cook R.G., and Allis C.D. 1999. Methylation of

histone H3 at lysine 4 is highly conserved and correlates with transcriptionally active nuclei in *Tetrahymena*. *Proc. Natl. Acad. Sci.* **96:** 14967–14972.

Su G.H., Sohn T.A., Ryu B., and Kern S.E. 2000. A novel histone deacetylase inhibitor identified by high-throughput transcriptional screening of a compound library. *Cancer Res.* **60:** 3137–3142.

Sung S. and Amasino R.M. 2004. Vernalization in *Arabidopsis thaliana* is mediated by the PHD finger protein VIN3. *Nature* **427:** 159–164.

Surani M.A., Barton S.C., and Norris M.L. 1984. Development of reconstituted mouse eggs suggests imprinting of the genome during gametogenesis. *Nature* **308:** 548–550.

Tagami H., Ray-Gallet D., Almouzni G., and Nakatani Y. 2004. Histone H3.1 and H3.3 complexes mediate nucleosome assembly pathways dependent or independent of DNA synthesis. *Cell* **116:** 51–61.

Tamaru H. and Selker E.U. 2001. A histone H3 methyltransferase controls DNA methylation in *Neurospora crassa*. *Nature* **414:** 277–283.

Tanaka E.M. 2003. Regeneration: If they can do it, why can't we? *Cell* **113:** 559–562.

Thomas J.O. 1999. Histone H1: Location and role. *Curr. Opin. Cell Biol.* **11:** 312–317.

Tsukada Y., Fang J., Erdjument-Bromage H., Warren M.E., Borchers C.H., Tempst P., and Zhang Y. 2006. Histone demethylation by a family of JmjC domain-containing proteins. *Nature* **439:** 811–816.

Tsukiyama T., Daniel C., Tamkun J., and Wu C. 1995. ISWI, a member of the SWI2/SNF2 ATPase family, encodes the 140 kDa subunit of the nucleosome remodeling factor. *Cell* **83:** 1021–1026.

Turner B.M. 2000. Histone acetylation and an epigenetic code. *BioEssays* **22:** 836–845.

Valk-Lingbeek M.E., Bruggeman S.W., and van Lohuizen M. 2004. Stem cells and cancer; the polycomb connection. *Cell* **118:** 409–418.

van Attikum H. and Gasser S.M. 2005. The histone code at DNA breaks: A guide to repair? *Nat. Rev. Mol. Cell Biol.* **6:** 757–765.

van der Heijden G.W., Dieker J.W., Derijck A.A., Muller S., Berden J.H., Braat D.D., van der Vlag J., and de Boer P. 2005. Asymmetry in histone H3 variants and lysine methylation between paternal and maternal chromatin of the early mouse zygote. *Mech. Dev.* **122:** 1008–1022.

Vaquero A., Loyola A., and Reinberg D. 2003. The constantly changing face of chromatin. *Sci. Aging Knowledge Environ.* **2003:** RE4.

Varga-Weisz P.D., Wilm M., Bonte E., Dumas K., Mann M., and Becker P.B. 1997. Chromatin-remodeling factor CHRAC contains the ATPases ISWI and topoisomerase II. *Nature* **388:** 598–602.

Verdel A., Jia S., Gerber S., Sugiyama T., Gygi S., Grewal S.I., and Moazed D. 2004. RNAi-mediated targeting of heterochromatin by the RITS complex. *Science* **303:** 672–676.

Vidanes G.M., Bonilla C.Y., and Toczyski D.P. 2005. Complicated tails: Histone modifications and the DNA damage response. *Cell* **121:** 973–976.

Vignali M., Hassan A.H., Neely K.E., and Workman J.L.. 2000. ATP-dependent chromatin-remodeling complexes. *Mol. Cell. Biol.* **20:** 1899–1910.

Vire E., Brenner C., Deplus R., Blanchon L., Fraga M., Didelot C., Morey L., Van E.A., Bernard D., Vanderwinden J.M., et al. 2005. The Polycomb group protein EZH2 directly controls DNA methylation. *Nature* **439:** 861–874.

Volpe T.A., Kidner C., Hall I.M., Teng G., Grewal S.I., and Martienssen R.A. 2002. Regulation of heterochromatic silencing and histone H3 lysine-9 methylation by RNAi. *Science* **297:** 1833–1837.

Waddington C.H. 1957. *The strategy of the genes.* MacMillan, New York.

Wade P.A., Gegonne A., Jones P.L., Ballestar E., Aubry F., and Wolffe A.P. 1999. Mi-2 complex couples DNA methylation to chromatin remodeling and histone deacetylation. *Nat. Genet.* **23:** 62–66.

Walsh C.P., Chaillet J.R., and Bestor T.H. 1998. Transcription of IAP endogenous retroviruses is constrained by cytosine methylation. *Nat. Genet.* **20:** 116–117.

Watanabe Y., Yokobayashi S., Yamamoto M., and Nurse P. 2001. Pre-meiotic S phase is linked to reductional chromosome segregation and recombination. *Nature* **409:** 359–363.

Watson J.D. 2003. Celebrating the genetic jubilee: A conversation with James D. Watson. Interviewed by John Rennie. *Sci. Am.* **288:** 66–69.

Wei Y., Yu L., Bowen J., Gorovsky M.A., and Allis C.D. 1999. Phosphorylation of histone H3 is required for proper chromosome condensation and segregation. *Cell* **97:** 99–109.

Whetstine J.R., Nottke A., Lan F., Huarte M., Smolikov S., Chen Z., Spooner E., Li E., Zhang G., Colaiacovo M., and Shi Y. 2006. Reversal of histone lysine trimethylation by the JMJD2 family of histone demethylases. *Cell* **125:** 467–481.

Wolffe A.P. and Matzke M.A. 1999. Epigenetics: Regulation through repression. *Science* **286:** 481–486.

Wysocka J., Swigut T., Milne T.A., Dou Y., Zhang X., Burlingame A.L., Roeder R.G., Brivanlou A.H., and Allis C.D. 2005. WDR5 associates with histone H3 methylated at K4 and is essential for H3 K4 methylation and vertebrate development. *Cell* **121:** 859–872.

Yan Q., Huang J., Fan T., Zhu H., and Muegge K. 2003. Lsh, a modulator of CpG methylation, is crucial for normal histone methylation. *EMBO J.* **22:** 5154–5162.

Yu B., Yang Z., Li J., Minakhina S., Yang M., Padgett R.W., Steward R., and Chen X. 2005. Methylation as a crucial step in plant microRNA biogenesis. *Science* **307:** 932–935.

Zhang Y. and Reinberg D. 2001. Transcription regulation by histone methylation: Interplay between different covalent modifications of the core histone tails. *Genes Dev.* **15:** 2343–2360.

Zhang Y., LeRoy G., Seelig H.P., Lane W.S., and Reinberg D. 1998. The dermatomyositis-specific autoantigen Mi2 is a component of a complex containing histone deacetylase and nucleosome remodeling activities. *Cell* **95:** 279–289.

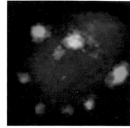

Epigenetics in *Saccharomyces cerevisiae*

Michael Grunstein[1] and Susan M. Gasser[2]

[1]*University of California, Los Angeles, California 90095-1570*
[2]*Friedrich Miescher Institute for Biomedical Research, 4058 Basel, Switzerland*

CONTENTS

1. The Genetic and Molecular Tools of Yeast, 65

2. The Life Cycle of Yeast, 66

3. Yeast Heterochromatin Is Present at the Silent *HM* Mating Loci and at Telomeres, 67

4. Heterochromatin Is Distinguished by a Repressive Structure That Spreads through the Entire Silent Domain, 69

5. Distinct Steps in Heterochromatin Assembly, 70
   5.1 *HM Heterochromatin, 70*
   5.2 *Telomeric Heterochromatin, 71*

6. Histone Deacetylation by Sir2 Provides Binding Sites for the Spread of SIR Complexes, 71

7. Sir2 Deacetylates Histone H4 at Lysine 16, 72

8. Histone Acetylation in Euchromatin Restricts SIR Complex Spreading, 73

9. Telomere Looping, 73

10. Discontinuity of Repression at Natural Subtelomeric Elements by Telomere Looping, 74

11. Trans-interaction of Telomeres, and Perinuclear Attachment of Heterochromatin, 74

12. Inheritance of Epigenetic States, 75

13. Aging and Sir2: Linked by rDNA Repeat Instability, 76

14. Summary, 77

References, 78

GENERAL SUMMARY

The fraction of chromatin in a eukaryotic nucleus that bears its active genes is termed euchromatin. This chromatin condenses in mitosis to allow chromosomal segregation and decondenses in interphase of the cell cycle to allow transcription to occur. However, some chromosomal domains were observed by cytological criteria to remain condensed in interphase, and this constitutively compacted chromatin was called heterochromatin. With the development of new techniques, molecular rather than cytological features have been used to define this portion of the genome, and the constitutively compacted chromatin found at centromeres and telomeres was shown to contain many thousands of simple repeat sequences. Such heterochromatin tends to replicate late in S phase of the cell cycle and is found clustered at the nuclear periphery or near the nucleolus. Importantly, its characteristic nuclease-resistant chromatin structure can spread and repress nearby genes in a stochastic manner. In the case of the fly locus *white*, a gene that determines red eye color, epigenetic repression yields a red and white sectored eye due to a phenomenon called position-effect variegation (PEV). Mechanistically, PEV reflects the recognition of methylated histone H3K9 by heterochromatin protein 1 (HP1) and the spreading of this mark along the chromosomal arm. In *Saccharomyces cerevisiae*, also known as budding yeast, a distinct mechanism of heterochromatin formation has evolved, yet it achieves a very similar result.

S. cerevisiae is a microorganism commonly used in making beer and baking bread. However, unlike bacteria, it is a eukaryote. The chromosomes of budding yeast, like those of more complex eukaryotes, are complexed with histones, enclosed in a nucleus, and replicated from multiple origins during S phase of the cell cycle. Still, the yeast genome is tiny, with only 14 megabase pairs of genomic DNA divided among 16 chromosomes, some not much larger than certain bacteriophage genomes. There are approximately 6000 genes in the yeast genome, closely packed along chromosomal arms with generally less than 2 kb spacing between them. The vast majority of yeast genes are in an open chromatin state, meaning that they are either actively transcribed or can be very rapidly induced. This, coupled with a very limited amount of simple repeat DNA, makes the detection of heterochromatin by cytological techniques virtually impossible in budding yeast.

Nonetheless, using molecular tools, it has been determined that yeast has distinct heterochromatin-like regions adjacent to the telomeres on all 16 chromosomes and at two silent mating loci on chromosome III. Transcriptional repression of these latter two loci is essential for maintaining a mating-competent haploid state. Both the subtelomeric regions and the silent mating-type loci repress integrated reporter genes in a position-dependent, epigenetic manner; they replicate late in S phase and are present at the nuclear periphery. Thus, these loci bear many of the characteristic features of heterochromatin, other than the cytologically visible condensation in interphase. Indeed, for the scientist studying heterochromatin, yeast combines the advantages of a small genome and the genetic and biochemical tools available in microorganisms with important aspects of higher eukaryotic chromosomes.

1 The Genetic and Molecular Tools of Yeast

Yeast provides a flexible and rapid genetic system for studying cellular events. With an approximate generation time of 90 minutes, colonies containing millions of cells are produced after just 2 days of growth. In addition, yeast can propagate in both haploid and diploid forms—greatly facilitating genetic analyses. Like bacteria, haploid yeast cells can be mutated to produce specific nutritional requirements or auxotrophic genetic phenotypes, and recessive lethal mutations can be maintained either in haploids bearing conditional lethal alleles (e.g., temperature-sensitive mutants) or in heterozygous diploids (bearing both wild-type and mutant alleles). The highly efficient system of homologous recombination in yeast allows the alteration of any chosen chromosomal sequence at will. In addition, portions of chromosomes can be manipulated by recombinant means on plasmids that can be stably maintained in dividing yeast cells by including short sequences that provide centromere and origin of DNA replication function. Even linear plasmids, or minichromosomes, which carry telomeric repeats to cap their ends, propagate stably in yeast.

PEV using the fly *white* gene as a reporter has been important in defining epigenetic gene regulation and the genes that affect this unique form of gene repression (see Chapter 5 for more detail). The discovery and characterization of a similar phenomenon near yeast telomeres, called telomere position effect (TPE), has been analogously aided by the use of *Ura3* and *Ade2* reporter genes (Fig. 1). In the presence of 5-fluoroorotic acid (5-FOA), the Ura3 protein converts 5-FOA to 5-fluorouracil (5-FU), an inhibitor of DNA synthesis that causes cell death. However, when *Ura3* is integrated into regions of heterochromatin, the *Ura3* gene is repressed in some, but not all, cells, and only the cells that silence *Ura3* are able to grow in the presence of 5-FOA. Thus, by scoring the efficiency of growth on 5-FOA with a serial dilution drop assay (Fig. 1a), one can quantify the repression of this reporter gene over a very large range (e.g., $10-10^6$-fold). Moreover, mutations that disrupt TPE can be readily identified by monitoring for increased sensitivity to 5-FOA.

Similarly, when the *Ade2* gene is targeted for integration into a region of heterochromatin, the gene is repressed and a precursor in adenine biosynthesis accumulates in the cell, turning it a reddish color. Importantly, the epigenetic nature of *Ade2* repression is visible within a single colony of genetically identical cells: The gene can be "on" in some cells and "off" in others, pro-

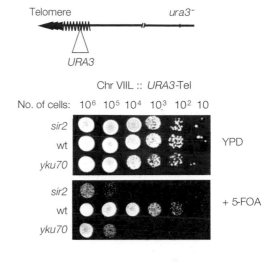

a TPE of *URA3* expression in *S. cerevisiae*

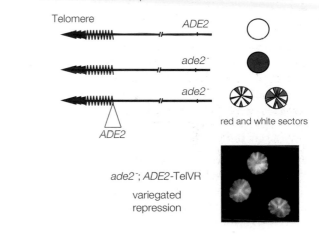

b TPE of *ADE2* expression in *S. cerevisiae*

Figure 1. Silencing and TPE in Yeast

(*a*) The *Ura3* gene, inserted near the telomeric simple TG-rich repeat at the left arm of chromosome VII, is silenced by telomeric heterochromatin in this yeast strain. In normal rich medium (YPD), no growth difference can be detected between wild-type (wt) cells that repress the subtelomeric *Ura3* gene and silencing mutants that lose telomeric heterochromatin and express *Ura3*. In media containing 5-FOA (*lower panel*), on the other hand, cells that repress *Ura3* (e.g., wt cells) can grow, whereas cells that express it (*sir2* and *yku70* mutants) cannot. This is because the *Ura3* gene product converts 5-FOA to the toxic intermediate 5-fluorouracil. The serial dilution/drop assay allows detection of silencing in as few as 1 in 10^6 cells. (*b*) Cells containing the wt *Ade2* gene produce a colony that is "white," whereas those containing mutant ade2 appear red, due to the accumulation of a reddish intermediate in adenine biosynthesis. When the *Ade2* gene is inserted near the telomere at the right arm of chromosome V, it is silenced in an epigenetic manner. The silent *Ade2* state and the active *Ade2* state in genetically identical cells are both inherited, creating red and white sectors in a colony (much like PEV).

ducing red sectors in a white colony background or vice versa (Fig. 1b). Unlike the *Ura3* assay, there is no selection against cells that fail to repress *Ade2*, and therefore, the phenotype of the *Ade2* reporter inserted in subtelomeric heterochromatin demonstrates the switching rate as well as the heritability of the epigenetic state. The *Ade2* color assay provides a striking illustration of the semi-stable nature of both repressed and derepressed states.

Combined with these genetic approaches, biochemical techniques are readily applied to protease-deficient strains grown either synchronously or asynchronously in large cultures. Recently, the battery of tools available has broadened to include sophisticated microarray and protein network techniques that easily accommodate the small genome of yeast. These methods have enabled genome-wide analyses of transcription, transcription factor binding, histone modifications, and protein–protein interactions. This broad range of sophisticated tools has allowed scientists to explore the mechanisms that regulate both the establishment of heterochromatin and its physiological roles in budding yeast. However, before describing these discoveries further, it is necessary to review the life cycle of yeast in more detail.

2 The Life Cycle of Yeast

S. cerevisiae multiplies through mitotic division in either a haploid or a diploid state, by producing a bud that enlarges and eventually separates from the mother cell (Fig. 2a). Haploid yeast cells can mate with each other (i.e., conjugate), since they exist in one of two mating types, termed **a** or α, reminiscent of the two sexes in mammals. Yeast cells of each mating type produce a distinct pheromone that attracts the cells of the opposite mating type: **a** cells produce a peptide of 12 amino acids called **a** factor, which binds to a membrane-spanning **a**-factor receptor on the surface of an α cell. Conversely, α cells produce a 13 aa peptide that binds to the α-factor receptor on the surface of **a** cells. These interactions result in the arrest of the cells in mid-to-late G$_1$ phase of the cell cycle. The arrested cells assume "shmoo"-like shapes (named after the pear-shaped Al Capp cartoon character; Fig. 2b), and the shmoos of opposite mating type fuse at their tips, producing an **a**/α diploid. The mating response is repressed in diploid cells, which propagate vegetatively (i.e., by mitotic division) just like haploid cells. On the other hand, exposure to starvation conditions will induce a meiotic program that results in the formation of an ascus containing four spores, two of

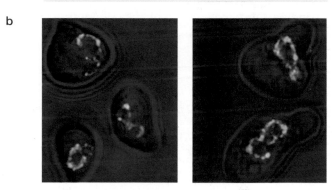

Figure 2. The Life Cycle of Budding Yeast

(*a*) Yeast cells divide mitotically in both haploid and diploid forms. Sporulation is induced in a diploid by starvation, whereas mating occurs spontaneously when haploids of opposite mating type are in the vicinity of each other. This occurs by pheromone secretion, which arrests the cell cycle in G$_1$ of a cell of the opposite mating type, and after sufficient exposure to pheromone, the mating pathway is induced. The diploid state represses the mating pathway. (*b*) In response to pheromone, haploid cells distort toward cells of the opposite mating type. These are called shmoos. The nuclear envelope is visible as green fluorescence.

each mating type. Given sufficient nutrients, the haploid spores grow into cells that are again capable of mating, starting the life cycle over again.

Although haploid yeast cells in the laboratory are usually designated as one mating type or the other, in the wild, yeast switch their mating type nearly each cell cycle (Fig. 3a). Mating-type switching is provoked by an endonuclease activity (HO) that induces a site-specific double-strand break at the *MAT* locus. A gene conversion event then transposes the opposite mating-type

a   Yeast life cycle

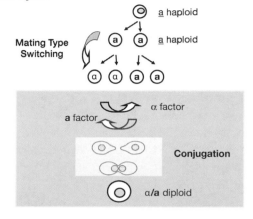

b   Chromosome III

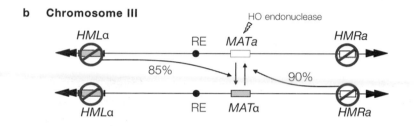

c   Transcriptionally silent domains and silencer elements

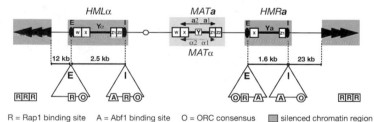

R = Rap1 binding site   A = Abf1 binding site   O = ORC consensus   ▪ silenced chromatin region

Figure 3. Mating Type Switching in Yeast

(*a*) Homothallic yeast strains are able to switch mating type after one division cycle. The switch occurs before DNA replication so that both mother and daughter cells assume the new mating type. (*b*) The position of the silent and expressed mating-type loci on chromosome III are shown here. The active *MAT* locus is able to switch through gene conversion roughly once per cell cycle, due to a double-strand break induced by the HO endonuclease. The percentages indicated show the frequency with which the gene conversion event replaced the *MAT* locus with the opposite mating-type information. The directionality of switching is guaranteed by the recombination enhancer (RE) on the left arm of chromosome III. (*c*) Repression at the silent mating-type loci *HMR* and *HML* is mediated by two silencer DNA elements that flank the silent genes. These silencers are termed E (for essential) or I (for important) (Brand et al. 1997) and provide binding sites for Rap1 (R), Abf1 (A), and ORC (O). Artificial silencers can be created using various combinations of the redundant binding sites, although their efficiency is less than that of the native silencers. *HML*α and *HMR***a** are 12 kb and 23 kb, respectively, from the telomeres of chromosome III. Telomeric heterochromatin domains at chromosome III are silenced independently from the HM loci in a process that is initiated at the telomeres through multiple binding sites for Rap1 (R).

information from a constitutively silent donor locus, *HML*α or *HMR***a**, to the active *MAT* locus. Such strains are called *homothallic*. This means that a vegetatively growing *MAT***a** cell will rapidly produce *MAT*α progeny, and vice versa. Because in the laboratory it is desirable to have cells with stable mating types, laboratory strains are usually constructed to contain a mutant *HO* endonuclease gene, which eliminates cleavage at the *MAT* locus. The loss of *HO* endonuclease activity prevents mating-type switching, producing a *heterothallic* strain. These strains contain silent *HM* loci and an active *MAT* locus whose mating type information is stably either **a** or α. Two silent mating loci (Fig. 3b), one for each "sex," are maintained constitutively silent in an epigenetic manner and have become a classic system for the study of heterochromatin.

3   Yeast Heterochromatin Is Present at the Silent *HM* Mating Loci and at Telomeres

The three mating-type loci, *HML*α, *MAT*, and *HMR***a** are located on chromosome III and contain the information that determines α or **a** mating type in yeast. *HML*α (~11 kb from the left telomere) and *HMR***a** (~23 kb from the right telomere; Fig. 3b,c) are situated between short DNA elements called E and I silencers. Only when either of the silent cassettes is copied and integrated into the active *MAT* locus is it capable of transcription in a normal cell. The transfer of *HML*α information into *MAT* results in an α mating type (*MAT*α) cell, whereas the transfer of *HMR***a** information into *MAT* results in the **a** mating type (*MAT***a**)(Fig. 3b). This shows that the genes and promoters at the *HM* loci are completely intact, although they remain stably repressed when they are positioned at *HMR* and *HML*. This is essential for the maintenance of mating potential, because the combined expression of **a** and α transcripts in the same cell results in a non-mating sterile state. The scoring of sterility as a phenotype proved very useful for identifying mutations that impair silencing at the *HM* loci. In this manner, the silent information regulatory proteins, *SIR1*, *SIR2*, *SIR3*, and *SIR4*, were identified as being essential for the full repression of silent *HM* loci (for review, see Rusche et al. 2003). Mutations in *sir2*, *sir3*, or *sir4* caused a complete loss of silencing, whereas in *sir1* mutants, only a fraction of *MATa* cells were unable to mate due to a loss of *HM* repression. Taking advantage of the partial phenotype of *sir1*-deficient cells, it could be shown that the two alternative states (mating and non-mating) are heritable through successive cell divisions in genetically identical cells (Pillus and Rine 1989). This provided a clear demonstration that mating-type repression displays the hallmark characteristic of epigenetically controlled repression. In addition, it was shown from other studies that the amino termini of histones H3 and H4, repressor activator protein 1 (Rap1), and the origin recognition complex (ORC) are also involved as structural components of heterochromatin (for review, see Rusche et al. 2003).

Heterochromatin is also present immediately adjacent to the yeast telomeric repeat DNA ($C_{1-3}A/TG_{1-3}$). As mentioned above, when reporter genes such as *Ura3* or *Ade2* were integrated adjacent to these telomeric repeats, they were repressed in a variegated and epigenetic manner (Gottschling et al. 1990). This TPE shared the *HM* requirement for Rap1, Sir2, Sir3, Sir4, and the histone amino termini (Kayne et al. 1988; Aparicio et al. 1991). Genetics argued strongly that with the exception of *Sir1*, similar mechanisms silence genes at the *HM* mating loci and at telomere-adjacent sites. Moreover, given that the subtelomeric reporters could switch at detectable rates between silent and expressed states, the gene repression appeared to be very similar to fly PEV.

In yeast, the four Sir proteins that mediate repression share no extensive homology among themselves, and the Sir1, Sir3, and Sir4 proteins appear to be conserved only in *S. cerevisiae* and closely related budding yeasts. Sir2, on the other hand, is the founding member of a large family of NAD-dependent histone deacetylases, which is conserved from bacteria to man (Fig. 4). A role for Sir2-like histone deacetylases in transcriptional repression is observed even in organisms such as fission yeast and flies, which lack the other Sir proteins. The *Schizosaccharomyces pombe* Sir2 activity is required for transcriptional silencing near telomeres, and *Drosophila* Sir2 affects the stability of PEV (for review, see Chopra and Mishra 2005). The coupling of NAD hydrolysis with deacetyla-

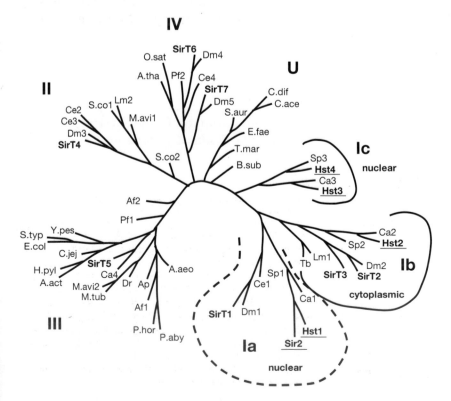

Figure 4. Sir2 Family of Deacetylases

Sir2 is the founding member of a large family of NAD-dependent deacetylases. The Sir2 family of proteins is unusually conserved and is found in organisms that range from bacteria to humans, and contains both nuclear and cytoplasmic branches of the evolutionary tree. This phylogenetic unrooted tree of Sir2 homologs was generated using CLUSTAL W® and TREEVIEW® programs to compare the core domain sequences of homologs identified in cDNA and unique libraries. The six subclasses and unlinked group (U) are described in Frye (2000). The mammalian homologs are labeled SirT1–7 and are in bold, and the budding yeast proteins are underlined. Other species are indicated by the species name. (Modified, with permission, from Frye 2000 [© Elsevier].)

tion by Sir2 produces O-acetyl ADP ribose, an intermediate that may have a function of its own (Tanner et al. 2000; also see Section 13). It is important to note that the Sir2 family of enzymes modifies many substrates other than histones, with a large branch of the Sir2 family actually being cytoplasmic enzymes (Fig. 4). The diversity of Sir2 functions is illustrated by the fact that mammalian Sir2 deacetylates the transcription factors FOXO and p53 in response to stress and DNA damage, altering their interaction. In budding yeast, Sir2 has an important role in addition to gene silencing, which is to suppress nonreciprocal recombination in the highly repetitive genes of the rDNA locus that is found within the nucleolus (Gottlieb and Esposito 1989).

4 Heterochromatin Is Distinguished by a Repressive Structure That Spreads through the Entire Silent Domain

Repression of gene activity in euchromatin can occur due to the presence of a repressive protein or complex that recognizes a specific sequence in the promoter of a gene, thus preventing movement or engagement of the transcription machinery. Heterochromatic repression occurs through a different mechanism that is not promoter-specific: Repression initiates at specific sites, yet spreads continuously throughout the domain, silencing any and all promoters in the region (Fig. 5) (Renauld et al. 1993). This was most clearly demonstrated by the use of chro-

matin-immunoprecipitation techniques, which showed that Sir2, Sir3, and Sir4 proteins interact physically with chromatin throughout the subtelomeric domain of silent chromatin (Hecht et al. 1996; Strahl-Bolsinger et al. 1997). Evidence that this induces a repressive, less accessible chromatin structure comes from other approaches. For instance, it was shown that the DNA of silenced chromatin was not methylated efficiently in yeast cells that express a bacterial *dam* methylase, although the enzyme readily methylated sequences outside the silent region. This suggested that heterochromatin can restrict access to macromolecules like *dam* methyltransferase (Gottschling 1992). Similarly, the approximately 3-kb *HMR* locus in isolated nuclei is preferentially resistant to certain restriction endonucleases (Loo and Rine 1994), and nucleosomes were shown to be tightly positioned between two silencer elements, creating nuclease-resistant domains at silent, but not active, *HM* loci (Weiss and Simpson 1998). Thus, yeast heterochromatin clearly assumes a distinct chromatin structure.

The extent to which either yeast or metazoan heterochromatin is hyper-condensed, and condensation sterically hinders access to transcription factors, is less certain. Surprisingly, the repressive complex formed by the interaction of Sir proteins and histones appears to be dynamic, because Sir proteins can be incorporated into *HM* silent chromatin even when cells are arrested at a stage in the cell cycle when heterochromatin assembly generally does not occur (Cheng and Gartenberg 2000).

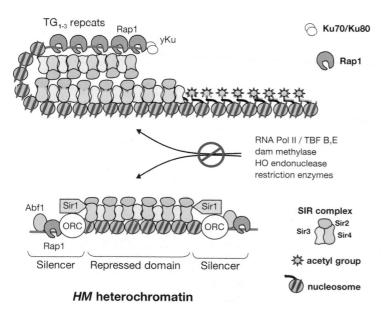

Telomeric heterochromatin

TG$_{1-3}$ repeats   Rap1   yKu   Ku70/Ku80   Rap1

RNA Pol II / TBF B,E
dam methylase
HO endonuclease
restriction enzymes

Abf1   Sir1   Sir1   ORC   ORC   Rap1

SIR complex   Sir2   Sir3   Sir4   acetyl group   nucleosome

Silencer   Repressed domain   Silencer

***HM* heterochromatin**

Figure 5. Model for Yeast Heterochromatin at Telomeres and the *HM* Loci

The telomere and HM silencer mechanisms for nucleating SIR complex spreading both use Rap1, Sir2, Sir3, and Sir4, yet they differ in that telomeres also rely on yKu whereas the HM silencer elements use the factors ORC, Abf1, and Sir1. Telomeric heterochromatin is thought to fold back onto itself to form a cap that protects the telomere from degradation and whose condensation and folding silences genes. In the case of HM heterochromatin, the repressed domain between the silencer elements consists of closely spaced nucleosomes that form a condensed structure. Both the telomeric and HM silent regions are inaccessible to the transcription machinery and degradative enzymes.

This may explain why Sir-bound heterochromatin can serve as a binding site for certain transcription factors even in its repressed state (Sekinger and Gross 1999). Although such studies argue that heterochromatin does not act by sterically hindering all non-histone-protein access, no obvious transcription occurs, and engaged RNA polymerases cannot be detected experimentally. Experiments by Chen and Widom (2005) argue that the step which is specifically prevented by heterochromatin is the recruitment of complexes containing RNA polymerase II and the promoter-binding transcription factors TFIIB and TFIIE. Thus, although the silent yeast chromatin is dynamic, allowing SIR factors and possibly some transcription factors to exchange, it selectively impedes the binding of the basal transcription machinery, and thereby blocks the production of mRNA (for more detail, see Chapter 10).

5 Distinct Steps in Heterochromatin Assembly

The assembly of heterochromatin in all species involves a series of molecular steps, several of which have been identified in budding yeast. One of the best characterized is the site-specific nucleation of heterochromatin, an event that requires sequence-specific DNA-binding factors. Next, heterochromatin spreads from the initiation site. Its spreading is limited by specific boundary mechanisms that are well characterized in yeast. Finally, yeast has been useful for demonstrating a role for subnuclear compartments in heterochromatin-mediated repression. The assembly of heterochromatin at telomeres varies in some ways from its assembly at *HM* loci, but both reflect a very similar principle: the presence of specific DNA-binding factors that nucleate the spread of general repressors. We describe these mechanisms in detail below (Fig. 6).

5.1 HM Heterochromatin

The silent mating loci *HML* and *HMR* are bracketed by short DNA elements termed silencers, labeled E (for essential) and I (for important; Fig. 3b,c). Silencer elements provide binding sites for at least two multifunctional nuclear factors, namely Rap1 and Abf1, as well as the origin recognition complex (ORC) (Brand et al. 1987). Although the deletion of *HMR*-E, which has all three recognition sites, has a much stronger effect on silencing than deletion of *HMR*-I, which has two, each silencer at *HMR* and at *HML* can serve as a specific nucleation site for Sir silencing complex recruitment, to promote subsequent Sir protein spreading along the intervening nucleosomes. Contact in vitro between Rap1 molecules at two separate binding sites argued that direct interaction may occur between the factors bound at E and I silencers through looping of the repressed domain, which would explain the cooperative effects by E and I on the initiation of repression (Hofmann et al. 1989).

Redundancy of silencer element function is a hallmark of repression by heterochromatin and is also true within a silencer element. Rap1 and Abf1, which are general transcription factors, and ORC, which seeds the pre-replication complex at origins of replication, function in a redundant manner. This was demonstrated by deletion experiments: DNA-binding sites for any two of these factors are sufficient to allow silencing (Brand et al. 1987), despite the fact that Rap1, Abf1, and ORC share little structural similarity. The explanation for their redundancy comes from an analysis of the proteins they recruit. For example, Rap1 recruits Sir4 both at *HM* silencers and at telomeres, whereas Abf1 interacts with Sir3, and ORC has high affinity for Sir1, a SIR factor specific for *HM* repression (for review, see Rusche et al. 2003). Sir1 itself interacts directly with the amino terminus of Sir4, providing the bridge between ORC and the SIR2-3-4 complex. Thus, the various silencer binding factors all lead to the recruitment of Sir4 and, in turn, the SIR2-3-4 complex, which is required in all cases for repression. The apparent redundancy among Rap1, Abf1, and ORC (at silencers), as well as the Ku heterodimer (at telomeres, see below), can be attributed to the ability of each to nucleate repression by direct contact with different components of the SIR complex.

It should be noted that Sir1 is involved primarily in the establishment rather than the maintenance of heterochromatic repression. Once Sir1 helps establish silencing, it is no longer needed for the stable maintenance of the repressed state (Pillus and Rine 1989). The important role played by Sir1 in establishment was shown by tethering the protein artificially through a Gal4 DNA-binding domain to Gal4-binding sites, which replaced the *HMR*-E silencer. In this context, GBD-Sir1 can efficiently nucleate repression, rendering the silencer and its binding factors unnecessary (Chien et al. 1993). Nonetheless, the Sir1-targeted repression still required all the other Sir proteins and intact histone tails. This argues that one of the primary roles of the silencer-binding factors is to attract Sir1, which in turn nucleates repression by recruiting the other Sir proteins to interact with adjacent nucleosomes. In support of this is the fact that, unlike the other Sir proteins, Sir1 does not spread with the SIR complex beyond the silencers (Fig. 5) (Rusche et al. 2002).

STEP 1)   Recruitment of Sir4, then Sir2 and Sir3 to telomere-bound Rap1

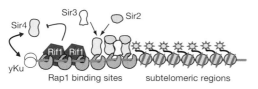

STEP 2)   Sir2-mediated deacetylation of histone H4K16

STEP 3)   Spreading of the SIR complex along nucleosomes

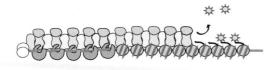

STEP 4)   Folding of a silent telomere into a higher-order structure

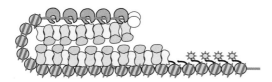

Figure 6. Steps in the Assembly of Telomeric Heterochromatin

(*Step 1*) At telomeres, Rap1 and yKu recruit Sir4 even in the absence of Sir2 or Sir3. Only Sir4 can be recruited, in the absence of the other Sir proteins, and its binding is antagonized by Rif1 and Rif2 (Mishra and Shore 1999). (*Step 2*) Sir4-Sir2 and Sir4-Sir3 interact strongly, creating Sir complexes along the TG repeats. Sir2 NAD-dependent histone deacetylase activity is stimulated by complex formation, and Sir2 deacetylates the acetylated histone H4 K16 residue in nearby nucleosomes. (*Step 3*) SIR complexes spread along the nucleosomes, perhaps making use of the O-acetyl ADP ribose intermediate produced by NAD hydrolysis (Liou et al. 2005). Sir3 and Sir4 bind the deacetylated histone H4 tails. Although the deacetylated histone H3 amino-terminal tail also binds Sir3 and Sir4 proteins, it is not shown here. (*Step 4*) The silent chromatin "matures" at the end of M phase to create an inaccessible structure. This may entail higher-order folding and sequestering at the nuclear envelope.

5.2 Telomeric Heterochromatin

At telomeres, an RNA-based enzyme called telomerase maintains a simple but irregular TG-rich repeat of 300–350 bp in length, which provides 16–20 consensus sites for Rap1. The array of Rap1-binding sites forms a non-nucleosomal cap on the chromosomal end and plays a critical role in telomere length maintenance (Marcand et al. 1997). Along the telomeric repeat, Rap1

binds its consensus through a core DNA-binding domain and binds Sir4 through its carboxy-terminal domain, even in the absence of the other SIR proteins or the H4 amino terminus. Since the disruption of Sir4 prevents other proteins from binding to telomeric chromatin (Luo et al. 2002), Sir4 appears to be a crucial link between nucleating events and the ensuing silent chromatin structure (Fig. 6).

The DNA end-binding complex yKu70/yKu80 helps Rap1 recruit Sir4 to chromosome ends. Indeed, loss of yKu strongly derepresses telomeric silencing, and a targeted GBD-yKu fusion can efficiently nucleate repression at silencer-compromised reporter genes. The requirement for yKu can be bypassed by elimination of the Rap1-binding factor Rif1, which competes for the interaction of Sir4 with the Rap1 carboxy-terminal domain (Fig. 6) (Mishra and Shore 1999). The cooperative effects of yKu and Rap1 in the nucleation of heterochromatin are demonstrated by the observation that 600 bp of telomeric repeat sequence, which provides more than 30 Rap1-binding sites, is not sufficient to nucleate repression at internal chromosomal loci, although insertion of 900 bp or about 45 Rap1-binding sites can (Stavenhagen and Zakian 1994). It should be noted that at promoters dispersed throughout the yeast genome, Rap1 serves as a general transcription factor contributing to the activation of many genes, particularly those encoding ribosomal proteins. Why Rap1 recruits activators to these promoters rather than nucleating heterochromatin, is presently unknown.

6 Histone Deacetylation by Sir2 Provides Binding Sites for the Spread of SIR Complexes

The molecular interactions of the SIR proteins have been well-characterized, with Sir4 playing a key scaffolding role for their assembly. Sir4 interacts strongly with Sir2 in vitro. Sir4 also interacts independently with Sir3, whereas Sir3 and Sir2 appear to interact very weakly (Moazed et al. 1997; Strahl-Bolsinger et al. 1997; Hoppe et al. 2002). Sir3 and Sir4 also homodimerize (Moretti et al. 1994). Nonetheless, when coordinately expressed in insect cells, Sir2, Sir3, and Sir4 are readily isolated as a stable 360-kD complex with a 1:1:1 stoichiometry of SIR proteins (Cubizolles et al. 2006). Consistent with a functional heterotrimeric complex of SIR2-3-4, it was shown by chromatin immunoprecipitation that the three SIR components spread to equal extents throughout a heterochromatic domain (Hecht et al. 1996; Strahl-Bolsinger et al. 1997). It is evident, nonetheless, that Sir3 has a special role in this process, because Sir3 overexpression can extend the silent domain, coincident with the spreading of Sir3 from its normal ~3 kb to ~15 kb (Renauld et al. 1993; Hecht et al. 1996). Imbalanced expression of Sir2 or Sir4 alone, or even expression of subdomains of either protein, has precisely the opposite effect: Overexpression of either Sir2 or Sir4 disrupts TPE, although the coordinated ectopic expression of Sir3 and Sir4 counteracts this imbalance and again restores silencing (Maillet et al. 1996). This underscores the importance of dosage within the SIR complex for its repressive function, which is also true for Polycomb complexes in flies. Consistent with a unique ability of overexpressed Sir3 to spread along chromatin, it was demonstrated that Sir3 can form a stable multimer in vitro (Liou et al. 2005).

The platform upon which the SIR complex spreads comprises nucleosomes with deacetylated histone H3 and H4 amino termini (Braunstein et al. 1996; Suka et al. 2001). The manner in which SIR proteins interact with histones helps explain how spreading occurs (Fig. 7). Sir3 and Sir4 proteins bind deacetylated histone H3 and H4 amino termini in vitro and in vivo (Hecht et al. 1995, 1996), and neither the H2A nor H2B tail is required for this interaction. The most important histone region in this regard is contained in residues 16–29 of histone H4, of which lysine 16 in particular must be deacetylated or positively charged for Sir3 to bind (Johnson et al. 1990, 1992). Unlike mutations at other acetylation sites, mutation of H4K16 by even a conservative change disrupts telomeric silencing completely. The histone H3/H4 tails, in particular the region of H4 (residues 16–24), also promote nucleosome array compaction in vitro, in which case the acetylation state of H4K16 is likely to regulate higher-order folding of the nucleosomal fiber. How is deacetylation of H4K16 regulated in vivo?

7 Sir2 Deacetylates Histone H4 at Lysine 16

Sir2 is an NAD-dependent histone deacetylase whose activity is enhanced by association with Sir4. Sir2 activity links deacetylation with the conversion of NAD to O-acetyl-ADP-ribose using an ADP-ribosyl transferase activity (Tanner et al. 2000). Given that a positively charged H4 lysine 16 is critical for forming heterochromatin, it is striking that Sir2 can deacetylate in vitro and in vivo H4 lysine 16, although the enzyme also deacetylates other lysines within the H4 amino terminus and H3 lysine 9 and lysine 14 as well (Imai et al. 2000; Suka et al. 2002; Cubizolles et al. 2006). All these target sites are within domains of H3 and H4 that are required for silencing. Interestingly, O-acetyl-ADP-ribose itself promotes not only the multimerization of Sir3, but also the interaction of Sir3 with Sir4-Sir2 in vitro (Liou et al. 2005). Together these data argue that histone deacetylation by Sir2 promotes the formation and multimerization of the Sir complex, as well as preparing deacetylated binding sites on adjacent nucleosomes for SIR protein binding.

We summarize the different steps for the initiation and spreading of heterochromatin in telomeric regions and at the *HM* loci in Figure 6. At telomeres, Rap1 and yKu recruit Sir4, Sir4 recruits Sir2 to deacetylate histone H4 and H3 amino-terminal tails. Sir4 also recruits Sir3. The deacetylation of the histone tails produces Sir3/Sir4-binding sites and nucleates binding of the SIR2-3-4 complex. The mutual interaction of Sir3 with Sir4, and of both with histone amino termini, is thought to stabilize the Sir complex on the nucleosomal fiber, allowing it to spread along the histone tails. Finally, the folding of the chromatin fiber (discussed below) may stabilize the repressed state. Most of these events are likely to be very similar at *HM* loci, although the initial recruitment of Sir4 is mediated by Rap1, Abf1, ORC, and Sir1. What then, causes spreading to stop?

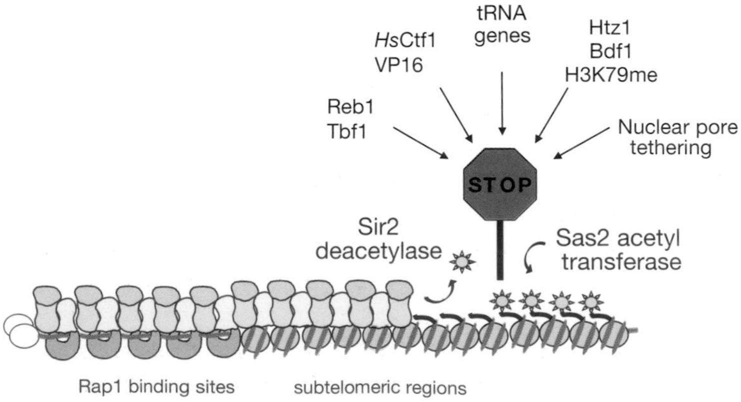

Figure 7. Heterochromatin Boundary Function in Budding Yeast

Spreading of heterochromatin through deacetylation of histone H4 K16 by Sir2 is limited by the competing activity of Sas2 histone acetyltransferase which acetylates H4K16 in adjacent euchromatin, thus preventing Sir3 binding. Methylation of H3K79 in adjacent euchromatin also affects the spreading of heterochromatin. In addition, factors such as Reb1, Tbf1, and mammalian or viral factors Ctf1 or VP16, nuclear pore tethering, and the presence of tRNA genes may also mediate boundary function. It is conceivable that several of these factors function through the recruitment of histone acetyltransferases.

8 Histone Acetylation in Euchromatin Restricts SIR Complex Spreading

Since the deacetylation of histone H4 lysine 16 by Sir2 is crucial to the formation of heterochromatin, it is not surprising that modification of this site also plays a key role in providing a barrier to heterochromatin propagation. Interestingly, of all histone acetylation sites, only H4K16 is modified in monoacetylated H4 of euchromatin (Clarke et al. 1993). One of the enzymes that contributes to H4K16 acetylation in subtelomeric regions in yeast is Sas2, a member of the highly conserved MYST class of histone acetyltransferases (HATs). If *Sas2* is deleted, preventing the acetylation of H4K16, or if H4K16 is changed to arginine to simulate the deacetylated state, the SIR complex spreads at low levels approximately fivefold farther at the right telomere of chromosome VI than in a wild-type cell. This argues that the spreading of subtelomeric heterochromatin is controlled, at least in part, by the opposing activities of Sir2 and Sas2 on H4K16 (Fig. 7) (Kimura et al. 2002; Suka et al. 2002).

At the *HM* loci, restricting the spread of silent chromatin is perhaps even more critical than at telomeres, since genes important for growth are found along the arm of chromosome III, and silencers were shown to function bidirectionally, promoting repression of flanking DNA sequences. One boundary that prevents further spreading of silencing toward the telomere from *HMR* is a tRNA gene (Donze and Kamakaka 2001). This boundary function is likely to require the HAT activity that is associated with transcription or transcriptional potential of this locus. It is significant that one of these HATs is Sas2, although the histone H3 HAT, Gcn5, also promotes the boundary function of tRNA genes. This suggests that transcriptional activators can generally restrict SIR complex propagation by recruiting HATs. Consistently, boundary activity has also been attributed to the transcription factors, Reb1, Tbf1, to a mammalian factor CTCF, as well as to the acidic *trans*-activating domain of VP16 (Fourel et al. 1999, 2001). Each of these may also promote hyperacetylation of histones, thereby attenuating SIR complex propagation by impairing its association with nucleosomes (Fig. 7).

Finally, coupled with the mechanism described above, it was reported that the presence in euchromatin of the variant histone H2A.Z and the RNA polymerase-associated factor Bdf1 (Meneghini et al. 2003), the methylation of histone H3 lysine 79 (van Leeuwen et al. 2002), and the tethering of DNA to nuclear pores (Ishii et al. 2002) all help limit the spread of silent chromatin. Although the mechanisms by which these factors affect heterochromatin spreading are unknown, it is interesting to note that some active genes are associated with nuclear pores (Ishii et al. 2002; Brickner and Walter 2004). Thus, the common characteristic of boundary factors in yeast may be that of a strong transcriptional activator or nucleosome remodeler that directly or indirectly disrupts histone interactions with heterochromatin proteins.

9 Telomere Looping

Several lines of evidence support the notion that long-range interactions enable chromosomal ends to loop back, bypassing subtelomeric boundary elements and stabilizing repressed chromatin at subtelomeric genes (Figs. 5, 6). For instance, despite the presence of Rap1-binding sites only within the first ~300 bp of TG repeat DNA on the end of a telomere, chromatin immunoprecipitation showed that Rap1 is associated with nucleosomes as far as ~3 kb away from the TG repeat (Strahl-Bolsinger et al. 1997). Similarly, yKu is recovered for ~3 kb from the chromosomal end to which it binds (Martin et al. 1999). Furthermore, when silencing is disrupted by mutation of *SIR* genes, both Rap1 and yKu are lost exclusively from the more internal sequences and not from the terminal TG repeats (Martin et al. 1999). To account for the recovery of the 3 kb of heterochromatin with Rap1 and/or yKu after shearing the chromatin into fragments of <500 bp, it was proposed that the truncated telomere folds back, enabling TG-bound Rap1 and yKu to bind SIR proteins across the chromosome in *trans* (Figs. 5, 6). This structure might contribute to the "capping" function of telomere-bound proteins.

Supporting evidence for telomere looping comes from the work of de Bruin et al. (2001), who have exploited the inability of transcriptional activators such as Gal4 to function from a site downstream of the gene whose promoter they are meant to activate. Strains were constructed in which the Gal4 upstream activating sequence (UAS) element was placed downstream of the reporter, and the construct was inserted either at an internal chromosomal location or near a telomere. At an internal site, this construct could not support galactose-inducible transcription. However, in a subtelomeric context, the Gal4 UAS could activate the promoter from a site 1.9 kb downstream of the promoter, in a Sir3-dependent manner. It was argued that the telomeric end can fold back in the presence, but not in the absence, of Sir3 to allow the Gal4 UAS to position itself proximal to the transcription start site (de Bruin et al. 2001).

10 Discontinuity of Repression at Natural Subtelomeric Elements by Telomere Looping

We have set forth here a simplistic view of continuous silent chromatin emanating from the telomeric Rap1-binding sites, yet the situation at native telomeres is significantly more complex, largely due to the presence of natural boundary elements found in subtelomeric repeat sequences. Generally, when reporter constructs for telomeric repression are integrated, the subtelomeric repeat elements called X and Y′ at telomeres are deleted, placing the reporter gene and unique sequence immediately adjacent to TG repeats. All native telomeres, on the other hand, contain a core subtelomeric repeat element, X, which is positioned between the TG repeat and the most telomere-proximal gene, and 50–70% of native telomeres also contain at least one copy of a larger subtelomeric element called Y′ (Fig. 8). Both X and Y′ elements contain binding sites for the transcriptional regulators Tbf1 and Reb1, and these have been shown to reduce the spread of silent chromatin (Fourel et al. 1999). However, X elements also contain the consensus for ORC and binding sites for Abf1, which have the opposite effect: These reinitiate or boost the repression of reporters placed on the centromere-proximal side of these elements. The result is one of discontinuity in silencing at native telomeres, which differs from the model of continuous spreading outlined in Figure 6. To explain this, Pryde and Louis (1999) have also proposed that telomeres loop back to allow a region of unrepressed chromatin to intervene between two repressed domains, leading to discontinuity in silent domains without eliminating the need for nucleation and spreading from the TG repeats.

11 Trans-interaction of Telomeres, and Perinuclear Attachment of Heterochromatin

One of the most universally conserved aspects of heterochromatin is that it occurs in discrete nuclear subcompartments. This is also true in budding yeast, where telomeres cluster into groups during interphase, remaining closely associated with the nuclear periphery. This clustering was initially observed as prominent foci of Rap1 and SIR proteins that were detected above a diffuse nuclear background of these factors by immunostaining (Fig. 9). Disruption of silencing by histone H4 K16 mutation, or interference in Rap1 or yKu function, led to the dispersion of the SIR proteins from these clusters (Hecht et al. 1995; Laroche et al. 1998). Later it was shown that not only telomeres, but also the *HML* and *HMR* loci, are closely associated with the nuclear envelope. This association is mediated by redundant pathways that depend either on the telomere-bound yKu factor, or on the forma-

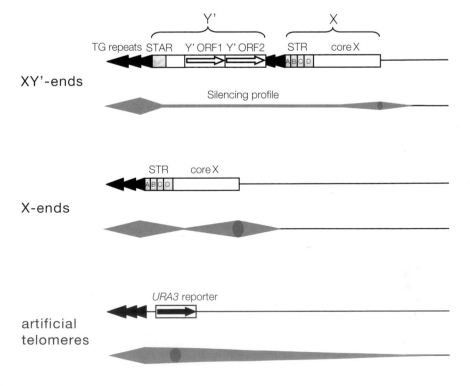

Figure 8. The Organization of Native Telomeres and Their Silencing Patterns

Subtelomeric elements are shown with their major protein-binding sites. Telomeres fall into the two general classes: X-containing or X+Y′-containing ends. The STAR and STR elements block the propagation of repression and leave a region of reduced repression within the Y′ or X element. This is not the case at artificially truncated telomeres where there is a gradient of repression that extends 3–4 kb from the TG repeat. Looping similar to that in Fig. 6 is proposed for native telomeres so that repressed regions contact each other, leaving unrepressed chromatin in between areas of contact. (Adapted from Pryde and Louis 1999.)

tion of silent chromatin itself (Hediger et al. 2002). Within silent chromatin, the anchoring function has been assigned to a subdomain of Sir4 that binds a nuclear envelope-associated protein called Esc1 (enhances silent chromatin 1; Taddei et al. 2004). Sir4–Esc1 interactions tether the SIR-repressed chromatin domain at the nuclear envelope, at sites distinct from pores. Even in the absence of a yKu anchoring pathway, the association of telomeres with the nuclear envelope can be achieved through the Sir4–Esc1 association as long as repression is maintained. Moreover, excised rings of silent chromatin separated from their adjacent telomeres by recombination remain tightly associated with the nuclear periphery in a SIR-dependent manner (Gartenberg et al. 2004).

The initial recruitment of telomeres to the nuclear envelope is probably mediated by yKu, since this functions even in the absence of silencing. This anchoring, together with interactions in *trans* between telomeres, allows a nuclear subcompartment to form that in turn sequesters SIR proteins (Fig. 10). This compartment is critical for creating a gradient of silencing at telomeres. Even silencer-flanked *HM* constructs repress more efficiently when they are integrated near telomeres (Thompson et al. 1994; Maillet et al. 1996) or when they are artificially tethered at the nuclear envelope by a trans-membrane factor (Andrulis et al. 1998). Importantly, the ability to improve repression due to telomere proximity is lost when Sir3 and Sir4 are no longer sequestered in foci or are overexpressed (Maillet et al. 1996; Marcand et al. 1996). This argues that the SIR protein concentration gradient is the feature of telomere clustering that is critical for promoting repression. Finally, it is proposed that the sequestering of general repressors which are in limiting concentrations helps a cell ensure epigenetic inheritance of the silent state as outlined in Figure 10. In brief, the model proposes that the assembly of newly replicated DNA into heterochromatin is likely to be favored if DNA is replicated within a subcompartment that is enriched for silencing factors.

12 Inheritance of Epigenetic States

A universal characteristic of heterochromatin is that its silent state is passed from one generation to the next. This requires that the reassembly of a heterochromatic structure on daughter strands occur soon after replication of the DNA template. Pioneering work on the question of how the cell cycle affects the establishment or inheritance of chromatin states was performed by Miller and Nasmyth (1984), who studied the onset and loss of silencing with a temperature-sensitive *sir3^{ts}* mutant. A shift from permissive temperature to nonpermissive temperature caused silencing to be lost immediately, indicating that Sir3 was required for maintenance of the repressed state. However, in the reciprocal experiment, shifting from nonpermissive temperature to permissive temperature did not lead to immediate restoration of repression: Passage through the cell cycle was required. They concluded that an event in S phase was required for establishment of heritably repressed chromatin. This requirement was later shown to

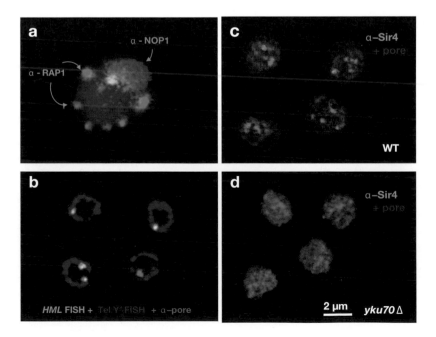

Figure 9. SIR Proteins and Rap1 Are Found in Foci at the Nuclear Periphery

In panel *a*, Rap1 (*green*) identifies 7 clusters representing all 64 telomeres in this diploid cell. They are either perinuclear or adjacent to the nucleolus (*blue*, anti-Nop1). DNA is in red. In panel *b*, telomeric DNA (*red*) is identified by fluorescent in situ hybridization (FISH), and *HML* is visualized in green. The two colocalize in about 70% of the cases, and both are adjacent to the nuclear envelope (*blue*) (Heun et al. 2001). Panel *c* shows the focal distribution of Sir4 (*green*) adjacent to the nuclear envelope (Mab414, *red*). This pattern is lost in a yKu70 deletion strain, coincident with the loss of telomeric silencing (Laroche et al. 1998).

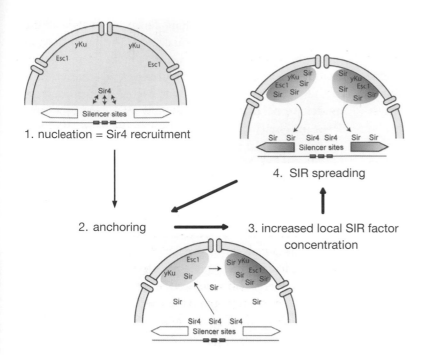

1. nucleation = Sir4 recruitment

2. anchoring → 3. increased local SIR factor concentration

4. SIR spreading

Figure 10. Spontaneous Formation of Silencing Subcompartments

A simple model for the formation of subnuclear compartments is shown. (*1*) Sir4 is first recruited at the nucleation center by DNA-binding proteins that can bind Sir4. These include Rap1, ORC, Abf1, and yKu. (*2*) The presence of Sir4 at the locus will then bring it to the nuclear periphery through one of the two Sir4-anchoring pathways (yKu or Esc1). (*3*) At the nuclear envelope, the high local concentrations of SIR proteins will help silencing complexes assemble and spread. (*4*) The ability of silent loci to remain attached at the periphery increases the local concentration of SIR proteins and reinforces the silencing of other loci within this region. Importantly, telomere-bound yKu can independently recruit telomeres to the nuclear envelope just as Sir4 recruits silencer sequences.

involve events in both S and G_2/M phases (Lau et al. 2002).

Initially, it was thought that origin firing from the silencer-linked ARS elements might be a critical event in the establishment or inheritance of silent chromatin, but because initiation could not be detected at the origins of the *HML* locus, this seemed an unlikely explanation. Indeed, the experiment showing that ORC can be efficiently replaced by a targeted GDB–Sir1 fusion protein put to rest the notion that origin firing contributes in an essential manner to the inheritance of silent chromatin. On the other hand, several lines of evidence indicate that passage through S phase is necessary for heterochromatin assembly. This was widely interpreted as a requirement for DNA replication and its associated reassembly of nucleosomal structure, yet recent experiments have shown that establishment of repression can occur on DNA that does not replicate (Kirchmaier and Rine 2001; Li et al. 2001). Candidates for the missing factor(s) needed for establishing the repressed chromatin state are thus proteins that might be specifically activated in late S phase, or have a specific S–G_2 phase function. Among these may be de novo synthesized histones or histone-modifying enzymes. They may also include chaperones, such as the chromatin assembly factor 1(CAF1) complex, which might ensure a critical histone assembly step.

Other studies have shown that robust silencing is not achieved until telophase, well beyond the S-phase window of nucleosome assembly. It appears that a cohesin subunit, Scc1, inhibits stable repression unless it is destroyed at the metaphase/anaphase transition (Lau et al. 2002). This correlates with findings that the targeting of transcription factors can efficiently disrupt or compete with the establishment of silent chromatin in G_2/M phase, but not after cells have passed M and entered G_1 (Aparicio and Gottschling 1994). Together, these findings argue that in addition to a critical S-phase component or event, there is an additional step that requires passage through mitosis and depends either directly or indirectly on the loss of sister chromatid cohesion.

13 Aging and Sir2: Linked by rDNA Repeat Instability

In *Drosophila*, highly active rDNA repeats are adjacent to centromeric heterochromatin, and in many higher eukaryotic species, nucleoli and condensed heterochromatin are spatially juxtaposed. It is significant, therefore, that in yeast, Sir2, independent of the other SIR proteins, is genetically and physically associated with the highly transcribed rDNA repeats (Gotta et al. 1997). Indeed, rDNA recombination is suppressed by Sir2 (Gottlieb and Esposito 1989), as is the repression of RNA pol-II-dependent reporters that are introduced into the rDNA array. Due to their tandemly repeated nature, the rDNA repeats are prone to unequal recombination events that can lead to either a reduction or increase of the rDNA array. Such instability has been also correlated with the accumulation of extrachromosomal rDNA circles (Fig.

11) (Sinclair and Guarente 1997). Sir2 is required both to repress reporter genes inserted in the rDNA array and to prevent aberrant recombination that leads to a loss of rDNA repeats, perhaps either by positioning nucleosomes (Fritze et al. 1997) or by aligning repeats between sister chromatids to preclude unequal exchange events (Kobayashi et al. 2004).

The most surprising phenotype of the yeast *sir2* null allele is a reduction in life span, which in yeast has no direct link to the cell's loss of telomeric repression or to the length of the TG tract. Indeed, the short-lived phenotype of *sir2*-deficient yeast cells means that these cells divide on average less than 12 times, rather than 20–25 times as observed for wild-type cells (Kaeberlein et al. 1999). It is now convincingly established that the production of extrachromosomal rDNA repeat circles (ERC) due to unequal recombination events in the rDNA, and their accumulation in mother cells, correlates with yeast senescence (Fig. 11). Importantly, life span can be not only shortened by the loss of Sir2, but also lengthened by Sir2 overexpression, which increases the amount of Sir2 bound to rDNA. Other mutations that reduce the efficiency of rDNA excision, for instance, elimination of the replication fork barrier protein Fob1 (Defossez et al. 1999), also extend life span in yeast, just as the artificial production of ERC is sufficient to cause cellular aging (Sinclair and Guarente 1997). Thus, rDNA instability is clearly correlated with aging in yeast, although its contribution to senescence may be indirect. One model suggests that the high levels of ERC titrate DNA repair or replication proteins from other genomic loci, leading to increased genomic damage or reduced replication of the rest of the yeast genome.

Because Sir2 is an NAD-dependent deacetylase, and because NAD levels act as a metabolic thermostat, it was proposed that the effect of yeast Sir2 on life span might be related to the extension of life span by caloric restriction, a conserved pathway that attenuates replicative aging in many species. Although this view is supported by studies showing that caloric restriction increases Sir2 activity in yeast, flies, and mammals, yeast life span is extended by growth on low glucose (caloric restriction) in a manner that is independent of and additive to the role of Sir2 (Kaeberlein et al. 2004). Thus, Sir2 and caloric restriction increase life span through independent pathways.

The accumulation of excised rDNA rings has not been detected in any other species, yet it has been proposed for both *Caenorhabditis elegans* and rodents that other types of genomic instability are associated with shortened life span in these species. Analogous to the events in budding yeast where loss of Sir2 leads to unequal inter-sister recombination, it is possible that the loss of heterochromatin at mammalian telomeres leads to end-to-end chromosomal fusions, which restrict the division potential of cells. Although it is not yet known whether mammalian Sir2 influences these mechanisms, it is nonetheless likely that genomic instability will be a common factor in aging, and that the loss of heterochromatin structure may well contribute specifically to these events.

14 Summary

Combined genetic, biochemical, and cytological techniques have been exploited in budding yeast to demonstrate fundamental principles at work during heterochromatin-mediated gene silencing. These princi

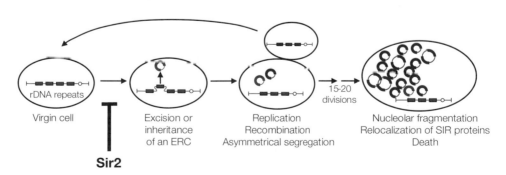

Figure 11. rDNA Recombination Leads to Cellular Senescence in Yeast

The rDNA is organized in an array of 140–200 direct repeats of a 9.1-kb unit (*red block*). These encode the 18S, 5.8S, 25S, and 5S rRNAs, and contain two Sir2-responsive elements downstream of the 5S gene and within the 18S gene. The rDNA repeats tend to be excised in aging yeast cells, and the circles accumulate in the mother cell (Kaeberlein et al. 1999). This correlates with premature senescence and can be antagonized by Sir2, which helps suppress unequal recombination and ring excision.

ples include the mechanism of initiation, spreading and barriers to spreading of heterochromatin; the balance of heterochromatin factors and their distribution within a subnuclear environment; and heterochromatin looping and cell cycle involvement in its formation. Moreover, in vitro systems are being developed for the reconstitution of yeast heterochromatin. These in vivo and in vitro studies provide a strong mechanistic basis for our understanding of the assembly of heterochromatin from chromatin fibers in all eukaryotes.

References

Andrulis E.D., Neiman A.M., Zappulla D.C., and Sternglanz R. 1998. Perinuclear localization of chromatin facilitates transcriptional silencing. *Nature* **394:** 592–595.

Aparicio O.M. and Gottschling D.E. 1994. Overcoming telomeric silencing: A trans-activator competes to establish gene expression in a cell cycle-dependent way. *Genes Dev.* **8:** 1133–1146.

Aparicio O.M., Billington B.L., and Gottschling D.E. 1991. Modifiers of position effect are shared between telomeric and silent mating-type loci in *S. cerevisiae. Cell* **66:** 1279–1287.

Boscheron C., Maillet L., Marcand S., Tsai-Pflugfelder M., Gasser S.M., and Gilson E. 1996. Cooperation at a distance between silencers and proto-silencers at the yeast *HML* locus. *EMBO J.* **15:** 2184–2195.

Brand A.H., Micklem G., and Nasmyth K. 1987. A yeast silencer contains sequences that can promote autonomous plasmid replication and transcriptional activation. *Cell* **51:** 709–719.

Braunstein M., Sobel R.E., Allis C.D., Turner B.M., and Broach J.R. 1996. Efficient transcriptional silencing in *Saccharomyces cerevisiae* requires a heterochromatin histone acetylation pattern. *Mol. Cell. Biol.* **16:** 4349–4356.

Brickner J.H. and Walter P. 2004. Gene recruitment of the activated INO1 locus to the nuclear membrane. *PLoS Biol.* **2:** e342.

Chen L. and Widom J. 2005. Mechanism of transcriptional silencing in yeast. *Cell* **120:** 37–48.

Cheng T.H. and Gartenberg M.R. 2000. Yeast heterochromatin is a dynamic structure that requires silencers continuously. *Genes Dev.* **14:** 452–463.

Chien C.T., Buck S., Sternglanz R., and Shore D. 1993. Targeting of SIR1 protein establishes transcriptional silencing at HM loci and telomeres in yeast. *Cell* **75:** 531–541.

Chopra V.S. and Mishra R.K. 2005. To SIR with Polycomb: Linking silencing mechanisms. *Bioessays* **27:** 119–121.

Clarke D.J., O'Neill L.P., and Turner B.M. 1993. Selective use of H4 acetylation sites in the yeast *S. cerevisiae. Biochem. J.* **294:** 557–561.

Cubizolles F., Martino F., Perrod S., and Gasser S.M. 2006. A homotrimer-heterotrimer switch in Sir2 structure differentiates rDNA and telomeric silencing. *Mol. Cell* **21:** 825–836.

de Bruin D., Zaman Z., Liberatore R.A., and Ptashne M. 2001. Telomere looping permits gene activation by a downstream UAS in yeast. *Nature* **409:** 109–113.

Defossez P.A., Prusty R., Kaeberlein M., Lin S.J., Ferrigno P., Silver P.A., Keil R.L., and Guarente L. 1999. Elimination of replication block protein Fob1 extends the life span of yeast mother cells. *Mol. Cell* **3:** 447–455.

Donze D. and Kamakaka R.T. 2001. RNA polymerase III and RNA

polymerase II promoter complexes are heterochromatin barriers in *S. cerevisiae. EMBO J.* **20:** 520–531.

Fourel G., Revardel E., Koering C.E., and Gilson E. 1999. Cohabitation of insulators and silencing elements in yeast subtelomeric regions. *EMBO J.* **18:** 2522–2537.

Fourel G., Boscheron C., Revardel E., Lebrun E., Hu Y.F., Simmen K.C., Muller K., Li R., Mermod N., and Gilson E. 2001. An activation-independent role of transcription factors in insulator function. *EMBO Rep.* **2:** 124–132.

Fritze C.E., Verschueren K., Strich R., and Easton Esposito R. 1997. Direct evidence for SIR2 modulation of chromatin structure in yeast rDNA. *EMBO J.* **16:** 6495–6509.

Frye R.A. 2000. Phylogenetic classification of prokaryotic and eukaryotic Sir2-like proteins. *Biochem. Biophys. Res. Comm.* **273:** 793–798.

Gartenberg M.R., Neumann F.R., Laroche T., Baszczyk M., and Gasser S.M. 2004. Sir-mediated repression can occur independently of chromosomal and subnuclear contexts. *Cell* **119:** 955–967.

Gotta M., Strahl-Bolsinger S., Renauld H., Laroche T., Kennedy B.K., Grunstein M., and Gasser S.M. 1997. Localization of Sir2p: The nucleolus as a compartment for silent information regulators. *EMBO J.* **16:** 3243–5503.

Gottlieb S. and Esposito R.E. 1989. A new role for a yeast transcriptional silencer gene, SIR2, in regulation of recombination in ribosomal DNA. *Cell* **56:** 771–776.

Gottschling D.E. 1992. Telomere-proximal DNA in *Saccharomyces cerevisiae* is refractory to methyltransferase activity in vivo. *Proc. Natl. Acad. Sci.* **89:** 4062–4065.

Gottschling D.E., Aparicio O.M., Billington B.L., and Zakian V.A. 1990. Position effect at *S. cerevisiae* telomeres: Reversible repression of PolII transcription. *Cell* **63:** 751–762.

Hecht A., Strahl-Bolsinger S., and Grunstein M. 1996. Spreading of transcriptional repressor SIR3 from telomeric heterochromatin. *Nature* **383:** 92–96.

Hecht A., Laroche T., Strahl-Bolsinger S., Gasser S.M., and Grunstein M. 1995. Histone H3 and H4 N-termini interact with SIR3 and SIR4 proteins: A molecular model for the formation of heterochromatin in yeast. *Cell* **80:** 583–592.

Hediger F., Neumann F.R., Van Houwe G., Dubrana K., and Gasser S.M. 2002. Live imaging of telomeres: yKu and Sir proteins define redundant telomere-anchoring pathways in yeast. *Curr. Biol.* **12:** 2076–2089.

Heun P., Laroche T., Raghuraman M.K., and Gasser S.M. 2001. The positioning and dynamics of origins of replication in the budding yeast nucleus. *J. Cell Biol.* **152:** 385–400.

Hoppe G.J., Tanny J.C., Rudner A.D., Gerber S.A., Danaie S., Gygi S.P., and Moazed D. 2002. Steps in assembly of silent chromatin in yeast: Sir3-independent binding of a Sir2/Sir4 complex to silencers and role for Sir2-dependent deacetylation. *Mol. Cell. Biol.* **12:** 4167–4180.

Imai S.I., Armstrong C., Kaeberlein M., and Guarente L. 2000. Transcriptional silencing and longevity protein Sir2 is an NAD-dependent histone deacetylase. *Nature* **403:** 795–800.

Ishii K., Arib G., Lin C., Van Houwe G., and Laemmli U.K. 2002. Chromatin boundaries in budding yeast: The nuclear pore connection. *Cell* **109:** 551–562.

Johnson L.M., Fisher-Adams G., and Grunstein M. 1992. Identification of a non-basic domain in the histone H4 N-terminus required for repression of the yeast silent mating loci. *EMBO J.* **11:** 2201–2209.

Johnson L.M., Kayne P.S., Kahn E.S., and Grunstein M. 1990. Genetic evidence for an interaction between SIR3 and histone H4 in the repression of the silent mating loci in S. cerevisiae. Proc. Natl. Acad. Sci. 87: 6286–6290.

Kaeberlein M., McVey M., and Guarente L. 1999. The SIR2/3/4 complex and SIR2 alone promote longevity in S. cerevisiae by two different mechanisms. Genes Dev. 13: 2570–2580.

Kaeberlein M., Kirkland K.T., Fields S., and Kennedy B.K. 2004. Sir2-independent life span extension by calorie restriction in yeast. PLoS Biol. 2: 296–307.

Kayne P.S., Kim U.J., Han M., Mullen J.R., Yoshizaki F., and Grunstein M. 1988. Extremely conserved histone H4 N terminus is dispensable for growth but essential for repressing the silent mating loci in yeast. Cell 55: 27–39.

Kimura A., Umehara T., and Horikoshi M. 2002. Chromosomal gradient of histone acetylation established by Sas2p and Sir2p functions as a shield against gene silencing. Nat. Genet. 3: 370–377.

Kirchmaier A.L. and Rine J. 2001. DNA replication-independent silencing in S. cerevisiae. Science 291: 646–650.

Kobayashi T., Horiuchi T., Tongaonkar P., Vu L., and Nomura M. 2004. Sir2 regulates recombination between different rDNA repeats, but not recombination within individual rRNA genes in yeast. Cell 117: 441–453.

Laroche T., Martin S.G., Gotta M., Gorham H.C., Pryde F.E., Louis E.J., and Gasser S.M. 1998. Mutation of yeast Ku genes disrupts the subnuclear organization of telomeres. Curr. Biol. 8: 653–656.

Lau A., Blitzblau H., and Bell S.P. 2002. Cell-cycle control of the establishment of mating-type silencing in S. cerevisiae. Genes Dev. 16: 2935-2945.

Li Y.C., Cheng T.H., and Gartenberg M.R. 2001. Establishment of transcriptional silencing in the absence of DNA replication. Science 291: 650–653.

Liou G.G., Tanny J.C., Kruger R.G., Walz T., and Moazed D. 2005. Assembly of the SIR complex and its regulation by O-acetyl-ADP-ribose, a product of NAD-dependent histone deacetylation. Cell 121: 515–527.

Loo S. and Rine J. 1994. Silencers and domains of generalized repression. Science 264: 1768–1771.

Luo K., Vega-Palas M.A., and Grunstein M. 2002. Rap1-Sir4 binding independent of other Sir, yKu, or histone interactions initiates the assembly of telomeric heterochromatin in yeast. Genes Dev. 12: 1528–1539.

Maillet L., Boscheron C., Gotta M., Marcand S., Gilson E., and Gasser S.M. 1996. Evidence for silencing compartments within the yeast nucleus: A role for telomere proximity and Sir protein concentration in silencer-mediated repression. Genes Dev. 10: 1796–1811.

Marcand S., Gilson E., and Shore D. 1997. A protein-counting mechanism for telomere length regulation in yeast. Science 275: 986–990.

Marcand S., Buck S.W., Moretti P., Gilson E., and Shore D. 1996. Silencing of genes at nontelomeric sites in yeast is controlled by sequestration of silencing factors at telomeres by Rap1 protein. Genes Dev. 10: 1297–1309.

Martin S.G., Laroche T., Suka N., Grunstein M., and Gasser S.M. 1999. Relocalization of telomeric Ku and SIR proteins in response to DNA strand breaks in yeast. Cell 97: 621–633.

Meneghini M.D., Wu M., and Madhani H.D. 2003. Conserved histone variant H2A.Z protects euchromatin from the ectopic spread of silent heterochromatin. Cell 112: 725–736.

Miller A.M. and Nasmyth K.A. 1984. Role of DNA replication in the repression of silent mating type loci in yeast. Nature 312: 247–251.

Mishra K. and Shore D. 1999. Yeast Ku protein plays a direct role in telomeric silencing and counteracts inhibition by Rif proteins. Curr. Biol. 9: 1123–1126.

Moazed D., Kistler A., Axelrod A., Rine J., and Johnson A.D. 1997. Silent information regulator protein complexes in S. cerevisiae: A SIR2/SIR4 complex and evidence for a regulatory domain in SIR4 that inhibits its interaction with SIR3. Proc. Natl. Acad. Sci. 94: 2186–2191.

Moretti P., Freeman K., Coodly L., and Shore D. 1994. Evidence that a complex of SIR proteins interacts with the silencer and telomere-binding protein RAP1. Genes Dev. 8: 2257–2269.

Palladino F., Laroche T., Gilson E., Axelrod A., Pillus L., and Gasser S.M. 1993. SIR3 and SIR4 proteins are required for the positioning and integrity of yeast telomeres. Cell 75: 542–555.

Pillus L. and Rine J. 1989. Epigenetic inheritance of transcriptional states in S. cerevisiae. Cell 59: 637–647.

Pryde F.E. and Louis E.J. 1999. Limitations of silencing at native yeast telomeres. EMBO J. 18: 2538–2550.

Renauld H., Aparicio O.M., Zierath P.D., Billington B.L., Chhablani S.K., and Gottschling D.E. 1993. Silent domains are assembled continuously from the telomere and are defined by promoter distance and strength, and by SIR3 dosage. Genes Dev. 7: 1133–1145.

Rusche L.N., Kirchmaier A.L., and Rine J. 2002. Ordered nucleation and spreading of silenced chromatin in Saccharomyces cerevisiae. Mol. Biol. Cell 7: 2207–2222.

———. 2003. The establishment, inheritance, and function of silenced chromatin in Saccharomyces cerevisiae. Annu. Rev. Biochem. 72: 481–516.

Sekinger E.A. and Gross D.S. 1999. SIR repression of a yeast heat shock gene: UAS and TATA footprints persist within heterochromatin. EMBO J. 18: 7041–7055.

Sinclair D.A. and Guarente L. 1997. Extrachromosomal rDNA circles—A cause of aging in yeast. Cell 91: 1033–1042.

Stavenhagen J.B. and Zakian V.A. 1994. Internal tracts of telomeric DNA act as silencers in Saccharomyces cerevisiae. Genes Dev. 8: 1411–1422.

Strahl-Bolsinger S., Hecht A., Luo K., and Grunstein M. 1997. SIR2 and SIR4 interactions differ in core and extended telomeric heterochromatin in yeast. Genes Dev. 11: 83–93.

Suka N., Luo K., and Grunstein M. 2002. Sir2p and Sas2p opposingly regulate acetylation of yeast histone H4 lysine16 and spreading of heterochromatin. Nat. Genet. 3: 378–383.

Suka N., Suka Y., Carmen A.A., Wu J., and Grunstein M. 2001. Highly specific antibodies determine histone acetylation site usage in yeast heterochromatin and euchromatin. Mol. Cell 8: 473–479.

Taddei A., Hediger F., Neumann F.R., Bauer C., and Gasser S.M. 2004. Separation of silencing from perinuclear anchoring functions in yeast Ku80 Sir4 and Esc1 proteins. EMBO J. 23: 1301–1312.

Tanner K.G., Landry J., Sternglanz R., and Denu J.M. 2000. Silent information regulator 2 family of NAD dependent histone/protein deacetylases generates a unique product, 1-O-acetyl-ADP-ribose. Proc. Natl Acad. Sci. 97: 14178–14182.

Thompson J.S., Johnson L.M., and Grunstein M. 1994. Specific repression of the yeast silent mating locus HMR by an adjacent telomere. Mol. Cell. Biol. 14: 446–455.

van Leeuwen F., Gafken P.R., and Gottschling D.E. 2002. Dot1p modulates silencing in yeast by methylation of the nucleosome core. Cell 109: 745–756.

Weiss K. and Simpson R.T. 1998. High-resolution structural analysis of chromatin at specific loci: S. cerevisiae silent mating type locus HMLα. Mol. Cell. Biol. 18: 5392–5403.

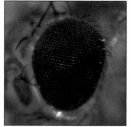
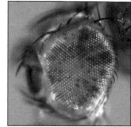

Position-Effect Variegation, Heterochromatin Formation, and Gene Silencing in *Drosophila*

Sarah C.R. Elgin[1] and Gunter Reuter[2]

[1]*Department of Biology, Washington University, St. Louis, Missouri 63130*
[2]*Institute of Genetics, Biologicum, Martin Luther University Halle, D-06120 Halle, Germany*

CONTENTS

1. Genes Abnormally Juxtaposed with Heterochromatin Exhibit a Variegating Phenotype, 83

2. Screens for Suppressors and Enhancers of PEV Have Identified Chromosomal Proteins and Modifiers of Chromosomal Proteins, 85

3. Immunofluorescent Staining of Polytene Chromosomes Has Identified Proteins Specifically Associated with Heterochromatin, 88

4. Histone Modification Plays a Key Role in Heterochromatin Silencing, 88

5. Chromosomal Proteins Form Mutually Dependent Complexes to Maintain and Spread Heterochromatic Structure, 91

6. How Is Heterochromatin Formation Targeted in *Drosophila*?, 93

7. Not All Heterochromatin Is Identical, 95

8. PEV, Heterochromatin Formation, and Gene Silencing in Different Organisms, 96

9. Summing Up: There Is Much That We Do Not Know about Heterochromatin, 97

Acknowledgments, 98

References, 98

GENERAL SUMMARY

Genes that are abnormally juxtaposed with heterochromatin, either by rearrangement or by transposition, exhibit a variegating phenotype, indicating that the gene has been silenced in some of the cells in which it is normally active (position-effect variegation, PEV). The silencing that occurs in PEV can be attributed to packaging of the reporter gene in a heterochromatic form, indicating that heterochromatin formation, once initiated, can spread to encompass nearby genes. Genetic, cytological, and biochemical analyses are all possible in *Drosophila melanogaster*, and in this chapter we show how these different approaches have converged to identify many potential contributors to this system, leading to characterization of several proteins that play key roles in establishing and maintaining heterochromatin. Heterochromatin formation depends critically on methylation of histone H3 at lysine 9, with concomitant association of Heterochromatin Protein 1 (HP1) and other interacting proteins, including H3K9 methyltransferases; the multiple interactions of these proteins are required for maintenance and spreading of heterochromatin. Targeting of heterochromatin formation, including accumulation of H3K9me, appears to involve the RNA interference (RNAi) machinery, although specific protein–DNA interactions may also play a role. Although heterochromatic regions (pericentromeric regions, telomeres, and the small fourth chromosome) share a common biochemistry, each is distinct, and the pericentromeric regions are mosaic. Heterochromatin in *Drosophila* is gene-poor, but it is not devoid of genes, and those genes that reside in heterochromatin are dependent on this environment for full expression. The final model for heterochromatin formation and maintenance (including targeting and spreading) will need to take into account the different responses of different genes to this chromatin environment.

1 Genes Abnormally Juxtaposed with Heterochromatin Exhibit a Variegating Phenotype

Large segments of the eukaryotic genome, primarily repetitious sequences, are packaged in a permanently inactive form as constitutive heterochromatin. This chromatin fraction was originally identified as that portion of the genome that remains condensed and deeply staining (heteropycnotic) as the cell makes the transition from metaphase to interphase; such material is generally associated with the telomeres and pericentromeric regions of the chromosomes. Heterochromatic regions tend to be late replicating and show little or no meiotic recombination. These regions are gene-poor, but they are not devoid of genes, and those genes that are present frequently are dependent on that environment for optimal expression. About one-third of the *Drosophila* genome is considered heterochromatic, including the entire Y chromosome, most of the small fourth chromosome, the pericentromeric 40% of the X chromosome, and the pericentromeric 20% of the large autosomes. During the last few decades, we have learned a great deal about the biochemistry of heterochromatin, and much of that understanding derives from our studies with *Drosophila* (Richards and Elgin 2002; Schotta et al. 2003).

One of the first mutations identified in *D. melanogaster* was *white*, a mutation that results in a fly with a white eye, rather than the characteristic red pigmentation. Using X rays as a mutagen, Muller (1930) observed an unusual phenotype, in which the eye was variegating, with some patches of red and some patches of white facets (Fig. 1). This phenotype suggested that the *white* gene itself was not damaged—after all, some facets remained red, and flies with entirely red eyes could be recovered as revertants, again using X rays as the mutagen. However, the *white* gene had clearly been silenced in some of the cells in which it is normally expressed. Subsequent examination of the polytene chromosomes (shown below, see Fig. 4) indicated that such phenotypes were the consequence of an inversion or rearrangement, with one breakpoint within the pericentromeric heterochromatin and one breakpoint adjacent to the *white* gene (see Fig. 1). Because the variegating phenotype is caused by a change in the position of the gene within the chromosome, this phenomenon is referred to as position-effect variegation (PEV). In *Drosophila*, virtually every gene that has been examined in an appropriate rearrangement has been shown to variegate, and rearrangements involving the pericentromeric heterochromatin of any chromosome

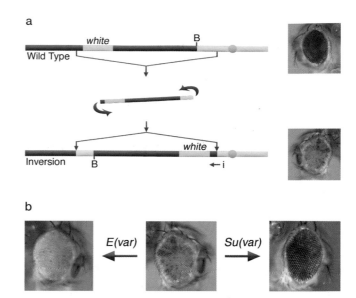

Figure 1. Schematic Illustration of *white* Variegation in the X-Chromosome Inversion *In(1)w^{m4}*

The *white* locus, normally located in the distal euchromatin (*blue*), is now placed within 25 kb of a breakpoint within the pericentromeric heterochromatin (*pink*) of the X chromosome due to an X-ray-induced inversion. Spreading of heterochromatin packaging into the euchromatic domain results in silencing; loss of silencing in some cells during differentiation results in a variegating phenotype. Given a fly exhibiting PEV, one can select for second-site mutations that either suppress the phenotype (*Su(var)* mutations; resulting in a loss of silencing) or enhance the phenotype (*E(var)* mutations; causing an increase in silencing).

can lead to PEV. PEV has been observed in a variety of organisms, including yeasts, flies, and mammals, but has been used as a tool to study heterochromatin formation primarily in *Drosophila*.

PEV indicates that such rearrangements allow packaging in a heterochromatic configuration to "spread" along the chromosome. Apparently, the rearrangement has removed a normally existing barrier or buffer zone. The consequence is an altered packaging and silencing of genes normally arranged in a euchromatic form. Visual inspection of the polytene chromosomes of larvae carrying such a rearrangement shows that the region carrying the reporter gene is now packaged in a dense block of heterochromatin, but only in the cells in which the gene is inactive (Zhimulev et al. 1986). Patterns observed as a consequence of rearrangement of *white* can vary in the number of pigmented cells, the size of the pigmented patches, and the level of pigment in the two different cell types observed (Fig. 1). In a system using an inducible *lac-Z* gene as a reporter, investigators observed that silencing occurs in embryogenesis, when heterochromatin is first observed cytologically, and is epigenetically inherited in both somatic and germ-line lineages; the mosaic phenotype was determined during differentiation by variegated relaxation of silencing in third-instar larvae (Lu et al. 1996). However, not all variegating genes remain silent until after differentiation, and the balance of factors leading to the "ON/OFF" decision no doubt differs for different genes. (See Ashburner et al. 2005b, for a more detailed discussion.)

Given a fly exhibiting a PEV phenotype, it is straightforward to screen for dominant second-site mutations (induced by chemical mutagens that cause point mutations or small insertions/deletions) that are either suppressors of PEV (denoted *Suppressor of variegation, Su[var]*), resulting in a loss of silencing, or enhancers of PEV (denoted *Enhancer of variegation, E[var]*), resulting in an increase in silencing (Fig. 1). About 30 modifiers of PEV have been isolated and characterized, but many more candidates are predicted from such screens. Where the gene has been cloned and the product characterized, one generally finds a chromosomal protein or a modifier of a chromosomal protein (see below). A small subset of these loci cause both a haplo-abnormal and an opposite triplo-abnormal phenotype; i.e., if one copy of the gene results in suppression of PEV, three copies result in enhancement of PEV. Identification of such loci has led to the suggestion that the protein products of these genes play a structural role in heterochromatin, and that the spread of heterochromatic packaging can be driven by the dosage of these proteins in a stochastic manner (Fig. 2) (Locke et al. 1988). However, "spreading" is a complex process, not a simple linear continuum—which most likely is dependent on the organization of the DNA in the region being silenced (see below).

The results observed on rearrangement of chromosomes suggest that a euchromatic gene inserted into a heterochromatic domain by transposition will also show a variegating phenotype, and this has been found to be the case. The *P* element, a DNA transposon found in many strains of *Drosophila* in the wild, can be engineered for this purpose. A natural *P* element has distinctive inverted repeat sequences at each end, and codes for just one enzyme, the *P*-specific DNA transposase. Reporter constructs lacking the DNA transposase but containing other genes of interest can be inserted into the *Drosophila* genome in the presence of active transposase by co-injection into *Drosophila* embryos. A *P*-based transposable element such as that shown in Figure 3a, carrying an *hsp70*-driven copy of *white*, can be used in a fly with no endogenous copy of *white* to identify domains of heterochromatin. When the *P* element is inserted into euchro-

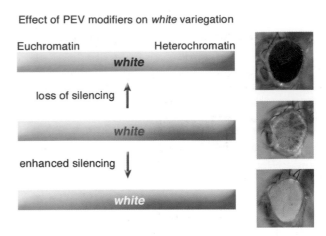

Effect of PEV modifiers on *white* variegation

Figure 2. Dosage-dependent Effects of Some Modifiers of PEV

The modifiers of PEV that have a dosage-dependent effect are thought to be structural proteins of heterochromatin. Whereas a variegating phenotype (exhibited here by a *white* reporter gene) is seen when the wild-type modifier gene is present in two copies (*middle chromosome, middle fly eye*), the presence of three wild-type copies of the modifier gene will drive more extensive heterochromatin formation, resulting in an enhancement of reporter gene silencing (*lower chromosome, lower fly eye*). Conversely, the presence of only one wild-type copy of the modifier gene will result in less heterochromatin formation and more expression from the reporter gene (*upper chromosome, upper fly eye*).

matin, the fly has a red eye. When this *P* is mobilized (by crossing in the gene encoding the transposase), approximately 1% of the lines recovered show a variegating eye phenotype. In situ hybridization shows that in these cases, the *P* element has jumped into the pericentromeric heterochromatin, the telomeres, or the small fourth chromosome (Wallrath and Elgin 1995). This identification of heterochromatic domains is in agreement with earlier cytological studies.

The use of such *P* elements has allowed comparison of the packaging of the same reporter gene in heterochromatic and euchromatic environments. Heterochromatin is relatively resistant to cleavage by nucleases, whether nonspecific (e.g., DNase I) or specific (restriction enzymes), and is less accessible to other exogenous probes, such as *dam* methyltransferase. Analysis of the same *hsp26* transgene (marked with a fragment of unique plant DNA, Fig. 3a) in euchromatin and pericentromeric heterochromatin using micrococcal nuclease (MNase) reveals a shift to a more ordered nucleosome array, indicating regular spacing of the nucleosomes in heterochromatin (Fig. 3b,c). The MNase cleavage fragments are well-defined, suggesting a smaller MNase target than usual in the linker region. The ordered nucleosome array extends across the 5′ regulatory region of the gene, a shift

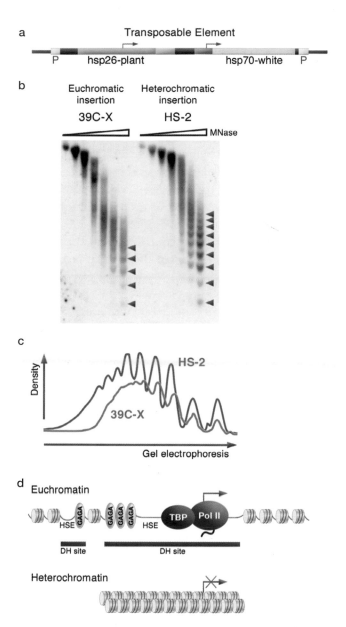

a Transposable Element

P   hsp26-plant   hsp70-white   P

b
Euchromatic   Heterochromatic
insertion   insertion

39C-X   HS-2

MNase

c
Density

HS-2

39C-X

Gel electrophoresis

d Euchromatin

GAGA GAGA GAGA GAGA TBP Pol II

HSE   GAGA GAGA HSE

DH site   DH site

Heterochromatin

Figure 3. Heterochromatin Is Packaged into a Regular Nucleosome Array

A transposable element such as that shown (*a*), carrying a marked copy of a heat shock gene for study and an *hsp70*-driven copy of *white* as a visual marker, can be used to examine the same gene in different chromatin domains. Nuclei from *Drosophila* embryos from a line carrying this transgene in a euchromatic domain (39C-X; red eye) and a line carrying the same transgene in a heterochromatic domain (HS-2; variegating eye) were digested with increasing amounts of MNase, the DNA purified and run out on an agarose gel, and a Southern blot hybridized with a probe unique to the transgene (*b*). Linker sites cleaved by MNase are marked with arrows. (*c*) Densitometer scans from the last lane of each sample are compared (top to bottom is left to right). An array of 9–10 nucleosomes can be detected in heterochromatin (*red line*), compared to 5–6 in euchromatin (*blue line*), indicating more regular spacing in the former case. (*d*) A diagrammatic representation of the results. (*b, c,* Adapted, with permission, from Sun et al. 2001 [© American Society for Microbiology].)

that no doubt contributes to the observed loss of 5′ hypersensitive (HS) sites (Sun et al. 2001). Indeed, although the mechanism of silencing is as yet incompletely understood, there is abundant evidence of transcriptional repression of strongly variegating genes, including loss of binding of TFIID and other transcription factors (Cryderman et al. 1999b).

2 Screens for Suppressors and Enhancers of PEV Have Identified Chromosomal Proteins and Modifiers of Chromosomal Proteins

PEV can be modified by a variety of factors. The temperature during development and the amount of heterochromatin within the genome were the first factors shown to affect the extent of variegation. As a rule, an increase in developmental temperature (from 25°C to 29°C) results in suppression of variegation (loss of silencing), whereas lower temperatures (e.g., 18°C) cause enhancement of variegation (increase in silencing). Other changes in culture conditions that accelerate or slow the rate of development can have similar effects. Strong suppression is found in flies carrying an additional Y chromosome (XXY females and XYY males), whereas strong enhancement is shown in males without a Y chromosome (X0). In general, duplication of heterochromatic material suppresses, whereas deletions of heterochromatic material enhance, variegation. These effects may be due to the titration of a fixed amount of key proteins required for heterochromatin packaging. The first second-site mutations to suppress or enhance PEV were identified by Schultz (1950) and Spofford (1967). At present, approximately 150 genes are implicated as modifiers of PEV loci.

The *Su(var)* and *E(var)* mutations identify genes causally connected with the onset of heterochromatic gene silencing in PEV. Molecular analysis of these genes has been essential in developing an understanding of the mechanisms leading to heterochromatin formation and gene silencing. In most cases, the modifying effect of the mutations on PEV is dominant, and *Su(var)*/+ or *E(var)*/+ heterozygotes show a suppressed or enhanced PEV phenotype (Fig. 1). Efficient isolation and thorough genetic analysis of *Su(var)* and *E(var)* mutations depend on the availability of an experimentally suitable PEV rearrangement. Although a large number of PEV rearrangements have been described (Flybase 2005), only a few can be readily used for efficient genetic screens to isolate dominant modifier mutations. One of the most useful PEV rearrangements for such experimental work is *In(1)w^{m4}* (Muller 1930). This rearrangement variegates for *white*, a phenotype easily recognizable in the eye of adult flies, as shown in Figure 1. Penetrance of white variegation in w^{m4} is 100%, so every fly in the starting stock shows an eye with a white variegated phenotype. Inactivation of the *white* gene does not affect viability or fertility, allowing unlimited work with flies homozygous for w^{m4}.

In the w^{m4} rearrangement, an inversion results in juxtaposition of the *white* gene with heterochromatic material of the X chromosome located at the distal border of the nucleolus organizer (Cooper 1959). This region contains tandem arrays of R1 type mobile elements; the heterochromatic breakpoint of $In(1)w^{m4}$ is found within an R1 repeat unit (A. Ebert and G. Reuter, unpubl.). Phenotypic w^+ revertants of w^{m4} have been isolated after X-ray or EMS (ethane methyl sulfonate, a chemical mutagen) treatment (Tartof et al. 1984; Reuter et al. 1985). Analysis of a series of more than 50 of the w^+ revertant chromosomes (all exhibiting reinversion or translocation of the *white* gene to a euchromatic neighborhood) has suggested that the heterochromatic material immediately flanking the breakpoint causes the inactivation of the *white* gene in w^{m4}. Most of the revertants show white variegation again if strong *E(var)* mutations are introduced, suggesting that some heterochromatic sequences remain associated with the *white* gene after relocation (Reuter et al. 1985), which is not surprising, given that the breakpoint in the flanking DNA is randomly introduced. These studies implicate repetitious DNA (in this case the R1 repeat units) as a target for heterochromatin formation.

Most of the PEV modifier mutations known have been isolated using a sensitized genetic background. For isolation of dominant suppressor mutations, the test genotype contains a dominant enhancer, whereas a dominant suppressor is used in schemes for the isolation of enhancer mutations (Dorn et al. 1993b). If the test genotype contains an enhancer of variegation, all of the w^{m4} flies have white eyes, and exceptions with variegated or red eyes indicate newly induced dominant *Su(var)* mutations. Correspondingly, with a dominant suppressor in the test line, all w^{m4} flies have red eyes, and exceptional flies with a variegated phenotype indicate newly induced *E(var)* mutations. These sensitized genetic schemes favor isolation of strongly dominant *Su(var)* and *E(var)* mutations, which have been found to be very useful for detailed genetic analysis.

More than one million flies have been inspected in different screens using this approach, and more than 140 *Su(var)* and 230 *E(var)* mutations have been isolated (Schotta et al. 2003). Mutations have been induced by EMS, by X-ray treatment, or by remobilization of *P* elements. Another set of *Su(var)* mutations has been isolated in a direct screen with w^{m4} (Sinclair et al. 1983). Screens with a *Df(1;f)* chromosome, which shows strong variegation for the *yellow* gene, a body-color marker, resulted in isolation of 70 PEV modifier mutations (Donaldson et al. 2002). In addition, screens for dominant modifiers of transposon reporter gene expression have identified several mutations with a *Su(var)* effect (Birchler et al. 1994). A subset of critical regulatory genes is known to be down-regulated by the Polycomb group (*PcG*) genes, and up-regulated by the trithorax group (*trxG*) genes. In direct tests, relatively few mutations in *PcG* genes modify PEV (e.g., Sinclair et al. 1998). In contrast, many mutations in the *trxG* genes are enhancers of PEV (Dorn et al. 1993a; Farkas et al. 1994).

Altogether, these screens have identified a large number of dominant *Su(var)* and *E(var)* mutations. Based on the genetic analysis performed to date, the total number of *Su(var)* and *E(var)* mutations can be estimated to be around 150. The large number of *Su(var)* and *E(var)* genes with almost identical phenotypic effects has sometimes resulted in inconsistencies in the genetic nomenclature. Most frequently, the *Su(var)* and *E(var)* gene symbols are combined with numbers indicating the chromosome where the mutation is located, the gene number, and the number of the allele. Thus, $Su(var)3-9^{17}$ symbolizes allele 17 of the ninth *Su(var)* gene identified on the third chromosome. At present, only around 30 of the corresponding genes have been carefully mapped, and alleles have been identified (Table 1). Dosage-dependent effects have been observed for about one-third of the identified genes using a series of overlapping deficiencies and duplications. In these cases, reduction in the amount of the gene products, due to loss of one copy of the gene, consistently results in modification of the variegating phenotype. Deletions of these *Su(var)* or *E(var)* loci suppress or enhance gene silencing, respectively. The duplication studies identified a few modifier loci that show an opposite (antipodal) effect on PEV if an extra copy of the gene is introduced by a duplication or by a transgene insertion. The total number of PEV modifier genes showing dosage-dependent effects is estimated to be about 15–20 (Schotta et al 2003).

If loss of one copy of a gene results in suppression of PEV, and the presence of three copies of a gene leads to an enhancement of PEV, this suggests that the encoded gene product is required in stoichiometric amounts for the establishment of heterochromatin, with concomitant gene silencing (see Fig. 2). Three such loci, *Su(var)2-5* (encoding HP1), *Su(var)3-7* (encoding a zinc finger protein), and

Table 1. Genetically defined *Su(var)* and *E(var)* genes and their molecular functions

Su(var)/ E(var) gene	Cytological position	Molecular function, protein distribution, and phenotypic effects
Suv4-20 [Su(var)]	X; 1B13 14	HKMT, histone H4K20 trimethylation
Su(z)5 [Su(var)]	2L; 21B2	S-adenosylmethionine synthetase
chm (chameau) [Su(var)]	2L; 27F3-4	Myst domain HAT; suppresses PEV but enhances Polycomb-group mutations
Su(var)2-5 (HP1)	2L; 28F2-3	heterochromatin protein HP1, binding of di- and trimethyl H3K9; binding of SU(VAR)3-9
Su(var)2-HP2	2R; 51B6	heterochromatin-associated protein, binds HP1
Su(var)2-10	2R; 45A8-9	PIAS protein, negative regulators of JAK/STAT pathway
Su(var)3-64B (HDAC1=RPD3)	3L; 64B12	histone deacetylase HDAC1, deacetylation of H3K9
E(z) [Su(var)]	3L; 67E5	HKMT, H3K27 mono-, di-, and trimethylation; extra gene copy enhances PEV; in null mutation, all euchromatic and heterochromatic H3K27 methylation lost, H3K9 methylation not affected
SuUR [Su(var)]	3L; 68A4	suppresses heterochromatin underreplication; heterochromatin-associated protein
Su(var)3-1 (JIL1)	3L; 68A5-6	antimorphic JIL1 mutations, carboxy-terminal protein truncations do not affect kinase function; blocking of heterochromatin spreading
Dom (Domina) [Su(var)]	3R; 86B1-2	fork head winged-helix (FKH/WH) protein; heterochromatin-associated
Su(var)3-6	3R; 87B9-10	PP1 protein serine/threonine phosphatase
Su(var)3-7	3R; 87E3	zinc-finger protein, heterochromatin-associated; interacts with HP1 and SU(VAR)3-9
Su(var)3-9	3R; 89E6-8	HKMT, histone H3K9 methylation, heterochromatin-associated, interaction with HP1
mod (modulo) [Su(var)]	3R; 100E3	DNA- and RNA-binding protein, phosphorylated Mod binds rRNA
E(var)3-64E/ Ubp64[Evar1]	3L; 64E5-6	putative ubiquitin-specific protease (Ubp46)
Trl (Trithorax-like) [E(var)]	3L; 70F4	GAGA factor, binding of repetitive DNA sequences
Mod(mdg4)/ E(var)3-93D	3R; 93D7	transcription regulator, more than 20 protein isoforms produced by trans-splicing
E(var)3-93E	3R; 93E9-F1	E2F transcription factor, haplo-enhancer and triplo-suppressor

See Flybase for original citations.

Su(var)3-9 (encoding a histone lysine methyltransferase), have been well characterized. *Su(var)2-5* was cloned by screening a cDNA expression library with a monoclonal antibody that recognizes heterochromatin (James and Elgin 1986). The encoded heterochromatin-associated protein was consequently designated HP1, (heterochromatin protein 1). In situ hybridization analysis using the isolated cloned DNA identified a gene in region 28–29 of the polytene chromosomes, where *Su(var)2-5* had been previously mapped. DNA sequence analysis of the mutant alleles confirmed that the *Su(var)2-5* locus at chromosome position 28F1-2 encodes HP1 (Eissenberg et al. 1990). HP1 contains two conserved domains, an amino-terminal chromodomain and a carboxy-terminal chromoshadow domain (Paro and Hogness 1991), and interacts with several other chromosomal proteins. *Su(var)3-7* was first cytogenetically mapped (using a series of overlapping deletions and dupli-

cations) to region 87E1-4 in the third chromosome. This region had been analyzed at the DNA level as part of the first chromosomal walk performed in *Drosophila* (Bender et al. 1983). Using a series of overlapping genomic clones, *Su(var)3-7* was defined within a DNA fragment of 7.8 kb which had a triplo-enhancer effect on a variegating reporter (Reuter et. al. 1990). *Su(var)3-7* encodes a protein with seven regularly spaced zinc fingers, domains that have been shown to function in DNA binding (Cleard and Spierer 2001). *Su(var)3-9* was cloned by P-element transposon tagging (Tschiersch et al. 1994). The *Su(var)3-9* gene in *Drosophila* (and in all other holometabolic insects studied to date) forms a bicistronic unit with the gene encoding eIF2γ (Krauss and Reuter 2000). Because the *Su(var)3-9* transcription unit has no introns, it is likely that *Su(var)3-9* was inserted into an intron of the *eIF2γ* gene via retrotransposition. The SU(VAR)3-9 protein contains a chromodomain in its amino-terminal region and the SET domain (identified first in the proteins SU(VAR)3-9, ENHANCER OF ZESTE [E(Z)], and TRITHORAX) (Jones and Gelbart 1993; Tschiersch et al. 1994) at its carboxyl terminus. This protein is a histone methyltransferase that specifically modifies histone H3 at lysine 9.

Seven different mutant alleles of Su(var)2-5 have been described, including missense mutations in the chromodomain, premature stop codons, and splicing errors (Eissenberg et al. 1992). Su(var)3-7 mutations have been generated with the help of homologous recombination (Seum et al. 2002); additional alleles have been recovered as suppressors of P-element-dependent silencing (Bushey and Locke 2004). Forty mutant alleles of Su(var)3-9 have been recovered and defined at the molecular level (Ebert et al. 2004). Immunocytological analyses using specific antibodies or transgene-expressed fusion proteins have demonstrated that all three proteins are preferentially associated with heterochromatin (see below and Fig. 4) (James et al. 1989; Cleard et al. 1997; Schotta et al. 2002). Strong colocalization is particularly evident for HP1 and SU(VAR)3-9. These proteins also bind to telomeres and at a number of euchromatic sites (Fanti et al. 1998; Schotta et al. 2002).

Several P-element insertions carrying the w^+ reporter gene into telomeric regions show white variegation. This phenomenon is called telomere position effect (TPE). Heterochromatin-like packaging is observed at telomere-associated satellite (TAS) sequences, clusters of repetitious DNA just proximal to the HeT-A and TART retroviral elements that make up Drosophila telomeres (Cryderman et al. 1999a). In general, TPE is not found to be modified by mutations in known modifier genes, although HP1 is important for telomere integrity. In cells deficient for this protein, the chromosomes frequently fuse at their telomeres (Fanti et al. 1998). No trans-acting dominant modifier of TPE was identified in Drosophila in a recent screen (Mason et al. 2004), suggesting that these regions are silenced by two (or more) independent mechanisms.

3 Immunofluorescent Staining of Polytene Chromosomes Has Identified Proteins Specifically Associated with Heterochromatin

One advantage of working with Drosophila is the ability to examine the polytene chromosomes, which provide a visual road map of the genome. During the larval stage, the chromosomes in many terminally differentiated cells are replicated but do not go through mitosis; rather, the chromatin strands remain paired, in perfect synapsis,

with all copies aligned. The most extreme case is found in the salivary glands, where the euchromatic arms of the chromosomes have undergone 10 rounds of replication, generating about 1000 copies. Replication is not uniform, however; many repetitious sequences are underreplicated, and satellite DNA sequences are not replicated at all. All of the chromosome arms fuse in a common chromocenter. Thus, in D. melanogaster, one observes five long arms (the X, second left [2L], second right [2R], third left [3L], third right [3R]), and the short fourth chromosome arm emanating from the condensed chromocenter made up of pericentromeric heterochromatin (see Fig. 4a) (for review, see Ashburner et al. 2005).

Although genetic analysis has identified many of the loci required for heterochromatin formation, it does not, in itself, allow us to determine whether the product of a given locus plays a direct or indirect role. Specific association of a protein with heterochromatin was initially observed in a screen of monoclonal antibodies (generated using a fraction of tight-binding nuclear proteins), analyzing the distribution patterns on polytene chromosomes. Antibodies specific for a 22-kD protein subsequently designated HP1 resulted in immunofluorescent "staining" of the pericentromeric heterochromatin, the telomeres, and the banded portion of the small fourth chromosome, all known sites of heterochromatin (Fig. 4a) (James and Elgin 1986). Subsequent analysis (described above) demonstrated that the HP1 protein is encoded by Su(var)2-5, a known suppressor of PEV (Eissenberg et al. 1990). Examining chromosomal localization with specific antibodies, using either mitotic chromosomes (Fanti and Pimpinelli 2004) or polytene chromosomes (which give more resolution, but are deficient in centromeric heterochromatin) (Silver and Elgin 1976), remains the best demonstration that the product of a Su(var) locus encodes a chromosomal protein. Approximately 10 such heterochromatin-specific proteins have been identified; if mutations in the genes encoding these proteins are available, one often observes dominant suppression of PEV (see Table 1) (Ashburner et al. 2005b). These proteins, including the recently identified HP2 (Fig. 4a) (Shaffer et al. 2002), are candidates to be structural components of heterochromatin.

4 Histone Modification Plays a Key Role in Heterochromatin Silencing

Analysis of SU(VAR)3-9 has identified a key function required for heterochromatic gene silencing (Tschiersch et al. 1994). The protein contains a SET domain that enzymatically functions in histone H3K9 methylation.

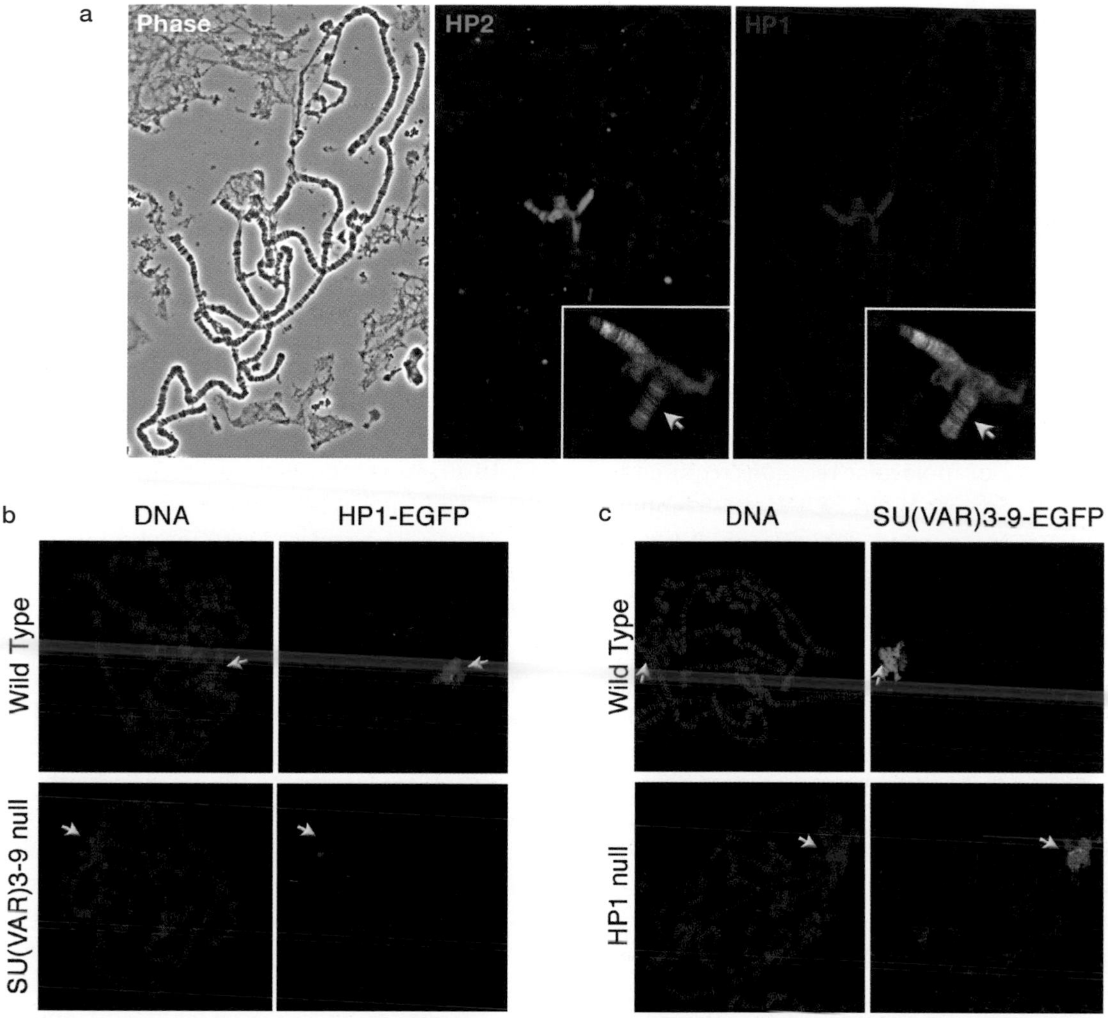

Figure 4. Immunofluorescent Staining of the Polytene Chromosomes Identifies Proteins Predominantly Associated with Heterochromatin

(a) The polytene chromosomes, prepared by fixation and squashing of the larval salivary gland (shown by phase contrast microscopy, left) are "stained" by incubating first with antibodies specific for a given chromosomal protein, and then with a secondary antibody coupled to a fluorescent tag. HP1 (right) and HP2 (center) have similar distribution patterns showing prominent association with the pericentromeric heterochromatin, small fourth chromosome (inset, arrow), and a small set of sites in the euchromatin arms. Note that the efficacy of any antibody can be affected by the choice of fixation protocol (see Stephens et al. 2003). (b, c) Association of HP1 and SU(VAR)3-9 with pericentromeric heterochromatin is interdependent. Mutations in Su(var)3-9 result in a loss of HP1 from the pericentromeric heterochromatin (but not the fourth chromosome, see text) (b), whereas mutations in Su(var)2-5 result in delocalization of SU(VAR)3-9 (c). (Adapted from Shaffer et al. 2002.)

That this protein is a histone lysine methyltransferase (HKMT) that targets H3K9 was first shown by characterization of the human SUV39H1 homolog (Rea et al. 2000). In Drosophila, SU(VAR)3-9 is the main, but not the only, H3K9 HKMT (Schotta et al. 2002; Ebert et al. 2004). SU(VAR)3-9 controls dimethylation of H3K9 in the bulk of the pericentromeric heterochromatin, but not at the fourth chromosome, the telomeres, or euchromatic sites. Trimethylation of H3K9, which in

Drosophila is observed primarily in the inner chromocenter, is also controlled by SU(VAR)3-9. Dimethylation of this inner region is independent of SU(VAR)3-9, as is monomethylation of H3K9 in pericentromeric heterochromatin (Ebert et al. 2004). The HKMTs responsible for these modifications are still unknown. The importance of H3K9 dimethylation in heterochromatic gene silencing is demonstrated by the strong dosage-dependent effect of SU(VAR)3-9 on PEV (discussed above), as well as by the finding that suppression of gene silencing by *Su(var)3-9* mutations correlates with their HKMT activity. The enzymatically hyperactive *Su(var)3-9^ptn* mutation is a strong enhancer of PEV and causes elevated H3K9me2 and H3K9me3 at the chromocenter, as well as generating prominent H3K9me2 signals at many euchromatic sites (ectopic heterochromatin). *S*-Adenosylmethionine functions as the methyl donor for all of these methylation reactions; consequently, mutations in the gene encoding *S*-adenosylmethionine synthase, *Su(z)5*, are dominant suppressors of PEV (Larsson et al. 1996).

Studies using mutations in *Su(var)* genes have begun to reveal the sequence of molecular reactions required to establish heterochromatic domains. SU(VAR)3-9 binding at heterochromatic sequences depends on both its chromo and its SET domains (Schotta et al. 2002). How SU(VAR)3-9 binding is controlled is not yet understood. Methylation of H3K9 by SU(VAR)3-9 establishes binding sites for HP1. The HP1 chromodomain specifically binds H3K9me2 and H3K9me3 (Bannister et al. 2001; Lachner et al. 2001). That SU(VAR)3-9 binds HP1 has been shown by yeast two-hybrid tests and by immunoprecipitation (Schotta et al. 2002). The region of SU(VAR)3-9 amino-terminal to its chromodomain interacts with the chromoshadow domain of HP1. This region of SU(VAR)3-9 also interacts with the carboxy-terminal domain of SU(VAR)3-7. SU(VAR)3-7 interacts at three different sites with the chromoshadow domain of HP1 (Delattre et al. 2000). Given this pattern of interactions, one can suggest that the three proteins—HP1, SU(VAR)3-7, and SU(VAR)3-9—physically associate in multimeric heterochromatin protein complexes.

Association of SU(VAR)3-9 and HP1 with pericentromeric heterochromatin is interdependent (Schotta et al. 2002). SU(VAR)3-9 causes H3K9 dimethylation, which is specifically recognized by the chromodomain of HP1 (Bannister et al. 2001; Lachner et al. 2001). Consequently, in *Su(var)3-9* null larvae, HP1 binding to pericentromeric heterochromatin is impaired (see Fig. 4b). This reflects the specific activity of HP1, which binds to H3K9me2 but not to H3K9me1; monomethylation is not affected by SU(VAR)3-9 (Ebert et al. 2004). H3K9 dimethylation in the inner chromocenter, the fourth chromosome, at telomeres, and at euchromatic sites does not depend on SU(VAR)3-9, and consequently, HP1 continues to be found at all of these sites in the mutant lines. SU(VAR)3-9 associates with these sites in wild-type cells, but appears to be inactive; an unknown HKMT controls H3K9 methylation in these regions.

Conversely, if HP1 is not present (having been depleted by mutations), SU(VAR)3-9 is no longer associated with the pericentromeric heterochromatin, but is also found along the euchromatic chromosome arms (Fig. 4b). It is now seen at almost all bands, where it causes ectopic mono- and dimethylation of H3K9 (H3K9me1 and H3K9me2) (Fig. 5). Thus, HP1 is essential for the restricted binding of SU(VAR)3-9 to pericentromeric heterochromatin. These data suggest a sequence of reactions starting with SU(VAR)3-9 association with heterochromatic domains and consequent generation of H3K9me2. This mark is recognized by the chromodomain of HP1; binding of SU(VAR)3-9 to the HP1 chromoshadow domain ensures its association with heterochromatin. A chimeric HP1-PC protein has been generated in which the chromodomain of HP1 is replaced with the chromodomain of the Polycomb (PC) protein (Platero et al. 1996). The chromodomain of PC binds strongly to H3K27me3 (Fischle et al. 2003). The HP1-PC chimeric protein therefore recognizes H3K27me3 Polycomb-binding sites in the euchromatic arms; in the presence of such a chimeric HP1-PC protein, the SU(VAR)3-9 protein is also found at PC-binding sites, demonstrating its strong association with the chromoshadow domain of HP1 (Schotta et al. 2002).

In SU(VAR)3-9 null cells, another heterochromatin-specific methylation mark, H4K20 trimethylation (H4K20me3), is strongly reduced (Schotta et al. 2004). The interdependence between H3K9 dimethylation and H4K20 trimethylation in heterochromatin has been shown to reflect an interaction between the SU(VAR)3-9, HP1, and SUV4-20 proteins. SUV4-20 is a HKMT that controls H4K20 methylation in heterochromatin. This heterochromatin-specific methylation mark is strongly impaired in SU(VAR)3-9 as well as in HP1 null cells, suggesting association of SU(VAR)3-9, HP1, and SUV4-20 in a mutually dependent protein complex. Mutations in the *Suv4-20* gene cause strong suppression of PEV-induced gene silencing, indicating that the H4K20me3 mark is required for this process.

A third histone methylation mark that is functionally connected with heterochromatin formation is H3K27 methylation catalyzed by the E(Z) HKMT. In *Drosophila*, E(Z) controls all mono-, di-, and trimethylation of H3K27 in both euchromatin and heterochromatin. Consequently, in *E(z)* null cells, all H3K27 methylation is lost (Ebert et al. 2004). A function of H3K27 methylation in heterochromatic gene silencing is indicated by both the Su(var) effect of *E(z)* loss-of-function mutations and the enhancer effect of additional *E(z)* gene copies (Laible et al. 1997). It is not clear whether this effect is direct or indirect. H3K27 methylation is critical for the Polycomb silencing system, which operates in euchromatic domains. Relatively little overlap has been observed between the distribution patterns, and functional roles, of PC and HP1. How H3K27 methylation might fit into the HP1-dependent heterochromatin complexes remains to be elucidated. The HP1 protein has a central linker function in heterochromatin formation and the associated gene silencing, binding H3K9me2 and H3K9me3, and interacting directly with SU(VAR)3-9 (the H3K9 HKMT), SUV4-20 (the H4K20 HKMT), and several additional proteins. Given the number of identified *Su(var)* loci, the model is certain to become more complex. In mammals and plants, histone H3K9 methylation and DNA methylation represent interrelated marks of repressed chromatin (Martienssen and Colot 2001; Bird 2002). Whether or not DNA methylation occurs at all in *Drosophila* has been a point of contention for many years. Recent reports of low levels of DNA methylation in the early embryo have renewed this discussion (Kunert et al. 2003). Analysis of the genome indicates that the only recognizable DNA methyltransferase present is Dnmt2. Mutations in this gene have little impact on the organism. Nonetheless, a role in early embryogenesis cannot be ruled out.

5 Chromosomal Proteins Form Mutually Dependent Complexes to Maintain and Spread Heterochromatic Structure

PEV reflects a change in gene expression, specifically a loss in expression of the reporter gene in some of the cells in which it is normally active, as a consequence of a genetic rearrangement. Several different models, not all mutually exclusive, have been suggested to explain PEV. One possibility originally considered was the random loss of the gene, perhaps as a consequence of late replication (Karpen and Spradling 1990). Quantitative Southern blot analysis has shown that this explanation is not generally applicable; variegating genes are generally fully replicated in diploid tissue (Wallrath et al. 1996). Other models have focused on the association of the variegating gene with a heterochromatic compartment in the nucleus, and/or on the spreading of heterochromatic structure from the newly adjacent heterochromatin. The spreading model, which is based on extensive genetic and cytological data, explains gene silencing as a consequence of heterochromatin packaging spreading across the breakpoint into normally euchromatic domains. In normal chromosomes, euchromatic and heterochromatic regions appear to be insulated from each other by specific sequences or buffer zones. Because these "insulating sequences" (never well-defined in *Drosophila*) are not present at the euchromatic–heterochromatic junction in PEV rearrangements (see Fig. 1), heterochromatinization of euchromatic sequences is variably induced. This heterochromatinization is cytologically visible in the polytene chromosomes

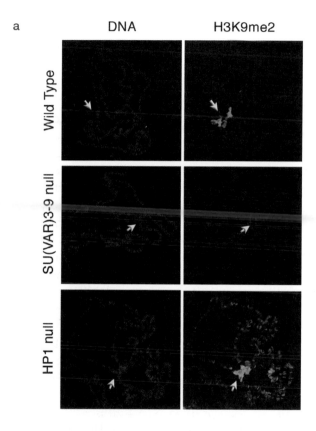

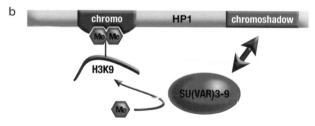

Figure 5. Interaction of SU(VAR)3-9 and HP1 in Setting the Distribution Pattern of H3K9me

(*a*) SU(VAR)3-9 is responsible for dimethylation of H3K9 (H3K9me2); loss of enzyme results in loss of this modification in the pericentromeric heterochromatin, as shown by loss of antibody staining of the polytene chromosomes (compare *middle panel* with *top panel*). Loss of HP1 results in a loss of targeting of SU(VAR)3-9; high levels of H3K9me2 are now seen throughout the chromosome arms (*bottom panel*). (*b*) HP1 interacts with H3K9me2 through its chromodomain, and with SU(VAR)3-9 through its chromoshadow domain. By recognizing both the histone modification and the enzyme responsible for that modification, HP1 provides a mechanism for heterochromatin spreading and epigenetic inheritance.

as a shift from a banded to an amorphous structure at the base of the chromosome arms (Hartmann-Goldstein 1967); the extent of this change can be modified by *Su(var)* and *E(var)* mutations (Reuter et al. 1982).

Inactivation of euchromatic genes over a distance along the chromosome can be genetically demonstrated (Demerec and Slizynska 1937). The affected regions become associated with HP1 (Belyaeva et al. 1993) and show H3K9me2, a typical mark of *Drosophila* heterochromatin (Ebert et al. 2004). Because the spreading model postulates a competition between packaging into euchromatin versus packaging into heterochromatin, PEV modifier genes could encode functions controlling either heterochromatin formation or euchromatin formation. The recovery of dosage-dependent modifiers, as discussed above, supports such a model (Locke et al. 1988; Henikoff 1996). Recently, *Su(var)* mutations controlling the balance between euchromatin and heterochromatin have indeed been identified (Ebert et al. 2004). PEV rearrangements have allowed us to visualize and study cases where heterochromatin packaging spreads into the flanking euchromatin domain. The spreading effect clearly depends on a series of molecular reactions within the euchromatic regions. Several histone modifications are known that are

mutually exclusive and that define these alternative chromatin states. Acetylation of H3K9, methylation of H3K4, and phosphorylation of H3S10 are typical marks of active euchromatin, whereas methylation of H3K9, H3K27, and H4K20 is a specific mark of silenced regions. Heterochromatinization of euchromatic regions therefore requires specific deacetylation, demethylation, and dephosphorylation reactions within euchromatin, as illustrated in Figure 6. This transition depends initially on H3K9 deacetylation by HDAC1. Mutations in the *rpd3* gene, encoding the histone H3K9-specific deacetylase HDAC1, are strong suppressors of PEV (Mottus et al. 2000), antagonizing the effect of SU(VAR)3-9 in gene silencing (Czermin et al. 2001). HDAC1 has been shown to be associated in vivo with the SU(VAR)3-9/HP1 complex; the two enzymes work cooperatively to methylate pre-acetylated histones.

It has recently been observed that spreading of heterochromatin into euchromatin is completely blocked in *Su(var)3-1* mutations (Ebert et al. 2004). *Su(var)3-1* mutations are frameshift mutations within the gene encoding JIL1 kinase that result in expression of a truncated JIL1 protein, lacking the carboxy-terminal region. The JIL1 protein contains two kinase domains and catalyzes H3S10 phosphorylation in euchromatin. The *JIL1*^{Su(var)3-1} muta-

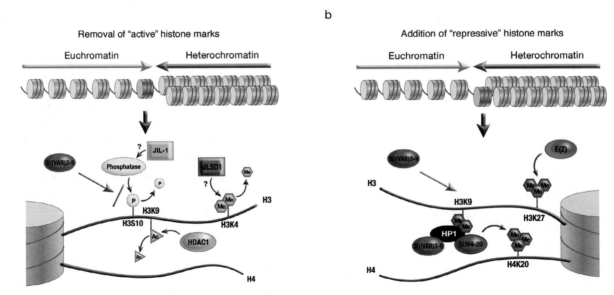

Figure 6. The Transition from a Euchromatic State to a Heterochromatic State Requires a Series of Changes in Histone Modification

(*a*) Active genes are marked by H3K4me2 and me3; if present, this mark must presumably be removed by LSD1 (not yet characterized in *Drosophila*). H3K9 is normally acetylated in euchromatin; this mark must be removed by a histone deacetylase, HDAC1. Phosphorylation of H3S10 can interfere with methylation of H3K9; dephosphorylation appears to involve a phosphatase targeted by interaction with the carboxyl terminus of the JIL1 kinase. These transitions set the stage for acquisition of the modifications associated with silencing, shown in *b*, including methylation of H3K9 by SU(VAR)3-9, binding of HP1, and subsequent methylation of H4K20 by SUV4-20, an enzyme recruited by HP1. Methylation of H3K27 by E(Z) may also occur.

tions do not affect H3S10 phosphorylation, but probably impair dephosphorylation of H3S10, effectively inhibiting methylation of H3K9. This suggests involvement of a phosphatase. Whether the PP1 enzyme (which has been identified with *Su(var)3-5* mutations) (Baksa et al. 1993) is directly involved in this reaction is not known. Demethylation of H3K4 appears to be another prerequisite for heterochromatinization of euchromatic regions. Recent work has shown that the LSD1 aminoxidase functions in mammalian systems as an H3K4 demethylase (Shi et al. 2005). The putative *Drosophila* LSD1 homolog SU(VAR)3-3 facilitates spreading of heterochromatin into euchromatic regions in all PEV rearrangements tested (S. Lein et al., unpubl.). Consistent with this, in *Su(var)3-3* null cells, lacking LSD1, the acquisition of H3K9 methylation in the euchromatin flanking a breakpoint is eliminated, although constitutively heterochromatic regions are not affected. These findings demonstrate that the coordinated function of several enzymes is required to remove euchromatin-specific histone modification marks before the transition to heterochromatin packaging can take place (see Fig. 6). It seems likely that the required enzymes will be found to form complexes with SU(VAR)3-9/HP1, as has already been shown for HDAC1.

6 How Is Heterochromatin Formation Targeted in *Drosophila*?

Although we have learned about many mechanistic aspects and the biochemistry of heterochromatin structure, as discussed above, this leaves open the question of how heterochromatin formation is targeted to selected regions of the genome in its normal configuration. All heterochromatic domains share certain features, and two of these features have been considered as essential inputs for assembling heterochromatin on a given DNA sequence: the position of the locus relative to spatially distinct subdomains of heterochromatin in the nucleus, and the presence of repetitious DNA.

In general, heterochromatic masses are seen at the nuclear periphery and around the nucleolus. In *Drosophila* embryos, this tendency is even more pronounced. Heterochromatic masses are first seen in early embryogenesis, as the nuclei move to the periphery of the egg. Early development in *Drosophila* is syncitial until nuclear division cycle 14, when cell walls form between the nuclei, creating the typical blastula, a ball of cells. The heterochromatic material (centromeres, chromosome four) is concentrated at one side of the nucleus, oriented to the exterior surface of the egg (Foe and Alberts 1985).

Such spatial subdivision of the nucleus persists during development, leading to the concept of heterochromatin "compartments" within the nucleus. These compartments might maintain a high concentration of factors required for heterochromatin formation (such as HP1 and HKMTs), while being depleted in factors required for euchromatin assembly and gene expression (such as HATs and RNA pol II). Indeed, proximity to heterochromatic masses, both in position along the chromosome and in three dimensions, has been shown to be a factor in PEV.

Proximity to the mass of centric heterochromatin has been shown to have an impact on variegation both for euchromatic genes (of which *white* is an example, see above), and for heterochromatic genes, the best-studied examples being *light* and *rolled*. Heterochromatic genes, mapped to those domains, can be observed to variegate when a rearrangement places them in juxtaposition with euchromatin; generally they show the opposite dependencies, requiring normal levels of HP1 for full expression, and showing an enhancement of variegation when HP1 is depleted.

Variegation of *light* depends not only on its juxtaposition to euchromatin, but also on the position of the breakpoint, specifically on the distance from heterochromatin measuring along the chromosome arm (Wakimoto and Hearn 1990). Similar results have been reported for *rolled*. Investigations of *brown dominant* (*bw^D*), a euchromatic gene induced to variegate by insertion of repetitious DNA, have shown that a shift in proximity of the locus to the centric heterochromatin can result in enhancement of silencing (if closer) or suppression of silencing (if farther away) (Henikoff et al. 1995). Similarly, translocation of a fourth chromosome carrying a *white* reporter to the distal half of chromosome arm 2L or 2R results in a dramatic loss of silencing; this was correlated with a change in nuclear disposition, to frequent occupancy of sites distant from the chromocenter in the salivary gland nucleus (Cryderman et al. 1999a).

A recent study using high-resolution microscopy examined both gene activity (using antibodies specific for the product) and nuclear location of a reporter (using FISH, fluorescence in situ hybridization) in the same cell during the normal time frame of expression. A *white* variegating inversion, *bw^D*, and a variegating *lacZ* transgene were studied in differentiating eye discs or adult eyes. This investigation found a strong correlation between the position of the reporter gene in the cell nucleus relative to pericentromeric heterochromatin and the level of expression, supporting the idea that a heterochromatic "compartment" exists, and that positioning within this compartment is correlated with gene silencing (Harmon and Sedat 2005). However, the correlation is not absolute. This is not surprising, given that studies with a *white* reporter indicate the presence of both euchromatic and heterochromatic domains interspersed on the small fourth chromosome (which is always close to the mass of pericentromeric heterochromatin). These latter observations point to other local determinants that contribute to packaging chromatin into one form or the other.

In *D. melanogaster*, it is estimated that one-third of the genome is heterochromatic by cytological criteria. This includes large blocks that flank the centromeres, smaller blocks associated with the telomeres, the whole of the Y chromosome, and most of the small fourth chromosome. The centromeric regions are made up of large (0.2–1 Mb) blocks of satellite DNA interspersed with "islands" of complex sequences, generally transposable elements (Le et al. 1995). Although gene-poor, these regions are not devoid of genes; the current estimate is that several hundred genes reside in the pericentromeric heterochromatin (Hoskins et al. 2002). The telomeres of *Drosophila* do not have the typical G-rich repeats seen elsewhere, but are composed of copies of HeT-A and TART retrotransposons. Telomere-associated sequences (TAS), blocks of 10^2–10^3 nucleotide repeats, are found just proximal, and *white* transgene reporters inserted in these regions display a variegating phenotype. Although the Y chromosome does carry the genes for a number of male fertility factors, the bulk of the chromosome is made up of satellite DNA, and it remains condensed in cells other than the male germ line. The small fourth chromosome is on the order of 4.3 Mb in size, with about 3 Mb made up of satellite DNA. The distal 1.2 Mb can be considered euchromatic in that it is polytenized in the salivary gland (see Fig. 4), but it appears heterochromatic by virtue of its late replication, its complete lack of meiotic exchange, and its association with HP1, HP2, and H3K9me2 (Fig. 4). This region has a six- to sevenfold higher density of transposon fragments than is found in the euchromatic arms, similar to regions at the junction of centric heterochromatin and euchromatin on the other chromosomes (Kaminker et al. 2002). Interestingly, an investigation of the fourth chromosome using the *white* reporter P element discussed above (Fig. 3a) found both euchromatic domains (resulting in a red-eye phenotype) and heterochromatic domains (resulting in a variegating phenotype) interspersed (Sun et al. 2004).

This finding suggests the presence of local elements in the DNA that can drive the formation of heterochromatin or euchromatin. Genetic screens for a switch in phenotype (from red to variegating or vice versa) have demonstrated that local deletions or duplications of 5–80 kb of DNA flanking a transposon reporter can lead to the loss or acquisition of variegation, pointing to short-range *cis*-acting determinants for silencing (see Fig. 7). This silencing is dependent on HP1 and correlates with a change in chromatin structure, as shown by a change in nuclease accessibility, pointing to a shift from a euchromatic to a heterochromatic state. Mapping data in one region of the fourth implicate the *1360* transposon as a target for heterochromatin formation and indicate that once heterochromatin formation is initiated at dispersed repetitive elements, it can spread along the fourth chromosome for about 10 kb, or until it encounters competition from a euchromatic determinant (Sun et al. 2004). Short-range *cis*-acting determinants related to copy number are also implied by the observation that tandem or inverted repeats of reporter *P* elements will result in heterochromatin formation and gene silencing (Dorer and Henikoff 1994).

Such *cis*-acting elements in the DNA might function by sequence-specific binding of a protein capable of triggering heterochromatin formation. Proteins that bind specifically to some of the satellite DNAs have been identified (e.g., D1, Aulner et al. 2002). The importance of these interactions has been inferred from the impact of satellite-specific DNA-binding drugs, which can suppress PEV (Janssen et al. 2000). However, the findings in yeast and plants (Elgin and Grewal 2003; Matzke and Birchler 2005; see Chapters 8 and 9) suggest a second model, specifically that an RNAi-based mechanism could be used to target heterochromatin formation to repetitious elements. Work from several labs has demonstrated that the RNAi system is present in *Drosophila* and plays an important role in developmental regulation via posttranscriptional gene silencing (PTGS). *D. melanogaster* has two genes encoding DICER proteins and numerous genes (*aubergine, AGO1, AGO2, spindleE* [aka *homeless*], *vasa intronic gene* [*VIG*], *armitage, Fmr1*) encoding components or proteins required for assembly of the RNA-induced silencing complex (RISC) (Sontheimer 2005). The system has been implicated in the PTGS of repetitious sequences, notably the tandemly repeated *Stellate* genes, several retrotransposons, and *Alcohol dehydrogenase* (*Adh*) transgenes, and in the transcriptional gene silencing (TGS) of *Adh* transgenes (Aravin et al. 2001; Pal-Bhadra et al. 2002). In a direct test, Pal-Bhadra et al. (2004) found that mutations in *piwi* (a member of the PAZ domain family) and *homeless* (a DEAD box helicase) suppress the PEV associated with tandem arrays

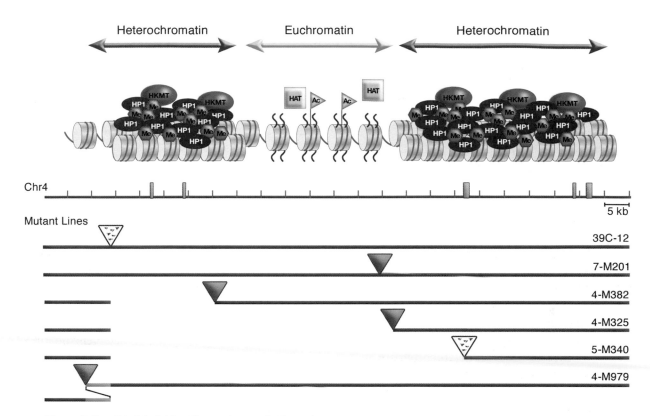

Figure 7. Possible Model for Heterochromatin Targeting

dsRNA from repetitious sequences is processed through RISC to generate a hypothetical "targeting complex," which directs either histone modification or HP1 association as an initial step in assembling heterochromatin at the site identified by the small ssRNA. Data from the fourth chromosome suggest that *1360* DNA transposon fragments (*orange bars*) are a target for heterochromatin formation; local deletions or duplications that shift the position of the *P* element reporter (*triangle*) away from a *1360* element lead to loss of silencing (*red triangle* indicates a red eye), whereas proximity to *1360* leads to silencing (*dotted triangle* indicates a variegating eye). (Based on data in Sun et al 2004.)

of the *white* gene, and that mutations in *piwi, aubergine,* and *homeless* suppress silencing of the *white* transgene *P[hsp70-w]* in pericentromeric heterochromatin or the fourth chromosome. This suppression of PEV was associated with a significant decrease in the levels of H3K9 methylation. Repeat-associated small interfering RNAs (rasiRNAs) have been identified from 40% of the known transposable elements (including *1360*) and other repeated sequences (Aravin et al. 2003).

Put together, the results discussed above suggest that heterochromatin formation may be dependent both on nuclear location (perhaps providing an abundant pool of required proteins) and on specific targeting based on RNAi recognition and processing of double-stranded RNA from repetitious elements, particularly some of the DNA transposons. Such targeting via a RISC could bring either a histone H3 methyltransferase or a complex including HP1 (or both) to a site to trigger the assembly process discussed in Section 4.

7 Not All Heterochromatin Is Identical

Although heterochromatin has been described above in general terms, it is clear that heterochromatic domains vary in detail. All heterochromatic domains are characterized by repetitious DNA (see above), but this can vary from a tandem array of short repeats (satellite DNA) found in blocks in centromeric regions, to a high density of interspersed repetitious sequences, as seen on the fourth chromosome. Whereas all heterochromatic regions appear to be associated with HP1 and H3K9me2, it is clear that the protein complexes involved must differ in other ways. Examination of the impact of 70 different modifiers on different variegating genes (including w^{m4}, bw^D, P-element reporters in pericentromeric heterochromatin or in a TAS array) showed that whereas there is substantial overlap in the targets of modifiers, there is also surprising complexity. This set of tests divided the modifiers into seven different groups in terms of their ability to affect silencing in a given compartment (Donaldson et al. 2002). Interestingly, the only modifer in this group to affect silencing in the TAS array was a new allele of *Su(var)3-9*.

These differences no doubt reflect changes in the local biochemistry, or in the enzymes used to achieve it. For example, cytological results indicate that whereas H3K9me2 is highly concentrated along the fourth chromosome, the enzyme responsible is not SU(VAR)3-9 (Schotta et al. 2002; K.A. Haynes et al., unpubl.). Even within the pericentromeric heterochromatin, one should anticipate a mosaic, given the differences in the underlying blocks of DNA, which vary from satellite DNA to clusters of interspersed repeats (Le et al. 1995), which might utilize a different mix of heterochromatin proteins. The consequences have been seen in studies that examine the impact of different blocks of pericentromeric heterochromatin on expression from a reporter, where one can observe that the severity of the phenotype does not depend simply on the amount of heterochromatin in *cis*, but varies depending on the local heterochromatin environment (Howe et al. 1995). Heterochromatin-associated proteins that might play a role in specific subdomains include the AT-hook protein D1, preferentially associated with the 1.688 g/cm^3 satellite III (Aulner et al. 2002), and DDP1, a multi-KH-domain protein homologous to vigilin that binds the pyrimidine-rich C strand of the dodeca satellite (Cortes and Azorin 2000).

8 PEV, Heterochromatin Formation, and Gene Silencing in Different Organisms

The phenomenon of position-effect variegation was initially detected in *Drosophila*, simply because this was one of the first organisms for which X irradiation was used to induce mutations. X irradiation is much more likely than other commonly used mutagens to induce chromosomal rearrangements, which can result in PEV. Similar mutations have been isolated from the mouse, where variegating coat color indicates PEV. Genetic analysis revealed an insertion of the autosomal region carrying wild-type alleles of the fur-color genes into the X chromosome (Cattanach 1961; Russel and Bangham 1961). Variegation is only observed in females carrying this insertion combined with a homozygous mutation in the original coat-color genes. In these females, the wild-type allele becomes inactivated as a consequence of X inactivation by heterochromatinization (see Chapter 17). In plants, the only unequivocal case of PEV that has been described was reported in *Oenothera blandina* (Catcheside 1939). In these cases, as in *Drosophila*, PEV silencing of euchro-

matic genes is connected with placement of those genes into a new heterochromatic neighborhood.

Transcriptional gene silencing has also been observed for repeated sequences (RIGS; repeat-induced gene silencing), particularly in plants. Analysis of the affected sequences has revealed the appearance of similar epigenetic marks (histone and DNA methylation) as found in heterochromatin and in regions silenced by PEV. If DNA fragments containing tandemly arranged luciferase genes are introduced into *Arabidopsis*, variegated luciferase expression is seen. Again, heterochromatin formation is responsible for the gene silencing observed. The underlying molecular mechanisms are conserved in higher eukaryotic organisms.

A central feature of heterochromatic gene silencing in *Drosophila* is the interaction of HP1 with H3K9me2 and the SU(VAR)3-9 HKMT. HP1 is conserved from the yeast *Schizosaccharomyces pombe* to man, and is consistently associated with pericentromeric heterochromatin. The human HP1 genes can be used to rescue the deficiency in *Drosophila* (Ma et al. 2001). However, HP1 has not been identified in plants as such. SU(VAR)3-9 is even more widely represented, having been identified in fission yeast (Clr4), *Neurospora* (DIM5), *Arabidopsis*, and mammals (SUV39H). All of the SU(VAR)3-9 homologs catalyze H3K9 methylation and function in heterochromatin formation. Again, a human SUV39H1 transgene can completely compensate for the loss of the endogenous *Drosophila* protein in mutant lines (Schotta et al. 2002). In higher plants (rice, *Arabidopsis*, and maize), several SU(VAR)3-9 homologous proteins (SUVH) are found (Baumbusch et al. 2001). The high number of HKMTs might reflect the plasticity of plant development or the need to respond to environmental factors (see Chapter 9 for further discussion). Four SUVH proteins, SUVH1, SUVH2, SUVH4 (KYP), and SUVH6, have been studied in detail (Jackson et al. 2002; Naumann et al. 2005). All are histone H3K9 methyltransferases. SUVH2 plays a pivotal role in control of heterochromatin states, exhibiting dosage-dependent effects on heterochromatin formation similar to those reported for *Drosophila* SU(VAR)3-9 (Naumann et al. 2005). SUVH2 loss of function strongly suppresses repeat-dependent silencing, and overexpression causes significant enhancement of such silencing in plants with luciferase transgenes.

Other genes identified by *Drosophila Su(var)* mutations encode proteins with conserved functions. The SUV4-20 HKMT has been characterized in mammals and in *Drosophila* (Schotta et al. 2004). In both organisms, it controls trimethylation of H4K20. Histone demethylases, acetylases, and deacetylases are also conserved (T. Rudolph et al., unpubl.). The evolutionary conservation of many of the key enzymes controlling histone modification supports the idea of a histone code (Jenuwein and Allis 2001). However, examination of the heterochromatin-specific histone modification marks observed in *Drosophila*, mammals, and plants (*Arabidopsis*) also identifies some genus-specific elements.

Significant hallmarks of constitutive heterochromatin in mammals include H3K9me3, H3K27me1, and H4K20me3 (Peters et al. 2003; Rice et al. 2003; Schotta et al. 2004). *Drosophila* heterochromatin is characterized by H3K9me1/me2, H3K27me1/me2/me3, and H4K20me3 (Schotta et al. 2002, 2004; Ebert et al. 2004). In contrast to mammals, H3K9me3 is underrepresented in *Drosophila*. In mammals, H3K9me1 is not a heterochromatic mark. In *Arabidopsis*, as in *Drosophila*, H3K9me1/me2 are heterochromatic marks, whereas H3K9me3 is euchromatic (Naumann et al. 2005). H3K27me1 and H3K27me2 are heterochromatic marks in *Arabidopsis,* whereas these marks in *Drosophila* are found in euchromatin and heterochromatin. H3K27me3 is exclusively euchromatic in *Arabidopsis*. H4K20me1 in *Arabidopsis* is heterochromatic, but H4K20me2 and H4K20me3 are euchromatic. Another striking difference between *Arabidopsis* and animals concerns the chromosomal distribution of H3S10 phosphorylation. This mark is heterochromatic in *Arabidopsis* (A. Fischer and G. Reuter, unpubl.) but euchromatic in *Drosophila* (Wang et al. 2001; Ebert et al. 2004). Similarities and differences in heterochromatin-specific histone modification marks between mammals, *Drosophila*, and *Arabidopsis* clearly indicate that the histone code is not completely universal, but rather exists in different dialects.

9 Summing Up: There Is Much That We Do Not Know about Heterochromatin

Although PEV has provided us with an extraordinary opportunity to study heterochromatin formation and gene silencing, the phenotype itself remains puzzling. Why do we observe a variegating pattern of silencing? What tips the balance, leading to a switch from the active to the silent state, or vice versa? Why does this appear to be clonally inherited? PEV is generally analyzed as a problem of maintaining the reporter gene "ON" or "OFF," but in many instances (particularly when using *P*-element-based reporters), one observes red facets on a yellow or pale orange background, suggesting that gene expression has been reduced uni-

formly, but that that down-regulation has been lost in some cells. Careful analysis of such lines might lead to identification of chromatin states with an intermediate impact on gene expression. Although the data support a crude model for loss or maintenance of silencing based on mass action, the final model will be complex, involving numerous interacting proteins (see, e.g., the proposal by Henikoff 1996). One is tempted to consider the nucleosome as a summation device, collecting modifications and displaying the results in terms of both particular protein-binding patterns and facility for remodeling in that region. The chromatin state might then reflect the results of competition for achieving different modifications. Such a model could be useful in sorting out the effects noted above. It is also compatible with observations demonstrating that the frequency of silencing of a GAL4-dependent reporter is sensitive to GAL4 levels (Ahmad and Henikoff 2001).

The RNAi system provides a plausible mechanism for targeting heterochromatin formation, presumably by targeting a complex including HP1, an H3K9 HKMT, or both. However, many questions remain. What is the source of the dsRNA? Must it be produced in *cis* (as implied by the results in *S. pombe*), or can it operate in *trans* (as suggested by results in plants); i.e., can the production of dsRNA from one *1360* site result in targeting of all *1360* sites? Are all repetitious elements potential targets? This seems unlikely from the fourth-chromosome analysis described above. If a subset of repetitious elements plays a key role, what determines that choice? The results obtained on the fourth chromosome argue that the density and distribution of critical repetitious elements will affect expression of the genes in the vicinity. This argues for the need to ascertain this characteristic when sequencing a genome.

How is spreading of heterochromatin accomplished, and what are the normal barriers to spreading? Note that there is no evidence for transitive RNAi in *Drosophila*; i.e., the spread of silencing to targets in a transcript that lie upstream of the dsRNA sequence (Celotto and Gravely 2002). This is in congruence with the lack of evidence for any RNA-dependent RNA polymerase in this system. An assembly system based on the interactions of HP1, H3K9me2, and an HKMT might well account for the spread of heterochromatin for approximately 10 kb, as observed on the fourth chromosome; this type of spreading could be limited by a site of histone acetylation. But what about the spreading that occurs in rearrangements, which has been found to extend for hundreds of kilobases? This form of spreading is not contiguous, but again appears to depend critically on chromatin proteins, notably JIL-1, in a role that does not depend on its kinase activity. These and other questions remain unanswered.

Acknowledgments

We thank Gabriella Farkas for creating the figures used here, Anja Ebert for immunocytological photos, and the members of our research groups for a critical review of this chapter. Our work is supported by Deutsche Forschungsgemeinschaft and the *Epigenome* Network of Excellence of the European Union (G.R.) and by grants from the National Institutes of Health (S.C.R.E.).

References

Ahmad K. and Henikoff S. 2001. Modulation of a transcription factor counteracts heterochromatin gene silencing in *Drosophila*. *Cell* **104**: 839–847.

Aravin A.A., Numova H. M., Tulin A. V., Vagin, V.V., Rozovsky Y.M., and Gvozdev V.A. 2001. Double-stranded RNA-mediated silencing of genomic tandem repeats and transposable elements in the *D. melanogaster* germline. *Curr. Biol.* **11**: 1017–1027.

Aravin A. A., Lagos-Quintana M., Yalcin A., Zavolan M., Marks D., Snyder B., Gaasterland T., Meyer J., and Tuschl T. 2003. The small RNA profile during *Drosophila melanogaster* development. *Dev. Cell* **5**: 337–350.

Ashburner M., Golic K.G., and Hawley R.S. 2005a. Chromosomes. In *Drosophila: A laboratory handbook*, 2nd edition. Cold Spring Harbor Laboratory Press, Cold Spring Harbor, New York, pp. 24–57.

———. 2005b. Position effect variegation. In *Drosophila: A laboratory handbook*, 2nd edition. Cold Spring Harbor Laboratory Press, Cold Spring Harbor, New York, pp.1007–1049.

Aulner N., Monod C., Mandicourt G., Jullien D., Cuvier O., Sall A., Janssen S., Laemmli U.K, and Kas E. 2002. The AT-hook protein D1 is essential for *Drosophila melanogaster* development and is implicated in position-effect variegation. *Mol. Cell. Biol.* **22**: 1218–1232.

Baksa K., Morawietz H., Dombradi V., Axton M., Taubert H., Szabo G., Török I., Gyurkovics H., Szöör B., Gloover D., et al. 1993. Mutations in the phosphatase 1 gene at 87B can differentially affect suppression of position-effect variegation and mitosis in *Drosophila melanogaster*. *Genetics* **135**: 117–125.

Bannister A.J., Zegermann P., Patridge J.F., Miska E.A., Thomas J.O., Allshire T.C., and Kouzarides T. 2001. Selective recognition of methylated lysine 9 on histone H3 by the HP1 chromo domain. *Nature* **410**: 120–124.

Baumbusch L.O., Thorstensen T., Krauss V., Fischer A., Naumann K., Assalkhou R., Schulz I., Reuter G., and Aalen R. 2001. The *Arabidopsis thaliana* genome contains at least 29 active genes encoding SET domain proteins that can be assigned to four evolutionary conserved classes. *Nucleic Acids Res.* **29**: 4319–4333.

Belyaeva E.S., Demakova O.V., Umbetova G.H., and Zhimulev I.F. 1993. Cytogenetic and molecular aspects of position-effect variegation in *Drosophila melanogaster*. V. Heterochromatin-associated protein HP1 appears in euchromatic chromosomal regions that are inactivated as a result of position-effect variegation. *Chromosoma* **102**: 583–590.

Bender W., Spierer P., and Hogness D.S. 1983. Chromosome walking and jumping to isolate DNA from the *Ace* and *rosy* loci and the bithorax complex of *Drosophila melanogaster*. *J. Mol. Biol.* **168**: 17–33.

Birchler J.A., Bhadra U., Rabinow L., Linsk R., and Nguyen-Huyuh A.T. 1994. *Weakener of white* (*Wow*), a gene that modifies the expression of white eye color locus and that suppresses position effect variegation in *Drosophila melanogaster*. *Genetics* **137**: 1057–1070.

Bird A. 2002. DNA methylation patterns and epigenetic memory. *Genes Dev.* **16**: 6–21.

Bushey D. and Locke J. 2004. Mutations in *Su(var)205* and *Su(var)3-7* suppress P-element-dependent silencing in *Drosophila melanogaster*. *Genetics* **168**: 1395–1411.

Catcheside D.G. 1939. A position effect in *Oenothera*. *J. Genet.* **38**: 345–352.

Cattanach B.M. 1961. A chemically-induced variegated-type position effect in the mouse. *Z. Vererbungsl.* **92**: 165–182.

Celotto A.M. and Graveley B.R. 2002 Exon-specific RNAi: A tool for dissecting the functional relevance of alternative splicing. *RNA* **8**: 718–724.

Cleard F. and Spierer P. 2001. Position-effect variegation in *Drosophila*: The modifier *Su(var)3-7* is a modular DNA-binding protein. *EMBO Rep.* **21**: 1095–1100.

Cleard F., Delattre M., and Spierer P. 1997. SU(VAR)3-7 a *Drosophila* heterochromatin-associated protein and companion of HP1 in the genomic silencing of position-effect variegation. *EMBO J.* **16**: 5280–5288.

Cooper K.W. 1959. Cytogenetic analysis of major heterochromatic elements (especially Xh and Y) in *Drosophila melanogaster* and the theory of "heterochromatin". *Chromosoma* **10**: 535–588.

Cortes A. and Azorin F. 2000. DDP1, a heterochromatin-asociated multi-KH-domain protein of *Drosophila melanogaster*, interacts specifically with centromeric satellite DNA sequences. *Mol. Cell. Biol.* **20**: 3860–3869.

Cryderman D.E., Morris E.J., Biessmann H., Elgin S.C.R., and Wallrath L.L. 1999a. Silencing at *Drosophila* telomeres: Nuclear organization and chromatin structure play critical roles. *EMBO J.* **18**: 3724–3735.

Cryderman D. E., Tang H., Bell C., Gilmour D.S., and Wallrath L.L. 1999b. Heterochromatic silencing of *Drosophila* heat shock genes acts at the level of promoter potentiation. *Nucleic Acids Res.* **27**: 3364–3370.

Czermin B., Schotta G., Hülsmann B.B., Brehm A., Becker P.B., Reuter G., and Imhof A. 2001. Physical and functional interaction of SU(VAR)3-9 and HDAC1 in *Drosophila*. *EMBO Rep.* **2**: 915–919.

Delattre M., Spierer A., Tonka C.H., and Spierer P. 2000. The genomic silencing of position-effect variegation in *Drosophila melanogaster*: Interaction between the heterochromatin-associated proteins Su(var)3-7 and HP1. *J. Cell Sci.* **113**: 4253–4261.

Demerec M. and Slizynska H. 1937. Mottled white 258-18 of *Drosophila melanogaster*. *Genetics* **22**: 641–649.

Donaldson K.M., Lui A., and Karpen G.H. 2002. Modifiers of terminal deficiency-associated position effect variegation in *Drosophila*. *Genetics* **160**: 995–1009.

Dorer D.R. and Henikoff S. 1994. Expansion of transgene repeats causes heterochromatin formation and gene silencing in *Drosophila*. *Cell* **77**: 993–1002.

Dorn R., Krauss V., Reuter G., and Saumweber H. 1993a. The enhancer of position-effect variegation *E(var)93D*, codes for a chromatin protein containing a conserved domain common to several transcriptional regulators. *Proc. Natl. Acad. Sci.* **90**: 11376–11380.

Dorn R., Szidonya J., Korge G., Sehnert M., Taubert H., Archoukieh I., Tschiersch B., Morawietz H., Wustmann G., Hoffmann G., and Reuter G. 1993b. P Transposon-induced dominant enhancer mutations of position-effect variegation in *Drosophila melanogaster*. *Genetics* **133**: 279–290.

Ebert A., Schotta G., Lein S., Kubicek S., Krauss V., Jenuwein T., and Reuter G. 2004. Su(var) genes regulate the balance between euchromatin and heterochromatin in *Drosophila*. *Genes Dev.* **18**: 2973–2983.

Eissenberg J.C., Morris G.D., Reuter G., and Hartnett T. 1992. The hetero-chromatin-associated protein HP-1 is an essential protein in *Drosophila* with dosage-dependent effects on position-effect variegation. *Genetics* **131**: 345–352.

Eissenberg J.C., James T.C., Foster-Hartnett D.M., Hartnett T., Ngan V., and Elgin S.C.R. 1990. A mutation in a heterochromatin-specific chromosomal protein is associated with suppression of position effect variegation in *Drosophila melanogaster*. *Proc. Natl. Acad. Sci.* **87**: 9923–9927.

Elgin S.C.R. and Grewal S.I.S. 2003. Heterochromatin: Silence is Golden. *Curr. Biol.* **13**: R895–R898.

Fanti L. and Pimpinelli S. 2004. Immunostaining of squash preparations of chromosomes of larval brains. *Methods Mol. Biol.* **247**: 353–361.

Fanti L., Giovinazzo G., Berloco M., and Pimpinelli S. 1998. The heterochromatin protein 1 prevents telomere fusions in *Drosophila*. *Mol. Cell* **2**. 527–538.

Farkas G., Gausz J., Galloni M., Reuter G., Gyurkovics H., and Krach F. 1994. The *trithorax-like* gene encodes the *Drosophila* GAGA factor. *Nature* **371**: 806–808.

Fischle W., Wang Y., Jacobs S.A., Kim Y., Allis C.D., and Khorasanizadeh S. 2003. Molecular basis for the discrimination of repressive methyl-lysine marks in histone H3 by Polycomb and HP1 chromodomains. *Genes Dev.* **17**: 1870–1881.

Flybase 2005. The *Drosophila* database. Available from World Wide Web at the URLs http://morgan/harvard.edu and http://www.ebi.ac.uk/flybase/

Foe V.E. and Alberts B.M. 1985. Reversible chromosome condensation induced in *Drosophila* embryos by anoxia: Visualization of interphase nuclear organization. *J. Cell Biol.* **100**: 1623–1636.

Harmon B. and Sedat J. 2005. Cell-by-cell dissection of gene expression and chromosomal interactions reveals consequences of nuclear reorganization. *PLoS Biol.* **3**: e67.

Hartmann-Goldstein I.J. 1967. On the relationship between heterochromatization and variegation in *Drosophila*, with special reference to temperature sensitive periods. *Genet. Res.* **10**: 143–159.

Henikoff S. 1996. Dosage-dependent modification of position-effect variegation in *Drosophila*. *BioEssays* **18**: 401–409.

Henikoff S., Jackson J.M., and Talbert P.B. 1995. Distance and pairing effects on the *brown*[Dominant] heterochromatic element in *Drosophila*. *Genetics* **140**: 1007–1017.

Hoskins R.A., Smith C.D., Carlson J.W., Carvalho A.B., Halpern A., Kaminker J.S., Kennedey C., Mungall C.J., Sullivan B.A., Sutton G.G., et al. 2002. Heterochromatic sequences in a *Drosophila* whole-genome shotgun assembly. *Genome Biol.* **3**: RESEARCH0085.

Howe M., Dimitri P., Berloco M., and Wakimoto B.T. 1995. *cis*-effects of heterochromatin on heterochromatic and euchromatic gene activity in *Drosophila melanogaster*. *Genetics*. **140**: 1033–1045.

Jackson J.P., Lindroth A.M., Cao X., and Jacobsen S.E. 2002. Control of CpNpG DNA methylation by the *KRYPONITE* histone H3 methyltransferase. *Nature* **416**: 556–560.

James T.C. and Elgin S.C.R. 1986. Identification of a nonhistone chromosomal protein associated with heterochromatin in *Drosophila melanogaster* and its gene. *Mol. Cell Biol.* **6**: 3862–3872.

James T.C., Eissenberg J.C., Craig C., Dietrich V., Hobson A., and Elgin S.C.R. 1989. Distribution patterns of HP1, a heterochromatin-associated nonhistone chromosomal protein of *Drosophila*. *Eur. J. Cell Biol.* **50**: 170–180.

Janssen S., Cuvier O., Muller M., and Laemmli U.K. 2000. Specific gain- and loss-of-function phenotypes induced by satellite-specific DNA-binding drugs fed to *Drosophila melanogaster*. *Mol. Cell* **6**: 1013–1024.

Jenuwein T. and Allis C.D. 2001. Translating the histone code. *Science* **293**: 1074–1080.

Jones R.S. and Gelbart W.M. 1993. The *Drosophila* Polycomb-group gene *Enhancer of zeste* contains a region with sequence similarity to trithorax. *Mol. Cell. Biol.* **13**: 6357–6366.

Kaminker J.S. Bergman C.M., Kronmiller B., Carlson J., Svirskas R., Patel S., Frise E., Wheeler D.A., Lewis S.E., Rubin G.M., et al. 2002. The transposable elements of the *Drosophila melanogaster* genome: A genomics perspective. *Genome Biol.* **3**: RESEARCH0084.

Karpen G.H. and Spradling A.C. 1990. Reduced DNA polytenization of a minichromosome region undergoing position-effect variegation in *Drosophila*. *Cell* **63**: 97–107.

Krauss V. and Reuter G. 2000. Two genes become one: The genes encoding heterochromatin protein SU(VAR)3-9 and translation initiation factor subunit eIF-2γ are joined to a dicistronic unit in holometabolic insects. *Genetics* **156**: 1157–1167.

Kunert N., Marhold J., Stanke J., Stach D., and Lyko F. 2003. A Dnmt2-like protein mediates DNA methylation in *Drosophila*. *Development* **130**: 5083–5090.

Lachner M., O'Carroll D., Rea S., Mechtler K., and Jenuwein T. 2001. Methylation of histone H3 lysine 9 creates a binding site for HP1 proteins. *Nature* **410**: 116–120.

Laible G., Wolf A., Dorn R., Reuter G., Nislow C., Lebesorger A., Popkin D., Pillus L., and Jenuwein T. 1997. Mammalian homologues of the Polycomb-group gene *Enhancer of zeste* mediate gene silencing in *Drosophila* heterochromatin and at *S. cerevisiae*. *EMBO J.* **16**: 3219–3232.

Larsson J., Zhang J., and Rasmuson-Lestander A. 1996. Mutations in the *Drosophila melanogaster* S-adenosylmethionine synthase suppress position-effect variegation. *Genetics* **143**: 887–896.

Le M.H., Duricka D., and Karpen G.H. 1995. Islands of complex DNA are widespread in *Drosophila* centric heterochromatin. *Genetics* **141**: 283–303.

Locke J., Kotarski M. A., and Tartof K.D. 1988. Dosage-dependent modifiers of position effect variegation in *Drosophila* and a mass action model that explains their effect. *Genetics* **120**: 181–198.

Lu B.Y., Bishop C.P., and Eissenberg J.C. 1996. Developmetal timing and tissue specificity of heterochromatin-mediated silencing. *EMBO J.* **15**: 1323–1332.

Ma J., Hwang K.K., Worman H.J., Courvalin J.C., and Eissenberg J.C. 2001. Expression and functional analysis of three isoforms of human heterochromatin-associated protein HP1 in *Drosophila*. *Chromosoma* **109**: 536–544.

Martienssen R.A. and Colot V. 2001. DNA methylation and epigenetic inheritance in plants and filamentous fungi. *Science* **293**: 1070–1074.

Mason J.M., Ransom J., and Konev A.Y. 2004. A deficiency screen for dominant suppressors of telomeric silencing in *Drosophila*. *Genetics* **168**: 1353–1370.

Matzke M.A. and Birchler J.A. 2005. RNAi-mediated pathways in the nucleus. *Nat. Rev. Genet.* **6**: 24–35.

Mottus R., Sobels R.E., and Grigliatti T.A. 2000. Mutational analysis of a histone deacetylase in *Drosophila melanogaster*: Missense mutations suppress gene silencing associated with position effect variegation. *Genetics* **154**: 657–668.

Muller H.J. 1930. Types of visible variations induced by X-rays in *Drosophila*. *J. Genet.* **22:** 299–334.

Naumann K., Fischer A., Hofmann I., Krauss V., Phalke S., Irmler K., Hause G., Aurich A.C., Dorn R., Jenuwein T., and Reuter G. 2005. Pivotal role of *At*SUVH2 in control of heterochromatic histone methylation and gene silencing in *Arabidopsis*. *EMBO J.* **24:** 1418–1429.

Pal-Bhadra M., Bhadra U., and Birchler J.A. 2002. RNAi related mechanisms affect both transcriptional and post-transcriptional transgene silencing in *Drosophila*. *Mol. Cell* **9:** 315–327.

Pal-Bhadra M., Leibovitch B.A., Gandhi S.G., Rao M., Bhadra U., Birchler J.A., and Elgin S.C.R. 2004. Heterochromatic silencing and HP1 localization in *Drosophila* are dependent on the RNAi machinery. *Science* **303:** 669–672.

Paro R. and Hogness D.S. 1991. The Polycomb protein shares a homologous domain with a heterochromatin-associated protein of *Drosophila*. *Proc. Natl. Acad. Sci.* **88:** 263–267.

Peters A.H.F.M., Kubicek S., Mechtler K., O'Sullivan J., Derijck A.A.H.A., Perez-Burgos L., Kohlmaier A., Opravil S., Tachibana M., Shinkai Y., et al. 2003. Partitioning and plasticity of repressive histone methylation states in mammalian chromatin. *Mol. Cell* **12:** 1577–1589.

Platero J.S., Sharp E.J., Adler P.N., and Eissenberg J.C. 1996. In vivo assay for protein-protein interaction using *Drosophila* chromosomes. *Chromosoma* **104:** 393–404.

Rea S., Eisenhaber F., O'Carroll D., Strahl B.D., Sun Z-W., Schmid M., Opravil S., Mechtler K., Ponting C.P., Allis C.D., and Jenuwein T. 2000. Regulation of chromatin structure by site-specific histone H3 methyltransferases. *Nature* **406:** 593–599.

Reuter G., Werner W., and Hofmann H.J. 1982. Mutants affecting position-effect heterochromatinization in *Drosophila melanogaster*. *Chromosoma* **85:** 539–551.

Reuter G., Wolff I., and Friede B. 1985. Functional properties of the heterochromatic sequences inducing w^{m4} position-effect variation in *Drosophila melanogaster*. *Chromosoma* **93:** 132–139.

Reuter G., Giarre N., Farah J., Gausz J., Spierer A., and Spierer P. 1990. Dependence of position-effect variegation in *Drosophila* on dose of a gene encoding an unusual zinc-finger protein. *Nature* **344:** 219–223.

Rice J.C., Briggs S.D., Ueberheide B., Barber C.M., Shabanowitz J., Hunt D.F., Shinkai Y., and Allis C.D. 2003. Histone methyltransferases direct different degrees of methylation to define distinct chromatin domains. *Mol. Cell* **12:** 1591–1598.

Richards E.J. and Elgin S.C.R. 2002. Epigenetic codes for heterochromatin formation and silencing: Rounding up the usual suspects. *Cell* **108:** 489–500.

Russel L.B. and Bangham J.W. 1961. Variegated type position effects in the mouse. *Genetics* **46:** 509–525.

Schotta G., Ebert A., Krauss V., Fischer A., Hoffmann J., Rea S., Jenuwein T., and Reuter G. 2002. Central role of *Drosophila* SU(VAR)3-9 in histone H3-K9 methylation and heterochromatic gene silencing. *EMBO J.* **21:** 1121–1131.

Schotta G., Ebert A., Dorn R., and Reuter G. 2003. Position-effect variegation and the genetic dissection of chromatin regulation in *Drosophila*. *Semin. Cell Dev. Biol.* **14:** 67–75.

Schotta G., Lachner M., Sarma K., Ebert A., Sengupta R., Reuter G., Reinberg D., and Jenuwein T. 2004. A silencing pathway to induce H3-K9 and H4-K20 trimethylation at constitutive heterochromatin. *Genes Dev.* **18:** 1251–1262.

Schultz J. 1950. Interrelations of factors affecting heterochromatin-induced variegation in *Drosophila*. *Genetics* **35:** 134.

Seum C., Pauli D., Delattre M., Jaquet Y., Spierer A., and Spierer P. 2002. Isolation of *Su(var)3-7* mutations by homologous recombination in *Drosophila melanogaster*. *Genetics* **161:** 1125–1136.

Shaffer C.D., Stephens G.E., Thompson B.A., Funches L., Bernat J.A., Craig C.A., and Elgin S.C.R. 2002. Heterochromatin protein 2 (HP2), a partner of HP1 in *Drosophila* heterochromatin. *Proc. Natl. Acad. Sci.* **99:** 14332–14337.

Shi Y., Lan F., Matson C., Mulligan P., Whetstine J.R., Cole P.A., Casero R.A., and Shi Y. 2005. Histone demethylation mediated by the nuclear amine oxidase homolog LSD1. *Cell* **119:** 941–953.

Silver L.M., and Elgin S.C.R. 1976. A method for determination of the in situ distribution of chromosomal proteins. *Proc. Natl. Acad. Sci.* **73:** 423–427.

Sinclair D.A.R., Mottus R.C., and Grigliatti T.A. 1983. Genes which suppress position effect variegation in *Drosophila melanogaster* are clustered. *Mol. Gen. Genet.* **191:** 326–333.

Sinclair D.A.R., Clegg N.J, Antonchuk J, Milner T.A., Stankunas K., Ruse C., Grigliatti T.A., Kassis J., and Brock H.W. 1998. *Enhancer of Polycomb* is a suppressor of position-effect variegation in *Drosophila melanogaster*. *Genetics* **148:** 211–220.

Sontheimer E.J. 2005. Assembly and function of RNA silencing complexes. *Nat. Rev. Mol. Cell Biol.* **6:** 127–138.

Spofford J.B. 1967. Single-locus modification of position-effect variegation in *Drosophila melanogaster*. I. white variegation. *Genetics* **57:** 751–766.

Stephens G.E., Craig C.A., Li Y., Wallrath L.L., and Elgin S.C.R. 2003. Immunofluorescent staining of polytene chromosomes: Exploiting genetic tools. *Methods Enzymol.* **376:** 372–393.

Sun F.-L., Cuaycong M.H., and Elgin S.C.R. 2001. Long-range nucleosome ordering is associated with gene silencing in *Drosophila melanogaster* pericentromeric heterochromatin. *Mol. Cell. Biol.* **21:** 2867–2879.

Sun F.-L., Haynes K., Simpson C.L., Lee S.D., Collins L., Wuller J., Eissenberg J.C., and Elgin S.C.R. 2004. cis-acting determinants of heterochromatic formation on *Drosophila melanogaster* chromosome four. *Mol. Cell. Biol.* **24:** 8210–8220.

Tartof K.D., Hobbs C., and Johnes M. 1984. A structural basis of variegating position effects. *Cell* **37:** 869–878.

Tschiersch B., Hofmann A., Krauss V., Dorn R., Korge G., and Reuter G. 1994. The protein encoded by the *Drosophila* position effect variegation suppressor gene *Su(var)3-9* combines domains of antagonistic regulators of homeotic gene complexes. *EMBO J.* **13:** 3822–3831.

Wakimoto B.T. and Hearn M.G. 1990. The effects of chromosome rearrangements on the expression of heterochromatic genes in chromosome 2L of *Drosophila melanogaster*. *Genetics* **125:** 141–154.

Wallrath L.L. and Elgin S.C.R. 1995. Position effect variegation in *Drosophila* is associated with an altered chromatin structure. *Genes Dev.* **9:** 1263–1277.

Wallrath L.L., Gunter V.P., Rosman L.E., and Elgin S.C.R. 1996. DNA representation of variegating heterochromatic P element inserts in diploid and polytene tissue of *Drosophila melanogaster*. *Chromosoma* **104:** 519–527.

Wang Y., Zhang W., Jin Y., Johansen J., and Johansen K.M. 2001. The JIL-1 tandem kinase mediates histone H3 phosphorylation and is required for maintenance of chromatin structure in *Drosophila*. *Cell* **105:** 433–443.

Zhimulev I.F., Belyaeva E.S., Fomina O.V., Protopopov M.O., and Bolshkov V.N. 1986. Cytogenetic and molecular aspects of position effect variegation in *Drosophila melanogaster*. *Chromosoma* **94:** 492–504.

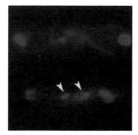
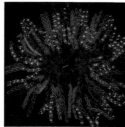

Fungal Models for Epigenetic Research: *Schizosaccharomyces pombe* and *Neurospora crassa*

Robin C. Allshire[1] and Eric U. Selker[2]

[1]Wellcome Trust Centre for Cell Biology, The University of Edinburgh, Edinburgh EH9 3JR, Scotland, United Kingdom
[2]Institute of Molecular Biology, University of Oregon, Eugene, Oregon 97403-1229

CONTENTS

1. *Schizosaccharomyces pombe:* The Organism, 103

   1.1 Chromatin Silencing in S. pombe *Is Different from That in* S. cerevisiae, 103

   1.2 Genes Placed in Fission Yeast Centromeres Are Silenced, 104

   1.3 Fission Yeast Centromeres Are Composed of Distinct Heterochromatin and Central Kinetochore Domains, 104

   1.4 Centromere Outer Repeats Alone Allow the Assembly of Silent Chromatin, 107

   1.5 RNA Interference Directs the Assembly of Silent Chromatin, 108

   1.6 Centromere Repeat Transcription by RNA Polymerase II Links RNAi to Chromatin Modification, 109

   1.7 Silent Chromatin at Centromeres Is Required to Mediate Sister-Centromere Cohesion and Normal Chromosome Segregation, 110

   1.8 Epigenetic Inheritance of the Functional Centromere State, 110

   1.9 Diverse Silencing Mechanisms in Fungi, 111

2. *Neurospora crassa:* History and Features of the Organism, 112

   2.1 DNA Methylation in Neurospora, 113

   2.2 RIP, a Genome Defense System with Both Genetic and Epigenetic Aspects, 115

   2.3 Studies of Relics of RIP Provided Insights into the Control of DNA Methylation, 116

   2.4 Quelling, 117

   2.5 Meiotic Silencing by Unpaired DNA (MSUD), 119

   2.6 Probable Functions and Practical Uses of RIP, Quelling, and MSUD, 120

3. Concluding Remarks, 121

Acknowledgments, 122

References, 122

GENERAL SUMMARY

Fungi provide excellent models for understanding the structure and function of chromatin both in actively transcribed regions (euchromatin) and in transcriptionally silent regions (heterochromatin). The budding yeast, *Saccharomyces cerevisiae*, has been an invaluable eukaryotic model for studying chromatin structure associated with transcription at euchromatic regions and for providing a paradigm for silent chromatin. The fission yeast, *Schizosaccharomyces pombe*, and the filamentous fungus, *Neurospora crassa*, on the other hand, have been instrumental for studying forms of silencing more closely related to those of higher eukaryotes. Heterochromatic regions are relatively small and not essential for viability in these fungi, making them easier to dissect and manipulate. Our understanding of heterochromatin around centromere and telomere regions is most advanced in the yeasts *S. cerevisiae* and *S. pombe*; however, the mechanism of chromatin silencing employed by *S. pombe* exhibits features that are conserved with heterochromatic regions of higher eukaryotes. Indeed, both fungi discussed in this chapter—*N. crassa* and *S. pombe*—contrast with *S. cerevisiae* in that they employ RNA interference (RNAi), histone H3 methylation of lysine 9 and heterochromatin protein 1 (HP1) type proteins to bring about silent chromatin formation in a manner that is conserved or similar to that in plants and metazoa. In addition, *N. crassa* sports DNA methylation, which is a characteristic feature of heterochromatin in many higher eukaryotes and which is a classic epigenetic phenomenon. The nature and function of heterochromatin is first discussed in *S. pombe* after a brief introduction to the organism. We then turn to *N. crassa* to demonstrate how this filamentous fungus has contributed to epigenetics research.

1 Schizosaccharomyces pombe: The Organism

Fission yeast, S. pombe, is found in the fermentations involved in the production of beer in subtropical regions; "pombe" is in fact the Swahili word for beer. S. pombe is primarily a haploid (1N) unicellular organism. In medium rich with nutrients, wild-type cells undergo a mitotic division approximately every 2 hours. But a variety of conditions, or conditional mutants, can be used to block cells at distinct stages of the cell cycle or to synchronize cell cultures in G_1/S, G_2, or at metaphase. This is particularly useful, since G_1 phase is very short in fast-growing cultures, and cells pass almost immediately into S phase following cytokinesis; the major portion of the cell cycle is spent in G_2 (Fig. 1) (Egel 2004).

Like S. cerevisiae, S. pombe can switch between opposite mating types, named Plus (+) and Minus (−). Mating types are equivalent to dimorphic sexes in higher eukaryotes, albeit they are haploid. The information for both mating types resides in the genome as epigenetically regulated silent cassettes—the mat2-P(+) or mat3-M(−) loci. These silent loci provide the genetic template for mating-type identity, but mating type itself is determined by which particular information (+ or −) resides at the active mat1 locus. Switching of information at the active locus mat1, and hence mating type, occurs by recombination between a silent locus and the mat1 locus according to a strict pattern (Egel 2004). When starved of nitrogen, cells stop dividing and arrest in G_1, which promotes the sexual phase of the life cycle through conjugation of pairs of + and − cells to form diploid zygotes (Fig. 1). After mating and nuclear fusion, premeiotic replication occurs (increasing DNA content from 2N to 4N), pairing and recombination of homologous chromosomes then occurs, and this is followed by the reductional meiosis I division and the equational meiosis II division. This produces four separate haploid nuclei (1N) that become encapsulated into spores enclosed in an ascus. The subsequent provision of a rich nutrient source allows germination and resumption of vegetative growth and mitotic cell division.

Non-switching derivatives have been isolated or constructed in which all cells are either + or − mating type. This facilitates controlled mating between strains of distinct genotypes. Although S. pombe is normally haploid, it is possible to select for diploid strains. Such diploid cells can then divide by vegetative mitotic growth until starved of nitrogen, when they too undergo meiosis and form "azygotic asci" (Fig. 1).

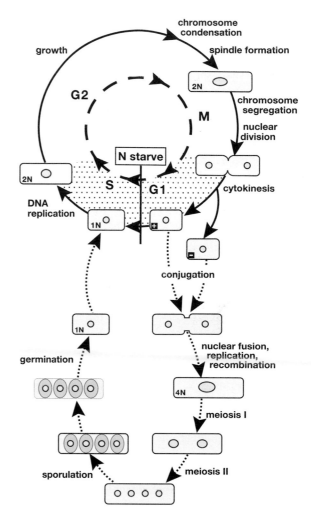

Figure 1. Life Cycle of the Fission Yeast, S. pombe

Fission yeast has a short G_1 taking less than 10% of the cell cycle (stippled area is expanded to aid representation). In rich medium, G_1 cells proceed into S phase followed by a long G_2 (~70% of the cell cycle), mitosis, and cytokinesis. When starved of nitrogen, cells of opposite mating type (+ and −) conjugate, after which nuclei fuse in a process known as karyogamy. Premeiotic replication and recombination allow meiosis I and II to proceed, resulting in four haploid nuclei that are separated into four spores in an ascus. Provision of rich medium allows germination of spores and resumption of the vegetative cell cycle.

1.1 Chromatin Silencing in S. pombe Is Different from That in S. cerevisiae

Fission yeast is a particularly useful model organism in which to study silent chromatin and related epigenetic effects. It is unlike S. cerevisiae, but more akin to N. crassa, plants, and metazoa in the mechanisms that it employs to achieve silencing via heterochromatin. Telomeres and the mating-type loci regions of the genome are subject to silencing in both S. cerevisiae and S. pombe and, in addition, centromeres exhibit silencing in *S. pombe*. However, as described in Chapter 4, *S. cerevisiae* employs a unique set of proteins—the Sir proteins—to achieve chromatin silencing. Fission yeast and other eukaryotes, on the other hand, appear to use a combination of distinct histone modifications (in particular, histone H3K9 methylation) and RNA interference (RNAi) proteins to silence chromatin. As discussed in detail below, studies in *S. pombe* have revealed that this silent chromatin is essential for the function of each specialized region of heterochromatin (i.e., centromeres and mating-type loci). At centromeres, heterochromatin is necessary to ensure normal chromosome segregation (Allshire et al. 1995; Ekwall et al. 1995) whereas at the mating-type loci, it facilitates and regulates mating-type switching (for review, see Egel 2004). In addition, silent chromatin is formed adjacent to telomeres, although a function has yet to be ascribed to this telomeric silent chromatin (Nimmo et al. 1994; Kanoh et al. 2005). Marker genes inserted within rDNA are also silenced (Thon and Verhein-Hansen 2000; Cam et al. 2005).

In contrast to *N. crassa* and higher eukaryotes, *S. pombe* appears to lack any detectable DNA methylation (Wilkinson et al. 1995), a common mechanism used for silencing chromatin in many eukaryotes (see Chapters 9 and 18); thus, silencing in fission yeast is mediated primarily by chromatin modification. As discussed below, the establishment of these modifications, and thus silent chromatin, employs the RNAi machinery.

1.2 Genes Placed in Fission Yeast Centromeres Are Silenced

Heterochromatin formation at fission yeast centromeres is essential to allow normal chromosome segregation during nuclear division. Studies have shown that the centromere in fact consists of two distinct chromatin structures: heterochromatin and CENP-A containing kinetochore chromatin. This has been demonstrated by studying the variable silencing of reporter genes inserted into different centromere regions.

At the DNA level, centromere regions in fission yeast are composed of outer repeats (subdivided into elements known as dg and dh, or K and L) which flank the central domain that includes the inner repeats (imr or B), and a central core (cnt or CC) (Fig. 2a). The three centromeres, *cen1*, *cen2*, and *cen3*, occupy ~40, ~60, and ~120 kb on chromosomes I, II, and III, respectively (for reviews, see Egel 2004; Pidoux and Allshire 2004). The repetitive nature of fission yeast centromere DNA resembles the larger, more complex repeated structures associated with many metazoan centromeres, but they are more amenable to manipu-

lation (Takahashi et al. 1992; Steiner et al. 1993; Ngan and Clarke 1997). Because repetitive DNA frequently correlates with the presence of heterochromatin in other eukaryotes (see also Chapter 5), the presence of repetitive DNA at fission yeast centromeres suggested that they might have heterochromatic properties such as the ability to hinder gene expression. As described below, two blocks of heterochromatin flank the central domain of each fission yeast centromere. The central domain itself is assembled in a distinct type of chromatin (CENP-A chromatin) that differs from the neighboring heterochromatin.

It is well known that the type of chromatin surrounding a gene can strongly influence its expression. This was originally demonstrated in the fruit fly, *Drosophila melanogaster*, where chromosomal rearrangements that move the *white* gene close to centromeric heterochromatin lead to its variable expression in eye facets and, thus, variegation in eye coloration (see Fig. 1 in Chapter 5). It is now apparent that transgenes in many organisms can be influenced by the environment into which they are placed.

In fission yeast, gene silencing can be monitored by phenotypic assays similar to those used in *S. cerevisiae* that assess expression of reporter genes. For example, when the *ura4*⁺ reporter is silenced, 5-fluoroorotic acid-resistant colonies are formed. Alternatively, silencing of the *ade6*⁺ reporter results in red rather than white colony color (Fig. 3a). Placement of a normally expressed gene, such as *ura4*⁺ or *ade6*⁺, within the centromere (as defined by the outer repeat and central domain elements) results in its transcriptional silencing. Silencing is robust in the outer repeats, such that most cells form colonies in which repression of markers is stably maintained (i.e., for the *ade6*⁺ reporter, red colonies are formed). Within the central domain, however, silencing of *ade6*⁺ is comparatively unstable, resulting in variegated colonies, manifested as either red, white, or red-white, sectored colonies (Fig. 3a). However, no silencing occurs just 1 kb distal to the outer repeats (Allshire et al. 1994, 1995), indicating that this transcriptional repression is confined to the centromere as defined by the central domain and flanking outer repeats.

1.3 Fission Yeast Centromeres Are Composed of Distinct Heterochromatin and Central Kinetochore Domains

The difference in the quality of silencing across fission yeast centromeres reflects the fact that repression of transcription is a result of different chromatin structures, including the associations of different non-histone proteins, at outer repeats and the central domain (Partridge et al. 2000). There are two distinct chromatin structures that have been characterized in centromeric regions of

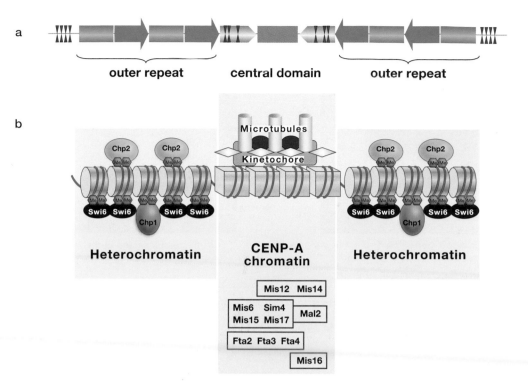

Figure 2. Distinct Outer Repeat Heterochromatin and Central Kinetochore Domains at Fission Yeast Centromeres

(*a*) Representation of a fission yeast centromere. The central domain (*pink*, kinetochore) is composed of *imr* and *cnt* elements, the outer repeats contain transcribed *dg* and *dh* repeats (*green*, heterochromatin). All three centromeres have a similar overall arrangement; however, the number of outer repeats differs: *cen1* (40 kb) has two, *cen2* (65 kb) has three, and *cen3* (110 kb) has approximately thirteen. Clusters of tRNA genes (*double arrowheads*) occur in the imr region and at the extremities of all three centromeres. Transcription of marker genes placed within the outer repeats or central domain is silenced. (*b*) *Heterochromatin*: Outer repeats are packaged in nucleosomes which are methylated on H3K9me2, allowing binding of the chromodomain proteins Chp1, Chp2, and Swi6. *Central "kinetochore" chromatin*: CENP-A is found in the central domain where it probably replaces the majority of H3 to form specialized nucleosomes (*pink squares*). In addition to CENP-A, several kinetochore proteins (those indicated) have been shown to associate with central domain sequences but not the outer repeats. Kinetochore assembly within the central domain mediates attachment to microtubules upon spindle formation and chromosome segregation. Mutation of heterochromatin components alleviates silencing of marker genes in the outer repeats but not the central domain. Defects in some kinetochore components allow expression of marker genes in the central domain but not the outer repeats.

fission yeast: heterochromatin over the outer repeat regions, and "CENP-A" chromatin coating the central domain where the kinetochore is assembled. Different proteins associate with, and are required for, the silencing of reporter genes in the two domains.

HETEROCHROMATIN

In chromatin, the amino-terminal tails of histones H3 and H4 are subject to a range of posttranslational modifications, which generally correlate with active or repressed states (see Chapter 3). Centromeric heterochromatin at outer repeats is associated with the histone H3 lysine 9 di- and trimethyl states (H3K9me2 and H3K9me3) (Nakayama et al. 2001; Yamada et al. 2005). The formation of centromeric heterochromatin requires the action of several proteins that modify chromatin and

thereby promote other factors to bind. Heterochromatin formation first requires the histone deacetylases (HDACs; such as Clr3, Clr6, and Sir2) to deacetylate histone H3. This subsequently allows the histone lysine methyltransferase (HKMT) Clr4 to methylate histone H3 on lysine 9 over the centromeric outer repeats. This modification creates a specific binding site that is recognized by the chromodomain motif present in Swi6 and Chp2 (homologs of heterochromatin protein 1 [HP1] described in Chapter 5) and another chromodomain protein Chp1 (Fig. 2b). These proteins all contribute to the formation of silent chromatin over the outer repeats (Allshire et al. 1995; Cowieson et al. 2000; Partridge et al. 2000; Bannister et al. 2001; Nakayama et al. 2001; Shankaranarayana et al. 2003; Sadaie et al. 2004).

The chromatin associated with reporter gene insertions at the outer repeats (e.g., the *ura4*+ gene) is notably

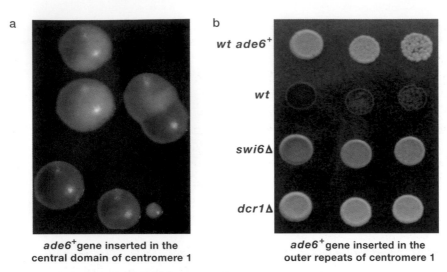

a

b

wt ade6⁺

wt

swi6Δ

dcr1Δ

ade6⁺gene inserted in the
central domain of centromere 1

ade6⁺gene inserted in the
outer repeats of centromere 1

Figure 3. Variegation and Alleviation of Marker Gene Silencing

(*a*) Expression of *ade6⁺* from the central domain is variegated, resulting in red, white, and sectored colonies. Cells expressing *ade6⁺* form white colonies, whereas when *ade6⁺* is repressed, red colonies are formed. (*b*) *ade6⁺* inserted in the outer repeats is robustly silenced, resulting in uniform red colonies. This silencing requires heterochromatin proteins such as Swi6 (which binds methylated H3K9) and RNAi components such as Dcr1 (Dicer, an RNase III ribonuclease).

enriched in H3K9me2 and Swi6 proteins. This indicates that chromatin modification and Swi6 can spread from neighboring centromeric repeat DNA into interposing sequences (Cowieson et al. 2000; Nakayama et al. 2001). Swi6 localization and silencing are dependent on H3K9 methylation, as illustrated by disruption of Swi6 localization in cells lacking the H3K9-specific HKMT, Clr4, or in H3 mutants where lysine 9 is replaced with arginine (Ekwall et al. 1996; Mellone et al. 2003). The Swi6 protein dimerizes via its chromoshadow domain (Cowieson et al. 2000), and this probably facilitates its spreading along chromatin fibers, aided by the sequential action of HDACs and the Clr4 H3K9 HKMT.

In addition to Swi6, the Chp1 and Chp2 chromodomain proteins also associate with outer repeat chromatin at centromeres by binding histone H3 methylated on lysine 9. Chp1 has been shown to be a component of the RNAi effector complex RITS (see Chapter 8) and is required for complete methylation of histone H3K9 over the outer repeats and inserted reporter genes (Partridge et al. 2002; Motamedi et al. 2004; Sadaie et al. 2004).

CENTRAL KINETOCHORE DOMAIN CHROMATIN

Before discussing the details of how heterochromatin is formed on the outer repeats at centromeres, it is important to appreciate that central domain chromatin, where the kinetochore is assembled, is very distinct from the flanking outer repeat heterochromatin, because this is

where the kinetochore is assembled. In contrast to silencing at the outer repeats, the silencing of reporter genes in the approximately 10-kb central domain of *cen1* is essentially independent of Clr4 and therefore does not involve methylation of histone H3 on lysine 9. In fact, the central domain has been shown to have a distinct chromatin composition. This was initially demonstrated by micrococcal nuclease analysis (for explanation of MNase digestion, see Chapter 5), which revealed a smear in contrast to the regular 150-bp ladder characteristic of flanking outer repeat chromatin (Polizzi and Clarke 1991; Takahashi et al. 1992). This distinct pattern differentiates central domain chromatin from heterochromatin and euchromatin, and is related to its assembly in distinctive CENP-A chromatin and the assembly of the kinetochore over this region. In all eukaryotes examined, a histone H3-like protein, known as CENP-A (or cenH3), associates specifically with active centromeres (Cleveland et al. 2003), and CENP-A chromatin is critical for specifying the site of kinetochore assembly (Fig. 2b).

In the central domain chromatin of fission yeast centromeres, most histone H3 is replaced by the CENP-A ortholog, known as Cnp1 (Fig. 4b) (Takahashi et al. 2000). CENP-A^{Cnp1} deposition can occur in a replication-dependent manner at S phase or in a replication-independent manner during G$_2$ (for more detail, see Chapter 13). Kinetochore proteins themselves govern the localization and assembly of CENP-A^{Cnp1} specifically within the centromere central domain (Goshima et al. 1999; Takahashi et al. 2000; Pidoux et al. 2003). Ams2 (a GATA factor), for instance, directs replication-coupled CENP-A^{cnp1} deposition in S phase. In its absence, replication-independent deposition by Mis6 in G$_2$ can compensate, allowing the levels of CENP-A^{cnp1} to be topped up in interphase (Chen et al. 2003; Takahashi et al. 2005). If CENP-A^{cnp1} chromatin structure is disrupted (as in the case of *cnp1*, *mal2*, *mis6*, *sim4*, and other mutants), the specific smeared micrococcal nuclease digestion pattern reverts to a pattern more typical of bulk chromatin (i.e., a nucleosomal ladder). Mutants that affect central domain chromatin have no detectable effects on silencing in the outer repeats (Cowieson et al. 2000; Jin et al. 2002; Pidoux et al. 2003; Hayashi et al. 2004). Furthermore, the fact that CENP-A^{Cnp1}, Mal2, Mis6, Sim4, and other kinetochore proteins only associate with the central domain demonstrates that the central kinetochore domain is structurally complex and functionally distinct from outer repeat silent chromatin (Fig. 2b).

Not all kinetochore domain proteins have been tested, but it appears that silencing within the central domain results from the assembly of an intact kinetochore which, as in *S. cerevisiae*, involves at least 50 proteins (Measday and Hieter 2004). This large complex of proteins presumably restricts access of RNA polymerase II to reporter genes placed within this region and thereby impedes their transcription. In mutants such as *cnp1*, *mal2*, *mis6*, and *sim4*, kinetochore integrity is clearly partially defective at the permissive temperature,

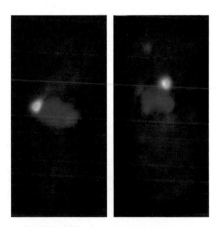

Figure 4. Silent Chromatin in *S. pombe* Nuclei

Two interphase nuclei with heterochromatin (centromeres, telomeres, and the silent *mat2-mat3* loci) decorated by red fluorescent immunolocalization of Swi6, and kinetochore chromatin (centromeres only) decorated by green fluorescent immunolocalization of CENP-A^{cnp1}. Red signals not in close proximity to green represent telomeres or *mat2-mat3*. All centromeres are clustered at the nuclear periphery adjacent to the spindle pole body.

and this allows increased transcription of reporter genes. A spin-off of this is that a normally silent reporter gene has been used to assay for defects in central core chromatin, leading to the identification of novel kinetochore proteins (Pidoux et al. 2003).

Centromere regions are not completely devoid of genes and, intriguingly, several tRNA genes reside between the outer repeats and the central kinetochore domain (Fig. 2a) (Kuhn et al. 1991; Takahashi et al. 1991). Recently, these have been shown to act as a barrier preventing heterochromatin from encroaching into the central domain (Scott et al. 2006).

1.4 Centromere Outer Repeats Alone Allow the Assembly of Silent Chromatin

Clr4-dependent silent chromatin is assembled not only at centromeres, but also over a region of about 20 kb containing the silent mating-type loci (*mat2-mat3*) (Noma et al. 2001) and adjacent to the terminal telomeric repeats (consensus TTACAGG) added by telomerase (Nimmo et al. 1994, 1998; Allshire et al. 1995; Kanoh et al. 2005). The cenH (for centromere homologous) region that resides between *mat2* and *mat3* shares a high degree of similarity, over ~7 kb, with the outer repeats found at centromeres (Grewal and Klar 1997). In addition, at least 0.5 kb of DNA sequences with >84% identity to cenH are located within the telomere-associated sequences (TAS) that occupy up to 40 kb proximal to the telomeres of chromosomes I and II (Kanoh et al. 2005). This suggests that the outer centromeric and related cenH repeats might act in *cis* to bring about silent chromatin assembly.

Assembly of Clr4-dependent silent chromatin occurs even on adjacent marker genes when centromeric outer repeat (*dg*) or *mat2-mat3* (cenH) DNA sequences are inserted in regions of the genome where silencing does not normally occur ("ectopic" silencing; Fig. 5) (Ayoub et al. 2000; Partridge et al. 2002; Volpe et al. 2003). A simple explanation would be that DNA-binding proteins recognize these repeats, and when bound, these proteins recruit HDACs and the HKMT Clr4, resulting in H3K9 methylation, the binding of chromodomain proteins, and the formation of heterochromatin. However, the situation is more complicated than this. It is now known that the centromeric outer repeats are transcribed. Remarkably, this transcription results in the production of a double-stranded RNA (dsRNA) substrate for the RNAi machinery, and this then recruits the Clr4 HKMT to trigger the assembly of silent chromatin (Volpe et al. 2002; Sadaie et al. 2004). RNAi also acts on the related cenH repeats

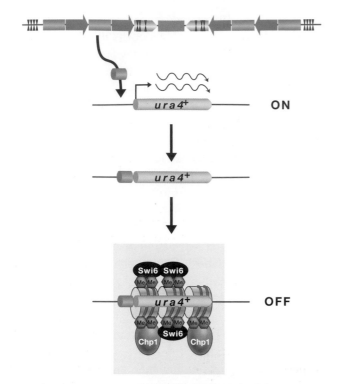

Figure 5. Centromere Repeat Sequences Mediate Silencing

Insertion of fragments (1–2 kb) adjacent to an expressed *ura4*⁺ gene at a locus that is not normally prone to silencing results in H3K9 methylation, binding of Swi6 and Chp1, and transcriptional silencing.

found in the *mat2-mat3* region and in the TAS repeats at telomeres (Cam et al. 2005; Kanoh et al. 2005).

However, the picture is further complicated by the demonstration that at *mat2-mat3*, RNAi acts to establish silent chromatin while the DNA-binding proteins Atf1 and Pcr1 bind near cenH and maintain silent chromatin over *mat2-mat3* even in the absence of key RNAi components (Jia et al. 2004; Kim et al. 2004). Similarly, overlapping mechanisms of silencing also operate at telomeres; terminal repeats alone can recruit Clr4 HKMT and thus Swi6 via the telomere repeat-binding protein Taz1, but RNAi also acts via the cenH part of the TAS elements to form an extended region of silent chromatin at telomeres (Nimmo et al. 1994; Allshire et al. 1995; Kanoh et al. 2005). Is there a comparable overlapping mechanism to maintain silent chromatin at centromeres? Although silencing of outer repeat reporter genes and centromeric outer repeat transcripts themselves is defective in cells lacking RNAi, H3K9me2 is retained on the outer repeat chromatin in the absence of RNAi components (Sadaie et al. 2004). What factors are responsible for maintaining this methylation? One possibility is the CENP-B-related proteins (Abp1, Cbh1, Cbh2), which contain a conserved DNA-binding

domain, bind centromeric repeat DNA, and are required for effective silent chromatin formation (Irelan et al. 2001; Nakagawa et al. 2002). Other observations indicate that the Clr3 HDAC also acts independently of RNAi to maintain heterochromatin integrity (Yamada et al. 2005).

1.5 RNA Interference Directs the Assembly of Silent Chromatin

The phenomenon of RNAi was first discovered in *Caenorhabditis elegans*, where it was found that expression of dsRNA results in loss of expression of a homologous gene. It soon became apparent that this form of RNAi is related to the process of transcriptional gene silencing (TGS) described in plants and quelling in *N. crassa* (described below).

It is known in plants and metazoa that dsRNA, when processed by RNAi machinery, yields small RNAs that bring about DNA and/or chromatin modifications on homologous chromatin. The presence of fission yeast orthologs of the main components of the RNAi pathway, i.e., Argonaute (Ago1), Dicer (Dcr1), and RNA-dependent RNA polymerase (Rdp1), provoked their investigation in fission yeast (Volpe et al. 2002), which has led to significant advances in understanding RNAi-mediated chromatin modification and silencing.

The phenotypic consequences resulting from loss of Ago1, Dcr1, or Rdp1 function are similar to those of *swi6* mutants: reduced H3K9me2 and loss of silencing over the outer repeats of centromeres (Fig. 3b). In RNAi mutants, however, overlapping noncoding RNA (ncRNA) transcripts of a discrete size were detected, originating from centromeric outer repeats. These ncRNAs are homologous to naturally occurring small RNAs called short interfering RNAs (siRNAs; ~21 nt) that have been isolated and sequenced from *S. pombe* (Reinhart and Bartel 2002; Cam et al. 2005). These siRNAs are generated by Dicer (an RNAi component) mediated by cleavage of the double-stranded derivatives of the long centromere repeat homologous noncoding transcripts.

The chromodomain protein Chp1 also turns out to be a component of the RNAi machinery. Chp1 binds to H3K9me2 chromatin and is required for reporter gene silencing at the outer repeats of centromeres, but is also required for the generation of siRNAs homologous to centromere repeats (Noma et al. 2004) and for normal levels of H3K9 methylation on the centromere repeats (Partridge et al. 2002; Sadaie et al. 2004). The role of Chp1 in RNAi was further supported by the identification of the RITS (RNA-induced transcriptional silencing) effector complex, which contains Chp1, Ago1, Tas3, and siRNAs homologous to centromeric outer repeats (Motamedi et al. 2004). In addition to RITS, a complex containing Rdp1 (RDRC: RNA-directed RNA polymerase complex) has been identified which contains a predicted RNA helicase (Hrr1) and a putative poly(A) polymerase (Cid12). These also appear to be integral to the process of RNAi and the assembly of intact silent chromatin on the outer repeats at centromeres (Fig. 6) (Motamedi et al. 2004).

The presence of the RNAi-independent (Atf1/Pcr1-dependent) silencing mechanism operating at the silent mating-type locus to maintain heterochromatin was discovered by treatment of cells lacking RNAi with the HDAC inhibitor trichostatin A (TSA). This resulted in the complete loss of silent chromatin from the *mat2-mat3* region mimicking the effect of deleting *atf1* or *pcr1* in RNAi mutants (Hall et al. 2002; Jia et al. 2004). However, as noted above, Atf1 and Pcr1 are not required for the formation of silent chromatin at the centromeric outer repeats (Kim et al. 2004). Transient treatment with TSA

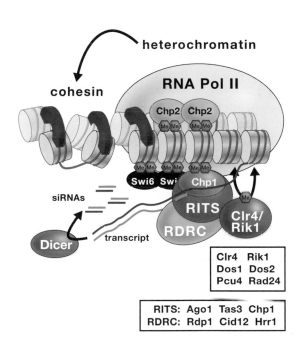

Figure 6. Centromere Repeat Transcription Links RNAi, Heterochromatin Formation, and Cohesion

Transcription of outer repeats by RNA pol II provides an initial substrate for RNAi and Dicer-dependent siRNA generation. Loading of Ago1 in the RITS complex (Ago1, Tas3, Chp1) with siRNA allows targeting of the homologous transcript. The action of the RDRC (Rdp1, Cid12, and Hrr1) would allow dsRNA production providing more substrate for Dcr1 to produce siRNA and perhaps amplify the signal. Interactions between the transcript, RNA pol II subunits, and RNAi components recruit Clr4, which methylates histone H3 on lysine 9, allowing binding of chromodomain proteins, recruitment of cohesin, and sister-centromere cohesion.

also causes loss of silencing at centromeres, resulting in hyperacetylation of histones coupled with defective centromere function (Ekwall et al. 1997). Although this TSA-induced "epistate" is metastable, it can be propagated through several rounds of cell divison and even meiosis. The most likely explanation is that heterochromatin is difficult to reestablish once this abnormal hyperacetylated state is attained. Epistates can also be established at a compromised *mat2-mat3* locus, and these too are propagated through meiosis (Nakayama et al. 2000).

1.6 Centromere Repeat Transcription by RNA Polymerase II Links RNAi to Chromatin Modification

The previous section demonstrated that RNAi and histone modifications are required to assemble centromeric heterochromatin, which can spread into adjacent inserted reporter genes. These observations raised obvious questions, such as, How is RNAi linked with the covalent modification of histones to form silent chromatin? and How are particular regions of the genome targeted for such RNAi-directed chromatin modification and silencing?

In many organisms, the expression of a specific dsRNA, homologous to a gene of interest, results in either transcriptional (DNA/chromatin modification) gene silencing (TGS) or posttranscriptional (mRNA degradation) gene silencing (PTGS). Can any dsRNA be processed to form siRNAs in fission yeast, and do such siRNAs induce only posttranscriptional silencing (RNA knockdown), or can they also bring about chromatin modifications (e.g., H3K9me2) that silence transcription from the homologous gene? Expression of dsRNAs homologous to a GFP reporter produces siRNAs that cause a reduction in GFP transcripts, but the transcriptional activity of the GFP reporter gene does not decline. Thus, although GFP-siRNAs reduce GFP mRNA levels, they are unable to bring about the chromatin modifications that result in transcriptional silencing. These GFP-siRNAs must therefore only act posttranscriptionally to destroy GFP mRNA (Sigova et al. 2004). It is unclear why GFP-siRNAs do not induce chromatin modification on the homologous gene, but it may be related to the nature of the nascent transcript or the strength of the RNA pol II promoter driving GFP expression.

Which RNA polymerase is responsible for transcription of centromere repeats? Mutation of either of two RNA pol II subunits (Rpb2 and Rpb7) results in defective centromere silencing (Djupedal et al. 2005; Kato et al. 2005), although these mutations show very different phenotypes. The *rpb7-1* mutant shows reduced levels of centromere repeat transcription, resulting in less ncRNA and, consequently, less siRNA production and a loss of silent chromatin. This implies that RNA pol II is required to transcribe centromere repeats to provide the primary substrate for RNAi. In contrast, in the *rpb2-m203* mutant, centromeric transcripts are produced, but they are not processed to siRNA, and H3K9me at centromeres is reduced. These studies indicate that RNAi not only requires an RNA pol II transcript, but also that, like other RNA processing events, the production of centromeric siRNA may be coupled to transcription by interactions between RNAi machinery, chromatin, modifying enzymes, and RNA pol II (Fig. 6). It is possible that RITS-associated siRNAs may home in on nascent transcripts as they emerge from RNA pol II engaged with the homologous locus. Once recognition has taken place, the RITS–siRNA complex might be stabilized on these transcripts, resulting in recruitment of chromatin-modifying activities such as Clr4. Surprisingly, centromeric siRNAs are also lost in cells lacking Clr4 HKMT activity (Noma et al. 2004; Hong et al. 2005). It is possible that the absence of Clr4 affects siRNA production by destabilizing associations between various components at the interface between transcription, RNAi, and chromatin modification (Motamedi et al. 2004). Alternatively, H3K9 methylation may be required to allow the generation of siRNAs in *cis* via the action of various RNA processing activities (e.g., RdRP) on primary centromeric transcripts (Fig. 6) (Noma et al. 2004; Sugiyama et al. 2005; see Chapter 8).

1.7 Silent Chromatin at Centromeres Is Required to Mediate Sister-Centromere Cohesion and Normal Chromosome Segregation

How do outer repeat centromeric heterochromatin and CENP-A^{cnp1} kinetochore chromatin affect the overall function of chromosome segregation? Clr4-dependent silent chromatin assembles on outer repeats at centromeres, at a related repeat at the mating-type locus, and adjacent to telomeres. Experiments with naked DNA plasmid constructs have shown that outer repeats contribute in some way to the assembly of a functional centromere, imparting the ability for these Cen-plasmids to segregate on mitotic and meiotic spindles. But neither the outer repeats nor the central domain alone is sufficient to assemble a functional centromere (Clarke and Baum 1990; Takahashi et al. 1992; Baum et al. 1994; Ngan and Clarke 1997; Pidoux and Allshire 2004).

Mutants that cause loss of silencing at the outer repeats (i.e., those defective in Clr4, RNAi components,

or Swi6) have elevated rates of mitotic chromosome loss and a high incidence of lagging chromosomes on late anaphase spindles (Fig. 7a) (Allshire et al. 1995; Ekwall et al. 1996; Bernard et al. 2001; Nonaka et al. 2002; Hall et al. 2003; Volpe et al. 2003). Cells lacking Swi6 are defective in cohesion at centromeres but retain cohesion along the chromosome arms (Bernard et al. 2001; Nonaka et al. 2002). The formation of a properly bioriented spindle requires that sister kinetochores attach to microtubule fibers emanating from opposite spindle poles. The forces exerted on bioriented kinetochores require that sister kinetochores be held together tightly (Fig. 7b). Swi6 is required to recruit cohesin to outer repeat chromatin and thereby mediate tight physical cohesion between sister centromeres (Fig. 6). Thus, one function of silent chromatin at centromeres is to mediate cohesion.

Cohesin is also strongly associated with telomeres and the *mat2-mat3* region (Bernard et al. 2001; Nonaka et al. 2002). In addition, cohesin is also recruited to silent chromatin formed on a *ura4⁺* gene in response to an adjacent ectopic centromere repeat (Fig. 5), underscoring the link between silent chromatin and cohesion (Partridge et al. 2002). Thus, the recruitment of cohesin seems to be a general property of Swi6-associated silent chromatin. How Swi6 chromatin brings about cohesin recruitment is not known, but Swi6 does interact with the Psc3 cohesin subunit (Nonaka et al. 2002). In addition, Dfp1, the regulatory subunit of the conserved kinase Hsk1 (Cdc7), interacts with Swi6 and is required to recruit cohesin to centromeres (Bailis et al. 2003). This functional link between heterochromatin and chromatid cohesion appears to be conserved in other organisms, since depletion of the RNAi component, Dicer, appears to affect heterochromatin integrity and sister-centromere cohesion in vertebrate cells (Fukagawa et al. 2004).

1.8 Epigenetic Inheritance of the Functional Centromere State

An interesting epigenetic phenomenon has been described with respect to the assembly of functional centromeres in fission yeast on plasmids containing minimal regions for centromere function. Although constructs retaining only part of an outer repeat and most of the central domain inefficiently assemble a functional centromere, surprisingly, once this functional centromere active state has been established, it can be propagated through many mitotic divisions and even through meiosis (Steiner and Clarke 1994; Ngan and

a

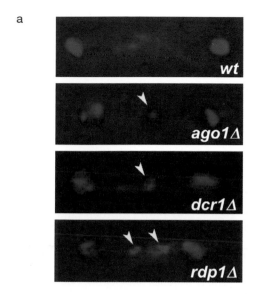

b

Wildtype

Bi-oriented sister centromeres

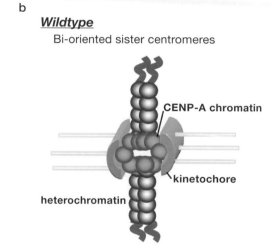

Defective heterochromatin

Merotelically oriented single centromere

Figure 7. Loss of Heterochromatin Results in Defective Chromosome Segregation

(*a*) Cells lacking RNAi or heterochromatin components display elevated rates of chromosome loss and lagging chromosomes on late anaphase spindles. (*b*) Lagging chromosomes in cells with defective heterochromatin may result from disorganized kinetochores so that one centromere can attach to microtubules from opposite poles. Such merotelic orientation could persist into anaphase; breakage of attachment with one pole or other would lead to random segregation and result in chromosome loss/gain events.

Clarke 1997). One interpretation is that the outer repeats provide an environment that is favorable for kinetochore assembly (Pidoux and Allshire 2005), but once assembled, CENP-A^{Cnp1} chromatin, and thus the kinetochore, is propagated at this position by a templating mechanism that may be coupled to replication (Takahashi et al. 2005). It is possible that heterochromatin somehow induces or aids the deposition of CENP-A^{Cnp1} in the central domain (Fig. 8) and that only one block of heterochromatin does not permit efficient kinetochore assembly. An alternative explanation is that one outer repeat is insufficient to recruit enough cohesin, and this leads to defective centromeric cohesion and elevated rates of chromosome loss. Such centromere constructs may stochastically accumulate sufficient cohesin after several cell divisions, resulting in increased mitotic stability. Once attained, this stabilized state must be somehow duplicated on daughter molecules to allow its propagation through subsequent divisions (Fig. 8).

As mentioned above, TSA can also set up a centromere-defective state caused by loss of silencing (Ekwall et al. 1997). Given the connection between silent chromatin formation and cohesion at centromeres, it seems likely that TSA-induced hyperacetylation blocks the efficient reestablishment of silent chromatin via RNAi and that defective centromere function is propagated due to the loss of silencing and thus sister-centromere cohesion (for more detail, see Chapter 14).

1.9 Diverse Silencing Mechanisms in Fungi

In the first half of this chapter, we described how in a single, relatively simple eukaryote, the fission yeast *S. pombe*, reporter genes are silenced by two distinct types of chromatin at centromeres. CENP-A chromatin resides in the middle of the centromere and is flanked on both sides by blocks of heterochromatin. CENP-A chromatin marks the region over which the kinetochore forms and presumably attaches to microtubules. RNAi is utilized in the formation of flanking heterochromatin to target noncoding transcripts emanating from the outer repeats and deliver chromatin-modifying enzymes such as Clr4 histone H3K9 methyltransferase, which creates binding sites for chromodomain proteins and, thus, robust silencing. This heterochromatin contributes to centromere function by providing tight physical cohesion and perhaps aiding kinetochore organization. The use of RNAi in forming silent chromatin at centromeres may be derived from its role in genome defense against RNA viruses and transposable elements, as has been described in plants. Indeed,

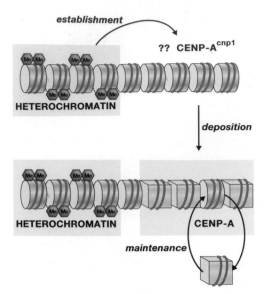

Figure 8. Establishment and Maintenance of CENP-A Chromatin

Central domain DNA alone is unable to establish a functional centromere; outer repeats are required. Loss of heterochromatin from established centromeres does not affect CENP-A^{cnp1} or kinetochore assembly on the central domain. This suggests that heterochromatin may in some way direct the site of CENP-A^{cnp1} chromatin and thus kinetochore assembly. It is not known how CENP-A^{cnp1} is deposited in nucleosomes or how this specialized chromatin is maintained in the central domain.

it is possible that centromere repeats themselves are the remnants of ancient mobile elements.

This possibility is strengthened by early studies of centromeres of the filamentous fungus, *N. crassa*, the subject of the second half of this chapter. Although there are clearly common epigenetic mechanisms at work in the diverse pair of fungi presented in this chapter, it is clear that these fungi show dramatic differences, as described below. For example, unlike the well-studied budding and fission yeasts, *Neurospora* uses DNA methylation, a classic epigenetic process common in higher eukaryotes such as mammals and flowering plants. In addition, studies in *Neurospora* have revealed several independent silencing systems operating at distinct stages of its life cycle. The first such mechanism, named repeat-induced point mutation (RIP), has both epigenetic and genetic aspects and clearly serves as a genome defense system. The second, named quelling, is an RNAi-based mechanism that results in silencing of transgenes and their native homologs. The third, named meiotic silencing by unpaired DNA (MSUD), is also RNAi-based but is distinct from quelling in its time of action, targets, and apparent purpose. Although we are still in the early days of epigenetic studies in all organisms, including the model fungi presented in this chapter, it is already clear that *S. pombe* and *N. crassa* will continue to serve as rich sources of information on epigenetic mechanisms operative in eukaryotes.

2 *Neurospora crassa:* History and Features of the Organism

The filamentous fungus *Neurospora crassa* (see Figs. 9 and 10) was first developed into an experimental organism by Dodge in the late 1920s and, about 10 years later, was adopted by Beadle and Tatum for their famous "one gene–one protein" studies linking biochemistry and genetics (Davis 2000). Beadle and Tatum selected *Neurospora* in part because this organism grows fast and is easy to propagate on defined growth media and because genetic manipulations, such as mutagenesis, complementation tests, and mapping, are simple with *Neurospora*. Although not as widely studied as some model eukaryotes, *Neurospora* continues to attract researchers because of its moderate complexity and because it is well suited for a variety of genetic, developmental, and subcellular studies (Borkovich et al. 2004). *Neurospora* has been especially useful for studies of photobiology, circadian rhythm, population biology, morphogenesis, mitochondrial import, DNA repair and recombination, DNA methylation, and other epigenetic processes.

The *N. crassa* life cycle is illustrated in Figure 9. The vegetative phase is initiated when either a sexual spore (ascospore) or an asexual spore (conidium) germinates, giving rise to multinucleate cells that form branched filaments (hyphae; Fig. 10C). The two mating types (*A* and *a*) are morphologically indistinguishable (Fig. 10B). Conveniently, ascospores require a heat shock to germinate, whereas conidia germinate spontaneously. The hyphal system spreads out rapidly (linear growth >5 mm per hour at 37°C) to form a "mycelium." After the mycelium is well established, aerial hyphae develop, leading to the production of the abundant orange conidia that are characteristic of the organism (Fig. 10A, B). The conidia, which contain one to several nuclei each, can either establish new vegetative cultures or fertilize crosses. If nutrients are limiting, *Neurospora* prepares to enter its sexual phase by producing nascent fruiting bodies ("protoperithecia"). When a specialized hypha ("trichogyne") projecting from the protoperithecium contacts tissue of the opposite mating type, a nucleus is picked up and transported back to the protoperithecium. The sexual phase of *Neurospora* and other filamentous ascomycetes differs from that of yeasts in having a prolonged heterokaryotic phase between fertilization

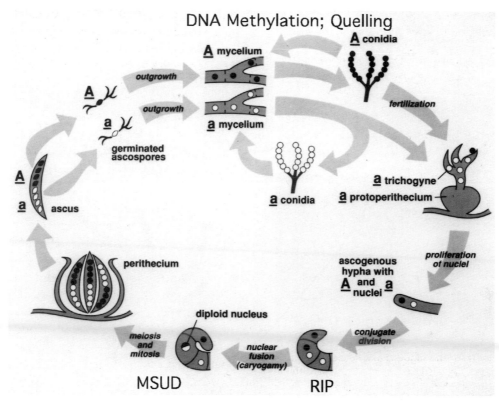

DNA Methylation; Quelling

Figure 9. Life Cycle of *N. crassa*

The stages in which epigenetic processes that are described in the text occur are indicated. (Adapted, with permission, from Shiu et al. 2001.)

and karyogamy (nuclear fusion). The heterokaryotic cells resulting from fertilization proliferate in the developing perithecium. In the final divisions, the cells are binucleate, containing one nucleus of each mating type. These cells bend to form hook-shaped cells ("croziers"), a final mitosis occurs to produce four nuclei, and septa are laid down to produce one binucleate cell at the crook of the crozier. This cell gives rise to an ascus. Genetic analyses have indicated that, in general, the ~100 or more asci of a perithecium are derived from a single maternal nucleus and a single paternal nucleus. When karyogamy occurs, the resulting diploid nucleus immediately enters into meiosis. Thus, the diploid phase of the life cycle is brief and limited to a single cell. The meiotic products undergo one mitotic division before being packaged as ascospores and undergo additional mitoses in the developing ascospores (see Fig. 9 and Davis 2000).

The ~40-megabase *N. crassa* genome consists of seven chromosomes with approximately 10,000 predicted protein-coding genes (Galagan et al. 2003) and a total genetic map length of roughly 1000 map units (Perkins et al. 2001). Only about 10% of the genome consists of repeti-

tive DNA and, aside from a tandem array of ~70 copies of the ~9-kb rDNA unit encoding the three large rRNAs, most of the repetitive DNA consists of inactivated transposable elements. That most strains of *Neurospora* lack active transposons and have very few close paralogs almost certainly reflects the operation of RIP, the first homology-dependent genome defense system discovered in eukaryotes (Selker 1990). We know that *Neurospora* has at least three gene-silencing processes which should serve to conserve the structure of the genome: RIP, quelling, and MSUD (Borkovich et al. 2004). All of these processes have epigenetic aspects and have direct or indirect connections with DNA methylation, a basic epigenetic mechanism found in *Neurospora* and many other eukaryotes. We discuss DNA methylation and then RIP, quelling, and MSUD.

2.1 DNA Methylation in Neurospora

Since its discovery decades ago, DNA methylation in eukaryotes has remained remarkably enigmatic. Basic questions are still debated, such as, What determines which chromosomal regions are methylated? and What is

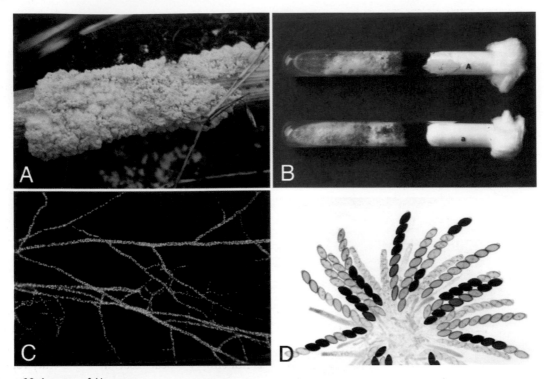

Figure 10. Images of *N. crassa*

(*A*) Vegetative growth in the wild on sugarcane (photo by D. Jacobson; Stanford University). (*B*) Slants of vegetative cultures of *N. crassa* in the laboratory (photo by N. B. Raju, Stanford University). (*C*) Hyphae of *N. crassa* stained with DAPI to show abundant nuclei (photo by M. Springer, Stanford University). (*D*) Rosette of maturing asci showing ascospore patterns. (*D*, Reprinted, with permission, from Raju 1980 [©Elsevier].)

the function of DNA methylation? *Neurospora* revealed itself to be an excellent system to study the control and function of DNA methylation. Some model eukaryotes, including the nematode *C. elegans* and the yeasts *S. cerevisiae* and *S. pombe*, lack detectable DNA methylation, and reports of DNA methylation in another model organism, *D. melanogaster*, remain controversial. DNA methylation is essential for viability in some organisms, such as mammals, complicating certain analyses. In *Neurospora* DNA, about 1.5% of the cytosines are methylated, but this methylation is dispensable, facilitating genetic studies. Although one must be cautious when extrapolating from one system to another, at least some aspects of DNA methylation appear conserved. For example, all known DNA methyltransferases (DMTs), including those from both prokaryotes and eukaryotes, show striking homology in their catalytic domains (Grace Goll and Bestor 2005). Findings from *Neurospora*, *Arabidopsis*, mice, and other systems in the last decade have revealed both important similarities and interesting differences in the control and function of DNA methylation, demonstrating the value of carrying out investigations in multiple model systems.

Discovery of DNA methylation in *Neurospora* initially attracted interest because it was not limited to symmetrical sites, such as CpG dinucleotides or CpNpG trinucleotides. Riggs, and Holliday and Pugh, had proposed an attractive model for the "inheritance" or "maintenance" of methylation patterns that relied on the symmetrical nature of methylated sites observed in animals. Although results of a variety of in vitro and in vivo studies have supported the "maintenance methylase" model (see Chapter 18), mechanisms for maintenance methylation that do not rely on faithful copying at symmetrical sites can be imagined and may be operative in a variety of organisms (see, e.g., Selker et al. 2002). The possibility that the observed methylation at asymmetric sites represented "de novo methylation" was exciting because mechanisms that blindly propagate methylation patterns can complicate determination of which sequences are methylated in the first place. Indeed, results of DNA-mediated transformation and methylation inhibitor studies with *Neurospora* demonstrated reproducible de novo methylation (see, e.g., Singer et al. 1995). Additional studies defined, in part, the underlying signals for de novo methylation (see, e.g., Tamaru and Selker 2003).

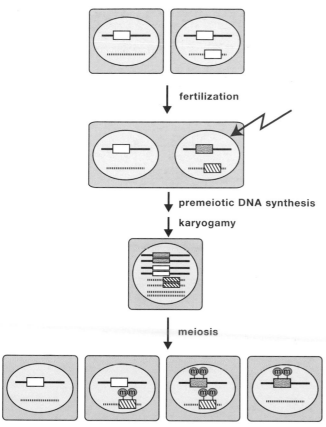

Figure 11. Repeat-induced Point Mutation (RIP)

For clarity, only two chromosomes are illustrated. The open box represents a gene, or any chromosomal segment, which when duplicated (e.g., in the strain indicated on the *top right*) is subject to RIP (symbolized by *lightning bolt*) between fertilization and karyogamy. Results of genetic experiments reveal that duplications can be repeatedly subjected to volleys of C-to-T transitions (symbolized by *filled boxes*) during this period of ~10 mitoses, right up to the final premeiotic DNA synthesis (Selker et al. 1987; Watters et al. 1999). The four possible combinations of chromosomes in progeny are indicated, and the red "m" represents DNA methylation, which is frequently (although not always) associated with products of RIP.

The first methylated patch characterized in detail was the 1.6-kb ζ-η region, which consists of a diverged tandem duplication of a 0.8-kb segment of DNA, including a 5S rRNA gene. Comparison of this region with the corresponding chromosomal region of strains lacking the duplication initially led to the idea that repeated sequences can somehow induce DNA methylation and ultimately led to the discovery of the genome defense system named RIP. Elucidation of RIP revealed that repeated sequences do not directly trigger DNA methylation, at least in *Neurospora*; instead, repeats trigger RIP, which is closely tied to DNA methylation, as described below. Both the ζ-η region and the ψ63 region, the second methylated region discovered in *Neurospora*, are products of RIP. Moreover, subsequent

genome-wide analyses of DNA methylation revealed that the majority of methylated regions in *Neurospora* are relics of transposons inactivated by RIP (Galagan et al. 2003; Selker et al. 2003). Indeed, the only DNA methylation in *Neurospora* that may not have resulted from RIP is that in the tandemly arranged rDNA (Perkins et al. 1986).

2.2 RIP, a Genome Defense System with Both Genetic and Epigenetic Aspects

RIP was discovered as a result of a detailed analysis of progeny from crosses of *Neurospora* transformants (Selker 1990). It was noticed that duplicated sequences, whether native or foreign, and whether linked or unlinked, were subjected to numerous polarized transition mutations (G:C to A:T) in the haploid genomes of the special binucleate cells resulting from fertilization (Fig. 12). Experiments in which the stability of a gene was tested when it was unique in the genome or else combined with an unlinked homolog demonstrated that RIP is not simply repeat-*associated*; it is truly repeat-*induced*. In a single passage through the sexual cycle, up to about 30% of the G:C pairs in duplicated sequences can be mutated. Frequently (but not invariably), sequences altered by RIP become methylated de novo. It is likely that the mutations from RIP occur by enzymatic deamination of 5-meCs or by deamination of Cs followed by DNA replication (Selker 1990). Cytosine methylation involves a reaction intermediate that is prone to spontaneous deamination, suggesting that the putative deamination step of RIP might be catalyzed by a DMT or DMT-like enzyme. Consistent with this possibility, one of the two DMT homologs predicted from the *Neurospora* genome sequence is involved in RIP (Freitag et al. 2002). Progeny from homozygous crosses of mutants with defects in this gene, *rid* (*RIP* defective), do not show new instances of RIP. *Rid* mutants do not display any noticeable defects in DNA methylation, fertility, growth, or development. In contrast, the second *Neurospora* DMT homolog (DIM-2), which was identified genetically and shown to be necessary for all known DNA methylation, is not required for RIP (Kouzminova and Selker 2001).

All indications are that every sizable duplication (greater than ~400 bp for tandem duplication or ~1000 bp for unlinked duplication) is subject to RIP in some fraction of the special binucleate cells. Nevertheless, duplications escape RIP at some frequency (typically less than 1% for a tandem duplication or ~50% for an unlinked duplication). Even duplications of chromosomal segments containing numerous genes are sensitive to RIP (Perkins et al.

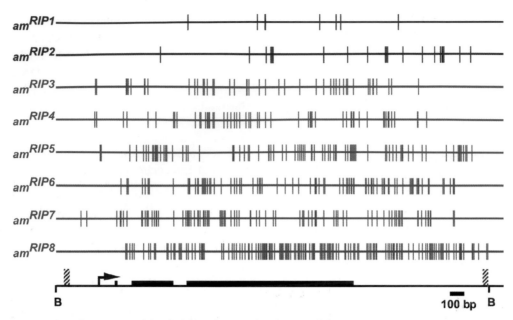

Figure 12. Mutations from RIP and Methylation Status of Eight *am* Alleles

Vertical bars indicate mutations. Alleles shown in black were not methylated, alleles in blue were initially methylated but, after loss of methylation was induced with 5-azacytidine, or by cloning and gene replacement, did not become remethylated. Alleles shown in red were not only initially methylated, but also triggered methylation de novo. (Adapted from Singer et al. 1995.)

1997; Bhat and Kasbekar 2001). Although RIP is limited to the sexual phase of the life cycle, the existence of this process raised the question of whether *Neurospora* can utilize gene duplications to evolve. The genome sequence revealed gene families, but tellingly, virtually all paralogs were found to be sufficiently divergent that they should not trigger RIP (Galagan and Selker 2004). We conclude that RIP may indeed limit evolution in *Neurospora*. Interestingly, some fungi show what appear to be milder genome defense systems that are similar to RIP. The most notable example is methylation induced premeiotically (MIP), a process that detects linked and unlinked sequence duplications during the period between fertilization and karyogamy, like RIP, but which relies exclusively on DNA methylation for inactivation; no evidence of mutations has been found in sequences inactivated by MIP (Rossignol and Faugeron 1994).

2.3 Studies of Relics of RIP Provided Insights into the Control of DNA Methylation

NONCANONICAL MAINTENANCE METHYLATION

The finding that a single DMT, DIM-2, is responsible for all detected DNA methylation was as surprising as the initial finding of rampant methylation at nonsymmetrical sites in *Neurospora* because no previously identified DMT was

known to methylate cytosines in a variety of sequence contexts. An obvious but important question was, Does methylation at nonsymmetrical sites necessarily reflect the potential of the corresponding sequences to induce methylation de novo? Early transformation experiments were consistent with this possibility; methylated sequences that were stripped of their methylation (e.g., by cloning) regained their normal methylation when reintroduced into vegetative cells. A surprise came, however, when eight alleles of the *am* gene that were generated by RIP were tested for their capacity to induce methylation de novo (Singer et al. 1995). Some products of RIP with relatively few mutations (Fig. 12, *am*^{RIP3} and *am*^{RIP4}) did not become re-methylated, even at their normal locus, suggesting that the observed methylation represented propagation of methylation established earlier. Importantly, their methylation, like other observed methylation in *Neurospora*, was not limited to symmetrical sites, did not significantly spread with time, and was "heterogeneous" in the sense that the pattern of methylated residues was not invariant within a clonal population of cells. Thus, although dependent on preexisting methylation established in the sexual phase (perhaps by RIP), this methylation could not reflect the action of a "maintenance methylase" of the type envisioned in the original model for inheritance of methylation patterns. It is noteworthy that MIP in *Ascobolus*, which also results in heterogeneous methylation, provided the first evidence for propagation of DNA methylation in fungi (Rossignol and Faugeron 1994). The capacity of *Neurospora* to perform maintenance methylation was confirmed experimentally (Selker et al. 2002). Interestingly, propagation of methylation was found to be sequence-specific (i.e., it did not work on all sequences), adding a new dimension to the maintenance methylation concept. Although a number of potential schemes that would result in propagation of DNA methylation can be imagined, the actual mechanism operative in *Neurospora* remains unknown. In principle, maintenance of methylation at nonsymmetrical sites could depend on methylation of nearby symmetrical sites, but the observed heterogeneous methylation, including at CpG sites, renders this possibility unlikely. Feedback mechanisms involving proteins associated with the methylated DNA could result in methylation that depends on preexisting methylation, i.e., maintenance methylation. As discussed below, findings from *Neurospora* (and other organisms) implicate histone modifications in the control of DNA methylation, raising the possibility that histones play a role in maintenance methylation.

INVOLVEMENT OF HISTONES IN DNA METHYLATION

The first indication of a role of histones in DNA methylation came from the observation that treatment of *Neurospora* with the histone deacetylase inhibitor trichostatin A (TSA) reduced methylation in some chromosomal regions (Selker 1998). The selectivity of demethylation by TSA could reflect differential access to histone acetyltransferases, but this has not been thoroughly investigated (Selker et al. 2002). Chromatin was unambiguously tied to the control of DNA methylation through investigations with the *Neurospora* mutant *dim-5*, which, like *dim-2* strains, shows a complete loss of DNA methylation. The SET domain protein DIM-5 was found to be a histone H3 methyltransferase that specifically trimethylates lysine 9 (Tamaru and Selker 2001; Tamaru et al. 2003). Confirmation that histone H3 is the physiologically relevant substrate of DIM-5 came from two demonstrations: (1) Replacement of lysine 9 in H3 with other amino acids caused loss of DNA methylation and (2) trimethyl-lysine 9 was found specifically at methylated chromosomal regions.

The discovery that histone methylation controls DNA methylation, at least in *Neurospora*, led to two important questions: (1) What tells DIM-5 which nucleosomes to methylate? (2) What reads the trimethyl mark and transmits this information to the DMT, DIM-2? The

factors that control DIM-5 have not been definitively identified, but there are strong suggestions that it is controlled by one or more proteins that recognize products of RIP and by the modification state of amino acid residues in the neighborhood of its action (Selker et al. 2002). The latter should allow DIM-5 to integrate information relevant to whether DNA in a particular region should be methylated and provides a possible explanation for the observation that TSA can inhibit DNA methylation in certain regions.

Findings in other systems led to discovery of what reads the trimethyl mark on lysine 9 of histone H3 in *Neurospora*. Knowledge that HP1, a protein first identified in *Drosophila* (see Chapter 5), binds methylated lysine 9 of histone H3 in vitro motivated a search for an HP1 homolog in *Neurospora*. A likely homolog was found, and its involvement in DNA methylation was tested by gene disruption (Freitag et al. 2004a). The gene, named *hpo* (HPone) was indeed found to be essential for DNA methylation. As another test of whether *Neurospora* HP1 reads the mark generated by DIM-5, its subcellular localization was examined in wild type and *dim-5* strains. In wild type, HP1-GFP localized to heterochromatic foci, but this localization was lost in *dim-5*, confirming that *Neurospora* HP1 is recruited by the trimethyl-lysine 9 mark generated by DIM-5.

Evidence that RNA interference (RNAi) is important for heterochromatin formation and maintenance in *S. pombe* raised the question of whether the RNAi machinery of *Neurospora* is involved in HP1 localization and/or DNA methylation. *Neurospora* has homologs of a variety of genes implicated in RNAi (Borkovich et al. 2004). Studies of mutants with null mutations in all three RNA-dependent RNA polymerase (RdRP) genes, in both Dicer genes, or in other presumptive RNAi genes revealed no evidence that RNAi is involved in methylation of H3K9, heterochromatin formation, or DNA methylation in *Neurospora* (Chicas et al. 2004; Freitag et al. 2004c). However, as discussed below, the *Neurospora* RNAi genes are involved in at least two other silencing mechanisms with epigenetic aspects, quelling and meiotic silencing.

2.4 Quelling

Soon after transformation techniques were established for *Neurospora*, researchers in several laboratories noticed that a sizable fraction (e.g., ~30%) of *Neurospora* transformants showed silencing of transforming DNA and, more surprisingly, silencing of native sequences homologous to those of the transforming DNA. The latter form of vegetative phase silencing was named "quelling" by the Macino laboratory, which has carried out most of the research on this phenomenon to date (Pickford et al. 2002). Quelling is most apparent with visible markers such as the *albino* genes, which encode enzymes required for carotenoid biosynthesis (Fig. 13), and is thought to be comparable to "cosuppression" or posttranscriptional gene silencing (PTGS) in plants. Interestingly, genes seem to vary in their sensitivity to quelling. For genes that are sensitive, quelling seems most common in transformants bearing multiple copies of transforming DNA in a tight array. Nuclei flow freely in hyphae of *Neurospora*, allowing "heterokaryosis" in which genetically distinct nuclei share a common cytoplasm. Thus, it was easy to demonstrate that quelling is "dominant"; i.e., a transformed nucleus can silence homologous sequences in nearby nuclei (Cogoni et al. 1996). This implicated a cytoplasmic silencing factor, perhaps an RNA species. Consistent with this possibility, identification of genes involved in quelling revealed that quelling is closely related to RNAi in other systems (Pickford et al. 2002). Specifically, *qde-1*, *qde-2*, *qde-3*, *dcl-1*, and *dcl-2* encode, respectively,

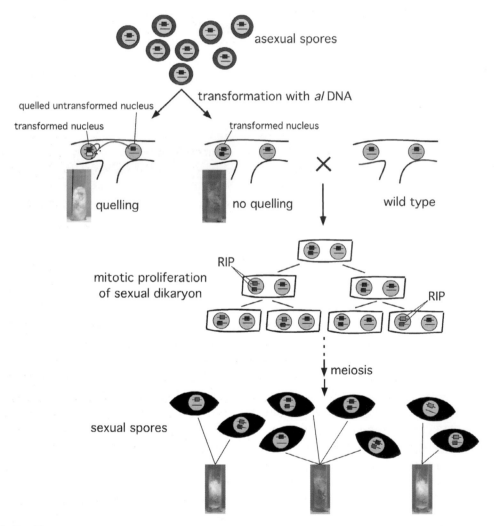

Figure 13. Quelling

For simplicity, only two of the seven chromosomes are diagrammed (*straight line segments in gray circles* representing nuclei). The native *albino* gene (*al*) is indicated by the dark orange rectangle on the top chromosome; the other (*dark orange* or *yellow*) rectangle represents ectopic *al* sequences introduced by transformation. Since transformed cells are often multinucleate, transformants are often heterokaryotic, as illustrated. Whether or not the transforming DNA includes the entire coding region, in some transformants it silences ("quells") the native *al*⁺ gene in both transformed and non-transformed nuclei through an undefined *trans*-acting molecule (*red lines* emanating from the transforming DNA indicated by the *yellow rectangle*). This results in poorly pigmented or albino (Al⁻) tissue in some transformants, as shown.

an RdRP, an "Argonaute"-like protein thought to be involved in small interfering RNA (siRNA)-guided mRNA cleavage (Baumberger and Baulcombe 2005), a RecQ-like presumptive DNA helicase (*qde-3*) and two "Dicers," which are presumably involved in generating siRNAs (Catalanotto et al. 2004). Although critical details are lacking, the developing model is that, in some cases, transforming DNA generates an "aberrant" transcript that somehow triggers the RNAi machinery, leading to degradation of homologous mRNAs. Although DNA methylation is frequently associated with transforming DNA, neither the DNA methyltransferase, DIM-2, nor the histone H3 lysine 9 methyltransferase, DIM-5, is required for quelling (Cogoni et al. 1996; Chicas et al. 2005).

2.5 Meiotic Silencing by Unpaired DNA (MSUD)

MSUD, the most recent addition to the list of silencing mechanisms operative in *Neurospora*, was discovered by Metzenberg and colleagues (Aramayo and Metzenberg 1996; Shiu et al. 2001; Shiu and Metzenberg 2002) while they were investigating the curious observation that a deletion mutation in the *Asm-1* (ascospore maturation) gene is functionally dominant. An elegant series of experiments, cartooned in Figure 14, led to the conclusion that sequences that lack a pairing partner in meiotic prophase cause meiotic silencing of identical, or nearly identical, sequences. A possible explanation for the observation that a deletion of *Asm-1* is dominant (Fig. 14b) was that the remaining single dose of the gene produced inadequate gene product, but this possibility was rendered unlikely by the observation that a functional copy of the gene at an ectopic location failed to complement the defect (Fig. 14c). Conversely, only one functional copy of the gene was required in meiosis if the functional allele had a pairing partner; i.e., its partner could harbor mutations rendering it unable to produce a functional product (Fig. 14d). Normal meiotic expression of the gene was observed in a strain having paired ectopic alleles and deletions of the native gene on both homologs, showing that the deletion was not itself "toxic" and that the ectopic copies were indeed functional (Fig. 14e). Interestingly, some alleles generated by RIP elicit MSUD if the strains are proficient for DNA methylation but fail to elicit MSUD in *dim-2* strains, suggesting that the DNA methylation frequently associated with such alleles can inhibit pairing (compare d and f in Fig. 14) (Pratt et al. 2004). (Incidentally, this observation also provided the first evidence that DIM-2 is functional in the sexual phase of *Neurospora*.) Alleles with

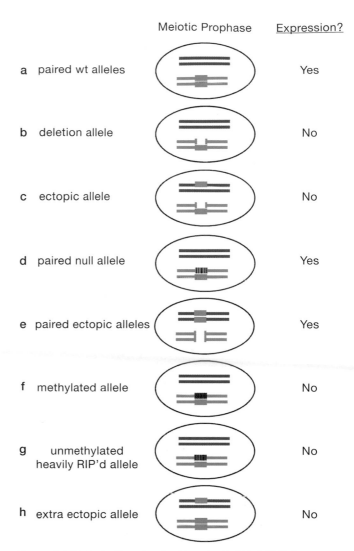

Meiotic Prophase | Expression?

a paired wt alleles — Yes

b deletion allele — No

c ectopic allele — No

d paired null allele — Yes

e paired ectopic alleles — Yes

f methylated allele — No

g unmethylated heavily RIP'd allele — No

h extra ectopic allele — No

Figure 14. Meiotic Silencing by Unpaired DNA (MSUD)

Cartoon of tests conducted by the Metzenberg and Aramayo laboratories (Aramayo and Metzenberg 1996; Shiu et al. 2001; Pratt et al. 2004) that defined the phenomenon. Only two of the seven *Neurospora* chromosomes are indicated, with their sequences illustrated in green and blue, respectively. The rectangular box signifies a gene normally functional during meiosis. Vertical lines indicate mutations, and DNA methylation is indicated by red coloring. Deletions are diagramed as gaps. See text for interpretations.

high densities of mutations by RIP were found to be dominant, like deletions (Fig. 14g) (Shiu et al. 2001; Shiu and Metzenberg 2002; Pratt et al. 2004), consistent with the hypothesis that they are unable to pair with the wild-type allele. To distinguish between the possibility that MSUD is due to absence of pairing and the possibility that it is due to presence of an unpaired allele, the researchers analyzed a cross in which the meiotic nucleus would have three copies of a gene: two wild-type alleles (which should pair) and an ectopic copy (which should be unpaired). Silencing was observed, implying that MSUD results from the presence of unpaired alleles rather than from the absence of paired alleles (Fig. 14h). MSUD can be observed cytologically in crosses heterozygous for a gene encoding a GFP-tagged protein, as illustrated in Figure 15.

A hunt for mutants defective in MSUD resulted in the identification of a telling member of the MSUD machinery. The *Sad-1* (Suppressor of ascus dominance) gene, identified by selection for mutants that were able to pass through a cross in which *asm-1* is not paired, encodes an RdRP (Shiu et al. 2001). This suggested that MSUD is related to quelling in *Neurospora*, and to RNAi generally. Interestingly, the *Neurospora* genome contains genes predicted to encode three putative RdRPs (one required for quelling, one required for MSUD, and one of unknown function), two Argonaute-like proteins (one required for quelling and one required for MSUD), and two Dicer-like proteins. It will be fascinating to learn the detailed mechanism of MSUD.

2.6 Probable Functions and Practical Uses of RIP, Quelling, and MSUD

RIP seems custom-made to limit the expression of "selfish DNA" such as transposons that direct the production of copies of themselves in a genome. Consistent with this possibility, the vast majority of relics of RIP are recognizably similar to transposons known from other organisms,

and most strains of *Neurospora* lack active transposons (Galagan et al. 2003; Selker et al. 2003). Nevertheless, because RIP is limited to the premeiotic dikaryotic cells, this process should neither prevent the spread of a new (e.g., horizontally acquired) transposon in vegetative cells nor prevent the duplication of a single-copy transposon in meiotic cells. Quelling and MSUD should deal with such eventualities, however. Although quelling does not completely suppress the spread of transposons in vegetative cells, as evidenced by the proliferation of an introduced copy of the LINE-like transposon, Tad, it does appear to partially silence such transposons (Nolan et al. 2005). Information about the action of MSUD suggests that this process will silence any transposed sequence in meiotic cells, even if it is only present as a single copy in the genome (Shiu et al. 2001). In addition to dealing with errant transposons in meiosis, MSUD also appears to play an important role in the process of speciation, as shown by the observation that mutants defective in MSUD relieve the sterility of strains bearing large duplications of chromosome segments and allow closely related species to mate with *N. crassa* (Shiu et al. 2001).

Although RIP, quelling, and MSUD can all be a nuisance for some genetic experiments, all have been exploited for research purposes. RIP provided the first simple method to knock out genes in *Neurospora* and is still the preferred method for generating partial-function mutants. Quelling has also been used to reduce, if not eliminate, gene function, much as RNAi is exploited in a

Figure 15. *Neurospora crassa*

Fluorescent image of a rosette of maturing asci from a heterozygous cross of a transformant engineered to express GFP-tagged histone H1. Four ascospores of each ascus show glowing nuclei (Freitag et al. 2004b). (Photo courtesy of N. B. Raju, Stanford University.)

variety of organisms. And MSUD provides a simple assay to test whether particular genes are required to function in (or immediately after) meiosis; if a gene is found to cause sterility when duplicated, or when at an ectopic location, and the sterility is rescued by a mutation blocking MSUD, it is safe to assume that it plays an important function in meiosis.

In addition to the postulated evolutionary roles of RIP, quelling, and MSUD, and to their utility in the laboratory, it is worth considering the possibility that these processes serve in other ways. For example, the fact that *Sad-1* function is required for full fertility suggests that MSUD is directly or indirectly required for meiosis (Shiu and Metzenberg 2002). In the case of RIP, although this process is nonessential, the distribution of products of RIP in the *Neurospora* genome suggests that junked transposons can serve the organism as substrates for kinetochore formation, much as repeated sequences do in *S. pombe* and other organisms. Analyses of DNA sequences from chromosome 7 illustrate that sequences around the genetically mapped centromeres of *Neurospora* consist primarily of relics of transposons heavily mutated by RIP

(Fig. 16). Relics of RIP are also found adjacent to telomere sequences of *Neurospora*. Interestingly, transposons and relics of transposons are also commonly found in heterochromatic sequences of other organisms, such as *Drosophila* (Chapter 5), mammals (Chapter 17), plants (Chapter 9), and other fungi.

3 Concluding Remarks

The fungi *S. pombe* and *N. crassa* have emerged as powerful systems to discover and elucidate epigenetic phenomena. The field of epigenetics is still in its infancy, with epigenetic mechanisms continuing to come to light. Therefore, it is not surprising that the depth and breadth of our current understanding of epigenetic processes, such as those described in this chapter, vary between organisms. It is too early to know how general the various epigenetic mechanisms described are. For example, it is possible that some organisms, like *S. pombe*, rely primarily on RNAi for silencing and heterochromatin formation, whereas others, such as *N. crassa*, rely more heavily on DNA methylation. It is already clear that even these two model eukaryotes have

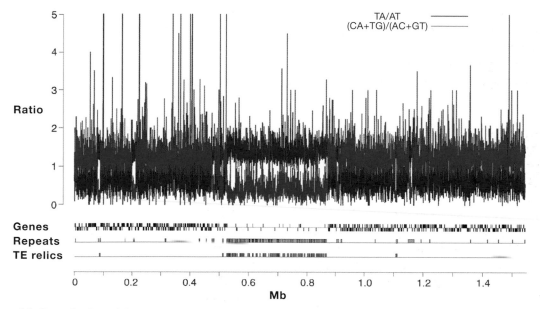

Figure 16. Organization of Centromere 7 Region of *N. crassa*

Contigs 249, 255, and 21 of genomic sequence release 7.0 (http://www.broad.mit.edu/annotation/fungi/neurospora_crassa_7/index.html), and all except the first 400 kb of sequence contig 10, were assembled, and the combined sequence file was analyzed in 200-bp increments for the "RIP indices" (TpA/ApT [*blue*] and CpA+TpG/ApC+GpT [*red*]) (Galagan et al. 2003; Selker et al. 2003; Galagan and Selker 2004). An ~360-kb region with a high density of transposable elements (TE) inactivated by RIP (retrotransposon relics in *blue*; DNA transposon relic in *violet*) was found between markers flanking the centromere, which was mapped genetically. The ~1.5-Mb segment shown includes 383 annotated genes (above and below line to indicate genes in opposite orientations), of which only 20 short predicted genes are within the predicted centromere region. The sizes of sequence gaps between the contigs (positions 0.5466, 0.6956, and 0.9058 Mb in the figure) are unknown.

both important differences and striking similarities. Both organisms utilize histone H3K9 methylation and RNAi, neither of which is found in the budding yeast, *S. cerevisiae*. Of these three fungi, however, only *Neurospora* sports DNA methylation. It is also noteworthy that a given process may be functionally rather different in two organisms. For instance, in *Neurospora*, RNAi components have been implicated in quelling and meiotic silencing, but not in heterochromatin formation, whereas in fission yeast, RNAi components contribute to heterochromatin formation, but other roles are not established. Finally, it is worth noting that even shared features, such as heterochromatin associated with centromeres of both fission yeast and *Neurospora*, may have important differences. An important goal for the future is to discover the extent to which information gleaned from one organism is applicable to others. Further exploration of epigenetic processes in various model organisms, including *S. pombe* and *N. crassa*, will provide this information. We anticipate that the richly diverse fungi will continue to serve as useful systems for epigenetic research for many years.

Acknowledgments

R.C.A. thanks Alison L. Pidoux and Sharon A. White for Figure 4 and Figure 3, respectively; and Alison L. Pidoux and Elizabeth H. Bayne for comments on the manuscript. Research in the Allshire lab is supported by the Wellcome Trust, MRC, and the EC FP6 *Epigenome* Network of Excellence. R.C.A. is a Wellcome Trust Principal Research Fellow. E.U.S. thanks N.B. Raju for help assembling figures of *Neurospora*, and Robert L. Metzenberg for comments on the manuscript. Research in the Selker laboratory is supported by U.S. Public Health Service grant GM35690 from the National Institutes of Health and by National Science Foundation grant MCB0131383.

References

Allshire R.C., Javerzat J.P., Redhead N.J., and Cranston G. 1994. Position effect variegation at fission yeast centromeres. *Cell* **76:** 157–169.

Allshire R.C., Nimmo E.R., Ekwall K., Javerzat J.P., and Cranston G. 1995. Mutations derepressing silent centromeric domains in fission yeast disrupt chromosome segregation. *Genes Dev.* **9:** 218–233.

Aramayo R. and Metzenberg R.L. 1996. Meiotic transvection in fungi. *Cell* **86:** 103–113.

Ayoub N., Goldshmidt I., Lyakhovetsky R., and Cohen A. 2000. A fission yeast repression element cooperates with centromere-like sequences and defines a mat silent domain boundary. *Genetics* **156:** 983–994.

Bailis J.M., Bernard P., Antonelli R., Allshire R.C., and Forsburg S.L. 2003. Hsk1-Dfp1 is required for heterochromatin-mediated cohesion at centromeres. *Nat. Cell Biol.* **5:** 1111–1116.

Bannister A.J., Zegerman P., Partridge J.F., Miska E.A., Thomas J.O., Allshire R.C., and Kouzarides T. 2001. Selective recognition of methylated lysine 9 on histone H3 by the HP1 chromo domain. *Nature* **410:** 120–124.

Baum M., Ngan V.K., and Clarke L. 1994. The centromeric K-type repeat and the central core are together sufficient to establish a functional *Schizosaccharomyces pombe* centromere. *Mol. Biol. Cell* **5:** 747–761.

Baumberger N. and Baulcombe D.C. 2005. Arabidopsis ARGONAUTE1 is an RNA Slicer that selectively recruits microRNAs and short interfering RNAs. *Proc. Natl. Acad. Sci.* **102:** 11928–11933.

Bernard P., Maure J.F., Partridge J.F., Genier S., Javerzat J.P., and Allshire R.C. 2001. Requirement of heterochromatin for cohesion at centromeres. *Science* **294:** 2539–2542.

Bhat A. and Kasbekar D.P. 2001. Escape from repeat-induced point mutation of a gene-sized duplication in *Neurospora crassa* crosses that are heterozygous for a larger chromosome segment duplication. *Genetics* **157:** 1581–1590.

Borkovich K.A., Alex L.A., Yarden O., Freitag M., Turner G.E., Read N.D., Seiler S., Bell-Pedersen D., Paietta J., Plesofsky N., et al. 2004. Lessons from the genome sequence of *Neurospora crassa*: Tracing the path from genomic blueprint to multicellular organism. *Microbiol. Mol. Biol. Rev.* **68:** 1–108.

Cam H.P., Sugiyama T., Chen E.S., Chen X., FitzGerald P.C., and Grewal S.I. 2005. Comprehensive analysis of heterochromatin- and RNAi-mediated epigenetic control of the fission yeast genome. *Nat. Genet.* **37:** 809–819.

Catalanotto C., Pallotta M., ReFalo P., Sachs M.S., Vayssie L., Macino G., and Cogoni C. 2004. Redundancy of the two dicer genes in transgene-induced posttranscriptional gene silencing in *Neurospora crassa*. *Mol. Cell. Biol.* **24:** 2536–2545.

Chen E.S., Saitoh S., Yanagida M., and Takahashi K. 2003. A cell cycle-regulated GATA factor promotes centromeric localization of CENP-A in fission yeast. *Mol. Cell* **11:** 175–187.

Chicas A., Cogoni C., and Macino G. 2004. RNAi-dependent and RNAi-independent mechanisms contribute to the silencing of RIPed sequences in *Neurospora crassa*. *Nucleic Acids Res.* **32:** 4237–4243.

Chicas A., Forrest E.C., Sepich S., Cogoni C., and Macino G. 2005. Small interfering RNAs that trigger posttranscriptional gene silencing are not required for the histone H3 Lys9 methylation necessary for transgenic tandem repeat stabilization in *Neurospora crassa*. *Mol. Cell. Biol.* **25:** 3793–3801.

Clarke, L. and M.P. Baum. 1990. Functional analysis of a centromere from fission yeast: A role for centromere-specific repeated DNA sequences. *Mol. Cell. Biol.* **10:** 1863–1872.

Cleveland D.W., Mao Y., and Sullivan K.F. 2003. Centromeres and kinetochores: From epigenetics to mitotic checkpoint signaling. *Cell* **112:** 407–421.

Cogoni C., Irelan J.T., Schumacher M., Schmidhauser T.J., Selker E.U., and Macino G. 1996. Transgene silencing of the *al-1* gene in vegetative cells of *Neurospora* is mediated by a cytoplasmic effector and does not depend on DNA-DNA interactions or DNA methylation. *EMBO J.* **15:** 3153–3163.

Cowieson N.P., Partridge J.F., Allshire R.C., and McLaughlin P.J. 2000. Dimerisation of a chromo shadow domain and distinctions from the chromodomain as revealed by structural analysis. *Curr. Biol.* **10:** 517–525.

Davis R.H. 2000. Neurospora: *Contributions of a model organism.* Oxford University Press, New York.

Djupedal I., Portoso M., Spahr H., Bonilla C., Gustafsson C.M., Allshire R.C., and Ekwall K. 2005. RNA Pol II subunit Rpb7 promotes centromeric transcription and RNAi-directed chromatin silencing. *Genes Dev.* **19:** 2301–2306.

Egel R. 2004. *The molecular biology of* Schizosaccharomyces pombe: *Genetics, genomics and beyond.* Springer-Verlag, Heidelberg, Germany.

Ekwall K., Javerzat J.P., Lorentz A., Schmidt H., Cranston G., and Allshire R. 1995. The chromodomain protein Swi6: A key component at fission yeast centromeres. *Science* **269:** 1429–1431.

Ekwall K., Nimmo E.R., Javerzat J.P., Borgstrom B., Egel R., Cranston G., and Allshire R. 1996. Mutations in the fission yeast silencing factors clr4⁺ and rik1⁺ disrupt the localisation of the chromo domain protein Swi6p and impair centromere function. *J. Cell Sci.* **109:** 2637–2648.

Ekwall K., Olsson T., Turner B.M., Cranston G., and Allshire R.C. 1997. Transient inhibition of histone deacetylation alters the structural and functional imprint at fission yeast centromeres. *Cell* **91:** 1021–1032.

Freitag M., Hickey P.C., Khlafallah T.K., Read N.D., and Selker E.U. 2004a. HP1 is essential for DNA methylation in *Neurospora. Mol. Cell* **13:** 427–434.

Freitag M., Hickey P.C., Raju N.B., Selker E.U., and Read N.D. 2004b. GFP as a tool to analyze the organization, dynamics and function of nuclei and microtubules in *Neurospora crassa. Fungal Genet. Biol.* **41:** 897–910.

Freitag M., Lee D.W., Kothe G.O., Pratt R.J., Aramayo R., and Selker E.U. 2004c. DNA methylation is independent of RNA interference in *Neurospora. Science* **304:** 1939.

Freitag M., Williams R.L., Kothe G.O., and Selker E.U. 2002. A cytosine methyltransferase homologue is essential for repeat-induced point mutation in *Neurospora crassa. Proc. Natl. Acad. Sci.* **99:** 8802–8807.

Fukagawa T., Nogami M., Yoshikawa M., Ikeno M., Okazaki T., Takami Y., Nakayama T., and Oshimura M. 2004. Dicer is essential for formation of the heterochromatin structure in vertebrate cells. *Nat. Cell Biol.* **6:** 784–791.

Galagan J.E., Calvo S.E., Borkovich K.A., Selker E.U., Read N.D., Jaffe D., FitzHugh W., Ma L.J., Smirnov S., Purcell S., et al. 2003. The genome sequence of the filamentous fungus *Neurospora crassa. Nature* **422:** 859–868.

Galagan J.E. and Selker E.U. 2004. RIP: The evolutionary cost of genome defense. *Trends Genet.* **20:** 417–423.

Goshima G., Saitoh S., and Yanagida M. 1999. Proper metaphase spindle length is determined by centromere proteins Mis12 and Mis6 required for faithful chromosome segregation. *Genes Dev.* **13:** 1664–1677.

Grace Goll M. and Bestor T.H. 2005. Eukaryotic cytosine methyltransferases. *Annu. Rev. Biochem.* **74:** 481–514.

Grewal S.I. and Klar A.J. 1997. A recombinationally repressed region between mat2 and mat3 loci shares homology to centromeric repeats and regulates directionality of mating-type switching in fission yeast. *Genetics* **146:** 1221–1238.

Hall I.M., Noma K., and Grewal S.I. 2003. RNA interference machinery regulates chromosome dynamics during mitosis and meiosis in fission yeast. *Proc. Natl. Acad. Sci.* **100:** 193–198.

Hall I.M., Shankaranarayana G.D., Noma K., Ayoub N., Cohen A., and Grewal S.I. 2002. Establishment and maintenance of a heterochromatin domain. *Science* **297:** 2232–2237.

Hayashi T., Fujita Y., Iwasaki O., Adachi Y., Takahashi K., and Yanagida M. 2004. Mis16 and Mis18 are required for CENP-A loading and histone deacetylation at centromeres. *Cell* **118:** 715–729.

Hong E.-J.E., Villen J., Gerace E.L., Gygi S., and Moazed D. 2005. A cullin E3 ubiquitin ligase complex associates with Rik1 and the Clr4 histone H3-K9 methyltransferase and is required for RNAi-mediated heterochromatin formation. *RNA Biol.* **2:** 106–111.

Irelan J.T., Gutkin G.I., and Clarke L. 2001. Functional redundancies, distinct localizations and interactions among three fission yeast homologs of centromere protein-B. *Genetics* **157:** 1191–1203.

Jia S., Noma K., and Grewal S.I. 2004. RNAi-independent heterochromatin nucleation by the stress-activated ATF/CREB family proteins. *Science* **304:** 1971–1976.

Jin Q.W., Pidoux A.L., Decker C., Allshire R.C., and Fleig U. 2002. The mal2p protein is an essential component of the fission yeast centromere. *Mol. Cell Biol.* **22:** 7168–7183.

Kanoh J., Sadaie M., Urano T., and Ishikawa F. 2005. Telomere binding protein Taz1 establishes Swi6 heterochromatin independently of RNAi at telomeres. *Curr. Biol.* **15:** 1808–1819.

Kato H., Goto D.B., Martienssen R.A., Urano T., Furukawa K., and Murakami Y. 2005. RNA polymerase II is required for RNAi-dependent heterochromatin assembly. *Science* **309:** 467–469.

Kim H.S., Choi E.S., Shin J.A., Jang Y.K., and Park S.D. 2004. Regulation of Swi6/HP1-dependent heterochromatin assembly by cooperation of components of the mitogen-activated protein kinase pathway and a histone deacetylase Clr6. *J. Biol. Chem.* **279:** 42850–42859.

Kouzminova E.A. and Selker E.U. 2001. *Dim-2* encodes a DNA-methyltransferase responsible for all known cytosine methylation in *Neurospora. EMBO J.* **20:** 4309–4323.

Kuhn R.M., Clarke L., and Carbon J. 1991. Clustered tRNA genes in *Schizosaccharomyces pombe* centromeric DNA sequence repeats. *Proc. Natl. Acad. Sci.* **88:** 1306–1310.

Measday V. and Hieter P. 2004. Kinetochore sub-structure comes to MIND. *Nat. Cell Biol.* **6:** 94–95.

Mellone B.G., Ball L., Suka N., Grunstein M.R., Partridge J.F., and Allshire R.C. 2003. Centromere silencing and function in fission yeast is governed by the amino terminus of histone H3. *Curr Biol* **13:** 1748–1757.

Motamedi M.R., Verdel A., Colmenares S.U., Gerber S.A., Gygi S.P., and Moazed D. 2004. Two RNAi complexes, RITS and RDRC, physically interact and localize to noncoding centromeric RNAs. *Cell* **119:** 789–802.

Nakagawa H., Lee J.K., Hurwitz J., Allshire R.C., Nakayama J., Grewal S.I., Tanaka K., and Murakami Y. 2002. Fission yeast CENP-B homologs nucleate centromeric heterochromatin by promoting heterochromatin-specific histone tail modifications. *Genes Dev.* **16:** 1766–1778.

Nakayama J., Klar A.J., and Grewal S.I. 2000. A chromodomain protein, Swi6, performs imprinting functions in fission yeast during mitosis and meiosis. *Cell* **101:** 307–317.

Nakayama J., Rice J.C., Strahl B.D., Allis C.D., and Grewal S.I. 2001. Role of histone H3 lysine 9 methylation in epigenetic control of heterochromatin assembly. *Science* **292:** 110–113.

Ngan V.K. and Clarke L. 1997. The centromere enhancer mediates centromere activation in *Schizosaccharomyces pombe. Mol. Cell. Biol.* **17:** 3305–3314.

Nimmo E.R., Cranston G., and Allshire R.C. 1994. Telomere-associated chromosome breakage in fission yeast results in variegated expression of adjacent genes. *EMBO J.* **13:** 3801–3811.

Nimmo E.R., Pidoux A.L., Perry P.E., and Allshire R.C. 1998. Defective meiosis in telomere-silencing mutants of *Schizosaccharomyces pombe. Nature* **392:** 825–828.

Nolan T., Braccini L., Azzalin G., De Toni A., Macino G., and Cogoni C. 2005. The post-transcriptional gene silencing machinery functions independently of DNA methylation to repress a LINE1-like retrotransposon in *Neurospora crassa. Nucleic Acids Res.* **33:** 1564–1573.

Noma K., Allis C.D., and Grewal S.I. 2001. Transitions in distinct histone H3 methylation patterns at the heterochromatin domain boundaries. *Science* **293:** 1150–1155.

Noma K., Sugiyama T., Cam H., Verdel A., Zofall M., Jia S., Moazed D., and Grewal S.I. 2004. RITS acts in cis to promote RNA interference-mediated transcriptional and post-transcriptional silencing. *Nat. Genet.* **36:** 1174–1180.

Nonaka N., Kitajima T., Yokobayashi S., Xiao G., Yamamoto M., Grewal S.I., and Watanabe Y. 2002. Recruitment of cohesin to heterochromatic regions by Swi6/HP1 in fission yeast. *Nat. Cell Biol.* **4:** 89–93.

Partridge J.F., Borgstrom B., and Allshire R.C. 2000. Distinct protein interaction domains and protein spreading in a complex centromere. *Genes Dev.* **14:** 783–791.

Partridge J.F., Scott K.S., Bannister A.J., Kouzarides T., and Allshire R.C. 2002. cis-acting DNA from fission yeast centromeres mediates histone H3 methylation and recruitment of silencing factors and cohesin to an ectopic site. *Curr. Biol.* **12:** 1652–1660.

Perkins D.D., Margolin B.S., Selker E.U., and Haedo S.D. 1997. Occurrence of repeat induced point mutation in long segmental duplications of *Neurospora. Genetics* **147:** 125–136.

Perkins D.D., Metzenberg R.L., Raju N.B., Selker E.U., and Barry E.G. 1986. Reversal of a *Neurospora* translocation by crossing over involving displaced rDNA, and methylation of the rDNA segments that result from recombination. *Genetics* **114:** 791–817.

Perkins D.D., Radford A., and Sachs M.S. 2001. *The* Neurospora *compendium; Chromosomal loci.* Academic Press, San Diego, California.

Pickford A.S., Catalanotto C., Cogoni C., and Macino G. 2002. Quelling in *Neurospora crassa. Adv. Genet.* **46:** 277–303.

Pidoux A.L. and Allshire R.C. 2004. Kinetochore and heterochromatin domains of the fission yeast centromere. *Chromosome Res.* **12:** 521–534.

———. 2005. The role of heterochromatin in centromere function. *Philos. Trans. R. Soc. Lond. B Biol. Sci.* **360:** 569–579.

Pidoux A.L., Richardson W., and Allshire R.C. 2003. Sim4: A novel fission yeast kinetochore protein required for centromeric silencing and chromosome segregation. *J. Cell Biol.* **161:** 295–307.

Polizzi C. and Clarke L. 1991. The chromatin structure of centromeres from fission yeast: Differentiation of the central core that correlates with function. *J. Cell Biol.* **112:** 191–201.

Pratt R.J., Lee D.W., and Aramayo R. 2004. DNA methylation affects meiotic trans-sensing, not meiotic silencing, in *Neurospora. Genetics* **168:** 1925–1935.

Raju N.B. 1980. Meiosis and ascospore genesis in *Neurospora. Eur. J. Cell Biol.* **23:** 208–223.

Reinhart B.J. and Bartel D.P. 2002. Small RNAs correspond to centromere heterochromatic repeats. *Science* **297:** 1831.

Rossignol J.-L. and Faugeron G. 1994. Gene inactivation triggered by recognition between DNA repeats. *Experientia* **50:** 307–317.

Sadaie M., Iida T., Urano T., and Nakayama J. 2004. A chromodomain protein, Chp1, is required for the establishment of heterochromatin in fission yeast. *EMBO J.* **23:** 3825–3835.

Scott K.C., Merrett S.L., and Willard H.F. 2006. A heterochromatin barrier partitions the fission yeast centromere into discrete chromatin domains. *Curr. Biol.* **16:** 119–129.

Selker E.U. 1990. Premeiotic instability of repeated sequences in *Neurospora crassa. Annu. Rev. Genet.* **24:** 579–613.

———. 1998. Trichostatin A causes selective loss of DNA methylation in *Neurospora. Proc. Natl. Acad. Sci.* **95:** 9430–9435.

Selker E.U., Cambareri E.B., Jensen B.C., and Haack K.R. 1987. Rearrangement of duplicated DNA in specialized cells of *Neurospora. Cell* **51:** 741–752.

Selker E.U., Freitag M., Kothe G.O., Margolin B.S., Rountree M.R., Allis C.D., and Tamaru H. 2002. Induction and maintenance of nonsymmetrical DNA methylation in *Neurospora. Proc. Natl. Acad. Sci.* (suppl. 4) **99:** 16485–16490.

Selker E.U., Tountas N.A., Cross S.H., Margolin B.S., Murphy J.G., Bird A.P., and Freitag M. 2003. The methylated component of the *Neurospora crassa* genome. *Nature* **422:** 893–897.

Shankaranarayana G.D., Motamedi M.R., Moazed D., and Grewal S.I. 2003. Sir2 regulates histone H3 lysine 9 methylation and heterochromatin assembly in fission yeast. *Curr. Biol.* **13:** 1240–1246.

Shiu P.K. and Metzenberg R.L. 2002. Meiotic silencing by unpaired DNA: Properties, regulation and suppression. *Genetics* **161:** 1483–1495.

Shiu P.K., Raju N.B., Zickler D., and Metzenberg R.L. 2001. Meiotic silencing by unpaired DNA. *Cell* **107:** 905–916.

Sigova A., Rhind N., and Zamore P.D. 2004. A single Argonaute protein mediates both transcriptional and posttranscriptional silencing in *Schizosaccharomyces pombe. Genes Dev.* **18:** 2359–2367.

Singer M.J., Marcotte B.A., and Selker E.U. 1995. DNA methylation associated with repeat-induced point mutation in *Neurospora crassa. Mol. Cell. Biol.* **15:** 5586–5597.

Steiner N.C. and Clarke L. 1994. A novel epigenetic effect can alter centromere function in fission yeast. *Cell* **79:** 865–874.

Steiner N.C., Hahnenberger K.M., and Clarke L. 1993. Centromeres of the fission yeast *Schizosaccharomyces pombe* are highly variable genetic loci. *Mol. Cell. Biol.* **13:** 4578–4587.

Sugiyama T., Cam H., Verdel A., Moazed D., and Grewal S.I. 2005. RNA-dependent RNA polymerase is an essential component of a self-enforcing loop coupling heterochromatin assembly to siRNA production. *Proc. Natl. Acad. Sci.* **102:** 152–157.

Takahashi K., Chen E.S., and Yanagida M. 2000. Requirement of Mis6 centromere connector for localizing a CENP-A-like protein in fission yeast. *Science* **288:** 2215–2219.

Takahashi K., Murakami S., Chikashige Y., Funabiki H., Niwa O., and Yanagida M. 1992. A low copy number central sequence with strict symmetry and unusual chromatin structure in fission yeast centromere. *Mol. Biol. Cell* **3:** 819–835.

Takahashi K., Murakami S., Chikashige Y., Niwa O., and Yanagida M. 1991. A large number of tRNA genes are symmetrically located in fission yeast centromeres. *J. Mol. Biol.* **218:** 13–17.

Takahashi K., Takayama Y., Masuda F., Kobayashi Y., and Saitoh S. 2005. Two distinct pathways responsible for the loading of CENP-A to centromeres in the fission yeast cell cycle. *Philos. Trans. R. Soc. Lond. B Biol. Sci.* **360:** 595–607.

Tamaru H. and Selker E.U. 2001. A histone H3 methyltransferase controls DNA methylation in *Neurospora crassa. Nature* **414:** 277–283.

———. 2003. Synthesis of signals for de novo DNA methylation in *Neurospora crassa. Mol. Cell. Biol.* **23:** 2379–2394.

Tamaru H., Zhang X., McMillen D., Singh P.B., Nakayama J., Grewal S.I., Allis C.D., Cheng X., and Selker E.U 2003. Trimethylated lysine 9 of histone H3 is a mark for DNA methylation in *Neurospora crassa. Nat. Genet.* **34:** 75–79.

Thon G. and Verhein-Hansen J. 2000. Four chromo-domain proteins of *Schizosaccharomyces pombe* differentially repress transcription at various chromosomal locations. *Genetics* **155:** 551–568.

Volpe T., Schramke V., Hamilton G.L., White S.A., Teng G., Martienssen R.A., and Allshire R.C. 2003. RNA interference is required for normal centromere function in fission yeast. *Chromosome Res.* **11:** 137–146.

Volpe T.A., Kidner C., Hall I.M., Teng G., Grewal S.I., and Martienssen R.A. 2002. Regulation of heterochromatic silencing and histone H3 lysine-9 methylation by RNAi. *Science* **297:** 1833–1837.

Watters M.K., Randall T.A., Margolin B.S., Selker E.U., and Stadler D.R. 1999. Action of repeat-induced point mutation on both strands of a duplex and on tandem duplications of various sizes in *Neurospora. Genetics* **153:** 705–714.

Wilkinson C.R., Bartlett R., Nurse P., and Bird A.P. 1995. The fission yeast gene pmt1+ encodes a DNA methyltransferase homologue. *Nucleic Acids Res.* **23:** 203–210.

Yamada T., Fischle W., Sugiyama T., Allis C.D., and Grewal S.I. 2005. The nucleation and maintenance of heterochromatin by a histone deacetylase in fission yeast. *Mol. Cell* **20:** 173–185.

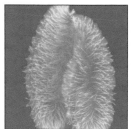
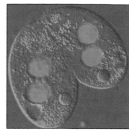

Epigenetics of Ciliates

Eric Meyer[1] and Douglas L. Chalker[2]

[1]Laboratoire de Genetique Moleculaire, CNRS UMR 8541 Ecole Normale Superieure, 75005 Paris, France
[2]Deptartment of Biology, Washington University, St. Louis, Missouri 63130

CONTENTS

1. Ciliates: Single Cells with Two Different Genomes, 129

2. Conjugation: Reciprocal Fertilization Reveals non-Mendelian Inheritance, 129

3. Cytoplasmic Inheritance in Ciliates, 130

4. Cortical Patterning: A Case of Structural Inheritance, 131

5. Macronuclei and Micronuclei: A Model for Active Versus Silent Chromatin, 132

   5.1 Separation of Micro- and Macronuclear Histones Reveals Distinct Roles for Histone Variants, 132

   5.2 Chromatin Modifications Correlate with Activity States, 133

6. Genome-wide Rearrangements Occur during Macronuclear Development, 134

   6.1 Internal DNA Deletion: Precise (Intragenic IESs) and Imprecise (Intergenic Repeats) Events, 135

   6.2 Chromosome Fragmentation, 135

7. Mechanisms of Genome Rearrangements, 136

8. Homology-dependent Gene Silencing in Ciliates, 136

   8.1 Transgene-induced Silencing, 137

   8.2 Silencing Is Induced by Double-stranded RNA, 137

9. Genome Rearrangements Are Guided by Homology-dependent Mechanisms, 137

9.1. Experimental Induction of Specific Deletions in the Developing Macronucleus, 137

10. Rearrangement Patterns Are Likely Determined by a Comparison of Germ-line and Somatic Genomes, 138

   10.1 The d48 Paradigm: Epigenetic Inheritance of Alternative Rearrangements, 138

   10.2 Epigenetic Inheritance of Experimentally Induced Deletions, 139

   10.3 "Spontaneous" Elimination of Foreign Sequences Introduced into the Micronuclear Genome, 140

   10.4 Experimental Rescue of Inherited Macronuclear Deletions, 140

   10.5 Experimental Inhibition of IES Elimination in the Developing Macronucleus, 140

11. A trans-Nuclear Comparison of Whole Genomes Mediated by RNA Interference, 142

   11.1 Linking Short RNAs to DNA Elimination, 142

   11.2 Transporting RNA from Maternal to Zygotic Macronuclei in Paramecium, 144

12. Conclusions: DNA Elimination as a Genome Defense Mechanism, 144

13. Future Contributions of Ciliate Research to Epigenetics, 146

References, 147

GENERAL SUMMARY

Anyone watching ciliates under a microscope is sure to be fascinated by these complex little animals that use their hair-like cilia to swim, eat, and find a mate. Vegetative growing cells duplicate by simple binary fission; yet periodically, ciliates will mate with a partner or, in some species, undergo self-fertilization, resulting in sexual progeny with a different genotype. What uniquely distinguishes these single-celled eukaryotes is that they maintain two functionally distinct genomes, carried in separate nuclei, within a common cytoplasm. The smaller of these, the micronucleus, contains the germ-line genome. It is transcriptionally silent during growth, but stores the genetic information that is passed to progeny at each sexual generation. The larger macronucleus performs somatic functions because it is responsible for all gene expression and thus governs the cell's phenotype. It is discarded at the end of each vegetative cycle when a new macronucleus differentiates from the germ line. During macronuclear development, massive DNA rearrangements generate a streamlined version of the genome ready for expression. A portion of the germ-line genome, including all the repetitive DNA that has long been considered "junk," is eliminated, while all the genes needed for the organism's survival throughout the life cycle are amplified to a high ploidy level.

The ciliates' genetic oddities have made them very useful model organisms with which to discover and understand epigenetic mechanisms. In some species, the two sibling progeny that develop upon mating begin with identical genomes, but their somatic nuclei differentiate within the context of two different parental cells. This permits easy detection of hereditary characters that are not solely determined by the nuclear genome. Genetic experiments conducted with *Paramecium tetraurelia*, primarily those of Tracy Sonneborn, provided some of the earliest descriptions of non-Mendelian inheritance in any eukaryote. Among cases in which genetically identical siblings expressed different variants of specific traits, some were true cases of cytoplasmic inheritance (maternal inheritance of organelle DNA), but others, such as the inheritance of an individual's mating type, were of a different kind. A progeny's mating type is not determined by its genotype, but rather is specified by the preexisting type of the parental cell in which its somatic genome developed. In simple terms, different environments (the parental cytoplasm) direct expression of alternative traits from identical DNA complements. This is the hallmark of epigenetics.

The peculiar genetic organization of ciliates also implies mechanisms that differentially regulate homologous sequences contained within the distinct nuclei. Early studies aimed to elucidate the means by which the germ line was kept silent and the somatic genome transcriptionally active. The compartmentalization of gene expression states offered researchers an opportunity to investigate the role of chromatin proteins and their modifications in epigenetic regulation. They could readily correlate specific histones and their modification with transcriptional activity or cell cycle stage. For instance, by comparing chromatin proteins from germ-line and somatic nuclei of *Tetrahymena thermophila*, some of the first histone variants were identified. Furthermore, new chromatin regulators, such as the first histone acetyltransferase (HAT), were identified in this ciliate, in part by taking advantage of the fact that only the macronucleus contains acetylated histones.

Although ciliate genetics may seem unconventional, the underlying mechanisms are widely used for epigenetic regulation in eukaryotes, as illustrated by the role of RNA interference in whole-genome rearrangements. The extent and form of these rearrangements are remarkably diverse among ciliate species, yet one common feature is that they normally direct the elimination of transposon-like elements and other repetitive sequences. In both *Paramecium* and *Tetrahymena*, short RNAs are generated from the germ-line genome during meiosis. The discovery of these small RNAs, together with the demonstration in *Tetrahymena* that Argonaute and Dicer homologs are required for DNA rearrangements, has led to the realization that an RNAi-like mechanism is involved. The small RNAs are thought to target histone H3 lysine 9 methylation to homologous sequences, marking them for elimination. Thus, ciliate DNA rearrangements are mechanistically similar to the more broadly used RNA-directed establishment of heterochromatin. The use of RNAi to eliminate transposable elements further underscores the importance of this pathway as a genome defense mechanism. Furthermore, many experiments have shown that DNA rearrangement patterns are not strictly determined by the germ-line genome, but are controlled, at least in part, by preexisting rearrangements within the parental somatic genome. The implication is that the germ-line and somatic genomes are compared to each other during nuclear differentiation, a comparison that is likely mediated by homology-dependent interactions between germ-line and somatic RNAs. Fully understanding this process will undoubtedly provide new insight into the roles of RNA in the epigenetic programming of the genome.

1 Ciliates: Single Cells with Two Different Genomes

Ciliates, which comprise a monophyletic lineage that emerged about one billion years ago (Philippe et al. 2000), were among the first unicellular eukaryotes to be used as genetic models. In the late 1930s, when T.M. Sonneborn discovered the mating types of *Paramecium aurelia* (Sonneborn 1937), the chromosome theory of inheritance elaborated by T.H. Morgan was still unsatisfying to many researchers, in particular embryologists (for historical detail, see Chapter 2). Unable to envision how such static entities as genes could be the sole basis of heredity, they believed that the cytoplasm had to be involved, if only to coordinate gene action (see Harwood 1985). Whereas mainstream geneticists largely focused on gene action, Sonneborn's early genetic analyses showed that the transmission of many heritable characteristics could not be fully explained by Mendel's laws. Due to their unique biology, the study of ciliates revealed some of the first examples of cytoplasmic inheritance and continues to provide new insights into epigenetic mechanisms.

One of the most distinctive features of ciliates is nuclear dimorphism: Each cell contains two kinds of nuclei that differ in structure and function. The diploid micronuclei are transcriptionally silent during vegetative growth but contain the germ-line genome. These nuclei undergo meiosis to produce gametic nuclei that transmit the Mendelian genome to the next sexual generation (Fig. 1). In contrast, the highly polyploid macronuclei are responsible for gene expression during vegetative growth and thus govern the cell's phenotype, but they are lost during sexual development and can therefore be considered the equivalent of the soma (Fig. 1). The numbers of nuclei of each type vary in different species. For example, *P. aurelia* species have two micronuclei and one macronucleus, whereas *Tetrahymena thermophila* has just one of each.

Macro- and micronuclei divide by separate mechanisms. Micronuclei divide via conventional closed mitosis. Macronuclei, in contrast, divide by a poorly understood amitotic mechanism that does not involve spindle formation or visible condensation of the centromere-less, somatic chromosomes. After DNA synthesis, the macronucleus simply splits into two roughly equal halves. There does not appear to be any mechanism to ensure equal segregation of macronuclear chromosomes to the two daughter cells. Instead, it is likely that the high ploidy level ($\sim$800n in *P. tetraurelia*, $\sim$45n in *T. thermophila*) prevents lethal gene loss for a number of vegetative divisions. Most species have a finite vegetative life span, and clonal cell lines will eventually die if they do not engage in sexual reproduction before they become senescent.

2 Conjugation: Reciprocal Fertilization Reveals non-Mendelian Inheritance

Ciliates are hermaphroditic species capable of conjugation, a mating process that involves cross-fertilization between two parent cells. Mature cells of appropriate clonal age will become sexually reactive upon mild starvation and pair with cells of compatible mating types to initiate conjugation. If no compatible partner is available, some species will undergo a self-fertilization process called autogamy. In both cases, nuclear reorganization ensues, starting with meiosis of micronuclei. The sequence of nuclear events is similar in all species with some variations, and is depicted in Figure 1 for the *P. aurelia* and *T. thermophila* species (see Sonneborn 1975).

Postmeiotic development starts with the selection of a single haploid nucleus in each cell to pass on the genome. The selected nucleus undergoes an additional division that produces two genetically identical gametic nuclei. In the case of conjugation, the two mates exchange one of their two haploid nuclei, and subsequent karyogamy (i.e., the fusion of two haploid nuclei) therefore generates genetically identical zygotic nuclei in each conjugant (stages 3–5 in Fig. 1). In autogamy, the two gametic nuclei within the single cell fuse to produce an entirely homozygous diploid genome. In both cases, the resulting diploid zygotic nucleus (stage 5) divides twice more, and the four products differentiate, two into new micronuclei and two into new macronuclei (stages 6 and 7). Upon completion of nuclear development, either both new micronuclei are maintained in the new vegetative clones as occurs in *P. aurelia* species, or one of the two is degraded as in *T. thermophila*. In both species, the two new macronuclei do not divide during the first cellular division (stage 9) but are distributed to the two daughter cells; they start dividing only at the second vegetative division.

While the parental micronuclei give rise to the new micro- and macronuclei of the next generation, the parental macronucleus is lost. In *P. aurelia*, it is fragmented into about 30 pieces in which DNA replication is rapidly inhibited, although transcription continues actively throughout the differentiation of the new macronuclei. When vegetative growth resumes, the fragments are distributed randomly to daughter cells until none is left (stage 9). In *T. thermophila*, the parental macronucleus does not fragment, but becomes pycnotic and is degraded by an apoptosis-like mechanism before the first vegetative division (Davis et al. 1992) (stages 7 and 8).

3 Cytoplasmic Inheritance in Ciliates

The biology of *P. aurelia* conjugation is quite favorable for the detection of epigenetic phenomena associated with cytoplasmic effects. Whereas reciprocal fertilization generates genetically identical zygotic nuclei in the two conjugants, almost no cytoplasm is exchanged between them, which effectively distinguishes the action of nuclear genes from that of the most influential of environments, the cytoplasm of the mother cell (Fig. 2). Each genetic cross is

therefore equivalent to a study of monozygotic twins being born to different mothers. Sonneborn's studies revealed that phenotypic differences between two parental cells can be maintained in their respective progeny even though the latter have identical genotypes. In a few cases, the phenotypic differences were later found to be determined by extranuclear genes and would not today qualify as epigenetic phenomena. For instance, the deadly properties of *killer* strains of *P. aurelia* are due to endosymbiotic bacteria of the genus *Caedibacter*, harbored in the cytoplasm, that release a toxin killing sensitive strains (for reviews, see Preer et al. 1974; Pond et al. 1989). Other cases, such as the maternal inheritance of serotype, which requires mutually exclusive expression of one of several paralogous surface antigen genes, are clear cases of cytoplasmic influence on gene activity.

As with serotypes, the two complementary mating types of *P. tetraurelia*, which are called O and E, exhibit a cytoplasmic pattern of inheritance. The O and E traits are terminally differentiated phenotypes that are determined during development of the somatic macronucleus from a totipotent germ line. After conjugation, a vegetative clone descended from the O parent is almost always mating type O, whereas one arising from the E parent is almost always of type E, even though both exconjugants develop from identical zygotic genomes (Fig. 2b). Furthermore, when a large cytoplasmic bridge forms between the two conjugating cells, allowing a significant exchange of cytoplasm, the progeny of both parents usually develop as type E (Sonneborn 1977). Thus, a cytoplasmic factor

Paramecium

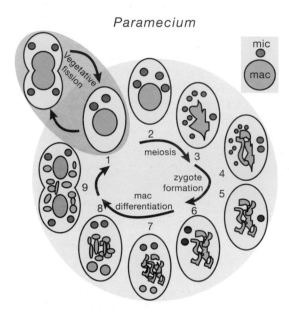

Ciliate Life Cycles

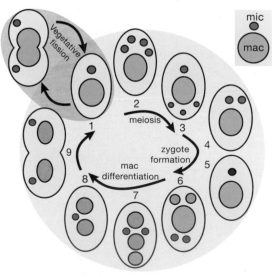

Tetrahymena

Figure 1. Life Cycles of *Paramecium* (*top*) and *Tetrahymena* (*bottom*)

(*Stage 1*) Vegetative cells multiply by binary fission. Sexual development, *stages 2–8*, will initiate upon conjugation of two cells or autogamy (in *Paramecium* only). (*Stages 2–3*) Micronuclear meiosis ends with selection of one of the haploid products as the gametic nucleus and degeneration of those remaining. In *Paramecium*, the parental macronucleus starts forming lobes. (*Stages 4–6*) Zygote formation. An additional division of the selected nucleus produces two genetically identical haploid nuclei. During conjugation, one of the two identical gametic nuclei is exchanged between the two mates and subsequent karyogamy produces the diploid zygotic nucleus (*red*). During autogamy, the two identical gametic nuclei simply fuse together. Two additional postzygotic divisions (*6*) produce the undifferentiated micro- and macronuclei. (*Stages 6–8*) Nuclear differentiation. After the second postzygotic division, two of the resulting nuclei become the new micronuclei, while the other two begin differentiating into new macronuclei (*pink*). In *Paramecium*, the maternal macronucleus is fragmented. In *Tetrahymena*, it becomes pycnotic. Also in *Tetrahymena*, one of the new micronuclei degenerates. (*Stage 9*) Caryonidal division: This first vegetative division is special, as new macronuclei are distributed to the daughter cells without division while micronuclei are segregated to progeny by mitosis. Finally, fragments of the *Paramecium* parental macronucleus are nondividing, but remain until lost through random distribution during subsequent fissions.

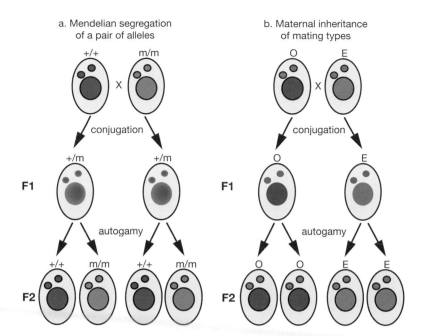

a. Mendelian segregation of a pair of alleles

b. Maternal inheritance of mating types

Figure 2. Mendelian vs. Cytoplasmic Inheritance

(*a*) Conjugation and autogamy are illustrated by a cross between two *Paramecium* cells, each homozygous for different alleles, M or m. Conjugation involves the reciprocal exchange of one of two identical gametic nuclei. This results in F₁ exconjugants with identical heterozygous genotypes. Autogamy, a self-fertilization process, generates an entirely homozygous genotype in just one sexual generation such that these F₂ individuals have a 50% chance of becoming M/M or m/m. (*b*) Phenotypic difference between F₁ clones reveals cytoplasmically inherited characteristics. In *Paramecium*, mating type (O or E) is irreversibly determined during the development of the somatic macronucleus (*large circle*) from the totipotent germ-line micronucleus (*small circle*); however, the parental macronucleus directs differentiation of each exconjugant toward maintaining the existing mating type.

must exist that directs development of the E type.

As introduced below, the germ-line genome undergoes extensive DNA rearrangements during macronuclear development, and although the putative mating-type gene has yet to be identified, one Mendelian mutation affecting mating-type determination has been shown to perturb these genome rearrangements at other loci (Meyer and Keller 1996). If regulated genome rearrangements determine mating type, the E-determining cytoplasmic factor should have the capacity to direct an alternative rearrangement of the mating-type gene, resulting in a macronuclear form of the gene that specifies type E. During conjugation, this form must also produce the E-determining cytoplasmic factor required for its further inheritance. As described in Section 10, alternative rearrangement patterns can be transmitted from maternal macronuclei to zygotic macronuclei, a *trans*-nuclear effect that appears to be mediated by RNA molecules acting in a homology-dependent manner. The cytoplasmic factor responsible for non-Mendelian inheritance is thus likely to be an RNA molecule that controls the developmental "mutation" of the mating-type gene.

4 Cortical Patterning: A Case of Structural Inheritance

Studies of the complex architecture of the cell cortex revealed another form of non-Mendelian inheritance. The *Paramecium* cell is covered with about 4000 cilia arranged in longitudinal rows of anchored units (Fig. 3a).

Each cilium is rooted in a basal body or kinetosome, a complex structure with both antero-posterior and left-right asymmetries. As cells divide, the duplication of basal bodies is constrained by the structure of ciliary units, so that the new basal bodies remain in the same orientation (Fig. 3b). However, surgical grafting of a small patch of cortex in the reverse polarity will direct the eventual formation of a complete inverted ciliary row as the grafted basal bodies duplicate (Fig. 3c). This antero-posterior inversion will be propagated for an indefinite number of vegetative cell divisions and will be maternally inherited during conjugation (Beisson and Sonneborn 1965).

These grafting experiments showed that genes are not responsible for the inheritance of such structural variation, and revealed the essential role of preexisting structures for the correct assembly of new structures. The oriented duplication of the centriole (Beisson and Wright 2003), and the propagation of flagellar shape upon cell division of trypanosomes (Moreira-Leite et al. 2001), represent other examples. Prions further exemplify self-propagating protein conformations that are responsible for cytoplasmic inheritance in yeast and mammals (Shorter and Lindquist 2005). Even the centromere of eukaryotic chromosomes behaves as a self-replicating protein complex that resides on DNA but is not determined by it (Cleveland et al. 2003; see Chapter 14). What these epigenetic phenomena tell us is that not all cellular structures can be assembled de novo by simply reading the information contained in genes. In a broader sense, replication of DNA itself is a case of structural inheritance, but the

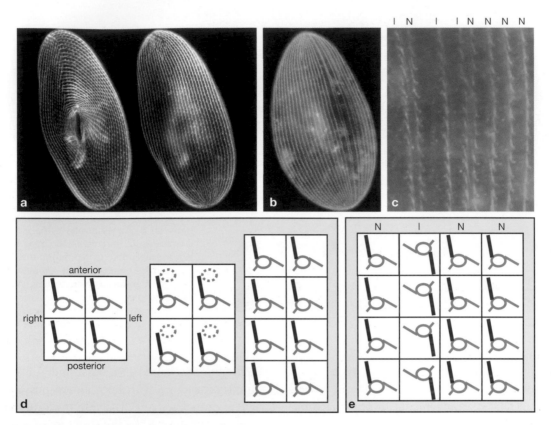

Figure 3. Structural Inheritance of Cortical Unit Polarity in *Paramecium*

(*a*) Immunolabeling of basal bodies and ciliary rootlets highlights the regular organization of parallel cortical rows of wild-type cells. Shown are ventral (*left*) and dorsal (*right*) views. (*b*) Dorsal view of a cell exhibiting disruption of the regular organization due to the reversed antero-posterior polarity of a few rows of cortical units. Basal bodies immunolabeled in red, and ciliary rootlets in green. (*c*) Enlargement of a patch of cortex shows the reversed orientation of ciliary rootlets in inverted rows (I) relative to normal rows (N). (*d*) Schematic of basal bodies (*green circles*) duplication during growth; each is shown flanked on its right side by an anteriorly oriented ciliary rootlet (*purple*) and two microtubular ribbons. Duplication occurs with a fixed geometry: Each new basal body is positioned anterior of its parent, ensuring identical polarity. (*e*) The repeated duplication of basal bodies within each row maintains homogeneous orientation indefinitely. (Photographs courtesy of Janine Beisson.)

genome is certainly not the only structure that dividing cells need to duplicate. Thus, far from being a rare curiosity, "epigenetic" structural inheritance may be viewed as one of the most fundamental mechanisms of life.

5 Macronuclei and Micronuclei: A Model for Active Versus Silent Chromatin

One basic concept in epigenetics is that individual copies of a DNA sequence can possess different activities and that differential states can be stably maintained. The nuclear dimorphism of ciliates is a natural example of homologous sequences that are maintained in a common cytoplasm, yet possess opposite activity states. The macronucleus serves as a model for the transcriptionally active state, the micronucleus for the repressed or silent state (Fig. 4). Early biochemical and immunohistochemi-

cal studies, primarily in *Tetrahymena*, aimed to compare the properties of these different nuclei, both in vegetative cells and during sexual development, in order to define how these different activity states might be determined, particularly at the level of chromatin structure.

5.1 Separation of Micro- and Macronuclear Histones Reveals Distinct Roles for Histone Variants

Isolation of histone proteins separately from the macronucleus and micronucleus of *Tetrahymena* led to the discovery of specific histone variants. Histone variants hv1 and hv2 that are now known to correspond to the H2A.Z and H3.3 variants of other eukaryotes, respectively (see Chapter 13), reside exclusively within macronuclei, which provided an early indication that these variants are important for maintaining transcrip-

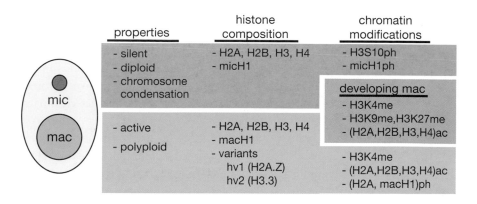

properties	histone composition	chromatin modifications

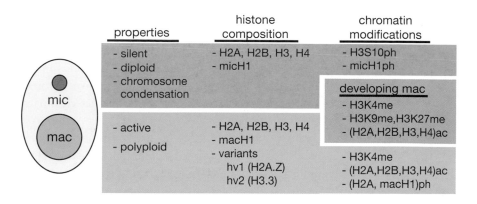

Figure 4. Nuclear Dimorphism of Ciliates

The germ-line micronucleus, the developing macronucleus, and the somatic macronucleus contain different histone complements and modifications. Those known to occur specifically in each or in the developing somatic genome are listed.

tional activity (Allis et al. 1980; Hayashi et al. 1984). The hv2 (H3.3) variant was shown to be constitutively expressed, a property critical for its deposition into chromatin outside of S phase, allowing this isoform of histone H3 to serve as a replacement histone (Yu and Gorovsky 1997).

In addition to the different complement of core histone variants, the macro- and micronucleus contain different linker histones. The macronuclear H1 has similar amino acid composition and biochemical properties to the linker histones of other eukaryotes but lacks a central globular domain (Wu et al. 1986; Hayashi et al. 1987). Neither linker histone gene is essential for cell viability. Interestingly, gene knock-outs of either cause the volume of their respective nuclei to increase, indicating that both perform roles in overall chromatin compaction, possibly by stabilizing higher-order chromatin structure (Shen et al. 1995). Loss of the macronuclear H1-like protein also leads to gene-specific changes in expression, implicating this linker histone in maintaining proper transcriptional regulation (Shen and Gorovsky 1996).

5.2 Chromatin Modifications Correlate with Activity States

ACETYLATION

The hyperacetylation of histones in the macronucleus and absence of this modification from the micronucleus have provided further evidence correlating this post-translational modification with gene activation (Vavra et al. 1982). The enzymes that performed histone acetylation in any organism remained unknown until the mid-1990s, when C. David Allis and coworkers purified the first type A (nuclear) histone acetyltransferase (HAT) (Brownell and Allis 1995; Brownell et al. 1996). These researchers started with highly purified macronuclei to separate this activity from type B cytoplasmic HAT activ-

ity and followed their purification using an in-gel assay. For this assay, purified histones were polymerized into the polyacrylamide matrix of the gel used to separate purified protein fractions. After electrophoresis, the proteins were renatured and incubated with radiolabeled acetyl-CoA to reveal a polypeptide with an apparent molecular weight of 55 kD that could incorporate the label into the histone matrix.

The real breakthrough came after microsequencing the purified protein and cloning the gene. This *Tetrahymena* HAT was found to be homologous to a well-characterized transcriptional regulator of baker's yeast, the Gcn5 protein. Before this discovery, transcriptional activators were primarily thought to act by recruiting RNA polymerase to promoters, but this work established that transcriptional activators may also possess enzymatic activity, modifying chromatin or other transcriptional regulators, thus changing the state of the template. The door was opened, and many known regulators were quickly thereafter shown to act as HATs.

METHYLATION

The nuclear dimorphism of ciliates again proved advantageous in elucidating roles of histone methylation. This modification is restricted to the macronuclei in growing *Tetrahymena* cells (Fig. 4). The histone lysine methyltransferase (HKMT) activity purified from these nuclei specifically modified histone H3 at lysine 4 (H3K4me), providing an early correlation between this specific modification and transcriptional activity (Strahl et al. 1999).

Histone H3 lysine 9 methylation is absent from vegetatively growing cells but occurs specifically during macronuclear development on germ-line-limited sequences that are eliminated from the somatic genome (Taverna et al. 2002). The developmentally regulated establishment of H3K9me2 (dimethylated) on these specific sequences provides a useful model with which to elucidate the targeting of this modification to the equivalent of heterochromatin (described in detail below).

PHOSPHORYLATION

Purification of ^{32}P-radiolabeled histones from micro- and macronuclei showed that H3, H2A, and linker histones were highly phosphorylated (Allis and Gorovsky 1981). Multiple sites of macronuclear H1 are phosphorylated, and this modification was shown to participate in the regulation of specific gene transcription (Mizzen et al. 1999). Using mutational analysis, Dou and Gorovsky found that this requirement for phosphorylation could be mimicked by the addition of charged amino acids into H1 (Dou et al. 1999). The charged residues, however, did not need to be present in the corresponding positions of the phosphorylated amino acid, but the complementary effect required a cluster of charged sites (Dou and Gorovsky 2000, 2002). These studies indicated that phosphorylation per se was not required, but that a critical charge density promoted proper transcription.

A single position, serine 10, is phosphorylated in histone H3 (H3S10ph) (Wei et al. 1998). This modification is cell-cycle-dependent and is correlated with mitosis in many eukaryotes. In *Tetrahymena*, it is restricted to micronuclei during mitosis and meiosis. Replacing the normal histone H3 gene with a mutant form containing an alanine substitution at serine 10 (S10A) causes defects in micronuclear division resulting in lagging chromosomes and aneuploidy (Wei et al. 1999). Macronuclear amitotic division, however, is not affected. These results demonstrated that H3 phosphorylation plays an important role in chromosome condensation and/or segregation. The unique nuclear dimorphism of the ciliate again revealed key insight into the role of a chromatin modification.

6 Genome-wide Rearrangements Occur during Macronuclear Development

The sequencing of the somatic (macronuclear) genomes of *P. tetraurelia* and *T. thermophila* revealed very high gene numbers (~40,000 and ~27,000, respectively) despite relatively small genome sizes (~72 Mb and ~104 Mb, respectively). This organization is consistent with a genome that is optimized for efficient gene expression (see Fig. 15 of Chapter 3 for relative comparisons with other eukaryotic organisms). This "streamlining" of the somatic genome is achieved by massive DNA rearrangements of the germ-line-derived chromosomes that occur

at specific stages of macronuclear development (for reviews, see Prescott 1994; Coyne et al. 1996; Klobutcher and Herrick 1997; Jahn and Klobutcher 2002). Thus, the nuclear differentiation of the germ-line and somatic genomes involves a physical reorganization of chromosomes in addition to establishing the distinct chromatin states just described.

Two types of rearrangements are commonly observed and are virtually unique to all ciliates that have been studied, internal DNA deletion and chromosome fragmentation (Fig. 5). Instances of this in other eukaryotes can be exemplified by localized VDJ locus recombination in the lymphocytes of mammals (see Chapter 21). Depending on the species, these events eliminate between 10% and 95% of the germ-line genome from each newly formed macronucleus. Virtually all repeated sequences, including transposable elements, are eliminated, which can in part account for the high macronuclear gene densities observed. The numbers of rearrangement sites also vary between species, from about 6,000 in *Tetrahymena* to perhaps as many as 100,000 in hypotrichous ciliates. The genome-wide distribution of these highly regulated and reproducible events provides an opportunity to investigate how cells identify and direct action on particular

Paramecium: precise and imprecise deletions

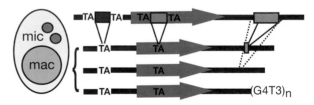

Tetrahymena: imprecise deletions and chromosome fragmentation

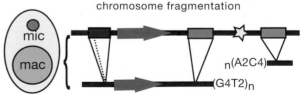

Figure 5. DNA Rearrangements of Ciliates

During development of a new macronucleus, extensive chromosome fragmentation and DNA elimination occur. (*Top*) DNA rearrangements occurring in *Paramecium* include both precise deletion of TA-bounded IESs (*colored bars* found in coding [*arrow*] and noncoding regions) and imprecise deletions (*orange bar*) that result alternatively in DNA deletion or fragmentation (G4T3)$_n$. (*Bottom*) In *Tetrahymena* imprecise deletion of IESs (*colored bars*) occurs at about 6000 loci, and chromosome fragmentation is specified by a conserved 15-bp sequence, the CBS (*star*).

DNA segments. Studies of ciliate DNA rearrangements have already revealed unique insights into epigenetic mechanisms, particularly regarding homology-dependent processes. The following description of DNA rearrangement events should provide the background sufficient for the subsequent discussion of the associated epigenetic regulation.

6.1 Internal DNA Deletion: Precise (Intragenic IESs) and Imprecise (Intergenic Repeats) Events

PRECISE DELETIONS

Precise deletions are those that occur at the same nucleotide positions in all copies of a macronuclear chromosome. The internal eliminated sequences (IESs) are short, single-copy DNA segments that are primarily removed from coding sequences, but are also found in intergenic or intronic regions of germ-line chromosomes (Fig. 5). Precisely excised IESs are bounded by short direct repeats, which typically vary in sequence between species. A prominent class, the so-called "TA" IESs found in *Paramecium* and some spirotrichs such as *Euplotes crassus*, are identified as having invariable 5′-TA-3′ repeats at their boundaries, one copy of which remains within the macronuclear locus after excision (Betermier 2004). The few nucleotide positions internal to the TA dinucleotides are not random and form a degenerate consensus that resembles the ends of Tc1/mariner transposons. Thus, these IESs may be evolutionarily derived from ancient insertions of such mobile elements (Klobutcher and Herrick 1997). Nevertheless, many IES ends conform poorly to the consensus sequence whereas many perfect matches can be found in macronuclear sequences that are not excised, indicating that the consensus does not contain sufficient information to specify excision of the approximately 60,000 IESs per haploid genome of *P. tetraurelia*, raising the question of how they are recognized.

An amazing variation to precise excision occurs in the stichotrichs, a subgroup of the spirotrichs, in which IES removal occurs simultaneously with gene "unscrambling" (Prescott 1999). In the micronuclear version of scrambled genes, the macronuclear destined sequences (MDSs, i.e., the DNA "exons") are not only separated by IESs, but are also disordered relative to the linear arrangement found in the reorganized macronuclear sequence. In the germ line, two MDSs that will be joined to form the expressed gene can be located far apart, sometimes in unlinked loci (Landweber et al. 2000), and may even be in an inverted orientation relative to each other. The precision of reordering appears to be guided by relatively long (11 bp

on average) homologous repeats, shared by cognate MDS ends, that are unrelated in sequence to those at other MDS ends. Although these long repeats certainly contribute to accurate unscrambling, it is not clear that they are sufficient, which has led to the proposal that a preexisting template may be involved (Prescott et al. 2003).

IMPRECISE DELETIONS

Whereas IESs are efficiently and reproducibly removed from the somatic genome, in some cases, the deletion boundaries formed by independent excision events in the polyploid nucleus vary in position (Fig. 5). This heterogeneity extends over tens of base pairs in *Tetrahymena* and up to several kilobase pairs in *Paramecium*. Imprecise deletion is characteristic of all studied *Tetrahymena* IESs and is primarily responsible for the removal of repeated sequences such as intergenic transposons or minisatellites of *Paramecium* (Le Mouël et al. 2003). Like precise IES excision, these deletions typically occur between short direct repeats, one of which is maintained in the macronuclear sequence. In *Tetrahymena*, the repeat sequences are highly variable among the known IESs, whereas in *Paramecium*, the repeats always contain at least one TA dinucleotide, suggesting that the mechanism involved may be related to that of precise IES excision.

6.2 Chromosome Fragmentation

During macronuclear development, the germ-line-derived chromosomes are fragmented into shorter molecules that are capped by de novo addition of telomeric repeats (Fig. 5). The resulting macronuclear chromosomes apparently lack centromeres, and thus, chromosome fragmentation may facilitate the equal distribution of these molecules during amitotic divisions of the macronucleus. The extent of fragmentation varies widely among species. In spirotrichs it is carried out to an extreme, producing tiny "nanochromosomes" that typically contain single genes, whereas in *Paramecium* and *Tetrahymena*, macronuclear chromosomes range in size from 20 kb to over 1 Mb and contain many genes.

The process is imprecise in most ciliates, so that the exact position of telomeric repeat addition is heterogeneous and often results in some loss of germ-line sequence. One exception is *Euplotes crassus*, in which telomeres are always added at the same nucleotide positions (Klobutcher 1999). In *Tetrahymena*, a conserved 15-bp chromosome breakage sequence (CBS) that is found in an estimated 280 loci within germ-line chromosomes

(Cassidy-Hanley et al. 2005; Hamilton et al. 2005) is both necessary and sufficient to direct fragmentation and new telomere addition (Fan and Yao 1996). In contrast, no analogous CBS has been identified in *P. aurelia* species. Rather, all evidence indicates that chromosome fragmentation is an alternative outcome of imprecise deletion (Fig. 5) such that DNA elimination is healed either by the rejoining of flanking sequences or by telomere addition (Le Mouël et al. 2003).

7 Mechanisms of Genome Rearrangements

Although ciliate genome rearrangements have been studied for more than 30 years, the molecular mechanisms that carry out these events remain largely unknown (Yao et al. 2002). Part of the challenge for researchers has been to explain the diversity of events that take place among the different ciliates. Excision intermediates as well as circular excision (by)products have been described for several ciliates (Jaraczewski and Jahn 1993; Klobutcher et al. 1993; Williams et al. 1993; Saveliev and Cox 1995, 1996; Betermier et al. 2000; Gratias and Betermier 2003). The data do not allow a unified excision mechanism to be deduced, even for those IESs of different ciliates that have similar consensus sequences at their termini (see Klobutcher and Herrick 1995; Gratias and Betermier 2001; Betermier 2004).

Despite the diversity of excision mechanisms, there still may exist significant overlap in the events that guide DNA rearrangements. One commonality is that transposon-like and repetitive sequences appear to be preferentially eliminated. As described elsewhere throughout this book, one role of epigenetic mechanisms is to suppress the activity of these potentially deleterious DNA elements. Allowing a transposon to escape from the silent germ line into the highly active macronucleus is potentially disastrous to the somatic genome. Many lines of evidence implicate mechanisms related to RNAi and heterochromatin formation in these processes of DNA rearrangement (Mochizuki and Gorovsky 2004; Yao and Chao 2005).

In most eukaryotes, methylated histone H3 at lysine 9 (H3K9me) is widely associated with the repressed DNA that is partitioned in the nucleus as heterochromatin (see Section 7 of Chapter 3). In *Tetrahymena*, this modification is not found in the transcriptionally silent micronucleus as one might presume, but is exclusively found in developing macronuclei immediately preceding and concurrent with DNA rearrangement (Taverna et al. 2002). Chromatin immunoprecipitation experi-

ments have shown that this modification is enriched on the histones associated with IESs. DNA elimination and heterochromatin formation were originally linked by the identification of the chromodomain-containing Programmed DNA Degradation 1 (Pdd1) protein, an abundant, developmentally expressed protein that colocalizes within foci containing germ-line-limited DNA to be eliminated (i.e., IESs) from the somatic genome (Madireddi et al. 1994, 1996). Chromodomains are protein motifs that have binding affinity to certain methylated histone residues. Perhaps the archetypal model of chromodomain protein involvement in chromatin regulation is the binding of heterochromatin protein 1 (HP1—chromodomain containing) to methylated H3K9 in *Drosophila*, involved in the formation of heterochromatin domains (for more details, see Chapter 5). The chromodomains of Pdd1 and Pdd3, two proteins required for DNA rearrangement (Coyne et al. 1999; Nikiforov et al. 2000), bind to H3K9me2 peptides (Taverna et al. 2002). To demonstrate that this chromatin modification is required for DNA rearrangement, Liu, Mochizuki, and Gorovsky used homologous gene replacement to substitute the major histone H3 genes with copies that contain a K9Q lysine 9 substitution mutation (Liu et al. 2004). These cells could not efficiently remove IESs during development despite the fact that Pdd1 localized appropriately within the precursors of the macronuclei, thus showing that H3K9me2 is required for DNA elimination. Because the establishment of chromatin states is a key determinant of epigenetic regulation, the important question to answer is, How is the H3K9me2 mark specifically targeted to the DNA segments destined for elimination? As described below, several experiments have demonstrated that homologous RNAs are involved in guiding DNA rearrangements.

8 Homology-dependent Gene Silencing in Ciliates

Homology-dependent, RNA-mediated silencing mechanisms are widely used in eukaryotes for epigenetic regulation. Evidence that such mechanisms are active in ciliates was first observed in *Paramecium* after transformation of the vegetative macronucleus with nonexpressible transgenes that produced phenocopy of Mendelian mutants in the endogenous genes. Similar effects were then reproduced in *Paramecium* and in spirotrichs by feeding cells double-stranded RNA, suggesting the involvement of RNAi pathways. One of these pathways leads to the ultimate in gene silencing, DNA elimination.

8.1 Transgene-induced Silencing

The *Paramecium* macronucleus is easy to transform by microinjection because any introduced DNA fragment can be maintained at a wide range of copy numbers, replicating autonomously without the need for any specific origin. Transformation with high-copy, nonexpressible transgenes can trigger posttranscriptional silencing of endogenous genes that possess sufficient sequence similarity (Ruiz et al. 1998; Galvani and Sperling 2001). Silencing is not observed if the 3′ UTR of the gene is present in the transgene (Galvani and Sperling 2001), which suggests that regulatory signals present in the RNA influence a construct's silencing capacity. Subsequently, silencing was found to correlate with the accumulation of homologous short RNAs approximately 23 nucleotides (nt) in length (Garnier et al. 2004), indicating that an RNAi pathway is involved. The ~23-nt short RNAs appear to be responsible for the targeted degradation of homologous mRNAs and may thus be called siRNAs (short interfering RNAs). A similar class of 23- to 24-nt RNAs has also been identified in vegetative *Tetrahymena* cells, and although there are no data about their possible roles, it is likely that they represent endogenous siRNAs (Lee and Collins 2006).

8.2 Silencing Is Induced by Double-stranded RNA

Double-stranded RNA (dsRNA) is likely the primary trigger for the transgene-induced silencing observed in *Paramecium*. The silencing efficiency of transgenes correlates with the production of aberrant RNA molecules that correspond to both the sense and the antisense strands of the injected sequence. Furthermore, the ability of dsRNA to promote gene silencing was demonstrated by feeding *Paramecium* cells *Escherichia coli* expressing dsRNA of a cloned gene using methodology developed for *Caenorhabditis elegans* (Timmons and Fire 1998; Timmons et al. 2001). Silencing of the endogenous gene can be observed phenotypically after as little as three vegetative divisions; i.e., less than 24 hours (Galvani and Sperling 2002). Feeding heat-killed *E. coli* to spirotrich species that normally feed on algae also promotes gene silencing (Paschka et al. 2003), suggesting that a wide variety of ciliates have this mechanism. In *Paramecium*, molecular analyses showed that feeding dsRNA leads to the accumulation of the same ~23-nt siRNAs as observed upon transgene-induced silencing (Nowacki et al. 2005), indicating that both phenomena rely on a common RNAi pathway.

Silencing induced by dsRNA feeding in *Paramecium* can be reversed immediately by replacing *E. coli* with the normal food bacterium in the culture medium; similarly, direct microinjection of dsRNA into the cytoplasm induces only transient silencing of the homologous genes, presumably because the injected dsRNA is rapidly diluted out during vegetative growth (Galvani and Sperling 2002). These observations suggest that dsRNA molecules cannot be amplified to any significant degree in the cytoplasm of vegetative cells, unlike the apparent fate of dsRNA in *C. elegans* that can lead to a heritable silent state. This further implies that RNAi in *Paramecium* does not lead to the establishment of stable transcriptional gene silencing in the macronucleus. Heritable silencing would likely require histone H3K9 methylation, and as mentioned above, this modification is apparently absent from the vegetative macronucleus, at least in *T. thermophila*.

9 Genome Rearrangements Are Guided by Homology-dependent Mechanisms

During ciliate development, three different genomes must be distinguished and channeled toward disparate fates: The germ-line micronuclear genome must be preserved intact while the developing somatic genome in the new macronucleus is directed to undergo extensive reorganization, and the maternal somatic genome is destined for destruction. Within this broader framework, ciliate researchers have aimed to understand the reproducibility of DNA rearrangement patterns. Initial efforts attempted to identify *cis*-acting DNA sequence motifs that could recruit the recombination proteins, but these searches have had few clear successes (see Yao et al. 2002; Betermier 2004). Primary DNA sequence is clearly not the sole determinant guiding reorganization of the macronuclear genome, a conclusion that is supported by evidence that H3K9 methylation marks DNA segments for elimination. Furthermore, many rearrangement patterns are sensitive to homology-dependent effects, as described below, that allow alternative patterns to be maternally inherited in subsequent sexual generations, independently from the Mendelian transmission of the wild-type germ-line genome. The inheritance of rearrangement patterns therefore satisfies the definition of an epigenetic phenomenon.

9.1 Experimental Induction of Specific Deletions in the Developing Macronucleus

Even before posttranscriptional gene silencing was described in *Paramecium*, the introduction of cloned sequences at high copy number into the vegetative macronucleus was found to alter the DNA rearrangements in the sexual progeny of transformed clones. Strikingly, the sequences homologous to the transgene were specifically deleted by the imprecise mechanism, while the micronuclear genome remained intact (Fig. 6) (Meyer 1992). This phenomenon appeared to be quite general because all tested DNA fragments could produce deletions (Meyer et al. 1997). These experiments indicated that sequence-specific information is transmitted through the cytoplasm during sexual events from the transformed maternal macronucleus to the developing zygotic macronucleus. The generality of the effect did not support interpretations that invoked a role for sequence-specific DNA-binding proteins produced from or titrated by the injected transgenes. The most parsimonious explanation that satisfies the observed specificity assumes that nucleic acids, presumably RNA molecules, are transferred between nuclei and recognize their targets by pairing interactions.

The constructs that efficiently induce postzygotic deletions are nonexpressible transgenes, such as ones that contain frameshift mutations or truncations of 5′ or 3′ UTRs (Garnier et al. 2004). Those that can produce stable, translatable mRNAs do not promote the elimination of the homologous genes from developing macronuclei. The same constructs that promote DNA deletion also caused the silencing of endogenous maternal genes during the vegetative growth of transformed clones. Thus, postzygotic deletions correlate with prezygotic silencing and with the accumulation of ~23-nt siRNAs. The siRNAs were further shown to persist in the cells throughout development, suggesting that they may be responsible for triggering these deletions.

If silencing-associated siRNAs direct DNA rearrangements, then introducing dsRNA prior to or during development should promote the elimination of the homologous sequence within the descending progeny. To test this, *Paramecium* cells fed an *E. coli* strain producing dsRNA homologous to the *ND7* coding sequence, a gene involved in the regulated exocytosis of secretory vesicles called trichocysts, were allowed to undergo autogamy and develop new macronuclei from wild-type micronuclei. A number of postautogamous progeny showed an *ND7* mutant phenotype due to elimination of the gene from their macronuclei (Garnier et al. 2004). Even phenotypically wild-type progeny showed partial elimination of *ND7* gene copies. Like those associated with transgene-induced silencing, the ~23-nt siRNAs associated with dsRNA feeding were shown to persist in the cells throughout autogamy (Nowacki et al. 2005), confirming their implication in the targeting of postzygotic deletions. Induced DNA deletion directed by dsRNA is not restricted to *Paramecium*. Microinjection of conjugating *Tetrahymena* with in vitro-transcribed sense and antisense RNA that is homologous to genomic loci normally retained in the macronucleus resulted in the imprecise deletion of these sequences (Yao et al. 2003).

10 Rearrangement Patterns Are Likely Determined by a Comparison of Germ-line and Somatic Genomes

10.1 The d48 Paradigm: Epigenetic Inheritance of Alternative Rearrangements

Although transgene and dsRNA-induced rearrangements illustrate the epigenetic nature of the DNA elimination process, what may be more striking are observations that induced rearrangement patterns can be inherited through subsequent rounds of macronuclear development. The first evidence for epigenetic regulation of rearrangements was uncovered when an aberrant deletion from the

a. Wild type autogamy

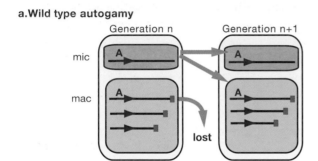

b. Transgene-induced deletion upon autogamy

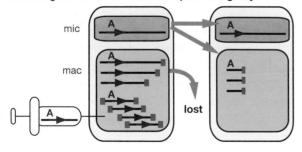

Figure 6. Transgene-induced Deletions in *Paramecium*

(*a*) In the wild type, the *A* gene (*red arrows*) sits near the heterogeneous ends of a macronuclear chromosome; the pink boxes = macronuclear telomeres. The rearrangement pattern of the micronuclear chromosome is faithfully reproduced from one generation to the next. (*b*) The introduction of a large copy number of *A* transgenes into the maternal macronucleus can induce the complete deletion of the endogenous *A* gene in sexual progeny, when a new macronucleus develops from the wild-type germ line. The new macronuclear telomeres are positioned just upstream of the *A* gene.

macronucleus of the gene encoding the *A* surface antigen of *P. tetraurelia* was found to be cytoplasmically inherited in crosses with the wild type (Fig. 7) (Epstein and Forney 1984). In wild-type macronuclei, the *A* gene is located near a chromosome end either 8, 13, or 26 kb away from the telomere, dictated by three alternative fragmentation sites. A variant cell line called d48 was found to lack *A*-gene expression because the gene itself was lost along with all downstream sequences as its telomere formed at the 5′ end of the gene (Forney and Blackburn 1988). Nuclear transplantation experiments confirmed that the d48 germ-line micronucleus carried the wild-type *A* gene. For example, replacement of the d48 micronucleus with one from a wild-type strain did not prevent maternal trans-

mission of the *A*-gene deletion to sexual progeny; and conversely, the d48 micronucleus, when transplanted into a wild-type cell, gave rise after autogamy to a new macronucleus that contained the *A* gene (Harumoto 1986; Kobayashi and Koizumi 1990). Similar experiments that focused on the maintenance or deletion of another telomere-proximal surface antigen gene, the *B* gene, showed that such maternal effects can be observed in other genomic regions (Scott et al. 1994). Together, these studies strongly suggest that a genomic region must be present in the maternal macronucleus to be effectively retained and amplified to a wild-type copy number in the developing macronucleus.

10.2 Epigenetic Inheritance of Experimentally Induced Deletions

Maternal influence on DNA rearrangements as observed for d48 appears to control the development of many, and possibly all, regions of the macronuclear genome of *Paramecium*. Indeed, the macronuclear deletions that are created experimentally in sexual progeny can be "spontaneously" reproduced in further sexual generations, following a maternal pattern of inheritance. This has been observed both for high-copy transgene-induced deletions of the *G* gene in *P. primaurelia*, another subtelomeric surface antigen gene (Meyer 1992), and for the macronuclear deletions of the *ND7* gene that were initially induced by dsRNA feeding (Garnier et al. 2004). In either case, the inducing transgene or introduced dsRNA was no longer needed to propagate the deletion to sexual progeny. In d48 and in these induced variant cell lines, the state of the somatic genome can occasionally revert to the wild-type rearrangement pattern after autogamy, confirming that the gene is still present in the micronuclear genome and highlighting the epigenetic mode of inheritance.

It is remarkable that gene silencing and the resulting developmental DNA deletion can be "remembered" in subsequent generations with the targeted genomic region treated like transposons and IESs during macronuclear development. The recurrent deletion of a gene in each sexual generation does not appear to be induced by ~23-nt siRNAs, since these have only been detected when silencing is experimentally induced in the first generation. Furthermore, genetic analyses of such cell lines have shown that the micronuclear gene does not carry any permanent imprint, since it can be normally amplified during macronuclear development when it is transferred by conjugation into a cell line with a wild-type macronucleus. It therefore appears that the

a. Wild type

b. d48

c. Rescued d48

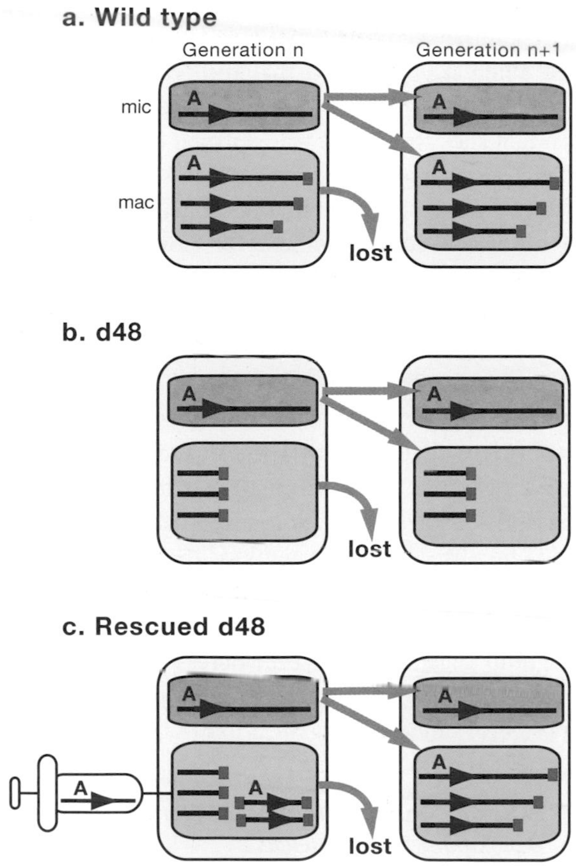

Figure 7. Epigenetic Inheritance and Experimental Rescue of Macronuclear *A*-gene Deletions

(*a*) Wild-type strain. (*b*) The d48 strain lacks the *A* gene in its macronucleus, but has a wild-type micronucleus. The *A* gene is reproducibly deleted during macronuclear development in each generation. (*c*) Transformation of the macronucleus of the d48 strain with *A*-gene sequences will specifically restore amplification of the germ-line *A* gene in the developing macronucleus of sexual progeny.

gene is deleted in the developing macronucleus simply because it is absent from the maternal macronucleus.

10.3 "Spontaneous" Elimination of Foreign Sequences Introduced into the Micronuclear Genome

The propagation of maternal rearrangement states suggests that the germ-line genome to be rearranged is compared to the existing rearranged genome, and the sequences that were absent in the previous generation are targeted for elimination from the newly forming somatic genome. If this were indeed the case, transgenes introduced into the germ-line micronucleus would be predicted to be frequently deleted during new macronuclear development. In *Tetrahymena*, where micronuclear transformation has been achieved, researchers have observed that integrated drug resistance markers used for gene disruption studies can be deleted from the macronuclear genome during successive rounds of conjugation (Yao et al. 2003; Liu et al. 2005). The genomic location of the transgene, as well as the number of copies introduced into the germ-line genome, significantly altered the efficiency of elimination (Liu et al. 2005), which indicates that the phenomenon in this ciliate is less generally induced. Nevertheless, these results demonstrate that a foreign sequence, the bacterial *neo* gene, could be recognized by the cell as an IES.

10.4 Experimental Rescue of Inherited Macronuclear Deletions

Is the mere absence of a genomic region from the maternal somatic macronucleus sufficient to direct its future elimination? If so, then reintroducing the *A* gene into the d48 macronucleus should rescue the defect in *A*-gene propagation during development. It was first shown that injection of either wild-type macronucleoplasm into the d48 vegetative macronucleus (Harumoto 1986), or cytoplasm from autogamous, wild-type cells into d48 cells early in development (Koizumi and Kobayashi 1989), resulted in a permanent reversion of d48 to wild type. Subsequently, direct microinjection of several nonoverlapping *A*-gene fragments, spread over most of the ~8-kb coding sequence, into the d48 macronucleus demonstrated that the *A*-gene sequence itself was sufficient to restore the wild-type macronuclear rearrangement pattern (Fig. 7) (Koizumi and Kobayashi 1989; Jessop-Murray et al. 1991; You et al. 1991).

The maternal genome's influence on DNA rearrangements shows marked sequence specificity. The *A* and *B* gene coding sequences are 74% identical overall. Nevertheless, injection of the *A*-gene sequences into the

macronucleus of a cell line that carried macronuclear deletions of both genes could only prevent *A*-gene deletion from the new macronucleus; and similarly, injection with the *B* gene could only prevent its own deletion (Scott et al. 1994). In addition, the d48 macronuclear deletion could not be rescued by transformation with the *G* gene from *P. primaurelia*, which shares 78% identity with the *A* gene. On the other hand, the macronuclear deletion of the *A* gene could be rescued by transformation with a different allele of the *A* gene, showing 97% identity (Forney et al. 1996). Thus, the maternal rescue of macronuclear deletions is a homology-dependent process that does not require any specific sequence within the genes, but requires a minimum level of sequence identity.

The d48 maternal effect rescue, at first glance, appears to be at odds with the transgene-induced deletions, since in these experiments injection of *A*-gene sequences into the maternal macronucleus had exactly opposite consequences on the zygotic *A* gene. This apparent paradox was solved when it was shown that the postzygotic deletion effect depends on the establishment of homology-dependent silencing in the transformed clones (Garnier et al. 2004). Conversely, the rescue effect is observed only in transformation conditions that do not elicit silencing (i.e., only moderate copy numbers for nonexpressible constructs, or any copy number for expressible transgenes) (Garnier et al. 2004). Thus, it appears that the accumulation of the ~23-nt siRNAs can prevent the rescue effect of *A*-gene sequences in the maternal macronucleus, which promote *A*-gene amplification in the developing macronucleus. This strongly suggests that the cytoplasmic factor mediating the *trans*-nuclear effect is an *A*-gene transcript. However, it is not necessarily the full-length mRNA, because even fragments of the coding sequence were shown to have rescue activity. Furthermore, clones containing the entire gene often express the mRNA throughout vegetative growth, whereas production of the rescuing cytoplasmic factor was shown to be restricted to the period of nuclear reorganization.

10.5 Experimental Inhibition of IES Elimination in the Developing Macronucleus

HOMOLOGY-DEPENDENT INHIBITION OF IES ELIMINATION IN PARAMECIUM

If the deletion or maintenance of cellular genes is controlled by a comparison of somatic and germ-line genome content, then even the normally efficient excision of IESs could perhaps be perturbed when copies are present in the maternal macronucleus. Examination of a Mendelian mutation, mtF^E, which has pleiotropic effects on macronuclear development, including an effect on mating-type determination (Brygoo and Keller 1981a,b), provided the opportunity to test this prediction. This mutation, when homozygous, was found to abolish the excision of an IES located within the *G* surface antigen gene. Surprisingly, when the wild-type allele of the *mtF* gene was reintroduced into the mutant strain by conjugation, the IES was still not excised during the subsequent macronuclear development (Meyer and Keller 1996). The genetic analysis of the resulting variant strain, which was called the IES⁺ strain, confirmed that it is genetically wild type and that the specific retention of this IES is maternally inherited in sexual progeny (Duharcourt et al. 1995).

Is the excision of the IES within developing macronuclei induced by maternal copies of the correctly rearranged *G* gene or inhibited by maternal copies of the IES-retaining *G* gene? To answer this question, the macronuclei of IES⁺ or IES⁻ cells were transformed by direct microinjection of plasmids containing a fragment of the *G* coding sequence in either its micronuclear (IES⁺) or macronuclear (IES⁻) versions (Fig. 8) (Duharcourt et al. 1995). After autogamy, transformed clones that contained the IES⁺ plasmid produced progeny lines that retained the IES in their newly formed macronuclei, while IES⁺ cells transformed with IES⁻ plasmid proved unable to induce excision, and their progeny remained in the IES⁺ state. A plasmid containing only the IES, without any flanking sequences, also caused IES retention, showing that the maternal IES copies alone inhibit excision from zygotic macronuclei (Duharcourt et al. 1995).

"MATERNALLY CONTROLLED" VERSUS "NON MATERNALLY CONTROLLED" IESS IN PARAMECIUM

Can the excision of other IESs be controlled by similar maternal effects? This question was addressed by transforming the macronuclei of wild type cells with large seg-

ments of micronuclear DNA (IES⁺) containing either the *G* or *A* surface antigen genes (Duharcourt et al. 1998). The injected segments contained 6 and 9 IESs, respectively. Excision of 13 of these IESs was examined, and 5 IESs were found to be retained in the macronuclei of postautogamous progeny of the transformed cells. The injection of plasmids containing single IESs showed that inhibition was strictly specific: Each of these 5 IESs induced the retention of only the homologous zygotic IES, but did not affect any other. A control DNA fragment containing most of the macronuclear (IES⁻) *G* gene had no effect on any of the IESs tested. Among the 13 *Paramecium* IESs tested, there is no obvious difference in size, base composition, or position within the genes between the 5 that show the maternal effect and the 8 that do not (Duharcourt et al. 1998).

HOMOLOGY-DEPENDENT INHIBITION OF IES ELIMINATION IN *TETRAHYMENA*

Experiments analogous to those performed in *Paramecium* revealed that the DNA content of the *Tetrahymena* parental macronucleus can regulate the elimination of the homologous sequences from the developing somatic genome. Two well-characterized IESs, the M and R deletion elements, were microinjected into the macronuclei of wild-type strains such that they were maintained on high-copy vectors. When these cells were induced to conjugate, the progeny of the cells containing maternal copies of the M element failed to efficiently eliminate the corresponding IESs during macronuclear development (Chalker and Yao 1996). Likewise, the cells whose parental macronuclei contained copies of the R element failed to excise the homologous IES. Significant inhibition of excision of nonhomologous elements was not observed. Thus, the inhibition of DNA elimination was sequence-specific. Sequences homologous to the IES itself were sufficient for this inhibition, and the immediately flanking DNA had no effect. Importantly, this

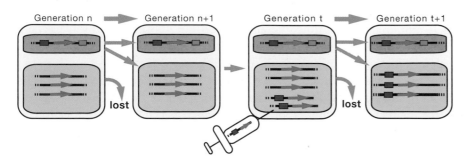

Figure 8. Homology-dependent Inhibition of IES Excision by the Maternal Macronucleus

During normal development, IESs (*red* and *green* bars) are excised efficiently. However, transformation of the maternal macronucleus with a given IES (initial transformants = Generation t) can inhibit the elimination of the homologous IES during the subsequent (Generation t + 1) (and future) rounds of new macronuclear differentiation.

induced failure of DNA elimination was heritable, as subsequent generations also retained genomic copies of the IES in their macronuclei.

Because *Tetrahymena* conjugating pairs extensively share cytoplasm during development, researchers were able to observe that the inhibition was transmitted through the cytoplasm such that only one mating partner need carry the IES in its parental somatic genome for the homologous sequences in all developing nuclei within a pair, including those in a wild-type partner, to be affected. Therefore, the DNA rearrangement state of the parental nuclei is transmitted through the cytoplasm to regulate the events that occur during the formation of the new somatic genome of the next generation. No exchange of germ-line nuclei is required to transmit the state of one mating partner to another, an observation that rules out that transmission occurs by imposing an imprint on the germ line during normal cell growth prior to entering development (Chalker et al. 2005). In fact, by physically separating mating pairs consisting of one wild-type cell and a partner containing copies of the M element in its maternal somatic nucleus, it was shown that transmission occurs after meiosis and very near the time that developing nuclei first begin differentiating into the new macronuclei and micronuclei. Therefore, the influence of the maternal somatic genome is actively established during development.

BIOLOGICAL IMPLICATIONS OF MATERNAL CONTROL

The ability of ciliates to alter DNA rearrangements to reflect a maternal pattern provides cells with a simple way to transmit alternative somatic versions of the genome to sexual progeny. This dynamic regulation implies that a stable equilibrium between two alternative genetic states, e.g., IES$^+$ (100% excision) and IES$^-$ (0% excision), can be reached and maintained over the course of many sexual generations. This has been shown for at least one *Paramecium* IES for which the maternal influence is sufficient to drive a smaller fraction of macronuclear IES$^+$ copies to a greater fraction in the macronucleus of the following generation (Duharcourt et al. 1995). Such influence parallels the stable, maternal inheritance of *O* and *E* mating types in *P. tetraurelia*. Comparable asymmetry is found in both systems, because both IES excision and *O*-type determination appear to be the default developmental pathways, whereas both IES retention and *E-type* determination require a cytoplasmic signal from the maternal macronucleus to alter their pathways. These systems may very well be related, because the Mendelian *mtF*^{E}

mutation, which impairs excision of one *G* gene IES, regardless of the parental state, similarly makes determination for *E* constitutive, an observation that links the alternative states of both examples. Elucidating the mechanism(s) that underlies these phenomena will undoubtedly reveal novel modes for maternal inheritance of epigenetic information.

11 A *trans*-Nuclear Comparison of Whole Genomes Mediated by RNA Interference

The homology-dependent effects described above demonstrate that a cross talk occurs between the maternal somatic and germ-line genomes during nuclear differentiation, which can profoundly alter DNA rearrangement patterns. The observation that only highly homologous sequences are affected suggests strongly that this cross talk is mediated by nucleic acids. This regulatory mechanism could possibly involve transcripts from the maternal macronucleus that are exported to the developing macronucleus where, if they contain an IES, they would prevent the elimination of the homologous sequences. However, it was shown that putative maternal transcripts do not participate as donor templates during the repair of double-stranded breaks induced by constitutive IES excision (Duharcourt et al. 1995). If protective maternal transcripts exist, the ~23-nt siRNAs that dictate gene silencing may target DNA deletion indirectly by promoting the destruction of those with sequence identity. This role for siRNAs alone cannot explain the maternal inheritance of rearrangement patterns in subsequent generations. As discussed below, it is likely the interplay between maternal macronuclear transcripts and a novel class of short RNAs derived from the meiotic micronucleus that ultimately directs genome reorganization.

11.1 Linking Short RNAs to DNA Elimination

In *Tetrahymena*, the canonical heterochromatin modification, H3K9 methylation, marks the chromatin associated with IESs just prior to their excision. How is this modification specifically targeted to these DNA segments destined for elimination? An RNAi-like pathway has been described that employs RNA guides to direct it to the proper loci (for review, see Mochizuki and Gorovsky 2004; Yao and Chao 2005). An initial indication that DNA deletion utilizes homologous RNA molecules was the observation that *Tetrahymena* IESs are bidirectionally transcribed early during conjugation (Chalker and Yao 2001). A real breakthrough came when Mochizuki et. al.

(2002) identified a PIWI/Argonaute family protein encoded by *TWI1* that is developmentally expressed and was required for DNA rearrangement (Mochizuki et al. 2002). Because Argonaute proteins are key players in RNAi-triggered processes, these researchers looked for and found a species of endogenous small (~28 nt) RNAs that are preferentially complementary to germ-line-limited sequences. Disruption of the *TWI1* gene destabilized these short RNAs and abolished H3K9 methylation in the developing somatic nucleus. This work, published in 2002, together with experiments in *Schizosaccharomyces pombe* (see Chapters 6 and 8), established a new paradigm that heterochromatin is generated by the action of short homologous RNAs targeting the H3K9me2 silencing mark to specific loci.

Characterization of the gene, *DCL1*, encoding the Dicer ribonuclease that processes the bidirectional transcripts into the ~28-nt small RNAs, has provided additional insight into the overall regulation of this process (Malone et al. 2005; Mochizuki and Gorovsky 2005).

The protein is expressed at high levels early in conjugation and localizes to the premeiotic micronuclei, which indicates that generation of the small RNAs is temporally and spatially compartmentalized within this germ-line nucleus. Disruption of the *DCL1* gene caused loss of small RNA production, accumulation of germ-line transcripts, and ultimately, failure of IES excision. Intriguingly, loss of the small RNAs did not abolish H3K9 methylation as was observed in the *TWI1* knockout lines. Nevertheless, chromatin immunoprecipitation analysis showed that this modification is no longer enriched on IESs; thus, the small RNAs are required to target this chromatin modification to the proper loci (Malone et al. 2005).

To explain the role of the maternal genome in regulating these events, the scan RNA model was proposed (Mochizuki et al. 2002). In the variation of this model shown in Figure 9, the 28-nt "scan" (scn)RNAs, which are generated in the micronucleus, assemble with a Twi1 containing RISC-like complex in the cytoplasm and are ini-

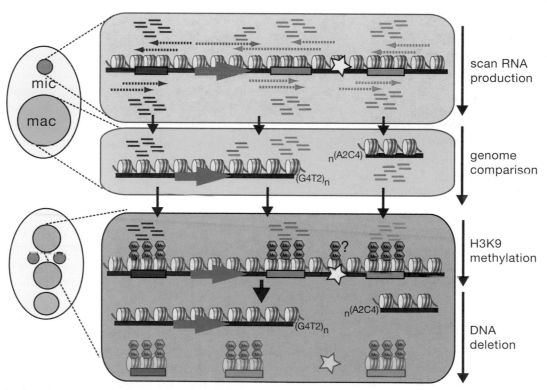

Figure 9. The Scan RNA Model for Control of DNA Deletion

Bidirectional transcription of a large portion of the germ-line genome occurs early in development and leads to the production of scnRNAs. These are then transported into the maternal macronucleus where any encounter with a homologous sequence will trigger their removal from the active pool. The remaining, micronucleus-specific RNAs are redirected to the developing macronucleus, where they target H3K9 methylation to homologous sequences, signaling their removal from the genome. Model adapted from Mochizuki et al. (2002).

tially channeled to the maternal macronucleus. There they scan the existing rearranged genome for homology. scnRNAs that pair with maternal sequences are removed from the pool of active complexes. The remaining Twi1-associated scnRNAs are then transported to the developing macronucleus where they target H3K9 methylation to the homologous sequences, marking them for excision by the DNA rearrangement machinery. This model is further supported by the observation that Twi1 localizes in the maternal macronucleus early in development after the bulk of small RNA production, but before the appearance of the new macronuclear precursors. Thus, the regulated trafficking of the small RNA protein complexes facilitates the comparison of somatic and germ-line genomes.

11.2 Transporting RNA from Maternal to Zygotic Macronuclei in Paramecium

The genome-wide comparison of germ-line and somatic sequences would require a highly sophisticated machinery, both to ensure the massive transport of RNA molecules between nuclei and to effect the very large number of pairing interactions implied by the scanning model. The novel nucleic-acid-binding protein Nowa1 appears to participate in this *trans*-nuclear cross talk in *Paramecium* (Nowacki et al. 2005). Nowa1 is synthesized shortly before meiosis and first accumulates in the maternal macronucleus and, like the *Tetrahymena* Twi1, relocalizes later to the developing zygotic macronucleus. Tagging Nowa1 with a photoactivatable GFP allowed researchers to conclusively demonstrate that this protein is transported from one nucleus type to the other. One domain of the protein can bind RNA, and a second domain is necessary and sufficient for internuclear transport; thus Nowa1 may be an RNA transporter.

Nowa1 is essential for the development of a viable new macronucleus, including the elimination of germ-line transposons and of a subset of IESs. Strikingly, only those IESs that are subject to maternal control are affected by Nowa1 knockdown, suggesting that the protein is involved in *trans*-nuclear genome comparison. IESs that are not sensitive to the presence of homologous sequences in the maternal macronucleus do not depend on Nowa1 for their excision, confirming the existence of mechanistically distinct classes of IESs in *Paramecium*. Although the nucleic acids bound by Nowa1 in vivo are not yet known, the effects of Nowa1 depletion are consistent with the scanning model proposed (Fig. 9) and suggest that the protein may carry RNAs that have been selected to target the elimination of maternally controlled IESs and transposons in the developing macronucleus.

12 Conclusions: DNA Elimination as a Genome Defense Mechanism

The study of RNAi, particularly in plants and nematodes, has led to the hypothesis that this pathway evolved as a defense mechanism that allows cells to control the proliferation of viruses and transposons by degrading mRNAs and targeting the formation of heterochromatin on these genomic parasites (Matzke and Birchler 2005). As already mentioned, transposable elements present in the germ-line genome of ciliates are eliminated during development of the somatic macronucleus, which will effectively negate their impact. The observation in *Tetrahymena* that H3K9 methylation marks genomic regions for developmental DNA deletion suggests that the use of RNAi in ciliates is fundamentally similar to its role in establishing heterochromatin in other eukaryotes; ciliates just go one step farther and eliminate heterochromatin from their somatic genome. Nevertheless, one original contribution of ciliate studies is the idea that the specific sequences that are targeted by RNAi for heterochromatin formation/elimination during early development are selected by a global comparison of maternal germ-line and somatic genomes, starting during meiosis. This mechanism would efficiently protect against the deleterious effects of transposition in the germ line: Any new transposon integrating into the maternal germ line would be recognized as alien by comparison with the somatic genome during sexual reproduction, leading to its removal from the transcribed somatic genome of progeny, thereby limiting its future spread.

The *Tetrahymena* scnRNAs that target DNA elimination likely mediate the *trans*-nuclear cross talk between the germ-line and somatic genomes. It is unclear whether the whole micronuclear genome produces scnRNAs, but unpublished evidence obtained in *Paramecium* suggests that at least a large fraction of it does. Northern blot analyses have revealed meiosis-specific, endogenous short RNAs that correspond to cellular genes, as well as to transposons and IESs (M. Nowacki et al., unpubl.). These short RNAs are ~25 nt in length and are clearly distinct from the ~23-nt siRNAs. They appear to be functionally equivalent to the *Tetrahymena* scnRNAs, because the inactivation of specific *Paramecium* Dicer-like genes suppresses their production during meiosis and abolishes the

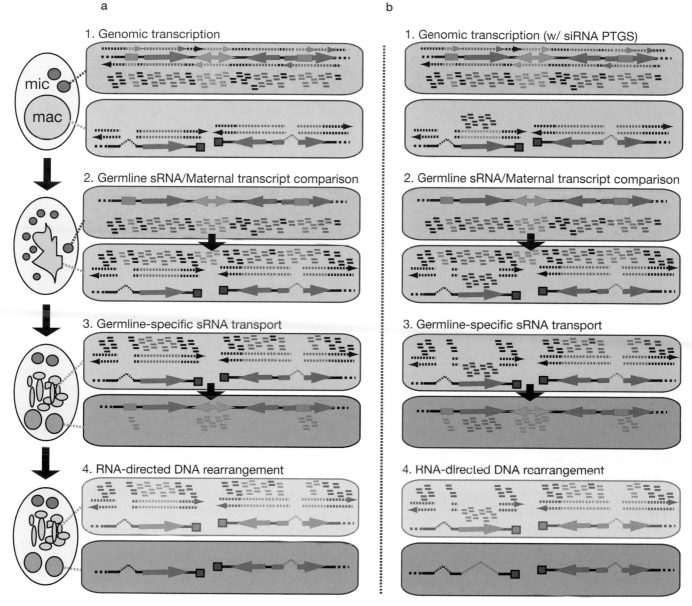

Figure 10. The Transcriptome Scanning Model for the Comparison of Germ-line and Somatic Genomes

(*a*) Default regulation. (*1*) Upon initiating meiosis, most or all of the macronuclear and micronuclear genomes is transcribed; the dashed lines in the macronucleus represent uncharacterized transcripts. In the micronucleus, transcription is bidirectional, resulting in the production of scnRNAs (short double-stranded molecules) for all types of sequences (cellular genes, *light purple arrows*; transposons, *orange double arrow*; IES, *green boxes*). (*2*) The scnRNAs are exported to the maternal macronucleus, where they may pair with homologous somatic transcripts. Pairing may also occur in the cytoplasm. (*3*) scnRNAs that pair with homologous transcripts are sequestered or destroyed, while the micronucleus-specific ones are re-exported to the developing zygotic macronucleus, where they pair with homologous sequences (DNA or nascent transcripts), thus targeting H3K9 methylation to micronuclear-specific sequences (transposons and IESs). (*4*) The marked sequences are eliminated. (*b*) Effects of posttranscriptional silencing of a gene in the maternal macronucleus. Experimental induction of posttranscriptional silencing by high-copy transgenes or dsRNA results in the production of double-stranded siRNAs homologous to that gene (in *red*). These siRNAs degrade the homologous maternal somatic transcripts, so that homologous scnRNAs will not be inactivated and will be free to target the deletion of the gene in the developing zygotic macronucleus.

elimination of transposons and maternally controlled IESs, as well as the maternal inheritance of macronuclear gene deletions (V. Serrano et al., unpubl.).

The original genome-scanning model proposed that scnRNAs are compared to the rearranged DNA in the maternal macronucleus, inactivating those matching macronuclear sequences and selecting for the micronucleus-specific pool (Fig. 10a). However, several lines of evidence suggest that scnRNAs may be compared with macronucleus-derived transcripts rather than with DNA itself. This would dispense with the need to open the DNA duplex along the entire genome to test for complementarity. IESs introduced into the maternal macronucleus of *Tetrahymena* that interfere with DNA deletion are transcribed, allowing for this possibility (Chalker and Yao 2001; Chalker et al. 2005). Transcripts arising from an IES in the maternal macronucleus would protect the zygotic IES from elimination, as initially postulated to explain IES inhibition in *Paramecium*, by titrating the homologous scnRNAs. The evidence for a cytoplasmic factor that can rescue the deletion of the *A* gene in the d48 strain also supports the existence of protective transcripts, and further suggests that scnRNA selection could occur in the cytoplasm as well as in the maternal macronucleus. Finally, this would explain how the posttranscriptional silencing of a given gene in *Paramecium* can cause its deletion in the next generation: If maternal macronuclear transcripts are degraded by the 23-nt siRNAs, homologous scnRNAs will be free to target deletions in the new macronucleus despite the presence of the gene in the maternal macronucleus.

One interesting aspect of this modified hypothesis, which may be called the transcriptome-scanning model, is that all pairing interactions may occur between RNA molecules. If scnRNAs finally target heterochromatin formation by interacting with nascent transcripts at the homologous locus, as is thought to occur in *S. pombe* and plants, these later pairing interactions need not be fundamentally different from those involved in the selection step. scnRNAs may simply pair with any available RNA after they leave the meiotic micronucleus. Pairing with the abundant macronuclear transcripts present in the cell early in development would efficiently remove the homologous scnRNAs from the pool by sequestering or destroying them. By the time the zygotic macronucleus forms and starts transcribing its unrearranged genome, only micronucleus-specific scnRNAs would be left to pair with nascent transcripts. These pairing interactions may lead to DNA deletion because they occur in the developing macronucleus, or because they occur at the correct developmental stage.

In this model, the recognition of self (macronuclear sequences) versus non-self (genomic parasites in the germ line) is achieved by a simple developmental switch that alters the outcome of similar pairing interactions. This is conceptually similar to the process by which the vertebrate immune system learns the distinction between self and non-self. A huge repertoire of lymphocytes expressing different antibodies is initially generated, but in early development all that recognize available antigens (likely self-antigens) are eliminated from the future pool. Once past this stage, the recognition of a cognate antigen (then likely to be alien) leads to the clonal expansion of the corresponding lymphocytes. The ciliate genomic immune system that utilizes RNAi thus has striking parallels to the cellular immune system of vertebrates.

13 Future Contributions of Ciliate Research to Epigenetics

The recent sequencing of the macronuclear genomes of *Tetrahymena* and *Paramecium* will sustain the utility of these facile unicellular models for investigation of epigenetic mechanisms. The biology of nuclear dimorphism has facilitated the discovery of chromatin regulators and will continue to provide an advantageous system for novel findings. With two distinct RNAi pathways, one used for posttranscriptional gene silencing and another for epigenetic modification of the genome, ciliates are also uniquely poised to unravel the complexity of RNA-guided, homology-dependent regulatory processes involved in development. Dicer-related ribonucleases, Argonaute-like proteins, and their partners have been or will be identified, and current studies promise to decipher their functional specialization. Given the phylogenetic position of ciliates on the tree of life, such information will undoubtedly shed light on the evolution of these mechanisms in eukaryotes.

It is the investigation of multiple epigenetic phenomena that led to an appreciation of the widespread use of RNA-mediated regulation in eukaryotes. Yet, even prior to this understanding, the observation that the DNA content of one nucleus imparts effects on homologous sequences in another forced ciliate researchers to postulate the existence of *trans*-acting RNA molecules directing homology-dependent cross talk. The discovery of RNAi and its role in DNA rearrangements has validated this speculation. What future insight of general relevance might be gained from uncovering the mechanisms that mediate *trans*-nuclear cross talk?

If the "transcriptome scanning" hypothesis (Fig. 10) is correct, maternal somatic transcripts protect homologous zygotic DNA sequences from elimination. If one equates

DNA elimination with heterochromatin formation, then the maternal transcripts would, by blocking elimination, enforce euchromatin-specific modifications on homologous sequences. The proposed positive role of these transcripts on homologous genes is thus indirect, arising from their capacity to inactivate homologous scnRNAs that would otherwise target the formation of heterochromatin. Nevertheless, this effect of maternal transcripts, which some experiments suggest are not necessarily protein-coding mRNAs, would represent a novel mechanism given that all known RNA-mediated, homology-dependent mechanisms lead to the down-regulation of the target gene. The degradation or sequestration of short RNAs by long transcripts is, in essence, the reverse of the demonstrated RNAi-based mechanisms, whereby short RNAs inactivate long ones.

If long and short RNAs can antagonize each other's action, their interaction must steer the system toward one of two alternative states, depending on their relative abundance. This suggests that the inheritance of the expression status of genes could depend on a constant feedback from an RNA pool, and not exclusively on semiconservative chromatin replication mechanisms. It is tempting to believe that such an RNA-mediated, *trans*-acting mechanism is responsible for the stability of surface antigen gene expression during vegetative growth in *Paramecium*, because this would also explain how, after sexual reproduction, the developing macronucleus can inherit the maternal serotype via the cytoplasm. Thus, the epigenetic inheritance of genome rearrangement patterns may be only one particular aspect of a comprehensive RNA-based inheritance system in ciliates.

As future ciliate research illuminates the players (RNAs and proteins) mediating this regulation, its existence in other organisms and its further connection to known epigenetic mechanisms should become apparent. It is not implausible that RNA-based inheritance, in its basic form, occurs widely. Recent studies of the transcriptome output of multicellular eukaryotes has revealed an unexpected abundance and complexity of noncoding transcripts (Mattick 2004; Meyers et al. 2004; Suzuki and Hayashizaki 2004; Cheng et al. 2005; Claverie 2005), as well as the frequent occurrence of short RNAs matching all types of genomic regions (Lu et al. 2005). Many homology-dependent effects, including meiotic silencing in fungi and paramutation in plants (see Chapters 6 and 9), remain incompletely understood. It will be the combined insight provided by future experiments in all eukaryotes, including ciliates, that will expose the full scope of epigenetic processes.

References

Allis C.D. and Gorovsky M.A. 1981. Histone phosphorylation in macro- and micronuclei of *Tetrahymena thermophila*. *Biochemistry* **20**: 3828–3833.

Allis C.D., Glover C.V., Bowen J.K., and Gorovsky M.A. 1980. Histone variants specific to the transcriptionally active, amitotically dividing macronucleus of the unicellular eucaryote, *Tetrahymena thermophila*. *Cell* **20**: 609–617.

Beisson J. and Sonneborn T.M. 1965. Cytoplasmic inheritance of the organization of the cell cortex in *Paramecium aurelia*. *Proc. Natl. Acad. Sci.* **53**: 275–282.

Beisson J. and Wright M. 2003. Basal body/centriole assembly and continuity. *Curr. Opin. Cell Biol.* **15**: 96–104.

Betermier M. 2004. Large-scale genome remodelling by the developmentally programmed elimination of germ line sequences in the ciliate *Paramecium*. *Res. Microbiol.* **155**: 399–408.

Betermier M., Duharcourt S., Seitz H., and Meyer E. 2000. Timing of developmentally programmed excision and circularization of *Paramecium* internal eliminated sequences. *Mol. Cell. Biol.* **20**: 1553–1561.

Brownell J.E. and Allis C.D. 1995. An activity gel assay detects a single, catalytically active histone acetyltransferase subunit in *Tetrahymena* macronuclei. *Proc. Natl. Acad. Sci.A* **92**: 6364–6368.

Brownell J.E., Zhou J., Ranalli T., Kobayashi R., Edmondson D.G., Roth S.Y., and Allis C.D. 1996. *Tetrahymena* histone acetyltransferase: A homolog to yeast Gcn5p linking histone acetylation to gene activation. *Cell* **84**: 843–851.

Brygoo Y. and Keller A.-M. 1981a. A mutation with pleitropic effects on macronuclearly differentiated functions in *Paramecium tetraurelia*. *Dev. Genet.* **2**: 23–34.

———. 1981b. Genetic analysis of mating type differentiation in *Paramecium tetraurelia*. III. A mutation restricted to mating type *E* and affecting the determination of mating type. *Dev. Genet.* **2**: 13–22.

Cassidy-Hanley D., Bisharyan Y., Fridman V., Gerber J., Lin C., Orias E., Orias J.D., Ryder H., Vong L., and Hamilton E.P. 2005. Genome-wide characterization of *Tetrahymena thermophila* chromosome breakage sites. II. Physical and genetic mapping. *Genetics* **170**: 1623–1631.

Chalker D.L. and Yao M.C. 1996. Non-Mendelian, heritable blocks to DNA rearrangement are induced by loading the somatic nucleus of *Tetrahymena thermophila* with germ line-limited DNA. *Mol. Cell. Biol.* **16**: 3658–3667.

———. 2001. Nongenic, bidirectional transcription precedes and may promote developmental DNA deletion in *Tetrahymena thermophila*. *Genes Dev.* **15**: 1287–1298.

Chalker D.L., Fuller P., and Yao M.C. 2005. Communication between parental and developing genomes during *Tetrahymena* nuclear differentiation is likely mediated by homologous RNAs. *Genetics* **169**: 149–160.

Cheng J., Kapranov P., Drenkow J., Dike S., Brubaker S., Patel S., Long J., Stern D., Tammana H., Helt G., et al. 2005. Transcriptional maps of 10 human chromosomes at 5-nucleotide resolution. *Science* **308**: 1149–1154.

Claverie J.M. 2005. Fewer genes, more noncoding RNA. *Science* **309**: 1529–1530.

Cleveland D.W., Mao, Y., and Sullivan K.F. 2003. Centromeres and kinetochores: From epigenetics to mitotic checkpoint signaling. *Cell* **112**: 407–421.

Coyne R.S., Chalker D.L., and Yao M.C. 1996. Genome downsizing during ciliate development: Nuclear division of labor through chromosome restructuring. *Annu. Rev. Genet.* **30**: 557–578.

Coyne R.S., Nikiforov M.A., Smothers J.F., Allis C.D., and Yao M.C. 1999. Parental expression of the chromodomain protein Pdd1p is required for completion of programmed DNA elimination and nuclear differentiation. *Mol. Cell* **4:** 865–872.

Davis M.C., Ward J.G., Herrick G., and Allis C.D. 1992. Programmed nuclear death: Apoptotic-like degradation of specific nuclei in conjugating *Tetrahymena*. *Dev. Biol.* **154:** 419–432.

Dou Y. and Gorovsky M.A. 2000. Phosphorylation of linker histone H1 regulates gene expression in vivo by creating a charge patch. *Mol. Cell* **6:** 225–231.

———. 2002. Regulation of transcription by H1 phosphorylation in *Tetrahymena* is position independent and requires clustered sites. *Proc. Natl. Acad. Sci.* **99:** 6142–6146.

Dou Y., Mizzen C.A., Abrams M., Allis C.D., and Gorovsky M.A. 1999. Phosphorylation of linker histone H1 regulates gene expression in vivo by mimicking H1 removal. *Mol. Cell* **4:** 641–647.

Duharcourt S., Butler A., and Meyer E. 1995. Epigenetic self-regulation of developmental excision of an internal eliminated sequence on *Paramecium tetraurelia*. *Genes Dev.* **9:** 2065–2077.

Duharcourt S., Keller A.M., and Meyer E. 1998. Homology-dependent maternal inhibition of developmental excision of internal eliminated sequences in *Paramecium tetraurelia*. *Mol. Cell. Biol.* **18:** 7075–7085.

Epstein L.M. and Forney J.D. 1984. Mendelian and non-Mendelian mutations affecting surface antigen expression in *Paramecium tetraurelia*. *Mol. Cell. Biol.* **4:** 1583–1590.

Fan Q. and Yao M.-C. 1996. New telomere formation coupled with site-specific chromosome breakage in *Tetrahymena thermophila*. *Mol. Cell. Biol.* **16:** 1267–1274.

Forney J.D. and Blackburn E.H. 1988. Developmentally controlled telomere addition in wild-type and mutant paramecia. *Mol. Cell. Biol.* **8:** 251–258.

Forney J.D., Yantiri F., and Mikami K. 1996. Developmentally controlled rearrangement of surface protein genes in *Paramecium tetraurelia*. *J. Eukaryot. Microbiol.* **43:** 462–467.

Galvani A. and Sperling L. 2001. Transgene-mediated post-transcriptional gene silencing is inhibited by 3' non-coding sequences in *Paramecium*. *Nucleic Acids Res.* **29:** 4387–4394.

———. 2002. RNA interference by feeding in *Paramecium*. *Trends Genet.* **18:** 11–12.

Garnier O., Serrano V., Duharcourt S., and Meyer E. 2004. RNA-mediated programming of developmental genome rearrangements in *Paramecium tetraurelia*. *Mol. Cell. Biol.* **24:** 7370–7379.

Gratias A. and Betermier M. 2001. Developmentally programmed excision of internal DNA sequences in *Paramecium aurelia*. *Biochimie* **83:** 1009–1022.

———. 2003. Processing of double-strand breaks is involved in the precise excision of *Paramecium* internal eliminated sequences. *Mol. Cell. Biol.* **23:** 7152–7162.

Hamilton E., Bruns P., Lin C., Merriam V., Orias E., Vong L., and Cassidy-Hanley D. 2005. Genome-wide characterization of *Tetrahymena thermophila* chromosome breakage sites. I. Cloning and identification of functional sites. *Genetics* **170:** 1611–1621.

Harumoto T. 1986. Induced change in a non-Mendelian determinant by transplantation of macronucleoplasm in *Paramecium tetraurelia*. *Mol. Cell. Biol.* **6:** 3498–3501.

Harwood J. 1985. The erratic career of cytoplasmic inheritance. *Trends Genet.* **1:** 298–300.

Hayashi T., Hayashi H., Fusauchi Y., and Iwai K. 1984. *Tetrahymena* histone H3. Purification and two variant sequences. *J. Biochem.* **95:** 1741–1749.

Hayashi T., Hayashi H., and Iwai K. 1987. *Tetrahymena* histone H1. Isolation and amino acid sequence lacking the central hydrophobic domain conserved in other H1 histones. *J. Biochem.* **102:** 369–376.

Jahn C.L. and Klobutcher L.A. 2002. Genome remodeling in ciliated protozoa. *Annu. Rev. Microbiol.* **56:** 489–520.

Jaraczewski J.W. and Jahn C.L. 1993. Elimination of Tec elements involves a novel excision process. *Genes Dev.* **7:** 95–105.

Jessop-Murray H., Martin L.D., Gilley D., Preer J.R., and Polisky B. 1991. Permanent rescue of a non-Mendelian mutation of *Paramecium* by microinjection of specific DNA sequences. *Genetics* **129:** 727–734.

Klobutcher L.A. 1999. Characterization of in vivo developmental chromosome fragmentation intermediates in *E. crassus*. *Mol. Cell* **4:** 695–704.

Klobutcher L.A. and Herrick G. 1995. Consensus inverted terminal repeat sequence of *Paramecium* IESs: Resemblance to termini of Tc1-related and *Euplotes* Tec transposons. *Nucleic Acids Res.* **23:** 2006–2013.

———. 1997. Developmental genome reorganization in ciliated protozoa: The transposon link. *Prog. Nucleic Acid Res. Mol. Biol.* **56:** 1–62.

Klobutcher L.A., Turner L.R., and LaPlante J. 1993. Circular forms of developmentally excised DNA in *Euplotes crassus* have a heteroduplex junction. *Genes Dev.* **7:** 84–94.

Kobayashi S. and Koizumi S. 1990. Characterization of non-Mendelian and Mendelian mutant strains by micronuclear transplantation in *Paramecium tetraurelia*. *J. Protozool.* **37:** 489–492.

Koizumi S. and Kobayashi S. 1989. Microinjection of plasmid DNA encoding the A surface antigen of *Paramecium tetraurelia* restores the ability to regenerate a wild-type macronucleus. *Mol. Cell. Biol.* **9:** 4398–4401.

Landweber L.F., Kuo T.C., and Curtis E.A. 2000. Evolution and assembly of an extremely scrambled gene. *Proc. Natl. Acad. Sci.* **97:** 3298–3303.

Le Mouël A., Butler A., Caron F., and Meyer E. 2003. Developmentally regulated chromosome fragmentation linked to imprecise elimination of repeated sequences in Paramecia. *Eukaryot. Cell* **2:** 1076–1090.

Lee S.R. and Collins K. 2006. Two classes of endogenous small RNAs in *Tetrahymena thermophila*. *Genes Dev.* **20:** 28–33.

Liu Y., Mochizuki K., and Gorovsky M.A. 2004. Histone H3 lysine 9 methylation is required for DNA elimination in developing macronuclei in *Tetrahymena*. *Proc. Natl. Acad. Sci.* **101:** 1679–1684.

Liu Y., Song X., Gorovsky M.A., and Karrer K.M. 2005. Elimination of foreign DNA during somatic differentiation in *Tetrahymena thermophila* shows position effect and is dosage dependent. *Eukaryot. Cell* **4:** 421–431.

Lu C., Tej S.S., Luo S., Haudenschild C.D., Meyers B.C., and Green P.J. 2005. Elucidation of the small RNA component of the transcriptome. *Science* **309:** 1567–1569.

Madireddi M.T., Davis M., and Allis D. 1994. Identification of a novel polypeptide involved in the formation of DNA-containing vesicles during macronuclear development in *Tetrahymena*. *Dev. Biol.* **165:** 418–431.

Madireddi M.T., Coyne R.S., Smothers J.F., Mickey K.M., Yao M.C., and Allis C.D. 1996. Pdd1p, a novel chromodomain-containing protein, links heterochromatin assembly and DNA elimination in *Tetrahymena*. *Cell* **87:** 75–84.

Malone C.D., Anderson A.M., Motl J.A., Rexer C.H., and Chalker D.L. 2005. Germ line transcripts are processed by a Dicer-like protein that is essential for developmentally programmed genome rearrangements of *Tetrahymena thermophila*. *Mol. Cell. Biol.* **25:** 9151–9164.

Mattick J.S. 2004. RNA regulation: A new genetics? *Nat. Rev. Genet.* **5:** 316–323.

Matzke M.A. and Birchler J.A. 2005. RNAi-mediated pathways in the nucleus. *Nat. Rev. Genet.* **6:** 24–35.

Meyer E. 1992. Induction of specific macronuclear developmental mutations by microinjection of a cloned telomeric gene in *Paramecium primaurelia. Genes Dev.* **6:** 211–222.

Meyer E. and Keller A.M. 1996. A Mendelian mutation affecting mating-type determination also affects developmental genomic rearrangements in *Paramecium tetraurelia. Genetics* **143:** 191–202.

Meyer E., Butler A., Dubrana K., Duharcourt S., and Caron F. 1997. Sequence-specific epigenetic effects of the maternal somatic genome on developmental rearrangements of the zygotic genome in *Paramecium primaurelia. Mol. Cell. Biol.* **17:** 3589–3599.

Meyers B.C., Vu T.H., Tej S.S., Ghazal H., Matvienko M., Agrawal V., Ning J., and Haudenschild C.D. 2004. Analysis of the transcriptional complexity of *Arabidopsis thaliana* by massively parallel signature sequencing. *Nat. Biotechnol.* **22:** 1006–1011.

Mizzen C.A., Dou Y., Liu Y., Cook R.G., Gorovsky M.A., and Allis C.D. 1999. Identification and mutation of phosphorylation sites in a linker histone. Phosphorylation of macronuclear H1 is not essential for viability in *Tetrahymena. J. Biol. Chem.* **274:** 14533–14536.

Mochizuki K. and Gorovsky M.A. 2004. Small RNAs in genome rearrangement in *Tetrahymena. Curr. Opin. Genet. Dev.* **14:** 181–187.

———. 2005. A Dicer-like protein in *Tetrahymena* has distinct functions in genome rearrangement, chromosome segregation, and meiotic prophase. *Genes Dev.* **19:** 77–89.

Mochizuki K., Fine N.A., Fujisawa T., and Gorovsky M.A. 2002. Analysis of a piwi-related gene implicates small RNAs in genome rearrangement in *Tetrahymena. Cell* **110:** 689–699.

Moreira-Leite F.F., Sherwin T., Kohl L., and Gull K. 2001. A trypanosome structure involved in transmitting cytoplasmic information during cell division. *Science* **294:** 610–612.

Nikiforov M.A., Gorovsky M.A., and Allis C.D. 2000. A novel chromodomain protein, pdd3p, associates with internal eliminated sequences during macronuclear development in *Tetrahymena thermophila. Mol. Cell. Biol.* **20:** 4128–4134.

Nowacki M., Zagorski-Ostoja W., and Meyer E. 2005. Nowa1p and Nowa2p: Novel putative RNA binding proteins involved in *trans*-nuclear crosstalk in *Paramecium tetraurelia. Curr. Biol.* **15:** 1616–1628.

Paschka A.G., Jönsson F., Maier V., Möllenbeck M., Paeschke K., Postberg J., Rupprecht S., and Lipps H.J. 2003. The use of RNAi to analyze gene function in spirotrichous ciliates. *Eur. J. Protistol.* **39:** 449–454.

Philippe H., Germot A., and Moreira D. 2000. The new phylogeny of eukaryotes. *Curr. Opin. Genet. Dev.* **10:** 596–601.

Pond F.R., Gibson I., Lalucat J., and Quackenbush R.L. 1989. R-body-producing bacteria. *Microbiol. Rev.* **53:** 25–67.

Preer J.R., Jr., Preer L.B., and Jurand A. 1974. Kappa and other endosymbionts in *Paramecium aurelia. Bacteriol. Rev.* **38:** 113–163.

Prescott D.M. 1994. The DNA of ciliated protozoa. *Microbiol. Rev.* **58:** 233–267.

———. 1999. The evolutionary scrambling and developmental unscrambling of germline genes in hypotrichous ciliates. *Nucleic Acids Res.* **27:** 1243–1250.

Prescott D.M., Ehrenfeucht A., and Rozenberg G. 2003. Template-guided recombination for IES elimination and unscrambling of genes in stichotrichous ciliates. *J. Theor. Biol.* **222:** 323–330.

Ruiz F., Vayssie L., Klotz C., Sperling L., and Madeddu L. 1998. Homol-ogy-dependent gene silencing in *Paramecium. Mol. Biol. Cell* **9:** 931–943.

Saveliev S.V. and Cox M.M. 1995. Transient DNA breaks associated with programmed genomic deletion events in conjugating cells of *Tetrahymena thermophila. Genes Dev.* **9:** 248–255.

———. 1996. Developmentally programmed DNA deletion in *Tetrahymena thermophila* by a transposition-like reaction pathway. *EMBO J.* **15:** 2858–2869.

Scott J.M., Mikami K., Leeck C.L., and Forney J.D. 1994. Non-Mendelian inheritance of macronuclear mutations is gene specific in *Paramecium tetraurelia. Mol. Cell. Biol.* **14:** 2479–2484.

Shen X. and Gorovsky M.A. 1996. Linker histone H1 regulates specific gene expression but not global transcription in vivo. *Cell* **86:** 475–483.

Shen X., Yu L., Weir J.W., and Gorovsky M.A. 1995. Linker histones are not essential and affect chromatin condensation in vivo. *Cell* **82:** 47–56.

Shorter J. and Lindquist S. 2005. Prions as adaptive conduits of memory and inheritance. *Nat. Rev. Genet.* **6:** 435–450.

Sonneborn T.M. 1937. Sex, sex inheritance and sex determination in *Paramecium aurelia. Proc. Natl. Acad. Sci.* **23:** 378–385.

———. 1975. *Paramecium aurelia.* In *Handbook of genetics* (ed. R. King), pp. 469–594. Plenum, New York.

———. 1977. Genetics of cellular differentiation: Stable nuclear differentiation in eucaryotic unicells. *Annu. Rev. Genet.* **11:** 349–367.

Strahl B.D., Ohba R., Cook R.G., and Allis C.D. 1999. Methylation of histone H3 at lysine 4 is highly conserved and correlates with transcriptionally active nuclei in *Tetrahymena. Proc. Natl. Acad. Sci.* **96:** 14967–14972.

Suzuki M. and Hayashizaki Y. 2004. Mouse-centric comparative transcriptomics of protein coding and non-coding RNAs. *Bioessays* **26:** 833–843.

Taverna S.D., Coyne R.S., and Allis C.D. 2002. Methylation of histone h3 at lysine 9 targets programmed DNA elimination in *Tetrahymena. Cell* **110:** 701–711.

Timmons L. and Fire A. 1998. Specific interference by ingested dsRNA. *Nature* **395:** 854.

Timmons L., Court D.L., and Fire A. 2001. Ingestion of bacterially expressed dsRNAs can produce specific and potent genetic interference in *Caenorhabditis elegans. Gene* **263:** 103–112.

Vavra K.J., Allis C.D., and Gorovsky M.A. 1982. Regulation of histone acetylation in *Tetrahymena* macro- and micronuclei. *J. Biol. Chem.* **257:** 2591–2598.

Wei Y., Yu L., Bowen J., Gorovsky M.A., and Allis C.D. 1999. Phosphorylation of histone H3 is required for proper chromosome condensation and segregation. *Cell* **97:** 99–109.

Wei Y., Mizzen C.A., Cook R.G., Gorovsky M.A., and Allis C.D. 1998. Phosphorylation of histone H3 at serine 10 is correlated with chromosome condensation during mitosis and meiosis in *Tetrahymena. Proc. Natl. Acad. Sci.* **95:** 7480–7484.

Williams K., Doak T.G., and Herrick G. 1993. Developmental precise excision of *Oxytricha trifallax* telomere-bearing elements and formation of circles closed by a copy of the flanking target duplication. *EMBO J.* **12:** 4593–4601.

Wu M., Allis C.D., Richman R., Cook R.G., and Gorovsky M.A. 1986. An intervening sequence in an unusual histone H1 gene of *Tetrahymena thermophila. Proc. Natl. Acad. Sci.* **83:** 8674–8678.

Yao M.C. and Chao J.L. 2005. RNA-guided DNA deletion in *Tetrahymena:* An RNAi-based mechanism for programmed genome rearrangements. *Annu. Rev. Genet.* **39:** 537–559.

Yao M.C., Duharcourt S., and Chalker D.L. 2002. Genome-wide

rearrangements of DNA in ciliates. in *Mobile DNA II* (ed. A. Lambowitz), pp. 730–758. Academic Press, New York.

Yao M.C., Fuller P., and Xi X. 2003. Programmed DNA deletion as an RNA-guided system of genome defense. *Science* **300:** 1581–1584.

You Y., Aufderheide K., Morand J., Rodkey K., and Forney J. 1991. Macronuclear transformation with specific DNA fragments controls the content of the new macronuclear genome in *Paramecium tetraurelia. Mol. Cell. Biol.* **11:** 1133–1137.

Yu L. and Gorovsky M.A. 1997. Constitutive expression, not a particular primary sequence, is the important feature of the H3 replacement variant hv2 in *Tetrahymena thermophila. Mol. Cell. Biol.* **17:** 6303–6310.

RNAi and Heterochromatin Assembly

Robert Martienssen[1] and Danesh Moazed[2]

[1]Cold Spring Harbor Laboratory, Cold Spring Harbor, New York 11724
[2]Department of Cell Biology, Harvard Medical School, Boston, Massachusetts 02115-5730

C O N T E N T S

1. Overview of the RNAi Pathway, 153

2. Early Evidence Implicating RNA as an Intermediate in Transcriptional Gene Silencing, 154

3. RNAi and Heterochromatin Assembly in *S. pombe*, 155

4. Small RNAs Initiate Heterochromatin Assembly in Association with an RNAi Effector Complex, 156

5. dsRNA Synthesis and siRNA Generation, 157

6. RNA–RNA Versus RNA–DNA Recognition Models, 159

7. How Does RNAi Recruit Chromatin-modifying Enzymes?, 159

8. RNAi-mediated Chromatin and DNA Modifications in *Arabidopsis*, 160

9. Conservation of RNAi-mediated Chromatin Modifications in Animals, 162

10. Concluding Remarks, 164

References, 164

GENERAL SUMMARY

The intersection of RNA interference (RNAi) and heterochromatin formation brought together two areas of gene regulation that had previously been thought to operate by different, perhaps even unrelated, mechanisms. Using cytological staining methods, heterochromatin was originally defined nearly 80 years ago as those chromosome regions that retained a condensed appearance throughout the cell cycle. Early investigators studying the relationship between chromosome structure and gene expression noticed that certain chromosome rearrangements resulted in the spreading of heterochromatin into adjacent genes, which then became silent. But the stochastic nature of spreading gave rise to genetically identical populations of cells that had different phenotypes, providing a striking example of epigenetic regulation. The term RNAi was first used to describe gene silencing when homologous antisense or double-stranded RNA (dsRNA) is introduced into the nematode *Caenorhabditis elegans*. It was soon recognized that a related mechanism accounted for posttranscriptional transgene silencing (PTGS) described earlier in petunia and other plants. In contrast, heterochromatin was widely believed to operate directly at the chromatin level to cause transcriptional repression, by a mechanism referred to as transcriptional gene silencing (TGS). This chapter focuses on the relationship between the RNAi pathway and the formation of epigenetically heritable heterochromatin at specific chromosome regions. It draws on recent examples that demonstrate this relationship in the fission yeast *Schizosaccharomyces pombe* and the mustard plant *Arabidopsis thaliana*.

The fission yeast nuclear genome is composed of three chromosomes that range from 3.5 Mb to 5.7 Mb in size. Each chromosome contains large blocks of repetitive DNA, particularly at centromeres, which are packaged into heterochromatin. In addition, the mating-type loci (which control cell type) and subtelomeric DNA regions also contain repetitive sequences that are packaged into heterochromatin. We now know that the assembly of DNA into heterochromatin plays both regulatory and structural roles. In the case of the mating-type loci in yeast, regulation of gene transcription by heterochromatin is important for cell-type identity. In the case of telomeres and centromeres, heterochromatin plays a structural role that is important for proper chromosome segregation during cell division. Moreover, repetitive DNA sequences and transposable elements account for a large fraction, in some cases more than half, of the genomes of many eukaryotic cells. Heterochromatin and associated mechanisms play a critical role in maintaining genome stability by regulating the activity of repeated sequences. Recent studies have uncovered a surprising requirement for components of the RNAi pathway in the process of heterochromatin formation in fission yeast and have provided insight into how these two pathways can work together at the chromatin level. Briefly, small interfering RNA (siRNA) molecules, which are a signature of RNAi and other dsRNA silencing mechanisms, assemble into the RNA-Induced Transcriptional Silencing (RITS) complex and direct epigenetic chromatin modifications and heterochromatin formation at complementary chromosome regions. RITS uses siRNA-dependent base-pairing to guide association with either DNA or nascent RNA sequences at the target locus destined to be silenced, an association that is stabilized by direct binding to methylated histone H3. The presence of these two activities in RITS triggers heterochromatin formation in concert with well-known heterochromatin-associated factors and directly links RNA silencing to heterochromatin modification.

In *A. thaliana* and other eukaryotes (with the exception of *Saccharomyces cerevisiae*), centromeric DNA regions are also composed of repetitive elements. These and other repeat sequences, such as retroelements and other transposons, are the source of siRNAs, attracting histone H3K9 and DNA methylation. Here again, several components of the RNAi pathway are required for the initiation and maintenance of these repressive methylation events. In this chapter, we discuss how heterochromatic siRNAs are produced and mediate DNA and/or chromatin modifications in fission yeast and *A. thaliana*.

1 Overview of the RNAi Pathway

Although the term RNAi was originally used to describe silencing that is mediated by exogenous dsRNA in *C. elegans* (Fire et al. 1998), it now broadly refers to gene silencing that is triggered by some kind of dsRNA. The steps involved in RNAi include the generation of dsRNA (which can be endogenous or exogenous such as viral RNA), processing into siRNA, and targeting of these molecules to either mRNAs (PTGS) or chromatin regions (TGS) to effect silencing. Therefore, before introducing the components of the RNAi machinery specific to TGS, we discuss the source of dsRNA that harnesses the RNAi machinery into action.

dsRNA may originate from bidirectional transcription of repetitive DNA elements, or transcription of RNA molecules that can base-pair internally to form dsRNA segments (see Fig. 1, a and b, respectively). For example, transcription through inverted repeat regions produces RNA molecules that fold back on themselves to produce hairpin structures. dsRNAs are then cleaved by Dicer, an RNase III class ribonuclease, which generates siRNAs. These are complementary duplexes, 21–27 nucleotides (nt) in size, that have a characteristic 2-nt overhang at each 3′ end of the duplex (Hamilton and Baulcombe 1999; Zamore et al. 2000; Bernstein et al. 2001; Elbashir et al. 2001; Hannon 2002; Zamore 2002; Bartel 2004; Baulcombe 2004). These duplexes are unwound into single-stranded siRNA to act as guides, through base-pairing interactions with complementary target sequences. They are therefore specificity factors and play a central role in all RNAi-mediated silencing mechanisms.

To date, two related complexes have been identified that incorporate siRNA: RISC and RITS. In the RNA-Induced Silencing Complex (RISC), siRNAs recognize target mRNAs and initiate their degradation by endonucleolytic cleavage within the mRNA region that is base-paired to the siRNA (Hannon 2002; Bartel 2004). The RNase H domain of the Argonaute/PIWI family protein (a subunit of RISC) carries out this initial mRNA cleavage event. In the nuclear RNA-Induced Transcriptional Silencing (RITS) complex (similar to the RISC), siRNAs target the complex to chromosome regions for chromatin modification (Verdel et al. 2004; Buhler et al. 2006). It is the RITS-mediated RNA pathway that is the focus of this chapter.

The central Argonaute and Dicer proteins are required for an additional type of RNA silencing mechanism involving microRNAs (miRNA). RNA, transcribed from endogenous noncoding genes that initially form hairpin RNA structures, due to extended dsRNA regions, is processed into miRNA through a series of steps (Bartel 2004; Filipowicz et al. 2005). Like siRNAs, miRNAs are 21–24 nt in size and form part of the RISC via the Argonaute proteins, to target specific mRNAs. This targeting can result in mRNA cleavage via the PIWI/RNAse H domain and translational repression involving interactions with the 7meG cap at the 5′-end of the mRNA. This may be coupled to sequestration of the mRNA to cytoplasmic RNA-processing organelles known as P bodies (Processing bodies). Thus, at least two different dsRNA-processing pathways result in the generation of siRNA or miRNA, yet these RNAs use a similar machinery to inactivate cognate mRNAs. The miRNA pathway distinguishes itself because miRNAs are all produced by endogenous noncoding genes that are largely developmentally regulated and, in turn, generally target and developmentally regulate the silencing of homologous genes.

Although dsRNAs can form by the annealing of forward and reverse RNAs that result from bidirectional

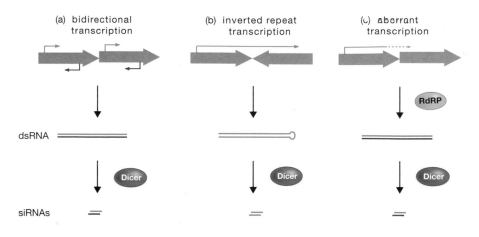

Figure 1. Sources of dsRNA, Which Act as a Substrate for Generation of siRNAs by the Dicer Ribonuclease, and Are the Trigger for RNA Silencing

(*a*) Bidirectional transcription has been observed at the *S. pombe* centromeric repeats and the *cenH* region of the silent mating-type locus. (*b*) Transcription through inverted repeats found in many plant and animal cells can potentially produce dsRNA. (*c*) Transcription of aberrant RNAs that may lack proper processing signals may trigger dsRNA synthesis by RdRPs.

transcription or are present in hairpin structures, in some cells, RNAi requires an additional enzyme to make dsRNA. This is the RNA-directed RNA polymerase (RdRP) found in plants and *C. elegans* (Dalmay et al. 2000; Sijen et al. 2001). It uses siRNAs as primers to generate more dsRNA, which can then be processed into additional siRNA by Dicer. The primary function of RdRP is thus thought to be in amplification of the RNAi response, but, as discussed later, RdRPs may have more specific roles in initiating dsRNA synthesis (see Section 5). Indeed, it seems to be involved in a process adapted for producing a better host defense response to the introduction of exogenous dsRNA. This idea is strengthened by the fact that RdRPs are not involved in the miRNA silencing pathways (Sijen et al. 2001). Interestingly, insects (including *Drosophila*) and vertebrates (including mammals) lack recognizable RdRP-like sequences in their genomes, but it remains possible that other polymerases carry out dsRNA synthesis in these organisms.

What then is the function of the various RNA silencing mechanisms? They are widely conserved in organisms ranging from fission yeast to plants to human, and they play central roles in the regulation of gene expression and genome stability (through stable heterochromatin formation at centromeres and telomeres). In addition, these silencing mechanisms are involved in defense against transposons and RNA viruses through degradation of their RNA transcripts (Plasterk 2002; Li and Ding 2005). Finally, transcription from some transposons generates aberrant RNAs that trigger RNAi by a mechanism thought to involve the conversion of aberrant transcripts to dsRNA by RdRPs (Fig. 1) (Baulcombe 2004).

2 Early Evidence Implicating RNA as an Intermediate in Transcriptional Gene Silencing

Before discussing the better-understood examples of RNAi-based chromatin modifications in fission yeast and *Arabidopsis*, we briefly discuss early experiments that suggested a role for RNA in mediating chromatin and DNA modifications. The earliest evidence for the role of an RNA intermediate in TGS came from studies of plant viroids. The potato spindle tuber viroid (PSTV) consists of a 359-nt RNA genome and replicates via an RNA–RNA pathway. The introduction of PSTV into the tobacco genome results in the DNA methylation of homologous nuclear sequences, albeit transgenic in origin (Wassenegger et al. 1994). However, these and integrated copies of PSTV DNA only become methylated in plants that sup-

port viroid RNA transcription, suggesting the involvement of an RNA intermediate that directs DNA methylation (Wassenegger et al. 1994). Furthermore, in *Arabidopsis*, the production of aberrant transcripts somehow results in the DNA methylation of all homologous promoter regions and transcriptional gene silencing (Mette et al. 1999). This, together with the finding that the replication of viral genomes in plants leads to the production of small RNAs that are 22 nt in size, suggests that RNAi-related mechanisms mediate DNA methylation (Mette et al. 2000). These observations, as well as repeat-induced silencing by transgenes, which was first discovered in petunia and in tobacco, are now widely recognized as the earliest examples of silencing by RNAi (Napoli et al. 1990; discussed in Chapter 9).

Further evidence for a link between RNAi and TGS comes from studies of repeat-induced gene silencing in *Drosophila* (see Chapter 5). The introduction of multiple tandem copies of a transgene results in the silencing of both the transgene and the endogenous copies (Pal-Bhadra et al. 1999). This silencing requires the chromodomain protein Polycomb, which is also involved in the packaging of homeotic regulatory genes into heterochromatin-like structure outside of their proper domains of action (Francis and Kingston 2001). In addition, this repeat-induced gene silencing requires Piwi, a *Drosophila* Argonaute family member required for RNAi (Pal-Bhadra et al. 2002). In *Tetrahymena*, another Piwi protein family member, Twi1, is required for small RNA accumulation and the massive DNA elimination that is observed in the somatic macronucleus of the protozoa (see Chapter 7). These and more recent results discussed in Section 8 suggest that the RNAi pathway is involved in the assembly of repressive chromatin structures in flies.

Other repeat-induced silencing mechanisms have been described in filamentous fungi, including Repeat-Induced Point mutation (RIP) in *Neurospora crassa* and Methylation Induced Pre-meiotically (MIP) in *Ascobolus immersus*, that do not appear to involve an RNA intermediate since they occur independently of the transcriptional state of the locus (Galagan and Selker 2004). Instead, RIP and MIP involve paired loci, where (for example) two out of three gene copies are silenced, suggesting some kind of DNA–DNA interaction mechanism involving homologous loci to induce silencing. Conversely, silencing of unpaired DNA in meiosis (MSUD), which also occurs in *Neurospora*, requires the RNAi pathway (Shiu et al. 2001; discussed in Chapter 6) and may have parallels in other organisms, including *C. elegans* (Maine et al. 2005; see Chapter 15).

3 RNAi and Heterochromatin Assembly in *S. pombe*

S. pombe chromosomes contain extensive heterochromatic DNA regions that are associated with underlying repetitive DNA elements at the centromeres and the silent mating-type loci (mat2/3) (Grewal 2000; Pidoux and Allshire 2004). Each fission yeast centromere contains a unique central core region (*cnt*) that is flanked by two types of repeats, called the *innermost* (*imr*) and *outermost* (*otr*) repeats (Fig. 2). The *otr* region itself is composed of *dh* and *dg* repeats.

Heterochromatin formation in *S. pombe* involves the concerted action of a number of *trans*-acting factors. These include histone deacetylases (HDACs), Clr4, a histone H3 lysine 9 methyltransferase (HKMT), and the histone H3K9-methyl binding proteins, Swi6 (an HP1 homolog) and Chp1. The initial recruitment of Swi6 and Clr4 to chromatin has been proposed to result in the spreading of H3K9 methylation and heterochromatin formation through sequential cycles of Clr4-catalyzed

H3K9 methylation coupled to Swi6-mediated spreading to adjacent nucleosomes through its self-association (Grewal and Moazed 2003).

Mutation in components of the RNAi pathway surprisingly results in a loss of centromeric heterochromatin and the accumulation of noncoding forward and reverse transcripts from bidirectional promoters within each *dg* and *dh* repeat (Fig. 2) (Volpe et al. 2002). Fission yeast contains a single gene for each of the RNAi proteins, Dicer, Argonaute, and RdRP (*dcr1*⁺, *ago1*⁺, and *rdp1*⁺, respectively). Deleting any of these genes results in the loss of histone H3K9 methylation, and mutants display defects in chromosome segregation, which are generally associated with defects in heterochromatin assembly (Provost et al. 2002; Volpe et al. 2003). Moreover, sequencing of a library of fission yeast small RNAs identified ~22-nt RNAs that mapped exclusively to centromeric repeat regions and ribosomal DNA repeats, suggesting that *cen* RNAs can produce dsRNAs that are processed into siRNAs (Reinhart and Bartel 2002). Thus, it was suggested that the RNAi pathway

S.pombe centromere

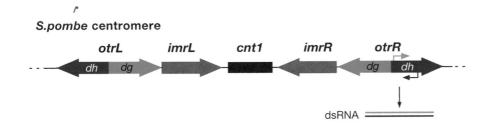

S.pombe silent mating type

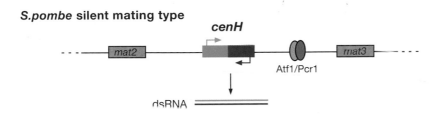

A.thaliana centromere

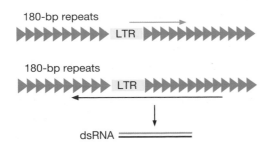

Figure 2. Organization of Heterochromatic Chromosome Regions in *S. pombe* and *A. thaliana*

The centromere of *S. pombe* chromosome 1 is shown as an example. The unique central core (*cnt1*) region is flanked by innermost (*imrL* and *imrR*) and outermost (*otrL* and *otrR*) repeats. The *otr* region is transcribed in both directions, giving rise to forward (*blue*) and reverse (*red*) transcripts. The region between the *mat2* and *mat3* genes contains a domain that is homologous to the centromeric *dg* and *dh* repeats (*cenH*) and is also bidirectionally transcribed. Atf1 and Pcr1 are DNA-binding proteins that act in parallel with RNAi in mating-type silencing. *Arabidopsis* centromeres are composed of 180-bp repeats (*green*) interspersed with retrotransposable elements (*yellow*). Forward transcripts initiating within the long terminal repeat (LTR) of the retroelement and reverse transcripts initiating within the 180-bp repeats are indicated.

could recruit Swi6 and Clr4 to chromatin to initiate and/or maintain heterochromatin formation at each of the above loci (Fig. 3) (Hall et al. 2002; Volpe et al. 2002).

Interestingly, both TGS and PTGS mechanisms appear to contribute to the down-regulation of *cen* RNAs. The forward strand transcript is primarily silenced at the transcriptional level, as demonstrated in RNAi mutants (Volpe et al. 2002). The reverse strand of *cen* transcripts, however, is not affected by Swi6 mutants (Volpe et al. 2002), and silencing of this cen-reverse transcript occurs primarily at the posttranscriptional level.

RNAi also plays a role in silencing the mating-type locus (*mat2/3*) (Hall et al. 2002). *mat2/3* is interrupted by a region of DNA that is highly homologous to centromeric repeats (called *cenH*, *cenHomology*)(Fig. 2). Like the *cen* repeats, the *cenH* region is divergently transcribed to produce forward and reverse RNA (Noma et al. 2004). These *cenH* transcripts accumulate to high levels in RNAi mutants. In contrast, RNAi is not necessary for silencing of a reporter transgene inserted at *mat2/3* if silencing is not first somehow compromised. The reason for this difference is that a partially redundant silencing mechanism, involving two DNA-binding proteins, Pcr1 and Atf1, which bind to *mat2/3*, recruits the heterochromatin

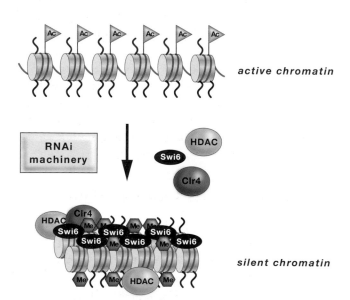

Figure 3. Assembly of Heterochromatin Involves the Concerted Action of Histone-modifying Enzymes (HDACs and Clr4) and Histone-binding Proteins (e.g., Swi6) and Can Be Directed by the RNAi Machinery

Deacetylation by HDACs is followed by recruitment of Clr4 and histone H3K9 methylation. Swi6 binds to H3K9-methylated histone tails, and spreading results from sequential cycles of H3K9 methylation that are coupled to Swi6 oligomerization.

machinery independently of RNAi (Jia et al. 2004). This is sufficient for silencing the reporter gene in the absence of RNAi but not for preventing the accumulation of noncoding *cenH* transcripts (Fig. 2).

4 Small RNAs Initiate Heterochromatin Assembly in Association with an RNAi Effector Complex

The discovery that the RNAi pathway is involved in heterochromatin formation in fission yeast and in transcriptional gene silencing in other systems raised the question of how it could directly regulate chromatin structure. Purification of Chp1, a chromodomain protein that is a structural component of heterochromatin, led to the identification of the RITS complex (Verdel et al. 2004). RITS contains the fission yeast Ago1 protein and Tas3, a protein of unknown function, in addition to Chp1. It also contains centromeric siRNAs, which are produced by the Dicer ribonuclease, and importantly, RITS associates with centromeric repeat regions in an siRNA-dependent fashion. RITS has therefore been proposed to use centromeric siRNAs to target specific chromosome regions for inactivation, and this provides a direct link between RNAi and heterochromatin assembly (Fig. 4).

Like RISC, which mediates PTGS, RITS uses siRNAs for target recognition. Unlike RISC, however, RITS associates with chromatin and initiates heterochromatin formation as opposed to mRNA inactivation. How can siRNAs target specific chromosome regions? Two possible mechanisms have been proposed. In the first model, siRNAs bound to Ago1 in the RITS complex must somehow base-pair with an unwound DNA double helix. In the second model, RITS-associated siRNAs base-pair with noncoding RNA transcripts at the target locus (Fig. 4).

According to either model, the association of RITS with chromatin via siRNA results in the recruitment of the Clr4 HKMT and subsequent histone H3K9 methylation. This is followed by Swi6 binding and the spreading of H3K9 methylation and heterochromatin. However, Clr4 is also required for the association of RITS with chromatin, suggesting that it provides methylated H3K9 to which the RITS complex can bind, thereby stabilizing its association with chromatin. The chromodomain of Chp1 was already known to bind specifically to methylated H3K9 residues (Partridge et al. 2002), and mutations in Clr4 or the chromodomain of Chp1 that are involved in this interaction result in a loss of RITS binding to chromatin (Partridge et al. 2002; Noma et al. 2004). Moreover, RITS can also bind to chromatin domains that are coated with methylated H3K9 through the chromo-

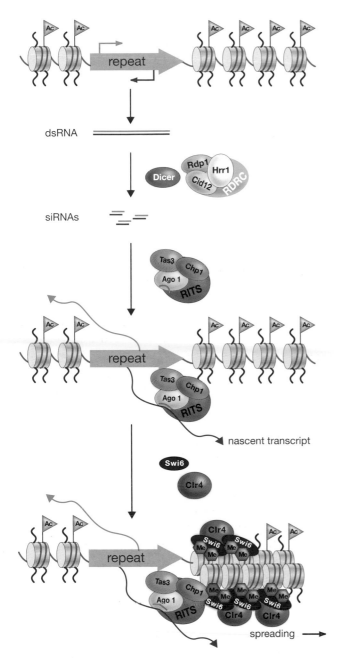

Figure 4. RNAi and siRNA-directed Assembly of Heterochromatin in *S. pombe*

Both Dicer and RDRC are required for siRNA generation. Initial targeting is proposed to involve RITS and siRNA-mediated recognition of cognate transcripts. The binding of RITS is stabilized by association of the chromodomain of Chp1 with H3K9 methylated histone. The recruitment of Clr4 and Swi6 mediates the spreading of H3K9 methylation.

domain of Chp1 at the *mat2/3* and telomeric regions in the absence of siRNAs (Noma et al. 2004; Petrie et al. 2005). In summary, the RITS complex shows affinity to chromatin via Chp1 binding to methylated H3K9 and

through base-pairing of siRNA with either DNA or RNA transcripts.

Recent evidence strongly supports a role for RITS and siRNAs in the initiation of heterochromatin assembly. Buhler et al. (2006) used a site-specific RNA-binding protein to artificially tether the RITS complex to the RNA transcript of the normally active *ura4+* gene. Remarkably, this tethering results in the generation of *ura4+* siRNAs and silencing of the *ura4+* gene in a manner that requires both RNAi and heterochromatin components. In addition, this system allowed a direct evaluation of the ability of newly generated siRNAs to initiate H3K9 methylation and Swi6 binding, which are molecular markers for heterochromatin formation. Interestingly, the newly generated *ura4+* siRNAs were found to be under negative control by the conserved siRNA ribonuclease, Eri1, which restricts them to the locus where they are produced. However, when the gene encoding Eri1 is deleted, *ura4+* siRNAs are able to act in *trans* to silence a second copy of the *ura4+*gene, which is inserted on a different chromosome in the same cell. This experiment therefore demonstrates that siRNAs can act as specificity factors that direct RITS and heterochromatin assembly to a previously active region of the genome.

The ability of siRNAs to initiate silencing in *S. pombe* has also been examined using a different method, which relies on the expression of a hairpin RNA to produce siRNAs homologous to a GFP transgene (Sigova et al. 2004). In this system, hairpin siRNAs promoted silencing of the GFP reporter gene at the PTGS, but not TGS, level (Sigova et al. 2004). It is unclear why the hairpin siRNAs cannot induce TGS and heterochromatin assembly at the chromosomal copy of GFP. One possible explanation is that heterochromatin is assembled at specific subnuclear locations, and assembly outside these locations occurs inefficiently (Gasser et al. 2004; Chapter 4).

5 dsRNA Synthesis and siRNA Generation

Bidirectional transcription of centromeric DNA repeats could in principle provide the initial source of dsRNA in fission yeast (Volpe et al. 2002). dsRNA resulting from the annealing of forward and reverse transcripts could then be a substrate for the Dicer ribonuclease. However, RNA-directed RNA polymerase (Rdp1) and its associated cofactors, as well as the Clr4 HKMT, are also required for siRNA production by Dicer (Hong et al. 2005; Li et al. 2005; Buhler et al. 2006). These observations indicate that the generation of heterochromatic siRNAs by Dicer is coupled to chromatin and Rdp1-dependent events (Fig. 4).

The Rdp1 enzyme resides in a multiprotein complex that also contains Hrr1, an RNA helicase, and Cid12, a member of the β family of DNA polymerases which includes poly(A) polymerase enzymes (Motamedi et al. 2004). This complex has been termed RNA-directed RNA polymerase complex (RDRC), and all of its subunits are required for heterochromatin formation at centromeric DNA regions (Motamedi et al. 2004). As expected from the presence of Rdp1, RDRC has RNA-directed RNA polymerase activity in vitro, and mutations that abolish this activity also abolish RNAi-dependent silencing in vivo (Motamedi et al. 2004; Sugiyama et al. 2005). The in vitro RNA synthesis activity of RDRC does not require an siRNA primer (Motamedi et al. 2004). RITS may therefore provide in vivo specificity by recruiting RDRC to selected RNA templates via siRNA. Consistent with this hypothesis, subunits of the RDRC are required for siRNA generation, and RITS complexes purified from cells that lack any subunit of the RDRC are devoid of siRNAs (Motamedi et al. 2004; Li et al. 2005; Sugiyama et al. 2005; Buhler et al. 2006).

The presence of Cid12 in the RDRC is intriguing and raises the possibility that another polymerase activity participates in chromosome-associated RNA silencing.

Because some members of this family have poly(A) polymerase activity, one possibility is that adenylation of Rdp1-produced dsRNA may be important for their further processing. Interestingly, Cid12-like proteins are conserved throughout eukaryotes (Table 1); mutations in Rde-3, a *C. elegans* member of this family, result in defective RNAi (Chen et al. 2005), corroborating a conserved role for these enzymes in the RNAi pathway.

There is evidence for dsRNA synthesis and processing associated with the generation of heterochromatic siRNAs occurring on the chromosome, at sites of transcription of noncoding centromeric RNAs (Fig. 4). Evidence includes, first, that Rdp1 can be cross-linked to centromeric DNA repeats (Volpe et al. 2002; Sugiyama et al. 2005), and to the forward and reverse RNA transcripts that originate from these regions (Motamedi et al. 2004). As is the case with cross-linking to DNA, cross-linking to centromeric RNAs requires Dicer and Clr4, and is therefore siRNA- and chromatin-dependent. Second, siRNA generation requires chromatin components, including Clr4, Swi6, and the HDAC Sir2 (Hong et al. 2005; Li et al. 2005; Buhler et al. 2006). Finally, the association of RDRC with RITS is dependent on siRNAs as well as Clr4, suggesting that it occurs on chromatin (Motamedi et al.

Table 1. Conservation of RNAi and heterochromatin proteins

S. pombe	*A. thaliana*	*C. elegans*	*Drosophila*	*H. sapiens*
Dcr1	DCL1 to 4	Dcr-1	Dcr1 and 2	DCR-1
Ago1	AGO1 to 10	Rde-1, Alg-1 and -2	Ago1 to 3, Piwi	AGO-1 to AGO-4
	PRG-1 and 2, and 19 others	Aubergine/Sting	Piwi-1 to Piwi-4	PIWI-1 to PIWI-4
Chp1[a]	CMT3	–	–	–
Tas3[b]	–	–	–	–
Rdp1	RDR1 to 6	Ego-1, Rrf-1 to -3	–	–
Hrr1	SGS2/SDE3	ZK1067.2	GH20028p	KIAA1404
Cid12	[c]	Rde-3, Trf-4[c]	CG11265[c]	POLS[c]
Swi6	LHP1 (TFL2)	Hpl-1, Hpl-2, F32E10.6	HP1	HP1α, β, γ
Clr4	SUVH2 to 6	[d]	Su(var)3-9	SUV39H1 and 2
Rik1[e]	DDB1	M18.5	Ddb1	DDB1
Cul4	CUL4	Cul4	Cul4	CUL4
Sir2	SIR2	Sir2-1	Sir2	SIRT1
Eri1	ERI1	Eri-1	CG6393	THEX1

[a] An obvious ortholog of the chromodomain protein, Chp1, has not been identified in the other model organisms listed here, but most eukaryotic cells contain multiple chromodomain proteins. CMT3 in *Arabidopsis* is a chromodomain DNA methyltransferase, which acts in the same pathway as AGO4 and may be analogous to Chp1.

[b] No obvious orthologs of Tas3 have been identified, but it shares weak sequence similarity with a mouse ovary testis specific protein (NP_035152).

[c] Cid12 belongs to a large family of conserved proteins that share sequence similarity with the classic poly(A) polymerase as well as 2′-5′-oligoadenylate enzymes.

[d] *C. elegans* have about 20 SET domain proteins, but an H3K9 HKMT has not yet been identified in this organism.

[e] *S. pombe* contains another Rik1-like protein, Ddb1, which is involved in DNA damage repair. Metazoans and plants appear to contain only a single Rik1-like gene, called Ddb1, which has been shown to be involved in DNA damage repair, but it is unknown whether it also participates in heterochromatin formation.

2004). Thus, the generation of dsRNA and heterochromatic siRNAs may involve the recruitment of RDRC to chromatin-associated nascent pre-mRNA transcripts as illustrated in Figure 5 (Martienssen et al. 2005; Verdel and Moazed 2005). The fact that transcription and siRNA generation are likely to occur simultaneously reinforces the difference between RNA silencing mechanisms that mediate chromatin modifications and PTGS. However, this distinction is unlikely to be absolute. For example, in *C. elegans*, mutations in several chromatin components, similar to *S. pombe*, result in defects in RNAi and transposon-induced RNA silencing (see Table 1) (Sijen and Plasterk 2003; Grishok et al. 2005; Kim et al. 2005), raising the possibility that in some cases dsRNA synthesis and processing may occur on the chromosome regardless of whether silencing occurs at the TGS or PTGS level.

6 RNA–RNA Versus RNA–DNA Recognition Models

An outstanding question in working out the role of RNAi in heterochromatin assembly is whether RITS/RDRC associates with DNA or nascent RNA. The observation that tethering components of the RNAi machinery to a gene transcript can induce heterochromatin-dependent gene silencing in *cis* clearly demonstrates that this process can be promoted via initial interactions with nascent RNA transcripts (Buhler et al. 2006). Importantly, *cis*-restriction rules out the possibility that the initial events of dsRNA synthesis and siRNA generation occur on mature transcripts where mRNA

products from different alleles cannot be distinguished. Furthermore, a direct prediction of the RNA–RNA interaction model is that transcription at the target locus should be required for RNAi-mediated heterochromatin assembly. Although the requirement for transcription has not been directly tested, mutations in two different subunits of RNA polymerase II (RNA pol II), denoted Rpb2 and Rpb7, have specific defects in siRNA generation and heterochromatin assembly, but not on general transcription (Djupedal et al. 2005; Kato et al. 2005). This is reminiscent of Rbp1 mutants, which have defects in histone modifications (i.e., H3K4 methylation and H2B ubiquitination) coupled to transcriptional elongation (Hampsey and Reinberg 2003), and provides a precedent for the hypothesis that RNAi-mediated H3K9 methylation and heterochromatin formation could be coupled to transcriptional elongation via the association of RNAi complexes with RNA pol II. In fact, contrary to the widely held view that heterochromatin is an inaccessible structure that inhibits transcription, RNAi-mediated heterochromatin assembly has little or no effect on the association of RNA pol II with *S. pombe* centromeric repeats (Volpe et al. 2002; Djupedal et al. 2005; Kato et al. 2005; Buhler et al. 2006). Therefore, nascent RNA transcripts, which act as templates for RITS in the RNA–RNA recognition model, are present in heterochromatic domains (Fig. 4).

The RNA–RNA targeting model is also supported by the observation that components of both the RITS and RDRC complexes can be localized to noncoding centromeric RNAs using in vivo cross-linking experiments (Motamedi et al. 2004). This localization is siRNA-dependent, which suggests that it involves base-pairing interactions with the noncoding RNA. In addition, it requires the Clr4 HKMT, suggesting that it is coupled to binding of RITS to methylated H3K9 and occurs on chromatin. Nonetheless, the possibility that siRNAs can also recognize DNA directly through base-pairing interactions cannot be ruled out. For example, in plants, siRNAs that are complementary to promoter regions that are (presumably) not transcribed can still direct DNA methylation, another modification which takes place during heterochromatin formation within these regions (see Chapter 9).

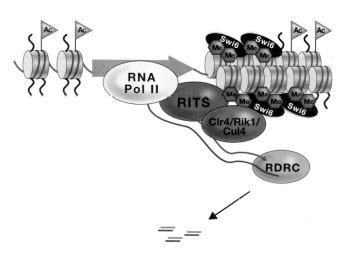

Figure 5. Model for Co-transcriptional dsRNA and siRNA Generation, and Recruitment of the Clr4-Rik1-Cul4 Histone Methyltransferase Complex in *S. pombe*

7 How Does RNAi Recruit Chromatin-modifying Enzymes?

The recruitment of Clr4 and Swi6 is a key step in initiating histone H3K9 methylation and heterochromatin assembly, through an autoregulatory modification-binding model (Figs. 3 and 4) (Grewal and Moazed 2003). However, because RITS association to chromatin and Clr4-catalyzed histone H3K9 methylation are interdependent processes, it has been difficult to determine the event that provides the initial trigger for RNAi-dependent heterochromatin assembly. One solution to this chicken-and-egg problem is that siRNA-dependent base-pairing interactions could provide the initial signal for heterochromatin assembly (Fig. 4). Consistent with this hypothesis, de novo generation of $ura4^+$ siRNAs promotes silencing of a previously active copy of the $ura4^+$ gene that is coupled to the recruitment of RITS and Swi6 to chromatin (Buhler et al. 2006). The initial binding of RITS may, however, be transient and difficult to detect; stable binding of RITS to chromatin would require dual interactions between (1) RITS-bound siRNAs and the nascent transcript and (2) the chromodomain binding of Chp1 to methylated H3K9. In this model, RITS itself directly recruits Clr4. Alternatively, Clr4 may be recruited by a parallel pathway that involves one or more DNA-binding proteins, as is the case at the silent mating type and telomeric regions (Jia et al. 2004; Kanoh et al. 2005). In either scenario, Clr4-mediated H3K9 methylation would be required to stabilize RITS association with chromatin, which then leads to the recruitment of RDRC, dsRNA synthesis, and siRNA generation (Fig. 5).

Clr4 has recently been found to be a component of a multiprotein complex that contains the heterochromatin protein Rik1, a Cullin E3 ubiquitin ligase, Cul4, and several other proteins (Hong et al. 2005; Horn et al. 2005; Jia et al. 2005; Li et al. 2005). These Clr4-associated proteins further strengthen the link between RNA and heterochromatin formation. The Rik1 protein is a member of a large family of β propeller WD repeat proteins that have been implicated in RNA or DNA binding. Members of this protein family include the Cleavage Polyadenylation Specificity Factor A (CPSF-A) involved in pre-mRNA splicing, and the DNA damage binding 1 (Ddb1) protein involved in binding UV-damaged DNA. CPSF-A is of particular interest because Rik1 shares sequence similarity with its putative RNA-binding domain involved in the recognition of mRNA polyadenylation sequences (Barabino et al. 2000). The Ddb1 protein, like Rik1, is a component of a Cul4 E3 ubiquitin ligase complex and is involved in the recognition and repair of UV-damaged DNA (Higa et al. 2003; Zhong et al. 2003). An exciting possibility is that Rik1 acts in a fashion that is similar to CPSF-A and Ddb1, binding to an RNAi-generated product during heterochromatin assembly (Fig. 5).

8 RNAi-mediated Chromatin and DNA Modifications in *Arabidopsis*

The mechanism by which RNAi guides heterochromatic modifications in plants is similar to the mechanism in fission yeast, but there are also many differences. The most important difference is that plants have methylated DNA at many repressive heterochromatin regions: In this respect they resemble vertebrates, but differ from worms and *Drosophila* (Lippman and Martienssen 2004). Four genetic screens for mutants that relieve RNA-mediated TGS have recovered mutants in H3K9-specific HKMTs, and in RNAi components, but they have also uncovered the required function of DNA methyltransferases, SWI/SNF remodeling complexes, and a novel RNA polymerase (Baulcombe 2004). These screens are described in detail in Chapter 9, but here we briefly compare the mechanism in fission yeast and plants.

Each of the silencing mutant screens used inverted repeats introduced in *trans* to induce the silencing of endogenous or transgenic reporter genes. Relief from silencing indicates a mutation that has arisen in a necessary component of the silencing pathway. The endogenous genes used were *PAI2* (involved in amino acid biosynthesis) (Mathieu and Bender 2004) and *SUPERMAN* (a transcription factor that regulates flower development) (Chan et al. 2004), and the reporter genes used were driven by either a strong viral promoter or a strong seed-specific promoter (Matzke et al. 2004). In each case, the promoter was targeted for silencing, in some cases along with the rest of the gene. A number of the genes found through these screens are illustrated in Figure 6 (see also Table 1 of Chapter 9). Only one RNAi mutant was identified, in only one of the screens, and this was the argonaute gene *AGO4*. However, three of the screens recovered mutants in DNA methyltransferases, including *MET1* and *CMT3*. A third DNA methyltransferase related to the mammalian DNMT3 was identified by reverse genetics, as this activity is encoded by *DRM1* and *DRM2*, two redundant genes unable to be determined in single mutant screens (see Chapter 9). Indeed, redundancy may account for the failure to recover additional components of the RNAi apparatus: for example, although *DCL3 (DICER-LIKE 3)* and *RNA-DEPENDENT RNA POLYMERASE 2 (RDRP)* are predominantly required for production of the 24-nt siRNA associated with transposons and repeats, at least two other *DCL* genes in *Arabidopsis* can substitute for *DCL3* to some extent (Gasciolli et al. 2005).

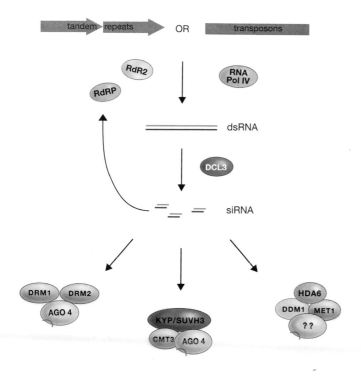

Figure 6. Summary of RNAi and Chromatin Proteins Required for RNAi-mediated DNA and Histone Methylation in *Arabidopsis*

Synthesis of dsRNA from repeated DNA elements provides a substrate for Dicer-mediated cleavage and siRNA generation (DCL3 and other Dicers). RNA-directed RNA polymerases (RdRP, RdR2) and RNA polymerase IV (RNA Pol IV) may be directly involved in the synthesis of dsRNA or its amplification. siRNAs then load onto Argonaute proteins (e.g., AGO4), which is likely to help target cognate repeat sequences for DNA and H3K9 methylation in association with other factors.

Other mutants found in the *PAI2* screen included mutants in the H3K9 methyltransferase gene *KYP/SUVH4* and the chromodomain-containing DNA methyltransferase gene *CMT3*. The parallels with fission yeast in this case are striking, as the RITS complex contains both an Argonaute protein and the chromodomain protein Chp1, which depends on H3K9 methylation for its association with the chromosome. Unlike fission yeast, however, loss of CMT3 or of H3K9me2 does not result in loss of siRNA in *Arabidopsis* (Lippman et al. 2003), and it is not yet clear whether these proteins form a complex with AGO4. There are, however, several other H3K9-specific HKMTs in *Arabidopsis*, at least three of which have genetic function, so redundancy may be part of the explanation here as well (Ebbs et al. 2005).

Mutants in the other DNA methyltransferases, *MET1* and *DRM1/2*, in contrast to mutants in *CMT3*, do result in loss of siRNA accumulation, at least from a subset of transposons and from tandem repeats, which generally

produce siRNA if they are transcribed (Martienssen 2003). In these cases, loss of siRNA is correlated with the loss of H3K9me2 (Cao et al. 2003; Lippman et al. 2003). Mutants in the SWI2/SNF2 chromatin-remodeling ATPase DDM1 (decreased DNA methylation) also abolish siRNA and H3K9me2 accumulation from a wide range of transposons, although when siRNA is retained, so is H3K9me2 (Lippman et al. 2003). It is possible, therefore, that siRNA in plants is bound to the chromosome via methylated DNA instead of, or in addition to, binding via methylated histones as is the case in *S. pombe* (Fig. 7).

DDM1 has an exquisite specificity for transposons and repeats, and must somehow recognize these as being different from genes. siRNA, perhaps bound to the chromosome by methyl-binding proteins, would have the required specificity to make this distinction. Transposons and repeats in *Arabidopsis* are a major source of 24-nt, and some 21-nt, siRNA, consistent with this idea (Lippman et al. 2004). Centromeric satellite repeats, which are arranged in tens of thousands of tandem copies on either side of each centromere, are also transcribed and processed by RNAi (Fig. 6). This processing depends on DCL3, RDR2, and DDM1. Silencing also depends on H3K9me2 and CMT3. However, silencing is more complex than in fission yeast, as retrotransposon insertions into the repeats can silence them, and this depends on other mechanisms including MET1, DDM1, and the histone deacetylase HDA6 (May et al. 2005).

As mentioned earlier, in fission yeast, subunits of RNA pol II are required for silencing and siRNA production, supporting the idea that the RNAi- and chromatin-modification apparatus is recruited to the chromosome by nascent transcripts (Fig. 5). In *Arabidopsis*, two subunits of a novel RNA polymerase (RNA pol IV) were recovered in one of the four screens mentioned above (Kanno et al. 2005) but were first isolated as weak mutants in PTGS, along with mutants in RNA-dependent RNA polymerase (Herr et al. 2005). It is not yet known what template is used by RNA pol IV, but both methylated DNA (Onodera et al. 2005) and double-stranded RNA have been proposed (Vaughn and Martienssen 2005). Only the largest subunits are unique to RNA pol IV, which presumably uses the same complement of small subunits as RNA pol II. Additional SWI2/SNF2 chromatin remodelers that were also recovered in these screens may alter local chromatin structure to facilitate processivity of RNA polymerases. It is therefore likely that they facilitate transcription by RNA pol IV (Kanno et al. 2004). A similar role can be proposed for DDM1, although the require-

a

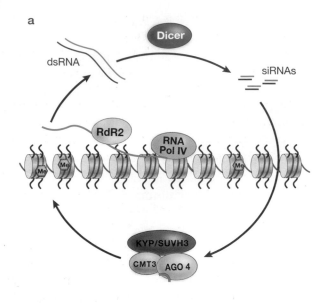

b

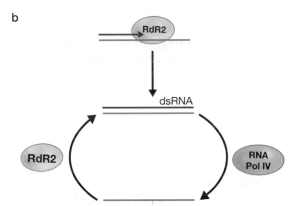

Figure 7. Hypothetical Models for the Role of RNA Pol IV in RNAi-directed DNA and/or Histone H3 Methylation

(*a*) RNA pol IV transcribes methylated DNA; RdR2 synthesizes dsRNA using the RNA pol IV product. siRNAs then direct a methyltransferase complex to the chromosome. (*b*) RNA pol IV uses a dsRNA template synthesized by RdR2 to produce more ssRNA template.

ment for DDM1 (also a chromatin remodeler) in silencing transposons is far more severe than that of RNA pol IV or the other SWI2/SNF2 proteins.

Genes can be silenced epigenetically by nearby transposons, and an important example in *Arabidopsis* is the imprinted homeobox gene *FWA* (Kinoshita et al. 2004). The first two exons of this gene are noncoding and form a tandem repeat due to the ancient integration of a SINE element at this site (Lippman et al. 2004). siRNAs from the SINE element are lost in *met1* (i.e., DNA methyltransferase) mutants, and the gene becomes strongly up-regulated in the inflorescence meristem, resulting in late

flowering. No such phenotype is observed in mutants of RNAi, even when siRNA is lost. Instead, RNAi may play a role in initiating *FWA* silencing, because *FWA* transgenes are rapidly silenced when first introduced into a plant, and this silencing depends on *DCL3*, *RDR2*, and *AGO4* (Chan et al. 2004). Silencing might then be maintained by DNA methylation, regulated by *MET1*. Similarly, transposons that lose siRNA in *met1* mutants cannot be re-silenced in backcrosses, but those that do not lose siRNA can be re-silenced, implicating siRNA in reestablishing silencing in *cis* rather than in *trans* (Lippman et al. 2003). Similarly, late-flowering *FWA* alleles are stably inherited in backcrosses after being removed from *met1* or *ddm1* mutant backgrounds because maintenance of epialleles is heritable (Soppe et al. 2000).

Finally, it is possible that miRNA may guide DNA methylation of genes in some circumstances. mRNA from the *PHABULOSA* gene is targeted for cleavage by miRNA 165 and 166 in *Arabidopsis*, and the gene itself is methylated downstream from the region that matches the miRNA. Interestingly, this match spans an exon junction, so that the spliced RNA must interact with the miRNA if this guides methylation (Bao et al. 2004). However, other members of the same gene family are not methylated in this way, and neither are most other miRNA target genes (Martienssen et al. 2004; Ronemus and Martienssen 2005). Conversely, several other genes are methylated in the *Arabidopsis* genome, and typically at their 3′ end, in a mechanism that requires MET1 but not DDM1 (Lippman et al. 2004; Tran et al. 2005). It remains to be seen whether RNA is involved in these cases.

9 Conservation of RNAi-mediated Chromatin Modifications in Animals

Perhaps the most widely studied examples of epigenetic silencing are found in animals, including *Drosophila* and *C. elegans*, as well as the mouse. The role of RNA and RNA interference in transcriptional silencing and heterochromatic modifications appears to be conserved in some model animals as well as in protists and plants. In *Drosophila*, both PIWI and the PIWI class Argonaute homolog, Aubergine (Sting), are required for epigenetic and heterochromatic silencing (see also Chapter 5). *Gypsy* retrotransposons are the target of silencing in ovary follicle cells and female gonads by PIWI itself (Sarot et al. 2004). This is mediated by the heterochromatic gene *Flamenco* (with as-yet-unknown function), and requires the 5′UTR of the *Gypsy* polyprotein gene. The detection of 25–27-nt small RNAs from this region suggests it occurs via an RNAi-mediated mechanism. Cut-and-paste DNA transposons are also affected by RNAi. For example, certain telomeric P elements (a type of DNA transposon) can suppress transposition to elsewhere in the genome when inherited through the female germ line, resulting in a strongly repressive "cytotype." This repression is completely dependent on the PIWI homolog, Aubergine, as well as the Swi6 homolog HP1 (Reiss et al. 2004). However, not all P-repressive cytotypes such as those mediated by other, nontelomeric P elements are dependent on Aubergine or HP1.

Unlinked transgenes in *Drosophila* are silenced posttranscriptionally when present in many copies (Pal-Bhadra et al. 1997, 2002). Silencing is associated with large amounts of 21-nt siRNA and depends on PIWI. Transgene fusions can also silence each other transcriptionally, in a manner that requires the Polycomb chromatin repressor. This silencing is not associated with increased levels of siRNA from the transgene transcript but is (largely) dependent on PIWI. Involvement of Polycomb in this example, and HP1 in other examples, of PIWI-dependent silencing, implicates the RNAi pathway and histone methylation in the silencing process. Tandem transgene arrays also exhibit position-effect variegation in *Drosophila*, and this variegation is strongly suppressed by mutants in HP1 as well as in piwi, aubergine, and the putative RNA helicase Spindle-E (homeless) (Pal-Bhadra et al. 2004). Transgenes inserted within centric heterochromatin are also affected, and heterochromatic levels of H3K9me2 are reduced in spindle-E mutant cells. These observations strongly support a role for both chromatin proteins and components of the RNAi pathway in gene silencing within *Drosophila* heterochromatin.

In the *Drosophila* male germ line, the heterochromatic *Suppressor of Stellate* repeats (*Su(ste)*), located on the Y chromosome, are transcribed first on the antisense strand, and then on both strands during spermatocyte development, possibly following the insertion of a nearby transposon (Aravin et al. 2001). These nuclear transcripts are required to silence sense transcripts of the closely related X-linked *Stellate* gene, whose overexpression results in defects in spermatogenesis. Although heterochromatic sequences are involved, silencing in this case appears to be posttranscriptional, is associated with 25–27-nt siRNA, and depends on both *Aubergine* and *Spindle-E*.

In *C. elegans*, examples of TGS in somatic cells have been reported. This depends on the RNAi pathway genes *rde-1*, *dcr-1*, *rde-4*, and *rrf-1*, as well as HP1 homologs and the histone modification apparatus (Grishok et al. 2005).

Somatic heterochromatin is not widespread in *C. elegans*, but an example of naturally occurring RNAi-dependent heterochromatic silencing has been described in the germ line (Sijen and Plasterk 2003). During meiosis, unpaired sequences, such as the X chromosome in males, are silenced via H3K9me2, and this silencing depends on RNA-dependent RNA polymerase (Maine et al. 2005; see Chapter 15), reminiscent of meiotic silencing of unpaired DNA (MSUD) in *Neurospora* (see Shiu et al. 2001; Chapter 6). However, other components of the RNAi apparatus have not yet been implicated in this process, and it is not known whether it is related mechanistically to RNAi-mediated heterochromatin assembly in fission yeast.

Finally, like *Drosophila*, mammalian cells lack genes related to RNA-dependent RNA polymerases found in plants, worms, and fungi. Nonetheless, antisense RNA has been implicated in the most widely studied epigenetic phenomena of all, imprinting and X inactivation (see Chapters 19 and 17, respectively). In the case of X inactivation, a 17-kb spliced and polyadenylated noncoding RNA known as Xist is required to silence the inactive X chromosome from which it is expressed. Conversely, *Xist* itself is silenced on the active X chromosome, a process that depends in part on the antisense RNA *Tsix*. Silencing is accompanied by modification of histones associated with upstream chromatin regions, which are marked with H3K9me2 and H3K27me3 (see Chapter 17). Silencing of other imprinted loci in the mouse, including *Igf2r* and the *Dlk1-Gtl2* region, is also maintained by antisense transcripts from the paternal or maternal allele, respectively. In the case of *Dlk1-Gtl2*, this noncoding RNA is specifically processed into miRNA that targets the antisense transcript from the paternal allele, encoding a *sushi* (*gypsy*) class retrotransposon (Davis et al. 2005).

Although the parallels with forward and reverse transcription from heterochromatic repeats in *S. pombe* are many, a role for RNAi itself in imprinting and X inactivation has so far proved elusive. Nonetheless, introduction of siRNA into cancer cell lines can result in chromatin being marked with H3K9me2 at homologous promoters (Ting et al. 2005). In some cases, it can also result in DNA methylation (Morris et al. 2004), perhaps mediated by direct binding of small RNA with DNA methyltransferases and DNA methylation binding proteins (Jeffery and Nakielny 2004). Finally, *Dicer* knockout vertebrate cell lines have chromosome segregation defects reminiscent of those found in fission yeast mutants, accompanied by changes in heterochromatic morphology, expression of satellite repeats, and mislocalization of cohesin (Fukagawa et al. 2004; Kanellopoulou et al. 2005).

10 Concluding Remarks

The possibility that genes may be regulated by small RNA molecules was suggested over 40 years ago (Jacob and Monod 1961), as well as the notion that "control RNA" might be related to repeats (Britten and Davidson 1969). Since the identification of the lambda and lac repressors as site-specific DNA-binding proteins in *Escherichia coli* and the infecting bacteriophage *lambda* (Gilbert and Muller-Hill 1966; Ptashne 1967), studies of gene regulation have focused almost exclusively on the role of nucleic-acid-binding proteins as specificity factors. The discovery of small RNA molecules as specificity agents in diverse RNA silencing mechanisms now clearly establishes a role for RNA as a sequence-specific regulator of genes and their RNA products. Studies in fission yeast, *Arabidopsis*, and other model organisms have revealed a surprisingly direct role for small RNAs in mediating epigenetic modifications of the genome that direct gene silencing and contribute to heterochromatic domains necessary for genome stability and nuclear division. Many important mechanistic questions remain at large, and future studies are likely to provide more surprises about how RNA regulates gene expression.

References

Aravin A.A., Naumova N.M., Tulin A.V., Vagin V.V., Rozovsky Y.M., and Gvozdev V.A. 2001. Double-stranded RNA-mediated silencing of genomic tandem repeats and transposable elements in the *D. melanogaster* germline. *Curr. Biol.* **11**: 1017–1027.

Bao N., Lye K.W., and Barton M.K. 2004. MicroRNA binding sites in *Arabidopsis* class III HD-ZIP mRNAs are required for methylation of the template chromosome. *Dev. Cell* **7**: 653–662.

Barabino S.M., Ohnacker M., and Keller W. 2000. Distinct roles of two Yth1p domains in 3′-end cleavage and polyadenylation of yeast pre-mRNAs. *EMBO J.* **19**: 3778–3787.

Bartel D.P. 2004. MicroRNAs: Genomics, biogenesis, mechanism, and function. *Cell* **116**: 281–297.

Baulcombe D. 2004. RNA silencing in plants. *Nature* **431**: 356–363.

Bernstein E., Caudy A.A., Hammond S.M., and Hannon G.J. 2001. Role for a bidentate ribonuclease in the initiation step of RNA interference. *Nature* **409**: 363–366.

Britten R.J. and Davidson E.H. 1969. Gene regulation for higher cells: A theory. *Science* **165**: 349–357.

Buhler M., Verdel A., and Moazed D. 2006. Tethering RITS to a nascent transcript initiates RNAi- and heterochromatin-dependent gene silencing. *Cell* **125**: 873–886.

Cao X., Aufsatz W., Zilberman D., Mette M.F., Huang M.S., Matzke M., and Jacobsen S.E. 2003. Role of the DRM and CMT3 methyltransferases in RNA-directed DNA methylation. *Curr. Biol.* **13**: 2212–2217.

Chan S.W., Zilberman D., Xie Z., Johansen L.K., Carrington, J.C., and Jacobsen S.E. 2004. RNA silencing genes control de novo DNA methylation. *Science* **303**: 1336.

Chen C.C., Simard M.J., Tabara H., Brownell D.R., McCollough J.A., and Mello C.C. 2005. A member of the polymerase β nucleotidyl-transferase superfamily is required for RNA interference in *C. elegans*. *Curr. Biol.* **15**: 378–383.

Dalmay T., Hamilton A., Mueller E., and Baulcombe D.C. 2000. *Potato virus X* amplicons in *Arabidopsis* mediate genetic and epigenetic gene silencing. *Plant Cell* **12**: 369–379.

Davis E., Caiment F., Tordoir X., Cavaille J., Ferguson-Smith A., Cockett N., Georges M., and Charlier C. 2005. RNAi-mediated allelic *trans*-interaction at the imprinted *Rtl1/Peg11* locus. *Curr. Biol.* **15**: 743–749.

Djupedal I., Portoso M., Spahr H., Bonilla C., Gustafsson C.M., Allshire R.C., and Ekwall K. 2005. RNA Pol II subunit Rpb7 promotes centromeric transcription and RNAi-directed chromatin silencing. *Genes Dev.* **19**: 2301–2306.

Ebbs M.L., Bartee L., and Bender J. 2005. H3 lysine 9 methylation is maintained on a transcribed inverted repeat by combined action of SUVH6 and SUVH4 methyltransferases. *Mol. Cell. Biol.* **25**: 10507–10515.

Elbashir S.M., Lendeckel W., and Tuschl T. 2001. RNA interference is mediated by 21- and 22-nucleotide RNAs. *Genes Dev.* **15**: 188–200.

Filipowicz W., Jaskiewicz L., Kolb F.A., and Pillai R.S. 2005. Post-transcriptional gene silencing by siRNAs and miRNAs. *Curr. Opin. Struct. Biol.* **15**: 331–341.

Fire A., Xu S., Montgomery M.K., Kostas S.A., Driver S.E., and Mello C.C. 1998. Potent and specific genetic interference by double-stranded RNA in *Caenorhabditis elegans*. *Nature* **391**: 806–811.

Francis N.J. and Kingston R.E. 2001. Mechanisms of transcriptional memory. *Nat. Rev. Mol. Biol.* **2**: 409–421.

Fukagawa T., Nogami M., Yoshikawa M., Ikeno M., Okazaki T., Takami Y., Nakayama T., and Oshimura M. 2004. Dicer is essential for formation of the heterochromatin structure in vertebrate cells. *Nat. Cell Biol.* **6**: 784–791.

Galagan J.E. and Selker E.U. 2004. RIP: The evolutionary cost of genome defense. *Trends Genet.* **20**: 417–423.

Gasciolli V., Mallory A.C., Bartel D.P., and Vaucheret H. 2005. Partially redundant functions of *Arabidopsis* DICER-like enzymes and a role for DCL4 in producing *trans*-acting siRNAs. *Curr. Biol.* **15**: 1494–1500.

Gasser S.M., Hediger F., Taddei A., Neumann F.R., and Gartenberg M.R. 2004. The function of telomere clustering in yeast: The circe effect. *Cold Spring Harbor Symp. Quant. Biol.* **69**: 327–337.

Gilbert W. and Muller-Hill B. 1966. Isolation of the Lac repressor. *Proc. Natl. Acad. Sci.* **56**: 1891–1898.

Grewal S.I. 2000. Transcriptional silencing in fission yeast. *J. Cell. Physiol.* **184**: 311–318.

Grewal S.I. and Moazed D. 2003. Heterochromatin and epigenetic control of gene expression. *Science* **30**: 798–802.

Grishok A., Sinskey J.L., and Sharp P.A. 2005. Transcriptional silencing of a transgene by RNAi in the soma of *C. elegans*. *Genes Dev.* **19**: 683–696.

Hall I.M., Shankaranarayana G.D., Noma K., Ayoub N., Cohen A., and Grewal S.I. 2002. Establishment and maintenance of a heterochromatin domain. *Science* **297**: 2232–2237.

Hamilton A.J. and Baulcombe D.C. 1999. A species of small antisense RNA in posttranscriptional gene silencing in plants. *Science* **286**: 950–952.

Hampsey M. and Reinberg D. 2003. Tails of intrigue: Phosphorylation of RNA polymerase II mediates histone methylation. *Cell* **113**: 429–432.

Hannon G.J. 2002. RNA interference. *Nature* **418**: 244–251.

Herr A.J., Jensen M.B., Dalmay T., and Baulcombe D.C. 2005. RNA polymerase IV directs silencing of endogenous DNA. *Science* **308**: 118–120.

Higa L.A., Mihaylov I.S., Banks D.P., Zheng J., and Zhang H. 2003. Radiation-mediated proteolysis of CDT1 by CUL4-ROC1 and CSN complexes constitutes a new checkpoint. *Nat. Cell Biol.* **5:** 1008–1015.

Hong E.E., Villen J., Gerace E.L., Gygi, S.P., and Moazed D. 2005. A Cullin E3 ubiquitin ligase complex associates with Rik1 and the Clr4 histone H3-K9 methyltransferase and is required for RNAi-mediated heterochromatin formation. *RNA Biol.* **2:** 106–111.

Horn P.J., Bastie J.N., and Peterson C.L. 2005. A Rik1-associated, cullin-dependent E3 ubiquitin ligase is essential for heterochromatin formation. *Genes Dev.* **19:** 1705–1714.

Jacob F. and Monod J. 1961. Genetic regulatory mechanisms in the synthesis of proteins. *J. Mol. Biol.* **3:** 318–356.

Jeffery L. and Nakielny S. 2004. Components of the DNA methylation system of chromatin control are RNA-binding proteins. *J. Biol. Chem.* **279:** 49479–49487.

Jia S., Kobayashi R., and Grewal S.I. 2005. Ubiquitin ligase component Cul4 associates with Clr4 histone methyltransferase to assemble heterochromatin. *Nat. Cell Biol.* **7:** 1007–1013.

Jia S., Noma K., and Grewal S.I. 2004. RNAi-independent heterochromatin nucleation by the stress-activated ATF/CREB family proteins. *Science* **304:** 1971–1976.

Kanellopoulou C., Muljo S.A., Kung A.L., Ganesan S., Drapkin R., Jenuwein T., Livingston D.M., and Rajewsky K. 2005. Dicer-deficient mouse embryonic stem cells are defective in differentiation and centromeric silencing. *Genes Dev.* **19:** 489–501.

Kanno T., Huettel B., Mette M.F., Aufsatz W., Jaligot E., Daxinger L., Kreil D.P., Matzke M., and Matzke A.J. 2005. Atypical RNA polymerase subunits required for RNA-directed DNA methylation. *Nat. Genet.* **37:** 761–765.

Kanno T., Mette M.F., Kreil D.P., Aufsatz W., Matzke M., and Matzke A.J. 2004. Involvement of putative SNF2 chromatin remodeling protein DRD1 in RNA-directed DNA methylation. *Curr. Biol.* **14:** 801–805.

Kanoh J., Sadaie M., Urano T., and Ishikawa F. 2005. Telomere binding protein Taz1 establishes Swi6 heterochromatin independently of RNAi at telomeres. *Curr. Biol.* **15:** 1808–1819.

Kato H., Goto D.B., Martienssen R.A., Urano T., Furukawa K., and Murakami Y. 2005. RNA polymerase II is required for RNAi-dependent heterochromatin assembly. *Science* **309:** 467–469.

Kim J.K., Gabel H.W., Kamath R.S., Tewari M., Pasquinelli A., Rual J.F., Kennedy S., Dybbs M., Bertin N., Kaplan J.M., et al. 2005. Functional genomic analysis of RNA interference in *C. elegans*. *Science* **308:** 1164–1167.

Kinoshita T., Miura A., Choi Y., Kinoshita Y., Cao X., Jacobsen S.E., Fischer R.L., and Kakutani T. 2004. One-way control of FWA imprinting in *Arabidopsis* endosperm by DNA methylation. *Science* **303:** 521–523.

Li F., Goto D.B., Zaratiegui M., Tang X., Martienssen R., and Cande W.Z. 2005. Two novel proteins, Dos1 and Dos2, interact with Rik1 to regulate heterochromatic RNA interference and histone modification. *Curr. Biol.* **15:** 1448–1457.

Li H.W. and Ding S.W. 2005. Antiviral silencing in animals. *FEBS Lett.* **579:** 5965–5973.

Lippman Z., Gendrel A.V., Black M., Vaughn M.W., Dedhia N., McCombie W.R., Lavine, K., Mittal V., May B., Kasschau K.D., et al. 2004. Role of transposable elements in heterochromatin and epigenetic control. *Nature* **430:** 471–476.

Lippman Z. and Martienssen R. 2004. The role of RNA interference in heterochromatic silencing. *Nature* **431:** 364–370.

Lippman Z., May B., Yordan C., Singer T., and Martienssen R. 2003. Distinct mechanisms determine transposon inheritance and methyla-

tion via small interfering RNA and histone modification. *PLoS Biol.* **1:** E67.

Maine E.M., Hauth J., Ratliff T., Vough V.E., She X., and Kelly W.G. 2005. EGO-1, a putative RNA-dependent RNA polymerase, is required for heterochromatin assembly on unpaired dna during *C. elegans* meiosis. *Curr. Biol.* **15:** 1972–1978.

Martienssen R.A. 2003. Maintenance of heterochromatin by RNA interference of tandem repeats. *Nat. Genet.* **35:** 213–214.

Martienssen R.A., Zaratiegui M., and Goto D.B. 2005. RNA interference and heterochromatin in the fission yeast *Schizosaccharomyces pombe*. *Trends Genet.* **21:** 450–456.

Martienssen R., Lippman Z., May B., Ronemus M., and Vaughn M. 2004. Transposons, tandem repeats, and the silencing of imprinted genes. *Cold Spring Harbor Symp. Quant. Biol.* **69:** 371–379.

Mathieu O. and Bender J. 2004. RNA-directed DNA methylation. *J. Cell Sci.* **117:** 4881–4888.

Matzke M., Aufsatz W., Kanno T., Daxinger L., Papp I., Mette M.F., and Matzke A.J. 2004. Genetic analysis of RNA-mediated transcriptional gene silencing. *Biochim. Biophys. Acta* **1677:** 129–141.

May B.P., Lippman Z.B., Fang Y., Spector D.L., and Martienssen R.A. 2005. Differential regulation of strand-specific transcripts from *Arabidopsis* centromeric satellite repeats. *PLoS Genet.* **1:** e79.

Mette M.F., van der Winden J., Matzke M.A., and Matzke A.J. 1999. Production of aberrant promoter transcripts contributes to methylation and silencing of unlinked homologous promoters in trans. *EMBO J.* **18:** 241–248.

Mette M.F., Aufsatz W., van der Winden J., Matzke M.A., and Matzke A.J. 2000. Transcriptional silencing and promoter methylation triggered by double-stranded RNA. *EMBO J.* **19:** 5194–5201.

Morris K.V., Chan S.W., Jacobsen S.E., and Looney D.J. 2004. Small interfering RNA-induced transcriptional gene silencing in human cells. *Science* **305:** 1289–1292.

Motamedi M.R., Verdel A., Colmenares S.U., Gerber S.A., Gygi S.P., and Moazed D. 2004. Two RNAi complexes, RITS and RDRC, physically interact and localize to noncoding centromeric RNAs. *Cell* **119:** 789–802.

Napoli C., Lemieux C., and Jorgensen R. 1990. Introduction of a chimeric chalcone synthase gene into Petunia results in reversible co-suppression of homologous genes *in trans*. *Plant Cell* **2:** 279–289.

Noma K., Sugiyama T., Cam H., Verdel A., Zofall M., Jia S., Moazed D., and Grewal S.I. 2004. RITS acts in *cis* to promote RNA interference-mediated transcriptional and post-transcriptional silencing. *Nat. Genet.* **36:** 1174–1180.

Onodera Y., Haag J.R., Ream T., Nunes P.C., Pontes O., and Pikaard C.S. 2005. Plant nuclear RNA polymerase IV mediates siRNA and DNA methylation-dependent heterochromatin formation. *Cell* **120:** 613–622.

Pal-Bhadra M., Bhadra U., and Birchler J.A. 1997. Cosuppression in *Drosophila*: Gene silencing of *Alcohol dehydrogenase* by *white-Adh* transgenes is Polycomb dependent. *Cell* **90:** 479–490.

———. 1999. Cosuppression of nonhomologous transgenes in *Drosophila* involves mutually related endogenous sequences. *Cell* **99:** 35–46.

———. 2002. RNAi related mechanisms affect both transcriptional and posttranscriptional transgene silencing in *Drosophila*. *Mol. Cell.* **9:** 315–327.

Pal-Bhadra M., Leibovitch B.A., Gandhi S.G., Rao M., Bhadra U., Birchler J.A., and Elgin S.C. 2004. Heterochromatic silencing and HP1 localization in Drosophila are dependent on the RNAi machinery. *Science* **303:** 669–672.

Partridge J.F., Scott K.S., Bannister A.J., Kouzarides T., and Allshire R.C. 2002. *cis*-acting DNA from fission yeast centromeres mediates histone H3 methylation and recruitment of silencing factors and cohesin to an ectopic site. *Curr. Biol.* **12:** 1652–1660.

Petrie V.J., Wuitschick J.D., Givens C.D., Kosinski A.M., and Partridge J.F. 2005. RNA interference (RNAi)-dependent and RNAi-independent association of the Chp1 chromodomain protein with distinct heterochromatic loci in fission yeast. *Mol. Cell. Biol.* **25:** 2331–2346.

Pidoux A.L. and Allshire R.C. 2004. Kinetochore and heterochromatin domains of the fission yeast centromere. *Chromosome Res.* **12:** 521–534.

Plasterk R.H. 2002. RNA silencing: The genome's immune system. *Science* **296:** 1263–1265.

Provost P., Silverstein R.A., Dishart D., Walfridsson J., Djupedal I., Kniola B., Wright A., Samuelsson B., Radmark O., and Ekwall K. 2002. Dicer is required for chromosome segregation and gene silencing in fission yeast cells. *Proc. Natl. Acad. Sci.* **99:** 16648–16653.

Ptashne M. 1967. Specific binding of the lambda phage repressor to lambda DNA. *Nature* **214:** 232–234.

Reinhart B.J. and Bartel D.P. 2002. Small RNAs correspond to centromere heterochromatic repeats. *Science* **297:** 1831.

Reiss D., Josse T., Anxolabehere D., and Ronsseray S. 2004. *aubergine* mutations in *Drosophila melanogaster* impair P cytotype determination by telomeric P elements inserted in heterochromatin. *Mol. Genet. Genomics.* **272:** 336–343.

Ronemus M. and Martienssen R. 2005. RNA interference: Methylation mystery. *Nature* **433:** 472–473.

Sarot E., Payen-Groschene G., Bucheton A., and Pelisson A. 2004. Evidence for a *piwi*-dependent RNA silencing of the *gypsy* endogenous retrovirus by the *Drosophila melanogaster flamenco* gene. *Genetics* **166:** 1313–1321.

Shiu P.K., Raju N.B., Zickler D., and Metzenberg R.L. 2001. Meiotic silencing by unpaired DNA. *Cell* **107:** 905–916.

Sigova A., Rhind N., and Zamore P.D. 2004. A single Argonaute protein mediates both transcriptional and posttranscriptional silencing in *Schizosaccharomyces pombe*. *Genes Dev.* **18:** 2359–2367.

Sijen T. and Plasterk R.H. 2003. Transposon silencing in the *Caenorhabditis elegans* germ line by natural RNAi. *Nature* **426:** 310–314.

Sijen T., Fleenor J., Simmer F., Thijssen K.L., Parrish S., Timmons L., Plasterk R.H., and Fire A. 2001. On the role of RNA amplification in dsRNA-triggered gene silencing. *Cell* **107:** 465–476.

Soppe W.J., Jacobsen S.E., Alonso-Blanco C., Jackson J.P., Kakutani T.,

Koornneef M., and Peeters A.J. 2000. The late flowering phenotype of *fwa* mutants is caused by gain-of-function epigenetic alleles of a homeodomain gene. *Mol. Cell.* **6:** 791–802.

Sugiyama T., Cam H., Verdel A., Moazed D., and Grewal S.I. 2005. RNA-dependent RNA polymerase is an essential component of a self-enforcing loop coupling heterochromatin assembly to siRNA production. *Proc. Natl. Acad. Sci.* **102:** 152–157.

Ting A.H., Schuebel K.E., Herman J.G., and Baylin S.B. 2005. Short double-stranded RNA induces transcriptional gene silencing in human cancer cells in the absence of DNA methylation. *Nat. Genet.* **37:** 906–910.

Tran R.K., Zilberman D., de Bustos C., Ditt R.F., Henikoff J.G., Lindroth A.M., Delrow J., Boyle T., Kwong S., Bryson T.D., et al. 2005. Chromatin and siRNA pathways cooperate to maintain DNA methylation of small transposable elements in *Arabidopsis*. *Genome Biol.* **6:** R90.

Vaughn M.W. and Martienssen R.A. 2005. Finding the right template: RNA Pol IV, a plant-specific RNA polymerase. *Mol. Cell* **17:** 754–756.

Verdel A., Jia S., Gerber S., Sugiyama T., Gygi S., Grewal S.I., and Moazed D. 2004. RNAi-mediated targeting of heterochromatin by the RITS complex. *Science* **303:** 672–676.

Verdel A. and Moazed D. 2005. RNAi-directed assembly of heterochromatin in fission yeast. *FEBS Lett.* **579:** 5872–5878.

Volpe T.A., Kidner C., Hall I.M., Teng G., Grewal S.I., and Martienssen R.A. 2002. Regulation of heterochromatic silencing and histone H3 lysine-9 methylation by RNAi. *Science* **297:** 1833–1837.

Volpe T., Schramke V., Hamilton G.L., White S.A., Teng G., Martienssen R.A., and Allshire R.C. 2003. RNA interference is required for normal centromere function in fission yeast. *Chromosome Res.* **11:** 137–146.

Wassenegger M., Heimes S., Riedel L., and Sanger H.L. 1994. RNA-directed de novo methylation of genomic sequences in plants. *Cell* **76:** 567–576.

Zamore P.D. 2002. Ancient pathways programmed by small RNAs. *Science* **296:** 1265–1269.

Zamore P.D., Tuschl T., Sharp P.A., and Bartel D.P. 2000. RNAi: Double-stranded RNA directs the ATP-dependent cleavage of mRNA at 21 to 23 nucleotide intervals. *Cell* **101:** 25–33.

Zhong W., Feng H., Santiago F.E., and Kipreos E.T. 2003. CUL-4 ubiquitin ligase maintains genome stability by restraining DNA-replication licensing. *Nature* **423:** 885–889.

Epigenetic Regulation in Plants

Marjori Matzke and Ortrun Mittelsten Scheid

Gregor Mendel Institute of Molecular Plant Biology, Austrian Academy of Sciences, A-1030 Vienna, Austria

CONTENTS

1. Benefits of Plants in Epigenetic Research, 169

    1.1 *Plants and Mammals Are Similar in Terms of (Epi)Genome Organization, 169*

    1.2 *Plants Provide Additional Topics for Epigenetic Research, 169*

    1.3 *Plants Tolerate Methodological Approaches That Are Difficult in Mammals, 170*

    1.4 *Plants Have a Proven Record of Contributing to Epigenetic Research, 174*

2. Molecular Components of Chromatin in Plants, 175

    2.1 *Regulators of DNA Methylation in Plants, 175*

    2.2 *Histone-modifying Enzymes, 177*

    2.3 *Other Chromatin Proteins, 178*

3. Molecular Components of RNAi-mediated Gene Silencing Pathways, 180

    3.1 *Elaboration of RNAi-mediated Silencing in Plants, 180*

    3.2 *Pathway 1: Transgene-related Posttranscriptional and Virus-induced Silencing (PTGS/VIGS), 181*

    3.3 *Pathway 2: Regulation of Plant Development by RNAs and Trans-acting siRNAs, 183*

    3.4 *Pathway 3: Transgene-related Transcriptional Silencing, RNA-directed DNA Methylation, and Heterochromatin Formation, 184*

4. Epigenetic Regulation without RNA Involvement, 186

5. Outlook, 186

References, 187

GENERAL SUMMARY

Plants are masters at epigenetic regulation. All major epigenetic mechanisms present in eukaryotes are used by plants and often elaborated to a degree unsurpassed in other kingdoms. DNA methylation, commonly associated with gene silencing, is found in CpG, CpNpG, and CpNpN nucleotide groups in plant genomes and relies on a number of plant-specific proteins, including several that might be specialized for active demethylation. Histone-modifying enzymes that modulate chromatin structure in plants are generally conserved within the catalytic domains, but they are frequently encoded by comparatively large gene families, which allows more extensive diversification or redundancy of gene function. RNAi-mediated gene-silencing pathways have also diversified in plants to combat viruses, tame transposons, orchestrate development, and organize the genome. Although the interplay between DNA methylation and histone modifications has been recognized for some time, the recent discovery that these modifications can be targeted to specific regions of the genome by the RNAi machinery has added a new dimension to epigenetics research. The intersections and overlaps among these silencing pathways provide plants with a multilayered and robust epigenetic circuitry.

The prominence of epigenetic regulation in plants reflects their evolutionary history, mode of development, and "lifestyle." Polyploidization—an increase in the number of sets of chromosomes—is a recurring event in plant lineages, amplifying gene families and fostering functional specialization of duplicated genes. Unlike mammals, where organ and tissue formation is largely completed during embryonic development, plants grow by continuously producing new aerial and underground parts from self-sustaining stem cell populations called meristems. Consequently, postembryonic development of plants is shaped by environmental influences and is characterized by a high degree of plasticity and variability. Because plants are unable to escape their surroundings, they are forced to cope with changeable and often unfavorable growth conditions. The inherent flexibility of epigenetic regulatory mechanisms can facilitate rapid changes in gene activity and fine-tune gene expression patterns, enabling plants to survive and reproduce successfully in unpredictable environments.

Historically, plants have provided excellent systems for discovering and analyzing epigenetic phenomena. A change from bilateral to radial symmetry in some variants of the plant *Linaria vulgaris* (see title figure), observed by Carl von Linné in the 18th century, was pinpointed to an epigenetic modification of the *cycloidea* gene, regulating flower development. Progress has been particularly impressive in the past 5 years, owing to the availability of the genome sequence of *Arabidopsis thaliana*—a "useful weed" that is highly amenable to genetic analyses—and to the synergy created by parallel studies of epigenetic phenomena in animal and fungal systems.

1 Benefits of Plants in Epigenetic Research

1.1 Plants and Mammals Are Similar in Terms of (Epi)Genome Organization

Soon after biology was established as an independent scientific discipline, animals and plants were grouped into separate kingdoms, and this view became traditionally manifest by training biologists in either zoology or botany. Of course, there are good arguments for this partition, including heterotrophic (i.e., requiring organic matter for growth) versus mainly autotrophic (i.e., self-sustaining) energy generation, mobile versus sessile lifestyle, potentially migrating and flexible cells versus motionless and rigid cells. However, geneticists and molecular biologists have uncovered in recent decades a degree of congruence between animals and plants that was surprising in the light of their long evolutionary separation. Common principles include sexual propagation via meiosis and fertilization, the regulation of individual development by a few master genes, the control of cell division and proliferation by related factors, and the reception of environmental factors through similar signaling cascades. This similarity extends to many aspects of genome and epigenome organization.

The resemblance is particularly striking between plants and mammals, which have comparable genome sizes, genome complexities, and ratios of heterochromatin. As in many other eukaryotes, euchromatin and heterochromatin are characterized by specific acetylation and methylation of histones, but heterochromatin of plants and mammals is specified additionally by significant DNA cytosine methylation. A comparison of components participating in genome organization and epigenetic regulation across different model systems reveals that there are more common features between plants and mammals than there are within the animal kingdom itself (Table 1). Therefore, even if interest is driven by an anthropocentric focus, similar questions can be addressed in plants and mammals, and basic information can often be shuttled between both systems.

1.2 Plants Provide Additional Topics for Epigenetic Research

In addition to elements shared with mammals, plants have acquired some specialties that are potentially relevant for epigenetic phenomena. Whereas in mammals fertilization is achieved by fusion of two haploid cells that are direct products of the preceding meiosis, plants have a haploid (gametophyte) growth stage between meiosis and fertilization (Fig. 1). The gametophytes correspond to the germinating pollen grain (male) and the embryo sac (female), each with several haploid nuclei that originate from the meiotic products by two or three subsequent mitotic divisions, respectively. In the gametophytic phase, any loss of genetic or epigenetic information cannot be compensated by information on homologous chromosomes or alleles. Although extensive studies have not yet been undertaken, there is no evidence for a massive programmed erasure of epigenetic marks during plant gametogenesis as occurs in mammals, and this might explain why epigenetic changes are often transmissible through meiosis in plants.

Another distinctive feature of plants is the less well-defined germ line and its separation from somatic cells only late in development (Fig. 1). Plants have apical meristems, which are growth points at shoot and root tips that generate new tissues and organs. The shoot apical meristem eventually forms the flowers that generate the gametes for sexual propagation, but additional lateral meristems can also grow out and form flowers, and many plants have developed specialized organs like rhizomes, tubers, or bulbs that contain meristems. These mechanisms of vegetative propagation can be even more common or successful than seed dissemination. Embryos can be formed not only by development of a fertilized egg, but also from somatic tissue (somatic embryogenesis). Upon manipulation in tissue culture, some differentiated somatic cells can undergo dedifferentiation and be reprogrammed toward alternative differentiation. This means that somatic cloning, still with low success rates in mammals, is routine in many plant species, and countless "green Dollies" have been produced over the years. Nevertheless, a surprising amount of phenotypic variability has been observed in supposedly genetically uniform populations of cloned plants. This so-called "somaclonal variation" has a strong epigenetic basis and is potentially useful for plant breeding and adaptation (for review, see Kaeppler et al. 2000).

Another plant-specific feature is the existence of plasmodesmata, cytoplasmatic bridges between individual cells, which are permeable to small molecules, some proteins, and RNAs, and viral genomic information. Despite the high degree of interconnection, plant shoots can be cut and grafted as scions on top of genetically different stocks (Fig. 1). This permits the production of chimeras in which vegetative tissue, and tissue that gives rise to progeny, are genetically different. Therefore, whereas epigenetic marks are transmitted through the germ line, they seem to be more flexible and reversible in plants relative to animals.

Table 1. Compilation of genomic and epigenetic components in epigenetic model systems

Feature	Saccharomyces cerevisiae	Schizosaccharomyces pombe	Neurospora crassa	Caenorhabditis elegans	Drosophila melanogaster	Mammals	Plants
Genome size	12 Mb	14 Mb	40 Mb	100 Mb	180 Mb	3,400 Mb	150–5,000 Mb
Number of genes	6,000	5,000	10,000	20,000	14,000	25,000–30,000	25,000–40,000
Average size of genes	1.45 kb	1.45 kb	1.7	2 kb	5 kb	35–46 kb	2 kb
Average number of introns/gene	≤ 1 (4% of genes with introns)	2 (40% of genes with introns)	2	5	3	6–8	4–5
% Genome as protein coding	70	60	44	25	13	4 (Hs)	26 (At) 10 (Os)
Transposon silencing	(+)	+	+ (+ RIP)[a]	+	+	+	+
Imprinting	–	–	–	–	+[b]	+	+
RNAi mechanisms	–	+	+	+	+	+	+
Repressive histone methylation	–	H3K9 (+) H3K27 –	H3K9 + H3K27 +	H3K9 + H3K27 +	H3K9 + H3K27 +	H3K9 + H3K27 +	H3K9 + H3K27 +
DNA methylation at CG	–	–	+	–	–	+	+
at CNG/CNN	–	–	+	–	(+)	(–)	+
Genes related to DNA methylation and recognition	–	+[c]	+	+[d]	+[e]	+	+
HP1-like protein	–	+	+	+	+	+	+
Polycomb proteins	–	–	–	+	+	+	+

(At) *Arabidopsis thaliana*, (Cb) *Caenorhabditis briggsae*, (Ce) *Caenorhabditis elegans*, (Dm) *Drosophila melanogaster*, (Hs) *Homo sapiens*, (Os) *Oryza sativa*, (Pp) *Pristionchus pacificus*.

[a] Repeat-induced point mutation, see Chapter 6.

[b] Chromosome- or genome-wide rather than gene-specific.

[c] Mutated Dnmt2.

[d] Dnmt2 (Pp) and MBD-domain proteins (Ce, Cb, Pp).

[e] Dnmt2 and MBD-domain proteins (Dm).

Plants have a higher tolerance toward polyploidy (the multiplication of the whole-chromosome complement) than mammals. The numerous wild polyploid species and cultivated polyploid plants—such as wheat, cotton, potato, peanut, sugarcane, and tobacco—suggest that polyploidy offers certain competitive advantages. Inspection of many plant genome sequences, including the small genome of *Arabidopsis thaliana*, provided clear evidence for ancient genome and gene duplication events. Even diploid plants can contain polyploid cells, which arise much more frequently than the few examples of highly specialized polyploid cells in mammals. The formation of polyploids is often associated with significant genomic and epigenetic changes (for review, see Adams and Wendel 2005). Some of these changes occur within one or a few generations and can contribute to rapid adaptation and evolution in plants.

1.3 Plants Tolerate Methodological Approaches That Are Difficult in Mammals

In mammals, genetic approaches are limited by demanding procedures for generating mutations and by the requirement for mating in order to establish homozygous genotypes, which are mandatory for revealing recessive traits. In contrast, plants allow efficient mutagenesis, either by chemical or physical treatments or by largely random insertion of sequence-tagged transgenes

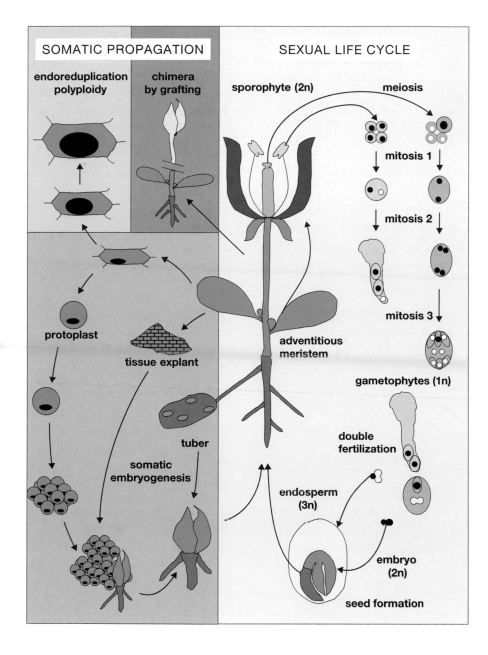

SOMATIC PROPAGATION

endoreduplication polyploidy

chimera by grafting

protoplast

tissue explant

tuber

somatic embryogenesis

SEXUAL LIFE CYCLE

sporophyte (2n)

meiosis

mitosis 1

mitosis 2

mitosis 3

adventitious meristem

gametophytes (1n)

double fertilization

endosperm (3n)

embryo (2n)

seed formation

Figure 1. Specialties of the Plant Life Cycle

Plants can propagate sexually (gametogenesis, fertilization, and seed formation, *right*) as well as somatically (vegetative sprigs, de- and re-differentiation or embryogenesis, *left*). The body of higher plants, with roots, stem, leaves, and flowers, is the diploid sporophyte. During meiosis, the chromosome number is reduced to half. Whereas in animals the meiotic products form the gametes without further division and fuse directly to produce the diploid embryo, plants form haploid male or female gametophytes by two or three mitotic divisions, respectively. The pollen tube ultimately contains one vegetative (*white*) and two generative (*black*) nuclei. The two generative nuclei fertilize the egg cell (*black*) and the central cell, which has a diploid nucleus derived from fusion of the two polar nuclei (*yellow*). This double fertilization gives rise to the diploid embryo and the triploid endosperm, which provides a nutrient source for the developing embryo. After seed germination, the embryo will grow into a new sporophyte. In addition, most plants have the potential for vegetative propagation through activation of quiescent lateral meristems, outgrowth of specialized root structures such as tubers, amplification in tissue culture, and even regeneration from individual somatic cells after removal of the cell wall (protoplasts). Endoreduplication is frequent in plants, producing polyploid cells or tissues. Plants can be grafted to produce chimeras. In summary, genetic and epigenetic information in plants therefore passes a much less well-defined germ line than in animals.

or transposable elements. Individuals with homozygous mutations are easily generated in Mendelian ratios in the following generation by self-pollination. Screens for mutations in epigenetic regulators were based on the recovery of gene expression from epigenetically inactivated marker genes or for epigenetic down-regulation of stably active reporter genes. In addition to such forward-directed unbiased methodology, the rapidly growing collections of insertion mutants within defined genomic integration sites permit reverse genetic approaches by analyzing the effects of mutations in chosen, defined genes orthologous to epigenetic regulators in other model organisms (Table 2).

The number of members within families of homologous genes can differ significantly between plants and mammals. As a consequence of functional redundancy, some mutations are less severe in either plants or mammals, and this can be important if a complete loss of function would eliminate the corresponding individuals prior to analysis in early development. Nonessential genes in pathways that determine coloration of plant tissues permit easy and inexpensive gene expression readouts in vivo (Fig. 2a–e). Other epigenetic changes can be followed by scoring for morphological defects, tolerated by many plants without lethal consequences (Fig. 2f). Additionally, thousands of individual plants can be

Table 2. Components of epigenetic regulation in the model plant *Arabidopsis thaliana* identified in forward or reverse genetic screens

Gene or mutant acronym	Gene or mutant name	Type of protein, confirmed or putative function	Mutant screen
		DNA methylation	
MET1 DDM2	Methyltransferase Decreased DNA methylation	DNA methyltransferase (CG)	reactivation of endogenous repeats reactivation of transgenes interference with RdDM hypomethylation of centromeric repeats
CMT3	Chromomethyltransferase	DNA methyltransferase (non CG)	reactivation of SUP-clk reactivation of PAI
DRM1	Domain-rearranged methyltransferase	de novo DNA methyltransferase	search for insertion mutants
DRM2	Domain-rearranged methyltransferase	de novo DNA methyltransferase	search for insertion mutants
HOG1	Homology-dependent gene silencing	S-adenosyl-L-homocysteine hydrolase	reactivation of repetitive transgene
ROS1	Repressor of silencing	DNA glycosylase-domain protein	inactivation of transgene
DME	Demeter	DNA glycosylase-domain protein	seed abortion
		Histone modification	
HDA1	Histone deacetylase	histone deacetylase	search for insertion mutants antisense expression
HDA6 SIL1 AXE1 RTS1	Histone deacetylase Modifier of silencing Auxin-gene repression RNA-mediated transcriptional silencing	histone deacetylase	reactivation of repetitive transgene interference with RdDM
SUVH2	Su(var)3–9 homolog	histone methyltransferase	search for insertion mutants antisense expression
SUVH4 KYP1	Su(var)3–9 homolog Kryptonite	histone methyltransferase	reactivation of PAI reactivation of SUP-clk
		Chromatin formation/remodeling	
DDM1 SOM	Decreased DNA methylation Somniferous	SWI2/SNF2 ATPase	hypomethylation of centromeric repeats reactivation of repetitive transgene
DRD1	Defective in RNA-directed DNA methylation	SWI2/SNF2 ATPase	interference with RdDM
SPD	Splayed	SWI2/SNF2 ATPase	altered meristem maintenance
PIE	Photoperiod-independent early flowering	ATP-dependent chromatin-remodeling protein	change of flowering time
PKL	Pickle	CHD3 chromatin-remodeling factor	abnormal root development
FAS1	Fasciated	chromatin assembly factor subunit	altered morphology
FAS2	Fasciated	chromatin assembly factor subunit	altered morphology
BRU1	Brushy	uncharacterized protein	DNA damage sensitivity
LHP1	Like heterochromatin protein	formation of repressive chromatin	early flowering and altered morphology
MOM1	Morpheus' molecule	incomplete SWI2/SNF2 ATPase	reactivation of repetitive transgene
RPA2	Replication protein A	subunit of the ssDNA-binding replication protein complex	suppressor screen in *ros1* reactivation of repetitive transgene
		RNAi-mediated silencing	
DCL1 CAF1 SIN1 EMB76 SUS1	Dicer-like Carpel factory Short integuments Embryo-defective Suspensor	RNase III (dsRNase)	search for insertion mutants abnormal flower development abnormal ovule development arrested embryo development suspensor proliferation

Table 2. (*continued*)

Gene or mutant acronym	Gene or mutant name	Type of protein, confirmed or putative function	Mutant screen
RNAi-mediated silencing (*continued*)			
DCL2	Dicer-like	RNase III (dsRNase)	search for insertion mutants
DCL3	Dicer-like	RNase III (dsRNase)	search for insertion mutants
AGO1	Argonaute	PAZ-PIWI domain protein	altered morphology reactivation of transgenes
AGO4	Argonaute	PAZ-PIWI domain protein	reactivation of SUP-clk
AGO7 ZIP	Argonaute Zippy	PAZ-PIWI domain protein	timing and type of trichome development
AGO10 PNH/ZLL	Argonaute	PAZ-PIWI domain protein	meristem defects
RDR1	RNA-dependent RNA polymerase	RNA-dependent RNA polymerase	search for insertion mutants
RDR2	RNA-dependent RNA polymerase	RNA-dependent RNA polymerase	search for insertion mutants
RDR6 SDE1 SGS2	RNA-dependent RNA polymerase Silencing-defective Suppressor of gene silencing	RNA-dependent RNA polymerase	reactivation of transgene altered trichome development
NRPD1a SDE4	RNA polymerase Silencing-defective	RNA polymerase IV subunit	reactivation of transgene
NRPD1b DRD3	RNA polymerase Defective in RNA-directed DNA methylation	RNA polymerase IV subunit	interference with RdDM
NRPD2a DRD2	RNA polymerase Defective in RNA-directed DNA methylation	RNA polymerase IV subunit	search for insertion mutants interference with RdDM
SDE3	Silencing-defective	RNA helicase	reactivation of transgene
SGS3	Suppressor of gene silencing	coiled-coil protein	reactivation of transgene altered trichome development
HEN1	HUA enhancer	dsRNA binding, methyltransferase	reactivation of transgene altered morphology
HYL1	Hyponastic leaves	nuclear dsRNA-binding protein	altered morphology search for insertion mutants
WEX	Werner syndrome-like exonuclease	RNase D exonuclease	search for insertion mutants
XRN4	XRN homolog	exoribonuclease	search for insertion mutants
HS1	Hasty	miRNA export receptor	timing and type of trichome development
Polycomb group proteins			
MEA FIS1	Medea Fertilization-independent seeds	Polycomb group protein	seed abortion fertilization-independent seed set
CLF	Curly leaf	Polycomb group protein	altered leaf morphology
FIE FIS3	Fertilization-independent endosperm Fertilization-independent seeds	Polycomb group protein	fertilization-independent seed set
MSI1	Multicopy suppressor of IRA homolog	Polycomb group protein chromatin assembly factor subunit	search for insertion mutants defects in endosperm patterning
SWN	Swinger	Polycomb group protein	search for insertion mutants
EMF2	Embryonic flower	Polycomb group protein	lack of vegetative stage
VRN2	Vernalization	Polycomb group protein	late flowering in spite of vernalization

Figure 2. Assays for Epigenetic Control in Plants

Genes determining coloration of plant tissue allow easy and inexpensive gene expression readout in vivo. (*a*) Expression of the dihydroflavonol reductase (*DFR*) gene is required for dark purple petunia flowers, whereas silencing of the DFR promoter gives rise to variegated, light coloration (courtesy of Jan Kooter). (*b*) Seeds from *Arabidopsis* expressing the chalcone synthase gene (*CHS*) have dark seed coats, whereas silencing of *CHS* after integration of a homologous transgene results in yellow seeds (courtesy of Ian Furner). (*c*) Maize plants with the *B-I* gene have purple pigmentation, in contrast to green plants with a paramutagenic, inactive *B′* gene. (*d*) Maize ear, segregating a transposon insertion (*Spm*) in the *B-Peru* gene required for anthocyanin pigment. Purple kernels represent revertants in which the *Spm* element excised from the gene in the germ line. The heavily spotted kernels contain the *Spm* element in the active form that induces frequent somatic excision sectors during kernel development. The kernels with rare, small purple sectors represent kernels in which the *Spm* element has been epigenetically silenced (courtesy of Vicky Chandler). (*e*) The dark color of soybeans (*middle*) is extinguished in cultivated varieties (*left*) due to natural posttranscriptional silencing of the *CHS* gene and can be partially reversed by infection of the parental plant with a virus possessing a PTGS suppressor protein, producing a mottled pattern (*right*). (*f*) Epigenetic regulation can become manifest also in plant morphology: Reduced function of a chromatin assembly factor subunit leads to a "fasciated" *Arabidopsis* stem. (*g*) Release of silencing from a transgenic resistance marker in *Arabidopsis* can be scored by growth on selective medium. (*c*, Reprinted, with permission, from Chandler et al. 2000 [Springer Science and Business Media]; *e*, reprinted, with permission, from Senda et al. 2004 [© American Society of Plant Biologists].)

screened in selection experiments scoring for the expression of endogenous or transgenic traits (Fig. 2g).

The sequence of the *A. thaliana* genome, together with several large collections of insertion mutants and facile mutagenesis, has made this species the current organism of choice for plant epigenetic research. The following discussion includes epigenetic modifiers for which functional information is available, obtained through screens for mutants deficient in transgene silencing, by the bioinformatic identification of homology with genes required for silencing in other organisms, and by recognizing similar developmental phenotypes to other silencing-defective

mutants in *Arabidopsis* forward or reverse genetic screens (Table 2). For a comprehensive list of chromatin genes in *Arabidopsis*, rice, and maize, the reader is referred to the Plant Chromatin Database (http://www.chromdb.org).

1.4 Plants Have a Proven Record of Contributing to Epigenetic Research

The best argument for appreciating the value of plant model systems in epigenetic research is the long list of seminal contributions that plant scientists have made to this field. The distinction between euchromatin and heterochromatin was first made after cytological analysis in mosses (Heitz 1928). The observation of heritable but reversible changes in gene expression after allelic interaction in tomato and maize, later termed paramutation, was early evidence for non-Mendelian genetics (for review, see Chandler and Stam 2004), now also apparent in mammalian systems. Likewise, parental imprinting of individual genes in plants was first observed in maize (for review, see Alleman and Doctor 2000). The repeated occurrence of individuals with altered flower symmetry, already described by Carl von Linné as "peloria" (monster) (see title figure), could now be explained by the formation of an epiallele, a stable epigenetic modification of a regulatory gene with the same sequence as the expressed version (Cubas et al. 1999). Cytological analysis in plants revealed changes in secondary chromosome constrictions that later were linked with nucleolar dominance, the silencing of one parental set of rRNA genes in interspecific hybrids, and shown to depend on epigenetic regulation (for review, see Pikaard 2000). The pioneering work on transposable elements in maize, by Barbara McClintock and other workers, revealed numerous links between their genetic behavior and epigenetic regulation (for review, see Fedoroff and Chandler 1994). Indeed, extant transposons and their degenerate remains provide the foundation for establishing epigenetic modifications throughout plant genomes (Section 3.4).

More recently, when transgenic technology became routine in the late 1980s for plants such as tobacco, petunia, and *Arabidopsis*, a major advance in epigenetic research arose from the unexpected results obtained in the course of introducing marker genes (for review, see Jorgensen 2003; Matzke and Matzke 2004). The concept of homology-dependent gene silencing was formulated as it became evident that silencing was often correlated with multiple copies of linked or unlinked transgenes. Different cases of homology-dependent gene silencing were due to either enhanced turnover of mRNA (posttranscriptional gene silencing, PTGS) or repression of transcription (transcriptional gene silencing, TGS), both of which were correlated with increased cytosine methylation of silenced genes. A striking example of PTGS in transgenic petunia was initially termed "cosuppression": Attempts to modify floral coloration by overexpression of chalcone synthase (CHS) genes that condition purple petals often produced variegated or even completely white flowers. The lack of pigmentation was shown to result from coordinate gene silencing of both the *CHS* transgene and the endogenous *CHS* gene (Jorgensen 2003). PTGS is now considered the plant equivalent of RNA interference

(RNAi) later described in *Caenorhabditis elegans* and other organisms (see Section 3.2).

By the mid-1990s, links between PTGS and virus resistance had been forged. PTGS was shown to naturally protect plants from unchecked replication of viruses, which can be both inducers and targets of PTGS. This principle was exploited in plants to experimentally down-regulate plant genes by constructing viral vectors containing sequences homologous to a target gene, resulting in virus-induced gene silencing (VIGS; for review, see Burch-Smith et al. 2004). In addition, RNA-directed DNA methylation (RdDM) was discovered in viroid-infected plants, providing the first demonstration that RNA could feed back on DNA to elicit epigenetic modifications (Wassenegger et al. 1994). This principle has been successfully used to transcriptionally silence and methylate promoters by intentionally generating homologous double-stranded RNA (see Section 3.4, RNA-directed DNA methylation).

2 Molecular Components of Chromatin in Plants

A number of molecular components of epigenetic regulation in plants were identified by the mutational approaches in *Arabidopsis* mentioned above (Table 2). However, mutant screens have probably not yet revealed a complete list of epigenetic modifiers because of either functional redundancy in large gene families or the lethal consequences of losing essential components.

2.1 Regulators of DNA Methylation in Plants

Methylation of carbon 5 of cytosines in DNA is a hallmark of epigenetic inactivation and heterochromatin in both plants and mammals (Table 1) (Chapter 18). In plants, however, DNA methylation has a number of unique features with respect to the pattern of methylation, proteins of the methylation machinery, and the possibility to reverse methylation in nondividing cells (for review, see Chan et al. 2005). In this section, we discuss the proteins required to establish, maintain, interpret, and erase DNA methylation. Special components needed for the process of RNA-directed DNA methylation are presented in Section 3.4.

DNA METHYLTRANSFERASES

DNA methylation can be divided into two steps: de novo methylation and maintenance methylation. De novo methylation refers to the modification of a previously unmethylated DNA sequence (Fig. 1) (Chapter 18). In plants, de novo methylation can alter CpG, CpNpG, and

CpNpN nucleotide groups (where N is A, T, or C). In contrast, methylation in mammals is largely restricted to CpG dinucleotides, and there is no evidence for extensive methylation in asymmetric CpNpN nucleotide groups. Although the signals that trigger de novo methylation are largely unknown, double-stranded RNA can fulfill this role in plants (Section 3.4). Maintenance methylation perpetuates methylation patterns during DNA replication and occurs most efficiently at CpG and CpNpG nucleotide groups with their palindromic symmetry. Maintenance of methylation occurs on a hemimethylated substrate after replication or repair, guided by the modification still present on the parental DNA strand. Although it is usually assumed that distinct DNA cytosine methyltransferase enzymes contribute to either de novo or maintenance methylation, an emerging view in plants is that enzymes with different site specificities (CpG or non-CpG) frequently cooperate to catalyze both steps.

The three conserved families of DNA methyltransferase are all present in plants. Members of the methyltransferase (MET1) family, which are homologs of the mammalian Dnmt1 type (see Chapter 18), are considered CpG maintenance methyltransferases, although one has also been assigned a role in CpG de novo methylation in the RdDM pathway (Section 3.4). The Dnmt2 class, of which one member is encoded in the *Arabidopsis* genome, comprises the most widespread and highly conserved DNA methyltransferase family (Table 2), but its function remains obscure. The plant Domains-rearranged methyltransferases (DRM) and their mammalian homologs, the Dnmt3 group, are usually considered de novo methyltransferases. The DRM enzymes catalyze methylation of cytosines in all sequence contexts and are prominent in the RdDM pathway (Section 4.4). As their name implies, the DRM proteins have rearranged domains (VI–X, followed by I–V) compared to Dnmt3 (I–X). This might give them the ability to methylate asymmetric CpNpN nucleotide groups, which are not methylated in mammalian cells. The plant-specific chromomethylase CMT3 modifies CpNpG trinucleotides. Similarly to MET1, CMT3 has been implicated in both de novo and maintenance methylation. The exact function of CMT3 is not entirely clear, although loss-of-function mutants reactivate certain silent transposons (for review, see Chan et al. 2005).

In contrast to mammals, where *dnmt1* and *dnmt3* mutants die during embryonic development or shortly after birth, *met1*, *cmt3*, and *drm* mutants are viable and usually fertile. The nonlethality of DNA methyltransferase mutations in plants has permitted more extensive analyses of deficiency mutants during development and sexual reproduction than is possible in mammals (for review, see Chan et al. 2005).

ACTIVE CpG DEMETHYLATION AND DNA GLYCOSYLASES

Epigenetic regulation implies that marks corresponding to active or inactive genetic states are potentially reversible. DNA methylation permits such reversibility, because it can be lost through passive or active means. Passive loss occurs when methylation fails to be maintained during multiple rounds of DNA replication. In contrast, active demethylation can occur in nondividing cells and requires enzymatic activities. Early reports from animal systems suggested that active demethylation can result from the action of DNA glycosylases, which are normally involved in base excision repair (for review, see Kress et al. 2001). Interest in this idea has been rekindled by the discovery in *Arabidopsis* of Demeter (DME) and Repressor of silencing (ROS1), which are large proteins containing DNA glycosylase domains. The *ROS1* gene was identified in a screen for epigenetic down-regulation and hypermethylation of a stably expressed reporter gene (Gong et al. 2002). The ROS1 protein displays nicking activity on methylated but not unmethylated DNA, which is consistent with a role in removing methylated cytosines from DNA in a pathway related to base excision repair. ROS1 is expressed constitutively and hence could potentially contribute to loss of DNA methylation in nondividing cells at all stages of development (Kapoor et al. 2005a). In contrast, DME activity is restricted to the female gametophyte, where it activates the imprinting factor Medea (MEA) in a manner that is dependent on a functional DNA glycosylase domain (Choi et al. 2002). The CG methyltransferase MET1 acts antagonistically to DME, suggesting that DME is indeed required for demethylation of CG dinucleotides (Hsieh and Fisher 2005). In *Arabidopsis*, there are two additional uncharacterized members of the *DME/ROS1* family that are unique to plants. The expansion of this gene family suggests that reversible gene silencing by active demethylation is important for plant physiology, development, or adaptation to the environment.

METHYL-DNA-BINDING PROTEINS

Methyl-CG-Binding Domain (MBD) proteins are thought to provide a means to transduce DNA methylation patterns into altered transcriptional activity. In mammals, MBD proteins bind methylated DNA and perform various functions, such as recruiting histone deacetylases, to reinforce transcriptional silencing. *Arabidopsis* has 12 MBD-containing genes, compared to 11 in mammals, 5 in *Drosophila*, 2 in *C. elegans*, and none in sequenced fungal genomes (Hung and Shen 2003). Little is known about the functions of *Arabidopsis* MBD proteins, although RNAi-knockdown of one, AtMBD11, was associated with pleiotropic effects on development (Springer and Kaeppler 2005). None of the *Arabidopsis* MBD proteins has been identified in forward genetic screens, perhaps because of functional redundancy. In addition, despite the amino acid conservation of DNA methyltransferases among plants and mammals, the MBD-containing proteins in the two kingdoms diverge completely outside of the methyl-CG-binding domain. Thus, even though plants and mammals establish and maintain DNA methylation patterns using related enzymes, they might have evolved different ways of interpreting these patterns by means of distinct MBD proteins (Springer and Kaeppler 2005).

COMPONENTS OF THE METHYL GROUP DONOR SYNTHESIS

Methylating enzymes require an activated methyl group, usually in the form of S-adenosyl-methionine. Therefore, it is surprising that the biochemical pathways providing this cofactor were not linked with epigenetic regulation earlier. Only recently, however, has a mutation (*hog1*) in the *Arabidopsis* gene encoding S-adenosyl-L-homocysteine hydrolase been found to be responsible for epigenetic defects (Rocha et al. 2005).

2.2 Histone-modifying Enzymes

Like other organisms (Table 1), plants contain enzymes that posttranslationally modify the amino-terminal tails of histones, thus establishing a putative histone code (for review, see Loidl 2004). In plants, histone-modifying enzymes are often encoded by comparatively large gene families. Functional information about most family members is still limited. The two most common modifications are histone acetylation/deacetylation and histone methylation.

HISTONE DEACETYLASES AND HISTONE ACETYLTRANSFERASES

The opposing functions of histone acetyltransferases (HATs) and deacetylases (HDACs) ensure reversibility of this epigenetic mark. The potential for reversibility is reinforced by the frequent coexistence at silent genes of histone hypoacetylation and CpG methylation, the latter of which can potentially be actively removed by DNA glycosylases (Section 2.1, Active CG demethylation and DNA glycosylases). *Arabidopsis* has 18 putative HDACs and 12 putative HATs (Pandey et al. 2002), which is around the same number found in mammals, but more than in other non-plant eukaryotes. The putative HDACs are generally conserved in all eukaryotes, but there is one plant-specific family, HD2, whose function remains obscure. Genetic screens have identified only two members of a conserved family: *HDA1* and *HDA6* (Table 2). HDA6 has roles in maintaining CpG methylation induced by RNA and in repeated sequences, but contributes minimally to development, as indicated by the normal phenotype of deficiency mutants. In contrast, reduced expression of HDA1 results in pleiotropic effects on development. None of the *Arabidopsis* HATs has been identified in forward genetic screens, which might reflect functional redundancy or the direction of most screens toward activation of silent genes.

HISTONE METHYLTRANSFERASES

Proteins that are able to methylate lysine residues in histones (referred to in this book as histone lysine methyltransferases or HKMTs) and other proteins contain a common SET domain (*SU(VAR)/E(Z)/TRX*). Through their ability to methylate histone H3 or H4 at various lysine residues, different complexes containing SET domain proteins play roles in promoting or inhibiting the transcription of specific genes and in forming heterochromatin. Some SET domain proteins are members of the Polycomb group (PcG) or trithorax group (trxG), which maintain transcriptionally repressed or active states, respectively, of homeotic genes during plant and animal development (see Chapters 11 and 12). Other SET domain proteins, such as SU(VAR)3-9, participate in maintaining condensed heterochromatin, often in repetitive regions, by methylating H3 at lysine 9 (H3K9).

The *Arabidopsis* genome encodes 32 SET domain proteins, 30 of which are expressed. They can be grouped into four conserved families: E(Z), TRX, ASH1, and SU(VAR)3-9, as well as a small fifth family present only in yeast and plants (Baumbusch et al. 2001; Springer et al. 2003). The number of expressed SET domain proteins in *Arabidopsis* is relatively high compared to the 14 in *Drosophila* and 4 in fission yeast, although there are 50 SET domain proteins in mice. In addition to expansion of the SET domain protein family by polyploidy, retrotransposition has also played a role in the amplification of SU(VAR)3-9 members in *Arabidopsis*. Outside of the SET domain, the plant and animal proteins are not always well conserved. The divergent regions are predicted to mediate protein–protein interactions, suggesting that plant SET domain proteins might act in complexes distinct from those in animals.

Although incomplete, the functional information available for *Arabidopsis* SET domain proteins implicates them in chromatin regulation and epigenetic inheritance. The first two SET domain proteins to be identified in genetic screens were Curly leaf (CLF) and Medea/Fertilization independent seed formation (MEA/FIS1), which are negative regulators related to *Drosophila* E(Z). In addition to being SET domain proteins, MEA, CLF, and E(Z) are also PcG proteins (Section 2.3, Other polycomb proteins). Mutations in *CLF* result in altered leaf morphology and homeotic changes in flower development. MEA/FIS1 regulates gametophyte-specific gene expression and is an imprinting factor that inhibits endosperm development in the absence of fertilization (for review, see Schubert et al. 2005). In contrast, the TRX family member *Arabidopsis* trithorax1 (ATX1) acts as an activator of floral homeotic genes, presumably by means of its ability to catalyze histone H3 lysine 4 (H3K4) methylation, a mark often associated with transcriptionally active chromatin (for review, see Hsieh and Fischer 2005).

Kryptonite/Suppressor of variegation 3-9 homolog 4 (*KYP/SUVH4*) was identified in screens for suppressors of epigenetic silencing at two endogenous genes (Jackson et al. 2002; Malagnac et al. 2002). KYP/SUVH4 catalyzes mono- and dimethylation of H3 at lysine 9 (H3K9me2/me3) and acts together with CMT3 to maintain CpNpG methylation of a subset of sequences in *Arabidopsis*. KYP/SUVH4 appears to play only a minor role in heterochromatin formation (Chan et al. 2005). In contrast, Suppressor of Variegation 3-9 homolog 2 (SUVH2), identified in a screen for reactivation of a silent transgene, appears to be the major activity responsible for methylation of H3 at lysines 9 (H3K9) and 27 (H3K27) in heterochromatin in *Arabidopsis* (Naumann et al. 2005).

Lysines in histones H3 and H4 can be mono-, di-, or trimethylated, which increases the combinatorial complexity of these modifications. Specific states define heterochromatin in different organisms. For example, H3K9me3 is a prominent feature of heterochromatin in animals and fungi, whereas this epigenetic mark is associated with euchromatin in *Arabidopsis*. Conversely, H3K9me1 and H3K9me2 are the predominant marks for

silenced heterochromatin in *Arabidopsis*, whereas they are euchromatic modifications in mammals. The origins of these differences and how they relate to the postulated histone code remain to be determined. In addition, the intricate relationships between specific histone modifications and DNA methylation patterns in plants remain to be fully elucidated (Tariq and Paszkowski 2004).

In contrast to histone acetylation, which can be dynamically regulated by the opposing activities of HDACs and HATs, histone methylation was thought until recently to be a more permanent epigenetic mark. Recent work in mammals, however, has identified a lysine demethylase, LSD1, that can remove H3K4me1 and H3K4me2 but not H3K4me3 (see Chapter 10). Four putative *LSD* homologs are encoded in the *Arabidopsis* genome, suggesting that at least some histone methylation is reversible in plants.

2.3 Other Chromatin Proteins

OTHER POLYCOMB PROTEINS

PcG proteins were initially identified in *Drosophila* as factors required to maintain repression of homeotic genes (see Chapter 11). In animals, structurally disparate PcG proteins act together in multiprotein complexes to repress gene expression. The PRC1 complex is absent in plants and *C. elegans* but present in *Drosophila* and mammals. The PRC2 complex, however, is found in plants and animals, where it has been shown to methylate predominantly H3 at lysine 27 (H3K27) through the histone methyltransferase activity of the SET domain and PcG protein E(Z).

Arabidopsis homologs of the core components of PRC2 have been identified in mutant screens designed to dissect various developmental pathways. In *Drosophila*, PRC2 components are encoded by single-copy genes. In contrast, genes encoding these proteins in *Arabidopsis* show functional diversification of at least three PRC2 complexes—FIS (fertilization independent seeds), EMF (embryonic flower), and VRN (vernalization)—that differ in their target gene specificity (Schubert et al. 2005; see also Fig. 2 in Chapter 11).

FIS genes were identified in screens for mutants showing partial seed development in the absence of fertilization. A major target is the MADS-box transcription factor PHERES (Köhler et al. 2005). Components of the EMF complex were identified by their common role in repressing floral homeotic genes, such as *Agamous* and *Apetala3*. A member of the VRN complex, VRN2, was identified on the basis of its contribution to epigenetic memory of vernalization, which is defined as the break of seed dormancy by cold treatment. Plants have to program their reproduction to occur during the proper season, and they do this in temperate climates by flowering only after extended periods of cold temperatures. The epigenetic memory of winter requires VRN2, which maintains cold-induced transcriptional repression of the gene encoding the flowering inhibitor FLC during later periods of growth at warmer temperatures. H3K27me2 is lost from FLC in vrn2 mutants, which is consistent with a role for PRC2 complexes in facilitating histone methylation (Schubert et al. 2005).

COMPONENTS OF IMPRINTING

Flowering plants and mammals are the only groups of organisms that have parental imprinting (Table 1), an epigenetic phenomenon in which a gene is differentially expressed depending on the parent from which it was inherited. In view of the parental conflict theory for the evolution of imprinting (for further discussion, see Chapter 19), the occurrence of parental imprinting in flowering plants and mammals likely reflects the fact that both taxa have a special maternal tissue that provides a nutrient source for the developing embryo. In mammals, this tissue is the placenta, and in plants it is the triploid endosperm, a terminally differentiated tissue that contains one paternal and two maternal genomes (Fig. 1). Indeed, the first example of parental imprinting of a single gene in any organism was observed in maize endosperm (for review, see Alleman and Doctor 2000). In Arabidopsis, two genes expressed in the endosperm, MEA and FWA (a flowering time control gene), are imprinted. In these cases, the two maternal copies are activated, presumably by DME-catalyzed active demethylation of CpGs in the female gametophyte (see Section 2.1, Active CpG demethylation and DNA glycosylases), whereas the paternal copy remains silent (for review, see Autran et al. 2005). Intriguingly, even though imprinting evolved independently in plants and mammals, DNA methylation and PcG proteins are required in both cases (Köhler et al. 2005).

CHROMATIN-REMODELING PROTEINS

Switch2/Sucrose Non-Fermentable2 (SWI2/SNF2) chromatin-remodeling factors constitute a conserved family of ATP-dependent chromatin remodelers that are able to displace nucleosomes or loosen histone/DNA contacts. Genetic screens have provided functional information for only a handful of the approximately 40 SWI2/SNF2 homologs encoded in the Arabidopsis

genome (Plant Chromatin Database). So far, only two—Decreased DNA methylation 1 (DDM1, Jeddeloh et al. 1999) and Defective in RNA-directed DNA methylation 1 (DRD1, Kanno et al. 2004)—have been implicated in regulating DNA methylation. Deficiency mutants of DDM1, which undergo genome-wide reduction of DNA methylation and transcriptionally reactivate a number of silent transposons and repeats, display severe developmental and morphological defects. These appear only after several generations of inbreeding homozygous ddm1 plants and appear to be due to the accumulation of epimutations and to insertional mutagenesis by transposons that are reactivated in the mutant. DDM1 has an ortholog in mammals, Lymphoid-Specific Helicase (LSH), which is likewise important for global CpG methylation and embryonic development. In contrast, DRD1 is unique to the plant kingdom and probably has a specialized role in RdDM (Section 3.4).

No phenotypic alteration other than a release of certain repetitive targets from silencing is caused by mutations of Morpheus' Molecule (MOM, Amedeo et al. 2000), a plant-specific gene with an incomplete ATP-dependent helicase motif. MOM acts synergistically with, but independently of, the DDM1/DNA methylation pathway, indicating multiple layers of transcriptional regulation in plants (Tariq and Paszkowski 2004).

Three more proteins with putative chromatin-remodeling function, Splayed (SPD), Photoperiod-independent early flowering (PIE), and Pickle (PKL), which were each identified by developmental effects in deficiency mutants, have not yet been implicated in specific chromatin modifications (Wagner 2003).

CHROMATIN ASSEMBLY FACTORS

Whereas the SWI2/SNF2 proteins probably act on assembled chromatin, other components are required to reestablish chromatin after replication and repair-associated DNA synthesis. The Chromatin Assembly Factor (CAF) complex, composed of three subunits, helps to bring semi-assembled nucleosomes to the replication fork. Mutations in genes of the two larger CAF subunits in Arabidopsis (fas1, fas2) cause characteristic morphological anomalies (fasciation, Fig. 2F), deficiencies in DNA repair, and derepression of repetitive targets (Takeda et al. 2004). This suggests that correct nucleosome deposition is essential for development and epigenetic control. Whereas the lack of CAF subunits does not interfere with maintenance of DNA methylation, it could lead to the erasure of other epigenetic marks, such as histone modifications. Reduced levels of the third CAF unit MSI1 do not reiterate fasciation but lead to distorted seed development and several morphological changes (for review, see Hennig et al. 2005). A mutation in the *BRU* gene that is unrelated to any known chromatin assembly protein, but results in a phenotype very similar to that of the *fas* mutants, makes it likely that additional factors are involved in maintaining the epigenetic information and genetic integrity during postreplicative chromatin assembly (Takeda et al. 2004). Finally, lack of RPA2, a subunit of the Replication Protein A complex, results in DNA damage sensitivity and release of transcriptional silencing, changing histone modification marks but not DNA methylation patterns (Elmayan et al. 2005; Kapoor et al. 2005b).

Heterochromatin-like Proteins

HP1 (heterochromatin protein 1) in *Drosophila* and mammals, and their homologs in fungi, are important components of silenced heterochromatin. The binding of HP1 through its chromodomain to methylated histone H3 at lysine 9 (H3K9me) promotes spreading of the silenced state to establish heterochromatic domains. The *Arabidopsis* genome encodes a single protein with homology to *Drosophila* HP1. Mutations in this gene, termed Like heterochromatin protein (*LHP1*) (Gaudin et al. 2001) or Terminal flower 2 (*TFL2*) (Kotake et al. 2003), result in changes in plant architecture, altered leaf development, and early onset of flowering. Although this mutant phenotype suggests an important role in regulating plant gene expression, it is unlikely that LHP1 acts through the formation of repressive chromatin complexes similarly to HP1 in other organisms. Instead, LHP1 in *Arabidopsis* regulates loci in euchromatin that are not targets of DNA methylation (Kotake et al. 2003; Tariq and Paszkowski 2004). Thus, LHP1 in plants and HP1 in other organisms appear to have evolved different modes of action.

3 Molecular Components of RNAi-mediated Gene Silencing Pathways

Modern epigenetics research has traditionally focused on DNA methylation and histone modifications. During the past several years, it has become evident that these alterations can be targeted to specific regions of the genome by the RNA interference pathway. Indeed, it is impossible nowadays to consider epigenetic regulation in many eukaryotes without integrating components of the RNAi machinery (Matzke and Birchler 2005). This is particularly true for plants, where the proliferation of RNAi-mediated gene-silencing pathways exceeds that present in any other type of organism.

3.1 Elaboration of RNAi-mediated Silencing in Plants

RNAi and related types of gene silencing represent cellular responses to double-stranded RNA (dsRNA). In these pathways, the dsRNA is processed by the RNase III-like endonuclease, Dicer, to produce small RNAs which determine the specificity of silencing by base-pairing to complementary target nucleic acids. Small RNAs incorporate into multiprotein silencing effector complexes to direct mRNA degradation, repress translation (PTGS), or guide chromatin modifications (TGS) in a sequence-specific manner. A key component of silencing effector complexes is an Argonaute protein, which binds small RNAs through its PAZ domain. Individual members of the Argonaute protein family, which comprises the largest group of proteins important for RNAi-mediated silencing, confer functional specificity to different silencing effector complexes (for review, see Carmell et al. 2002). In addition to participating in viral defense and transposon control, RNAi-mediated gene silencing plays essential roles in plant and animal development.

The elaboration of RNAi-mediated silencing in plants reflects in part their co-evolution with pathogens that generate dsRNA during replication, such as RNA viruses and viroids. Indeed, together with transgenes—another type of "foreign" nucleic acid—these RNA pathogens have been invaluable for detecting and studying various forms of RNAi-mediated gene silencing in plants.

The proliferation of RNAi-mediated gene-silencing pathways in plants is illustrated by

1. the expansion and functional diversification of gene families encoding core components of RNAi: the *Arabidopsis* genome encodes four DICER-LIKE (DCL) proteins and ten Argonaute (AGO) proteins

2. the heterogeneity in length and functional diversity of small RNAs, including the 21-nucleotide short interfering RNAs (siRNA) derived from transgenes and viruses, and several types of endogenous small RNAs, such as 21- to 24-nucleotide microRNAs; 21-nucleotide trans-acting siRNAs, and 24- to 26-nucleotide heterochromatic siRNAs

3. the various modes of gene silencing elicited by different small RNAs: PTGS involves mRNA degradation or repression of translation, and TGS is

associated with epigenetic modifications such as DNA cytosine methylation and histone methylation

4. the importance of PTGS in antiviral defense, which can be countered by a variety of plant viral proteins that repress silencing at different steps of the pathway

5. the existence of processes, such as non-cell-autonomous silencing and transitivity (see Section 3.2, Non-cell-autonomous silencing and transitivity), that rely on RNA-dependent RNA polymerases, six of which are encoded in the *Arabidopsis* genome

These aspects will be discussed in the framework of three major pathways of RNAi-mediated gene silencing in plants (Fig. 3a–c). However, it should be kept in mind

that the pathways feed into each other at various points. Components with assigned functions are listed in Table 2.

3.2 Pathway 1: Transgene-related Posttranscriptional and Virus-induced Gene Silencing (PTGS/VIGS)

RNAi-mediated gene silencing induced by transgenes and viruses appears to function primarily as a host defense to foreign or invasive nucleic acids, including viruses, transposons, and transgenes.

ORIGIN AND PROCESSING OF DsRNA

Transgene constructs can be introduced into plant genomes in sense or antisense orientations or as inverted DNA repeats. Viruses can have single-stranded

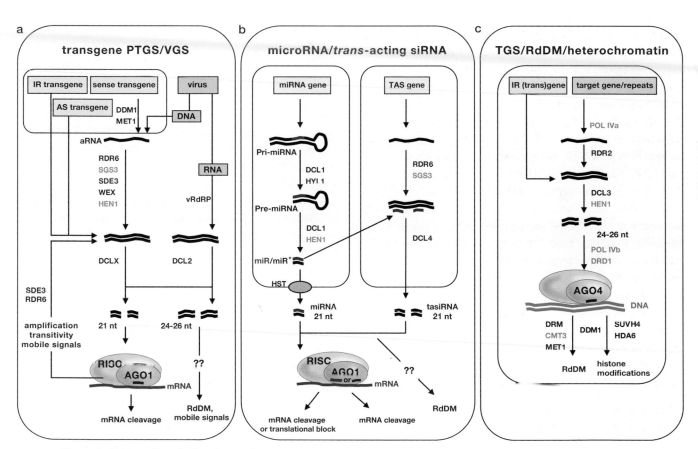

Figure 3. RNA-mediated Silencing Pathways in Plants

Although there are some overlaps and shared components, three major pathways can be distinguished by the source of dsRNA, class of small RNA, nature of the target sequence, and the mode of silencing evoked. Silencing effector complexes containing an Argonaute protein are shown as light gray spheres. Yellow boxes mark processes known to occur within the nucleus. See text for details and Table 2 for the names of regulatory components. Plant-specific proteins are labeled in green. (PTGS) Posttranscriptional gene silencing, (VIGS) virus-induced gene silencing, (TGS) transcriptional gene silencing, (RdDM) RNA-directed DNA methylation, (IR) inverted repeats, (AS) antisense, (vRdRP) virally encoded RNA-dependent RNA polymerase, (aRNA) aberrant RNA, (siRNA) short interfering RNA, (RISC) RNA-induced silencing complex. (Modified from Meins et al. 2005.)

or double-stranded DNA or RNA genomes. Therefore, in this pathway, dsRNA can be produced by a variety of routes. In principle, antisense transcripts can base-pair directly to target mRNAs to form dsRNA. Transcription through inverted DNA repeats can produce hairpin RNAs. RNA viruses, which encode their own RNA-dependent RNA polymerase (vRdRP) and replicate via dsRNA intermediates, enter the pathway directly at the level of dsRNA. In contrast, sense transgenes and DNA viruses, such as geminiviruses, require the cellular RNA-dependent RNA polymerase RDR6 for dsRNA synthesis as well as several other factors identified genetically (SDE3, SGS3, and WEX; Table 2). To render them substrates for RDR6, transcripts of sense transgenes and DNA viruses are presumed to be aberrant in some way; for example, by lacking a 5′ cap or a polyadenylated tail (for review, see Meins et al. 2005).

The DCL activity required to process dsRNA into siRNAs in the PTGS pathway has not yet been identified (DCLX). Tests of *dcl1* partial loss-of-function mutants indicated that DCL1 is unlikely to be involved in this processing step. The plant-specific protein HEN1 adds a methyl group to the 3′-most nucleotide of small RNAs, thus protecting them from uridylation and subsequent degradation (Li et al. 2005). DCL2 has been implicated in generating siRNAs from some, but not all, RNA viruses (Xie et al. 2004).

PTGS and VIGS result in the production of two distinct size classes of siRNA, 21–22 nucleotides and 24–26 nucleotides, that have been implicated in diverse functions (Baulcombe 2004). In general, the 21-nucleotide siRNAs are thought to guide mRNA cleavage, whereas the 24- to 26-nucleotide size class, termed heterochromatic siRNA, directs epigenetic modifications to homologous DNA sequences (i.e., TGS; see Section 3.4).

Following DCL processing, the siRNA duplex is unwound and the antisense strand associates with a member of the Argonaute protein family, as part of the assembly into the RNA-induced silencing complex (RISC). The siRNA-programmed RISC can then direct endonucleolytic cleavage of target mRNAs at a single site near the center of siRNA–mRNA complementarity. For the mammalian equivalent, cleavage is catalyzed by the Ago2 "slicer" activity (see Chapter 8). The *Arabidopsis* protein carrying out this function in the transgene PTGS pathway is AGO1 (Baumberger and Baulcombe 2005). Following endonucleolytic cleavage, the severed 3′ segment of the mRNA is degraded in the 5′ to 3′ direction by the exonuclease AtXRN4 (Souret et al. 2004); the 5′ portion is probably degraded by the exosome in a 3′ to 5′ direction.

NON-CELL-AUTONOMOUS SILENCING AND TRANSITIVITY

PTGS in plants has two special properties that rely on the activity of the RNA-dependent RNA polymerase RDR6: non-cell-autonomous silencing and transitivity (Fig. 3a). In the former, RNA signals that induce PTGS move from the cell of origin into neighboring cells through plasmodesmata or—as originally shown in grafting experiments—through the vascular system to induce sequence-specific gene silencing at distant sites (for review, see Voinnet 2005). Mobile small RNAs, providing a systemic silencing signal, thus might play the dual function of influencing plant development by facilitating communication between cells, and coordinating activities in remote parts of the plant. This proposal is supported by the finding of microRNAs (miRNAs; important for development, Section 3.3) and a small RNA-binding protein in phloem sap, which is the main transporter of metabolites through the plant vascular system (Yoo et al. 2004).

Transitivity refers to the generation of secondary siRNAs corresponding to sequences located outside the primarily targeted regions. To make these, RNA-dependent RNA polymerase catalyzes synthesis of secondary dsRNAs from transgene or viral template RNAs using primary siRNAs as primers. Dicer processing yields secondary siRNAs, which amplify the silencing reaction and, when viral RNAs are involved, strengthen virus resistance (Voinnet 2005).

The only other organism in which both non-cell-autonomous silencing and transitivity have been observed is *C. elegans* (see Chapter 8), which has putative RNA-dependent RNA polymerase activities that are absent in mammals and *Drosophila*.

VIRAL SUPPRESSORS OF SILENCING

Plant viruses are not only inducers and targets of silencing; they also encode proteins that can suppress silencing (for review, see Voinnet 2005). This reinforces the idea that PTGS is a natural defense to viruses, since these suppressor proteins constitute a counter-defense "strategy" of the pathogen. Most plant viruses encode at least one silencing suppressor protein that acts at a distinct step of the PTGS pathway, typically downstream of dsRNA processing. Suppression of PTGS by a virus is strikingly revealed in mottled soybeans, where the dark color is the result of reversal of natural PTGS (i.e., reactivation) of a pigment gene (Fig. 2e) (Senda et al. 2004). Viral suppressors of RNAi have recently also been found in an insect virus and a mammalian retrovirus (Lecellier et al. 2005; Voinnet 2005).

3.3 Pathway 2: Regulation of Plant Development by miRNAs and Trans-acting siRNAs

The discovery of endogenous populations of miRNAs in plants and animals opened a new era in research of developmental biology and RNAi-mediated gene silencing (for review, see Bartel 2004). miRNAs silence gene expression by base-pairing to target messenger RNAs (mRNAs) and inducing either mRNA cleavage or translation repression. The importance of miRNAs in plant development is illustrated by the fact that many genes needed for miRNA biogenesis and silencing—including *DCL1*, *AGO1*, *HEN1*, *HYL1*, and *HST*—were identified in screens for developmental mutants and only later shown to be important for miRNA accumulation. The phenotypes of mutants defective in these proteins suggest diverse roles for miRNAs in meristem function, organ polarity, vascular development, floral patterning, and stress/hormone responses (for review, see Kidner and Martienssen 2005). miRNAs have recently been implicated in the biogenesis of a new type of small RNA, the trans-acting siRNAs.

ROLES AND BIOGENESIS OF MIRNAS

miRNAs were initially recovered by cloning size-fractionated small RNAs ranging from about 18 to 28 nucleotides in length. Their high degree of complementarity to target mRNAs in plants facilitated identification of additional miRNAs by computational approaches. So far, 92 loci in *Arabidopsis* that encode 27 distinct miRNAs have been discovered, and there are a similar number in rice. The expression of many miRNA genes is developmentally or environmentally regulated. About 50% of their known targets in *Arabidopsis* are transcription factors, many of which were known modulators of meristem formation and identity, prior to their identification as miRNA targets. In contrast, animal miRNAs do not preferentially target transcription factors but regulate diverse genes that operate at many levels in the cell. Two essential proteins of the miRNA pathway in *Arabidopsis*, DCL1 and AGO1, are themselves regulated by miRNAs, providing a means for negative modulation by feedback control (Kidner and Martienssen 2005).

Many miRNAs are evolutionarily conserved among eukaryotes (Axtell and Bartel 2005), in some cases over extended periods of time. Remarkably, in flowering plants, gymnosperms, and more primitive plants, mRNAs of a group of transcription factors that regulate meristem formation and lateral organ asymmetry have maintained perfect complementarity to the cognate miRNA. This indicates conservation of function for at least 400 million years (Floyd and Bowman 2004).

miRNAs are encoded in regions between protein-coding genes or in introns. They originate from imperfect RNA hairpin precursors, ranging from 70 bp to more than 300 bp in length, that are transcribed by DNA-dependent RNA polymerase II. Processing of plant miRNA precursors occurs in multiple steps in the nucleus. First, the ends of the pri-miRNA are removed by nuclear DCL1. This step requires the dsRNA-binding protein HYL1, originally identified by the hormone response defects of its mutant phenotype (Han et al. 2004; Vasquez et al. 2004a). The second step involves release of the miRNA duplex (miR/miR*, Fig. 3b), again by DCL1, and 3'-end methylation by HEN1 (see Section 3.2, Origin and processing of dsRNA). Transport of the miR/miR* duplex from the nucleus to the cytoplasm requires HASTY (HST), a homolog of mammalian Exportin 5 (Park et al. 2005). Mature miRNAs are also found in nuclear fractions, suggesting that some may function in the nucleus to direct epigenetic modifications. Indeed, a miRNA that is complementary to the spliced, nascent transcript of a transcription factor induces cytosine methylation of DNA sequences downstream of the target gene, by an unknown mechanism (Schubert et al. 2005).

miRNA biogenesis differs somewhat in mammals, which have a single Dicer that is located in the cytoplasm and a second RNase III-type activity, Drosha, in the nucleus. Drosha, together with the dsRNA-binding protein Pasha—neither of which has a homolog in plants—cleaves the ends of the pri-miRNA. The resulting pre-miRNA is then transported to the cytoplasm by an Exportin5-mediated pathway to undergo final processing to mature miRNAs by Dicer (Du and Zamore 2005; Kim 2005).

PLANT MIRNAS GUIDE MRNA CLEAVAGE

In general, animal miRNAs show imperfect complementarity to target mRNAs and repress translation by binding to multiple sites in 3'UTRs. In contrast, the nearly perfect complementarity of plant miRNAs to the coding regions of target mRNAs favors mRNA cleavage, presumably in a manner similar to siRNAs. However, there are increasing exceptions to both of these "rules." For example, plant miR172 is able to block translation, and certain mammalian miRNAs may direct cleavage of target mRNAs (for review, see Du and Zamore 2005).

AGO1 is the founding member of the Argonaute family of proteins and the mRNA "slicer" component of miRNA-programmed RISC in *Arabidopsis* (Baumberger and Baulcombe 2005). AGO1 was identified prior to the discovery of miRNAs in a screen for *Arabidopsis* mutants defective in leaf development (for review, see Carmell et al. 2002). The name Argonaute was inspired by the phenotype of *ago1* mutants, which resemble a small squid because of their narrow, filamentous leaves. *Ago1* mutants display shoot apical meristem defects similar to mutants deficient in PNH/ZLL/AGO10 (Table 2), which is similar to AGO1 but not yet shown to be needed for PTGS (Vaucheret et al. 2004). The essential function of AGO proteins in plant meristems is consistent with a conserved function of these proteins in stem cell maintenance (Carmell et al. 2002; Kidner and Martienssen 2005).

TRANS-ACTING SIRNAS

Endogenous trans-acting siRNAs (ta-siRNAs) are a new type of small RNA that have been discovered recently in *Arabidopsis*. The ta-siRNAs, which elicit cleavage of their target mRNAs, share features with both siRNAs and miRNAs. Similarly to siRNAs, the synthesis of the dsRNA precursor of ta-siRNAs depends on RDR6 and SGS3. Similarly to miRNAs, ta-siRNAs originate from genomic regions that have little overall resemblance to their target mRNA. To ensure formation of the correct ta-siRNA with complementarity to the target mRNA, a miRNA sets the phased cleavage of the dsRNA precursor by DCL4 (Fig. 3b) (Allen et al. 2005; Gasciolli et al. 2005).

ta-siRNAs have been assigned a role in developmental timing. During development, the shoot of flowering plants undergoes two phase changes: the vegetative phase change, comprising the juvenile-to-adult transition, and the reproductive phase change, which results in growth of flower-containing branches instead of vegetative shoots (see Fig. 1). Although genetic analysis of floral induction is well advanced, less is known about the vegetative phase change. An Argonaute protein, ZIPPY/AGO7, however, has a specialized role in this transition. A screen for mutants undergoing precocious vegetative phase change similar to *zip/ago7* mutants identified *RDR6* and *SGS3*, two genes important for PTGS (Fig. 3b) (Peregrine et al. 2004). Further analysis showed that several genes that are up-regulated in *rdr6* and *sgs3* mutants are silenced post-transcriptionally by ta-siRNAs (Vazquez et al. 2004b). These findings imply that components of the PTGS machinery are important not only for viral defense and transgene silencing, but also for temporal control of developmental switches. It is not yet known whether ta-siRNAs have counterparts in animals.

3.4 Pathway 3: Transgene-related Transcriptional Silencing, RNA-directed DNA Methylation, and Heterochromatin Formation

Current concepts of RNAi-mediated transcriptional gene silencing grew out of early plant work on homology-dependent gene silencing triggered by multiple copies of promoter regions and on RNA-directed DNA methylation (for review, see Matzke and Matzke 2004). More recent studies on RNAi-mediated heterochromatin formation in fission yeast (see Chapters 6 and 8) and on siRNA-mediated TGS in mammalian cells have expanded the phylogenetic scope of this process and confirmed mechanistic overlaps to RNAi.

RNA-DIRECTED DNA METHYLATION

RdDM was first observed in tobacco plants infected with viroids (Wassenegger et al. 1994). Viroids are minute plant pathogens consisting solely of a non-protein-coding, circular RNA several hundred bases in length. In the original experiments, replicating viroids were found to trigger de novo methylation of viroid cDNAs integrated as transgenes into the tobacco genome. Transgene systems were subsequently used to establish that RdDM requires a dsRNA that is processed to small RNAs, a hallmark of RNAi. RNA viruses that replicate exclusively in the cytoplasm were shown to elicit methylation of homologous nuclear DNA, indicating that small RNAs produced in the cytoplasm as a consequence of PTGS are able to enter the nucleus and induce epigenetic changes. dsRNAs containing promoter sequences can direct DNA methylation and transcriptional silencing of cognate target promoters (for review, see Mathieu and Bender 2004; Matzke et al. 2005).

In plants, RNA induces a distinctive pattern of de novo methylation that is typified by the modification of cytosines in all sequence contexts, largely within the region of RNA–DNA sequence identity. This characteristic pattern hints that RNA–DNA base-pairing provides a substrate for de novo methylation, but this remains to be experimentally verified. Whereas asymmetrical CpNpN methylation is not efficiently retained after withdrawing the trigger RNA, symmetrical CpG and CpNpG methylation can be maintained to varying extents without RNA at different promoters. Differences in the efficiency of maintenance methylation might reflect differences in sequence composition or patterns of histone modifications.

Combined data from genetic screens using transgene and endogenous gene systems are revealing the molecular components needed for RdDM and TGS. Transgene systems rely on transcribed inverted repeats or viruses to produce dsRNA that is homologous to target DNA loci. Endogenous genes that have been informative in forward genetic screens include the phosphoribosyl anthranilate isomerase (*PAI*) gene family (Mathieu and Bender 2004) and the *SUPERMAN* gene (Chan et al. 2005). These genes have features that render them targets or inducers of RdDM and TGS. For example, the *PAI* gene family contains four members, two of which are arranged as an inverted repeat. Transcription through the inverted repeats from an unrelated upstream promoter produces a dsRNA that targets the singlet copies of the *PAI* gene for methylation and silencing.

PLANT-SPECIFIC MACHINERY FOR RDDM

For the most part, conserved DNA methyltransferases and histone-modifying enzymes are required for RdDM (Sections 2.1 and 2.2). De novo methylation of cytosines in all sequence contexts is catalyzed by the conserved DRM class of DNA methyltransferase. The conserved MET1 and plant-specific CMT3 function primarily to maintain methylation of CpG and CpNpG nucleotide groups, respectively, although minor contributions to de novo methylation have been reported. The conserved histone deacetylase HDA6 and the SWI2/SNF2 protein DDM1 help to maintain CpG methylation at some loci. The histone methyltransferase KYP/SUVH4 is involved in locus-specific maintenance of CpNpG methylation induced by RNA (for review, see Chan et al. 2005).

A recent, surprising finding is that RdDM requires a plant-specific RNA polymerase, termed pol IV. In all eukaryotes examined so far, there are three DNA-dependent RNA polymerases—pol I, pol II, and pol III—that contain multiple subunits encoded by distinct genes. The first hint of the existence of pol IV came from analyzing the *Arabidopsis* genome sequence, which revealed genes encoding the largest and second-largest subunits of an atypical RNA polymerase unique to plants. There appear to be two functionally diversified pol IV complexes that are specified by unique largest subunits that each act with a common second-largest subunit. pol IVa is needed to generate siRNAs, presumably by initially transcribing target genes (Herr et al. 2005; Onodera et al. 2005). The initial transcript is used as a template by RDR2 to synthesize dsRNA, which is processed by DCL3, a nuclear activity that is specialized for producing 24-nucleotide heterochromatic siRNAs from transposons and repeats (Xie et al. 2004). Downstream of siRNA production, pol IVb acts together with the plant-specific

SWI2/SNF2-like protein DRD1 to signal DNA methylation (Kanno et al. 2005), probably in cooperation with AGO4 (Fig. 3c) (Chan et al. 2005). Whether pol IVb actually transcribes RNA is not known, but its net effect is to create a chromatin structure that permits DNA methyltransferases to catalyze de novo cytosine methylation at the siRNA-targeted site. Even though other eukaryotes do not contain pol IV subunits, two subunits of pol II, which transcribes mRNA precursors, are required for RNAi-mediated heterochromatin formation in fission yeast (see Chapters 6 and 8).

Although promoter-directed siRNAs can induce TGS in human cells, there are conflicting reports about whether this is accompanied by detectable DNA methylation (Kawasaki et al. 2005; Ting et al. 2005). Many proteins required for RdDM in plants are found only in that kingdom (Fig. 3c). Thus, if RdDM occurs regularly in mammals, the mechanism or protein machinery differs from those in plants.

SILENCING OF ENDOGENOUS GENES BY RNAI-MEDIATED TGS

Many transposons and repetitive DNA sequences, such as the 5S rDNA arrays and the transposon-rich heterochromatic knob on *Arabidopsis* chromosome 4, are transcriptionally silenced and methylated by an RNAi-mediated mechanism (Lippman and Martienssen 2004; Chan et al. 2005). The endogenous DNA targets reflect the natural roles of RNAi-mediated TGS in repressing transposition and in packaging repeats into heterochromatin. However, plant genes containing transposon insertions can themselves become targets of RNAi-mediated silencing and methylation. For example, transposon-derived repeats in the promoter of the *Arabidopsis* floral gene *FWA* are targeted for methylation by cognate siRNAs (Lippman and Martienssen 2004), thus silencing the gene in vegetative tissues where it is not required. In some *Arabidopsis* accessions, a *Mu* element in an intron of the *FLC* gene, a repressor of flowering, renders the gene susceptible to repressive chromatin modifications directed by siRNAs originating from dispersed copies of *Mu* (Liu et al. 2004). The resulting lowered expression of *FLC* can accelerate flowering time, which might have adaptive significance in certain environments. Since many plant genes have transposon insertions in the vicinity of promoters or in introns, this mode of regulation might be common in the plant kingdom. Indeed, as more is learned about epigenetic regulation, McClintock's idea of transposons acting as elements that control host genes and development is gaining increasing support (McClintock 1956).

4 Epigenetic Regulation without RNA Involvement

Despite the specificity provided by small RNAs, they probably do not induce all epigenetic modifications in plants. For example, MOM, a protein with a partial SNF2 domain, has not yet been implicated in RNAi-mediated TGS. There is also no evidence that PcG proteins in plants are directed to their target genes by small RNAs. Other types of signal, such as homologous pairing of non-transcribed repetitive sequences or special sequence compositions, might nucleate heterochromatin formation or attract DNA methyltransferases. The RNAi machinery, for instance, is dispensable for DNA methylation and histone methylation in *Neurospora*, where TA-rich segments are preferentially targeted for modification (see Chapter 6). Moreover, there are pathways for heterochromatin formation in fission yeast that are independent of RNAi (see Chapters 6 and 8).

An unusual epigenetic phenomenon in plants that has not yet been shown to involve RNAi is paramutation. Paramutation occurs when certain alleles, termed paramutagenic, impose an epigenetic imprint on susceptible (paramutable) alleles. The epigenetic imprint is inherited through meiosis and persists even after the two interacting alleles segregate in progeny. Paramutation represents a violation of Mendel's law, which stipulates that alleles segregate unchanged from a heterozygote. Paramutation was first observed decades ago in maize and tomato, but the mechanism(s) has remained enigmatic. Paramutation-like phenomena have been observed recently in mammals, suggesting that it is not limited to the plant kingdom (for review, see Chandler and Stam 2004). The B locus in maize, one of the most intensively studied cases of paramutation, contains a series of direct repeats almost 100 kb from the transcription start site that mediate paramutation in an unknown manner. Although RNA-based silencing has not been fully ruled out, alternate mechanisms relying on pairing of alleles are still under consideration (for review, see Stam and Mittelsten Scheid 2005).

5 Outlook

In this chapter, we have discussed what is known about basic epigenetic principles in plants and their relationship to epigenetic regulation in other organisms. Plants clearly share a number of features of epigenetic control with other organisms, yet they have also evolved a number of plant-specific variations and innovations. These likely underpin the unique aspects of plant development and their extraordinary ability to survive and reproduce successfully in unpredictable environments.

Prominent among the plant-specific innovations is a built-in system for reversible epigenetic modifications, which likely makes a key contribution to plant developmental plasticity and adaptability. The capacity to induce or erase repressive modifications in nondividing cells—the former through RdDM and histone modifications, and the latter through the activity of DNA glycosylases such as DME and ROS1—allows epigenetic reprogramming without intervening cycles of DNA replication. The facile erasure of epigenetic marks from plant genomes probably accounts for the relative ease of cloning whole plants from single somatic cells. Nevertheless, induction as well as removal of epigenetic marks is likely neither perfect nor uniform throughout an individual, which creates epigenetic variability in populations of supposedly genetically identical cloned plants. Such somaclonal variation can be exploited in plant breeding programs.

Similarly, the differential inheritance of epigenetic marks during sexual reproduction can lead to epigenetic variation in natural populations. Selection can act on this variability by fixing specific epialleles that might have adaptive significance. As we have described for the process of vernalization, environmental cues can trigger epigenetic modifications in plants and alter physiological responses. Thus, plants can "learn" if environmentally or stress-induced epigenetic modifications in shoot meristem cells enter the germ line and are adaptive. Defining the full range of conditions under which epigenetic changes are likely to occur spontaneously or are programmed will reveal more about the biological functions of these modifications. Likewise, unraveling the mechanisms of meiotic inheritance of epigenetic marks in plants could eventually permit scientists to manipulate this feature for improvements in horticulture and agriculture.

In addition to responding appropriately to environmental stimuli, plants have confronted a variety of genomic challenges during their evolutionary and breeding histories. Polyploidization or hybridization can have a significant impact on epigenetic modifications owing to the still ill-defined process of genome shock, a response to an unusual stress leading to widespread mobilization of transposons (McClintock 1983). The origin of heterosis, the superior performance of hybrids compared to that of inbred parent lines, is still unknown, but it is likely to involve epigenetic alterations triggered by combining two related but distinct genomes. Similarly, polyploidization combines and/or multiplies whole genomes, with innumerable possibilities for epigenetic changes. Learning the epigenetic consequences of polyploidization in plants would also help to understand our own evolutionary history, which is increasingly thought to involve two whole-genome duplications (Furlong and Holland 2004). Clearly, even at this scale of inquiry, plant epigenetics can be informative for human biology, justifying their reputation as "masters of epigenetic regulation."

References

Adams K.L. and Wendel J.F. 2005. Polyploidy and genome evolution in plants. *Curr. Opin. Plant Biol.* **8:** 135–141.

Alleman M. and Doctor J. 2000. Genomic imprinting in plants: Observations and evolutionary implications. *Plant Mol. Biol.* **43:** 147–161.

Allen E., Xie Z., Gustafson A.M., and Carrington J.C. 2005. microRNA-directed phasing during *trans*-acting siRNA biogenesis in plants. *Cell* **121:** 207–221.

Amedeo P., Habu Y., Afsar K., Mittelsten Scheid O., and Paszkowski J. 2000. Disruption of the plant gene MOM releases transcriptional silencing of methylated genes. *Nature* **405:** 203–206.

Autran D., Huanca-Mamani W., and Vielle-Calzada J.P. 2005. Genomic imprinting in plants: The epigenetic version of an Oedipus complex. *Curr. Opin. Plant Biol.* **8:** 19–25.

Axtell M.J. and Bartel D.P. 2005. Antiquity of microRNAs and their targets in land plants. *Plant Cell* **17:** 1658–1673.

Bartel D.P. 2004. MicroRNAs: Genomics, biogenesis, mechanism, and function. *Cell* **116:** 281–297.

Baulcombe D.C. 2004. RNA silencing in plants. *Nature* **431:** 356–363.

Baumberger N. and Baulcombe D.C. 2005. *Arabidopsis* ARGONAUTE1 is an RNA slicer that selectively recruits microRNAs and short interfering RNAs. *Proc. Natl. Acad. Sci.* **102:** 11928–11933.

Baumbusch L.O., Thorstensen T., Krauss V., Fischer A., Naumann K., Assalkhou R., Schulz I., Reuter G., and Aalen R.B. 2001. The *Arabidopsis thaliana* genome contains at least 29 active genes encoding SET domain proteins that can be assigned to four evolutionarily conserved classes. *Nucleic Acids Res.* **29:** 4319–4333.

Burch-Smith T.M., Anderson J.C., Martin G.B., and Dinesh-Kumar S.P. 2004. Applications and advantages of virus-induced gene silencing for gene function studies in plants. *Plant J.* **39:** 734–746.

Carmell M.A., Xuan Z., Zhang M.Q., and Hannon G.J. 2002. The Argonaute family: Tentacles that reach into RNAi, developmental control, stem cell maintenance, and tumorigenesis. *Genes Dev.* **16:** 2733–2742.

Chan S.W.-L., Henderson I.R., and Jacobsen S.E. 2005. Gardening the genome: DNA methylation in *Arabidopsis thaliana*. *Nat. Rev. Genet.* **6:** 351–360.

Chandler V.L., and Stam M. 2004. Chromatin conversations: Mechanisms and implications of paramutation. *Nat. Rev. Genet.* **5:** 532–544.

Chandler V.L., Eggleston W.B., and Dorweiler J.E. 2000. Paramutation in maize. *Plant Mol. Biol.* **43:** 121–145.

Choi Y., Gehring M., Johnson L., Hannon M., Harada J.J., Goldberg R.B., Jacobsen S.E., and Fischer R.L. 2002. DEMETER, a DNA glycosylase domain protein, is required for endosperm gene imprinting and seed viability in *Arabidopsis*. *Cell* **110:** 33–42.

Cubas P., Vincent C., and Coen E. 1999. An epigenetic mutation responsible for natural variation in floral symmetry. *Nature* **401:** 157–161.

Du T. and Zamore P.D. 2005. microRNAPrimer: The biogenesis and function of microRNA. *Development* **132:** 4645–4652.

Elmayan T., Proux F., and Vaucheret H. 2005. *Arabidopsis* RPA2: A genetic link among transcriptional gene silencing, DNA repair, and DNA replication. *Curr. Biol.* **15:** 1919–1925.

Fedoroff N.V. and Chandler V. 1994. Inactivation of maize transposable elements. In *Homologous recombination and gene silencing in plants* (ed J. Paszkowski), pp. 349-385. Kluwer Academic Publishers, Dordrecht, The Netherlands.

Floyd S.K. and Bowman J.L. 2004. Gene regulation: Ancient microRNA target sequences in plants. *Nature* **428:** 485–486.

Furlong R.F. and Holland P.W. 2004. Polyploidy in vertebrate ancestry: Ohno and beyond. *Biol. J. Linnean Soc.* **82:** 425–430.

Gasciolli V., Mallory A.C., Bartel D.P., and Vaucheret H. 2005. Partially redundant functions of *Arabidopsis* DICER-like enzymes and a role for DCL4 in producing *trans*-acting siRNAs. *Curr. Biol.* **15:** 1494–1500.

Gaudin V., Libault M., Pouteau S., Juul T., Zhao G., Lefebvre D., and Grandjean O. 2001. Mutations in *LIKE HETEROCHROMATIN PROTEIN 1* affect flowering time and plant architecture in *Arabidopsis*. *Development* **128:** 4847–4858.

Gong Z., Morales-Ruiz T., Ariza R.R., Roldan-Arjona T., David L., and Zhu J.K. 2002. *ROS1*, a repressor of transcriptional gene silencing in *Arabidopsis*, encodes a DNA glycosylase/lyase. *Cell* **111:** 803–814.

Han M.H., Goud S., Song L., and Fedoroff N. 2004. The *Arabidopsis* double-stranded RNA-binding protein HYL1 plays a role in microRNA-mediated gene regulation. *Proc. Natl. Acad. Sci.* **101:** 1093–1098.

Heitz E. 1928. Das Heterochromatin der Moose. I. *Jahrb. Wiss. Bot.* **69:** 762–818.

Hennig L., Bouveret R., and Gruissem W. 2005. MSI1-like proteins: An escort service for chromatin assembly and remodeling complexes. *Trends Cell Biol.* **15:** 295–302.

Herr A.J., Jensen M.B., Dalmay T., and Baulcombe D.C. 2005. RNA polymerase IV directs silencing of endogenous DNA. *Science* **308:** 118–120.

Hsieh T.-F. and Fischer R.L. 2005. Biology of chromatin dynamics. *Annu. Rev. Plant Biol.* **56:** 327–351.

Hung M.-S. and Shen C.-K. 2003. Eukaryotic methyl-CpG-binding domain proteins and chromatin modification. *Eukaryot. Cell* **2:** 841–846.

Jackson J.P., Lindroth A.M., Cao X., and Jacobsen S.E. 2002. Control of CpNpG DNA methylation by the KRYPTONITE histone H3 methyltransferase. *Nature* **416:** 556–556.

Jeddeloh J.A., Stokes T.L., and Richards E.J. 1999. Maintenance of genomic methylation requires a SWI2/SNFs-like protein. *Nat. Genet.* **22:** 94–97.

Jorgensen R.A. 2003. Sense cosuppression in plants: Past, present, and future. In *RNAi: A guide to gene silencing* (ed. G.J. Hannon), pp. 5–22. Cold Spring Harbor Laboratory Press, Cold Spring Harbor, New York.

Kaeppler S.M., Kaeppler H.F., and Rhee Y. 2000. Epigenetic aspects of somaclonal variation in plants. *Plant Mol. Biol.* **43:** 59–68.

Kanno T., Mette M.F., Kreil D.P., Aufsatz W., Matzke M., and Matzke A.J. 2004. Involvement of putative SNF2 chromatin remodeling protein DRD1 in RNA-directed DNA methylation. *Curr. Biol.* **14:** 801–805.

Kanno T., Huettel B., Mette M.F., Aufsatz W., Jaligot E., Daxinger L., Kreil D.P., Matzke M., and Matzke A.J.M. 2005. Atypical RNA polymerase subunits required for RNA-directed DNA methylation. *Nat. Genet.* **37:** 761–765.

Kapoor A., Agius F., and Zhu J.K. 2005a. Preventing transcriptional gene silencing by active DNA demethylation. *FEBS Lett.* **579:** 5889–5898.

Kapoor A., Agarwal M., Andreucci A., Zheng X., Gong Z., Hasegawa P.M., Bressan R.A., and Zhu J.K. 2005b. Mutations in a conserved replication protein suppress transcriptional gene silencing in a DNA-methylation-independent manner in *Arabidopsis*. *Curr. Biol.* **15:** 1912–1918.

Kawasaki H., Taira K., and Morris K.V. 2005. siRNA induced transcriptional gene silencing in mammalian cells. *Cell Cycle* **4:** 442–448.

Kidner C.A. and Martienssen R.A. 2005. The developmental role of microRNA in plants. *Curr. Opin. Plant Biol.* **8:** 1–7.

Kim V.N. 2005. MicroRNA biogenesis: Coordinated cropping and dicing. *Nat. Rev. Genet.* **6:** 376–385.

Köhler C., Page D.R., Gagliardini V., and Grossniklaus U. 2005. The *Arabidopsis thaliana* MEDEA polycomb group protein controls expression of PHERES1 by parental imprinting. *Nat. Genet.* **37:** 28–30.

Kotake T., Takada S., Nakahigashi K., Ohto M., and Goto K. 2003. *Arabidopsis TERMINAL FLOWER 2* gene encodes a heterochromatin protein 1 homolog and represses both *FLOWERING LOCUS T* to regulate flowering time and several floral homeotic genes. *Plant Cell Physiol.* **44:** 555–564.

Kress C., Thomassin H., and Grange T. 2001. Local DNA demethylation in vertebrates: How could it be performed and targeted? *FEBS Lett.* **494:** 135–140.

Lecellier C.-H., Dunoyer P., Arar K., Lehmann-Che J., Eyquem S., Himber C., Saib A., and Voinnet O. 2005. A cellular microRNA mediates antiviral defense in human cells. *Science* **308:** 557–560.

Li J., Yang Z., Yu B., Liu J., and Chen X. 2005. Methylation protects miRNAs and siRNAs from 3′-end uridylation activity in *Arabidopsis*. *Curr. Biol.* **15:** 1501–1507.

Lippman Z. and Martienssen R. 2004. The role of RNA interference in heterochromatic silencing. *Nature* **431:** 364–370.

Liu J., He Y., Amasino R., and Chen X. 2004. siRNAs targeting an intronic transposon in the regulation of natural flowering behavior in *Arabidopsis*. *Genes Dev.* **18:** 2873–2878.

Loidl P. 2004. A plant dialect of the histone language. *Trends Plant Sci.* **9:** 84–90.

Malagnac F., Bartee L., and Bender J. 2002. An *Arabidopsis* SET domain protein required for maintenance but not establishment of DNA methylation. *EMBO J.* **21:** 6842–6852.

Mathieu O. and Bender J. 2004. RNA-directed DNA methylation. *J. Cell Sci.* **117:** 4881–4888.

Matzke M.A. and Birchler J.A. 2005. RNAi-mediated pathways in the nucleus. *Nat. Rev. Genet.* **6:** 24–35.

Matzke M. and Matzke A.J.M. 2004. Planting the seeds of a new paradigm. *PLoS Biol.* **2:** 528–586.

Matzke M., Kanno T., Huettel B., Jaligot E., Mette M.F., Kreil D.P., Daxinger L., Rovina P., Aufsatz W., and Matzke A.J.M. 2005. RNA-directed DNA methylation. In *Plant epigenetics* (ed. P. Meyer), pp. 69–96. Blackwell Publishing, Oxford, United Kingdom.

McClintock B. 1956. Intranuclear systems controlling gene action and mutation. *Brookhaven Symp. Biol.* **8:** 58–74.

———. 1983. The significance of responses of the genome to challenge. Nobel lecture. http://nobelprize.org/medicine/laureates/1983/mcclintock-lecture.pdf

Meins F., Si-Ammour A., and Blevins T. 2005. RNA silencing systems

and their relevance to plant development. *Annu. Rev. Cell Dev. Biol.* **21:** 297–318.

Naumann K., Fischer A., Hofmann I., Krauss V., Phalke S., Irmler K., Hause G., Aurich A.-C., Dorn R., Jenuwein T., and Reuter G. 2005. Pivotal role of AtSUVH2 in heterochromatic histone methylation and gene silencing in *Arabidopsis*. *EMBO J.* **24:** 1418–1429.

Onodera Y., Haag J.R., Ream T., Nunes P.C., Pontes O., and Pikaard C. 2005. Plant nuclear RNA polymerase IV mediates siRNA and DNA methylation-dependent heterochromatin formation. *Cell* **120:** 613–622.

Pandey R., Müller A., Napoli C.A., Selinger D.A., Pikaard C.S., Richards E.J., Bender J., Mount D.W., and Jorgensen R.A. 2002. Analysis of histone acetyltransferase and histone deacetylase families of *Arabidopsis thaliana* suggests functional diversification of chromatin modification among multicellular eukaryotes. *Nucl. Acids Res.* **30:** 5036–5055.

Park M.Y., Wu G., Gonzalez-Sulser A., Vaucheret H., and Poethig R.S. 2005. Nuclear processing and export of microRNAs in *Arabidopsis*. *Proc. Natl. Acad. Sci.* **102:** 3691–3696.

Peregrine A., Yoshikawa M., Wu G., Albrecht H.L., and Poethig R.S. 2004. SGS3 and SGS2/SDE1/RDR6 are required for juvenile development and the production of *trans*-acting siRNAs in *Arabidopsis*. *Genes. Dev.* **18:** 2368–2379.

Pikaard C.S. 2000. The epigenetics of nucleolar dominance. *Trends Genet.* **16:** 495–500.

Rocha P.S., Sheikh M., Melchiorre R., Fagard M., Boutet S., Loach R., Moffatt B., Wagner C., Vaucheret H., and Furner I. 2005. The *Arabidopsis HOMOLOGY-DEPENDENT GENE SILENCING1* gene codes for an S-adenosyl-L-homocysteine hydrolase required for DNA methylation-dependent silencing. *Plant Cell* **17:** 404–417.

Schubert D., Clarenz O., and Goodrich J. 2005. Epigenetic control of plant development by Polycomb-group proteins. *Curr. Opin. Plant Biol.* **8:** 553–561.

Senda M., Masuta C., Ohnishi S., Goto K., Kasai A., Sano T., Hong J.-S., and MacFarlane S. 2004. Patterning of virus-infected *Glycine max* seed coat is associated with suppression of endogenous silencing of chalcone synthase genes. *Plant Cell* **16:** 807–818.

Souret F.F., Kastenmayer J.P., and Green P.J. 2004. AtXRN4 degrades mRNA in *Arabidopsis* and its substrates include selected miRNA targets. *Mol. Cell* **15:** 173–183.

Springer N.M. and Kaeppler S.M. 2005. Evolutionary divergence of monocot and dicot methyl-CpG-binding domain proteins. *Plant Physiol.* **138:** 92–104.

Springer N.M., Napoli C.A., Selinger D.A., Pandey R., Cone K.C., Chandler V.L., Kaeppler H.F., and Kaeppler S.M. 2003. Comparative analysis of SET domain proteins in maize and *Arabidopsis* reveals multiple duplications preceding the divergence of monocots and dicots. *Plant Physiol.* **132:** 907–925.

Stam M. and Mittelsten Scheid O. 2005. Paramutation: An encounter leaving a lasting impression. *Trends Plant Sci.* **10:** 283–290.

Takeda S., Tadele Z., Hofmann I., Probst A.V., Angelis K.J., Kaya H., Araki T., Mengiste T., Mittelsten Scheid O., Shibahara K., et al. 2004. BRU1, a novel link between responses to DNA damage and epigenetic gene silencing in *Arabidopsis*. *Genes Dev.* **18:** 782–793.

Tariq M. and Paszkowski J. 2004. DNA and histone methylation in plants. *Trends Genet.* **20:** 244–251.

Ting A.H., Schuebel K.E., Herman J.G., and Baylin S.B. 2005. Short double-stranded RNA induces transcriptional gene silencing in human cancer cells in the absence of DNA methylation. *Nat. Genet.* **37:** 906–910.

Vaucheret H., Vazquez F., Crété P., and Bartel D.P. 2004. The action of

ARGONAUTE1 in the miRNA pathway and its regulation by the miRNA pathway are crucial for plant development. *Genes Dev.* **18:** 1187–1197.

Vazquez F., Gasciolli V., Crete P., and Vaucheret H. 2004a. The nuclear dsRNA binding protein HYL1 is required for microRNA accumulation and plant development, but not posttranscriptional transgene silencing. *Curr. Biol.* **14:** 346–345.

Vazquez F., Vaucheret H., Rajagopalan R., Lepers C., Gasciolli V., Mallory A.C., Hilbert J.-L., Bartel D.P., and Crété P. 2004b. Endogenous *trans*-acting siRNAs regulate the accumulation of *Arabidopsis* mRNAs. *Mol. Cell* **16:** 69–79.

Voinnet O. 2005. Induction and suppression of RNA silencing: Insights from viral infections. *Nat. Rev. Genet.* **6:** 206–220.

Wagner D. 2003. Chromatin regulation of plant development. *Curr. Opin. Plant Biol.* **6:** 20–28.

Wang X.J., Gaasterland T., and Chua N.H. 2005. Genome-wide prediction and identification of *cis*-natural antisense transcripts in *Arabidopsis thaliana*. *Genome Biol.* **6:** R30.

Wassenegger M., Heimes S., Riedel L., and Sanger H.L. 1994. RNA-directed de novo methylation of genomic sequences in plants. *Cell* **76:** 567–576.

Xie Z., Johansen L.K., Gustafson A.M., Kasschau K.D., Lellis A.D., Zilberman D., Jacobsen S.E., and Carrington J.C. 2004. Genetic and functional diversification of small RNA pathways in plants. *PLoS Biol.* **2:** E104.

Yoo B.-C., Kragler F., Varkonyi-Gasic E., Haywood V., Archer-Evans S., Lee Y.M., Lough T.J., and Lucas W.J. 2004. A systemic small RNA signalling system in plants. *Plant Cell* **16:** 1979–2000.

WWW Resources

http://asrp.cgrb.oregonstate.edu. *Arabidopsis thaliana* small RNA project

http://mpss.dbi.udel.edu. MPSS (Massively parallel signature sequencing)

http://www.arabidopsis.org/abrc. *Arabidopsis* Biological Resource Center Stocks

http://www.chromdb.org. Plant Chromatin Database

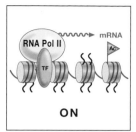
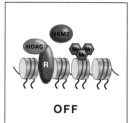

ON   OFF

Chromatin Modifications and Their Mechanism of Action

Tony Kouzarides[1] and Shelley L. Berger[2]

[1]*The Gurdon Institute, University of Cambridge, Cambridge, United Kingdom*
[2]*The Wistar Institute, Philadelphia, Pennsylvania 19104*

CONTENTS

1. Histones and Acetylation Are Regulatory to Transcription, 193

2. Acetylation and Deacetylation, 194

3. Phosphorylation, 196

4. Methylation, 197
   4.1 *Methylation of Lysines, 197*
   4.2 *Demethylation of Lysines, 201*
   4.3 *Methylation of Arginines, 201*

5. Deimination, 202

6. Ubiquitylation/Deubiquitylation and Sumoylation, 203

7. Themes in Modifications, 204
   7.1 *Histone Code, 204*
   7.2 *Modification Patterns, 204*
   7.3 *Changes in Chromatin Structure Associated with Transcription Activation and Elongation, 205*

8. Concluding Remarks, 206

References, 206

GENERAL SUMMARY

Histones are the building blocks of nucleosomes, making an octameric structure that packages DNA in eukaryotes forming a structure known as chromatin. Chromatin is not a uniform structure, however, and in recent years, an explosion in our knowledge of the variations in chromatin structure has occurred. This, in turn, has enhanced our understanding of the mechanisms that regulate genome templated processes, the posttranslational modifications of histone proteins (a central feature of this genomic regulation). There are, in fact, a large number of histone posttranslational modifications (HPTMs), and they divide into two groups. First, there are the small chemical groups, including acetylation, phosphorylation, and methylation. Second, there are the much larger peptides, including ubiquitylation and sumoylation.

How are HPTMs thought to affect genome regulation and function? Three mechanisms are commonly considered, and it is helpful to keep these mechanisms in mind as the wealth of information and history of HPTMs is presented in this chapter. First, HPTMs may somehow affect the structure of chromatin; for example, by preventing crucial contacts that facilitate certain chromatin conformations or higher-order structures (which can be considered as *cis*-modifying effects). In contrast, two other mechanisms are considered to operate in *trans*. HPTMs may disrupt the binding of proteins that associate with chromatin or histones. Alternatively, HPTMs may provide altered binding surfaces that attract certain effector proteins. This third mechanism has been characterized in the most detail, and such recruitment of proteins is defining with regard to the functional consequence: That is, it may have an activating or repressive outcome on transcription. The large number of HPTMs that have been discovered and their various combinations have led to the idea that HPTMs regulate via combinatorial patterns, in temporal sequences, and can be established over short- and long-range distances. These varied mechanisms establish different functional outcomes—some transient, others stable and epigenetically heritable.

It was during the 1960s that Vincent Allfrey identified acetylation, methylation, and phosphorylation of histones purified from many eukaryotes. Histones were also the first recognized ubiquitylated protein substrates. However, although Allfrey observed certain correlations between modifications and transcriptionally active chromatin sources, genetic and functional evidence to support a role for HPTMs in gene regulation did not emerge until much later. In fact, many scientists studying the biochemistry and genetics of gene regulation during the 1980s and 1990s were skeptical that HPTMs had a causal role in gene regulation.

1 Histones and Acetylation Are Regulatory to Transcription

As shown in this chapter, histones are subject to many different posttranslational modifications (PTMs), and time will undoubtedly reveal new, as-yet-unknown HPTMs. The known modifications can be categorized as either small chemical groups discussed in Sections 2–4 of this chapter, or as larger peptide changes to histones as discussed in Section 5 (see Table 1). The mechanism by which HPTMs affect the chromatin template and related processes such as gene transcription or repression are considered in the context of three conceptual models illustrated in Figure 1. Model 1 proposes that posttranslationally modified histones may, in some way, alter chromatin structure. In Model 2, an HPTM may inhibit the binding of a factor to the chromatin template, whereas Model 3 proposes that an HPTM creates a binding site for a particular protein (see also Section 5 of Chapter 3).

From a historical perspective, what changed the mainstream view that chromatin was largely inert packing material for DNA? Early evidence that HPTMs regulated transcriptional activation and silencing came from experiments in *Saccharomyces cerevisiae* during the late 1980s. This budding yeast provides an efficient organism to carry out genetic experiments (both forward and reverse genetics) to examine the importance of histones. The reason is that, unlike higher eukaryotes where there are multiple copies of each histone gene, the single-copy yeast

Model 1: Chromatin structural change

Model 2: Inhibit binding of negative-acting factor

Model 3: Recruit positive-acting factor

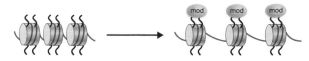

Figure 1. Models Showing How Histone Posttranslational Modifications Affect the Chromatin Template

Model 1 proposes that changes to chromatin structure are mediated by the *cis* effects of covalent histone modifications, such as histone acetylation or phosphorylation. Model 2 illustrates the inhibitory effect of an HPTM for the binding of a chromatin-associated factor (CF), as exemplified by H3S10 phosphorylation occluding HP1 binding at methylated H3K9. In Model 3, an HPTM may provide binding specificity for a chromatin-associated factor. A classic example is HP1 binding through its chromodomain to methylated H3K9.

histone genes can easily be genetically manipulated. For instance, in a background where all the histone genes have been deleted, a copy of each gene can be introduced, encoded on an episome that carries a selectable marker, such as the *URA3* gene, to maintain the episome. A second copy of the histones can be introduced on a second episome, carrying a different selectable marker. This second copy can be mutated by site-directed mutagenesis, and then the first, wild-type copy can be selectively lost from the cell using the 5-FOA (5-fluoroorotic acid) drug, which causes the *URA3* gene product to be toxic to the cell. The end result is that the only copy present in the cell is the altered second episomal copy, which contains any number of mutations to be tested.

In *S. cerevisiae*, the histone genes are arranged in pairs of H3/H4 and H2A/H2B, and their transcription is highly coordinated to coincide with S phase. Each nucleosome is assembled from an H3/H4 tetramer and two dimers of H2A/H2B; when one pair of either duo is under- or overtranscribed, nucleosomes are depleted. This alteration of histone dosage by genetic means provided some of the initial evidence that chromatin structure is crucial for regulating expression. One such

Table 1. Types of covalent histone posttranslational modifications

	Role in transcription	Histone-modified sites
GROUP 1		
Acetylation	activation	H3 (K9,K14,K18,K56) H4 (K5,K8K12,K16) H2A H2B (K6,K7,K16,K17)
Phosphorylation	activation	H3 (S10)
Methylation	activation	H3 (K4,K36,K79)
	repression	H3 (K9,K27) H4 (K20)
GROUP 2		
Ubiquitylation	activation	H2B (K123)
	repression	H2A (K119)
Sumoylation	repression	H3 (?) H4 (K5,K8,K12,K16) H2A (K126) H2B (K6,K7,K16,K17)

HPTMs are categorized into two groups: Group 1 represents small chemical group modifications, whereas Group 2 includes larger chemical modifications.

approach utilized "forward" genetics, where random mutations were selected that result in gene activation of a marker (Clark-Adams et al. 1988). These mutations were found to alter the amount of histone pairs. A second approach used "reverse" genetics, where directed depletion of histone genes provided clear evidence that histones regulate gene transcription (Han and Grunstein 1988). The next step was to direct deletion of only the histone amino-terminal tails (the sites of many HPTM) or to carry out substitution mutations of acetylation sites in histones. These more surgical changes also caused decreases in gene activation, suggesting that acetylation is required for gene transcription (Durrin et al. 1991).

Other approaches investigated whether nucleosomes are naturally altered during gene activation. Biochemical experiments had shown that nucleosomes were repressive to transcription on DNA templates in vitro (Workman and Roeder 1987), but whether this was true in vivo was under debate. Some promoters have naturally positioned nucleosomes upstream of transcriptional start sites, and these positioned nucleosomes became altered when the gene was activated (Svaren et al. 1994; Shim et al. 1998). In the case of PHO5, nucleosome alteration required an activator, showing that without transcription the nucleosomes were not changed. However, it was unclear whether this alteration was a cause or an effect of transcription. To address this, the TATA box was deleted, which abrogated transcription in yeast. Nucleosome position nevertheless changed, strongly suggesting that the alteration of nucleosomes preceded transcription.

Taken together, these experiments began to provide strong evidence that both nucleosome repositioning and acetylation of specific residues within the histone tails may be required for transcriptional activation.

2 Acetylation and Deacetylation

Additional experimental approaches continued to provide evidence that acetylation (versus the absence of acetylation) correlates with transcription. Regions that are transcriptionally active, or are poised for transcription, tend to have an "open" chromatin configuration and therefore are accessible to enzymes such as DNase and MNase, which, when added to isolated but intact nuclei, can digest DNA. In the early 1990s, researchers began to use chromatin immunoprecipitation (ChIP), a powerful technique for analyzing what proteins are bound to particular DNA sequences in vivo. This involves cross-linking proteins that are bound to DNA using a cell-permeable chemical such as formaldehyde, followed by sonication to break up the

DNA:protein complexes into smaller fragments. The DNA:protein complexes of interest are then immunoprecipitated using a specific antibody as a probe. The cross-links are then reversed in order to isolate and identify the DNA sequences that associated with the antibody-bound protein, by analysis using either radioactively labeled DNA probes or PCR. One group used this method to investigate the correlation between DNA sites around the active globin genes that are hypersensitive to DNase digestion and associate with acetylated histones in chicken erythrocytes; the correlation was remarkably close (Hebbes et al. 1994). In S. cerevisiae, similar approaches were employed within transcriptionally silenced regions of the genome, and they showed very low levels of histone acetylation (Braunstein et al. 1993). Conversely, genetic disruption of silencing correlated with increased acetylation.

All of these experiments were slowly revealing that histones and, in particular, sites of reversible acetylation play a role in gene regulation. However, it was not until the mid-1990s that the first nuclear histone acetylation and deacetylation enzymes were identified, and these provided the "smoking gun"—the most direct evidence that these enzymes play a role during transcription. The first nuclear histone acetyltransferase (HAT) was isolated from the so-called macronucleus (the very large transcriptionally active nucleus, as distinct from the meiotic micronucleus) from Tetrahymena, which has high transcription rates (Brownell et al. 1996). The key approach was the "in-gel" assay to detect HAT activity: A complex mixture of proteins from cell extracts was separated on a histone-permeated SDS gel, the peptides were then subjected to renaturation, and proteins with HAT enzymatic activity labeled the gel by the transfer of radiolabeled cofactor, acetyl coenzyme A, onto localized histones. This allowed further biochemical fractionation and purification of the polypeptide. The HAT enzyme that was identified was homologous to a previously isolated transcriptional coactivator in S. cerevisiae, called Gcn5, known to interact with transcriptional activators. Contemporaneously, the first histone deacetylase (HDAC) enzyme was isolated via biochemical purification (Taunton et al. 1996). In this case, the enzyme was purified from cell extracts using an inhibitor bound to an insoluble matrix, which physically bound to the catalytic site of the enzyme. The enzyme was homologous to a previously isolated gene, which has a cofactor role in gene repression. These remarkable parallel findings for the first enzymes found to metabolize acetyl groups on histones led to a model that is now the paradigm for gene-specific histone PTMs: DNA-bound activators recruit HATs to acetylate nucleosomal histones, while repressors recruit HDACs to deacetylate histones. These changes lead to alterations of the nucleosome and up- or down-regulation of the gene, respectively (Fig. 2).

Many other well-known coactivators and corepressors were shown to possess HAT or HDAC activity, or to associate with such enzymes (Sterner and Berger 2000; Roth et al. 2001). Moreover, the enzymatic activities of the HATs and HDACs are critical for their role in gene activation and repression. The enzymes are often components of large complexes that are modular in structure and function; histone-modifying enzymatic activity is just one function, and others include, for instance, the recruitment of the TATA-binding protein (TBP) (Grant et al. 1998). Interestingly, certain nuclear hormone receptors function both as DNA-binding transcriptional repressors (when not bound to hormone ligand) and as transcriptional activators (when bound to hormone ligand); the receptors do this partly through the PTM of target chromatin regions, by recruiting HDACs when unliganded and HATs when liganded (Baek and Rosenfeld 2004).

HAT proteins can acetylate lysine residues on all four core histones, but different enzymes possess distinct specificities in their substrate of choice (Fig. 3; Table 1), although each enzyme rarely targets just a single site. One major HAT family—GNAT (for Gcn5 related acetyltrans-

ferase)—targets histone H3 as its main substrate. A second major HAT family, the MYST family, targets histone H4 as its main substrate. A third major family— CBP/p300—targets both H3 and H4, and is the most promiscuous. Structural analyses have been carried out for the catalytic domains of the first two major families (GNAT and MYST), and they are distinct; the structure of the CBP/p300 family has not yet been solved. Incidentally, each of these acetyltransferase families is also able to acetylate non-histone substrates (Glozak et al. 2005).

As discussed above, there are three models for the role of HPTMs in regulating chromatin structure (Fig. 1). The first model considers structural changes to chromatin induced by the direct effects of HPTMs, such as changes in charge. In this case, the neutralization of positively charged lysine by acetylation reduces the strength of binding of the strongly basic histones or histone tails to negatively charged DNA, and thus opens DNA-binding sites (Vettese-Dadey et al. 1996). Still focusing on the first model, there is also evidence that acetylation can decompact nucleosome arrays, consistent with a role in opening chromatin for gene activation (Shogren-Knaak et al. 2006). The third model proposes that HPTMs provide a binding surface for proteins to associate with chromatin and regulate DNA-templated processes; this was first shown for acetylation. A specialized protein domain called a bromodomain, commonly found in chromatin-associated proteins, specifically binds to acetylated lysines (Fig. 3) (Dhalluin et al. 1999). Bromodomains are present in many HATs, such as Gcn5 and CBP/p300. Proteins with this motif, when part of large chromatin-associating/altering complexes such as the ATP-dependent remodeling complex, Swi/Snf, promote its binding to chromatin (Hassan et al. 2002). Other examples of proteins containing bromodomains with binding specificity to acetylated histone include Taf1 and Bdf1 in the TFIID complex, Rsc4 in the Rsc remodeling complex, and Brd2 in a large family of bromodomain proteins.

There are numerous HDAC enzymes that remove acetyl groups (Kurdistani and Grunstein 2003; Yang and Seto 2003). They fall into three catalytic groups, which are conserved through evolution from *S. cerevisiae* to mammals, and referred to as Type I, Type II, and Type III or Sir2-related enzymes. Type I and Type II have a related mechanism of deacetylation, which does not involve a cofactor, whereas the Sir2-related enzymes require the cofactor NAD as part of their catalytic mechanism. The structures of representatives for all three families have been solved. Many of the HDACs are found within large multisubunit complexes, components of which serve to target the enzymes to genes, leading to transcriptional

Gene activator recruits histone acetyltransferase

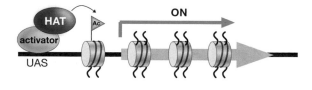

Gene repressor recruits histone deacetylase

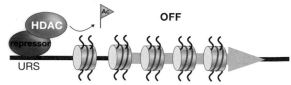

Figure 2. Histone-modifying Enzymes Are Recruited to Promoters by DNA-binding Transcription Factors

Histone acetyltransferases (HAT) are recruited by activators that bind to specific upstream activating sequences (UAS). This enzyme catalyzes the acetylation of local histones, known to contribute to transcriptional activation. Histone deacetylases (HDAC) are recruited by repressors of transcription that bind to upstream repressive sequences (URS) and deacetylate local histones. This contributes to transcriptional repression.

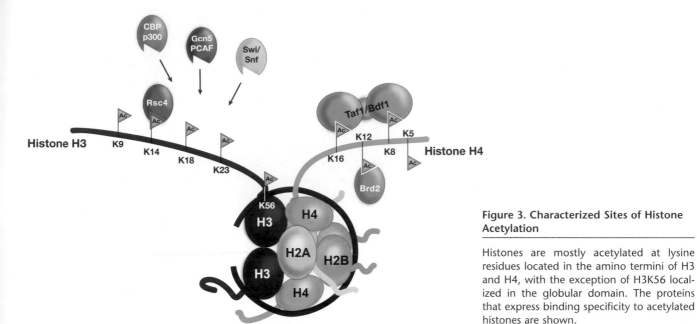

Figure 3. Characterized Sites of Histone Acetylation

Histones are mostly acetylated at lysine residues located in the amino termini of H3 and H4, with the exception of H3K56 localized in the globular domain. The proteins that express binding specificity to acetylated histones are shown.

repression. For example, Rpd3 is part of a large complex including the HDAC Sin3, which interacts with DNA-bound repressors (Kurdistani and Grunstein 2003; Yang and Seto 2003). Rpd3 is also part of a small complex, which is targeted to gene open reading frames (ORFs) via chromodomain association with H3K36me (see Section 4 for further discussion of chromodomains). This results in histone deacetylation, in part to suppress internal RNA polymerase II (pol II) initiation, and also to regulate different steps during the transcription cycle (Carrozza et al. 2005; Joshi and Struhl 2005).

3 Phosphorylation

Phosphorylation is the most well known PTM because it has long been understood that kinases regulate signal transduction from the cell surface, through the cytoplasm, and into the nucleus, leading to changes in gene expression. Histones were among the first proteins found to be phosphorylated. By 1991, it was discovered that when cells were stimulated to proliferate, the so-called "immediate-early" genes were induced to become transcriptionally active and to function in stimulating the cell cycle. This increased gene expression correlates with histone H3 phosphorylation (Mahadevan et al. 1991). The histone H3 Serine 10 residue (H3S10) has turned out to be an important phosphorylation site for transcription from yeast to humans, and appears to be especially important in *Drosophila* (Nowak and Corces 2004). Many kinases have been identified that target this site, including Msk1/2 and the related Rsk2 in

mammals, and SNF1 in *S. cerevisiae* (Sassone-Corsi et al. 1999; Lo et al. 2001; Soloaga et al. 2003). Studies of linker histone H1 in *Tetrahymena* have revealed that phosphorylation of this histone may also affect transcriptional control.

Perhaps counterintuitively, phosphorylation of certain residues correlates with chromosome condensation, during both mitosis and meiosis. It is unclear how phosphorylation contributes to the process, but recently, H3S10 phosphorylation acts like a temporal switch, ejecting HP1 bound to the adjacently methylated H3K9 residue, referred to as the methyl-phos binary switch (Fischle et al. 2005; Hirota et al. 2005). It remains to be seen whether this, perhaps in concert with the phosphorylation of H3S28 and H3T11, may effect chromatin condensation by recruiting the condensin complex and the mitotic spindle (Nowak and Corces 2004).

Less is known about the precise mechanistic role of histone phosphorylation. There is evidence to support all three models for the role of HPTMs. First, histone phosphorylation alters chromatin compaction in vivo (Model 1). Indeed, work in *Tetrahymena* demonstrated that the collective negative "charge patch" resulting from phosphorylation of clusters of nearby residues within linker histone H1 affects the affinity of its binding to DNA, positively increasing the transcriptional potential of the local chromatin environment (Dou and Gorovsky 2002). In support of Model 2, proteins bound to chromatin can be dislodged by phosphorylation. This was recently demonstrated by the lowered binding affinity of HP1 during mitosis subsequent to mitosis-specific H3S10 phosphorylation (Fischle et al. 2005; Hirota et al. 2005). In support of Model 3, the 14-3-3 adapter protein, a known phospho-binding protein, recognizes H3S10ph at promoters of inducible genes (Macdonald et al. 2005).

4 Methylation

Methylation as a histone covalent modification is more complex than any other, since it can occur on either lysines or arginines. Additionally, unlike any other modification in Group 1, the consequence of methylation can be either positive or negative toward transcriptional expression, depending on the position of the residue within the histone (Table 1). A further level of complexity lies in the fact that there can be multiple methylated states on each residue. Lysines can be mono- (me1), di- (me2), or tri- (me3) methylated, whereas arginines can be mono- (me1) or di- (me2) methylated. Given that there are at least 24 identified sites of lysine and arginine methylation on H3, H4, H2A, and H2B, the number of distinct nucleosomal methylated states is enormous. Such combinational potential of methylated nucleosomes may be necessary, at least partly, to allow for the regulation of complex and dynamic processes such as transcription, which requires sequential and precisely timed events (Jenuwein and Allis 2001; Y. Zhang and Reinberg 2001; Lee et al. 2005; Martin and Zhang 2005; Wysocka et al. 2006a).

4.1 Methylation of Lysines

The fact that lysine residues within histones are methylated has been known for many decades. The biological significance of this modification has only come to light recently, however, following the identification of the first lysine methyltransferase that uses histones as its substrate (Rea et al. 2000). Now, more histone lysine methyltransferases (HKMTs) have been identified, and their sites of modification on histones are defined (Martin and Zhang 2005). All of these enzymes, except Dot 1, share the SET domain, which contains the catalytically active site and allows binding to the S-adenosyl-L-methionine cofactor.

Of the many known methylated sites, six have been well characterized to date: five on H3 (K4, K9, K27, K36, K79) and one on H4 (K20). Methylation at H3K4, H3K36, and H3K79 has, in general, been linked to activation of transcription, and the rest to repression (Table 1). In addition, two of these sites—H3K79me and H4K20me—have been implicated in the process of DNA repair. Specific protein binders have been identified that recognize each of the six characterized methylation sites (Fig. 4). These proteins have one of three distinct types of methyl lysine recognition domains: the chromo, tudor, and PHD repeat domains. Below, each of these characterized modifications is discussed in more detail.

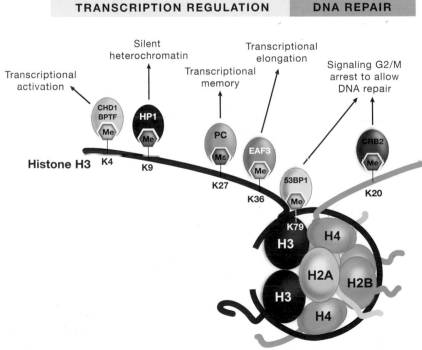

Figure 4. Sites of Histone Methylation, Their Protein Binders, and Functional Role in Genomic Processes

Methylation of histones occurs at lysine residues in histones H3 and H4. Certain methylated lysine residues are associated with activating transcription (*green Me flag*), whereas others are involved in repressive processes (*red Me flag*). Proteins that bind particular methylated lysine residues are indicated.

H3K4 METHYLATION

Methylation of H3K4 is associated with euchromatin and, specifically, with genes that are active or destined to be so. The demonstration that H3K4 methylation correlates with active chromatin came from analysis of the chicken β-globin locus and the budding yeast mating-type loci (Litt et al. 2001; Noma et al. 2001). ChIPs using antibodies specific for methylated H3K4 indicated that islands of the modified histones track active genes. Subsequent work in yeast established that different methyl states are important for activity and that the trimethyl state (H3K4me3) appears during the process of active transcription (Santos-Rosa et al. 2002).

H3K4me3 is observed at the 5′ ends of genes in yeast during activation of transcription. Three components of the transcriptional machinery are thought to be responsible for this mark. First, RNA pol II that has been phosphorylated at Ser-5 of the carboxy-terminal domain (CTD) can recruit the Set1 HKMT that methylates H3K4 in the vicinity of promoters (Fig. 5). Such phosphorylation normally releases RNA pol II from the transcription initiation complex into an early elongating complex (often referred to as promoter clearance or escape). The second component that recruits H3K4me3 is the PAF complex, which regulates different steps of RNA metabolism and also interacts with Set1. The third component important for the establishment of H3K4me3 is monoubiquitylation of H2B at Lys-123 (H2BK123ub1, or H2BK120ub1 in humans, discussed further in Section 6). What remains unclear is what transcriptional elongation processes H3K4me3 controls (Hampsey and Reinberg 2003); however, factors that bind specifically to methylated H3K4me3 are beginning to reveal its role.

Mechanistically, H3K4 methylation can lead to the recruitment of specific factors such as the CHD1 protein, shown to bind to H3K4me2 and me3 (Fig. 4), and the NURF complex, known to mobilize nucleosomes at active genes in *Drosophila*. The domains that mediate association with methylated H3K4 are a tandem set of chromodomains in Chd1 (Sims et al. 2006) and a PHD finger within NURF (Li et al. 2006). Other proteins recruited by H3K4 methylation include the ISWI ATPase, which binds indirectly via other protein(s). Conversely, there is evidence that the mammalian NuRD repressor complex no longer binds to methylated H3K4 tails (D.Y. Lee et al. 2005; Martin and Zhang 2005).

Methylation at H3K4 seems to communicate with other modifications. For instance, methylation of H3K9 by the SUV39H HKMT is prevented in vitro if H3K4 is methylated and H3S10 is phosphorylated. This may well be a way to occlude the repressive H3K9 modification on actively transcribed genes. In a more elaborate "trans-tail" form of communication, the monoubiquitylation of H2BK123 affects levels of H3K4me3. How this comes about is unclear, but one suggestion is that the Set1 complex cannot tri-methylate H3K4 unless the nucleosome(s) is in a certain conformational state defined by ubiquitylation of H2B (Zhang and Reinberg 2001; D.Y. Lee et al. 2005) .

The Set1/MLL/ALL1/HRX protein, which is the human homolog of Set1, can be recruited to *HOX* gene promoters. A distinct H3K4 HKMT, SMYD3, has been linked to transcriptional activation. Methylation by SMYD3 has also been linked to the induction of cell proliferation. Indeed, limited analysis of the human H3K4 methylating enzymes suggests that they are implicated in the genesis of cancer (D.Y. Lee et al. 2005).

H3K36 METHYLATION

Evidence has led to a proposal that H3K36 methylation is necessary for efficient elongation of RNA pol II through

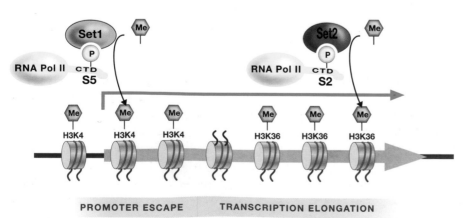

Figure 5. Role of Histone Lysine Methylation in Transcriptional Elongation

RNA polymerase II recruits distinct types of HKMTs, depending on the phosphorylation state of its carboxy-terminal domain (CTD). RNA pol II is activated for transcriptional initiation in the vicinity of the promoter, when Ser-5 is phosphorylated. This recruits the Set1 HKMT to methylate H3K4. Phosphorylation of Ser-2 occurs during transcriptional elongation, prompting H3K36 methylation as a result of Set2 HKMT recruitment to the chromatin template.

the coding region. This modification is highly enriched on the coding region of active genes, in contrast to the 5′ location of H3K4 methylation. The Set2 protein is the HKMT capable of methylating H3K36. The Set2 enzyme binds preferentially to RNA pol II that has been phosphorylated within its CTD at Ser-2 (Fig. 5). This form of RNA pol II, which, incidentally, is different from the phosphorylated state associated with promoter clearance, tends to accumulate within the transcribed regions as well as at the 3′ ends of the genes. This is consistent with the finding that H3K36me3 peaks at the 3′ ends of genes that are actively transcribed. The recruitment of Set2 to active genes also requires components of the PAF complex, as in the case for the recruitment of Set1. However, H2B monoubiquitylation has a negative repressive role on H3K36 methylation (Zhang and Reinberg 2001; Martin and Zhang 2005). Indeed, the SAGA complex, recruited to transcribed genes in yeast, contains Ubp8, a deubiquitinase that is specific for H2BK123. Further studies have suggested that ubiquitylation and deubiquitylation of H2BK123 is an active process during transcription elongation.

Processivity of RNA pol II through coding regions requires acetylation of nucleosomes. Transcriptional regulation also needs to suppress inappropriate internal initiation of transcription from cryptic start sites that occur inside coding regions. To suppress this process, methylation at H3K36 by Set2 creates a recognition site for the EAF3 protein through its chromodomain, which in turn mediates the recruitment of the Rpd3S HDAC complex. The deacetylase activity of Rpd3S then removes histone acetylation associated with elongation, thus suppressing internal initiation (Carrozza et al. 2005; Joshi and Struhl 2005; Keogh et al. 2005). Methylation of H3K36 has also been found at much lower levels in the promoter of inducible genes, but in this case, its effect appears to be repressive (Zhang and Reinberg 2001).

H3K79 METHYLATION

Methylation at H3K79 is unusual because the modification lies within the core of the nucleosome rather than in the tail, where most other characterized methylation sites are found. Global analysis has shown that H3K79 is methylated in euchromatic regions of yeast and associates primarily with the coding region of active genes. Limited analysis in higher eukaryotes shows the same profile.

The mammalian enzyme that methylates H3K79, hDOT1L, has been shown to mediate the leukemogenic functions of the MLL-AF10 fusion protein. There is, however, no protein to date that binds H3K79me and links it to

transcriptional events. The only mechanistic evidence of how H3K79 methylation functions in transcriptional activation comes from work in budding yeast. This shows that this modification somehow limits repressive proteins such as Sir2 and Sir3 at euchromatin, thus contributing to the regulation and maintenance of silent heterochromatin by enhancing their concentration at repressive chromatin regions. A distinct function ascribed to the H3K79 HKMT Dot1 in yeast is the mediation of DNA repair checkpoint. Consistent with this latter finding, a protein has been identified in human cells—P53BP1—that can bind to methylated H3K79 and has a role in DNA repair checkpoint function (Martin and Zhang 2005).

H3K9 METHYLATION

This has been the most studied of histone modifications to date, primarily because the enzyme that methylates H3K9—SUV39H1—was the first HKMT to be identified (Rea et al. 2000). The *Drosophila* homolog, Su(var)3-9, was initially identified as a suppressor of variegation, indicating that it was involved in the silencing mechanism of position-effect variegation (PEV), which involves the spreading of heterochromatin into adjacent euchromatic genes (for more detail, see Chapter 5). The realization that SUV39H1 had sequence similarity to a plant methyltransferase which had Rubisco as its substrate led to the identification of the Suv39 SET domain as the catalytic domain capable of methylating H3K9.

Progress has been made in defining the function of H3K9 methylation in pericentromeric heterochromatin formation, which is also discussed extensively in other chapters (for studies on *Drosophila*, see Chapter 5; for studies in *S. pombe*, see Chapter 6; and for studies on RNAi-mediated heterochromatin formation, see Chapter 8). The results have come largely from studies in fission yeast and mammals, where heterochromatic structures are thought to be reasonably well conserved (but note, H3K9 methylation has not been detected in budding yeast). To summarize, the first stage of our understanding emerged from studies on the factors involved in the establishment of heterochromatin. This involves the cooperation of two proteins: SUV39H (or Clr4 in fission yeast) and its binding partner HP1 (or Swi6 in fission yeast [Nakayama et al. 2001; Noma et al. 2001]). A model has been proposed whereby SUV39H methylates H3K9, creating a binding platform for HP1, through its chromodomain (Bannister et al. 2001; Lachner et al. 2001). Once HP1 binds, it can spread onto adjacent nucleosomes by its association with SUV39H, which further catalyzes neighboring histone methylation (Nakayama et al. 2001). In addition, HP1 self-associates via the chromoshadow domain facilitating the spread of heterochromatin. How HP1 spreading dictates the formation of the densely packed heterochromatic structures is, however, unknown.

The above model predicts that there should be a specific heterochromatin-based recruitment mechanism for the SUV39H HKMT enzyme, before HP1 can spread. The clue as to what this may be came from a series of experiments in fission yeast which showed a link between heterochromatin formation and the production of short interfering RNAs (siRNAs) (Hall et al. 2002; Volpe et al. 2002). These RNAs come from the bidirectional transcription of centrometric repeats which are processed into siRNAs by the enzyme dicer. The siRNAs are then packaged into the RITS complex, which contains the chromodomain-containing protein, Chp1, which binds methylated H3K9. Thus, the targeting of the RITS complex to chromatin forms the initiation stage of heterochromatin formation. The spreading and maintenance of heterochromatin over a 20-kb region, as described above, requires the methylation of histone H3K9 by the Clr4 HKMT and the binding of Swi6 to H3K9 methylated chromatin (Martin and Zhang 2005; for more detail, see Chapter 8).

The interdependence of different repressive epigenetic mechanisms has emerged from studies first in *Neurospora crassa*, but also in plants, notably demonstrating a link between H3K9 methylation and the process of DNA methylation (see Chapters 6 and 8). H3K9 methylation is necessary for DNA methylation to take place, and the reciprocal connection seems to be operational, whereby H3K9 methylation is dependent on DNA methylation. Moreover, recent studies in mammalian cancer cells lacking DNA methyltransferase enzymes (Dnmts) show reduced levels of H3K9 methylation, and this can be attributed to the fact that the methyl-CpG-binding protein 1 (MBD1) associates with the H3K9 HKMT, SETDB1 (Zhang and Reinberg 2001; Martin and Zhang 2005; for more detail, see Chapter 18).

Methylation at H3K9 also functions in the repression of euchromatic genes. ChIPs have detected this methylation at the promoter of mammalian genes when the genes are silent. The mechanism of this repression at euchromatic sites appears to be slightly different from those encountered at heterochromatic regions. The RB repressor protein delivers the SUV39H1 HKMT and HP1 to euchromatic genes such as the E2F-regulated cyclin E gene. Unlike heterochromatin, however, HP1 occupancy appears to be restricted to one or a few nucleosomes

around the initiation site, even though H3K9 methylation occurs elsewhere on the promoter. In another example, the KAP1-repressor brings the ESET/SETDB1 HKMT to the promoter of KAP1 regulated genes and silences transcription by methylation of H3K9 and HP1 recruitment. The special restriction of HP1 on these euchromatic promoters, and the prevention of spreading, suggest a distinct mechanism of action for HP1 relative to its heterochromatic role. One possible mode of action for HP1, which has some support, is that it acts as an anchor into heterochromatin-rich nuclear compartments. Movements have been observed during the repression of euchromatic genes which show that a silenced gene is displaced into a heterochromatic region, a movement which is dependent on the gamma isoform of HP1 (Martin and Zhang 2005).

Heterochromatin formation at telomeres, although involving HP1 and H3K9 methylation, varies from the aforementioned pericentromeric and silent euchromatic regions. In *Drosophila*, HP1 is not recruited to telomere ends through its chromo- or chromoshadow domain, and H3K9 methylation is catalyzed by an unknown HKMT. In mammals, distinct HP1 homologs, CBX1, CBX3, and CBX5, are involved in binding to methylated H3K9, transduced by the SUV39H1 and SUV39H2 proteins to form repressive chromatin domains at chromosome ends (for more detail, see Chapter 14).

H3K27 METHYLATION

H3K27 methylation is a repressive modification found in three distinct places in the cell: (1) euchromatic gene loci, predominantly where there are Polycomb Response Elements (PREs) in the case of *Drosophila*, (2) at pericentromeric heterochromatin, and (3) at the inactive X in mammals.

The enzyme that mediates H3K27 methylation is EZH2 in human cells, a homolog of the *Drosophila* ENHANCER OF ZESTE [E(Z)] protein. The EZH2 enzyme is found in a number of distinctive Polycomb repressive complexes (PRCs) which associate with specific repressive Polycomb DNA elements in promoters in *Drosophila* (see also Chapter 11). What targets the EZH2-containing complexes to specific genes in mammals is unknown, as Polycomb repressive elements have not been identified. However, targeting of these EZH2 complexes may be mediated by a variety of transcription factors, including GAGA and MYC. The mechanisms of repression by EZH2 involve methylation of H3K27 and the recruitment of the Polycomb (Pc) protein to this modified site (as in Model 3 in Fig. 1). An important aspect of the pathway that leads to H3K27 methylation is that it is implicated in cancer. The EZH2 H3K27 HKMT is found overexpressed in a number of cancer tissues, including breast and prostate (Martin and Zhang 2005).

H4K20 METHYLATION

Very little is known mechanistically about the role of this modification in transcriptional control. What is clear is that H4K20me2 and H4K20me3 are present at pericentromeric heterochromatin and that the HKMT enzymes that mediate these modifications are SUV4-20H1 and SUV4-20H2. Methylation of H3K9 seems to be required for methylation of H4K20.

Another enzyme that can mono-methylate H4K20 in higher eukaryotes is PR-Set7, which has been implicated in mitotic events. Last, there is functional evidence that H4K20 methylation has been linked to DNA repair via the binding of the DNA damage checkpoint protein CrB2 in budding yeast (Martin and Zhang 2005).

4.2 Demethylation of Lysines

Until recently, it was unclear whether histone lysine demethylation was taking place in the cell. The search for such enzymes had been fruitless, and evidence existed that methyl groups can be quite stable on heterochromatin regions. The discovery of LSD1 changed all that (Shi et al. 2004). This protein was shown to be an enzyme that removes methyl groups specifically from H3K4 and is involved in repression. LSD1 is present in a number of different repressor complexes, and some of these allow it to more efficiently demethylate lysine residues from nucleosomal histone H3 (M.G. Lee et al. 2005; Shi et al. 2005). The specificity of LSD1 can be changed if it binds a partner such as the androgen receptor (AR). An LSD1-AR complex demethylates H3K9 instead of H3K4, and under these conditions, activates, rather than represses, transcription (Metzger et al. 2005).

Recently, five new demethylases were identified that possess a common catalytic structure distinct from LSD1, called the JmjC-domain (Fig. 6). This domain was previously predicted to possess enzymatic activity (Trewick et al. 2005). These new demethylases are found to demethylate distinct methyl states of H3K9 and H3K36. JHDM1 demethylates H3K36me1 and me2, and JHDM2A demethylates H3K9me1 and me2 (Tsukada et al. 2006; Yamane et al. 2006). The tri-methyl state of these two modified residues is removed by a distinct set of enzymes. JHDM3A and JMJD2A can act on both H3K36me3 and H3K9me3 (Cloos et al. 2006; Fodor et al. 2006; Klose et al. 2006; Tsukada et al. 2006; Whetstine et al. 2006). It is perhaps surprising that enzymes exist which can simultaneously demethylate an active (e.g., H3K36me) and a repressive (e.g., H3K9me) mark. This may be explained by the recent finding that H3K9 methylation also associates with actively transcribed genes (Vakoc et al. 2005). In a more classic mechanism, the JHDM2A enzyme is recruited to promoters by AR, where it is involved in activating transcription via demethylation of H3K9 (Yamane et al. 2006). Structural analysis of JMJD2A has revealed that four distinct domains (JmjN, JmjC, an unusual Zing finger, and a carboxy-terminal domain) come together to form the catalytic core. A deep cleft is formed by these domains coming together, which demands a conformational change in the enzyme or substrate to accommodate the methyl group for demethylation. Such a conformational shift may explain the specificity of demethylation (Chen et al. 2006).

It is interesting to note that one of the newly discovered demethylases, JMJD2C, was previously known as GASC1, a gene amplified in squamous carcinoma. Consistent with a causative role for this enzyme in cancer development, the overexpression of GASC1 was shown to induce cell proliferation (Cloos et al. 2006). These results together imply that demethylases as well as HKMTs may be targets for anticancer drug development (see also Chapter 24).

4.3 Methylation of Arginines

The importance of histone arginine methylation in transcriptional control came after the identification of CARM1, an enzyme that can methylate arginines within H3 in vitro (Chen et al. 1999). In vivo, arginine methylation was subsequently demonstrated in experiments using specific antibodies to arginine-methylated sites (Strahl et al. 2001; Wang et al. 2001; Bauer et al. 2002).

Arginine methylation has been implicated in the positive and negative regulation of transcription. Two methyltransferases, PRMT1 (protein arginine methyltransferase) and PRMT4/CARM1, have been linked to transcriptional activation. PRMT1 has the ability to methylate H4R3 (Strahl et al. 2001; Wang et al. 2001), whereas PRMT4/CARM1 can catalyze the methylation of H3R2, H3R17, and H3R26 (Schurter et al. 2001; Bauer et al. 2002). Specific transcription factors (NR, p53, YY1, NFKB) recruit these enzymes to specific promoters where they activate transcription. In contrast, PRMT5 (which can methylate H3R8 and H4R3) acts as a repressor of numerous genes, including some regulated by MYC (Fabbrizio et al. 2002; Pal et al. 2003).

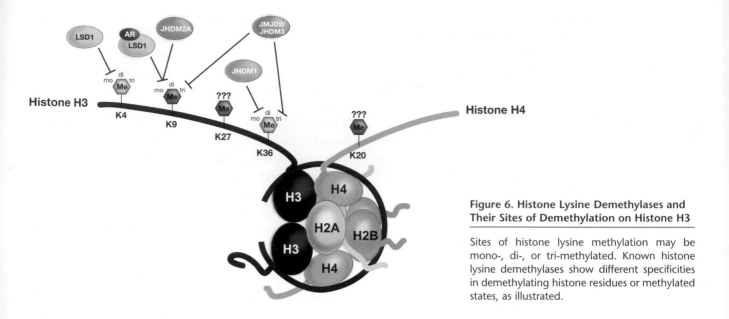

Figure 6. Histone Lysine Demethylases and Their Sites of Demethylation on Histone H3

Sites of histone lysine methylation may be mono-, di-, or tri-methylated. Known histone lysine demethylases show different specificities in demethylating histone residues or methylated states, as illustrated.

Most of our knowledge regarding arginine methylation comes from the analysis of the estrogen-signaling pathway that regulates the pS2 gene. ChIPs have indicated that a complex and cyclic set of events follows the stimulation of this gene by estrogen. The estrogen receptor is first recruited to the pS2 promoter within minutes of the stimulus and brings with it many protein complexes and enzymes that modify histones (Metivier et al. 2003). Relevant here is the recruitment of CARM1 and PRMT1, which can methylate arginine residues of histones H3 and H4 (Ma et al. 2001; Bauer et al. 2002). This methylation is detected very soon after the arrival of the enzymes and coincides with the appearance of active RNA pol II on the promoter. Surprisingly, however, minutes after these events, methylation at arginines is no longer detected by specific antibodies, and RNA pol II disappears. Soon after that, methylation of arginines and RNA pol II reappears (Metivier et al. 2003; Cuthbert et al. 2004; Wang et al. 2004). The reason for these cyclic events is not known. One possibility is that it provides a mechanism for rapid shutoff of transcription if estrogen signaling fails.

Experiments done on reconstituted chromatin templates have helped establish a direct role for arginine methylation in gene expression. Analysis of p53-mediated activation of transcription in vitro has shown that there is a synergistic effect of methylation transduced by PRMT1 and CARM1, and acetylation by CBP/p300 (An et al. 2004). Furthermore, these assays have confirmed the in vivo observations on the pS2 gene that a specific order of events takes place during activation in which the sequen-

tial activity of PRMT1, CBP/p300, and CARM1 is necessary (Metivier et al. 2003).

Given that arginine methylation is such a dynamic process, several ways have been described in which the effectiveness of arginine methyltransferase is controlled. First, the interaction of the enzyme with another protein can control its substrate specificity. Second, there is potential for competition between enzymes for a given arginine substrate. Both PRMT1 and PRMT5 can methylate H4R3, but the first enzyme is an activator and the second is a repressor of transcription. A third level of regulation of the methyl state may come from arginine demethylation. Such an activity has not yet been isolated, but there are clear indications of the rapid disappearance of methyl groups from arginines, making such an activity a very attractive possibility (Zhang and Reinberg 2001; D.Y. Lee et al. 2005; Wysocka et al. 2006a).

5 Deimination

The lack of an arginine demethylase prompted the suggestion that other types of enzymatic reactions may antagonize arginine methylation (Bannister et al. 2002). One such reaction is deimination, a process by which an arginine can be converted to citrulline via the removal of an imine group. If the arginine is mono-methylated, removal of methylamine would effectively result in the removal of the methyl group from the arginine. The presence of citrulline in histones has now been demonstrated, and the enzyme, PADI4, has been identified that can convert arginines within histones into citrulline (Cuthbert et al. 2004; Wang et al. 2004). Moreover, the appearance of citrulline on histones H3 and H4 correlates with the disappearance of arginine methylation in vivo. Additionally, analyses of estrogen-regulated promoters, where arginine methylation coincides with the active state of transcription, have shown that citrulline appears when the promoter is shut off. Many unanswered questions remain regarding this modification. Is the citrulline acting to suppress active methylation at arginines, or does it repress transcription by actively recruiting proteins? What about the reversal of citrulline deposition? This clearly takes place on the promoters at a very rapid pace, but is this an enzymatically driven reaction or is it merely due to the replacement of nucleosomal histones by histone variants, which contain arginine in place of citrulline?

6 Ubiquitylation/Deubiquitylation and Sumoylation

Ubiquitin and SUMO are quite distinct PTMs compared to acetylation, phosphorylation, and methylation. Whereas the latter PTMs are small chemical groups, Ub and SUMO are large polypeptides, which increase the size of the histone by approximately two-thirds. Ub and SUMO are 18% identical in sequence and share a three-dimensional structure, but are dissimilar in surface charge.

Histones were the first proteins shown to be monoubiquitylated, although precise positions of Ub were not identified until relatively recently (Robzyk et al. 2000; Wang et al. 2004). Like methylation, and unlike acetylation and phosphorylation (and, possibly, sumoylation), ubiquitylation can be either repressive or activating, depending on the specific sites. H2A and H2B are

monoubiquitylated, which contrasts with proteolysis-associated polyubiquitylation. The effects of monoubiquitylation on each core histone are opposite (Fig. 7). H2B monoubiquitylation is activating to transcription, transduced by Rad6/Bre1 (and the human counterpart RNF20/RNF40 + UbcH6) (Wood et al. 2003; Kim et al. 2005; Zhu et al. 2005), and leads to H3K4 methylation, as described in the previous section and in the next section (Henry et al. 2003; Kao et al. 2004). This sequence of events, although as yet not understood mechanistically, is conserved from yeast to human (Kim et al. 2005; Zhu et al. 2005). H2AK119ub1, on the other hand, is repressive to transcription in mammals and catalyzed by the Polycomb group Bmi1/Ring1A protein (Wang et al. 2004). There is no evidence for evolutionary conservation of repressive H2Aub in yeast.

To date, no histone-specific ubiquitin-binding proteins have been identified. However, because numerous ubiquitin interaction domains have been documented as binding to non-histone ubiquitylated substrates, it seems highly likely that effectors for ubiquitylated histones will be found. However, they may interact in a different manner than the chromatin interacting domains for acetylation and methylation; i.e., there are likely to be two simultaneous binding interactions, one to a surface on ubiquitin and a second interaction within histone sequences, to provide specificity of interaction.

Deubiquitylation of the H2BK123 site is involved in both gene activation and maintenance of heterochromatic silencing, through the action of two distinct proteases, Ubp8 and Ubp10. Ubp8 is a subunit of the SAGA histone acetylation complex (Sanders et al. 2002) and acts following ubiquitylation by Rad6 (Henry et al. 2003;

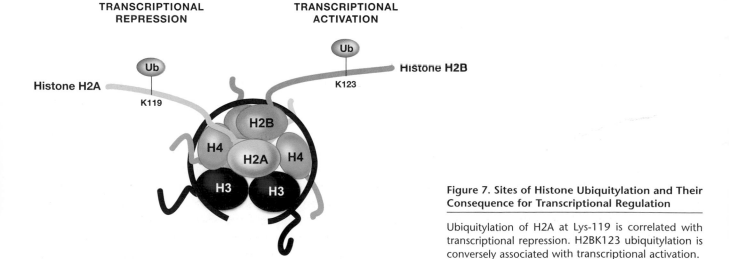

TRANSCRIPTIONAL REPRESSION

TRANSCRIPTIONAL ACTIVATION

Ub

Histone H2A K119

K123

Ub

Histone H2B

H2B

H4 H2A H4

H3 H3

Figure 7. Sites of Histone Ubiquitylation and Their Consequence for Transcriptional Regulation

Ubiquitylation of H2A at Lys-119 is correlated with transcriptional repression. H2BK123 ubiquitylation is conversely associated with transcriptional activation.

Daniel et al. 2004). This sequence of H2B ubiquitylation followed by deubiquitylation is required to establish the appropriate levels of H3K4 (H2Bub required) and H3K36 methylation (H2Bub not required) (Henry et al. 2003). Ubp10 functions at silenced regions to maintain low levels of H3K4me and H3/H4 lysine acetylation, and thus assists in preventing transcription (Emre et al. 2005; Gardner et al. 2005).

Sumoylation is the only HPTM described in yeast as repressive and is conserved in mammals (Shiio and Eisenman 2003). Its role may be generally negative-acting to prevent activating HPTMs. The inhibition of active HPTMs may occur through two mechanisms. First, SUMO-histone may directly block lysine substrate sites that are alternatively acetylated or sumoylated (as in Model 2 in Fig. 1). Second, sumoylated histones may recruit HDACs both to chromatin (Model 3) and via a SUMO group that occurs on DNA-bound repressors.

7 Themes in Modifications

The preceding discussion of the numerous types and sites of histone PTM occurring in transcription might lead to the conclusion that there are few overarching guiding principles or ideas. However, there do appear to be a number of broad themes that occur repeatedly, although the specifics may change depending on the histone, the sites of HPTM, and the binding proteins. Indeed, chromatin regulation may vary between promoters and distinct pathways.

7.1 Histone Code

One key question emerges after this lengthy discussion of the intricacies of HPTMs: Why are there so many modifications? Clearly, many of them correlate with transcription, and others occur during different DNA-templated processes. Thus, one hypothesis is that there is a histone "code," linking specific modifications with individual processes (Strahl and Allis 2000; Turner 2000). The simplest code would be a binary relationship between HPTMs and either gene activation or repression, and distinct HPTMs for other processes. The evidence supporting such a code is the observed tendency, as described above, for certain HPTMs to be positive-acting and others negative-acting. However, there are observations that are inconsistent with a simple binary code. For example, phosphorylation of H3S10 is both activating to transcription, which presumably involves opening the chromatin, and involved in chromosome condensation, making chromatin even more inaccessible (Nowak and Corces

2004). Similarly, H3K9me has recently been shown to increase during gene induction (Vakoc et al. 2005), in addition to its well-characterized role in heterochromatic silencing. Finally, many of the same HPTMs occur in both transcription and DNA repair, which are mechanistically distinct processes.

Based on some of these considerations, a more general hypothesis has been proposed where HPTMs serve as a nuclear DNA-associated signal transduction pathway, similar to cytoplasmic signal transduction that is generated and propagated largely through Ser/Thr phosphorylation (Schreiber and Bernstein 2002). In this model, there is not a strict histone code for specific processes, but rather HPTM recognition and binding via a plethora of protein-binding motifs. This model explains how any site could be *both* activating and repressing and involved in *more than one* process, because different binding effector proteins are cognates for the same HPTM for distinct processes.

7.2 Modification Patterns

Some experimental evidence points to the structural alteration of chromatin with certain HPTMs (Model 1). This can result from altering the charge of single or cluster of histone residues. This is particularly true when residues are acetylated or phosphorylated, which reduces the positive charge of histone regions (see Section 5 of Chapter 3). Such *cis* alterations can alter internucleosomal spacing and reduce the affinity of histones to negatively charged DNA, as exemplified by the negative charged patches that occur on linker histone (Dou and Gorovsky 2002). These types of HPTMs may be cumulative in their effect on, for example, transcriptional activation or for creating higher-order chromatin structures, rather than producing a binary ON/OFF effect (Kurdistani et al. 2004; Henikoff 2005).

Another model for the "output" of the myriad of HPTMs is that the code is complex and is read in patterns and often in temporal sequences. In this view, the intricacy of the patterns in three-dimensional space and over time during a process requiring many chromatin-associated steps, such as transcription, yields a meaningful mechanistic result. Two types of HPTM patterns have been identified. First, there are patterns on the same histone tail, or in "*cis*," and second, patterns on different histone tails, or in "*trans*." The most well-characterized *cis* pattern is between H3S10ph and H3K14ac on the H3 amino-terminal tail (Cheung et al. 2000; Lo et al. 2000), where H3S10ph leads to H3K14ac. The mechanism underlying the establishment of this pattern is understood in structural detail: The enzyme that acetylates binds to the previously phosphory-lated H3 tail with increased affinity due to a greater number of amino acid side-chain contact points (Clements et al. 2003). Other *cis* patterns are H3K23 acetylation and H3R17 methylation (Daujat et al. 2002) and H4R3 methylation and H4K8 acetylation (Wang et al. 2001).

As described above, one *trans* tail pattern has been identified, where initial ubiquitylation of H2BK123 leads to methylation of H3K4 (Briggs et al. 2002; Dover et al. 2002; Sun and Allis 2002). The mechanism linking these HPTMs has not been elucidated, although several possible hypotheses exist. Because the link is from one large modification (ubiquitin) to a nonadjacent HTPM, one model is that ubiquitin wedges the chromatin open, like a crowbar, to allow the methylating enzyme access to its site. A second general model is that H2BK123ub1 functions to recruit effector proteins, similar to the role of the other HPTMs. The noncatalytic portion of the proteosome requires H2B ubiquitylation for chromatin association (Ezhkova and Tansey 2004), and the function of the elongation complex FACT is stimulated by H2B ubiquitylation (Pavri et al. 2006), although neither complex has yet been shown to directly bind to ubiquitylated H2B.

7.3 Changes in Chromatin Structure Associated with Transcription Activation and Elongation

The transcriptionally active euchromatic regions contain nucleosomes, but in an "unfolded" state, denoted as "beads-on-a-string" or 11-nm fiber. The nucleosomes in this state still impose an intrinsic inhibition to the transcription machinery. Some transcription factors, be they activators or repressors, can gain access to their sites when contained in nucleosomes, but others cannot. Moreover, the machinery recruited by the DNA-bound regulators and responsible for delivering RNA pol II to promoters is constrained by the presence of nucleosomes. A number of distinct mechanisms serve to reconfigure the chromatin, poising genes for subsequent transcription, or promoting initiation or elongation. Some of these mechanisms are illustrated in Figure 8.

The nucleosome problem during transcription is solved in part by the recruitment of protein complexes to mobilize and/or alter the structure of the nucleosome. These complexes fall into two different families, one represented by SNF2H (or ISW1 and ISW2 in yeast), and one by the Brahma-Swi/Snf family (Narlikar et al. 2002; Peterson 2002; Flaus and Owen-Hughes 2004). The first family mobilizes nucleosomes, whereas Swi/Snf also transitorily alters the structure of the nucleosomes. Acetylated nucle-

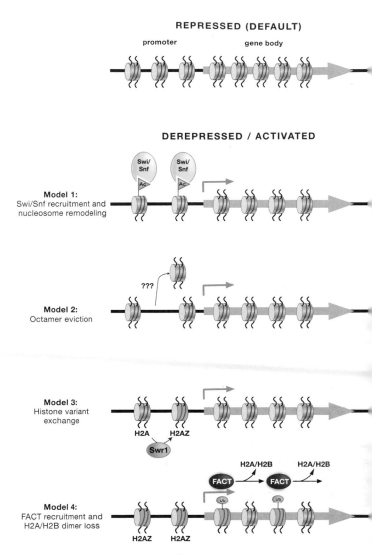

Figure 8. Models for the Involvement of Chromatin Remodeling and Histone Exchange in Transcriptional Processes

In Model 1, the Swi/Snf family of ATPase binds chromatin through bromodomain recognition of acetylated histones and acts to alter the local chromatin structure. Model 2 depicts the reported octamer eviction that occurs at certain loci such as *PHO5* by an unknown mechanism. In Model 3, the ATPase SWR1 catalyzes the replacement of histone H2A with H2AZ, which poises chromatin for transcription. Model 4 focuses on the involvement of FACT in transcriptional elongation, assisting in nucleosome unraveling by the displacement of an H2A/H2B dimer. Concomitantly, histone H3 may be exchanged with H3.3 during the process.

osomes are recognized by the Swi/Snf complex through bromodomain interaction (Model 1 in Fig. 8).

A second mechanism involved in gene activation is selective octamer loss at promoters. For example, histone octamers are evicted at the promoter of the *PHO5* gene in *S. cerevisiae* during transcriptional induction (Model 2 in Fig. 8) (Boeger et al. 2003; Reinke and Horz 2003). In addition, promoters of *S. cerevisiae* have a constitutively low density of nucleosomes, which allows access for transcription factors (Sekinger et al. 2005). It is not yet known whether or how ATP-dependent remodeling complexes assist in generating and maintaining this low occupancy.

A third major mechanism involved in setting up transcriptional states is the presence of histone variants. There are two types of histone variants associated with gene activity. First, a variant of H2A called H2AZ is found in nucleosomes around the promoter gap, and poises the gene for activation (Santisteban et al. 2000; Raisner et al. 2005; Zhang et al. 2005); a specific ATP-dependent remodeling complex, called Swr1, replaces H2A with H2AZ (Model 3 in Fig. 8) (Mizuguchi et al. 2004; for more detail, see Chapter 13). Second, one H3 isoform, called H3.1, is incorporated into chromatin during replication, whereas isoform H3.3 is incorporated in a replication-independent manner (Ahmad and Henikoff 2002) with the aid of the HIRA (histone regulator A) chaperone. This variant is predominantly found within gene ORFs (Mito et al. 2005), suggesting that its deposition is a transcription-coupled process.

There are additional mechanisms to overcome the nucleosomal barrier to elongating RNA pol II (and RNA pol I). A large number of factors have been isolated that affect transcription elongation (Sims et al. 2004). One of these factors was found to allow the RNA pol II to traverse nucleosomes. This factor is known as FACT (for FAcilitate Chromatin Transcription). Importantly, FACT functions exclusively through nucleosomes, binds to them, and then promotes the displacement of one H2A/H2B dimer (Model 4 in Fig. 8) (Belotserkovskaya et al. 2003). As transcription ceases, FACT also promotes the reconstitution of the nucleosome. Interestingly, FACT performs its functions in the absence of energy, but physically interacts with CHD1, a protein that hydrolyzes ATP to mobilize nucleosomes and bind to the active H3K4me mark. Moreover, FACT also interacts with NuA4, a complex that contains HAT activity. Although FACT can promote displacement of the H2A/H2B dimer in vitro in an ATP-independent manner, it is possible that this is promoted by its interaction with factors such as CHD1, which mobilize or alter the structure of nucleosomes in vivo and also the interplay with HPTMs (Reinberg and Sims 2006).

8 Concluding Remarks

We have come a long way in this "modern era" of histone modifications which covers the last 10 years. In this time, there have been six distinct types of histone modification pathways characterized and numerous sites of modifications identified. Yet this is clearly still the beginning of our understanding. Mechanistically we know that modifications affect the binding of proteins, but we are still unaware precisely how these proteins result in reorganization of chromatin structure. We still do not know whether there is a code or whether modifications are simply part of a signaling pathway. In addition, our knowledge is lacking on the many cellular processes, other than transcription, that modifications are involved in. Thus, in short, we have become aware of the complexity of the system, but we are a long way from making sense of the complexity. One thing is for sure, it is worth the effort to find out, since histone modifications play a fundamental role in both normal and diseased processes.

References

Ahmad K. and Henikoff S. 2002. The histone variant H3.3 marks active chromatin by replication-independent nucleosome assembly. *Mol. Cell* 9: 1191–1200.

An W., Kim J., and Roeder R.G. 2004. Ordered cooperative functions of PRMT1, p300, and CARM1 in transcriptional activation by p53. *Cell* 117: 735–748.

Baek S.H. and Rosenfeld M.G. 2004. Nuclear receptor coregulators: Their modification codes and regulatory mechanism by translocation. *Biochem. Biophys. Res. Commun.* 319: 707–714.

Bannister A.J., Schneider R., and Kouzarides T. 2002. Histone modification: Dynamic or static? *Cell* 109: 801–806.

Bannister A.J., Zegerman P., Partridge J.F., Miska E.A., Thomas J.O., Allshire R.C., and Kouzarides T. 2001. Selective recognition of methylated lysine 9 on histone H3 by the HP1 chromo domain. *Nature* 410: 120–124.

Bauer U.M., Daujat S., Nielsen S.J., Nightingale K., and Kouzarides T. 2002. Methylation at arginine 17 of histone H3 is linked to gene activation. *EMBO Rep.* 3: 39–44.

Belotserkovskaya R., Oh S., Bondarenko V.A., Orphanides G., Studitsky V.M., and Reinberg D. 2003. FACT facilitates transcription-dependent nucleosome alteration. *Science* 301: 1090–1093.

Boeger H., Griesenbeck J., Strattan J.S., and Kornberg R.D. 2003. Nucleosomes unfold completely at a transcriptionally active promoter. *Mol. Cell* 11: 1587–1598.

Braunstein M., Rose A.B., Holmes S.G., Allis C.D., and Broach J.R. 1993. Transcriptional silencing in yeast is associated with reduced nucleosome acetylation. *Genes Dev.* 7: 592–604.

Briggs S.D., Xiao T., Sun Z.W., Caldwell J.A., Shabanowitz J., Hunt D.F., Allis C.D., and Strahl B.D. 2002. Gene silencing: *Trans*-histone regulatory pathway in chromatin. *Nature* 418: 498.

Brownell J.E., Zhou J., Ranalli T., Kobayashi R., Edmondson D.G., Roth S.Y., and Allis C.D. 1996. *Tetrahymena* histone acetyltransferase A: A homolog to yeast Gcn5p linking histone acetylation to gene activation. *Cell* 84: 843–851.

Carrozza M.J., Li B., Florens L., Suganuma T., Swanson S.K., Lee K.K., Shia W.J., Anderson S., Yates J., Washburn M.P., and Workman J.L. 2005. Histone H3 methylation by Set2 directs deacetylation of coding regions by Rpd3S to suppress spurious intragenic transcription. *Cell* 123: 581–592.

Chen D., Ma H., Hong H., Koh S.S., Huang S.M., Schurter B.T., Aswad D.W., and Stallcup M.R. 1999. Regulation of transcription by a protein methyltransferase. *Science* 284: 2174–2177.

Chen Z., Zang J., Whetstine J., Hong X., Davrazou F., Kutateladze T.G.,

Simpson M., Mao Q., Pan C.H., Dai S., et al. 2006. Structural insights into histone demethylation by JMJD2 family members. *Cell* **125:** 691–702.

Cheung P., Tanner K.G., Cheung W.L., Sassone-Corsi P., Denu J.M., and Allis C.D. 2000. Synergistic coupling of histone H3 phosphorylation and acetylation in response to epidermal growth factor stimulation. *Mol. Cell* **5:** 905–915.

Clark-Adams C.D., Norris D., Osley M.A., Fassler J.S., and Winston F. 1988. Changes in histone gene dosage alter transcription in yeast. *Genes Dev.* **2:** 150–159.

Clements A., Poux A.N., Lo W.S., Pillus L., Berger S.L., and Marmorstein R. 2003. Structural basis for histone and phosphohistone binding by the GCN5 histone acetyltransferase. *Mol. Cell* **12:** 461–473.

Cloos P.A., Christensen J., Agger K., Maiolica A., Rappsilber J., Antal T., Hansen K.H., and Helin K. 2006. The putative oncogene GASC1 demethylates tri- and dimethylated lysine 9 on histone H3. *Nature* **442:** 307–311.

Cuthbert G.L., Daujat S., Snowden A.W., Erdjument-Bromage H., Hagiwara T., Yamada M., Schneider R., Gregory P.D., Tempst P., Bannister A.J., and Kouzarides T. 2004. Histone deimination antagonizes arginine methylation. *Cell* **118:** 545–553.

Daniel J.A., Torok M.S., Sun Z.W., Schieltz D., Allis C.D., Yates J.R., III, and Grant P.A. 2004. Deubiquitination of histone H2B by a yeast acetyltransferase complex regulates transcription. *J. Biol. Chem.* **279:** 1867–1871.

Daujat S., Bauer U.M., Shah V., Turner B., Berger S., and Kouzarides T. 2002. Crosstalk between CARM1 methylation and CBP acetylation on histone H3. *Curr. Biol.* **12:** 2090–2097.

Dhalluin C., Carlson J.E., Zeng L., He C., Aggarwal A.K., and Zhou M.M. 1999. Structure and ligand of a histone acetyltransferase bromodomain. *Nature* **399:** 491–496.

Dou Y. and Gorovsky M.A. 2002. Regulation of transcription by H1 phosphorylation in *Tetrahymena* is position independent and requires clustered sites. *Proc. Natl. Acad. Sci.* **99:** 6142–6146.

Dover J., Schneider J., Tawiah-Boateng M.A., Wood A., Dean K., Johnston M., and Shilatifard A. 2002. Methylation of histone H3 by COMPASS requires ubiquitination of histone H2B by Rad6. *J. Biol. Chem.* **277:** 28368–28371.

Durrin L.K., Mann R.K., Kayne P.S., and Grunstein M. 1991. Yeast histone H4 N-terminal sequence is required for promoter activation in vivo. *Cell* **65:** 1023–1031.

Emre N.C., Ingvarsdottir K., Wyce A., Wood A., Krogan N.J., Henry K.W., Li K., Marmorstein R., Greenblatt J.F., Shilatifard A., and Berger S.L. 2005. Maintenance of low histone ubiquitylation by Ubp10 correlates with telomere-proximal Sir2 association and gene silencing. *Mol. Cell* **17:** 585–594.

Ezhkova E. and Tansey W.P. 2004. Proteasomal ATPases link ubiquitylation of histone H2B to methylation of histone H3. *Mol. Cell* **13:** 435–442.

Fabbrizio E., El Messaoudi S., Polanowska J., Paul C., Cook J.R., Lee J.H., Negre V., Rousset M., Pestka S., Le Cam A., and Sardet C. 2002. Negative regulation of transcription by the type II arginine methyltransferase PRMT5. *EMBO Rep.* **3:** 641–645.

Fischle W., Tseng B.S., Dormann H.L., Ueberheide B.M., Garcia B.A., Shabanowitz J., Hunt D.F., Funabiki H., and Allis C.D. 2005. Regulation of HP1-chromatin binding by histone H3 methylation and phosphorylation. *Nature* **438:** 1116–1122.

Flaus A. and Owen-Hughes T. 2004. Mechanisms for ATP-dependent chromatin remodelling: Farewell to the tuna-can octamer? *Curr. Opin. Genet. Dev.* **14:** 165–173.

Fodor B.D., Kubicek S., Yonezawa M., O'Sullivan R.J., Sengupta R.,

Perez-Burgos L., Opravil S., Mechtler K., Schotta G., and Jenuwein T. 2006. Jmjd2b antagonizes H3K9 trimethylation at pericentric heterochromatin in mammalian cells. *Genes Dev.* **20:** 1557–1562.

Gardner R.G., Nelson Z.W., and Gottschling D.E. 2005. Ubp10/Dot4p regulates the persistence of ubiquitinated histone H2B: Distinct roles in telomeric silencing and general chromatin. *Mol. Cell. Biol.* **25:** 6123–6139.

Glozak M.A., Sengupta N., Zhang X., and Seto E. 2005. Acetylation and deacetylation of non-histone proteins. *Gene* **363:** 15–23.

Grant P.A., Sterner D.E., Duggan L.J., Workman J.L., and Berger S.L. 1998. The SAGA unfolds: Convergence of transcription regulators in chromatin-modifying complexes. *Trends Cell Biol.* **8:** 193–197.

Hall I.M., Shankaranarayana G.D., Noma K., Ayoub N., Cohen A., and Grewal S.I. 2002. Establishment and maintenance of a heterochromatin domain. *Science* **297:** 2232–2237.

Hampsey M. and Reinberg D. 2003. Tails of intrigue: Phosphorylation of RNA polymerase II mediates histone methylation. *Cell* **113:** 429–432.

Han M. and Grunstein M. 1988. Nucleosome loss activates yeast downstream promoters in vivo. *Cell* **55:** 1137–1145.

Hassan A.H., Prochasson P., Neely K.E., Galasinski S.C., Chandy M., Carrozza M.J., and Workman J.L. 2002. Function and selectivity of bromodomains in anchoring chromatin-modifying complexes to promoter nucleosomes. *Cell* **111:** 369–379.

Hebbes T.R., Clayton A.L., Thorne A.W., and Crane-Robinson C. 1994. Core histone hyperacetylation co-maps with generalized DNase I sensitivity in the chicken beta-globin chromosomal domain. *EMBO J.* **13:** 1823–1830.

Henikoff S. 2005. Histone modifications: Combinatorial complexity or cumulative simplicity? *Proc. Natl. Acad. Sci.* **102:** 5308–5309.

Henry K.W., Wyce A., Lo W.S., Duggan L.J., Emre N.C., Kao C.F., Pillus L., Shilatifard A., Osley M.A., and Berger S.L. 2003. Transcriptional activation via sequential histone H2B ubiquitylation and deubiquitylation, mediated by SAGA-associated Ubp8. *Genes Dev.* **17:** 2648–2663.

Hirota T., Lipp J.J., Toh B.H., and Peters J.M. 2005. Histone H3 serine 10 phosphorylation by Aurora B causes HP1 dissociation from heterochromatin. *Nature* **438:** 1176–1180.

Jenuwein T. and Allis C.D. 2001. Translating the histone code. *Science* **293:** 1074–1080.

Joshi A.A. and Struhl K. 2005. Eaf3 chromodomain interaction with methylated H3-K36 links histone deacetylation to Pol II elongation. *Mol. Cell* **20:** 971–978.

Kao C.F., Hillyer C., Tsukuda T., Henry K., Berger S., and Osley M.A. 2004. Rad6 plays a role in transcriptional activation through ubiquitylation of histone H2B. *Genes Dev.* **18:** 184–195.

Keogh M.C., Kurdistani S.K., Morris S.A., Ahn S.H., Podolny V., Collins S.R., Schuldiner M., Chin K., Punna T., Thompson N.J., et al. 2005. Cotranscriptional set2 methylation of histone H3 lysine 36 recruits a repressive Rpd3 complex. *Cell* **123:** 593–605.

Kim J., Hake S.B., and Roeder R.G. 2005. The human homolog of yeast BRE1 functions as a transcriptional coactivator through direct activator interactions. *Mol. Cell* **20:** 759–770.

Klose R.J., Yamane K., Bae Y., Zhang D., Erdjument-Bromage H., Tempst P., Wong J., and Zhang Y. 2006. The transcriptional repressor JHDM3A demethylates trimethyl histone H3 lysine 9 and lysine 36. *Nature* **442:** 312–316.

Kurdistani S.K. and Grunstein M. 2003. Histone acetylation and deacetylation in yeast. *Nat. Rev. Mol. Cell. Biol.* **4:** 276–284.

Kurdistani S.K., Tavazoie S., and Grunstein M. 2004. Mapping global histone acetylation patterns to gene expression. *Cell* **117:** 721–733.

Lachner M., O'Carroll D., Rea S., Mechtler K., and Jenuwein T. 2001.

Methylation of histone H3 lysine 9 creates a binding site for HP1 proteins. *Nature* **410:** 116–120.

Lee D.Y., Teyssier C., Strahl B.D., and Stallcup M.R. 2005. Role of protein methylation in regulation of transcription. *Endocr. Rev.* **26:** 147–170.

Lee M.G., Wynder C., Cooch N., and Shiekhattar R. 2005. An essential role for CoREST in nucleosomal histone 3 lysine 4 demethylation. *Nature* **437:** 432–435.

Li H., Ilin S., Wang W., Duncan E.M., Wysocka J., Allis C.D., and Patel D.J. 2006. Molecular basis for site-specific read-out of histone H3K4me3 by the BPTF PHD finger of NURF. *Nature* **442:** 91–95.

Litt M.D., Simpson M., Gaszner M., Allis C.D., and Felsenfeld G. 2001. Correlation between histone lysine methylation and developmental changes at the chicken beta-globin locus. *Science* **293:** 2453–2455.

Lo W.S., Duggan L., Emre N.C., Belotserkovskaya R., Lane W.S., Shiekhattar R., and Berger S.L. 2001. Snf1—A histone kinase that works in concert with the histone acetyltransferase Gcn5 to regulate transcription. *Science* **293:** 1142–1146.

Lo W.S., Trievel R.C., Rojas J.R., Duggan L., Hsu J.Y., Allis C.D., Marmorstein R., and Berger S.L. 2000. Phosphorylation of serine 10 in histone H3 is functionally linked in vitro and in vivo to Gcn5-mediated acetylation at lysine 14. *Mol. Cell* **5:** 917–926.

Ma H., Baumann C.T., Li H., Strahl B.D., Rice R., Jelinek M.A., Aswad D.W., Allis C.D., Hager G.L., and Stallcup M.R. 2001. Hormone-dependent, CARM1-directed, arginine-specific methylation of histone H3 on a steroid-regulated promoter. *Curr. Biol.* **11:** 1981–1985.

Macdonald N., Welburn J.P., Noble M.E., Nguyen A., Yaffe M.B., Clynes D., Moggs J.G., Orphanides G., Thomson S., Edmunds J.W., et al. 2005. Molecular basis for the recognition of phosphorylated and phosphoacetylated histone H3 by 14-3-3. *Mol. Cell* **20:** 199–211.

Mahadevan L.C., Willis A.C., and Barratt M.J. 1991. Rapid histone H3 phosphorylation in response to growth factors, phorbol esters, okadaic acid, and protein synthesis inhibitors. *Cell* **65:** 775–783.

Martin C. and Zhang Y. 2005. The diverse functions of histone lysine methylation. *Nat. Rev. Mol. Cell. Biol.* **6:** 838–849.

Metivier R., Penot G., Hubner M.R., Reid G., Brand H., Kos M., and Gannon F. 2003. Estrogen receptor-α directs ordered, cyclical, and combinatorial recruitment of cofactors on a natural target promoter. *Cell* **115:** 751–763.

Metzger E., Wissmann M., Yin N., Muller J.M., Schneider R., Peters A.H., Gunther T., Buettner R., and Schule R. 2005. LSD1 demethylates repressive histone marks to promote androgen-receptor-dependent transcription. *Nature* **437:** 436–439.

Mito Y., Henikoff J.G., and Henikoff S. 2005. Genome-scale profiling of histone H3.3 replacement patterns. *Nat. Genet.* **37:** 1090–1097.

Mizuguchi G., Shen X., Landry J., Wu W.H., Sen S., and Wu C. 2004. ATP-driven exchange of histone H2AZ variant catalyzed by SWR1 chromatin remodeling complex. *Science* **303:** 343–348.

Nakayama J., Rice J.C., Strahl B.D., Allis C.D., and Grewal S.I. 2001. Role of histone H3 lysine 9 methylation in epigenetic control of heterochromatin assembly. *Science* **292:** 110–113.

Narlikar G.J., Fan H.Y., and Kingston R.E. 2002. Cooperation between complexes that regulate chromatin structure and transcription. *Cell* **108:** 475–487.

Noma K., Allis C.D., and Grewal S.I. 2001. Transitions in distinct histone H3 methylation patterns at the heterochromatin domain boundaries. *Science* **293:** 1150–1155.

Nowak S.J. and Corces V.G. 2004. Phosphorylation of histone H3: A balancing act between chromosome condensation and transcriptional activation. *Trends Genet.* **20:** 214–220.

Pal S., Yun R., Datta A., Lacomis L., Erdjument-Bromage H., Kumar J., Tempst P., and Sif S. 2003. mSin3A/histone deacetylase 2- and PRMT5-containing Brg1 complex is involved in transcriptional repression of the Myc target gene cad. *Mol. Cell. Biol.* **21:** 7475–7487.

Pavri R., Zhu B., Li G., Trojer P., Mandal S., Shilatifard A., and Reinberg D. 2006. Histone H2B monoubiquitination functions cooperatively with FACT to regulate elongation by RNA polymerase II. *Cell* **125:** 703–717.

Peterson C.L. 2002. Chromatin remodeling: Nucleosomes bulging at the seams. *Curr. Biol.* **12:** R245–R247.

Raisner R.M., Hartley P.D., Meneghini M.D., Bao M.Z., Liu C.L., Schreiber S.L., Rando O.J., and Madhani H.D. 2005. Histone variant H2A.Z marks the 5′ ends of both active and inactive genes in euchromatin. *Cell* **123:** 233–248.

Rea S., Eisenhaber F., O'Carroll D., Strahl B.D., Sun Z.W., Schmid M., Opravil S., Mechtler K., Ponting C.P., Allis C.D., and Jenuwein T. 2000. Regulation of chromatin structure by site-specific histone H3 methyltransferases. *Nature* **406:** 593–599.

Reinberg D. and Sims R.J., III. 2006. de FACTo nucleosome dynamics. *J. Biol. Chem.* (in press).

Reinke H. and Horz W. 2003. Histones are first hyperacetylated and then lose contact with the activated *PHO5* promoter. *Mol. Cell* **11:** 1599–1607.

Robzyk K., Recht J., and Osley M.A. 2000. Rad6-dependent ubiquitination of histone H2B in yeast. *Science* **287:** 501–504.

Roth S.Y., Denu J.M., and Allis C.D. 2001. Histone acetyltransferases. *Annu. Rev. Biochem.* **70:** 81–120.

Sanders S.L., Jennings J., Canutescu A., Link A.J., and Weil P.A. 2002. Proteomics of the eukaryotic transcription machinery: Identification of proteins associated with components of yeast TFIID by multidimensional mass spectrometry. *Mol. Cell. Biol.* **22:** 4723–4738.

Santisteban M.S., Kalashnikova T., and Smith M.M. 2000. Histone H2A.Z regulates transcription and is partially redundant with nucleosome remodeling complexes. *Cell* **103:** 411–422.

Santos-Rosa H., Schneider R., Bannister A.J., Sherriff J., Bernstein B.E., Emre N.C., Schreiber S.L., Mellor J., and Kouzarides T. 2002. Active genes are tri-methylated at K4 of histone H3. *Nature* **419:** 407–411.

Sassone-Corsi P., Mizzen C.A., Cheung P., Crosio C., Monaco L., Jacquot S., Hanauer A., and Allis C.D. 1999. Requirement of Rsk-2 for epidermal growth factor-activated phosphorylation of histone H3. *Science* **285:** 886–891.

Schreiber S.L. and Bernstein B.E. 2002. Signaling network model of chromatin. *Cell* **111:** 771–778.

Schurter B.T., Koh S.S., Chen D., Bunick G.J., Harp J.M., Hanson B.L., Henschen-Edman A., Mackay D.R., Stallcup M.R., and Aswad D.W. 2001. Methylation of histone H3 by coactivator-associated arginine methyltransferase 1. *Biochemistry* **40:** 5747–5756.

Sekinger E.A., Moqtaderi Z., and Struhl K. 2005. Intrinsic histone-DNA interactions and low nucleosome density are important for preferential accessibility of promoter regions in yeast. *Mol. Cell* **18:** 735–748.

Shi Y., Lan F., Matson C., Mulligan P., Whetstine J.R., Cole P.A., Casero R.A., and Shi Y. 2004. Histone demethylation mediated by the nuclear amine oxidase homolog LSD1. *Cell* **119:** 941–953.

Shi Y.J., Matson C., Lan F., Iwase S., Baba T., and Shi Y. 2005. Regulation of LSD1 histone demethylase activity by its associated factors. *Mol. Cell.* **19:** 857–864.

Shiio Y. and Eisenman R.N. 2003. Histone sumoylation is associated with transcriptional repression. *Proc. Natl. Acad. Sci.* **100:** 13225–13230.

Shim E.Y., Woodcock C., and Zaret K.S. 1998. Nucleosome positioning by the winged helix transcription factor HNF3. *Genes Dev.* 12: 5–10.

Shogren-Knaak M., Ishii H., Sun J.M., Pazin M.J., Davie J.R., and Peterson C.L. 2006. Histone H4-K16 acetylation controls chromatin structure and protein interactions. *Science* 311: 844–847.

Sims. R.J., Belotserkovskaya R., and Reinberg D. 2004. Elongation by RNA polymerase II: The short and long of it. *Genes Dev.* 18: 2437–2468.

Sims R.J., Chen C.F., Santos-Rosa H., Kouzarides T., Patel S.S., and Reinberg D. 2006. Human but not yeast CHD1 binds directly and selectively to histone H3 methylated at lysine 4 via its tandem chromodomains. *J. Biol. Chem* 51: 41789–41792.

Soloaga A., Thomson S., Wiggin G.R., Rampersaud N., Dyson M.H., Hazzalin C.A., Mahadevan L.C., and Arthur J.S. 2003. MSK2 and MSK1 mediate the mitogen- and stress-induced phosphorylation of histone H3 and HMG-14. *EMBO J.* 22: 2788–2797.

Sterner D.E. and Berger S.L. 2000. Acetylation of histones and transcription-related factors. *Microbiol. Mol. Biol. Rev.* 64: 435–459.

Strahl B.D. and Allis C.D. 2000. The language of covalent histone modifications. *Nature* 403: 41–45.

Strahl B.D., Briggs S.D., Brame C.J., Caldwell J.A., Koh S.S., Ma H., Cook R.G., Shabanowitz J., Hunt D.F., Stallcup M.R., and Allis C.D. 2001. Methylation of histone H4 at arginine 3 occurs in vivo and is mediated by the nuclear receptor coactivator PRMT1. *Curr. Biol.* 26: 996–1000.

Sun Z.W. and Allis C.D. 2002. Ubiquitination of histone H2B regulates H3 methylation and gene silencing in yeast. *Nature* 418: 104–108.

Svaren J., Schmitz J., and Horz W. 1994. The transactivation domain of Pho4 is required for nucleosome disruption at the *PHO5* promoter. *EMBO J.* 13: 4856–4862.

Taunton J., Hassig C.A., and Schreiber S.L. 1996. A mammalian histone deacetylase related to the yeast transcriptional regulator Rpd3p. *Science* 272: 408–411.

Trewick S.C., McLaughlin P.J., and Allshire R.C. 2005. Methylation: Lost in hydroxylation? *EMBO Rep.* 6: 315–320.

Tsukada Y., Fang J., Erdjument-Bromage H., Warren M.E., Borchers C.H., Tempst P., and Zhang Y. 2006. Histone demethylation by a family of JmjC domain-containing proteins. *Nature* 439: 811–816.

Turner B.M. 2000. Histone acetylation and an epigenetic code. *Bioessays* 22: 836–45.

Vakoc C.R., Mandat S.A., Olenchock B.A., and Blobel G.A. 2005. Histone H3 lysine 9 methylation and HP1γ are associated with transcription elongation through mammalian chromatin. *Mol. Cell* 19: 381–391.

Vettese-Dadey M., Grant P.A., Hebbes T.R., Crane-Robinson C., Allis C.D., and Workman J.L. 1996. Acetylation of histone H4 plays a primary role in enhancing transcription factor binding to nucleo-

somal DNA in vitro. *EMBO J.* 15: 2508–2518.

Volpe T.A., Kidner C., Hall I.M., Teng G., Grewal S.I., and Martienssen R.A. 2002. Regulation of heterochromatic silencing and histone H3 lysine-9 methylation by RNAi. *Science* 297: 1833–1837.

Wang H., Wang L., Erdjument-Bromage H., Vidal M., Tempst P., Jones R.S., and Zhang Y. 2004. Role of histone H2A ubiquitination in Polycomb silencing. *Nature* 431: 873–878.

Wang H., Huang Z.Q., Xia L., Feng Q., Erdjument-Bromage H., Strahl B.D., Briggs S.D., Allis C.D., Wong J., Tempst P., and Zhang Y. 2001. Methylation of histone H4 at arginine 3 facilitating transcriptional activation by nuclear hormone receptor. *Science* 293: 853–857.

Wang Y., Wysocka J., Sayegh J., Lee Y.H., Perlin J.R., Leonelli L., Sonbuchner L.S., McDonald C.H., Cook R.G., Dou Y., et al. 2004. Human PAD4 regulates histone arginine methylation levels via demethylimination. *Science* 306: 279–283.

Whetstine J.R., Nottke A., Lan F., Huarte M., Smolikov S., Chen Z., Spooner E., Li E., Zhang G., Colaiacovo M., and Shi Y. 2006. Reversal of histone lysine trimethylation by the JMJD2 family of histone demethylases. *Cell* 125: 467–481.

Wood A., Krogan N.J., Dover J., Schneider J., Heidt J., Boateng M.A., Dean K., Golshani A., Zhang Y., Greenblatt J.F., et al. 2003. Bre1, an E3 ubiquitin ligase required for recruitment and substrate selection of Rad6 at a promoter. *Mol. Cell* 11: 267–274.

Workman J.L. and Roeder R.G. 1987. Binding of transcription factor TFIID to the major late promoter during in vitro nucleosome assembly potentiates subsequent initiation by RNA polymerase II. *Cell* 51: 613–622.

Wysocka J., Allis C.D., and Coonrod S. 2006a. Histone arginine methylation and its dynamic regulation. *Front. Biosci.* 11: 344–355.

Yamane K., Toumazou C., Tsukada Y., Erdjument-Bromage H., Tempst P., Wong J., and Zhang Y. 2006. JHDM2A, a JmjC-containing H3K9 demethylase, facilitates transcription activation by androgen receptor. *Cell* 125: 483–495.

Yang X.J. and Seto E. 2003. Collaborative spirit of histone deacetylases in regulating chromatin structure and gene expression. *Curr. Opin. Genet. Dev.* 13: 143–153.

Zhang H., Roberts D.N., and Cairns B.R. 2005. Genome-wide dynamics of Htz1, a histone H2A variant that poises repressed/basal promoters for activation through histone loss. *Cell* 123: 219–231.

Zhang Y. and Reinberg D. 2001. Transcription regulation by histone methylation: Interplay between different covalent modifications of the core histone tails. *Genes Dev.* 15: 2343–2360.

Zhu B., Zheng Y., Pham A.D., Mandal S.S., Erdjument-Bromage H., Tempst P., and Reinberg D. 2005. Monoubiquitination of human histone H2B: The factors involved and their roles in HOX gene regulation. *Mol. Cell* 20: 601–611.

C H A P T E R   1 1

Transcriptional Silencing by Polycomb Group Proteins

Ueli Grossniklaus[1] and Renato Paro[2]

[1]Institute of Plant Biology and Zürich-Basel Plant Science Center, University of Zürich, CH-8008 Zürich, Switzerland
[2]ZMBH, University of Heidelberg, D-69120 Heidelberg, Germany

CONTENTS

1. Introduction, 213
   1.1 The Concept of Cellular Memory, 213
   1.2 Genetic Identification of the Polycomb Group, 213

2. Establishing Silencing Marks on Chromatin, 215
   2.1 Components and Evolutionary Conservation of PRC2, 215
   2.2 Chromatin-modifying Activity of PRC2, 219
   2.3 Dynamic Function of PRC2 during Development, 219

3. Maintaining Transcriptional Silencing, 220
   3.1 Components of PRC1, 220

3.2 Targeting PRC1 to Silenced Genes, 222
3.3 Establishment of Repressive Functions by PRC1, 223
3.4 Preventing Heritable Repression by Anti-silencing, 224

4. PcG Repression in Mammalian Development, 224
   4.1 From Gene to Chromosome Repression, 224
   4.2 Consequences of Aberrant Transcriptional Activation, 225
   4.3 Maintaining Stem Cell Fate, 226

5. Conclusion and Outlook, 227

References, 228

GENERAL SUMMARY

The organs of humans, animals, and plants are constructed from a large pool of distinct cell types, each performing a specialized physiological or structural function. With very few exceptions, all cell types contain the same genetic information encoded in their DNA. Thus, the distinctiveness of a given cell type is achieved through specific gene expression programs. However, cell lineages need to have these programs of gene expression maintained during growth and cell division. This implies the existence of a memory system that ensures a faithful transmission of information for which gene has to be active or repressed from mother to daughter cells. The existence of such a system is illustrated by the fact that cultured tissues of plants and animals usually maintain their differentiated characters even if grown in a foreign environment. By way of example, ivy plants regenerated after tissue culture produce the type of leaf corresponding to the phase of development from which the original tissue was taken (i.e., juvenile or adult leaf).

The major question to be addressed in this and the following chapter concerns the molecular identity of factors contributing to the mechanism(s) which maintains determined states over many cell divisions (a process termed "cellular memory" or "transcriptional memory"). Genetic analyses in *Drosophila* have identified regulators crucial in maintaining the fate of individual body segments that are determined by the action of the *HOX* genes. In *Drosophila* males, the first thoracic segment has legs with sex combs. Legs on the second and third thoracic segment lack these structures (see the left panel of the title images). In the 1940s, *Drosophila* mutants were identified (*Polycomb* and *extra sex combs*) where males had sex combs on all legs (see the right panel of the title images). They correspond to homeotic transformations of the second and third leg identities into the first leg identity. Genetic and molecular studies showed that these mutations did not affect the products of the *HOX* genes themselves, but rather the way *HOX* gene activity was spatially controlled. Over the years, a large number of such regulatory genes were identified, which could be classified into two antagonistic groups, the *Polycomb* (PcG) and *trithorax* (trxG) groups. Whereas the PcG proteins are required to maintain the silenced state of developmental regulators such as the *HOX* genes, the trxG proteins are generally involved in maintaining the active state of gene expression. Thus, PcG and trxG proteins form the molecular basis for cellular memory.

Proteins of the PcG and trxG are organized into large multimeric complexes that act on their target genes by modulating chromatin structure. In this chapter, we focus on the molecular nature and function of two major *Polycomb* Repressive Complexes, PRC1 and PRC2; the molecular nature of the trxG complexes is described in the next chapter. PcG complexes are recruited to target genes through a DNA sequence called a PcG Response Element (PRE) in *Drosophila*. Once recruited, they establish a silent chromatin state that can be inherited over many cell divisions. Members of PRC2 are conserved between plants and animals, whereas PRC1 proteins are only present in *Drosophila* and vertebrates. This implies conservation but also diversity in the basic building blocks of the cellular memory system. In addition to their function in the maintenance of cell types, PcG complexes may also play important roles in stem cell plasticity. Their deregulation can lead to neoplastic transformation and cancer in vertebrates. Thus, PcG proteins are crucial for many fundamental processes of normal development and disease in multicellular eukaryotes.

1 Introduction

All multicellular organisms start from a single cell, the zygote, which during development gives rise to a multitude of distinct cell types with specialized functions. This poses the problem of how, once determined, cell types can be maintained over many cell divisions occurring during growth phases.

1.1 The Concept of Cellular Memory

An adult animal has 200–300 structurally distinct cell types, whereas a plant has between 30 and 40. If complex gene expression patterns are taken into consideration, an even larger number of different cell populations can be distinguished. The identity and function of a given cell type are determined by its characteristic gene expression profile, where appropriate sets of genes are either active or silent. During development and adult homeostasis, it is crucial to memorize and faithfully reproduce this state after each cell division. This is particularly important during the replication of genetic material (S phase) and the separation of chromosomes during mitosis (M phase), which are recurring events at each cell cycle that interrupt gene expression processes. Thus, how can differential gene expression patterns be maintained from one cell generation to the next?

We know from experiments done in the 1960s and 1970s that plant and animal tissues "remember" a determined state even after prolonged passage in culture (Hadorn 1968; Hackett et al. 1987). Hadorn and colleagues showed that imaginal disc cells found in *Drosophila* larvae have an inherent cellular memory, allowing them to remember determined states that are fixed in early

embryogenesis. Imaginal discs are clusters of epithelial cells set aside in the developing embryo as precursors for specific external structures and appendages during metamorphosis. For instance, of the two pairs of imaginal discs in the second thoracic segment, one forms a midleg and the other a wing (see Fig. 2 in Chapter 12). Imaginal discs can be cultured by transplantation into the hemocoel of adult females, where they continue to proliferate but do not differentiate. When transplanted back into a larva prior to metamorphosis, the disc will subsequently differentiate into the expected adult structures, even after successive transplantation passages. More recently, the PcG and trxG proteins were shown to be required for maintenance of the determined state of imaginal disc cells (Fig. 1). In the culture experiments described above, however, it was observed that in rare cases an imaginal disc can change its fate, a process called transdetermination. This process involves the down-regulation of PcG repression by the JNK signaling cascade in transdetermined cells (Fig. 2 a,b) (Lee et al. 2005). PcG mutants also have elevated frequencies of transdetermination, supporting a role for PcG proteins in maintaining imaginal disc cell fates (Klebes et al. 2005). Thus, PcG proteins seem to play a crucial role in the reprogramming of cellular fates both during normal development and in regeneration processes (see Section 4.3 for more detail).

1.2 Genetic Identification of the Polycomb Group

In all metazoans, the anterior–posterior axis is specified through defined expression patterns of *HOX* genes (see Fig. 2 in Chapter 12). During embryogenesis in

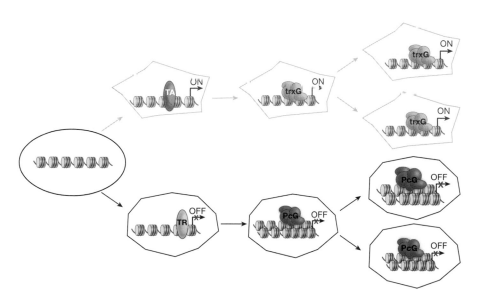

Figure 1. The Concept of Cellular Memory

Schematic illustration of the involvement of PcG and trxG complexes in the determination of active and repressed states of gene expression and, thereby, cellular differentiation, which is maintained over many cell divisions. (TA) Transcriptional activator; (TR) transrciptional repressor.

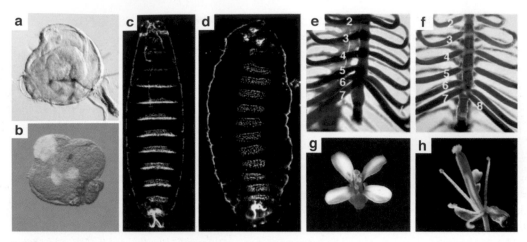

Figure 2. Homeotic Transformations in PcG Mutants of Various Species

(*a–d*) *Drosophila melanogaster*, (*e, f*) *Mus musculus*, (*g, h*) *Arabidopsis thaliana*. (*a, b*) Leg imaginal discs undergoing a trans-determination event as indicated by the expression of the wing-specific gene *vestigial* (which is marked by GFP). (*c, d*) Cuticles of a wild-type (*c*) and a *Su(z)12* mutant embryo (*d*). In the *Su(z)12* mutant embryo, all abdominal, thoracic, and several head segments (not all visible in this focal plane) are homeotically transformed into copies of the eighth abdominal segment due to misexpression of the *Abd-B* gene in every segment. (*e, f*) Axial skeleton of newborn wild-type (*e*) and *Ring1A*$^{-/-}$ mice (*f*). Views of the thoracic regions of cleared skeletons showing bone (*red*) and cartilage (*blue*). The mutant displays anterior transformation of the eighth thoracic vertebra as indicated by the presence of eight (1–8) vertebrosternal ribs, instead of seven (1–7) as in the wild type. (*g, h*) Wild-type (*g*) and *clf-2* mutant (*h*) flowers. The wild-type flower shows the normal arrangement of sepals, petals, stamens, and carpels. In the *clf-2* flower, petals are absent or reduced in number. (*a,b,* Courtesy of N. Lee and R. Paro; *c,d,* reprinted, with permission, from Birve et al. 2001 [©Company of Biologists Ltd.]; *e,f,* reprinted, with permission, from del Mar Lorente et al. 2000 [©Company of Biologists Ltd.]; *h,* courtesy of J. Goodrich.)

Drosophila, the activity of maternally (i.e., inherited through the oocyte) and zygotically produced transcription factors generates a specific combination of *HOX* expression required for each segment. This segment-specific profile of *HOX* gene activity is maintained throughout the development of the fly, long after the early transcriptional regulators have disappeared. When the function of *HOX* genes was genetically characterized, many *trans*-acting regulators were isolated. Among them, *Polycomb* (*Pc*) was identified and genetically analyzed by Pam and Ed Lewis (Lewis 1978). Heterozygous *Pc* mutant males have additional sex combs on the second and third legs, whereas wild-type males only carry sex combs on the first leg (see title figure). Homozygous mutants are embryonic lethal, exhibiting a transformation of all cuticular segments toward the most posterior abdominal segment (Fig. 2c,d). These classic PcG phenotypes were interpreted as being caused by ectopic expression of *HOX* genes. Thus, *Pc* and the other genes with similar phenotypes were defined as repressors of *HOX* gene activity. Detailed analyses subsequently uncovered the fact that the PcG proteins are only required for the maintenance of *HOX* repression, rather than the position-specific establishment of *HOX* activity

during pattern formation. This latter task is performed by the transcription factors encoded by the early acting segmentation genes. Based on their repressing or activating effect on *HOX* gene expression, these newly identified *trans*-acting regulators were divided into two antagonistic classes, the PcG and trxG, respectively (Fig. 1) (Kennison 1995).

The molecular isolation of *Drosophila* PcG genes has made it possible to study the function of vertebrate orthologs in mice, where they were also found to be key regulators of *HOX* gene expression (van der Lugt et al. 1994; Core et al. 1997). In mammals, mutations in PcG genes lead to homeotic transformations of the vertebrae (Fig. 2e,f). In addition, PcG genes play a crucial role in the control of cell proliferation, stem cell maintenance, and cancer (see Sections 4.2 and 4.3). The remarkable conservation of PcG genes between flies and mammals has facilitated biochemical analyses and led to the identification of some novel members of PcG complexes, e.g., the RING1 protein (Satijn and Otte 1999). Targeted mutation of RING1a in the mouse, for instance, led to the classic homeotic transformation phenotype. Only subsequently was it found to correspond to the PcG gene *Sex combs extra* in *Drosophila*.

In two other model organisms, namely the worm *Caenorhabditis elegans* and the flowering plant *Arabidopsis thaliana*, the molecular characterization of mutants isolated in various genetic screens revealed the existence of other PcG protein orthologs. In *C. elegans*, PcG members were identified in screens for *maternal effect sterile* (*mes*) mutants and were shown to be involved in X-chromosome silencing in the hermaphrodite germ line (Fong et al. 2002; see Chapter 15).

In *Arabidopsis*, PcG genes were identified in several genetic screens investigating distinct developmental processes (Hsieh et al. 2003). The first PcG gene in plants, *CURLY LEAF* (*CLF*), was identified as a mutant with homeotic transformations of floral organs (Fig. 2g,h) (Goodrich et al. 1997). Mutations in the *FERTILIZATION-INDEPENDENT SEED* (*FIS*) class of genes were found in screens for mutants showing maternal-effect seed abortion (Grossniklaus et al. 1998), or allowing aspects of seed development to occur in the absence of fertilization (Luo et al. 1999; Ohad et al. 1999). Finally, PcG genes were identified in screens for flowering time mutants, e.g., mutants that flower directly after germination (Yoshida et al. 2001) or that disrupt the vernalization response, i.e., the process rendering plants competent to flower after prolonged exposure to cold (Gendall et al. 2001).

The variety of processes regulated by PcG proteins illustrates the importance of maintaining the repressed state of key developmental regulators in different organisms. On the one hand, there is an amazing conservation of some biological functions from plants to mammals, e.g., the regulation of key developmental regulators such as homeotic genes or involvement in the tight regulation of cell proliferation. On the other hand, PcG complexes appear to be versatile and dynamic molecular modules that have been employed to control a large variety of developmental and cellular processes.

2 Establishing Silencing Marks on Chromatin

PcG proteins fall into two biochemically characterized classes, which form the *Polycomb* repressive complexes 1 and 2 (PRC1 and PRC2). The two complexes are required for consecutive steps in the repression of gene expression. First, PRC2 has histone-modifying activity and methylates H3K27 and/or H3K9 at genes targeted for silencing. PRC1 components can then recognize and bind to such modifications and induce appropriate structural changes in chromatin. Whereas PRC2 proteins are present in all multicellular model species, PRC1 components have not been identified in *C. elegans* and *Arabidopsis*.

2.1 Components and Evolutionary Conservation of PRC2

Several variants of PRC2 have been purified from *Drosophila* embryos, but all of these complexes contain four core proteins (Levine et al. 2004): the SET histone-methyltransferase Enhancer of Zeste (E(Z)), the WD40 protein ESC, the histone-binding protein p55, and Suppressor of Zeste12 (SU(Z)12) (Table 1 and Fig. 3). Based on this composition, PRC2 was originally referred to as the E(Z)-ESC complex. This section highlights the molecular and biochemical details known about the different PRC2 components identified to date in different model organisms.

The E(z) gene encodes a 760-amino acid protein, containing a SET domain that confers histone lysine methyltransferase (HKMT) activity. The SET domain is preceded by a CXC or Pre-SET domain (Tschiersch et al. 1994), which contains nine conserved cysteines that bind three zinc ions and is thought to stabilize the SET domain. Such a structural role is supported by the fact that several temperature-sensitive E(z) alleles affect one of the conserved cysteines (Carrington and Jones 1996). In addition, E(z) contains SANT domains implicated in histone binding, and a C5 domain required for the physical interaction with SU(Z)12.

ESC is a short protein of 425 amino acids that contains five WD40 repeats, shown to form a β propeller structure. This serves as a platform for protein–protein interactions, hence giving ESC a central role in PRC2, to physically interact with both E(z) and p55 in all model systems analyzed.

The SU(Z)12 protein is 900 amino acids long and characterized by a C_2H_2-type zinc finger and a carboxy-terminal VEFS domain. The VEFS domain was identified as a conserved region between SU(Z)12 and its three homologs in plants: *VRN2*, *EMF2*, *FIS2* (see Fig. 3). Several mutant *Su(z)12* alleles alter this domain, showing it is required for the interaction with the C5 domain of E(Z) (Chanvivattana et al. 2004; Yamamoto et al. 2004).

The p55 protein was not identified as a PcG member by genetic approaches, possibly because it takes part in a multitude of other protein complexes associated with chromatin (Hennig et al. 2005). The p55 protein was, however, identified biochemically as part of PRC2. It is 430 amino acids long and contains six WD40 repeats, which physically interact with ESC or its orthologs in mammals and plants (Tie et al. 2001; Köhler et al. 2003a).

In addition to the core PRC2 proteins, some variants of the complex contain the RPD3 histone deacetylase

(HDAC) or the Polycomb-like (PCL) protein. The interaction with RPD3 is noteworthy, because histone deacetylation is correlated with a repressed state of gene expression (see Chapter 10). The different compositions of PRC2 likely reflect dynamic changes during development or tissue-specific variants.

PRC2 is highly conserved in invertebrates, vertebrates, and plants (Fig. 3). In *C. elegans*, only homologs of E(Z) and ESC are present: MES-2 and MES-6. Together with another nonconserved protein, MES-3, they form a small complex of about 230 kD required for X-chromosome silencing in the hermaphrodite germ line (see Chapter 15). In plants and mammals, all four core proteins of PRC2 are present. As in *Drosophila*, the mammalian complex is about 600 kD and plays a role not only in regulating homeotic gene expression, but also in the control of cell proliferation, X-chromosome inactivation, and

imprinted gene expression (see Section 4 and relevant chapters for more detail).

In plants, several genes encoding PRC2 components have undergone duplications such that they now are present as small gene families. In *Arabidopsis* there is only one homolog of ESC, FERTILIZATION-INDEPENDENT ENDOSPERM (FIE), but three homologs of E(Z), three homologs of SU(Z)12, and five homologs of p55 (referred to as MSI1-5) (Table 1). Varying combinations of these proteins form at least three distinct complexes that control specific developmental processes (Figs. 3 and 4) (Reyes and Grossniklaus 2003; Chanvivattana et al. 2004).

The best studied of these complexes is formed by members of the *FERTILIZATION-INDEPENDENT SEED* (*FIS*) class, which play a crucial role in the control of cell proliferation in the seed (Grossniklaus et al. 2001). This FIS or MEA-FIE complex contains MEDEA, FIE, FIS2,

Table 1. Core PcG genes in model systems

	D. melanogaster		M. musculus	A. thaliana	C. elegans
		PcG DNA-binding proteins			
pho	Pleiohomeotic	zinc finger	*Yy1*		
phol	Pleiohomeotic-like	zinc finger			
Psq	Pipsqueak	BTB-POZ domain			
Dsp1	Dorsal Switch Protein 1	HMG domain protein	*HMGB2*		
		PRC2 core complex			
esc	Extra sex combs	WD 40 repeats	*Eed*	*FIE*	*MES-6*
E(z)	Enhancer of zeste	SET domain	*Ezh1*/Enx2, *Ezh2*/Enx1	*CLF* *MEA* *SWN*	*MES-2*
Su(z)12	Suppressor of zeste12	zinc finger VEFS box	*mSU(Z) 12*	*FIS2* *VRN2* *EMF2*	
p55	p55	histone-binding domain	*RbAp48* *RbAp46*	*MSI1* (MSI2,3,4,5)	
		PRC1 core complex			
Pc	Polycomb	chromodomain	*Cbx2*/M33 *Cbx4*/MPc2 *Cbx6* *Cbx8*/MPc3 *Cbx7*		
Ph	Polyhomeotic	zinc finger SAM/SPM domain	*Edr1*/Mph1/Rae28 *Edr2*/Mph2		*SOP-2*
Psc	Posterior Sex Combs	zinc finger HTH domain	*Bmi1* *Rnf110*/Zfp144/ Mel18		
dRing / Sce	dRing / Sex combs extra	RING zinc finger	*Ring1*/Ring1a *Rnf2*/ Ring1b		

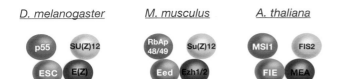

FIS complex

EMF complex

MSI1? VRN2

FIE? CLF/ SWN

VRN complex

Figure 3. Conserved PRC2 Core Complexes

The core members of PRC2 in *D. melanogaster*, *M. musculus*, *A. thaliana*, and *C. elegans* are shown. In *A. thaliana*, an ancestral complex is proposed to have diversified into three variants with discrete functions in development. In *C. elegans*, the PRC2 core complex contains only three proteins: MES-3 does not have homology with any other identified PRC2 protein. The colors indicate homology, the contacts indicate interactions. (Adapted from Reyes and Grossniklaus 2003 and Chanvivattana et al. 2004.)

and MSI1. The FIS complex was found to regulate the genes encoding PHERES1 (PHE1), a MADS domain transcription factor; and MEIDOS, a homolog of Skp1, which in yeast plays a crucial role in the control of cell proliferation (Köhler et al. 2003b). Interestingly, the paternal allele of *PHE1* is expressed at higher levels than the maternal allele. This regulation of gene expression by genomic imprinting is under the control of the FIS complex, which specifically represses the maternal allele (Köhler et al. 2005). Thus, as outlined below, the FIS complex shares with its mammalian counterpart functions in regulating cell proliferation as well as imprinted gene expression.

The EMF complex contains CLF and EMBRYONIC FLOWER2 (EMF2) (Chanvivattana et al. 2004). Mutations in either of these show weak homeotic transformations and an early flowering phenotype. The EMF complex is required to repress homeotic genes, whose combinatorial action determines the identity of floral organs (Goodrich et al. 1997). Thus, the EMF complex has a similar function in maintaining the repressed state of homeotic genes as PRC2 in *Drosophila* and vertebrates (Fig. 2). However, homeotic genes in plants do not encode homeodomain proteins, but rather other transcription factors belonging to the MADS-domain and the plant-specific AP2-domain families. Strong mutants of *EMF2*, however, have more severe phenotypes where their seedlings produce flowers directly after germination, bypassing the vegetative phase of development (Yoshida et al. 2001). Thus, the EMF complex plays a role both early in development, where it prevents immediate flowering, and later in floral organogenesis (Chanvivattana et al. 2004). At both stages, the

EMF complex represses floral homeotic genes such as *AG* and *APETALA3* (*AP3*) (Fig. 4). The FIS class proteins, FIE and MSI1, have also been implicated in the control of homeotic gene expression (Figs. 3 and 4). Because mutations in both cause maternal-effect embryo lethality, this function was only revealed when partial loss-of-function alleles could be studied at later stages of development (Kinoshita et al. 2001; Hennig et al. 2003).

Finally, the VRN complex plays a key role in a well-known epigenetic process: vernalization (extended exposure to low temperature). Vernalization induces flowering in winter annuals, but the effect is only seen after many cell divisions (Fig. 4). A plant cell will remember that it was vernalized for many months, or even years, after the cold period. This cellular memory is maintained through passages in cell culture but not from one generation to the next (Sung and Amasino 2004a). The *VERNALIZATION* (*VRN*) genes mediate the response to vernalization. *VRN2* was found to encode a SU(Z)12 homolog (Gendall et al. 2001), which interacts with the plant E(Z) homologs CLF and SWINGER (SWN) in yeast two-hybrid assays (Chanvivattana et al. 2004). The transition to flowering is not only controlled by vernalization, but involves the perception of endogenous (developmental stage and age) as well as exogenous factors (day length, light conditions, temperature). Four pathways have been defined by genetic analyses: (1) The autonomous pathway constitutively represses flowering, (2) the photoperiod pathway accelerates flowering under long days, (3) the vernalization pathway induces flowering in response to exposure to cold temperature, and (4) gibberellins promote flowering. The flowering time

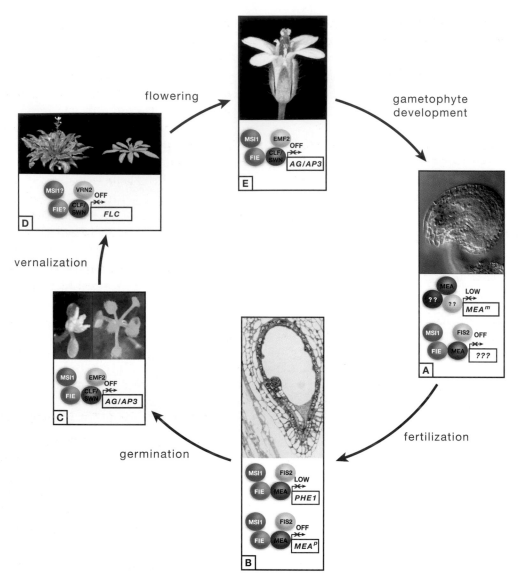

Figure 4. Involvement of Distinct PRC2s at Various Stages of Plant Development

During the plant life cycle, distinct variants of PRC2 (see Fig. 3) control developmental progression. (*A*) A cleared wild-type ovule harboring the female gametophyte in its center. The FIS complex represses target genes that control proliferation of the central cell; as in all *fis* class mutants, this cell proliferates in the absence of fertilization. Around fertilization, MEA is also required to maintain a low level of *MEA^m* expression, but this activity is independent of other components of the FIS complex. (*B*) Section of a wild-type seed harboring the embryo and endosperm, enclosed by the seed coat. After fertilization, the FIS complex is involved in the control of cell proliferation in embryo and endosperm. It maintains a low level of expression of *PHE1* and is required to keep the paternal *MEA^p* allele silent. (*C*) Wild-type (*right*) and *emf2* mutant (*left*) seedling 21 days after germination. The *emf2* seedling produced a flower with homeotic transformations but no leaves. The EMF complex prevents flowering and represses floral homeotic genes such as *AG*, *AP3*, and others. (*D*) Vernalized (*right*) and non-vernalized (*left*) plants, the latter being characterized by a prolonged vegetative phase and the production of many leaves. During the vegetative phase of development, exogenous and endogenous signals induce flowering. Vernalization leads to the repression of the floral repressor *FLC* and thus promotes flowering. The maintenance of this repression depends on the VRN complex. (*E*) Wild-type *Arabidopsis* flower. During flower organogenesis, the EMF complex regulates floral homeotic genes that determine the identity of floral organs. (*A*, Courtesy of J.M. Moore and U. Grossniklaus; *B*, courtesy of J.-P. Vielle-Calzada and U. Grossniklaus; *C*, reprinted, with permission, from Moon et al. 2003 [©ASPB]; *D*, reprinted, with permission, from Sung and Amasino 2004a [©Elsevier]; *E*, reprinted, with permission, from Page and Grossniklaus 2002 [©Macmillan].)

gene *FLC*, which contains a MADS box, is a key integrator of the flowering response: It represses flowering. *FLC* expression is reduced by both the vernalization and the autonomous pathway. Whereas the initial repression of *FLC* is independent of the VRN complex, the maintenance of repression requires *VRN2*, which alters chromatin organization at the *FLC* locus (Gendall et al. 2001). Note that one of the components of the autonomous pathway is a p55 homolog, *FVE* (or *MSI4*), which affects flowering time response but does not act in the vernalization pathway (Ausin et al. 2004; Kim et al. 2004). Because no biochemical studies on the VRN complex have been reported, its exact composition is currently unknown (Figs. 3 and 4).

2.2 Chromatin-modifying Activity of PRC2

How does PRC2 mediate its repressive effect? Several proteins of the PcG and trxG have SET domains, including the PRC2 component E(Z). The discovery that SET domain proteins possess HKMT activity (Rea et al. 2000) suggested an involvement of histone methylation in PcG function. Indeed, mammalian and *Drosophila* PRC2 complexes were shown to methylate histone H3 at lysine 27 (H3K27) and, to a lesser extent, H3K9 both in vivo and in vitro (Cao et al. 2002; Czermin et al. 2002; Kuzmichev et al. 2002; Müller et al. 2002). These histone marks are usually associated with a transcriptionally silent state. Furthermore, H3K9 and H3K27 methylation has been associated with repressed homeotic genes of the *bithorax* complex (Müller et al. 2002). However, only H3K27 methylation was lost in *E(z)* mutants, stressing the importance of H3K27 methylation in PcG silencing. Unlike the SU(VAR)3-9 protein, which methylates H3K9 on its own, E(Z) proteins on their own do not have H3K27 HKMT activity. The smallest complex acting as a HKMT also requires ESC and SU(Z)12, which may have modulating functions. It was recently shown that PRC2 complexes can also methylate H1K26 (Kuzmichev et al. 2004). Distinct isoforms of the mammalian ESC homolog, Eed, determine the specificity of mammalian PRC2 for H1K26 versus H3K27 methylation (Kuzmichev et al. 2004). However, the functional relevance of H1K26 methylation for PcG silencing remains unclear.

In plants, the HKMT activity of PRC2 complexes has not yet been demonstrated in vitro. However, studies of *FLC* regulation have shown that vernalization induces a loss of acetylation and an increase of H3K9 and H3K27 methylation, mainly in the first intron of the gene (Bastow et al. 2004; Sung and Amasino 2004b). Both methylation marks were lost in *vrn2* mutants, implicating the VRN complex in setting these repressive histone methyla-

tion marks. In two other mutants, *vrn1* and *vernalization insensitive3* (*vin3*), only the H3K9me2 mark is missing. *VRN1* and *VIN3* encode transcription factors of the B3-domain and homeodomain families, respectively, but the exact molecular mechanism of their involvement in modifying chromatin is currently unclear.

From numerous studies to date, the main function of PRC2 seems to involve HKMT activity, but there are other chromatin-modifying activities present in some PRC2 variants. The *Rpd3* gene encodes a HDAC that has been implicated in PcG silencing (Tie et al. 2001). However, although *rpd3* mutations enhance PcG phenotypes, they do not show the typical homeotic transformations by themselves. The fact that RPD3 is not present in all PRC2 preparations may thus reflect either a weak overall interaction, or a tissue- and stage-specific interaction with the PRC2 core components. The interaction of RPD3 with PRC2 represents an interesting partnership, as both HKMT and HDAC activities associate with silent chromatin, and in combination may reinforce transcriptionally silent states.

2.3 Dynamic Function of PRC2 during Development

As pointed out above, the PRC1 and PRC2 core complexes are associated with distinct factors that may play a role in recruiting PcG complexes to tissue-specific target loci or in modulating their activity (Otte and Kwaks 2003). The different steps of PcG repression shown in Figure 5 illustrate the stage-specific compositions of PcG complexes during *Drosophila* embryogenesis. So far, it has been difficult to characterize differences with respect to distinct tissues or cell types in flies because whole embryonic extracts are usually used for biochemical purifications. Studies performed in mammals and plants, however, clearly show that PcG complexes have distinct memberships in specific tissues and that their composition changes during cellular differentiation (Chanvivattana et al. 2004; Kuzmichev et al. 2005; Baroux et al. 2006). In mammals, expression levels of PcG genes differ tremendously from one cell line to the next. PcG complexes may even differ between target genes in the same cell, suggesting a highly dynamic behavior at different developmental stages.

In *Drosophila*, PcG proteins maintain repressed states of homeotic genes, established during early embryogenesis, thereby fixing developmental decisions. Once the silent state of a PcG target has been fixed, it will remain in that state for the remainder of an individual's life span. In plants, a similar situation may occur with the VRN com-

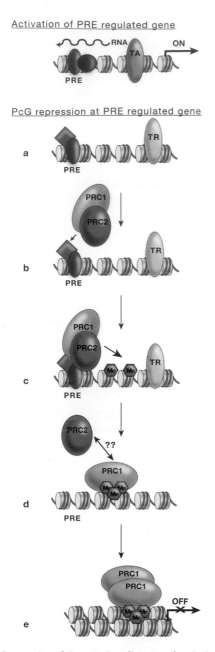

Figure 5. Sequence of Events Leading to the PcG-dependent Repressed State of Gene Expression in *Drosophila* Embryos

The original gene expression state of a PRE-regulated gene is determined by the activity of transcriptional regulators, either transcriptional repressors (TR) or activators (TA). Transcription through the PRE prevents the establishment of the "OFF" state and leads to the trxG-dependent "ON" state (for details, see Fig. 8 in Chapter 12). (*a–b*) A nontranscribed PRE binds specific DNA-binding proteins (e.g., PHO, PHOL, DSP1, or GAF) that are involved in the recruitment of the early PcG complex containing proteins of both PRC1 and PRC2. (*c*) This early PcG complex marks chromatin by E(Z)-dependent histone methylation. (*d*) Maintenance of the silent state occurs through interactions of the two distinct complexes, PRC1 and PRC2, in the absence of the original transcriptional repressor. Maintenance of PRC1 is stabilized through binding of H3K27me3 via the chromodomain of PC. (*e*) PRC1 can compact chromatin, further establishing tightly condensed, silent chromatin.

plex: Once vernalized, the target gene(s) will be permanently inactivated and only reset in the next generation. Other plant PRC2 complexes, however, seem to respond more quickly to developmental or environmental stimuli. For instance, one function of the FIS complex is to repress cell proliferation in the absence of fertilization. Upon fertilization, however, cell proliferation is rapidly induced, presumably through the derepression of PcG target genes. This indicates that PcG repression is the default state, which has to be overcome by some unknown mechanism to allow normal developmental progression. The inactivation of PcG complexes as part of the normal plant life cycle may explain the absence of PRC1 proteins in plants (Fig. 4). PRC1 plays an important role in the permanent, stable, and long-lasting inactivation of target genes. Such permanent inactivation would be detrimental to plant development, where often PcG repression is released upon appropriate stimuli.

3 Maintaining Transcriptional Silencing

3.1 Components of PRC1

The molecular analysis of the PcG gene products has revealed a structurally diverse group of chromatin-associated proteins. PRC1 contains four PcG proteins; Polycomb (PC), Polyhomeotic (PH), Posterior Sex Combs (PSC), and Ring 1 (dRing1/SCE) (see Table 1) (Francis et al. 2001). They occur in stoichiometric amounts, and additional partner proteins have been identified depending on the material used for purification. A related complex has been purified from mammalian cells, suggesting that these four subunits form the core of PRC1 (Levine et al. 2002). Immunostaining of *Drosophila* polytene chromosomes, using antibodies directed against PRC1 proteins, showed overlapping localization patterns, which indicated that these proteins cooperate at a defined and common set of target genes (Fig. 6a). Additionally, the approximately 100 bands observed on the chromosomes provided evidence that the *HOX* genes are just part of a larger regulatory network, including other gene targets subject to PcG silencing.

The *PC* gene encodes a 390-amino acid protein containing a chromodomain at its amino-terminal end. This conserved motif has homology with HP1, a *Drosophila* protein required for heterochromatin formation (Paro and Hogness 1991; see also Chapter 5). The chromodomain was subsequently found to bind to methyl moieties at H3K27 and H3K9 (Bannister et al. 2001; Fischle et al. 2003). Another conserved domain is present at the carboxy-terminal end. The conservation, as well as the occur-

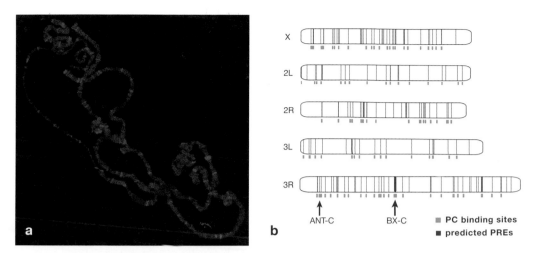

Figure 6. Targeting of PRC1 to PREs on Polytene Chromosomes

(*a*) Immunostaining of *Drosophila* polytene chromosomes to visualize the distribution of the PC protein. (*b*) Alignment of chromosome arms showing the overlap between predicted PRE sites on the *Drosophila* genome and the cytologically mapped PC-binding sites on polytene chromosomes. The two *HOX* gene clusters (ANT-C and BX-C) are prominent binding sites for PRC1s.

rence of several aberrations in mutant alleles, suggests an important but as-yet-unknown regulatory function in this part of the protein. The carboxyl terminus of PC is dispensable for targeting the protein to silenced genes (fulfilled by the chromodomain) but was found to interact in vitro with nucleosomes (Breiling et al. 1999). Whether this indicates an undiscovered recognition motif for another histone modification remains to be seen. For human Pc2, a SUMO E3 ligase activity has been demonstrated, pointing to SUMO modifications as important marks in the PcG silencing process (Kagey et al. 2003).

The amino-terminal part of the PSC protein is conserved in the vertebrate proto-oncogene *bmi-1* and the tumor suppressor gene *mel-18*. This region contains a C_3HC_4 ring finger motif, which may mediate protein–protein interactions. The ring finger motif has been implicated in subnuclear localization of Bmi-1/Mel-18, which is correlated with cellular transformations.

In *Drosophila*, the *polyhomeotic* (*ph*) locus is duplicated, consisting of a proximal (*ph-p*) and a distal (*ph-d*) gene sharing extensive homology. Homologous mouse PH proteins have been identified. All share a conserved single zinc finger and a SAM (also known as SEP or SPM) domain. This domain is also found in another PcG protein, Sex Combs on the Midleg (SCM). SAM domains are involved in protein–protein interactions, as it has been demonstrated that they participate in homo- or heterotypic interactions with other proteins. These findings support a possible function in generating large nuclear

complexes, required for silencing. Indeed, PcG proteins have been localized in subnuclear foci called PcG bodies, which might function as silencing compartments (Saurin et al. 1998).

As mentioned above, dRING1 was not initially recognized as a PcG member. Only biochemical purification uncovered the presence of this factor with a RING finger motif in PRC1, in which it is thought to play a structural role (Francis et al. 2001; Lavigne et al. 2004). The Ring1A and Ring1B proteins of mammalian PRC1 have been found to be associated with ubiquitylated H2A on the inactive X chromosome, and the maintenance of this histone mark was dependent on the Ring1 proteins (de Napoles et al. 2004; Fang et al. 2004; Cao et al. 2005; for more detail, see Section 4.1 and Chapter 17).

These four proteins comprise the core structure of PRC1. However, other PcG proteins like SCM or the Zeste protein were found to be associated with the complex (Otte and Kwaks 2003). Their molecular function in PRC1 remains unclear, as they seem to have additional roles in the nucleus; e.g., a transcriptional activator function of Zeste. Still other PcG genes were identified by virtue of their role as transcriptional regulators of the core PcG genes (Ali and Bender 2004). Namely, three PcG genes are upstream regulators of genes encoding PRC1 components. Negative feedback loops among PRC1 components, as well as positive regulation of PRC1 components by PRC2, further suggest a complicated cross-regulatory network among the PcG genes to ensure the fine-tuning of protein levels in the complexes (Fig. 7a). Similarly, complex regulatory interactions have been described for the genes of the FIS complex in *Arabidopsis* (Baroux et al. 2006).

3.2 Targeting PRC1 to Silenced Genes

Transgene analyses of *Drosophila* homeotic gene clusters uncovered regulatory elements that are required for the maintenance of appropriate segment-specific expression of the *HOX* genes. These DNA elements—called *Polycomb* Response Elements or PREs—maintain the segment-specific expression of *HOX* genes beyond the embryonic initiation phase. PREs attract proteins of the PRC1 when integrated at ectopic sites in the polytene chromosomes, suggesting that they define sequence specificity for the recognition and anchoring of PRC1s to target genes. However, the issue of PcG targeting appears to be a complex one. The size of functionally characterized PREs ranges from a few hundred to several thousand base pairs, containing consensus binding sites for many different DNA-binding proteins, and usually two or more PREs are found at a given target locus. So far, all characterized PREs come from *Drosophila*, and no PREs have yet been defined in mammals or plants. Despite the complexity of PREs, four

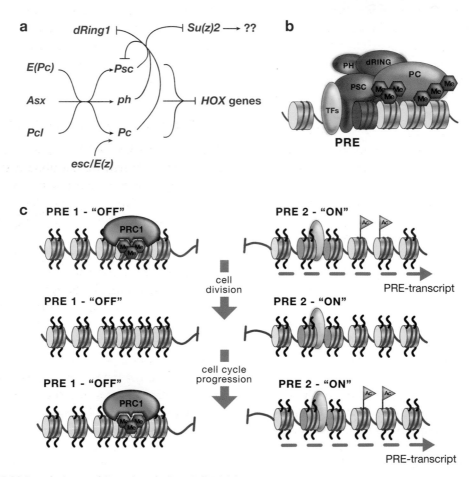

Figure 7. PRC1 Regulation and Function during Cell Division

(*a*) Cross-regulatory interactions among the PcG genes, as suggested from genetic evidence. *E(Pc)*, *Pcl*, and *Asx* are positive regulators of the core PRC1 members acting upstream. PRC2 members *Esc* and *E(z)* act as positive regulators of *Pc* transcription. A negative feedback by core PRC1 members on *Psc* and *dRing1*, as well as on *Su(z)2*, is observed. The fine-tuning of gene product level is probably required for well-balanced processes based on chemical equilibrium. (*b*) Sequence-specific transcription factors (TF) tether components of PRC1 to a PRE. A stable silencing complex requires anchoring of PRC1 via the chromodomain of PC to neighboring methylated histone tails. (*c*) Possible model for how differential gene expression states can be inherited. The process of intergenic transcription places positive epigenetic marks (e.g., acetylated histone tails, histone variants) at PREs that control active genes (PRE 2). All other PREs are silenced by default (PRE 1). During DNA replication and mitosis, only the positive epigenetic signal needs to be transmitted to the daughter cells, ensuring that in the next interphase intergenic transcription is restarted at PRE 2 before default silencing is reestablished at all other PREs. (*a*, Adapted from Ali and Bender 2004.)

consensus sequence motifs could be identified and were shown to play a role in *Drosophila* PRE function. One of these motifs (GCCAT) is bound by both the Pleiohomeotic (PHO) and Pleiohomeotic-like (PHOL) proteins, which have partially redundant functions. PHO and PHOL function in PcG targeting, as they are found in PcG complexes isolated from early embryonic extracts, coimmunprecipitate with members of both PRC1 and PRC2, and bind PREs in vitro (Fig. 5a) (Poux et al. 2001). Recently, a role in PcG recruitment was also demonstrated for DSP1, which binds the GAAA motif found in many PREs (Dejardin et al. 2005). Finally, the trxG proteins Zeste and GAF (encoded by the *Trithorax-like* gene) may help to recruit PcG proteins to their targets.

A newly developed algorithm, based on the finding that clustered pairs of GAF, Zeste, and PHO/PHOL sites characterize a PRE, predicts known PREs with high probability and thus can identify new potential PcG target genes in the *Drosophila* genome (Fig. 6b) (Ringrose et al. 2003). The family of PRE-controlled genes ranges from the well-known developmentally important transcription factor genes required for pattern formation to genes encoding factors involved in cell cycle regulation and senescence.

PRC1, once bound, interacts with neighboring histones to generate stable silencing complexes at PREs (Fig. 7b). The H3K27me3 marks provided by the PRC2 act as additional binding sites for the chromodomain of PC (Fig. 7c). In their absence, as shown by competition with a soluble methylated histone tail peptide, the PRC1s dissociate from their target genes (Czermin et al. 2002; Ringrose et al. 2004; Wang et al. 2004). The discovery of the HKMT activity of PRC2 and the associated histone marks typical of silent chromatin has suggested a new mechanism for the establishment of PcG repression. Following PRC2-catalyzed modification of H3K27me3, PRC1 binds through the chromodomain of the PC protein to stabilize silencing. This is corroborated by the findings that (1) H3K27me3 marks and PC colocalize on polytene chromosomes and (2) PC binding is lost in *E(z)* mutants, which lack HKMT activity that modifies H3 with H3K27me3 marks at PREs, serving to recruit PC to its targets (Fig. 5). Although such a model is certainly attractive, the situation at PREs seems more complex because PRC2s and PRC1s do not act sequentially, but rather are present together on PREs in early embryogenesis (Fig. 5b, c). Thus, it seems likely that H3K27 methylation is a downstream event after PcG recruitment, but plays a crucial role in establishing the silenced state.

The model described above shows parallels to heterochromatin formation, where the Heterochromatin Protein 1 (HP1) is recruited via its chromodomain to H3K9me marks generated by SU(VAR)3-9 (see Chapter 5). Thus, a productive silencing complex is targeted by transcription factors to defined DNA sequence elements but requires, in addition, an appropriately modified histone layer in the vicinity to generate a higher-order repressive chromatin structure (Fig. 5).

During evolution, PREs have retained remarkably little sequence conservation. Even within closely related *Drosophila* species, the number, position, and composition of PREs vary substantially (L. Ringrose and R. Paro, unpubl.). This suggests that the sequence requirements as well as the position of the PREs are flexible and may be adapted to species-specific requirements. Nevertheless, the components of PRC1 are highly conserved, and they presumably utilize the same basic molecular mechanism(s) to induce higher-order chromatin changes at silenced target genes.

3.3 Establishment of Repressive Functions by PRC1

The way in which PRE-bound PRC1s interact with the promoter to prevent transcription is still unknown. The anchoring of paused RNA polymerase complexes at promoters, preventing initiation, has been attributed to PRE–PRC1 interactions described for reporter constructs (Dellino et al. 2004). Additionally, PRC1 was shown to counteract remodeling of nucleosomes in vitro and to induce a compact chromatin structure. Thus, PRC1 potentially blocks the accessibility to DNA of transcription factors and other complexes required for transcription (Francis et al. 2004). Using the algorithm described above, PRE-like sequences are predicted to exist at almost all promoters of PcG-controlled *Drosophila* target genes. This suggests that PRC1 occupation at both promoter and regulatory sites might foster interactions between PREs and promoters, and establish stably repressed chromatin structures unfavorable for transcription (Ringrose et al. 2003).

The stability of silencing complexes, as demonstrated by anchoring via methylated histone tails, appears to be a hallmark of the long-term repressive function of the PcG proteins. However, when analyzed in vivo at the cellular level, a remarkably dynamic behavior is observed. PcG proteins cluster in PcG bodies, which vary in size and composition between cells, suggesting an interaction of silencing complexes in the nucleus in a developmentally regulated manner. Furthermore, dynamic in vivo analyses of GFP-marked PC and PH proteins uncovered a very high exchange rate of unbound proteins with their complexes at silenced targets (Ficz et al. 2005). These results suggest that long-term repression is primarily based on a chemical equilibrium between bound and unbound proteins rather than on high-affinity protection of DNA-binding sites.

3.4 Preventing Heritable Repression by Anti-silencing

The binding of PRC1s to PREs appears to be induced by default, as many of the anchoring PcG components and DNA-binding proteins are expressed in all cells, and PREs globally silence reporter genes in transgenic constructs. The counteracting proteins of the trxG do not, in fact, function as activators, but rather as anti-repressors (Klymenko and Müller 2004; see Chapter 12). Thus, to maintain active transcription of a PRE-controlled gene, the silencing at that PRE has to be prevented in a tissue- and stage-specific manner. In *Drosophila*, for example, the activation of *HOX* genes is controlled by the early cascade of transcription factors encoded by the segmentation genes. Interestingly, these factors induce transcription not only of the *HOX* genes, but also of intergenic, noncoding RNAs that are transcribed through the associated PREs often found upstream or downstream (Fig. 5). It was demonstrated that transcription through PREs is required to prevent silencing and to maintain the active state of a reporter gene using transgenic constructs (Schmitt et al. 2005). The process of transcription most probably remodels PRE chromatin to generate an active state characterized, for instance, by a lack of repressive histone methylation and the presence of histone acetylation. Thus, even though the DNA-binding proteins attract PRC1 to this particular activated PRE, the histone environment does not allow anchoring of PC via H3K27me3, and no stable silencing will be established. Since silencing is induced by default in the PcG system, epigenetic inheritance of a differential gene expression pattern only requires the transmission of the active PRE state during DNA replication and mitosis (Fig. 7c). How this is achieved at the molecular level, and which epigenetic mark(s) is responsible for maintaining an active PRE state, are still open questions. Interestingly, it was recently shown that at a *Drosophila* PRE of the homeotic *Ubx* gene, noncoding RNAs produced at the PREs stay associated with chromatin and recruit the trxG regulator Absent Small or Homeotic discs 1 (ASH1). Destruction of these RNAs by RNAi attenuates ASH1 recruitment to the PRE, suggesting that this interaction plays an important role in the epigenetic activation of the homeotic genes, by overriding default PcG-induced silencing (Sanchez-Elsner et al. 2006).

4 PcG Repression in Mammalian Development

4.1 From Gene to Chromosome Repression

Mutations in members of the murine PRC1 exhibit homeotic transformations of the axial skeleton. This can cause the appearance of additional vertebrae as a consequence of a derepression of *HOX* genes (Fig. 2e,f) (Core et al. 1997). In addition, the mutant mice display severe combined immunodeficiencies, caused by a lack of proliferative responses of hematopoietic cells (Raaphorst 2005). The role of PcG proteins has been particularly well studied in blood cells, in light of the fact that most blood-cell lineages are characterized by their well-described cell-type-specific transcription programs. However, lineage commitment and restriction somehow need to be faithfully maintained through cell division. It turns out that in PcG knockout mice, B- and T-cell precursor populations are produced normally, indicating that PcG control is not involved in establishing lineage-specific gene expression patterns. PcG proteins, however, contribute to the irreversibility of the lineage choice, rather than being involved in the decision to follow a particular developmental pathway.

Besides the control of the *HOX* genes, whose expression patterns characterize different blood-cell lineages, PcG proteins play a major role in controlling proliferation. The *bmi1* gene, an ortholog of *Drosophila Psc*, was initially identified as an oncogene that, in collaboration with *myc*, induces murine lymphomagenesis (van Lohuizen et al. 1991). The Bmi1 protein controls the cell cycle regulators p16^{INK4a} and p19^{ARF} (Jacobs et al. 1999). Both Bmi1 and the related protein Mel-18 are negative regulators of the INK4c-ARF locus required for normal lymphoid proliferation control. Misregulation of this important cell cycle checkpoint affects apoptosis and senescence in mice (Akasaka et al. 2001).

Mammalian PcG proteins are also associated with the classic epigenetic phenomenon of X-chromosome inactivation (see Chapter 17). The inactivation of one X chromosome in female XX cells is accompanied by a series of chromatin modifications that involve PcG proteins (Heard 2004). In particular, components of PRC2, like the ESC homolog, Eed (Embryonic ectoderm expression), or the E(Z) homolog, Enx1 (Table 1), play a major role in the establishment of histone marks associated with transcriptional silencing. Transient association of this PRC2 with the X chromosome, coated by Xist RNA, is accompanied by H3K27 methylation. In contrast, *eed* mutant mouse embryos show no recruitment of the Enx1 HKMT, nor can any H3K27 methylation be observed. However, the absence of these PRC2 components does not lead to a complete derepression of the entire inactive X chromosome; rather, the sporadic reexpression of X-linked genes and an increase in epigenetic marks associated with an active state (H3K9ac and H3K4me3) are observed in some cells. This is likely because other partially redundant epigenetic mechanisms are in place to ensure the maintenance of one inactive X chromosome.

Recruitment of PRC2 to the inactive X chromosome appears to be dependent on Xist RNA. Because association of PRC2 to the inactive X is only transient, it appears that the complex is only required to set epigenetic marks (i.e., H3K27me3) for the maintenance of silencing. Currently, it is not known whether PRC1 directly recognizes these marks and is required for the permanent silencing of the inactive X chromosome, but PRC1 components are found to be associated with the inactive X chromosome. However, DNA methylation is known to accompany the maintenance phase and is required for permanent X inactivation.

PRC2 is specifically involved in the regulation of monoallelic expression of the X chromosome both in the embryo, where X-chromosome inactivation is random, and in extraembryonic tissues, where the paternally inherited X chromosome is always inactivated (imprinted X-chromosome inactivation). In addition, it was recently found that PRC2 is involved in the regulation of some autosomal imprinted genes. For instance, an analysis of 14 imprinted loci from six unlinked imprinting clusters showed that four of these were biallelically expressed in *eed* mutant mice (Mager et al. 2003; for more detail, see Chapter 19). Interestingly, all loci that lost imprinted expression were normally repressed when paternally inherited, whereas none of the maternally repressed loci was affected. Because it was recently shown that Ezh2 directly interacts with the mammalian DNA methyltransferases and is required for their activity (Viré et al. 2006), it is possible that PRC2 plays a role in the regulation of these imprinted genes via DNA methylation (see Chapter 18).

An involvement of PRC2 in the regulation of imprinted gene expression has also been reported in *Arabidospis*, where the *PHE1* locus is expressed at much higher levels from the paternal allele (Köhler et al. 2005). In mutants for the *E(z)* homolog *MEA*, the maternal *PHE1* allele is specifically derepressed. Similarly, *MEA* also regulates its own imprinted expression: Early in reproductive development, the maternal *MEA* allele is strongly derepressed in *mea* mutants. This effect, however, is independent of the other components of the FIS complex (Baroux et al. 2006). In contrast, later in development, the FIS complex ensures the stable repression of the paternal *MEA* allele (Baroux et al. 2006; Gehring et al. 2006; Jullien et al. 2006). In this latter case, the FIS complex is involved in the silencing of paternally repressed imprinted genes similar to the situation in mammals. In addition, *MEA* also has a role in keeping expression of the maternal *PHE1* and *MEA* alleles at low levels as described above (Fig. 4).

Because PRC2 components are present in plants, invertebrates, and mammals, PRC2 represents an ancient molecular module suitable for gene repression that was already present in the unicellular ancestor of plants and animals, prior to the evolution of multicellularity. Thus, these examples suggest that PRC2 was recruited independently for the regulation of imprinted gene expression in plants and mammals, the two lineages where genomic imprinting evolved (Grossniklaus 2005).

4.2 Consequences of Aberrant Transcriptional Activation

The finding that *Bmi1* misregulation causes malignant lymphomas in mice raises the question of whether human BMI1 (a PRC1 component) itself contributes to the development of cancer in a similar fashion. There is accumulating evidence that altered PcG gene expression is widespread in human malignant lymphomas (Raaphorst 2005). For instance, the level of BMI1 overexpression in B-cell lymphomas correlates with the degree of malignancy, suggesting that PRC1 components do play a role in the development of human cancer. However, the target genes of BMI1 in human cells appear to be different from those of mouse lymphocytes, as no obvious down-regulation of p16^{INK4a} could be correlated to the overexpression of the oncogenes.

PcG gene overexpression is not only observed in hematological malignancies, but is also found in solid tumors, including medulloblastomas, and tumors originating from liver, colon, breast, lung, penis, and prostate (Fig. 8). The high expression of a PRC2 marker, Ezh2, is often found in early stages of highly proliferative lung carcinomas. This suggests that the well-known cascade of PRC2 initiation and PRC1 maintenance (Fig. 5) might also accompany the development of a tumor cell lineage.

Interestingly, PRC2 components also play a crucial role in the control of cell proliferation in *Arabidopsis*. Although aberrant growth does not lead to cancer and death in plants, a strict control of cell proliferation is

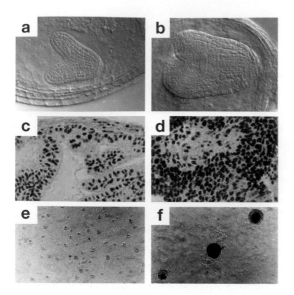

Figure 8. PRC2 Regulates Cell Proliferation in Mammals and Plants

(*a, b*) Plant embryos derived from wild-type and *mea* mutant egg cells. *MEA* encodes a protein of the FIS complex and regulates cell proliferation. The giant *mea* embryo is much larger than the corresponding wild-type embryo at the same stage of development (late heart stage). Mutant embryos develop more slowly and have approximately twice the number of cell layers. (*c, d*) Normal and cancerous prostate epithelium. In the cancerous epithelium, Ezh2 expression is highly increased (labeled with an anti-Ezh2 antibody). Thus, both loss of E(Z) function in plants and overexpression of E(Z) function in humans can lead to defects in cell proliferation. (*e, f*) Control and RING1 overexpressing rat 1a fibroblast cells. Overexpression of RING1 leads to anchorage-independent growth in soft agar, typical of neoplastically transformed cells. (*a,b*, Courtesy of J.-P. Vielle-Calzada and U. Grossniklaus; *c,d*, reprinted, with permission, from Kuzmichev et al. 2005 [©National Academy of Sciences]; *e,f*, reprinted, with permission, from Satijn and Otte 1999 [©American Society for Microbiology].)

essential for normal development. In mutants of the *fis* class, the two fertilization products of flowering plants, the embryo and endosperm, overproliferate, and the resulting seeds abort (Fig. 8) (Grossniklaus et al. 2001; Hsieh et al. 2003; Guitton and Berger 2005). Effects on cell proliferation are also observed in double mutants of *clf* and *swn*, two of the plant *E(z)* homologs. Such plants undergo normal seed development after germination but produce a mass of proliferating, undifferentiated tissue (callus) rather than leaves (Chanvivattana et al. 2004).

Although it is not known how PRC2 controls cell proliferation in plants, it is likely to involve interactions with RBR1, the plant homolog of the Retinoblastoma (Rb) protein (Ebel et al. 2004; Mosquna et al. 2004). Mutants in the *FIS* class of genes not only show proliferation defects during seed development after fertilization, but are also required to prevent proliferation of the endosperm in the absence of fertilization. This latter aspect of the pheno-

type is shared with *rbr1* mutants, providing a link to the Rb pathway. Remarkably, a connection between the Rb pathway and PRC2 has also been reported in mammals (Bracken et al. 2003), illustrating conserved regulatory networks between plants and mammals.

4.3 Maintaining Stem Cell Fate

Stem cells play an ever-increasing role in medicine. Their potential to provide progenitors for the healing of damaged tissue places them into a well-treasured toolbox of regenerative medicine. Not surprisingly, it is in the very well characterized blood-cell lineage where we know most about the identity and location of stem cells.

Hematopoietic stem cells (HSCs) maintain the pool of blood cells by self-renewing as well as by producing daughter cells that differentiate into the lymphoid, myeloid, and erythroid lineages. The stem cell niche in the adult bone marrow provides the cells with specific external signals to maintain their fate. On the other hand, cell-intrinsic cues for the maintenance of the "stem cellness" state seem to rely on the PcG system.

Mouse mutants affecting PRC1 genes (e.g., *bmi1/mel-18, mph1/rae28,* and *m33;* see Table 1) suffer from various defects in the hematopoietic system, such as hyperplasia (i.e., increased cell proliferation) in spleen and thymus, reduction in B and T cells, and an impaired proliferative response of lymphoid precursors to cytokines. The requirements for Bmi1 and Mel-18 in stem cell self-renewal during different stages of development suggest a changing pool of target genes between embryonic and adult stem cells.

The PcG system is also required for neural stem cells (NSCs) as indicated by the neuronal defects observed in *bmi1* mouse mutants (Bruggeman et al. 2005; Zencak et al. 2005). In particular, the mice are depleted of cerebral NSCs postnatally, indicating an in vivo requirement of Bmi1 in NSC renewal. As was found for the hematopoietic system, it appears that embryonal NSC maintenance is under a different PcG network control than adult NSC self-renewal.

External signals like the sonic-Hedgehog signaling cascade modulate the Bmi1 response in NSCs and ensure a proliferative/self-renewal capacity (Leung et al. 2004). The identification of these external cues controlling PcG repression came through the analysis of the development of cerebellar granule neuron progenitors (CGNPs). A postnatal wave of proliferation is induced by the signaling factor Sonic hedgehog (Shh), secreted by the Purkinje cells. The Shh signal branches to control N-Myc and Bmi1 levels (Fig. 9). Thus, Bmi1-deficient CGNPs have a

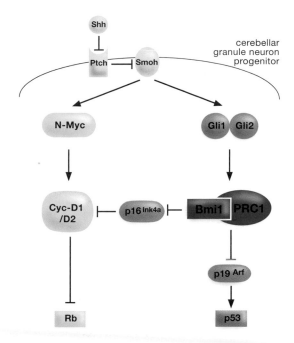

cerebellar
granule neuron
progenitor

proliferation / self-renewal pathway in stem cells

Figure 9. Sonic Hedgehog Signaling Maintains Proliferation/Self-renewal of Cerebellar Progenitor Cells

The Shh signaling cascade regulates both the Rb pathway and the p53 pathway via Bmi1 control of the p16/p19 proliferation checkpoint. Inhibition of Smoothened (Smoh) by the Shh receptor Patched (Ptch) results in downstream signaling in the nucleus. One part of the signal induces N-Myc, Cyclin D1, and D2, whereas the other part activates Bmi1 via the Gli effectors. (Adapted from Valk-Lingbeek et al. 2004.)

defective proliferative response upon Shh stimulation. The Shh signal is able to control proliferation of these stem cells ultimately by modulating both the downstream Rb pathway (via N-myc and Bmi1/p16^{INK4a}) and the p53 pathway (via Bmi1/p19^{ARF}). This mechanism explains why hyperactivation of Shh signaling leads to the development of medulloblastomas (Leung et al. 2004). HSCs are regulated by a similar Indian hedgehog-controlled pathway. In NSCs, expression of the *Hoxd8*, *Hoxd9*, and *Hoxc9* loci is under the control of Bmi1. The appropriate *HOX* expression profile confers the necessary stem cell fate.

Indeed, because stem cells represent a defined and committed cellular fate, it is not surprising that the PcG system maintains this particular fate in a mitotically heritable fashion. In the future, it will be interesting to identify the pool of targets of the PcG system in the different stem cell populations, and to learn how to influence the maintenance system to allow a controlled reprogramming of stem cell fates. At the moment, it is not clear whether the PcG plays a role in stem cell maintenance in plants.

Conceivably, however, the reprogramming of plant cells, which are totipotent and have the potential to form a complete new organism under appropriate conditions, could involve PcG regulation. Indeed, plants lacking the *E(z)* homologs *CLF* and *SWN* produce a mass of undifferentiated cells after germination, suggesting that PcG genes are required to maintain a differentiated state (Chanvivattana et al. 2004).

5 Conclusion and Outlook

It has been remarkable to follow the development of our understanding of PcG epigenetic regulation from the initial genetic identification of a *Drosophila* mutant possessing additional sex combs on the second and third leg. This eventually led to the discovery of a new class of regulators found to be required for fundamental epigenetic processes such as vernalization in plants and silencing of the mammalian X chromosome. Control of genetic information is highly influenced by chromatin structure and composition of histones in their various modified forms. The proteins of the PcG are directly involved in generating epigenetic marks, for instance, H3K27me3, as a consequence of developmental decisions. The same group "reads" (i.e., shows high affinity to), through the action of the PRC1 proteins, these epigenetic marks and translates them into a stable, transcriptionally repressed state. In the model organism *Drosophila*, we have a relatively clear picture of how PcG complexes are anchored at PREs, for a defined group of target genes that are subject to long-term repression. However, to date, no PREs have been identified in other organisms. Although the basic function of PcG proteins remains the same, it is unclear which part of the plant and vertebrate genomes is subjected to their repression and how they are targeted to their site of action. Additionally, we need to get a better understanding of how an apparently dynamic group of proteins can impose a stable state of transcriptional repression through a chemical equilibrium.

The other major question of the PcG research focuses on the heritability of the repressed state, the very essence of epigenetics. What is the identity of the molecular marks required to transmit a state of gene expression through DNA replication and mitosis? We know that the cooperation of trxG and PcG proteins maintains active or silent states of gene expression. Do both states need a corresponding epigenetic mark that is transmitted to daughter cells, or is only one sufficient, while the other represents the default state? The mechanism by which PcG proteins impose silencing on transcription during the interphase of the cell cycle has become increasingly clear. In the future, the focus of research will be on how the information regarding a state of gene expression endures the DNA replication process and is faithfully transmitted to the daughter cells following mitosis.

References

Akasaka T., van Lohuizen M., van der Lugt N., Mizutani-Koseki Y., Kanno M., Taniguchi M., Vidal M., Alkema M., Berns A., and Koseki H. 2001. Mice doubly deficient for the *Polycomb* Group genes *Mel18* and *Bmi1* reveal synergy and requirement for maintenance but not initiation of *HOX* gene expression. *Development* **128:** 1587–1597.

Ali J.Y. and Bender W. 2004. Cross-regulation among the *Polycomb* group genes in *Drosophila melanogaster*. *Mol. Cell. Biol.* **24:** 7737–7747.

Ausin I., Alonso-Blanco C., Jarillo J.A., Ruiz-Garcia L., and Martinez-Zapater J.M. 2004. Regulation of flowering time by FVE, a retinoblastoma-associated protein. *Nat. Genet.* **36:** 162–166.

Bannister A.J., Zegerman P., Partridge J.F., Miska E.A., Thomas J.O., Allshire R.C., and Kouzarides T. 2001. Selective recognition of methylated lysine 9 on histone H3 by the HP1 chromodomain. *Nature* **410:** 120–124.

Baroux C., Gagliardini V., Page D., and Grossniklaus U. 2006. Dynamic regulatory interactions of *Polycomb* group genes: *MEDEA* autoregulation is required for imprinted gene expression in *Arabidopsis*. *Genes Dev.* **20:** 1081–1086.

Bastow R., Mylne J.S., Lister C., Lippman Z., Martienssen R.A., and Dean C. 2004. Vernalization requires epigenetic silencing of *FLC* by histone methylation. *Nature* **427:** 164–167.

Birve A., Sengupta A.K., Beuchle D., Larsson J., Kennison J.A., Rasmuson-Lestander A., and Müller J. 2001. *Su(z)12*, a novel *Drosophila Polycomb* group gene that is conserved in vertebrates and plants. *Development* **128:** 3371–3379.

Bracken A.P., Pasini D., Capra M., Prosperini E., Colli E., and Helin K. 2003. *EZH2* is downstream of the pRB-E2F pathway, essential for proliferation and amplified in cancer. *EMBO J.* **22:** 5323–5335.

Breiling A., Bonte E., Ferrari S., Becker P.B., and Paro R. 1999. The *Drosophila Polycomb* protein interacts with nucleosomal core particles in vitro via its repression domain. *Mol. Cell. Biol.* **19:** 8451–8460.

Bruggeman S.W.M., Valk-Lingbeek M.E., van der Stoop P.P.M., Jacobs J.J.L., Kieboom K., Tanger E., Hulsman D., Leung C., Arsenijevic Y., Marino S., and van Lohuizen M. 2005. *Ink4a* and *Arf* differentially affect cell proliferation and neural stem cell self-renewal in *Bmi1*-deficient mice. *Genes Dev.* **19:** 1438–1443.

Cao R., Tsukada Y., and Zhang Y. 2005. Role of Bmi-1 and Ring1A in H2A ubiquitylation and *HOX* gene silencing. *Mol. Cell* **20:** 845–854.

Cao R., Wang L.J., Wang H.B., Xia L., Erdjument-Bromage H., Tempst P., Jones R.S., and Zhang Y. 2002. Role of histone H3 lysine 27 methylation in *Polycomb*-group silencing. *Science* **298:** 1039–1043.

Carrington E.A. and Jones R.S. 1996. The *Drosophila Enhancer of zeste* gene encodes a chromosomal protein: Examination of wild-type and mutant protein distribution. *Development* **122:** 4073–4083.

Chanvivattana Y., Bishopp A., Schubert D., Stock C., Moon Y.H., Sung Z.R., and Goodrich J. 2004. Interaction of *Polycomb*-group pro-

teins controlling flowering in *Arabidopsis*. *Development* **131:** 5263–5276.

Core N., Charroux B., McCormick A., Vola C., Fasano L., Scott M.P., and Kerridge S. 1997. Transcriptional regulation of the *Drosophila* homeotic gene *teashirt* by the homeodomain protein Fushi tarazu. *Mech. Dev.* **68:** 157–172.

Czermin B., Melfi R., McCabe D., Seitz V., Imhof A., and Pirrotta V. 2002. *Drosophila* enhancer of Zeste/ESC complexes have a histone H3 methyltransferase activity that marks chromosomal *Polycomb* sites. *Cell* **111:** 185–196.

Dejardin J., Rappailles A., Cuvier O., Grimaud C., Decoville M., Locker D., and Cavalli G. 2005. Recruitment of *Drosophila Polycomb* group proteins to chromatin by DSP1. *Nature* **434:** 533–538.

Dellino G.I., Schwartz Y.B., Farkas G., McCabe D., Elgin S.C., and Pirrotta V. 2004. *Polycomb* silencing blocks transcription initiation. *Mol. Cell* **13:** 887–893.

del Mar Lorente D., Marcos-Gutierrez C., Perez C., Schoorlemmer J., Ramirez A., Magin T., and Vidal M. 2000. Loss- and gain-of-function mutations show a *Polycomb* group function for Ring1A in mice. *Development* **127:** 5093–5100.

de Napoles M., Mermoud J.E., Wakao R., Tang Y.A., Endoh M., Appanah R., Nesterova T.B., Silva J., Otte A.P., Vidal M., et al. 2004. *Polycomb* group proteins Ring1A/B link ubiquitylation of histone H2A to heritable gene silencing and X inactivation. *Dev. Cell* **7:** 663–676.

Ebel C., Mariconti L., and Gruissem W. 2004. Plant retinoblastoma homologues control nuclear proliferation in the female gametophyte. *Nature* **429:** 776–780.

Fang J., Chen T.P., Chadwick B., Li E., and Zhang Y. 2004. Ring1b-mediated H2A ubiquitination associates with inactive X chromosomes and is involved in initiation of X inactivation. *J. Biol. Chem.* **279:** 52812–52815.

Ficz G., Heintzmann R., and Arndt Jovin D.J. 2005. *Polycomb* group protein complexes exchange rapidly in living *Drosophila*. *Development* **132:** 3963–3976.

Fischle W., Wang Y., Jacobs S.A., Kim Y., Allis C.D., and Khorasanizadeh S. 2003. Molecular basis for the discrimination of repressive methyl-lysine marks in histone H3 by Polycomb and HP1 chromodomains. *Genes Dev.* **17:** 1870–1881.

Fong Y., Bender L., Wang W., and Strome S. 2002. Regulation of the different chromatin states of autosomes and X chromosomes in the germ line of *C. elegans*. *Science* **296:** 2235–2238.

Francis N.J., Kingston R.E., and Woodcock C.L. 2004. Chromatin compaction by a *Polycomb* group protein complex. *Science* **306:** 1574–1577.

Francis N.J., Saurin A.J., Shao Z., and Kingston R.E. 2001. Reconstitution of a functional core *Polycomb* repressive complex. *Mol. Cell* **8:** 545–556.

Gehring M., Huh J.H., Hsieh T.F., Penterman J., Choi Y., Harada J.J., Goldberg R.B., and Fischer R.L. 2006. DEMETER DNA glycosylase establishes *MEDEA Polycomb* gene self-imprinting by allele-specific demethylation. *Cell* **124:** 495–506.

Gendall A.R., Levy Y.Y., Wilson A., and Dean C. 2001. The *VERNALIZATION2* gene mediates the epigenetic regulation of vernalization in *Arabidopsis*. *Cell* **107:** 525–535.

Goodrich J., Puangsomlee P., Martin M., Long D., Meyerowitz E.M., and Coupland G. 1997. A *Polycomb*-group gene regulates homeotic gene expression in *Arabidopsis*. *Nature* **386:** 44–51.

Grossniklaus U. 2005. Genomic imprinting in plants: A predominantly maternal affair. In *Annual plant reviews: Plant epigenetics* (ed. P. Meyer), pp. 174–200. Blackwell, Sheffield, United Kingdom.

Grossniklaus U., Spillane C., Page D.R., and Köhler C. 2001. Genomic imprinting and seed development: Endosperm formation with and without sex. *Curr. Opin. Plant Biol.* **4:** 21–27.

Grossniklaus U., Vielle-Calzada J.P., Hoeppner M.A., and Gagliano W.B. 1998. Maternal control of embryogenesis by *MEDEA*, a *Polycomb* group gene in *Arabidopsis*. *Science* **280:** 446–450.

Guitton A.E. and Berger F. 2005. Control of reproduction by *Polycomb* Group complexes in animals and plants. *Int. J. Dev. Biol.* **49:** 707–716.

Hackett W.P., Cordero R.E., and Sinivasan C. 1987. Apical meristem characteristics and activity in relation to juvenility in *Hedera*. In *Manipulation of flowering* (ed. J.G. Atherton), pp. 93–99. Butterworth, London.

Hadorn E. 1968. Transdetermination in cells. *Sci. Am.* **219:** 110.

Heard E. 2004. Recent advances in X-chromosome inactivation. *Curr. Opin. Cell Biol.* **16:** 247–255.

Hennig L., Bouveret R., and Gruissem W. 2005. MSI1-like proteins: An escort service for chromatin assembly and remodeling complexes. *Trends Cell Biol.* **15:** 295–302.

Hennig L., Taranto P., Walser M., Schonrock N., and Gruissem W. 2003. *Arabidopsis* MSI1 is required for epigenetic maintenance of reproductive development. *Development* **130:** 2555–2565.

Hsieh T.F., Hakim O., Ohad N., and Fischer R.L. 2003. From flour to flower: How *Polycomb* group proteins influence multiple aspects of plant development. *Trends Plant Sci.* **8:** 439–445.

Jacobs J.J.L., Scheijen B., Voncken J.W., Kieboom K., Berns A., and van Lohuizen M. 1999. Bmi-1 collaborates with c-Myc in tumorigenesis by inhibiting c-Myc-induced apoptosis via INK4a/ARF. *Genes Dev.* **13:** 2678–2690.

Jullien P.E., Katz A., Oliva M., Ohad N., and Berger F. 2006. *Polycomb* group complexes self-regulate imprinting of the *Polycomb* group gene *MEDEA* in *Arabidopsis*. *Curr. Biol.* **16:** 486–492.

Kagey M.H., Melhuish T.A., and Wotton D. 2003. The *Polycomb* protein Pc2 is a SUMO E3. *Cell* **113:** 127–137.

Kennison J.A. 1995. The *Polycomb* and *trithorax* group proteins of *Drosophila*: Trans-regulators of homeotic gene function. *Annu. Rev. Genet.* **29:** 289–303.

Kim H.J., Hyun Y., Park J.Y., Park M.J., Park M.K., Kim M.D., Kim H.J., Lee M.H., Moon J., Lee I., and Kim J. 2004. A genetic link between cold responses and flowering time through *FVE* in *Arabidopsis thaliana*. *Nat. Genet.* **36:** 167–171.

Kinoshita T., Harada J.J., Goldberg R.B., and Fischer R.L. 2001. *Polycomb* repression of flowering during early plant development. *Proc. Natl. Acad. Sci.* **98:** 14156–14161.

Klebes A., Sustar A., Kechris K., Li H., Schubiger G., and Kornberg T.B. 2005. Regulation of cellular plasticity in *Drosophila* imaginal disc cells by the *Polycomb* group, *trithorax* group and *lama* genes. *Development* **132:** 3753–3765.

Klymenko T. and Muller J. 2004. The histone methyltransferases Trithorax and Ash1 prevent transcriptional silencing by *Polycomb* group proteins. *EMBO Rep.* **5:** 373–377.

Köhler C., Page D.R., Gagliardini V., and Grossniklaus U. 2005. The *Arabidopsis thaliana* MEDEA *Polycomb* group protein controls expression of *PHERES1* by parental imprinting. *Nat. Genet.* **37:** 28–30.

Köhler C., Hennig L., Bouveret R., Gheyselinck J., Grossniklaus U., and Gruissem W. 2003a. *Arabidopsis* MSI1 is a component of the MEA/FIE *Polycomb* group complex and required for seed development. *EMBO J.* **22:** 4804–4814.

Köhler C., Hennig L., Spillane C., Pien S., Gruissem W., and Grossniklaus U. 2003b. The *Polycomb* group protein MEDEA regulates seed development by controlling expression of the MADS-box gene *PHERES1*. *Genes Dev.* **17:** 1540–1553.

Kuzmichev A., Jenuwein T., Tempst P., and Reinberg D. 2004. Different EZH2-containing complexes target methylation of histone H1 or nucleosomal histone H3. *Mol. Cell* **14:** 183–193.

Kuzmichev A., Nishioka K., Erdjument-Bromage H., Tempst P., and Reinberg D. 2002. Histone methyltransferase activity associated with a human multiprotein complex containing the Enhancer of Zeste protein. *Genes Dev.* **16:** 2893–2905.

Kuzmichev A., Margueron R., Vaquero A., Preissner T.S., Scher M., Kirmizis A., Ouyang X., Brockdorff N., Abate Shen C., Farnham P., and Reinberg D. 2005. Composition and histone substrates of *Polycomb* repressive group complexes change during cellular differentiation. *Proc. Natl. Acad. Sci.* **102:** 1859–1864.

Lavigne M., Francis N.J., King I.F., and Kingston R.E. 2004. Propagation of silencing; recruitment and repression of naive chromatin in *trans* by *Polycomb* repressed chromatin. *Mol. Cell* **13:** 415–425.

Lee N., Maurange C., Ringrose L., and Paro R. 2005. Suppression of *Polycomb* group proteins by JNK signalling induces transdetermination in *Drosophila* imaginal discs. *Nature* **438:** 234–237.

Leung C., Lingbeek M., Shakhova O., Liu J., Tanger E., Saremaslani P., van Lohuizen M., and Marino S. 2004. Bmi1 is essential for cerebellar development and is overexpressed in human medulloblastomas. *Nature* **428:** 337–341.

Levine S.S., King I.F., and Kingston R.E. 2004. Division of labor in *Polycomb* group repression. *Trends Biochem. Sci.* **29:** 478–485.

Levine S.S., Weiss A., Erdjument Bromage H., Shao Z., Tempst P., and Kingston R.E. 2002. The core of the *Polycomb* Repressive Complex is compositionally and functionally conserved in flies and humans. *Mol. Cell. Biol.* **22:** 6070–6078.

Lewis E.B. 1978. A gene complex controlling segmentation in *Drosophila*. *Nature* **276:** 565–570.

Luo M., Bilodeau P., Koltunow A., Dennis E.S., Peacock W.J., and Chaudhury A.M. 1999. Genes controlling fertilization-independent seed development in *Arabidopsis thaliana*. *Proc. Natl. Acad. Sci.* **96:** 296–301.

Mager J., Montgomery N.D., de Villena F.P.M., and Magnuson T. 2003. Genome imprinting regulated by the mouse *Polycomb* group protein Eed. *Nat. Genet.* **33:** 502–507.

Marx J. 2005. Developmental biology—Combing over the *Polycomb* group proteins. *Science* **308:** 624–626.

Moon Y.H., Chen L., Pan R.L., Chang H.S., Zhu T., Maffeo D.M., and Sung Z.R. 2003. EMF genes maintain vegetative development by repressing the flower program in *Arabidopsis*. *Plant Cell* **15:** 681–693.

Mosquna A., Katz A., Shochat S., Grafi G., and Ohad N. 2004. Interaction of FIE, a *Polycomb* protein, with pRb: a possible mechanism regulating endosperm development. *Mol. Genet. Genomics* **271:** 651–657.

Müller J., Hart C.M., Francis N.J., Vargas M.L., Sengupta A., Wild B., Miller E.L., O'Connor M.B., Kingston R.E., and Simon J.A. 2002. Histone methyltransferase activity of a *Drosophila* polycomb group repressor complex. *Cell* **111:** 197–208.

Ohad N., Yadegari R., Margossian L., Hannon M., Michaeli D., Harada J.J., Goldberg R.B., and Fischer R.L. 1999. Mutations in *FIE*, a WD *Polycomb* group gene, allow endosperm development without fertilization. *Plant Cell* **11:** 407–415.

Otte A.P. and Kwaks T.H. 2003. Gene repression by *Polycomb* group protein complexes: A distinct complex for every occasion? *Curr. Opin. Genet. Dev.* **13:** 448–454.

Page D.R. and Grossniklaus U. 2002. The art and design of genetic screens: *Arabidopsis thaliana*. *Nat. Rev. Genet.* **3:** 124–136.

Paro R. and Hogness D.S. 1991. The *Polycomb* protein shares a homologous domain with a heterochromatin-associated protein of *Drosophila*. *Proc. Natl. Acad. Sci.* **88:** 263–267.

Poux S., McCabe D., and Pirrotta V. 2001. Recruitment of components of *Polycomb* group chromatin complexes in *Drosophila*. *Development* **128:** 75–85.

Raaphorst F.M. 2005. Deregulated expression of *Polycomb*-group oncogenes in human malignant lymphomas and epithelial tumors. *Hum. Mol. Genet.* **14:** R93–100.

Rea S., Eisenhaber F., O'Carroll N., Strahl B.D., Sun Z.W., Schmid M., Opravil S., Mechtler K., Ponting C.P., Allis C.D., and Jenuwein T. 2000. Regulation of chromatin structure by site-specific histone H3 methyltransferases. *Nature* **406:** 593–599.

Reyes J.C. and Grossniklaus U. 2003. Diverse functions of *Polycomb* group proteins during plant development. *Semin. Cell Dev. Biol.* **14:** 77–84.

Ringrose L., Ehret H., and Paro R. 2004. Distinct contributions of histone H3 lysine 9 and 27 methylation to locus-specific stability of *Polycomb* complexes. *Mol. Cell* **16:** 641–653.

Ringrose L., Rehmsmeier M., Dura J.M., and Paro R. 2003. Genome-wide prediction of Polycomb/Trithorax response elements in *Drosophila melanogaster*. *Dev. Cell* **5:** 759–771.

Sanchez-Elsner T., Gou D., Kremmer E., and Sauer F. 2006. Noncoding RNAs of trithorax response elements recruit *Drosophila* Ash1 to *Ultrabithorax*. *Science* **311:** 1118–1123.

Satijn D.P. and Otte A.P. 1999. RING1 interacts with multiple *Polycomb*-group proteins and displays tumorigenic activity. *Mol. Cell. Biol.* **19:** 57–68.

Saurin A.J., Shiels C., Williamson J., Satijn D.P.E., Otte A.P., Sheer D., and Freemont P.S. 1998. The human polycomb group complex associates with pericentromeric heterochromatin to form a novel nuclear domain. *J. Cell Biol.* **142:** 887–898.

Schmitt S., Prestel M., and Paro R. 2005. Intergenic transcription through a *Polycomb* group response element counteracts silencing. *Genes Dev.* **19:** 697–708.

Sung S. and Amasino R.M. 2004a. Vernalization and epigenetics: How plants remember winter. *Curr. Opin. Plant Biol.* **7:** 4–10.

———. 2004b. Vernalization in *Arabidopsis thaliana* is mediated by the PHD finger protein VIN3. *Nature* **427:** 159–164.

Tie F., Furuyama T., Prasad-Sinha J., Jane E., and Harte P.J. 2001. The *Drosophila Polycomb* group proteins ESC and E(Z) are present in a complex containing the histone-binding protein p55 and the histone deacetylase RPD3. *Development* **128:** 275–286.

Tschiersch B., Hofmann A., Krauss V., Dorn R., Korge G., and Reuter G. 1994. The protein encoded by the *Drosophila* position-effect variegation suppressor gene *Su(var)3-9* combines domains of antagonistic regulators of homeotic gene complexes. *EMBO J.* **13:** 3822–3831.

Valk-Lingbeek M.E., Bruggeman S.W.M., and van Lohuizen M. 2004. Stem cells and cancer: The *Polycomb* connection. *Cell* **118:** 409–418.

van der Lugt N.M., Domen J., Linders K., van Roon M., Robanus-Maandag E., te Riele H., van der Valk M., Deschamps J., Sofroniew M., van Lohuizen M., et al. 1994. Posterior transformation, neurological abnormalities, and severe hematopoietic defects in mice with a targeted deletion of the *bmi-1* proto-oncogene. *Genes Dev.* **8:** 757–769.

van Lohuizen M., Verbeek S., Scheijen B., Wientjens E., van der Gulden H., and Berns A. 1991. Identification of cooperating oncogenes in Eμ-myc transgenic mice by provirus tagging. *Cell* **65:** 737–752.

Viré E., Brenner C., Deplus R., Blanchon L., Fraga M., Didelot C., Morey L., van Eynde A., Bernhard D., Vanderwinden J.M., et al. 2006. The *Polycomb* group protein EZH2 directly controls DNA methylation. *Nature* **439:** 871–874.

Wang L., Brown J.L., Cao R., Zhang Y., Kassis J.A., and Jones R.S. 2004. Hierarchical recruitment of *Polycomb* group silencing complexes. *Mol. Cell* **14:** 637–646.

Yamamoto Y., Girard F., Bello B., Affolter M., and Gehring W.J. 1997. The *cramped* gene of *Drosophila* is a member of the *Polycomb*-group, and interacts with *mus209*, the gene encoding proliferating cell nuclear antigen. *Development* **124:** 3385–3394.

Yamamoto K., Sonoda M., Inokuchi J., Shirasawa S., and Sasazuki T. 2004. *Polycomb* group Suppressor of zeste 12 links Heterochromatin Protein 1α and Enhancer of zeste 2. *J. Biol. Chem.* **279:** 401–406.

Yoshida N., Yanai Y., Chen L.J., Kato Y., Hiratsuka J., Miwa T., Sung Z.R., and Takahashi S. 2001. EMBRYONIC FLOWER2, a novel *Polycomb* group protein homolog, mediates shoot development and flowering in *Arabidopsis*. *Plant Cell* **13:** 2471–2481.

Zencak D., Lingbeek M., Kostic C., Tekaya M., Tanger E., Hornfeld D., Jaquet M., Munier F.L., Schorderet D.F., van Lohuizen M., and Arsenijevic Y. 2005. Bmi1 loss produces an increase in astroglial cells and a decrease in neural stem cell population and proliferation. *J. Neurosci.* **25:** 5774–5783.

C H A P T E R   12

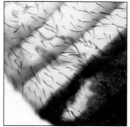

Transcriptional Regulation by Trithorax Group Proteins

Robert E. Kingston[1] and John W. Tamkun[2]

[1]Department of Molecular Biology, Massachusetts General Hospital, Boston, Massachusetts 02114
[2]Department of Molecular, Cell and Developmental Biology, University of California, Santa Cruz, California 95064

C O N T E N T S

1. Introduction, 233
   1.1 Identification of Genes Involved in the Maintenance of the Determined State, 233
   1.2 trxG Proteins in Other Organisms, 236
   1.3. trxG Proteins Play Diverse Roles in Eukaryotic Transcription, 237

2. Connections between trxG Proteins and Chromatin, 237
   2.1. trxG Proteins Involved in 238
   2.2. trxG Proteins That Covalently Modify Nucleosomal Histones, 242

3. Connections between trxG Proteins and the General Transcription Machinery, 243

4. Biochemical Functions of Other trxG Proteins, 244

5. Functional Interactions between trxG Proteins, 244

6. trxG Proteins: Activators or Anti-repressors?, 244

7. Conclusion and Outlook, 245

References, 246

GENERAL SUMMARY

All cells in an organism must be able to "remember" what type of cell they are meant to be. This process, referred to as "cellular memory" or "transcriptional memory," requires two basic classes of mechanisms. The first class, discussed in the previous chapter, functions to maintain an "OFF" state for genes that, if turned on, would specify an inappropriate cell type. The Polycomb-Group (PcG) proteins have as their primary function this repressive role in cellular memory. The second class of mechanisms are those that are required to maintain key genes in an "ON" state. Any cell type requires the expression of master regulatory proteins that direct the specific functions required for that cell type. The genes that encode these master regulatory proteins must be maintained in an "ON" state throughout the lifetime of an organism in order to maintain the proper cell types within that organism.

The striking multiple-winged fly in the left title figure illustrates the dramatic phenotypes that can result from the failure to maintain the "ON" state of a master regulatory gene. The proteins that are involved in maintaining the "ON" state are called trithorax-Group (trxG) proteins in honor of the trithorax gene, the founding member of this group of regulatory proteins. A large group of proteins with diverse functions make up the trxG. The roles these proteins play in the epigenetic mechanisms that maintain the "ON" state appear more complex at this juncture than the roles for PcG proteins in repression. The first complexity is that a very large number of proteins and mechanisms are needed to actively transcribe RNA from any gene. Thus, in contrast to repression, which might be accomplished by comparatively simple mechanisms that block access of all proteins, activation of a gene requires numerous steps, any of which might play a role in maintaining an "ON" state. Thus, there are numerous possible stages at which a trxG protein might work.

A second complexity in thinking about trxG proteins is that proteins which function in activation can also, in different contexts, function in repression. This might appear counterintuitive, but, depending on the precise architecture of a gene, the same protein carrying out its function might in one case help a gene become activated, and in another case help a different gene become repressed. At this time, it does not appear that trxG proteins are dedicated solely to the maintenance of gene expression, but that these proteins can also play multiple roles in the cell. These complexities evoke several interesting unanswered questions. Why are only some of the proteins needed to activate transcription also critical for maintenance of transcription? Do these proteins have functions that are uniquely suited to maintaining the active state? Or are some of these proteins needed for maintenance, solely due to an evolutionary accident that made them key regulators of a gene(s) particularly important to development?

As shown below, some of the trxG proteins are involved in regulating chromatin structure in opposition to the mechanisms used by the PcG proteins. trxG proteins can place covalent modifications on chromatin or can alter chromatin by changing the structure and position of the nucleosomes that are the building blocks of chromatin. Other trxG proteins function as part of the transcription machinery. Thus, these proteins are found in a wider variety of complexes than the PcG proteins and are likely to play more complicated roles in epigenetic mechanism.

1 Introduction

Numerous developmental decisions—including the determination of cell fates—are made in response to transient positional information in the early embryo. These decisions are dependent on changes in gene expression. This allows cells with identical genetic blueprints to acquire unique identities and to follow distinct pathways of differentiation. The changes in gene expression underlying the determination of cell fates are heritable; a cell's fate rarely changes once it is determined, even after numerous cell divisions and lengthy periods of developmental time. Understanding the molecular mechanisms underlying the maintenance of the determined state has long been a goal of developmental and molecular biologists.

Many of the regulatory proteins involved in the maintenance of heritable states of gene expression were identified in studies of *Drosophila* homeotic (Hox) genes. Hox genes encode homeodomain transcription factors that regulate the transcription of batteries of downstream target genes, which in turn specify the identities of body segments (Gellon and McGinnis 1998). In *Drosophila*, Hox genes are found in two gene complexes: the Antennapedia complex (ANT-C), which contains the Hox genes *labial* (*lab*), *Deformed* (*Dfd*), *Sex combs reduced* (*Scr*), and *Antennapedia* (*Antp*); and the bithorax complex (BX-C), which contains the Hox genes *Ultrabithorax* (*Ubx*), *abdominalA* (*abdA*), and *AbdominalB* (*AbdB*) (Duncan 1987; Kaufman et al. 1990). Each Hox gene specifies the identity of a particular segment, or group of segments, along the anterior–posterior axis of the developing fly. For example, *Antp* specifies the identity of the second thoracic segment, including the second pair of legs, whereas *Ubx* specifies the identity of the third thoracic segment, including the balancer organs located behind the wings. Thus, the transcription factors encoded by Hox genes function as master regulatory switches that direct the choice between alternative pathways of development.

The transcription of Hox genes must be regulated precisely, because dramatic alterations in cell fates can result from their inappropriate expression (Simon 1995; Simon and Tamkun 2002). For example, the derepression of *Antp* in head segments transforms antennae into legs, and the inactivation of *Ubx* in thoracic segments transforms balancer organs into wings. In *Drosophila*, the initial patterns of Hox transcription are established early in embryogenesis by transcription factors encoded by segmentation genes. The proteins encoded by segmentation genes—including the gap, pair-rule, and segment polarity genes—subdivide the early embryo into 14 identical segments.

These proteins also establish the initial patterns of Hox transcription, the first step toward the development of segments with distinct identities and morphology. Once established, the segmentally restricted patterns of Hox transcription must be maintained throughout subsequent embryonic, larval, and pupal stages in order to maintain the identities of the individual body segments. Because the majority of segmentation genes are transiently expressed during early development, this function is carried out by two other groups of regulatory proteins: the Polycomb group of repressors (PcG) and the trithorax group of transcriptional regulators (trxG) (Fig. 1). The regulation of Hox transcription therefore consists of at least two distinct phases: establishment (by segmentation genes) and maintenance (by PcG and trxG genes) (Fig. 2).

1.1 Identification of Genes Involved in the Maintenance of the Determined State

Because of their roles in the maintenance of cell fates, *Drosophila* PcG and trxG genes have been the subject of intense study for decades. As discussed in the previous chapter, the majority of PcG genes were identified by mutations that cause homeotic transformations due to the failure to maintain repressed states of Hox transcription. A classic example of a phenotype associated with PcG mutations is the transformation of second and third legs to first legs. This homeotic transformation results from the derepression of the ANT-C gene *Scr* and is manifested by the appearance of first leg bristles known as sex comb teeth on the second and third legs of the adult. This "Polycomb" or "extra sex combs" phenotype—together with other homeotic transformations resulting from the failure to maintain repression of Hox genes—led to the identification of more than a dozen PcG genes in *Drosophila*. The majority of PcG genes encode subunits of two complexes involved in transcriptional repression: Polycomb Repressive Complex (PRC) 1 and PRC2 (Levine et al. 2004). PRC1 and PRC2 are targeted to the vicinity of Hox (and other) promoters via *cis*-regulatory elements known as Polycomb-response elements (PREs). A large body of evidence suggests that PcG complexes repress transcription by modulating chromatin structure (Francis and Kingston 2001; Ringrose and Paro 2004).

Members of the trxG, including *trithorax* (*trx*); *absent, small or homeotic 1* (*ash1*), *absent, small or homeotic 2* (*ash2*), and *female-sterile homeotic* (*fsh*), were initially identified by mutations that mimic loss-of-function Hox mutations in *Drosophila* (Fig. 3) (Kennison 1995). For example, mutations in *trx*—the founding member of the trxG—

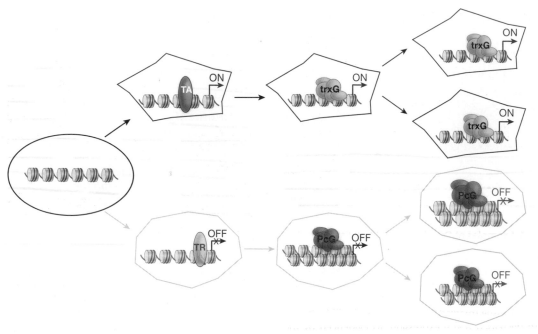

Figure 1. The Concept of Cellular Memory

Schematic illustration highlighting the role of trxG complexes in maintaining heritable states of active gene expression in contrast to heritable silencing by PcG complexes, as defined originally for the *Drosophila* Hox gene cluster.

cause the partial transformation of halteres to wings (due to decreased *Ubx* transcription); first legs to second legs (due to decreased *Scr* transcription); and posterior abdominal segments to more anterior identities (due to decreased *abdA* and *AbdB* transcription). Numerous other trxG members were identified in screens for extragenic suppressors of *Pc* (*Su(Pc)*) mutations (Kennison and Tamkun

1988). The rationale behind these genetic screens was that a reduction in the level of a protein that maintains an active state should compensate for a reduction in the level of a PcG repressor (Fig. 4). *brahma* (*brm*) and numerous other *Su(Pc)* loci were identified using this approach, bringing the total number of trxG members to more than 16 (Table 1). Many other proteins have been classified as trxG

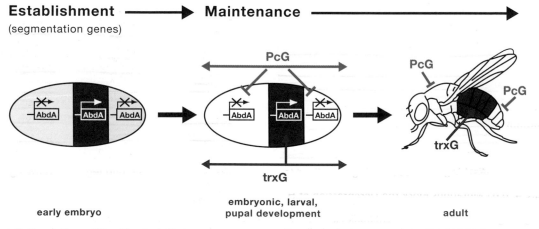

Figure 2. Regulation of Hox Transcription

The boundaries of *abd-A* transcription and other Hox genes are established by segmentation proteins. These include the products of gap and pair-rule genes, which subdivide the embryo into 14 identical segments. During subsequent development, the "OFF" or "ON" states of Hox transcription are maintained by the ubiquitously expressed members of the trxG of activators and the PcG of repressors via mechanisms that remain poorly understood.

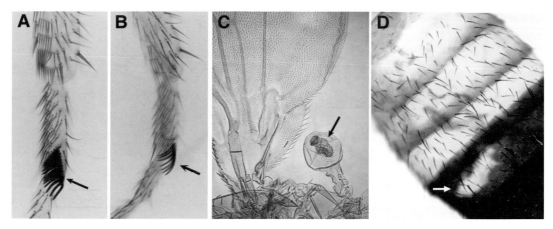

Figure 3. Examples of Developmental Cell Fate Transformations Associated with Mutations in *Drosophila* trxG Genes

(*A*) Wild-type first leg. The sex comb, unique to the first leg, is marked by an arrow. (*B*) A patch of *kis* mutant tissue (marked by an arrow) is partially transformed from the first leg to the second leg due to decreased *Scr* transcription, albeit incomplete, as evidenced by a reduction in the number of sex comb teeth. (*C*) A patch of *mor* mutant tissue (marked by an arrow) displays the partial transformation from balancer organ to wing, due to decreased Ubx expression. (*D*) A patch of *kis* mutant tissue (marked by an arrow) in the fifth abdominal segment is partially transformed to a more anterior identity due to decreased *Abd B* expression, as evidenced by the loss of the dark pigmentation characteristic of this segment. (*A,B,D*, Reprinted, with permission, from Daubresse et al. 1999.)

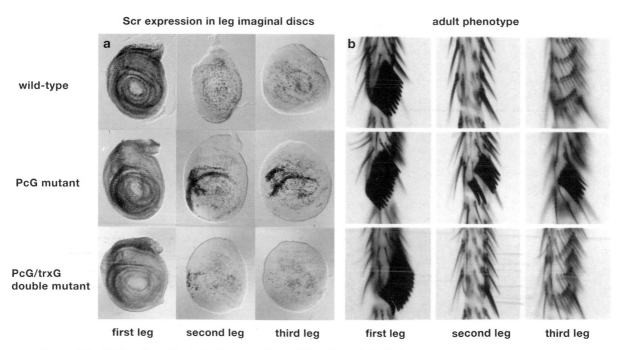

Figure 4. trxG Mutations Block the Derepression of Hox Genes in PcG Mutants

(*a*) Leg imaginal discs stained with antibodies against the protein encoded by the Hox gene, *Scr*, which specifies the identity of the labial and first thoracic segments, including the first leg. (*b*) Basitarsal segments of the legs of wild-type and mutant adults. Note the presence of sex comb teeth on the first leg, but not the second and third legs of wild-type adults. The *Scr* gene is partially derepressed in the second and third leg discs, where it is normally silent, in individuals heterozygous for mutations in PcG genes, leading to the appearance of ectopic sex comb teeth on the second and third legs. These phenotypes are suppressed by mutations in *brm* and many other trxG genes (*a*, Reprinted, with permission, from Tamkun et al. 1992 [©Elsevier]; *b*, portion modified, with permission, from Kennison 2003 [©Elsevier].).

Table 1. Biochemical functions of trxG proteins

| Known function | Organism | | | Complexed with non-trxG proteins? |
	Drosophila	human	yeast	
ATP-dependent chromatin remodeling	BRM	BRG1/HBRM	Swi2/Snf2, Sth1	yes (5–10)[a]
	OSA	BAF250	Swi1/Adr6	yes (5–10)
	MOR	BAF155, BAF170	Swi3, Rsc8	yes (5–10)
	SNR1	hSNF5/INI1	Snf5, Sfh1	yes (5–10)
	Kismet (KIS)	CHD7	–	not known
Histone methyltransferases	Trithorax (TRX)	MLL1, MLL2, hSET1	Set1	yes (5–20)
	Absent, small or homeotic 1(ASH1)	hASH1	–	not known
Mediator subunits	Kohtalo (KTO)	TRAP230	Srb8	yes (13–24)
	Skuld (SKD)	TRAP240	Srb9	yes (13–24)
Transcription factor	Trithorax-like (TRL)	BTBD14B	–	no
Growth factor receptor	Breathless (BTL)	FGFR3	–	not known
Other	Sallimus (SLS)	Titin		not known
	ASH2	hASH2L[b]	Bre2	yes (5–20)

[a]BRM, OSA, MOR, and SNR1 can all be found in stable association with each other in a single complex.
[b]Relatively low sequence similarity to ASH2.

members based on other, less stringent criteria, including sequence homology with known trxG proteins, physical association with trxG proteins, biochemical activity, or effects on Hox transcription in vitro or in vivo.

The functional relationship between members of the trxG, and the mechanistic connection between trxG function and maintenance of cell fate, is complicated. There are numerous mechanisms via which a protein might maintain an appropriately high level of expression of a homeotic gene (the genetic definition of a trxG protein) without being a devoted transcriptional activator, or a protein devoted to epigenetic control. Formal possibilities for trxG function (in addition to the ability to directly activate transcription) include the ability to increase function of direct activators, the ability to block function of PcG repressors, and the ability to create a "permissive" chromatin state that facilitates the function of numerous other regulatory complexes. Furthermore, as discussed below, some trxG proteins play complicated mechanistic roles that on some genes contribute to activation and that on other genes can contribute to repression.

Two brief examples illustrate the complexity of potential roles for trxG proteins. ATP-dependent remodeling complexes such as the one that contains trxG proteins BRM and MOR have been proposed to increase the ability of any sequence-specific DNA-binding protein to bind to chromatin. An unsettled issue is whether this ATP-depend-

ent remodeling complex can therefore use this ability to promote both activation of genes through increased binding of activators and repression through increased binding of repressors. Other studies have led to the hypothesis that some trxG protein complexes might function primarily by blocking the ability of a PcG repressor complex to function, and that repression by PcG proteins is the default state. Thus, in this latter instance, the role in maintaining an active state by some trxG proteins might reflect indirect, as opposed to direct, actions. The evolutionary conservation of this family, and the conserved functions of this family, offer hints concerning what types of mechanisms are needed to maintain the appropriate level of activation of master regulatory genes that determine cell fate.

1.2 trxG Proteins in Other Organisms

Functional counterparts of virtually all *Drosophila* trxG proteins are present in mammals, including humans (Table 1). Genetic and biochemical studies have shown that the fly and mammalian proteins play highly conserved roles in both gene expression and development. A good example of the functional conservation of trxG proteins is provided by MLL, the mammalian ortholog of *Drosophila trx*. Mutations in MLL cause homeotic transformations of the axial skeleton of mice due to the failure to maintain active transcription of Hox genes (Yu et al.

1995, 1998). Both MLL and trx function as histone lysine methyltransferases (HKMTs), and direct evidence of functional homology between the two proteins was provided by the use of human MLL to partially rescue developmental defects resulting from the loss of *trx* function in flies (Muyrers-Chen et al. 2004). Thus, the mechanisms underlying the maintenance of the determined state have been highly conserved during evolution.

Cancer and other human diseases can result from the failure to maintain a heritable state of gene expression. Not surprisingly, many human PcG and trxG genes function as proto-oncogenes or tumor suppressor genes. For example, the human trxG gene *MLL* was originally identified by 11q23 chromosome translocations associated with acute lymphoblastic (ALL) or myeloid (AML) leukemia. Mutations in other mammalian trxG genes are also associated with a variety of cancers (for more detail, see Chapter 23). For example, BRG1, the human counterpart of *Drosophila brm*, physically interacts with the retinoblastoma tumor suppressor protein; disruption of this interaction leads to increased cell division and malignant transformation in certain human tumor cell lines (Dunaief et al. 1994; Strober et al. 1996). Consistent with a role of BRG1 in tumor suppression, mice heterozygous for mutations in this gene are prone to develop a variety of tumors (Bultman et al. 2000). Mutations in INI1, the human counterpart of the *Drosophila* trxG gene *SNF5-related gene 1* (*SNR1*), also predispose individuals to cancers and have been identified in a large percentage of malignant rhabdoid tumors, an aggressive cancer of children (Versteege et al. 1998). These and other connections to human disease have provided researchers with additional motivation to understand the mechanism of action of trxG proteins.

1.3 trxG Proteins Play Diverse Roles in Eukaryotic Transcription

The trxG of activators is a large and functionally diverse group of regulatory proteins. This may reflect the complexity of eukaryotic transcription, which involves highly regulated interactions between gene-specific transcriptional activators, the numerous components of the general transcription machinery, and the DNA template that is transcribed. Transcriptional activation involves the binding of sequence-specific activating proteins, the recruitment of the general transcription machinery by those proteins, the formation of a pre-initiation complex in which RNA polymerase II is bound to the promoter, the opening of the DNA helix near the promoter, the effi-

cient escape of RNA polymerase from the promoter, and efficient elongation of RNA polymerase through the gene.

The ability to maintain an active transcriptional state might involve any of the numerous steps required for activation, because on any given gene, different steps might play a rate-determining role for transcriptional activity. The packaging of eukaryotic DNA into chromatin provides another level at which trxG proteins can regulate transcription. Nucleosomes and other components of chromatin tend to inhibit the binding of general and gene-specific transcription factors to DNA, as well as inhibit the elongation of RNA polymerase. Alterations in chromatin structure—including changes in the structure or positioning of nucleosomes—can influence virtually every step in the process of transcription.

Any protein that is required for transcription is required for the maintenance of the active state. Indeed, some trxG proteins play relatively general roles in transcription and are not dedicated solely to the maintenance of the determined state. Other trxG proteins, however, may play specialized roles in this process, either by directly counteracting PcG repression or by maintaining heritable states of gene activity through DNA replication and mitosis. The latter class of trxG proteins is of particular interest to developmental biologists.

2 Connections between trxG Proteins and Chromatin

Genetic studies indicating that trxG genes play key roles in transcription and development stimulated significant work to understand the biochemical function of their products. Many of these experiments have used, as their conceptual basis, the hypothesis that chromatin will be the biologically relevant substrate of trxG proteins. All genes are packaged into chromatin, and that packaging can create a compacted and inaccessible state or can be in an open and permissive state. Both the permissive and inaccessible states may conceivably be heritable. These considerations led to the simple hypothesis that trxG proteins might modulate chromatin structure to affect regulation. Furthermore, as trxG genes were cloned and sequenced, it became apparent that some of their products are related to proteins involved in ATP-dependent chromatin remodeling or the covalent modification of nucleosomal histones in other organisms, including the yeast *Saccharomyces cerevisiae*. Thus, although yeast lack either Hox genes or PcG repressors, this organism has provided valuable clues about potential roles for trxG proteins in eukaryotic transcription.

One of the first connections between the trxG and chromatin was provided by the discovery that the *Drosophila* trxG gene *brm* is highly related to yeast SWI2/SNF2 (Tamkun et al. 1992). SWI2/SNF2 was identified in screens for genes involved in mating-type switching (*switch* [*swi*] genes) and sucrose-fermentation (*sucrose-nonfermenting* [*snf*] genes). It was subsequently shown to be required for the activation of numerous inducible yeast genes (Holstege et al. 1998; Sudarsanam et al. 2000). The transcription defects observed in *swi2/snf2* mutants are suppressed by mutations in nucleosomal histones, an early observation which first suggested that SWI2/SNF2 activates transcription by counteracting chromatin repression (Kruger et al. 1995). Biochemical studies conducted in the early 1990s confirmed this hypothesis; SWI2/SNF2 and many of the other proteins identified in the *swi/snf* screens function as subunits of a large protein complex (SWI/SNF) that uses the energy of ATP hydrolysis to increase the ability of proteins to bind to nucleosomal DNA (Cote et al. 1994; Imbalzano et al. 1994; Kwon et al. 1994). SWI2/SNF2 functions as the ATPase subunit, or "engine," of this chromatin-remodeling machine; other subunits of the SWI/SNF complex mediate interactions with regulatory proteins or its chromatin substrate (Phelan et al. 1999).

Another connection between trxG and chromatin was suggested by the presence of SET domains in the trxG proteins Trithorax (TRX) and Absent, small or homeotic (ASH1). The SET domain was originally defined by a stretch of amino acids that shows homology between Su(var)3-9, Enhancer of zeste (E(z)), and TRX, the latter two proteins being, respectively, PcG and trxG members. In the late 1990s, the SET family of proteins was shown to have HKMT activity. Su(var)3-9 methylates H3K9, whereas (E(z)) methylates H3K27 (Rea et al. 2000; Levine et al. 2004; Ringrose and Paro 2004). As discussed elsewhere, H3K9 methylation promotes heterochromatin assembly, whereas H3K27 methylation appears to be required for PcG repression (for more detailed discussion, see Chapters 5 and 11, respectively). The presence of SET domains in trxG proteins suggested that the methylation of histone tails might also be important for the maintenance of active transcriptional states.

These findings, together with the growing realization that chromatin-remodeling and -modifying enzymes play key roles in transcriptional activation, motivated biochemists to identify protein complexes that contain trxG proteins and to examine the effect of these complexes on chromatin structure in vitro. Other experiments tested the hypothesis that trxG proteins might interact directly with the transcriptional machinery, another well-established method of affecting regulation. As described below, these studies revealed that some trxG proteins affect regulation by modifying chromatin structure whereas others function via direct interactions with components of the transcription machinery.

2.1 trxG Proteins Involved in ATP-dependent Chromatin Remodeling

Chromatin-remodeling complexes have been implicated in a wide variety of biological processes, including transcriptional repression and activation, chromatin assembly, the regulation of higher-order chromatin structure, and cellular differentiation. The most extensively studied trxG proteins involved in chromatin remodeling are BRM and its human counterparts, BRG1 and HBRM. As predicted, these proteins function as the ATPase subunits of complexes that are highly related to yeast SWI/SNF (Kwon et al. 1994; Wang et al. 1996). SWI/SNF complexes contain between 8 and 15 subunits and have been highly conserved during evolution (Fig. 5). The ATPase of each of these complexes is able to function as an isolated subunit. Although this family of proteins is historically referred to as containing a "helicase" motif, due to the similarity of their ATPase domain to that of true helicases, the proteins related to BRM have never been shown to possess helicase activity, but rather appear to use other mechanisms such as translocation along the DNA to effect changes in chromatin structure (Whitehouse et al. 2003; Saha et al. 2005). A second trxG gene identified in this screen, *moira* (*mor*), encodes another key member of this ATP-dependent remodeling complex in *Drosophila*, and homologs of BRM and MOR interact directly to form a functional core of SWI/SNF in humans (Phelan et al. 1999).

SWI/SNF and other chromatin-remodeling complexes use the energy of ATP hydrolysis to alter the structure or positioning of nucleosomes. By catalyzing ATP-dependent changes in chromatin structure, chromatin-remodeling complexes help transcription factors and other regulatory proteins gain access to DNA sequences that would normally be occluded by the histone proteins (Polach and Widom 1995; Logie and Peterson 1997). Models to create access to specific sites include "sliding" the histones along the DNA to move a site into a linker region, looping DNA away from the histone octamer or, most dramatically, evicting the entire histone octamer to a different place in the nucleus (Fig. 6). ATP-dependent remodeling can also lead to changes in the position of the nucleosome along a DNA

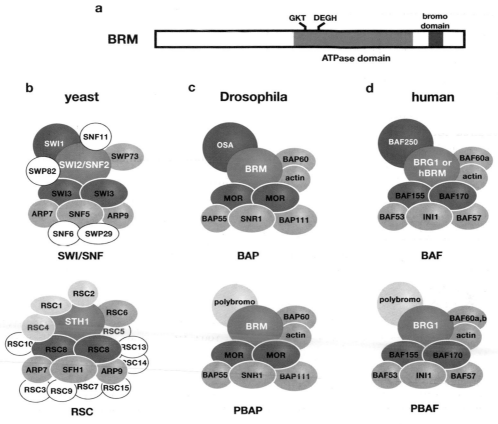

Figure 5. The SWI/SNF Family of Remodeling Complexes

Each complex contains a member of the SNF2/SWI2 family of ATPases and at least 8 other subunits. (*a*) Schematic diagram of the BRM protein, showing the location of the ATPase domain and carboxy-terminal bromodomain (which shows affinity to acetylated lysine residues in histone tails) that are conserved in all SNF2/SWI2 family members. SWI/SNF complexes in yeast (*b*), *Drosophila* (*c*), and human (*d*) are shown. *Drosophila* trxG proteins (BRM, MOR, and OSA) and their counterparts in other organisms are shown in color. Further information about these complexes and their subunits may be found in Mohrmann and Verrijzer (2005).

sequence, to changes in the spacing of nucleosomes, and to the exchange of histones into and out of the histone octamer that is the core of the nucleosome. Different remodeling complexes display different proclivities for each of these functions.

SWI/SNF complexes are abundant in higher eukaryotes; for example, each mammalian nucleus contains about 25,000 copies of SWI/SNF-family complexes. Biochemical analyses show that SWI/SNF complexes are able to create access to an unusually large spectrum of sites within the nucleosome when compared with other ATP-dependent chromatin-remodeling complexes (Fan et al. 2003). For example, SWI/SNF complexes are able to efficiently create access to sites at the center of a mononucleosome, which is energetically difficult because sites at the center of the nucleosome have approximately 70 base pairs of constrained nucleosomal DNA on both sides. Whether this is caused by an unusually potent ability to utilize the

energy of ATP hydrolysis relative to other remodelers, or instead represents a distinct mechanism of remodeling, is a topic of ongoing research (Kassabov et al. 2003). SWI/SNF complexes do not display measurable ability to evenly space nucleosomes, a hallmark of other chromatin-remodeling complexes. They also do not display the same degree of efficiency in "swapping" H2A/H2B dimers as some other chromatin-remodeling complexes, although they are able to do this and to evict octamers when tested in vitro (Lorch et al. 1999). Which of these abilities is related to the function of these complexes in the maintenance of the active state is not yet clear.

SWI/SNF complexes have been implicated in transcriptional activation in every species that has been examined. This family of complexes can be targeted to genes by interactions with transcriptional activators, can remodel nucleosomes to assist in the initial binding of general transcription factors and RNA polymerase II, and can

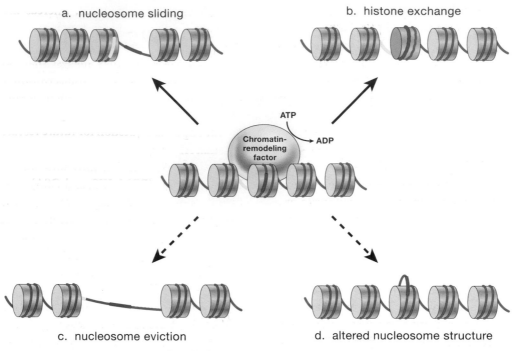

a. nucleosome sliding

b. histone exchange

c. nucleosome eviction

d. altered nucleosome structure

Figure 6. **Mechanisms for ATP-dependent Remodeling**

Models for chromatin remodeling are illustrated by showing the change in position or composition of nucleosomes relative to the DNA wrapped around it. The central panel indicates a starting chromatin region where linker DNA is indicated in yellow and nucleosomal DNA in red. (*a*) Movement of a nucleosome translationally along the DNA (sliding) to expose a region (marked in *red*) that was previously occluded; (*b*) exchange of a variant histone for a standard histone to create a variant nucleosome; (*c*) eviction of nucleosomes to open a large region of DNA. This mechanism might depend on other proteins, such as histone chaperones or DNA-binding factors, in addition to remodeling proteins; (*d*) creating a loop on the surface of the nucleosome. Remodelers in the SWI/SNF family have been hypothesized to use alternative mechanisms, such as creating stable loops of DNA on the surface of the nucleosome, to make sites available that are central to the nucleosome.

become targeted later in the activation process to assist with transcriptional elongation. Thus, SWI/SNF complexes appear to function at every step in the process of transcriptional activation, although there appears to be an emphasis on function at the early steps that lead to loading of RNA polymerase II. Microarray analysis in yeast shows that, in addition to these effects promoting activation, SWI/SNF complexes can also facilitate repression of some genes (Sudarsanam et al. 2000).

One simple hypothesis to explain these broad in vivo functions is that these remodeling complexes alter nucleosome structure in a manner that facilitates binding and function of a wide variety of regulatory factors and complexes. Thus, the potent remodeling characteristics observed in vitro might reflect an ability to significantly expand access to regulatory factors in vivo. It is possible that SWI/SNF complexes are uniquely able to broadly create access, which may account for their importance in the maintenance of the active state.

Each species studied has at least two distinct SWI/SNF complexes, all of which contain BRM or a highly related

chromatin-remodeling ATPase. Another trxG protein, OSA, provides distinction between the complexes, in that one class of complexes contains OSA and another evolutionarily conserved complex contains a Polybromo-domain protein (Fig. 6) (Mohrmann and Verrijzer 2005). The biochemical function of OSA is not clear. One attractive possibility is that it might target the SWI/SNF complex in which it resides to a specific set of genes.

SWI/SNF is not the only chromatin-remodeling factor that is present in eukaryotic cells. Dozens of different chromatin-remodeling complexes have been identified, including NURF, NURD, ACF, and CHRAC (Vignali et al. 2000). These complexes can be subdivided into several major groups based on the identities of their ATPase subunits. SWI/SNF complexes contain ATPases related to SWI2/SNF2; ISWI complexes (e.g., NURF, CHRAC, and ACF) contain ATPases related to Imitation-SWI (ISWI); and CHD complexes (e.g., NURD) contain ATPases related to CHD1 and Mi2.

Recent studies have implicated a *Drosophila* member of the CHD family of chromatin-remodeling factors—

kismet (*kis*)—in the maintenance of the active state. Like *brm*, *mor*, and *osa*, *kis* was identified in a screen for extragenic suppressors of *Pc*, suggesting that it acts antagonistically to PcG proteins to maintain active states of Hox transcription (Kennison and Tamkun 1988). Genetic studies revealed that *kis* is required for both segmentation and the maintenance of Hox transcription during *Drosophila* development (Daubresse et al. 1999). The molecular analysis of *kis* revealed that it encodes several large proteins, including an approximately 575-kD isoform (KIS-L) that contains an ATPase domain characteristic of chromatin-remodeling factors (Daubresse et al. 1999; Therrien et al. 2000). Conserved domains outside the ATPase domain (including bromodomains and chromodomains) contribute to the functional specificity of chromatin-remodeling factors by mediating interactions with nucleosomes or other proteins. BRM and other ATPase subunits of SWI/SNF complexes contain a single bromodomain (a protein motif associated with the binding of certain acetylated histones), whereas KIS-L contains two chromodomains (protein motifs that bind certain methylated histones) and is therefore more similar to Mi2 and other members of the CHD family of chromatin-remodeling factors. Although the large size of KIS-L (~575 kD) has made it difficult to analyze this protein biochemically, its sequence strongly suggests that it activates transcription by remodeling chromatin.

KIS-L is not physically associated with BRM and behaves chromatographically as if it is in a distinct protein complex (Srinivasan et al. 2005). The two proteins overlap extensively with each other and RNA polymerase II on polytene chromosomes, however, suggesting that both play relatively global roles in transcription (Fig. 7) (Armstrong et al. 2002; Srinivasan et al. 2005). Loss of BRM function blocks a relatively early step in transcrip-

tion (Armstrong et al. 2002), whereas the loss of KIS-L function leads to a decrease in the level of elongating, but not initiating, forms of RNA polymerase II (Srinivasan et al. 2005). These findings suggest that BRM and KIS-L facilitate distinct steps in transcription by RNA polymerase II by catalyzing ATP-dependent alterations in nucleosome structure or spacing.

An important question for future research concerns the role that ATP-dependent remodeling plays in maintenance of the activated state. It is intriguing that four trxG members are known (BRM, KIS, and OSA) or suspected (KIS) members of large ATP-dependent chromatin-remodeling complexes, but none of the other numerous ATP-dependent remodeling complexes has been identified in genetic screens for *Drosophila* trxG proteins. Two predominant hypotheses, not mutually exclusive, to explain this are that the BRM and KIS chromatin-remodeling complexes are targeted to genes important for developmental progression or that they have special remodeling characteristics which are uniquely required for maintenance. Thus, it is possible that generic ATP-dependent remodeling is required for all active states, and that maintenance of the active state of developmentally important genes happens to require these trxG members because they are targeted to these genes. It is also possible that maintenance requires special ATP-dependent functions that can only be carried out by the complexes that contain trxG members.

It is also intriguing to think about the mechanisms that remodelers might use to contribute to epigenetic regulation of the active state. At least three classes of mechanisms can be envisioned that might apply. First, remodeling functions might be required in a somewhat indirect manner to facilitate the binding (or re-binding following replication) of gene-specific activating proteins that are needed to maintain active transcription. In this case, the remodelers

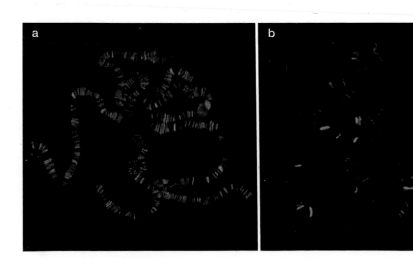

Figure 7. Chromosomal Distribution of trxG Proteins

The genome-wide distribution of trxG proteins was examined by staining *Drosophila* salivary gland polytene chromosomes with antibodies against BRM (*a*) or TRX (*b*). Consistent with a relatively global role in transcriptional activation, BRM is associated with hundreds of sites in a pattern that overlaps extensively with RNA pol II. In contrast, strong TRX signals are detected at a much smaller number of sites on polytene chromosomes.

would not be the "brains" of the epigenetic mechanism, but instead would act as a necessary tool to allow the proteins required to function efficiently. Second, remodelers could work alone or with histone chaperones to evict nucleosomes from a region, and this lack of occupancy by nucleosomes would hypothetically cause the region to remain non-nucleosomal following replication. As mentioned above, the ability of the replication/nucleosome deposition machinery to accurately recapitulate nucleosome modification or location is an important unanswered issue in epigenetics. Finally, remodeling machineries could reposition nucleosomes to create a structure that is amenable to activation. This latter mechanism has experimental support from studies of the albumin gene (Chaya et al. 2001; Cirillo et al. 2002). Several DNA-binding factors are required to maintain activity of this key gene in the liver. One of these factors, FoxA, binds to a site on a nucleosome, and the specific nucleosome-FoxA architecture is key to maintaining the active state of the albumin gene. Although it is not clear whether there is a required role for ATP-dependent remodeling to position this specific nucleosome in the liver, this example demonstrates the potential for specific nucleosome positioning to play a key epigenetic role.

2.2 trxG Proteins That Covalently Modify Nucleosomal Histones

A second common method of regulating gene expression involves covalent modification of the amino-terminal tails of the core histones that comprise the protein component of the nucleosome. These tails, which protrude from the surface of the nucleosome, can mediate interactions with other nucleosomes, as well as with a wide variety of structural and regulatory proteins. The covalent modification of histone tails by acetylation, methylation, or phosphorylation can help target regulatory complexes to chromatin and can also directly change the ability of nucleosomes to compact into repressive structures by changing the charge on the tails. Covalent modification might also provide a mark to help maintain a specific regulated state, as the covalently modified histones have the potential to divide to the two daughter strands and thereby propagate the information contained in the covalent mark to both mother and daughter cells following replication. Whether histones remain associated with one or both daughter strands following replication is an issue key to potential mechanisms of epigenetic regulation that remains controversial, in large part due to the challenge of finding techniques that will allow accurate tracking of individual histones in living cells.

Several trxG proteins are able to covalently modify histone tails, and these proteins are frequently found in complexes that are able to perform more than one type of modification reaction. For example, *Drosophila* TRX and its counterparts in other organisms methylate histone H3 at lysine 4 (H3K4): This covalent mark is tightly associated with active genes in a wide variety of organisms, including yeast, flies, and humans. A second trxG protein, ASH1 (see below), also has H3K4 methyltransferase activity (Beisel et al. 2002; Byrd and Shearn 2003). H3K4 methylation has been implicated in maintenance of active gene expression in yeast by the timing of its appearance and removal on active genes (Santos-Rosa et al. 2002; Pokholok et al. 2005). The finding that trxG members have this histone modification activity further ties the H3K4 mark to maintenance of the active state.

In yeast and in humans, counterparts of TRX are found in a complex that also contains a third trxG protein, Ash2, which is not related in sequence to Ash1. The yeast homolog of trithorax, Set1, is found in a complex (COMPASS or Set1C) that is approximately 400 kD in size and contains five other proteins in addition to Set1 and Ash2 (Miller et al. 2001; Roguev et al. 2001). The only known biochemical activity of this complex is methylation of H3K4; it is not yet clear what the function of each of the other proteins might be, although one component might help propagate the methylation mark (see below).

In humans, there are three TRX homologs, called MLL1, MLL2, and hSET1. The MLL1 protein has received the most attention in biochemical analyses and is found in a large complex (>10 members) that also contains the human homolog of ASH2 (Hughes et al. 2004; Yokoyama et al. 2004). This complex and the yeast complex both contain a WD40 repeat protein which is called WDR5 in humans (Dou et al. 2005; Wysocka et al. 2005). Recently, it has been shown that the WDR5 protein can bind to histone H3 that has been methylated at lysine 4 (Wysocka et al. 2005). Thus, binding of this protein to the mark created by the MLL1 complex in which it resides might provide a mechanism to facilitate spreading of the mark. This is similar to proposals made concerning the repressive complexes that methylate H3K9, which contain HP1, a protein that binds specifically to methylated K9 (for more details, see Chapters 5 and 6).

There is evidence from both *Drosophila* and humans that the complex containing TRX/MLL is also involved in acetylation. In humans, MLL is associated with the MOF acetyltransferase, which acetylates lysine 16 of histone H4, another modification normally linked to activation (Dou et al. 2005). In flies, TRX is associated with dCBP, an acetyltransferase with broad specificity that is involved in activation (Petruk et al. 2001). Acetylation might work synergistically with H3K4 methylation to direct an active state following function of these trxG complexes. Acetylation is also known to prevent the methylation of residues such as H3K9 and H3K27 that direct repression of the template.

The ASH1 protein, another trxG member, is also a histone methyltransferase that methylates H3K4 (Beisel et al. 2002; Byrd and Shearn 2003). The composition of any ASH1-containing complexes has not been established, nor is it understood how the activities of ASH1 and the complexes containing TRX/MLL1/SET1 are coordinated. However, ASH1 has also been seen to colocalize and associate with the CBP family of acetyltransferases (Bantignies et al. 2000), once again suggesting that methylation and acetylation go hand in hand.

There are numerous fascinating, as yet unanswered, questions concerning how covalent modification of histones might contribute to trxG function. What functional role do the marks play? Covalent modification can contribute to epigenetic regulation via a wide spectrum of mechanisms. Methylation and acetylation marks might serve to directly alter chromatin compaction (sometimes termed *cis*-effects, as in Fig. 8 of Chapter 3). The ability of chromatin to enter a compacted state, which is generally assumed to be repressive for transcription, is influenced by the charge distribution on the histone tails. Modifications that occur on lysine (e.g., acetylation) can eliminate the positive charge normally found with this residue, and therefore might directly decrease the ability of nucleosomes to form compacted structures, thus increasing the ability of the template to be transcribed.

Covalent marks have been proposed to create strong binding sites for complexes that direct transcriptional activation. These covalent modifications are able to create specific "knobs" on the surface of the nucleosome that fit into pockets on the complexes that promote activation, thus increasing binding energy and function of these complexes. For example, acetylation of histone tails increases binding by homologs of the BRM protein, thus promoting ATP-dependent remodeling of acetylated templates (Hassan et al. 2001). This type of mechanism, frequently referred to as the "histone code" or *trans*-effects of covalent histone modifications, has the potential to be a central epigenetic function. Further studies are needed to determine which marks created by trxG proteins enhance binding of which complexes, to determine the extent to which the energy of binding to a single modified residue can influence function and

targeting, and to determine the temporal order of addition of the marks and whether they are maintained across mitosis.

The flip side of this mechanism is that the marks could inhibit binding by repressive complexes. A covalent mark on a key residue required for optimal binding by a repressive complex could strongly inhibit binding by the repressive complex. For example, it is known that binding by repressive complexes is increased by methylation of histone H3 at K9 and K27 (Khorasanizadeh 2004). Acetylation of these residues would both block methylation and create an ill-shaped "knob" on the histone that impairs binding by the repressive complex. Thus, the ability of modifications to influence function of other complexes can cut in both directions, increasing the potency of this potential mode of epigenetic regulation.

These mechanisms not only are not mutually exclusive, but are likely to work together to help maintain an active state. Marks that chemically increase the ability to form a compacted state (a *cis*-effect) might also increase the ability of complexes to bind (a *trans*-effect), and further promote a compacted state. Conversely, marks that chemically decrease compaction might increase binding of complexes that also decompact nucleosomes. This mechanistically parsimonious use of covalent marks to alter several characteristics of chromatin structure and of the ability of regulatory complexes to bind could create a powerful means of maintaining an active state.

3 Connections between trxG Proteins and the General Transcription Machinery

The theme that trxG proteins frequently are found in the same complex is continued with the *skuld* (*skd*) and *kohtalo* (*kto*) proteins. These two proteins are homologs of the proteins identified biochemically as TRAP240 (Skuld) and TRAP 230 (Kohtalo), which are both members of the "Mediator" complex (Janody et al. 2003). The mediator complex is a large complex that functions at the interface between gene-specific activator proteins and formation of the pre-initiation complex that contains RNA polymerase II (Lewis and Reinberg 2003). Thus, these proteins are involved in general activation processes, much in the same way that the SWI/SNF-family remodelers are involved in general activation. SKD and KTO might have some special function involved in maintenance, because other components of the mediator complex were not identified in screens for trxG genes. The observation that SKD and KTO interact with each other, and that *skd kto* double mutants have the same phenotype as either single mutant, has led to the hypothesis that the two proteins together form a functional module that somehow alters mediator action (Janody et al. 2003).

4 Biochemical Functions of Other trxG Proteins

The biochemical activities of the majority of other trxG proteins remain relatively mysterious. Ring3, a human counterpart of the *Drosophila* trxG gene *female-sterile homeotic* (*fsh*), encodes a nuclear protein kinase with two bromodomains that has been implicated in cell cycle progression and leukemogenesis, but the substrates of this kinase are currently unknown (Denis and Green 1996). The trxG gene *Tonalli* (*Tna*) encodes a protein related to SP-RING finger proteins involved in sumoylation, suggesting that it may also regulate transcription via the covalent modification of other, as yet unidentified, proteins (Gutierrez et al. 2003). The trxG gene *sallimus* (*sls*) was identified in a screen for extragenic suppressors of *Pc* (Kennison and Tamkun 1988) and subsequently found to encode *Drosophila* Titin (Machado and Andrew 2000). Like its vertebrate counterpart, *Drosophila* Titin helps maintain the integrity and elasticity of the sarcomere. In addition, Titin is a chromosomal protein that is required for chromosome condensation and segregation (Machado and Andrew 2000). These intriguing findings suggest a potential role for trxG proteins in the regulation of higher-order chromatin structure.

5 Functional Interactions between trxG Proteins

Now that the basic biochemical activities of many trxG and PcG members have been identified, attention has shifted to the way in which their activities are coordinated to regulate transcription and maintain the active state. Despite the lack of in vitro systems for studying the maintenance of the determined state, good progress has been made toward addressing this issue. One popular hypothesis is that the trxG and PcG members facilitate a sequence of dependent events required for the maintenance of the active or repressed state. Support for this idea has come from recent studies of the PcG complexes PRC1 and PRC2; by methylating H3K27, the E(z) histone methyltransferase subunit of PRC2 creates a covalent mark that is directly recognized by the chromodomain of the Pc subunit of PRC1 (Jacobs and Khorasanizadeh 2002; Min et al. 2003). Thus, one PcG complex appears to directly promote the binding of another PcG complex to chromatin.

By analogy, it is possible that the covalent modification of nucleosomes by trxG members with histone methyltransferase or acetyltransferase activities (e.g.,

TRX or ASH1) directly regulates the targeting or activities of trxG members involved in ATP-dependent chromatin remodeling (e.g., BRM [SWI/SNF], or KIS) (Fig. 8). Consistent with this possibility, BRM and other subunits of SWI/SNF complexes contain bromodomains that can directly interact with acetylated histone tails, and KIS contains two chromodomains that may directly interact with methylated histone tails. This model, which is supported by recent studies of chromatin-remodeling factors in both yeast and mammals (Agalioti et al. 2000; Hassan et al. 2001), is particularly attractive because it provides a mechanism by which a heritable histone modification could perpetuate a constitutively "open" chromatin configuration that is permissive for active transcription.

6 trxG Proteins: Activators or Anti-repressors?

Another important issue concerns the functional relationship between PcG repressors and trxG activators. Do these regulatory proteins have independent roles in activation and repression, or do they act in direct opposition to maintain the heritable state? Recent genetic studies show that removal of PcG complexes will reactivate genes even in the absence of TRX and ASH1 (Klymenko and Muller 2004), suggesting that trxG proteins with histone methyltransferase activity may function as PcG antirepressors, as opposed to activators (Fig. 8).

Both biochemical and genetic analyses provide evidence that there might be direct connections between trxG function and PcG function. One interesting property of PcG proteins is that they are capable of repressing transcription when tethered near virtually any gene transcribed by RNA polymerase II. trxG members that play global roles in transcription—including BRM, KIS, and other trxG members involved in chromatin-remodeling—are thus excellent candidates for direct targets of PcG repressors (Fig. 8). One of the major PcG complexes, PRC1, blocks the function of SWI/SNF-family remodeling complexes, apparently by blocking access of this complex to the template (Francis et al. 2001). This is consistent with the notion that one mechanism for PcG repression might be to prevent ATP-dependent remodeling by trxG members. The Brahma complex and PRC1 are further connected by the fact that both directly interact with the Zeste protein, a protein which plays a complicated role in regulation of gene expression in *Drosophila* that might help direct cross talk between the two complexes.

A second protein that connects PcG proteins and trxG proteins is the GAGA factor, which is encoded by the *Trithorax-like* gene and is thus a trxG member (Farkas et al. 1994). This protein can function as a

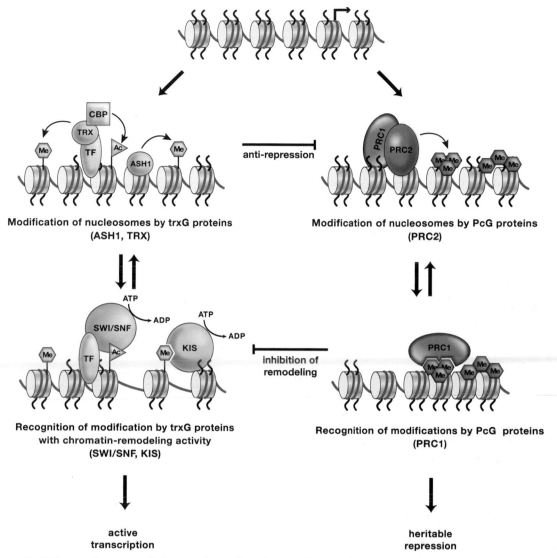

Figure 8. Trithorax Group and Polycomb Group Functions and Interactions

Both trxG and PcG families include proteins that covalently modify histones and proteins that noncovalently modify chromatin. Covalent modifications on histones can increase binding by noncovalent modifying complexes such as SWI/SNF, KIS, or PRC1. Binding by these latter complexes has the potential to lead to further covalent modification, thus leading to iterative cycles of covalent modification and recognition of the covalent marks.

sequence-specific activator protein at some promoters, but also is a prominent member of the proteins that bind to the Polycomb Repressive Element (PRE, see Chapter 8). PRE sequences direct PcG function, and at least one PRE can act as a memory module when affixed to a reporter construct, emphasizing the importance of these sequences. Sequences that bind the GAGA factor play an important role in PRE function, and tethering the GAGA protein to DNA has been proposed to enhance binding and function of PRC1 (Mahmoudi and Verrijzer 2001). Thus, the GAGA factor might play key roles in maintenance of activation (via its transcrip-

tional activating properties) and maintenance of repression (via interactions with the PcG proteins). An important issue for future research is to understand why proteins such as GAGA and Zeste appear to interact with both the activating and repressing machineries of maintenance.

7 Conclusion and Outlook

Two of the major issues regarding function of trxG proteins remain largely a matter for conjecture. First, why does a relatively small subset of the proteins required for transcriptional activation score genetically as being important for maintenance of the active state? Is this because these proteins play global roles in transcription but are expressed in limiting quantities or happen, by evolutionary serendipity, to be especially important for developmentally important genes? Second, how can the active state be maintained across replication and mitosis? Replication will create two daughter strands that must both be regulated identically, and mitosis requires condensation and thereby inhibition of transcription of most genes in a cell. What mechanisms create the epigenetic mark(s) that ensures reactivation of a gene on both daughter strands following mitosis?

The majority of trxG proteins are part of complexes that are broadly used in gene expression, and most of these complexes also contain many other proteins not in the trxG (see Table 1). This raises the important question as to whether there are special functions that are used for maintenance of active gene expression. It is possible that SWI/SNF remodelers are able to perform a special remodeling function, that H3K4 methylation targets special complexes and/or chromatin conformations, and that Skuld/Kohtalo alter function of Mediator in a specific manner important for maintenance. Alternatively, it is possible that each of these proteins performs a reaction that is normally used in activation of all types of genes, and that these complexes are among those that have emerged as being important for maintenance for a relatively uninteresting reason (e.g., because even relatively subtle changes in the expression of *Drosophila* Hox genes cause homeotic transformations). To resolve these issues, considerably more information is needed about the precise mechanisms that each of these proteins uses in activation. For example, do the SWI/SNF complexes harness the energy of ATP hydrolysis in the same manner as other ATP-dependent remodeling complexes, or do they differ in an important way in how this energy is used to alter nucleosome structure? Structural techniques including crystallography, biophysical techniques such as single-molecule analysis and FRET (fluorescence resonance energy transfer), and detailed imaging in vivo might help to shed light on whether there are mechanisms specially designed for epigenetic maintenance of activation. The initial functional studies that have been done with trxG complexes on simple model templates are just the beginning of the process for answering these important questions.

The epigenetic mechanisms that might maintain an active state are even less well understood. Are covalent marks distributed to help create an active mark? Are nucleosome positions maintained following replication to create "open" stretches of chromatin, or specially positioned nucleosomes, that increase binding of activators? Does trxG function cause active genes to compartmentalize within the nucleus to regions that favor active transcription? These are all viable hypotheses; more hypotheses exist, and others have not yet even been envisioned. The incredible complexity of the machinery that transcribes DNA offers numerous possibilities for regulation, and for the development of mechanisms that allow an epigenetic maintenance of active transcription. This intersection of two fields rich in intellectual history, transcriptional activation and epigenetic mechanism, will provide fertile ground for experimentalists for many years.

References

Agalioti T., Lomvardas S., Parekh B., Yie J., Maniatis T., and Thanos D. 2000. Ordered recruitment of chromatin modifying and general transcription factors to the IFN-β promoter. *Cell* **103**: 667–678.

Armstrong J.A., Papoulas O., Daubresse G., Sperling A.S., Lis J.T., Scott M.P., and Tamkun J.W. 2002. The *Drosophila* BRM complex facilitates global transcription by RNA polymerase II. *EMBO J.* **21**: 5245–5254.

Bantignies F., Goodman R.H., and Smolik S.M. 2000. Functional interaction between the coactivator *Drosophila* CREB-binding protein and ASH1, a member of the trithorax group of chromatin modifiers. *Mol. Cell. Biol.* **20**: 9317–9330.

Beisel C., Imhof A., Greene J., Kremmer E., and Sauer F. 2002. Histone methylation by the *Drosophila* epigenetic transcriptional regulator Ash1. *Nature* **419**: 857–862.

Bultman S., Gebuhr T., Yee D., La Mantia C., Nicholson J., Gilliam A., Randazzo F., Metzger D., Chambon P., Crabtree G., and Magnuson T. 2000. A Brg1 null mutation in the mouse reveals functional differences among mammalian SWI/SNF complexes. *Mol. Cell* **6**: 1287–1295.

Byrd K.N. and Shearn A. 2003. ASH1, a *Drosophila* trithorax group protein, is required for methylation of lysine 4 residues on histone H3. *Proc. Natl. Acad. Sci.* **100**: 11535–11540.

Chaya D., Hayamizu T., Bustin M., and Zaret K.S. 2001. Transcription factor FoxA (HNF3) on a nucleosome at an enhancer complex in liver chromatin. *J. Biol. Chem.* **276**: 44385–44389.

Cirillo L.A., Lin F.R., Cuesta I., Friedman D., Jarnik M., and Zaret K.S. 2002. Opening of compacted chromatin by early developmental transcription factors HNF3 (FoxA) and GATA-4. *Mol. Cell* **9**: 279–289.

Cote J., Quinn J., Workman J.L., and Peterson C.L. 1994. Stimulation of GAL4 derivative binding to nucleosomal DNA by the yeast SWI/SNF complex. *Science* **265**: 53–60.

Daubresse G., Deuring R., Moore L., Papoulas O., Zakrajsek I., Waldrip W.R., Scott M.P., Kennison J.A., and Tamkun J.W. 1999. The *Drosophila kismet* gene is related to chromatin-remodeling factors and is required for both segmentation and segment identity. *Development* **126**: 1175–1187.

Denis G.V. and Green M.R. 1996. A novel, mitogen-activated nuclear kinase is related to a *Drosophila* developmental regulator. *Genes Dev.* **10**: 261–271.

Dou Y., Milne T.A., Tackett A.J., Smith E.R., Fukuda A., Wysocka J., Allis C.D., Chait B.T., Hess J.L., and Roeder R.G. 2005. Physical association and coordinate function of the H3 K4 methyltransferase MLL1 and the H4 K16 acetyltransferase MOF. *Cell* **121:** 873–885.

Dunaief J.L., Strober B.E., Guha S., Khavari P.A., Alin K., Luban J., Begemann M., Crabtree G.R., and Goff S.P. 1994. The retinoblastoma protein and BRG1 form a complex and cooperate to induce cell cycle arrest. *Cell* **79:** 119–130.

Duncan I. 1987. The bithorax complex. *Annu. Rev. Genet.* **21:** 285–319.

Fan H.Y., He X., Kingston R.E., and Narlikar G.J. 2003. Distinct strategies to make nucleosomal DNA accessible. *Mol. Cell* **11:** 1311–1322.

Farkas G., Gausz J., Galloni M., Reuter G., Gyurkovics H., and Karch F. 1994. The *Trithorax-like* gene encodes the *Drosophila* GAGA factor. *Nature* **371:** 806–808.

Francis N.J. and Kingston R.E. 2001. Mechanisms of transcriptional memory. *Nat. Rev. Mol. Cell. Biol.* **2:** 409–421.

Francis N.J., Saurin A.J., Shao Z., and Kingston R.E. 2001. Reconstitution of a functional core polycomb repressive complex. *Mol. Cell* **8:** 545–556.

Gellon G. and McGinnis W. 1998. Shaping animal body plans in development and evolution by modulation of *Hox* expression patterns. *Bioessays* **20:** 116–125.

Gutierrez L., Zurita M., Kennison J.A., and Vazquez M. 2003. The *Drosophila* trithorax group gene *tonalli* (*tna*) interacts genetically with the Brahma remodeling complex and encodes an SP-RING finger protein. *Development* **130:** 343–354.

Hassan A.H., Neely K.E., and Workman J.L. 2001. Histone acetyltransferase complexes stabilize swi/snf binding to promoter nucleosomes. *Cell* **104:** 817–827.

Holstege F.C., Jennings E.G., Wyrick J.J., Lee T.I., Hengartner C.J., Green M.R., Golub T.R., Lander E.S., and Young R.A. 1998. Dissecting the regulatory circuitry of a eukaryotic genome. *Cell* **95:** 717–728.

Hughes C.M., Rozenblatt-Rosen O., Milne T.A., Copeland T.D., Levine S.S., Lee J.C., Hayes D.N., Shanmugam K.S., Bhattacharjee A., Biondi C.A., et al. 2004. Menin associates with a trithorax family histone methyltransferase complex and with the *hoxc8* locus. *Mol. Cell* **13:** 587–597.

Imbalzano A.N., Kwon H., Green M.R., and Kingston R.E. 1994. Facilitated binding of TATA-binding protein to nucleosomal DNA. *Nature* **370:** 481–485.

Jacobs S.A. and Khorasanizadeh S. 2002. Structure of HP1 chromodomain bound to a lysine 9-methylated histone H3 tail. *Science* **295:** 2080–2083.

Janody F., Martirosyan Z., Benlali A., and Treisman J.E. 2003. Two subunits of the *Drosophila* mediator complex act together to control cell affinity. *Development* **130:** 3691–3701.

Kassabov S.R., Zhang B., Persinger J., and Bartholomew B. 2003. SWI/SNF unwraps, slides, and rewraps the nucleosome. *Mol. Cell* **11:** 391–403.

Kaufman T.C., Seeger M.A., and Olsen G. 1990. Molecular and genetic organization of the antennapedia gene complex of *Drosophila melanogaster*. *Adv. Genet.* **27:** 309–362.

Kennison J.A. 1995. The Polycomb and trithorax group proteins of *Drosophila*: Trans-regulators of homeotic gene function. *Annu. Rev. Genet.* **29:** 289–303.

———. 2003. Introduction to Trx-G and Pc-G genes. *Methods Enzymol.* **377:** 61–70.

Kennison J.A. and Tamkun J.W. 1988. Dosage-dependent modifiers of Polycomb and Antennapedia mutations in *Drosophila*. *Proc. Natl. Acad. Sci.* **85:** 8136–8140.

Khorasanizadeh S. 2004. The nucleosome: From genomic organization to genomic regulation. *Cell* **116:** 259–272.

Klymenko T. and Muller J. 2004. The histone methyltransferases Trithorax and Ash1 prevent transcriptional silencing by Polycomb group proteins. *EMBO Rep.* **5:** 373–377.

Kruger W., Peterson C.L., Sil A., Coburn C., Arents G., Moudrianakis E.N., and Herskowitz I. 1995. Amino acid substitutions in the structured domains of histones H3 and H4 partially relieve the requirement of the yeast SWI/SNF complex for transcription. *Genes Dev.* **9:** 2770–2779.

Kwon H., Imbalzano A.N., Khavari P.A., Kingston R.E., and Green M.R. 1994. Nucleosome disruption and enhancement of activator binding by a human SWI/SNF complex. *Nature* **370:** 477–481.

Levine S.S., King I.F., and Kingston R.E. 2004. Division of labor in Polycomb group repression. *Trends Biochem. Sci.* **29:** 478–485.

Lewis B.A. and Reinberg D. 2003. The mediator coactivator complex: Functional and physical roles in transcriptional regulation. *J. Cell Sci.* **116:** 3667–3675.

Logie C. and Peterson C.L. 1997. Catalytic activity of the yeast SWI/SNF complex on reconstituted nucleosome arrays. *EMBO J.* **16:** 6772–6782.

Lorch Y., Zhang M., and Kornberg R.D. 1999. Histone octamer transfer by a chromatin-remodeling complex. *Cell* **96:** 389–392.

Machado C. and Andrew D.J. 2000. D-Titin: A giant protein with dual roles in chromosomes and muscles. *J. Cell Biol.* **151:** 639–652.

Mahmoudi T. and Verrijzer C.P. 2001. Chromatin silencing and activation by Polycomb and trithorax group proteins. *Oncogene* **20:** 3055–3066.

Miller T., Krogan N.J., Dover J., Erdjument-Bromage H., Tempst P., Johnston M., Greenblatt J.F., and Shilatifard A. 2001. COMPASS: A complex of proteins associated with a trithorax-related SET domain protein. *Proc. Natl. Acad. Sci.* **98:** 12902–12907.

Min J., Zhang Y., and Xu R.M. 2003. Structural basis for specific binding of Polycomb chromodomain to histone H3 methylated at Lys 27. *Genes Dev.* **17:** 1823–1828.

Mohrmann L. and Verrijzer C.P. 2005. Composition and functional specificity of SWI2/SNF2 class chromatin remodeling complexes. *Biochim. Biophys. Acta* **1681:** 59–73.

Muyrers-Chen I., Rozovskaia T., Lee N., Kersey J.H., Nakamura T., Canaani E., and Paro R. 2004. Expression of leukemic MLL fusion proteins in *Drosophila* affects cell cycle control and chromosome morphology. *Oncogene* **23:** 8639–8648.

Petruk S., Sedkov Y., Smith S., Tillib S., Kraevski V., Nakamura T., Canaani E., Croce C.M., and Mazo A. 2001. Trithorax and dCBP acting in a complex to maintain expression of a homeotic gene. *Science* **294:** 1331–1334.

Phelan M.L., Sif S., Narlikar G.J., and Kingston R.E. 1999. Reconstitution of a core chromatin remodeling complex from SWI/SNF subunits. *Mol. Cell* **3:** 247–253.

Pokholok D.K., Harbison C.T., Levine S., Cole M., Hannett N.M., Lee T.I., Bell G.W., Walker K., Rolfe P.A., Herbolsheimer E., et al. 2005. Genome-wide map of nucleosome acetylation and methylation in yeast. *Cell* **122:** 517–527.

Polach K.J. and Widom J. 1995. Mechanism of protein access to specific DNA sequences in chromatin: A dynamic equilibrium model for gene regulation. *J. Mol. Biol.* **254:** 130–149.

Rea S., Eisenhaber F., O'Carroll D., Strahl B.D., Sun Z.W., Schmid M., Opravil S., Mechtler K., Ponting C.P., Allis C.D., and Jenuwein T. 2000. Regulation of chromatin structure by site-specific histone H3 methyltransferases. *Nature* **406:** 593–599.

Ringrose L. and Paro R. 2004. Epigenetic regulation of cellular memory by the polycomb and trithorax group proteins. *Annu. Rev. Genet.* **38:** 413–443.

Roguev A., Schaft D., Shevchenko A., Pijnappel W.W., Wilm M., Aasland R., and Stewart A.F. 2001. The *Saccharomyces cerevisiae* Set1 complex includes an Ash2 homologue and methylates histone 3 lysine 4. *EMBO J.* **20:** 7137–7148.

Saha A., Wittmeyer J., and Cairns B.R. 2005. Chromatin remodeling through directional DNA translocation from an internal nucleosomal site. *Nat. Struct. Mol. Biol.* **12:** 747–755.

Santos-Rosa H., Schneider R., Bannister A.J., Sherriff J., Bernstein B.E., Emre N.C., Schreiber S.L., Mellor J., and Kouzarides T. 2002. Active genes are tri-methylated at K4 of histone H3. *Nature* **419:** 407–411.

Simon J. 1995. Locking in stable states of gene expression: Transcriptional control during *Drosophila* development. *Curr. Opin. Cell Biol.* **7:** 376–385.

Simon J.A. and Tamkun J.W. 2002. Programming off and on states in chromatin: Mechanisms of Polycomb and trithorax group complexes. *Curr. Opin. Genet. Dev.* **12:** 210–218.

Srinivasan S., Armstrong J.A., Deuring R., Dahlsveen I.K., McNeill H., and Tamkun J.W. 2005. The *Drosophila* trithorax group protein Kismet facilitates an early step in transcriptional elongation by RNA Polymerase II. *Development* **132:** 1623–1635.

Strober B.E., Dunaief J.L., Guha S, and Goff S.P. 1996. Functional interactions between the hBRM/hBRG1 transcriptional activators and the pRB family of proteins. *Mol. Cell. Biol.* **16:** 1576–1583.

Sudarsanam P., Iyer V.R., Brown P.O., and Winston F. 2000. Whole-genome expression analysis of snf/swi mutants of *Saccharomyces cerevisiae*. *Proc. Natl. Acad. Sci.* **97:** 3364–3369.

Tamkun J.W., Deuring R., Scott M.P., Kissinger M., Pattatucci A.M., Kaufman T.C., and Kennison J.A. 1992. *brahma*: A regulator of *Drosophila* homeotic genes structurally related to the yeast transcriptional activator SNF2/SWI2. *Cell* **68:** 561–572.

Therrien M., Morrison D.K., Wong A.M., and Rubin G.M. 2000. A genetic screen for modifiers of a *kinase suppressor of* Ras-dependent rough eye phenotype in *Drosophila*. *Genetics* **156:** 1231–1242.

Versteege I., Sevenet N., Lange J., Rousseau-Merck M.F., Ambros P., Handgretinger R., Aurias A., and Delattre O. 1998. Truncating mutations of hSNF5/INI1 in aggressive paediatric cancer. *Nature* **394:** 203–206.

Vignali M., Hassan A.H., Neely K.E., and Workman J.L. 2000. ATP-dependent chromatin-remodeling complexes. *Mol. Cell. Biol.* **20:** 1899–1910.

Wang W., Cote J., Xue Y., Zhou S., Khavari P.A., Biggar S.R., Muchardt C., Kalpana G.V., Goff S.P., Yaniv M., et al. 1996. Purification and biochemical heterogeneity of the mammalian SWI-SNF complex. *EMBO J.* **15:** 5370–5382.

Whitehouse I., Stockdale C., Flaus A., Szczelkun M.D., and Owen-Hughes T. 2003. Evidence for DNA translocation by the ISWI chromatin-remodeling enzyme. *Mol. Cell. Biol.* **23:** 1935–1945.

Wysocka J., Swigut T., Milne T.A., Dou Y., Zhang X., Burlingame A.L., Roeder R.G., Brivanlou A.H., and Allis C.D. 2005. WDR5 associates with histone H3 methylated at K4 and is essential for H3 K4 methylation and vertebrate development. *Cell* **121:** 859–872.

Yokoyama A., Wang Z., Wysocka J., Sanyal M., Aufiero D.J., Kitabayashi I., Herr W., and Cleary M.L. 2004. Leukemia proto-oncoprotein MLL forms a SET1-like histone methyltransferase complex with menin to regulate Hox gene expression. *Mol. Cell. Biol.* **24:** 5639–5649.

Yu B.D., Hanson R.D., Hess J.L., Horning S.E., and Korsmeyer S.J. 1998. MLL, a mammalian *trithorax*-group gene, functions as a transcriptional maintenance factor in morphogenesis. *Proc. Natl. Acad. Sci.* **95:** 10632–10636.

Yu B.D., Hess J.L., Horning S.E., Brown G.A., and Korsmeyer S.J. 1995. Altered Hox expression and segmental identity in *Mll*-mutant mice. *Nature* **378:** 505–508.

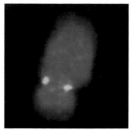

Histone Variants and Epigenetics

Steven Henikoff[1] and M. Mitchell Smith[2]

[1]Howard Hughes Medical Institute, Fred Hutchinson Cancer Research Center, Seattle, Washington 98109-1024
[2]Department of Microbiology, University of Virginia, Charlottesville, Virginia 22908

CONTENTS

1. DNA Is Packaged by Architectural Proteins in All Organisms, 251

2. Eukaryotic Core Histones Evolved from Archaeal Histones, 251

3. Bulk Histones Are Deposited after DNA Replication, 253

4. Variant Histones Are Deposited Throughout the Cell Cycle, 254

5. Centromeres Are Identified by a Special H3 Variant, 254

6. The Replacement Histone Variant H3.3 Is Found at Active Chromatin, 256

7. Phosphorylation of H2AX Functions in DNA Double-Strand Break Repair, 258

8. H2AZ Plays Roles in Transcriptional Regulation, 259

9. Protein Complexes for the Deposition and Replacement of H2A Variants, 261

10. Other H2A Variants Differentiate Chromatin but Their Functions Are As Yet Unknown, 262

11. Many Histones Have Evolved to More Tightly Package DNA, 262

12. Conclusions and Future Research, 263

References, 263

GENERAL SUMMARY

Histones package DNA by assembling into nucleosome core particles while the double helix wraps around. Over evolutionary time, histone fold domain proteins have diversified from archaeal ancestors into the four distinct subunits that comprise the familiar octamer of the eukaryotic nucleosome. Further diversification of histones into variants results in differentiation of chromatin that can have epigenetic consequences. Investigations into the evolution, structure, and metabolism of histone variants provide a foundation for understanding the participation of chromatin in important cellular processes and in epigenetic memory.

Most histones are synthesized at S phase for rapid deposition behind replication forks to fill in gaps resulting from the distribution of preexisting histones. In addition, the replacement of canonical S-phase histones by variants, independent of replication, can potentially differentiate chromatin.

The differentiation of chromatin by a histone variant is especially conspicuous at centromeres, where the H3 variant, CENP-A, is assembled into specialized nucleosomes that form the foundation for kinetochore assembly (see left panel of title figure). A centromeric H3 (CenH3) counterpart of CENP-A is found in all eukaryotes. In plants and animals, the faithful assembly of CenH3-containing nucleosomes at centromeres does not appear to require centromeric DNA sequences, a spectacular example of epigenetic inheritance. Some CenH3s have evolved adaptively in regions that contact DNA, which suggests that centromeres compete with each other, and that CenH3s and other centromere-specific DNA-binding proteins have adapted in response. This process could account for the large size and complexity of centromeres in plants and animals.

Chromatin can also be differentiated outside of centromeres by incorporation of a constitutively expressed form of H3, called H3.3, which is the substrate for repli-cation-independent nucleosome assembly. Replacement with H3.3 occurs at active genes (see right panel of title figure, showing H3.3 in green on a fruit fly chromosome), a dynamic process with potential epigenetic consequences. Differences between H3 and H3.3 in their complement of covalent modifications might underlie changes in the properties of chromatin at actively transcribed loci.

Several H2A variants can also differentiate or regulate chromatin. H2A.X is defined as a variant by a 4-amino acid carboxy-terminal motif whose serine residue is the site for phosphorylation at sites of DNA double-stranded breaks. Phosphorylation of H2AX is an early event in double-strand break repair, where it is thought to concentrate components of the repair machinery. H2AX phosphorylation also marks the inactive XY bivalent during mammalian spermatogenesis and is required for condensation, pairing, and fertility.

H2AZ is a structurally diverged variant that has long presented an enigma. Studies in yeast have implicated H2AZ in establishing transcriptional competence and in counteracting heterochromatic silencing. The biochemical complex that replaces H2A with H2AZ in nucleosomes is an ATP-dependent nucleosome remodeler, providing the first example of a specific function for a member of this diverse class of chromatin-associated machines.

Two vertebrate-specific variants, macroH2A and H2A^{Bbd}, display contrasting features when packaged into nucleosomes in vitro, with macroH2A impeding and H2A^{Bbd} facilitating transcription. These features are consistent with their localization patterns on the epigenetically inactivated mammalian X chromosome, with macroH2A showing enrichment and H2A^{Bbd} showing depletion.

The emerging view from these studies is that histone variants and the processes that deposit them into nucleosomes provide a primary differentiation of chromatin that might serve as the basis for epigenetic processes.

1 DNA Is Packaged by Architectural Proteins in All Organisms

The enormous length of the DNA double helix relative to the size of the organelle that contains it requires tight packaging, and architectural proteins have evolved for this purpose. The first level of packaging shortens the double helix and protects it from damage, while still allowing DNA polymerase to gain full access to each base pair every cell cycle. In addition, these architectural proteins facilitate higher-order folding to further reduce the length of a chromosome. Perhaps because of stringent requirements for packaging DNA, only two structural classes of architectural proteins are found in nearly all cellular life forms (Malik and Henikoff 2003). Bacterial DNA is packaged by HU proteins, eukaryotic DNA is packaged by histones, and archaeal DNA is packaged by either HU proteins or histones.

Histones package DNA into nucleosome particles, and this architectural role can account for the fact that histones comprise half of the mass of a eukaryotic chromosome. However, histones have also been found to play diverse roles in gene expression, chromosome segregation, DNA repair, and other basic chromosomal processes in eukaryotes. Specific requirements of these chromosomal processes have led to the evolution of distinct histone variants. The incorporation of a variant histone into a nucleosome represents a potentially profound alteration of chromatin. Indeed, recent work has revealed that some histone variants are deposited by distinct nucleosome assembly complexes, which suggests that chromatin is diversified, at least in part, by the incorporation and replacement of histone variants.

The four core histones differ with respect to their propensity to diversify into variants. For example, humans have only one H4 isotype but several H2A paralogs with different properties and functions. Evidently, the different positions of the core histones within the nucleosome particle have subjected them to different evolutionary forces, leading to important diversifications of H2A and H3 but not of H2B and H4. The availability of genomic sequences from a wide variety of eukaryotes allows us to conclude that these diversifications have occurred at various times during eukaryotic evolution. However, the evident diversification of an ancestral histone fold protein into the familiar four core histones must have occurred early in the evolution of the eukaryotic nucleus or perhaps before. By considering these ancient events, we gain insight into the forces that have resulted in subsequent diversification into present-day variants.

2 Eukaryotic Core Histones Evolved from Archaeal Histones

The eukaryotic nucleosome is a complex structure, consisting of an octamer of four core histones wrapped nearly twice by DNA, with histone tails and linker histones mediating a variety of packaging interactions outside the core particle (Arents et al. 1991; Wolffe 1992; Luger et al. 1997). Archaeal nucleosomes are much simpler, and it is evident that they resemble the ancestral particle from which eukaryotic nucleosomes evolved (Malik and Henikoff 2003). An archaeal nucleosome consists of histone fold domain proteins that lack tails and form a tetrameric particle that is wrapped only once by DNA. The kinship between archaeal and eukaryotic nucleosomes can be seen by comparing their structures: The backbone of the archaeal tetramer nearly superimposes over that of the (H3•H4)$_2$ tetramer (Fig. 1). When archaeal nucleosomes are reconstituted to form chromatin, the resulting fiber behaves similarly to "tetrasomes" of (H3•H4)$_2$. Therefore, it is thought that eukaryotic nucleosomes evolved from archaeal nucleosomes by addition of H2A•H2B dimers on either side of the tetrasome to allow a second DNA wrap, and by acqui-

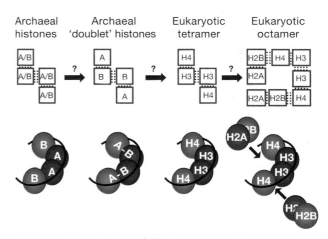

Figure 1. Model for the Evolution of the Eukaryotic Nucleosome from an Archaeal Doublet Histone Ancestor

An archaeal tetramer with interchangeable subunits A and B (A/B) may have evolved into a dimer of fused dimers ("doublet"). This could have been followed by a gene split to give rise to the eukaryotic tetramer of H3 and H4, forming an (H3•H4)$_2$ "tetrasome" that occupies a single turn of DNA. H2A and H2B may have arisen from a similar event, assembling above and below the tetramer as suggested in the cartoon so being able to accommodate two turns of DNA (not illustrated). Single dots in the top part of the diagram represent dimeric contacts and double dots represent four-helix bundles between adjacent dimers (Reprinted from Malik and Henikoff 2003).

sition of histone tails. In addition, DNA wraps into a right-handed superhelix around archaeal cores, but into a left-handed superhelix around eukaryotic cores.

Further insight into the origin of the eukaryotic nucleosomes comes from examination of the subunit structures of archaeal nucleosomes. Whereas most archaeal histones are undifferentiated monomers or are differentiated into structurally interchangeable variants that come together to form a tetramer, some are head-to-tail dimeric fusions that come together to form a dimer of fused dimers (Fig. 1). When two of these fused dimers assemble into a nucleosome particle, each member of the fused pair is in a structurally distinguishable position. By occupying distinct positions in the particle, each member of the archaeal fused dimer evolves independently, allowing it to adapt to a single position in the nucleosome particle. In contrast, monomers that occupy interchangeable positions are not free to adapt to particular positions. Indeed, the two members of archaeal dimers have diverged from one another in both independent lineages in which they are found. This process provides a possible scenario for the differentiation of an ancestral histone fold domain protein into four distinct subunits that occupy distinct positions in the eukaryotic nucleosome. Like their presumed archaeal ancestors, eukaryotic histones form dimers, where H2A dimerizes with H2B and H3 with H4 (which also stably tetramerizes in solution). The structural backbone of an archaeal histone dimer superimposes with those of H2A•H2B and H3•H4 at 2 Å resolution, with the first member of the dimeric repeat superimposing on H2A or H3 and the second member superimposing on H2B or H4. So, although all four eukaryotic histones lack significant sequence similarity to one another and to archaeal histones, the striking structural superposition of dimeric units suggests that eukaryotic histones evolved and differentiated from simpler archaeal ancestors.

The asymmetry of H2A•H2B and H3•H4 dimers, which appears to have originated from archaeal tandem dimers, could have led the way to subsequent diversification of eukaryotic histone variants. Both H2A and H3 correspond to the first member of archaeal tandem histone dimers, and both have subsequently diversified multiple times in eukaryotic evolution. In contrast, H2B and H4 correspond to the second member and have shown little (H2B) or no (H4) functional diversification. Both H3 and H2A make homodimeric contacts in the octamer (Fig. 2), whereas H4 and H2B only contact other histones. As a result, changes in the residues involved in homodimerization of either H2A or H3 can

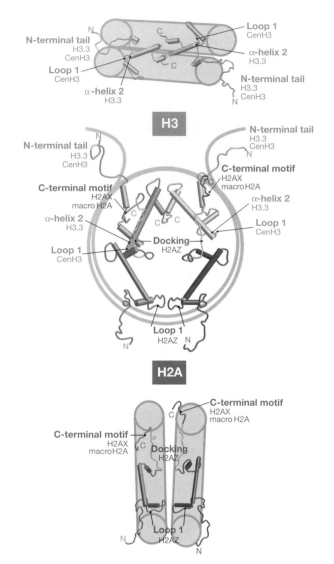

Figure 2. Location of Histone H3 (*blue*) and H2A (*brown*) in the Nucleosome Core Particle

The four residues that differ between H3 variants are indicated in yellow. (Reprinted, with permission, from Henikoff and Ahmad 2005.)

potentially resist formation of mixed octamers, allowing nucleosomes containing an H2A or H3 variant to evolve independently of parental nucleosomes. For example, the four-helix bundle comprising the interface between H3s determines the left-handed supercoiling of the DNA around the nucleosome (Arents et al. 1991; Luger et al. 1997), whereas DNA supercoils are right-handed in archaeal nucleosomes (Marc et al. 2002). Evidently, mutation of the four-helix bundle in an H3 ancestor was responsible for this reversal. In general, structural features that facilitated independent evolution of subunits may have been prerequisites for diversification of nucleosome particles.

Although we can rationalize the descent of the eukaryotic core histones from archaeal tandem dimers, other basic questions remain. Where did histone tails come from? When did H2A•H2B arrive on the scene? Did these events occur before, during, or after the evolution of the eukaryotic nucleus? Why do all known archaeal nucleosomes consist of tetramers with one wrap, whereas eukaryotic nucleosomes consist of octamers with two wraps? Why did the superhelical handedness switch? Perhaps the sequences of more archaea or of primitive eukaryotes will reveal intermediate forms that can answer these questions.

3 Bulk Histones Are Deposited after DNA Replication

The packaging of essentially all DNA in a eukaryotic cell into nucleosomes requires that chromatin is duplicated when DNA replicates (Fig. 3). Thus, canonical histones are produced during the DNA synthesis (S) phase of the cell cycle. S-phase coupling of histone synthesis to DNA synthesis is under tight cell cycle control (Marzluff and Duronio 2002). This is especially evident in animals, where special processing of histone transcripts by the U7 small

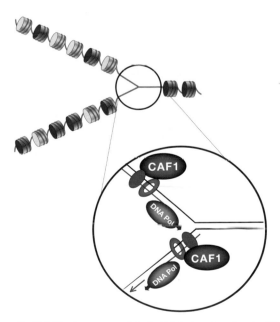

Figure 3. Old Nucleosomes (*dark disks*) Are Randomly Distributed behind the Replication Fork and New Nucleosomes (*light disks*) Are Deposited in the Gaps

CAF-1-mediated nucleosome assembly is depicted on the leading and lagging strand in magnification. DNA polymerase (*green*); replication processivity clamp, PCNA (*blue ring*); histone H3•H4 tetramers (*pink*); newly synthesized DNA (*red*).

nuclear ribonuclear protein complex, and mRNA stabilization by the stem-loop-binding protein (SLBP), contribute to the tight coordination of histone synthesis with DNA replication. The need for rapid and massive production of histones during S phase is very likely responsible for the fact that replication-coupled (RC) histones in animals are encoded in clusters that comprise many histone genes. For example, there are 14 H4 genes in the human genome, most of which are found in two major clusters, where these H4 genes are interspersed with other RC histone genes (Marzluff et al. 2002). In animals, RC histones are recognizable by the presence of a 26-bp 3′ sequence that forms a stem-loop for recognition by SLBP when transcribed into histone mRNA. Canonical plant histones are also encoded by multiple genes and are deposited during S phase, although plant histone transcripts are polyadenylated and there does not appear to be a counterpart to SLBP.

To the extent that epigenetic inheritance results from inheritance of a chromatin "state," the process of RC nucleosome assembly has been of intense interest. The biochemistry of the process was elucidated with the development of in vitro systems that could assemble nucleosomes onto replicating DNA. These studies revealed that a three-subunit complex, chromatin assembly factor 1 (CAF-1), acts as a histone chaperone that facilitates the incorporation of H3•H4 as a first step in nucleosome assembly (Loyola and Almouzni 2004). CAF-1 was shown to interact with the replication processivity clamp, PCNA, which implies that DNA replication and RC assembly occur in close proximity. Work in budding yeast revealed that none of the subunits of complexes involved in RC assembly in vitro is essential for growth, suggesting that, in vivo, there are redundant mechanisms for RC assembly. The fact that much of yeast chromatin is assembled in a replication-independent (RI) manner (Altheim and Schultz 1999) provides a rationale for this evident redundancy. As shown below, histone variants are typically deposited by RI nucleosome assembly.

RC assembly is not completely redundant in budding yeast. An intriguing finding is that absence of the large CAF-1 subunit leads to loss of epigenetic silencing at telomeres (Loyola and Almouzni 2004). The connection between RC assembly and epigenetic silencing has been extended to *Arabidopsis*, where loss of CAF-1 subunits results in a variety of defects attributable to loss of epigenetic memory. Although the mechanistic basis for these observations is unknown, it seems clear that the proper deposition of new nucleosomes behind the replication fork is important for maintaining an epigenetically silenced state.

A prerequisite for epigenetic inheritance of a nucleosome state is that preexisting nucleosomes must be distributed to daughter chromatids following replication (Fig. 3). Indeed, this is the case: Extensive studies have shown that old nucleosomes are inherited intact and evidently at random to daughter chromatids (Fig. 3) (Annunziato 2005). However, this process of inheritance is poorly understood, as is the process by which new histones might acquire epigenetic information. A popular model is that new nucleosomes are modified by their proximity to old nucleosomes (Jenuwein 2001); however, evidence for this hypothetical process is lacking, and alternative means of perpetuating an epigenetic state must be considered (Henikoff and Ahmad 2005). How epigenetic information is inherited to daughter cells remains a major unanswered question in biology, and the study of histone variants and the mechanisms of their deposition may provide clues.

4 Variant Histones Are Deposited Throughout the Cell Cycle

As we have seen, core histones can be classified on the basis of their ancestral sequence and position in the nucleosome. Linker histones are characterized by a winged helix domain, rather than a histone fold domain, and bind to the linker DNA that separates nucleosomes (Wolffe 1992). Although minor variants of these canonical histones exist, they appear to be interchangeable with the major form. For example, mammalian H3.1 and H3.2 differ by a single amino acid that is not known to impart different biological properties to the two isoforms. The existence of multiple genes that produce large amounts of canonical histones for S-phase deposition is typical of eukaryotic genomes. The near ubiquity and overwhelming abundance of canonical S-phase histones has resulted in relatively little attention being paid to histone variants until recently.

The renaissance of interest in histone variants came in part from the realization that they differ from canonical S-phase histones in ways that can lead to profound differentiation of chromatin. One way that they differ is in their mode of incorporation into chromatin. RC assembly incorporates new nucleosomes into gaps between old nucleosomes genome-wide, whereas RI assembly involves local replacement of an existing nucleosome or subunit (Marzluff et al. 2002). RI assembly therefore has the potential of switching a chromatin state by replacing a canonical histone with a variant. Replacing one histone with another also could erase or alter the pattern of post-translational modifications. Therefore, RI assembly can potentially reset epigenetic states that are thought to be

mediated by histones and their modifications. Recent progress in studying histone variants and the processes by which they are deposited has led to new insights into the basis for epigenetic inheritance and remodeling. Below, we discuss features of particular histone variants that contribute to chromatin differentiation and might be involved in propagating epigenetic information.

5 Centromeres Are Identified by a Special H3 Variant

A defining feature of the eukaryotic chromosome is the centromere, which is the site of attachment of spindle microtubules at mitosis. The first centromeres to be described in molecular detail were those of budding yeast (*Saccharomyces cerevisiae*), where a 125-bp sequence is necessary and sufficient for centromere formation (Amor et al. 2004a). However, centromeres of plants and animals are very different, typically consisting of megabase arrays of short tandem repeats. Unlike the situation for budding yeast, the role of DNA sequence at these complex centromeres is uncertain, because fully functional human neocentromeres are known to form spontaneously at ectopic sites that entirely lack sequences resembling centromeric repeats (Fig. 4). These and other observations argue against a direct role of

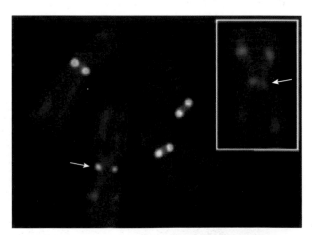

Figure 4. Human Neocentromeres (Indicated by an Arrow) Lack Centromeric α-Satellite DNA but Have CENP-A and Heterochromatin

Anti-CENP-A staining in green and anti-CENP-B staining in red (which marks α-satellite DNA) identify a Chromosome 4 neocentromere that lacks α-satellite (*main panel*). This Chromosome 4 is otherwise normal, having been transmitted for at least three meiotic generations in normal individuals. Inset shows anti-HP1 staining, which indicates that despite the lack of satellite DNA, heterochromatin forms around active neocentromeres. (Reprinted, with permission, from Amor et al. 2004b [©National Academy of Sciences].)

DNA sequence in determining the location of centromeres (see Chapter 6).

A key insight into the basis for centromere identity and inheritance came from the identification of a histone H3 variant, CENP-A (title figure), which was found to localize specifically to centromeres and to be incorporated into nucleosomal particles in place of H3 itself (Palmer et al. 1991). Remarkably, CENP-A remains associated with centromeres during the transition from histones to protamines during spermatogenesis, when essentially all other histones are lost (Palmer et al. 1990). This early observation in the study of CENP-A suggested that CENP-A contributes to centromere identity of the male genome. The generality of this insight was not fully appreciated until it was realized that CENP-A is a much better marker for centromeres than is DNA sequence (Amor et al. 2004a) and that counterparts of CENP-A can be found in the genomes of all eukaryotes (Fig. 5) (Malik and Henikoff 2003). Thus, although budding yeast centromeres are determined by a 125-bp consensus sequence, this is also the site of a centromeric nucleosome that contains the Cse4 centromeric H3 (CenH3) variant. In fission yeast (*Schizosaccharomyces pombe*), an array of CenH3-containing nucleosomes occupies the central core region of the centromere flanked by H3-containing nucleosomes that display heterochromatic features (Amor et al. 2004a). In flies and vertebrates, CenH3s are present in arrays which alternate with H3-containing arrays which display a unique pattern of histone modifications (Sullivan and Karpen 2004). Alternation can account for the fact that centromeres occupy only the outside edge of the cen-

tromeric constriction of metaphase chromosomes (title figure). This is consistent with the observation that in worm "holokinetic" chromosomes, microtubules attach throughout the length of each anaphase chromosome, and CenH3 occupies the leading edge all along its length (Fig. 5, right) (Malik and Henikoff 2003). Indeed, a unique CenH3 variant is found to precisely mark the centromere in all eukaryotes that have been examined. This apparent ubiquity, and the presence of centromeres to perform mitosis in all eukaryotes, raises the possibility that the first canonical H3 evolved from a CenH3.

Genetic experiments in a variety of eukaryotes have confirmed the essentiality of CenH3 for formation of the kinetochore and for chromosome segregation (Amor et al. 2004a). Because they remain in place throughout the cell cycle, CenH3-containing nucleosomes form the foundation for assembly of other kinetochore proteins during mitosis and meiosis (see Chapter 6). An outstanding question in chromosome research is just how these proteins interact to provide a linkage between the centromere and spindle microtubules that can hold up to the strong pulling forces exerted on kinetochores at anaphase. Several dozen kinetochore-specific proteins have been identified in yeast (for more detail, see Chapter 6), although how they interact with CenH3-containing nucleosomes and other foundation proteins, such as CENP-C, is currently unknown. An additional challenge is elucidation of the process that assembles CenH3 into nucleosomes. The fact that centromeres account for such a small proportion of chromatin overall has hampered biochemical approaches to this outstanding problem, but we expect

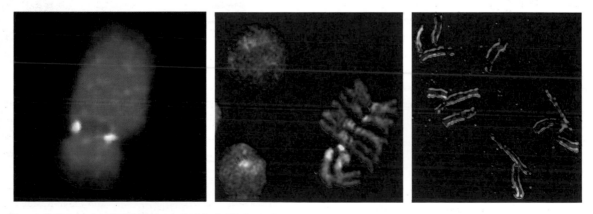

Figure 5. Centromeric H3 Variants in Model Eukaryotes

(*Left*) Human chromosome stained with an antibody against the centromere-specific histone H3 variant CENP-A (*green*) and anti-CENP-B (*red*) marking α-satellite DNA (image courtesy of Peter Warburton). (*Center*) *Drosophila melanogaster* anti-CenH3 antibody (*red*) stains centromeres in metaphase chromosomes and throughout interphase (image courtesy of Suso Platero). (*Right*) *Caenorhabditis elegans* anti-CenH3 antibody (*green*) stains the end-to-end holocentromeres of prophase chromosomes (*red*) (image courtesy of Landon Moore).

that improving technologies will lead to a better understanding of kinetochore structure and dynamics.

The evolution of CenH3s is unlike that of any other histone class. Whereas histone H3 is almost invariant in sequence, which reflects extraordinarily strong purifying selection on every residue, CenH3s are evolving rapidly, especially in plant and animal lineages (Malik and Henikoff 2003). This is most evident from the amino-terminal tails, which differ in length and sequence to such an extent that they cannot be aligned between the CenH3s of different taxonomic groups. Even the histone fold domain of CenH3 is evolving orders of magnitude faster than that of H3. What is the reason for this striking evolutionary difference between an H3 that functions at centromeres and an H3 that functions everywhere else?

Rapidly evolving regions of *Drosophila* and *Arabidopsis* CenH3 genes display an excess of replacement nucleotide substitutions over what would be expected from the rate of synonymous substitutions (Malik and Henikoff 2003). This excess is a hallmark of adaptive evolution. Adaptive evolution in plants and animals is also seen for another major centromere foundation protein, CENP-C (Talbert et al. 2004). Although adaptive evolution is well documented for genes involved in genetic conflicts, such as arms races between host and parasite interactions, these are the only known essential single-copy genes that are adaptively evolving in any organism. In the case of CenH3 and CENP-C, the regions of adaptive evolution correspond to regions of DNA binding and targeting. This suggests that the major centromere-binding proteins are adapting to the evolving centromeric DNA, thus allowing centromeric chromatin to interact with the conserved kinetochore machinery that connects the centromere to spindle microtubules. It has been proposed that centromeres compete during female meiosis to be included in the egg nucleus rather than being lost as polar bodies (Talbert et al. 2004). An arms race would develop leading to expansion of centromeres, probably by unequal crossing-over between sister chromatids. Host suppression of this meiotic drive process by CenH3 and CENP-C would lead to an excess of replacement changes in regions that interact with DNA. Organisms in which there is no opportunity for centromeres to compete, such as budding yeast, would not undergo centromere drive, and this might account for the fact that they have small centromeres and their CenH3 and CENP-C proteins are under strong purifying selection.

Thus, we see that a special region of the genome, the centromere, is distinguished by a single histone variant class, whose sequences reveal remnants of an arms race that may have led to the extraordinary complexity of centromeres. The RI assembly process that targets new CenH3-containing nucleosomes to centromeres every cell cycle remains unknown (Amor et al. 2004a). Centromeric nucleosomes show a remarkable lack of sequence specificity in that they not only can faithfully localize to neo-centromeres that are completely unlike native centromeres (Fig. 4), but also the yeast homolog Cse4 can functionally replace human CENP-A (Wieland et al. 2004) (neither of which is adaptively evolving; Talbert et al. 2004). It is extraordinary that our centromeres have remained in the same positions for tens of millions of years without any evident sequence determinants involved in the process that maintains them. To the extent that epigenetics refers to inheritance that does not depend on DNA sequence, the inheritance of centromeres on a geological timescale is the most extreme form imaginable. Yet, we are still seeking a mechanism to explain how they have maintained themselves for even a single cell cycle (topic discussed further in Chapter 14).

6 The Replacement Histone Variant H3.3 Is Found at Active Chromatin

Like centromeres, transcriptionally active chromatin is thought to be maintained epigenetically, and like centromeres, active chromatin is enriched in an H3 variant, called H3.3 (Henikoff and Ahmad 2005). H3.3 is very similar in sequence to the canonical forms of H3, differing by only four amino acids. With so few differences, it might be assumed that these two forms are interchangeable. However, in *Drosophila*, H3.3 is deposited by either RC or RI nucleosome assembly, whereas H3 is deposited only at replication foci in a RC manner. This difference between the two variants is encoded in the protein itself, with three of the four differences between H3 and H3.3 evidently involved in preventing H3 from being deposited by an RI pathway (in α-helix 2, Fig. 2). Purification of soluble human assembly complexes confirmed that these two forms participate in distinct assembly processes: H3.1 copurified with CAF-1 for RC assembly, and H3.3 copurified with other components, including HirA, and participated in RI assembly.

Although four-amino acid differences might seem practically insignificant, when one considers that humans, flies, and clams have precisely the same H3.3 sequence, these differences from H3 stand out. Phylogenetic analysis reveals that the H3/H3.3 pair evolved at least four separate times during eukaryotic evolution: in plants, animals/fungi, ciliates, and apicomplexans (Malik and Henikoff 2003). Despite having a separate origin from animals and fungi, the animal H3/H3.3 pair and the pair from plants (called H3.1 [RC] and H3.2 [RI]—to avoid confusion, we refer to all RC isoforms as H3 and all RI isoforms as H3.3) are strikingly similar. The same cluster of amino acids (positions 87–90) that prevents RI deposition of H3 in *Drosophila* is found to differ in plants, and the remaining difference in animals (position 31 is Ala for H3 and either Ser or Thr for H3.3) is also found in plants. Fungi are especially interesting. Ancestrally, they have both H3 and H3.3; however, ascomycetes, which include yeasts and molds, have lost the H3 form. Thus, the obligate RC form of histone 3 that has received the most attention in animals is not even present in yeast.

Studies of H3.3 in bulk chromatin showed that it is enriched in active fractions (Henikoff and Ahmad 2005). However, various factors contributed to the obscurity of this potential "mark" of active chromatin during a time of great excitement in the chromatin field when it was realized that histone modifications can distinguish active from silent chromatin. For one thing, no antibodies were available that could effectively distinguish H3 from H3.3 in chromatin (positions 87–90 are blocked by the DNA gyres in the nucleosome), whereas excellent antibodies against many different posttranslational modifications were readily available. In addition, the seemingly slight sequence differences between H3 and H3.3 did not suggest any fundamental distinctions in chromatin, whereas histone modifications were mostly on tail lysines that were known to affect chromatin interactions or to bind chromatin-associated proteins. This perception that the two histone-3 forms should be interchangeable was confirmed by the finding in *Tetrahymena* that the S-phase form can substitute for its replacement counterpart. Finally, the influential "histone code" hypothesis envisioned nucleosomes as fixed targets of modification enzymes during chromatin differentiation (Jenuwein and Allis 2001). However, it has become increasingly evident that chromatin is highly dynamic, and even heterochromatin-associated proteins bind with residence times of a minute or less (Phair et al. 2004). It appears that the chromatin of actively transcribed genes is in constant flux, characterized by continual histone replacement (Henikoff and Ahmad 2005). The three core differences that distinguish H3 and H3.3 make H3.3•H4 dimers the substrate for RI assembly, and RI assembly itself profoundly changes chromatin. As a result of this process, actively transcribed regions become marked by H3.3 (Fig. 6), and evidence for this process comes from the observation of RI replacement

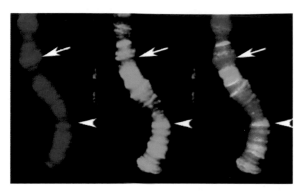

Figure 6. H3.3 Preferentially Localizes to Actively Transcribed Regions of *Drosophila* Polytene Chromosomes

DAPI staining (*red*) shows the DNA banding pattern (*left*), and H3.3-GFP (*green*) localizes to interbands (*middle*), which are sites of RNA polymerase II localization. Right shows the merge. (Reprinted from Schwartz and Ahmad 2005).

of H3 methylated on lysine 9 (H3K9me) with tagged H3.3 at RNA polymerase I and II (pol I and II) transcribed loci (Schwartz and Ahmad 2005).

The dynamic nature of chromatin at active loci results in the erasure of preexisting histone modifications. This provides a potential solution to the problem of how silent chromatin can become activated when it is hypermethylated on H3K9 and H3K27 (histone modifications commonly associated with repressive chromatin). Time-course studies showed that methyls on histones are as stable as the histones themselves (Waterborg 1993), although the recent discovery of a demethylase specific for mono- and di-methyl H3K4 (Shi et al. 2004) indicates that some methyls can be enzymatically removed from histones. In general, patterns of histone covalent modifications might result from modifications already present on the histones at the time that they are deposited. In this way, modification enzymes would track with the assembly machinery, perhaps facilitating the process (Henikoff and Ahmad 2005).

This dynamic assembly model predicts that histone modifications found to be enriched on active chromatin should be enriched on H3.3, and bulk measurements of modifications on H3 and H3.3 have shown this to be the case for both plants and animals. Furthermore, it is expected from this model that active lysine modifications such as acetylation of H3 and H4 and methylation of H3K4 and H3K79 will be strongly correlated with one another, as has been observed in diverse systems (O'Neill et al. 2003; Kurdistani et al. 2004; Schubeler et al. 2004). Finally, dynamic RI assembly at active genes can explain why CAF-1 mutations cause a loss of silencing (Loyola and

Almouzni 2004): Only about 10% of the yeast genome is considered to be in a silent state, and this may be the only chromatin that is not dynamically replaced in the yeast genome. In the absence of CAF-1-mediated RC assembly, RI assembly would occur over the entire yeast genome, activating previously silent regions. Perhaps the existence of an H3 variant dedicated to RC assembly in multicellular eukaryotes is an adaptation to keep the large majority of the chromatin in a cell in an epigenetically silent state.

Replacement by differentially modified H3.3·H4 dimers suggests a simple model for inheritance of active chromatin in dividing cells (Henikoff and Ahmad 2005). Active chromatin would remain active following dilution by ordinary nucleosomes after RC assembly if this random mixture of RI-deposited and RC-deposited nucleosomes does not obstruct active processes such as transcriptional initiation and elongation. Continuation of transcriptional activity as a result would restore chromatin in the next cell cycle, leading to perpetual maintenance of active chromatin throughout development. The possibility that a histone variant is perpetually maintained by an RI assembly process may also hold for CenH3s, which would incorporate into gaps caused by the unraveling of ordinary nucleosomes resulting from anaphase tension.

When cells exit the cell cycle and differentiate, they no longer produce or incorporate S-phase histones, and H3.3 accumulates as a result. For example, H3.3 accumulates in rat brains to a level of 87% of the histone 3 by the time that rats are 400 days old (Henikoff and Ahmad 2005). Whether or not this gradual replacement of chromatin is of functional significance is unknown. It is also unknown whether the active process that allows replacement to occur is the same as that seen at transcriptionally active loci. One possibility is that disruption of chromatin by a transiting RNA polymerase or chromatin-remodeling machine causes local unraveling of the nucleosome and occasional loss of an H3.3·H4 dimer (Fig. 7). This would be followed by reassembly of the nucleosome in the wake of the polymerase with replacement of the lost dimer with an H3.3·H4 dimer by the HirA complex. Only when polymerases are too densely packed for assembly to occur would nucleosomes completely unravel.

7 Phosphorylation of H2AX Functions in DNA Double-Strand Break Repair

The H2A histones also comprise a family of distinct variants found throughout eukaryotes. The H2AX variant is defined by the presence of a carboxy-terminal amino acid sequence motif, SQ(E or D)Θ, where Θ indicates a hydrophobic amino acid. The serine in this sequence motif is the site of phosphorylation producing a modified protein designated "γ-H2AX." The dynamic nature of chromatin, and H2AX phosphorylation, is especially evident when double-strand (ds) breaks occur in DNA (Morrison and Shen 2005). The lethality of even a single ds break requires immediate action to repair the lesion and restore the continuity of the double helix. The detection of a ds break normally occurs within a minute or so of its formation and this, in turn, triggers the rapid phosphorylation of H2AX in the immediate vicinity of a break site. This phosphorylation is carried out by members of the phosphoinositol 3-kinase-like kinase family. Following this initial event, H2AX phosphorylation then spreads quickly along the chromosome marking a relatively large chromatin domain surrounding the break. Finally, the ds break is eventually repaired by either homologous recombination or nonhomologous end-joining, and the phosphorylation mark is removed.

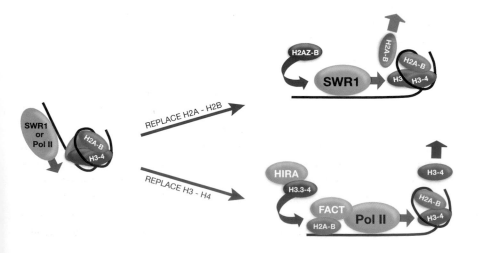

Figure 7. Model for Replication-independent Replacement or Exchange

A large molecular machine (either the SWR1 complex or RNA polymerase II) partially or completely unravels a nucleosome during transit. The result is either retention of heterodimeric subunits, such as the FACT-facilitated transfer of H2A·H2B from in front of RNA polymerase to behind (Formosa et al. 2002; Belotserkovskaya et al. 2003), or loss of a heterodimer. In the latter case, chromatin repair replaces the lost heterodimer with either H2AZ·H2B (*top*) or H3.3·H4 (*bottom*).

Phosphorylation of H2AX is not essential for detection or repair of ds breaks, because deletion of the gene or mutation of the target serine residue does not abolish repair. However, H2AX is not just a marker of damage, since such mutants have reduced efficiency of repair and are hypersensitive to radiation damage and genotoxic agents. Currently, H2AX is thought to function in ds break repair in at least two ways. First, it may help recruit or retain proteins required for repair at the site of the break (Morrison and Shen 2005). Second, it may stabilize the chromosome surrounding the broken ends, through the recruitment of cohesin, the protein complex responsible for keeping sister chromatids together (Lowndes and Toh 2005).

The evolution of H2AX is unlike that of other histone variants. Although a gene for H2AX is found in nearly all eukaryotes, it has had multiple relatively recent origins (Malik and Henikoff 2003). For example, the version of H2AX found in *Drosophila* is different from that found in another dipteran insect, *Anopheles*. Presumably, the ability to evolve a new H2AX from the canonical form of H2A is a consequence of the simple SQ(E or D)Θ motif. Evolving such a motif at the carboxyl terminus of the canonical H2A is expected to occur repeatedly over evolutionary time. Occasional loss of an existing H2AX with a newly minted version might be fueled by the need for H2AX to be very uniformly distributed, because ds breaks can occur anywhere in the genome. If mutations occur in an existing H2AX gene that reduce its similarity to the canonical H2A in such a way that its assembly becomes less efficient or uniform, there will be strong selection to replace it with a version that is more similar to canonical H2A. This rationale could help account for the exceptional case of *Drosophila* H2AX, which, unlike other eukaryotes, is not derived from its canonical H2A, but rather from the distant H2AZ lineage (described below). If all that is necessary to be an H2AX is to be in the H2A position in a nucleosome and to have the carboxy-terminal motif for phosphorylation, an H2AZ can evolve this capability.

ds break repair is clearly the universal function of H2AX phosphorylation, and there would seem to be no stable epigenetic aspect to this process. However, H2AX null mice are sterile, and cytological examination of mammalian spermatogenesis has revealed a striking epigenetic feature, in which H2AX is specifically phosphorylated on the XY bivalent (Fig. 8) (Fernandez-Capetillo et al. 2003). This chromosome pair occupies a distinct "sex body" during meiotic prophase which has been implicated in silencing of sex-linked genes during male meio-

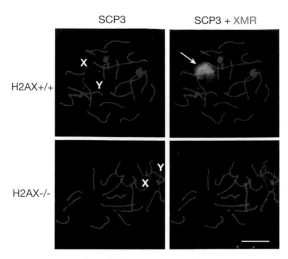

Figure 8. Pachytene Stage of Spermatogenesis Showing the Dependence of Sex Body Formation on H2AX

In normal mammalian spermatocytes, a nuclear structure, the sex body (*arrow, labeled green in right panels*), is seen to encompass the unpaired XY bivalent (*labeled in left panels*). The synaptonemal complex, which aligns paired chromosomes, is stained red. H2AX is normally enriched in the sex body (H2AX^+/+^). In H2AX^−/−^ spermatocytes, the sex body does not form and a sex body epitope becomes dispersed over autosomes (*lower right panel*). Bar, 10 μm. Images courtesy of Shantha Mahadevaiah and Paul Burgoyne (Fernandez-Capetillo et al. 2003).

sis. H2AX phosphorylation is essential for normal sex body formation, and H2AX-deficient spermatocytes fail to pair or condense and fail to inactivate X and Y genes during meiosis. H2AX phosphorylation of the XY bivalent is distinct from the process that occurs at ds breaks. XY phosphorylation in the sex body does not require breaks, but rather occurs most conspicuously at unpaired regions of the chromosomes. The mechanisms whereby H2AX phosphorylation is targeted to unpaired chromosomes, and how this event leads to condensation, pairing, and silencing, are currently unknown. However, it is interesting to speculate that this role may be related to its ability to interact with and recruit cohesin.

8 H2AZ Plays Roles in Transcriptional Regulation

The renaissance of interest in histone variants has been especially strong in the case of H2AZ (or H2A.Z) (Kamakaka and Biggins 2005). H2AZ is nearly ubiquitous, and it diverged from an ancestral H2A early in eukaryotic evolution. Consistent with this separate lineage, genetic experiments in budding yeast and flies have shown that histones H2A and H2AZ have evolved to perform separate nonoverlapping functions. H2AZ is an essential histone in most organisms, from ciliated protozoans to mammals. However, in budding and fission yeasts, deletion of the H2AZ gene produces viable cells, although the null mutants exhibit a variety of phenotypes. These properties have facilitated its genetic and biochemical characterization in yeast.

H2AZ makes up approximately 10% of the total H2A protein in most organisms tested to date. It is widely, but not uniformly, distributed throughout the chromosomes. This is most elegantly visualized in the case of Drosophila polytene chromosomes, where it produces a distinct banding pattern. The results of chromatin immunoprecipitation experiments using yeast and mouse cells are consistent with this pattern. Although H2AZ is preferentially localized to the promoter regions of yeast genes, this specificity is not true for all sites of deposition. In Drosophila, there is no discernible relationship between H2AZ localization and gene expression. Thus, although the mechanism of H2AZ deposition is known (discussed below), at present, the rules that determine where it is concentrated are not.

A variety of observations point to important roles for H2AZ in regulating gene expression (Kamakaka and Biggins 2005). Mutational analysis of budding yeast revealed that the function of H2AZ is partially redundant with two different classes of global transcription factors, the nucleosome-remodeling complex, Swi/Snf, and the histone modification complex, SAGA. Although the individual loss of function of H2AZ, Swi/Snf, or SAGA is viable, the simultaneous loss of any combination of two pathways is lethal. Additional genetic and biochemical experiments suggest that these roles include functions in both transcription initiation and elongation (for more detail, see Chapter 10). Moreover, the balance of H2AZ deposition is causally linked to epigenetics through its role as an anti-silencing factor. Deletion of the H2AZ gene results in extended spreading of silent chromatin inward from the telomeres, and this defect can be suppressed by the additional deletion of genes encoding the silencing factors themselves (Fig. 7) (see Chapter 4). The effect of deleting H2AZ on global gene expression has been assayed using yeast gene microarrays. Although the majority of regulated genes show decreased expression in the H2AZ null mutant, a substantial fraction show an increase in expression. Since it is not yet clear which changes reflect direct regulation and which are indirect, it may be that H2AZ nucleosomes function both positively and negatively to regulate gene transcription. Furthermore, it is not known whether the diverse roles of H2AZ in transcription and heterochromatin stem from a single unifying mechanism or a more complex combination of pathways.

In contrast to the current picture in budding yeast, H2AZ is preferentially located in heterochromatic regions of mammalian cells. Indeed, it has been shown to physically interact with Heterochromatin-associated Protein 1 (HP1) (Fan et al. 2004). Although this might suggest a role for H2AZ in silencing in metazoans, it is worth noting that the subset of expressed genes located in heterochromatin in Drosophila actually requires HP1 for expression (Weiler and Wakimoto 1995). If the location of H2AZ in mammalian cells reflects a similar process, then the clearly established roles for this variant in facilitating transcription and counteracting silencing in yeast would likely reflect general fundamental properties of this variant.

H2AZ may have an additional role in the epigenetics of chromosome segregation. One of the first phenotypes to be recognized for an H2AZ null mutant was a defect in mitotic chromosome segregation observed in fission yeast. More recent experiments have strengthened this connection. The experimental depletion of H2AZ in mammalian cells by RNA interference (RNAi) causes defects in pericentric HP1 association, genome instability, and chromosome mis-segregation (Kamakaka and Biggins 2005). Similarly, in budding yeast, H2AZ null mutants show increased mitotic chromosome loss and significant genetic interactions with genes encoding known components of the centromere and mitotic spindle (Krogan et al. 2004). It remains formally possible that the effect of H2AZ on chromosome segregation is an indirect consequence of its role in setting the program of gene transcription. However, an intriguing hypothesis is that mechanisms of chromosome segregation have evolved to exploit not only an H3 variant, but an H2A variant as well.

How does H2AZ affect transcriptional competence, silencing, heterochromatin, and perhaps chromosome segregation? The high-resolution structure of an H2AZ-containing nucleosome reveals several unique properties of the variant (Suto et al. 2000). Compared with H2A nucleosomes, H2AZ presents an extended acidic patch domain on the surface of the nucleosome, and this difference is likely to have functional significance. For example, it is part of the "docking domain" (Fig. 2) that interacts with histone H4 and defines the segment essential for function in Drosophila. Furthermore, the results of mutational studies and binding experiments in vitro argue that this extended acid patch makes a major contribution to the interaction of the nucleosome with HP1 (Dryhurst et al. 2004). Interestingly, HP1 contains a chromodomain, a protein motif that can bind to methylated H3 lysine 9 (see Chapters 3 and 4). Thus, H2AZ may act in synergy with histone H3 methylation to provide a binding platform for chromatin-associated proteins. In addition to its extended acid patch, H2AZ has a pair of histidine residues that coordinate an additional metal ion in the structure which, in vivo, might provide a unique physiological response that is unavailable to nucleosomes containing H2A. Finally, the crystal structure predicts that an asymmetric histone octamer, made up of one major H2A•H2B dimer plus one variant H2AZ•H2B dimer, would produce a clash of protein structures at Loop 1 (Fig. 2) and seems unlikely to occur in vivo.

Together, these novel features of H2AZ nucleosomes argue that the variant should confer unique physical properties to chromatin. This prediction is borne out experimentally. For example, H2AZ may stabilize dimer–tetramer interactions within the nucleosome, and nucleosome arrays composed of H2AZ nucleosomes can show enhanced higher-order folding and decreased intermolecular aggregation (Dryhurst et al. 2004). Thus, H2AZ is likely to modulate chromatin function in at least three ways. First, it undoubtedly alters the physical properties of its chromatin environment, thus influencing access or activity of *trans*-acting factors. Second, as is the case for other histones, posttranslational modifications within its amino-terminal and carboxy-terminal domains are likely to provide unique docking sites for chromatin-associated proteins (so-called *trans*-effects introduced in Chapter 3), or regulated changes in charge density (*cis*-effects). Third, its restricted and specific deposition in chromatin is likely to target unique functions to specific loci.

9 Protein Complexes for the Deposition and Replacement of H2A Variants

Although important questions still remain as to how H2A histone variants function, recent studies have elucidated the basis for their incorporation into chromatin. The first breakthrough came with the biochemical purification of the complex that catalyzes the transfer of H2AZ•H2B dimers into chromatin (Sarma and Reinberg 2005). This multisubunit complex, termed SWR1-C, contains as its catalytic subunit the protein Swr1, a member of the SWI/SNF family of ATP-dependent chromatin remodelers. In vivo, SWR1-C appears to be dedicated to this task, because the effects of deleting the gene *SWR1* are similar to the effects of deleting the gene encoding H2AZ itself. Furthermore, in a *swr1* null mutant, the preferential deposition of H2AZ at specific loci is completely lost. In vitro, when purified SWR1-C is presented with a nucleosomal array, it specifically replaces H2A•H2B dimers with H2AZ•H2B dimers in an ATP-dependent reaction (Fig. 7). An interesting aspect of this reaction stems from a prediction of the crystal structure mentioned above: Mixed nucleosomes containing both H2A and H2AZ should not be stable. Thus, the dimer replacement mediated by SWR1-C may be a concerted reaction in which the substitution of one H2AZ•H2B dimer facilitates the ejection and replacement of the remaining H2A•H2B dimer.

A second multisubunit protein complex carries out the replacement of phosphorylated H2AX with an unphosphorylated molecule in *Drosophila* (Morrison and Shen 2005). Remarkably, this single *Drosophila* complex, termed dTip60, is composed of proteins ordinarily found in two separate complexes: SWR1-C, the ATP-dependent chromatin-remodeling complex described above, and NuA4/Tip60, a histone modification complex with acetyltransferase activity. In vitro, the reaction requires both ATP and acetyl-CoA. Thus, this one complex integrates histone acetylation, nucleosome remodeling, and histone variant replacement. This combination likely reflects the fact that *Drosophila* H2AX is also its H2AZ, whereas H2AX in other eukaryotes evolved from canonical H2A. Despite this difference, there are reasons to expect that the basic pathway is conserved. In budding yeast and mammalian cells, SWR1-C, NuA4/Tip60, and another ATP-dependent nucleosome-remodeling complex, INO80-C, share common subunits. One of these is the actin-related protein Arp4. Interestingly, Arp4 has been shown to interact with phosphorylated H2AX in budding yeast and to result in the sequential recruitment of NuA4, SWR1, and INO80 complexes (Downs et al. 2004). This suggests that these complexes catalyze the replacement of both H2AX and H2AZ in this organism as well. This prediction remains to be demonstrated directly.

The discovery that chromatin-remodeling complexes are dedicated to RI nucleosome assembly is important not just for understanding how histone variants are incorporated, but also for providing the first specific in vivo functions for chromatin-remodeling machines. Prior to these discoveries, it was not clear why cells would have such an abundance of large machines that facilitate the movement of nucleosomes (Becker and Horz 2002). The diversity of SWI/SNF ATPases presented a puzzle that now can perhaps be better understood if some remodeling machines are dedicated to the assembly of different variants into nucleosomes. Perhaps nucleosome assembly is a concerted process in which histone-modifying enzymes act on their substrates while ATP-dependent chromatin remodelers provide the power stroke and specificity needed for RI replacement.

10 Other H2A Variants Differentiate Chromatin but Their Functions Are As Yet Unknown

Further diversification of H2A has occurred in vertebrates. In mammals, macroH2A and H2A[Bbd] (H2A Barr body deficient) represent unique lineages that appear to play roles in the epigenetic phenomenon of dosage compensation (discussed in detail in Chapter 17). macroH2A is so-called because in addition to the histone fold domain and amino- and carboxy-terminal tails, it contains a large carboxy-terminal globular domain (Ladurner 2003). Considering that the H2A carboxy-terminal tail exits near the linker DNA, it is possible that this globular domain interacts with linkers, H3 tails, or linker proteins such as H1 and High Mobility Group (HMG) proteins. Just what this interaction would be is unknown, although an intriguing possibility is that it has an enzymatic activity. This possibility is encouraged by the resemblance of the 200-amino acid globular domain to proteins with hydrolytic activities on polynucleotides and peptides. Alternatively, the globular domain might simply act as an impediment to transcriptional initiation, a role suggested by its ability to block transcription factors from binding in vitro (Sarma and Reinberg 2005). The histone fold domain of macroH2A also has distinct properties, as it is not acted upon by chromatin remodelers. These observations suggest that macroH2A-containing nucleosomes are less mobile and so may be resistant to active transcription. This might account for the enrichment of macroH2A in discrete regions of the facultatively inactive X chromosome of human females that alternate with regions of constitutive heterochromatin (Fig. 9a) (Chadwick and Willard 2004).

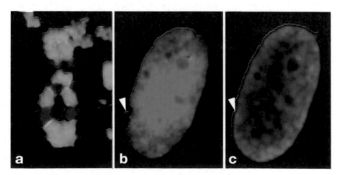

Figure 9. H2A Variants and the Inactive X Chromosome of Human Females

(*a*) macroH2A (*red*) stains discrete regions of the inactive X chromosome that alternate with a marker for heterochromatin (histone H3K9me3). (*b*) H2A[Bbd] (*green*) is excluded from the inactive X chromosome (*red dot with arrow pointing to it*). (*c*) Same nucleus as in *b*, but stained with DAPI to show chromatin. (*a*, Reprinted, with permission, from Chadwick and Willard 2004 [© National Academy of Sciences]; *b,c*, reprinted, with permission, from Chadwick and Willard 2001 [© The Rockefeller University Press].)

In contrast to macroH2A, H2A[Bbd] appears to be undetectable on the Barr body, but otherwise ubiquitous throughout the nucleus (Fig. 9b) (Chadwick and Willard 2001). The in vitro behavior of H2A[Bbd]-containing nucleosomes is consistent with its playing a role in facilitating transcription (Sarma and Reinberg 2005). H2A[Bbd] is rapidly evolving relative to other known H2A isoforms, although the reason for this accelerated evolution is not clear.

11 Many Histones Have Evolved to More Tightly Package DNA

When it is no longer necessary to gain access to DNA for replication and transcription, chromatin typically becomes further condensed, and this often involves replacement of canonical histones. This is obviously the case for sperm, and in some lineages, histone paralogs have evolved specialized packaging roles. For example, sea urchin sperm contains H1 and H2B variants with repeated tail motifs that bind to the minor grooves of DNA (Malik and Henikoff 2003), presumably an adaptation to tightly package chromosomes for inclusion into sperm heads. A similar adaptation is found in pollen-specific H2A variants in flowering plants. In vertebrates, sperm-specific specialized histone variants are found in mammalian testes, including an H2B paralog (SubH2Bv) that localizes to the acrosome and a testes-specific H3 variant (Witt et al. 1996).

The replacement of histones during sperm maturation by protamines and other proteins provides a potential means of erasing epigenetic information in the male germ line. However, evidence for trans-generational inheritance (Rakyan and Whitelaw 2003), especially in animals that lack DNA methylation, raises the possibility that a subset of nucleosomal histones survive this transition and transmit epigenetic information. As already pointed out, this is just what occurs for CENP-A at centromeres (Palmer et al. 1990), and it is possible that a small fraction of other variants, such as H3.3, remain with sperm for epigenetic inheritance of gene expression information. Although our understanding of the process that replaces histones during sperm development is rudimentary, we expect that much can be learned by understanding how CENP-A survives this transition.

Increased compaction also occurs in somatic cells that have finished dividing and undergo differentiation. In some cases, compaction involves quantitative and qualitative changes in linker histones. The stoichiometry of histone H1 relative to nucleosomes determines the average spacing within nucleosome arrays in vivo (Fan et al. 2003). In addition, the presence of H1 in chromatin promotes higher-order chromatin structure that generally inhibits transcription (Wolffe 1992). Linker histones are much more mobile than core histones in vivo. Residence times for H2A and H2B are hours in length, and cannot even be measured for H3 and H4, whereas the residence time of H1 is a few minutes (Phair et al. 2004). As a result, the incorporation of variant linker histones is unlikely to differentiate chromatin in a heritable manner. Rather, the role of H1 variants is thought to change the bulk properties of chromatin that can affect overall compaction (Wolffe 1992).

H1 variants share with core histones a distinction between RC and RI forms (Marzluff et al. 2002). RC variant forms of H1 appear to be interchangeable with one another, based on the fact that knock-out mice lacking one or two of the five RC H1 variants are phenotypically normal (Fan et al. 2003). In birds, the H5 linker histone variant is deposited during erythrocyte maturation, which accompanies extreme compaction of the nucleus. Another variant that is deposited at high levels in nondividing cells is H1^0, which is highly diverged from the canonical forms. Overexpression of H1^0 renders chromatin less accessible to nucleases than similar overexpression of a canonical form. The natural accumulation of H1^0 in nondividing cells might be a general mechanism for chromatin compaction as cells become quiescent.

12 Conclusions and Future Research

Histone variants provide the most fundamental level of differentiation of chromatin, and alternative mechanisms for depositing different variants can potentially establish and maintain epigenetic states. Histones H2A, H2B, H3, and H4 occupy distinct positions in the core particle as a result of an evolutionary process that began before the last common ancestor of eukaryotes. Key evolutionary innovations remain uncertain, including the emergence of an octamer from an ancestral H3•H4-like tetramer, and we look forward to the sequencing of more archaeal and primitive eukaryotic genomes that might provide missing links. Subsequent elaborations of the four core histones into distinct variants have provided the basis for epigenetic processes, including development and chromosome segregation. For a full understanding of epigenetic inheritance, we need a better understanding of the processes that incorporate variants by replacing canonical histones. An important recent development is the initial characterization of replication-independent assembly pathways dedicated to particular variants.

Centromeres are the most conspicuous examples of profoundly different chromatin that is attributable to special properties of a histone variant. Although it is clear that CenH3-containing nucleosomes form the foundation of the centromere, just how they are deposited in the same place every cell generation without any hint of sequence specificity is a major challenge for future research.

It is becoming evident that histone variants are also involved in epigenetic properties of active genes. Both H3.3 and H2AZ are enriched at transcriptionally active loci, and understanding the assembly processes that are responsible for their enrichment is an exciting area of current research. The dynamic behavior of chromatin leads to the realization that transcription, chromatin remodeling, and histone modification might be coupled to nucleosome assembly and disassembly. The study of dynamic processes coupled to histone turnover is only at an early stage, and we look forward to technological advances in molecular biology, cytogenetics, biochemistry, and structural biology that can be harnessed to better understand the dynamic nature of chromatin.

In addition to these universal processes, histone variants are also involved in particular epigenetic phenomena. In the case of the mammalian X chromosome, three different H2A variants, phospho-H2AX, macroH2A, and H2A^{Bbd}, have been recruited to participate in silencing or activation of genes for purposes of germ-line inactivation or dosage compensation. Understanding the function of these variants in epigenetic processes remains a major challenge for the future.

The availability of the first high-resolution structure of the nucleosome core particle (Luger et al. 1997) was a seminal advance in elucidating the properties of chromatin. By elaborating this basic structure in a way that has biological consequences, histone variants provide an opportunity to deepen our understanding of how these fascinating architectural proteins have evolved to play diverse roles in epigenetic processes.

References

Altheim B.A. and Schultz M.C. 1999. Histone modification governs the cell cycle regulation of a replication-independent chromatin assembly pathway in *Saccharomyces cerevisiae*. *Proc. Natl. Acad. Sci.* **96**: 1345–1350.

Amor D.J., Kalitsis P., Sumer H., and Choo K.H. 2004a. Building the centromere: From foundation proteins to 3D organization. *Trends Cell Biol.* **14**: 359.

Amor D.J., Bentley K., Ryan J., Perry J., Wong L., Slater H., and Choo K.H. 2004b. Human centromere repositioning "in progress". *Proc. Natl. Acad. Sci.* **101**: 6542–6547.

Annunziato A.T. 2005. Split decision: What happens to nucleosomes during DNA replication? *J. Biol. Chem.* **280**: 12065–12068.

Arents G., Burlingame R.W., Wang B.C., Love W.E., and Moudrianakis E.N. 1991. The nucleosomal core histone octamer at 3.1 Å resolution: A tripartite protein assembly and a left-handed superhelix.

Proc. Natl. Acad. Sci. **88:** 10148–10152.

Becker P.B. and Horz W. 2002. ATP-dependent nucleosome remodeling. *Annu. Rev. Biochem.* **71:** 247–273.

Belotserkovskaya R., Oh S., Bondarenko V.A., Orphanides G., Studitsky V.M., and Reinberg D. 2003. FACT facilitates transcription-dependent nucleosome alteration. *Science* **301:** 1090–1093.

Chadwick B.P. and Willard H.F. 2001. A novel chromatin protein, distantly related to histone H2A, is largely excluded from the inactive X chromosome. *J. Cell Biol.* **152:** 375–384.

———. 2004. Multiple spatially distinct types of facultative heterochromatin on the human inactive X chromosome. *Proc. Natl. Acad. Sci.* **101:** 17450–17455.

Downs J.A., Allard S., Jobin-Robitaille O., Javaheri A., Auger A., Bouchard N., Kron S.J., Jackson S.P., and Cote J. 2004. Binding of chromatin-modifying activities to phosphorylated histone H2A at DNA damage sites. *Mol. Cell* **16:** 979–990.

Dryhurst D., Thambirajah A.A., and Ausio J. 2004. New twists on H2A.Z: A histone variant with a controversial structural and functional past. *Biochem. Cell Biol.* **82:** 490–497.

Fan Y., Nikitina T., Morin-Kensicki E.M., Zhao J., Magnuson T.R., Woodcock C.L., and Skoultchi A.I. 2003. H1 linker histones are essential for mouse development and affect nucleosome spacing in vivo. *Mol. Cell. Biol.* **23:** 4559–4572.

Fernandez-Capetillo O., Mahadevaiah S.K., Celeste A., Romanienko P.J., Camerini-Otero R.D., Bonner W.M., Manova K., Burgoyne P., and Nussenzweig A. 2003. H2AX is required for chromatin remodeling and inactivation of sex chromosomes in male mouse meiosis. *Dev. Cell* **4:** 497–508.

Formosa T., Ruone S., Adams M.D., Olsen A.E., Eriksson P., Yu Y., Roades A.R., Kaufman P.D., and Stillman D.J. 2002. Defects in SPT16 or POB3 (yFACT) in *Saccharomyces cerevisiae* cause dependence on the Hir/Hpc pathway: Polymerase passage may degrade chromatin structure. *Genetics* **162:** 1557–1571.

Henikoff S. and Ahmad K. 2005. Assembly of variant histones into chromatin. *Annu. Rev. Cell Dev. Biol.* **21:** 133–153.

Jenuwein T. 2001. Re-SET-ting heterochromatin by histone methyltransferases. *Trends Cell Biol.* **11:** 266–273.

Jenuwein T. and Allis C.D. 2001. Translating the histone code. *Science* **293:** 1074–1080.

Kamakaka R.T. and Biggins S. 2005. Histone variants: Deviants? *Genes Dev.* **19:** 295–310.

Krogan N.J., Baetz K., Keogh M.C., Datta N., Sawa C., Kwok T.C., Thompson N.J., Davey M.G., Pootoolal J., Hughes T.R., et al. 2004. Regulation of chromosome stability by the histone H2A variant Htz1, the Swr1 chromatin remodeling complex, and the histone acetyltransferase NuA4. *Proc. Natl. Acad. Sci.* **101:** 13513–13518.

Kurdistani S.K., Tavazoie S., and Grunstein M. 2004. Mapping global histone acetylation patterns to gene expression. *Cell* **117:** 721–733.

Ladurner A.G. 2003. Inactivating chromosomes: A macro domain that minimizes transcription. *Mol. Cell* **12:** 1–3.

Lowndes N.F. and Toh G.W. 2005. DNA repair: The importance of phosphorylating histone H2AX. *Curr. Biol.* **15:** R99–R102.

Loyola A. and Almouzni G. 2004. Histone chaperones, a supporting role in the limelight. *Biochim. Biophys. Acta* **1677:** 3–11.

Luger K., Mader A.W., Richmond R.K., Sargent D.F., and Richmond T.J. 1997. Crystal structure of the nucleosome core particle at 2.8 Å resolution. *Nature* **389:** 251–260.

Malik H.S. and Henikoff S. 2003. Phylogenomics of the nucleosome. *Nat. Struct. Biol.* **10:** 882–891.

Marc F., Sandman K., Lurz R., and Reeve J.N. 2002. Archaeal histone tetramerization determines DNA affinity and the direction of DNA supercoiling. *J. Biol. Chem.* **277:** 30879–30886.

Marzluff W.F. and Duronio R.J. 2002. Histone mRNA expression: Multiple levels of cell cycle regulation and important developmental consequences. *Curr. Opin. Cell Biol.* **14:** 692–699.

Marzluff W.F., Gongidi P., Woods K.R., Jin J., and Maltais L.J. 2002. The human and mouse replication-dependent histone genes. *Genomics* **80:** 487–498.

Morrison A.J. and Shen X. 2005. DNA repair in the context of chromatin. *Cell Cycle* **4:** 568–571.

O'Neill L.P., Randall T.E., Lavender J., Spotswood H.T., Lee J.T., and Turner B.M. 2003. X-linked genes in female embryonic stem cells carry an epigenetic mark prior to the onset of X inactivation. *Hum. Mol. Genet.* **12:** 1783–1790.

Palmer D.K., O'Day K., and Margolis R.L. 1990. The centromere specific histone CENP-A is selectively retained in discrete foci in mammalian sperm nuclei. *Chromosoma* **100:** 32–36.

Palmer D.K., O'Day K., Trong H.L., Charbonneau H., and Margolis R.L. 1991. Purification of the centromere-specific protein CENP-A and demonstration that it is a distinctive histone. *Proc. Natl. Acad. Sci.* **88:** 3734–3738.

Phair R.D., Scaffidi P., Elbi C., Vecerova J., Dey A., Ozato K., Brown D.T., Hager G., Bustin M., and Misteli T. 2004. Global nature of dynamic protein-chromatin interactions in vivo: Three-dimensional genome scanning and dynamic interaction networks of chromatin proteins. *Mol. Cell. Biol.* **24:** 6393–6402.

Rakyan V. and Whitelaw E. 2003. Transgenerational epigenetic inheritance. *Curr. Biol.* **13:** R6.

Sarma K. and Reinberg D. 2005. Histone variants meet their match. *Nat. Rev. Mol. Cell Biol.* **6:** 139–149.

Schubeler D., MacAlpine D.M., Scalzo D., Wirbelauer C., Kooperberg C., van Leeuwen F., Gottschling D.E., O'Neill L.P., Turner B.M., Delrow J., et al. The histone modification pattern of active genes revealed through genome-wide chromatin analysis of a higher eukaryote. *Genes Dev.* **18:** 1263–1271.

Schwartz B.E. and Ahmad K. 2005. Transcriptional activation triggers deposition and removal of the histone variant H3.3. *Genes Dev.* **19:** 804–814.

Shi Y., Lan F., Matson C., Mulligan P., Whetstine J.R., Cole P.A., Casero R.A., and Shi Y. 2004. Histone demethylation mediated by the nuclear amine oxidase homolog LSD1. *Cell* **119:** 941–953.

Sullivan B.A. and Karpen G.H. 2004. Centromeric chromatin exhibits a histone modification pattern that is distinct from both euchromatin and heterochromatin. *Nat. Struct. Mol. Biol.* **11:** 1076–1083.

Suto R.K., Clarkson M.J., Tremethick D.J., and Luger K. 2000. Crystal structure of a nucleosome core particle containing the variant histone H2A.Z. *Nat. Struct. Biol.* **7:** 1121–1124.

Talbert P.B., Bryson T.D., and Henikoff S. 2004. Adaptive evolution of centromere proteins in plants and animals. *J. Biol.* **3:** 18.

Waterborg J.H. 1993. Dynamic methylation of alfalfa histone H3. *J. Biol. Chem.* **268:** 4918–4921.

Weiler K.S. and Wakimoto B.T. 1995. Heterochromatin and gene expression in *Drosophila*. *Annu. Rev. Genet.* **29:** 577–605.

Wieland G., Orthaus S., Ohndorf S., Diekmann S., and Hemmerich P. 2004. Functional complementation of human centromere protein A (CENP-A) by Cse4p from *Saccharomyces cerevisiae*. *Mol. Cell. Biol.* **24:** 6620–6630.

Witt O., Albig W., and Doenecke D. 1996. Testis-specific expression of a novel human H3 histone gene. *Exp. Cell Res.* **229:** 301–306.

Wolffe A.P. 1992. *Chromatin: Structure and function*. Academic Press, San Diego.

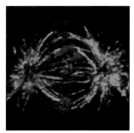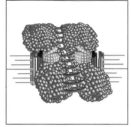

Epigenetic Regulation of Chromosome Inheritance

Gary H. Karpen[1] and R. Scott Hawley[2]

[1] Lawrence Berkeley National Laboratory, Department of Genome Biology, University of California at Berkeley, Department of Molecular and Cell Biology, Berkeley, California 94720
[2] Stowers Institute for Medical Research, Kansas City, Missouri 64110 and Department of Physiology, University of Kansas Medical Center, Kansas City, Missouri 66160

CONTENTS

1. Introduction, 267
   1.1   How Is Chromosome Inheritance Accomplished?, 267
   1.2   What Elements Are Required for Chromosome Inheritance?, 267

2. Epigenetic Regulation of DNA Replication, Repair, and Telomeres, 268
   2.1   Initiation of DNA Replication Is Controlled by Epigenetic Mechanisms, 268
   2.2   DNA Repair Involves Epigenetic Alterations in Chromatin Structure, 269
   2.3   Epigenetic Control of Telomere Structure and Function, 270

3. Epigenetic Regulation of Centromere Identity and Function, 273
   3.1   Centromere Structure and Function In Different Eukaryotes, 273
   3.2   Centromeric Sequences Are Not Necessary or Sufficient for Kinetochore Formation and Function, 274
   3.3   The Unusual Composition of Centromeric Chromatin, 275
   3.4   Models for Centromere Structure, Function, and Propagation, 278
   3.5   Epigenetics and Centromere Evolution, 279

4. Heterochromatin and Meiotic Pairing/Segregation, 280
   4.1   Discovery of a Heterochromatic Pairing Site in Drosophila Males, 280

4.2   Heterochromatin Pairing Facilitates Segregation in Drosophila Females, 281
4.3   Role of the Centromere in Facilitating Achiasmate Segregation in Budding Yeast, 282
4.4   The Heterochromatin-associated Ph1 Locus in Maize and Its Role in Mediating Homologous Versus Homeologous Pairing, 283

5. Heterochromatin and Meiotic Drive, 283
   5.1   Segregation Distorter in Drosophila Males, 283
   5.2   Paternal Chromosome Loss in Sciara and Mapping of the Response Element, 284
   5.3   Paternal Chromosome Loss in Nasonia, 284
   5.4   Knob 10 in Corn—The Role of Heterochromatic "Knob" Sequences in Facilitating Chromosome Segregation at Meiosis I, 284

6. The Silencing of Genes by Unpaired DNA during Meiosis, 285
   6.1   Meiotic Silencing of Unpaired DNA during Meiosis in Neurospora, 285
   6.2   Silencing of Unsynapsed Chromosomes in the Mouse, 286
   6.3   Sex Chromosome Dysfunction in Drosophila, 286

7. Perspectives and Conclusions, 286

References, 287

GENERAL SUMMARY

The duplication and transmission of genetic information are accomplished by two types of cell division, mitosis and meiosis, both of which are fundamental to life and evolution. Mitosis is the nuclear division that occurs in somatic cells, involving the identical partitioning of duplicated genetic material by way of chromosomes to daughter cells. Meiosis is a reductional nuclear division that occurs only in the germ cells of multicellular organisms or at particular stages of a unicellular life cycle to produce cells with a haploid genome prior to fertilization (or conjugation in some eukaryotes). Abnormal DNA replication or repair results in mutations and chromosome rearrangements. Perhaps more importantly, chromosome missegregation during nuclear division causes loss or gain of whole chromosomes (aneuploidy). These kinds of "genome instability" affect the viability of cells and fertility of all eukaryotes. Moreover, they play key roles in the etiology of human birth defects and cancer.

Early studies suggested that DNA sequence played a predominant role in specifying the sites and functions of chromosomal elements required for proper mitosis and meiosis, such as origins of DNA replication, sites of spindle attachment (centromeres and kinetochores), chromosome ends (telomeres), and meiotic pairing sites. However, in the last decade we have come to understand that epigenetic mechanisms regulate many key functions required for genome stability and chromosome inheritance. These include roles in the initiation of DNA replication, DNA repair and recombination, chromosome end protection (telomeres), chromosome movement (centromeres), and segregation of homologous chromosomes in meiosis. At first glance, epigenetic regulation appears to be at odds with the fact that these chromosomal functions are essential for cell and organismal viability, which implies that they should be "hard-wired" in the DNA sequence. However, when viewed through the lens of evolution, epigenetic "plasticity" of chromosomes during mitosis and meiosis appears to be important to compensate for the types of sequence changes and chromosome rearrangements associated with speciation. Understanding the molecular basis for epigenetic regulation of inheritance is fundamental to elucidating these basic biological processes, and for the diagnosis and treatment of human diseases.

1 Introduction

1.1 How Is Chromosome Inheritance Accomplished?

Mitosis is a basic type of cell division that produces identical, diploid daughter cells, and is utilized by somatic cells and premeiotic germ cells (Fig. 1a). There are four phases to the mitotic cell cycle, called G_1 (gap 1, a "resting" stage after mitosis), S (synthesis, DNA replication and gene expression), G_2 (gap 2, "resting" after S, preparation for mitosis), and M (mitosis, consisting of prophase, metaphase, anaphase, and telophase). During S phase, DNA is replicated, and the duplicated sister chromatids are held together by the establishment of cohesion. At the beginning of mitosis, chromosomes condense, and histone H3 becomes phosphorylated at Ser-10 (H3S10ph) (see Chapter 10). In addition, in most organisms, a single site (the centromere) on each sister chromatid forms a structure referred to as the kinetochore, which mediates attachment to spindle microtubules and serves as a cell cycle checkpoint (Fig. 1a). Pairs of sister chromatids congress to the metaphase plate in prometaphase, and segregate to the poles in anaphase. These movements are achieved by both the activities of kinetochore-associated microtubule motors (kinesins and dyneins) and regulation of microtubule assembly and disassembly, and also require destruction of sister chromatid cohesion during the metaphase–anaphase transition.

Meiosis occurs only in germ cells and is characterized by one round of replication followed by two divisions (meiosis I and II, Fig. 1b); this produces haploid eggs in the female germ line and haploid sperm in the male germ line of metazoans, rather than diploid daughter cells. In meiosis I, the replicated homologs are paired and segregate together. The sister chromatids of each homolog do not segregate from each other until meiosis II. Normal segregation during meiosis requires frequent recombination between homologs, as well as specialized cohesion in the centromere region that ensures the association of sister chromatids during meiosis I (Watanabe 2005).

1.2 What Elements Are Required for Chromosome Inheritance?

Both mitotic and meiotic cell divisions require the activities of specific chromosomal elements and binding proteins to accomplish accurate genome duplication and

a

S-PHASE ------ **MITOSIS** ------

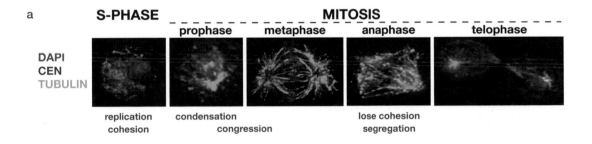

b

PREMEIOTIC ------ **MEIOSIS I PROPHASE** ------
S-PHASE

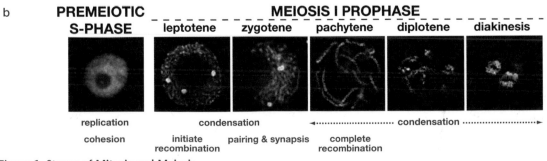

Figure 1. Stages of Mitosis and Meiosis

(*a*) Images from *Drosophila* cells indicate the behaviors of chromosomes (*blue*, text descriptions below), microtubules (*green*), and centromeres (*red*) in interphase and mitosis. (*b*) Chromosome behaviors are shown for maize meiosis I prophase, which is the stage in which homolog pairing, synapsis, and recombination occur (images supplied by Hank Bass and Shaun Murphy, Florida State University). Key chromosome functions that occur during each stage are indicated below (*blue text*). Subsequently, homologs segregate to opposite poles during meiosis I anaphase, completing a reductional division. Sister chromatids only separate during meiosis II (see Fig. 9).

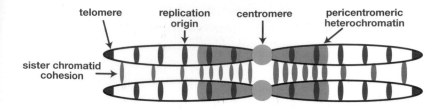

Figure 2. Chromosome Inheritance Elements

The diagram indicates the chromosomal elements essential for normal duplication (replication origins) and inheritance (centromeres, cohesion, telomeres) through mitosis and meiosis. Normal meiotic segregation also requires homolog pairing sites (not shown) and, in most cases, recombination.

chromosome segregation (Fig. 2). DNA replication is initiated at "origins," which in most eukaryotes are not strictly sequence dependent (discussed in Section 2). Sister chromatid cohesion then becomes visible along the entire length of the chromatids in mitotically dividing cells, although there is a higher concentration of cohesins in pericentromeric heterochromatin. Centromeres are large regions composed of DNA and specialized chromatin proteins that serve as the foundation for kinetochore formation and are critical for spindle attachments and normal meiotic and mitotic chromosome segregation (discussed in Section 3). In most eukaryotes, there is one and only one centromere per chromosome. Loss of the centromere results in spindle attachment failures and chromosome loss, and the presence of more than one centromere leads to attachments of the same chromatid to both poles, which causes chromosome bridges and fragmentation during anaphase. Rarely, organisms (e.g., the roundworm *Caenorhabditis elegans*) contain "polycentric" or "holocentric" chromosomes, in which kinetochores are present in multiple regions (for example, see Fig. 5 of Chapter 13). These chromosomes utilize special mechanisms to ensure attachment and segregation of sister chromatids to opposite poles. Telomeres are specialized chromatin structures found at the ends of chromosomes to protect them from degradation or recombination, and ensure complete DNA duplication. Meiotic segregation also requires centromeres, telomeres, cohesion, and origins of replication. However, additional elements and modification of centromere behavior are required to ensure homolog pairing and segregation in meiosis I (discussed in Section 4).

The essential nature of chromosome inheritance suggests that the specification and localization of inheritance elements should be "hard-wired" in the DNA sequence. Thus, it is surprising that many elements, including those outlined in this section, are instead regulated epigenetically, especially in multicellular eukaryotes. In summary, elements that are prone to epigenetic regulation to ensure faithful chromosome inheritance include DNA replication origins, telomeres, sister chromatid cohesion sites, centromeres, and homolog pairing sites.

2 Epigenetic Regulation of DNA Replication, Repair, and Telomeres

The first step in ensuring inheritance of genetic information involves the faithful duplication of the entire genome, which is accomplished by a process known as DNA replication. Unfortunately, errors occur during replication, causing changes in the DNA (mutations), which can be harmful to organismal viability. In addition, environmental agents such as radiation produce mutations, including DNA base changes, deletions, insertions, and rearrangements. Cells respond to DNA damage by activating DNA repair pathways, which do an amazing job of maintaining genome fidelity and stability. Finally, duplication of linear DNA molecules poses challenges that are overcome by the presence of specialized sequences and structures at chromosome ends, known as telomeres. Recent studies have shown that these basic processes required for accurate duplication and maintenance of DNA sequences are affected by epigenetic mechanisms that regulate chromatin.

2.1 Initiation of DNA Replication Is Controlled by Epigenetic Mechanisms

The faithful copying of DNA is accomplished by the 5′ to 3′ action of DNA polymerases, and starts at specific sites called "origins." A "bubble" consisting of two replication "forks" is formed at origins, and replication proceeds bidirectionally until forks generated by the next origins are met. Domains replicated from a single origin are usually quite large, covering hundreds of kilobases (Aladjem and Fanning 2004). Origins in the yeast *Saccharomyces cerevisiae* (also known as ARSs, for autonomously replicating sequences) function ectopically when cloned into plasmids, and usually upon integration into other chromosomal sites, indicating that initiation of replication is regulated by specific DNA sequences. ARSs are approximately 100–150 bp in size and contain one or more copies of an essential approximately 11-bp AT-rich sequence, plus other conserved elements (Weinreich et al. 2004). A protein complex known as the origin recognition complex (ORC) is required for initiation and is responsible for the recruitment of the prereplication complex (PRC) that includes minichromosome maintenance (MCM) proteins (Prasanth et al. 2004). For metazoans, replication origins can be identified in situ, but they are typically inactive upon cloning and reintroduction, suggesting that initiation in these organisms is regulated epigenetically, rather than by strict DNA sequence dependence (Aladjem and Fanning 2004). Although ORC and MCM proteins are conserved in metazoans, the factors and mechanisms responsible for regulating origin activity in these organisms remain mysterious. In addition, it is unclear whether chromatin structure affects the processivity of replication forks.

Clues about how metazoan origins might be regulated epigenetically come from detailed studies in *S. cerevisiae*. Although DNA sequences at origins are necessary and, for the most part, sufficient for replication initiation in this organism, there is clear evidence for chromatin structure effects on origin activity (Weinreich et al. 2004). Microarray analysis has shown that not all of the 332 sites of bidirectional replication, or the 429 sites bound by ORC and MCM proteins, are active in every cell cycle. Chromosomal context and chromatin structure affect the ability of a putative origin to be active in replication initiation. For example, an ARS located in the silenced mating-type loci cannot initiate replication unless moved to other chromosome locations. Another example of epigenetic regulation of origins involves the approximately 100–200 genes encoding ribosomal RNA (rDNA), which are present in a tandemly repeated array (Pasero et al. 2002). Each 9.1-kb rDNA unit contains an ARS, which initiates replication when inserted into a plasmid or elsewhere in the genome. However, only about 20% of rDNA origins are active during each S phase.

Origins can also be regulated to fire at different stages during S phase (Weinreich et al. 2004). For example, origins near telomeres are normally active in late S phase

(see Section 2.3), but insertion of the same sequences into circular plasmids results in replication early in S phase (Fig. 3) (Ferguson and Fangman 1992). Late replication can be recapitulated by linearization of the plasmid and telomere addition, which places the ARSs near the ends.

These results demonstrate that both origin activity and the timing of origin firing during S phase are regulated epigenetically. Further evidence comes from the observations that S-phase timing, and origin silencing at rDNA, telomeres, and mating-type loci, are regulated by many of the histone modifications and proteins responsible for epigenetic gene silencing, including the SIR proteins (see Chapter 4).

2.2 DNA Repair Involves Epigenetic Alterations in Chromatin Structure

Accumulation of DNA damage, due to replication errors or exposure to environmental agents, can lead to deleterious mutations, genome instability, cancer, cell senescence, and death. DNA damage is repaired by error-correction mechanisms during DNA replication, as well as independent pathways that act during G_2. Cells contain "checkpoints" that identify the presence of DNA damage and arrest or delay the cell cycle until repair is complete; these pathways also induce cell death (apoptosis) if the damage is not repaired, which contributes to organismal viability by removing defective cells. These processes are normally extremely efficient; for example, human skin cells exposed to the UV radiation present in sunlight contain a surprisingly large number of DNA lesions, the vast majority of which are properly repaired (Friedberg et al. 1995). Individuals who are deficient in repair, due to mutations in one or more components of the checkpoint or repair pathways, suffer from a variety of diseases, including predispositions to colon, breast, and skin cancers, and premature aging.

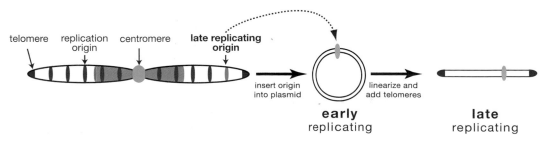

Figure 3. Epigenetic Regulation of Replication Timing in Yeast

One of the best examples of epigenetic effects on replication was demonstrated in the yeast *Saccharomyces cerevisiae*. Insertion of a late-replicating origin into a circular plasmid results in early replication in S phase, and late replication is restored upon linearization of the plasmid and addition of telomeres (Ferguson and Fangman 1992; Weinreich et al. 2004).

Early studies successfully identified molecules and pathways that recognize different types of lesions in DNA, such as double-strand breaks (DSBs) and pyrimidine dimers, and that recruit specific complexes to repair the damage. However, the packaging of DNA into chromatin could potentially block access of factors involved in recognizing sites of damage, or effecting repair, similar to the repressive impact of heterochromatin on gene expression (Hassa and Hottiger 2005). Repair, however, employs ATP-dependent chromatin-remodeling complexes, which presumably act to "expose" the defective DNA for repair.

More recent studies have shown that specific changes to the chromatin template, such as the presence of histone variants and posttranslational histone modifications, play key roles in the recognition of DNA lesions and the recruitment of the appropriate repair machinery (Hassa and Hottiger 2005). For example, the presence of DSBs results in rapid phosphorylation of the histone H2A variant H2AX at serines 136 and 139 (known as γH2AX). H2AX phosphorylation is required for the accumulation of repair proteins to large (megabase) regions that surround DSBs and for the assembly of repair "foci," rather than the initial recruitment of repair complexes to the primary sites of DNA damage (Bassing et al. 2002; Celeste et al. 2002). These observations suggest that γH2AX "spreading" from DSBs acts to amplify the signal emanating from DSBs, enhancing the recruitment and perhaps retention of repair factors (Fernandez-Capetillo et al. 2004). In addition to involvement in DSB repair, H2AX phosphorylation affects V(D)J recombination in mammalian lymphocytes and also acts as a suppressor of genome instability and tumors in mice (Fernandez-Capetillo et al. 2004).

Other types of chromatin changes, such as histone acetylation, SUMOylation, ubiquitination, and methylation have also demonstrated roles in successful repair of DNA lesions. For example, methylation of histone H3 at lysine 79 (H3K79me) is required for recruitment of the repair checkpoint protein 53BP1 to DSBs, and is mediated by the DOT1 histone lysine methyltransferase (HKMT) (Huyen et al. 2004). Interestingly, induced DNA damage does not change H3K79 methylation levels, suggesting that this modification is not added in response to DSBs. One possibility is that 53BP1 recruitment and "sensing" of DSBs involves exposure of preexisting H3K79 methylations in response to chromatin remodeling at sites of DNA lesions.

Although our current understanding of the impact of these and other chromatin factors on DNA repair suggests roles in signaling and recruitment of appropriate complexes to DNA lesions, it is likely that future studies will reveal more ways in which epigenetic mechanisms regulate pathways that maintain genome stability. It is important to note that the role of chromatin in DNA repair is dynamic and occurs in response to damage. Although histone modifications and other epigenetic regulatory proteins play key roles, the changes are not heritable through cell division, unlike the other examples discussed in this chapter.

2.3 Epigenetic Control of Telomere Structure and Function

The ends of linear eukaryotic chromosomes are specialized sites known as telomeres, which serve three essential functions. First, telomeres ensure that DNA replication includes the very ends of the chromosomes, overcoming the "end-replication problem" (Lue 2004). Second, telomeres protect the ends of the chromosomes from degradation and inhibit fusions with other chromosomes. Third, in many but not all organisms, telomeres facilitate chromosome pairing in meiosis. In most eukaryotes, telomeres are composed of simple, short repeats that are restored by the enzyme, telomerase. Telomere functions are regulated by both sequence-based and epigenetic mechanisms.

The end-replication problem arises because DNA polymerases require a primer to initiate 5′ to 3′ "lagging strand" synthesis; the consequence of this restricted enzyme activity is that replication cannot proceed all the way to the end of the chromosome (Lue 2004). Two mechanisms are utilized to overcome this problem. The predominant mechanism used by most organisms, including yeasts, mammals, and plants, involves an unusual enzyme complex known as telomerase. Telomeres in most eukaryotes are composed of simple, 6-bp repeats extending for tens to hundreds of kilobases. Telomerase complexes contain a reverse transcriptase-like enzymatic activity, as well as RNAs that have homology with the telomeric repeats. In essence, end replication is accomplished by targeting of the complex to telomeric repeats via the RNA component, followed by reverse transcription (3′ to 5′) to produce new repeats. Interestingly, loss of telomerase activity and shortened telomeres are correlated with cell senescence and aging, and conversely, cancer cells display enhanced telomerase activity and elongated telomeres (Blasco 2005).

There are also telomerase-independent mechanisms that maintain chromosome ends (Louis and Vershinin 2005). One well-studied alternative system appears to be restricted to *Drosophila* and other dipterans. These organisms lack an identified telomerase, and the ends do not contain the simple, short repeats found in most other eukaryotes. Instead, the ends of *Drosophila* chromosomes are composed of scrambled clusters of different non-LTR (long terminal repeat) retrotransposons ranging in size from 3 to 5 kb, and other repeats (TAS, for telomere-associated repeats) (Biessmann and Mason 2003). These transposons encode reverse transcriptase enzymes (hence *retro*transposon), suggesting that there may be an evolutionary relationship with the more standard telomerase mechanisms. The major difference, however, is that *Drosophila* chromosomes are not replicated to the very end; they lose about 70 bp per fly generation, roughly the amount expected from the end-replication problem. This loss does not cause deletion of essential genes, because the telomeric and subtelomeric repeat domains are about 50–100 kb in length, and it would take many generations to lose enough DNA to reach the genic regions. The loss of telomeric sequences, however, is compensated by infrequent addition of non-LTR retrotransposons (Biessmann and Mason 2003).

Epigenetic regulation affects telomere function, and gene expression of loci residing in the region. "Naked" telomeric DNA or internal DSBs both result in chromosome fusions and aneuploidy. Barbara McClintock first described a phenomenon known as the breakage-fusion-bridge cycle, in which fusions between broken chromosomes, or chromosome ends, produce dicentric chromosomes and anaphase bridges, which generate further breakage. Evidence for the epigenetic regulation of telomere end protection comes from studies in *Drosophila*, which showed it to be independent of DNA sequence. A broken chromosome end in *Drosophila* can behave as a DSB in one generation, but acts as a fully functional telomere subsequently, without any addition of retrotransposons or any sequence changes (Ahmad and Golic 1998). Furthermore, any end generated in *Drosophila* (known as terminal deletions) can be packaged as a telomere and protected against fusion events (Karpen and Spradling 1992). Additionally, telomere function in *Schizosaccharomyces pombe* depends on the Taz1 protein (Miller and Cooper 2003) and telomeric chromatin, in a manner that is independent of canonical telomeric repeats (Sadaie et al. 2003).

Telomeric regions contain chromatin modifications and properties that are similar to pericentromeric heterochromatin described in Section 3. Characterization of the epigenetic mechanisms that regulate telomeric and subtelomeric regions came from studies of gene expression in yeasts and *Drosophila*, but also occur in humans. Euchromatic genes inserted into telomeric regions are variably silenced. This is referred to as telomere position effect (TPE) and is similar to position-effect variegation (PEV) induced by adjacent centromeric heterochromatin

in flies and *S. pombe* (for more detail, see Chapters 5 and 6, respectively). In budding yeast, many of the distinct chromatin-related factors, such as the SIR proteins that affect mating-type silencing, also affect telomere-induced silencing (see Chapter 4). Surprisingly, almost none of the genes known to regulate PEV in *Drosophila* (*Suppressors* and *Enhancers of Variegation*, *Su(var)*s and *E(var)*s described in Chapter 5) have any effect on telomeric silencing. This suggests that PEV and TPE are mediated at least in part by different pathways (Cryderman et al. 1999; Donaldson et al. 2002).

Heterochromatin protein 1 (HP1, a Su(var) gene product) and H3K9 methylation, which are key components of heterochromatin-mediated silencing (see Chapter 8), are present at *Drosophila* telomeres and are required for telomere elongation (Fig. 4) (Perrini et al. 2004). Deletion of HP1 or its binding partner HOAP (for HP1/ORC-associated protein) results in a very high frequency of telomeric fusions (Cenci et al. 2003). HP1 typically is recruited to chromatin through its affinity to methylated H3K9 via the chromodomain. Interestingly, telomere capping by HP1 is independent of H3K9 methylation, suggesting that end protection is mediated by an alternative mechanism involving direct binding to telomeric DNA or non-telomeric sequences present in terminal deletions (Fig. 4a) (Perrini et al. 2004). One attractive model is that HP1 binds and protects ends independent of DNA sequence, then recruits an unknown H3K9 HKMT; local methylation of H3K9 would then recruit more HP1 to the region, which promotes the spreading of telomeric silencing (Fig. 4). This mechanism likely does not require the RNAi pathway, involved in establishing and silencing centromeric heterochromatin (see Chapter 8), but this component of the model needs to be tested directly.

Recent studies have shown that telomerase-dependent telomere elongation is also regulated epigenetically in mammals (Lai et al. 2005). For example, mice deleted for both copies of the H3K9 HKMTs, *Suvar39h1/2*, contain telomeres with reduced levels of H3K9me2 and H3K9me3 and exhibit abnormally long telomeres (Fig. 4b) (Garcia-Cao et al. 2004). These results suggest that Suv39h1/2 HKMT activity transduces the H3K9me modification into the di- and tri-methylated forms, facilitating the binding of HP1 homologs Cbx 3 and 5, which are required for the assembly of normal telomeric chromatin structure and regulation of telomere length.

Finally, meiotic recombination and chromosome transmission are also affected by the epigenetic modifications that occur at telomeres. For example, loss of Ndj1, a telomere protein necessary for both telomere

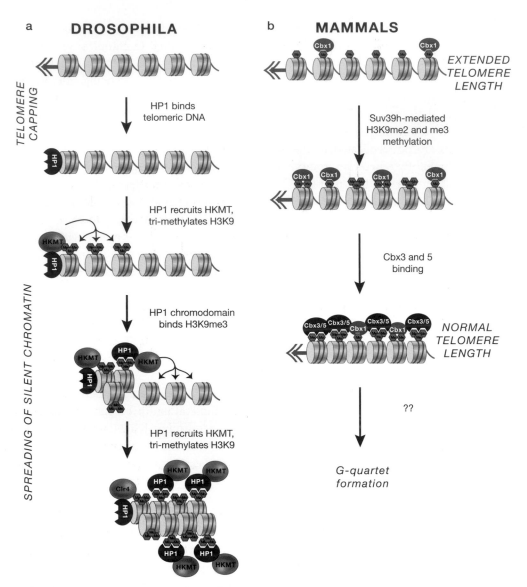

Figure 4. Telomere Function Is Epigenetically Regulated in Flies and Mammals

(*a*) In *Drosophila*, Heterochromatin Protein 1 (HP1) binds telomeric DNA independent of its chromodomain, and "caps" telomeres, which ensures normal segregation by blocking telomere fusions (Fanti et al. 1998; Perrini et al. 2004). HP1 then recruits an unknown histone methyltransferase (HKMT; not *Su(var)3-9*) that tri-methylates H3K9 on nearby nucleosomes; HP1 binds H3K9me3 through its chromodomain, which in turn recruits more HKMT, and successive rounds of HP1 binding/HKMT recruitment promote spreading of silent chromatin through subtelomeric regions. (*b*) In mice, knock-outs of both *Suv39* HKMT loci results in reduced levels of H3K9me3 and me2, and increased H3K9me modifications, altered chromatin structure, and changes in levels of proteins that bind di- and tri-methylated H3K9 (↓ Cbx 3 and 5), H3K9me (↑Cbx 1), and TERFs 1 and 2 (not shown) at telomeres (Garcia-Cao et al. 2004). These changes are correlated with extended telomere length, suggesting that tri-methylation of H3K9 by *Suv39h*s is required for normal telomerase function and regulation of telomere size.

bouquet formation (i.e., clustering) and meiotic recombination (Wu and Burgess 2006), confers a severe reduction in telomere deletion rates in the budding yeast (Joseph et al. 2005). Joseph et al. propose that Ndj1 facilitates telomere deletion "by promoting telomeric interactions during meiosis, resulting in an effective increase in the factors required for deletion." Similarly, mutants that are defective in the transcriptional silencing of genes placed near telomeres display severe defects in meiotic pairing and recombination, resulting in chromosome missegregation during meiosis (Nimmo et al. 1998). Thus, the epigenetic events that control both telomere length and transcriptional competence also appear to be employed in processes controlling chromosome behavior during meiosis.

3 Epigenetic Regulation of Centromere Identity and Function

Normal inheritance of genetic material requires that chromosomes segregate faithfully during mitosis and meiosis, after the genome is accurately duplicated and repaired during S phase. Centromeres were originally defined in 1880 by Flemming as a cytologically visible "primary" constriction in the chromosome. In the early 1900s, centromeres were also defined genetically as chromosomal sites essential for normal inheritance, and as regions of greatly reduced or absent meiotic recombination. We now define the centromere (CEN) as the DNA plus chromatin proteins responsible for kinetochore formation. The kinetochore is a proteinaceous structure facilitating the attachment to and travel along microtubules, plateward during prometaphase and poleward during anaphase of mitosis and meiosis. The kinetochore also serves as the site of action for a key cell cycle checkpoint, known as the spindle assembly checkpoint (SAC) or mitotic checkpoint (Cleveland et al. 2003).

A key question for organisms with mono-centric chromosomes concerns how one and only one site per chromosome is associated with centromere function (known as "centromere identity") and how this information is transmitted from one cell or organismal generation to the next ("centromere propagation"). Here we present the evidence that in most eukaryotes centromere identity and propagation are regulated epigenetically through chromatin structure, rather than by specific DNA sequences (Carroll and Straight 2006). A summary of the key pieces of data includes: (1) Centromeric sequences are not conserved between even closely related species, or even among chromosomes in a single species, (2) centromeric DNA is not necessary or sufficient for kinetochore formation, and (3) centromere positioning along a chromosome displays dramatic plasticity during evolution.

3.1 Centromere Structure and Function in Different Eukaryotes

Studies in the yeast *S. cerevisiae* during the 1980s led to the first cloning and analysis of a eukaryotic centromere. A 125-bp structure present on all 16 *S. cerevisiae* chromosomes was shown to be both necessary and sufficient for normal centromere function (Bloom et al. 1989); even

single-base changes in the highly conserved elements I and III resulted in complete loss of function. Thus, centromere identity and propagation in this single-cell eukaryote are determined by DNA sequence.

The hope that similar sequence-based mechanisms could regulate centromere identity in other eukaryotes was first dispelled by studies in another "simple" eukaryote, *S. pombe*. Centromeric sequences in this fission yeast are structurally larger and more complex than observed in *S. cerevisiae* (Clarke et al. 1986; Nakaseko et al. 1987). Nonhomologous 4- to 5-kb-long "central core" sequences, which are the sites of kinetochore formation, are flanked by various classes of inverted repeats that are shared among the three chromosomes. A minimum of 25 kb, containing the nonrepetitive central core, inner repeats, and a portion of the outer repeats is absolutely required for centromere function and stable chromosome transmission (Baum et al. 1994). Reasonable centromere function is observed for transfected plasmid constructs that carry a central core plus inner repeats (i.e., the central domain) and two flanking outer repeats. Interestingly, the deletion of inner repeats compromises meiotic sister chromatid segregation, demonstrating that centromeric regions play roles in processes other than kinetochore assembly. Indeed, both kinetochore and cohesion domains are closely linked and important for proper chromosome segregation.

Although centromeric regions in multicellular eukaryotes are even larger and more complex than in *S. pombe* (hundreds to thousands of kilobases of repeated DNAs), the overall organization and function of fission yeast centromeres has served as an excellent model for centromeres in mammals, plants, and insects. Centromeres in these organisms are embedded in the large heterochromatic blocks present on each chromosome, which are predominantly composed of satellite DNAs (simple, short repeats) and transposons. These centromeric regions are composed of subdomains responsible for different functions, most notably kinetochore formation and sister cohesion. Centromeric sequences, however, are not conserved among eukaryotes, or even among the different chromosomes in an individual species. It is the epigenetic composition of centromere functional subdomains that shows conservation, notably through histone variant composition and histone modification patterns, which appear to be epigenetically regulated.

In the nematode *C. elegans* and in other species, the holocentric chromosomes recruit and assemble centromeric proteins along the entire chromosome length (Dernburg 2001). Specific worm sequences are apparently not required, as concatemers of lambda and many other types of DNA are stably transmitted. Proteins are recruited in "bundles" in prophase, but by metaphase are spread evenly on the poleward face of chromosome arms, suggesting that many areas of the *C. elegans* genome can support kinetochore assembly in an epigenetic manner. Despite obvious differences with monocentric chromosomes, it is possible that organizational and structural attributes, such as 3D spiraling or looping of CEN DNA, are conserved (see Section 3.3).

3.2 Centromeric Sequences Are Not Necessary or Sufficient for Kinetochore Formation and Function

The large size and complexity of centromeric sequences in multicellular eukaryotes have made it difficult to analyze DNA sequence requirements with the kinds of defined constructs used so successfully in the yeast studies. Human artificial chromosomes (HACs) have nonetheless been generated at low frequency by transfecting tissue culture cells with arrays of satellite DNAs, but they exhibit a high rate of mitotic instability (Rudd et al. 2003). We know, however, that HACs are formed by concatemerization of the introduced satellite arrays, yet some alpha satellite arrays cannot form centromeres de novo, suggesting a requirement for multiple, unknown steps or factors. More recent studies have shown that the unique properties and components of centromeric chromatin (as explained in Section 3.3) are present on both the satellite arrays and non-centromeric sequences (e.g., plasmid vector and selectable marker sequences) in HACs (Lam et al. 2006). Thus, the sufficiency of specific DNA sequences in assembling and maintaining functional human centromeres is still unclear.

The first indication that centromere identity and propagation are regulated epigenetically resulted from studies of "minimal" centromere constructs in *S. pombe* (Steiner and Clarke 1994). A low frequency of the construct transformants exhibited a switch from reduced centromere function to high "active" centromere activity (0.6% of cells), which could subsequently be perpetuated in a lineage for many generations. Thus, the same DNA sequences can display two functionally different, heritable states, similar to observations of epigenetic effects seen for PEV (Chapter 5) or TPE (Chapter 4) on gene expression.

Other observations strongly suggest a primary role for epigenetic mechanisms in determining centromere identity and forming kinetochores in multicellular eukaryotes. First, DNA sequences normally associated with centromeres are not sufficient for function. For example, only a subset of mouse and human heterochromatic satellite

a Human

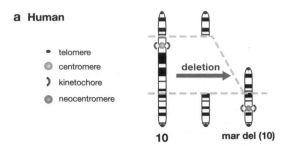

b Drosophila

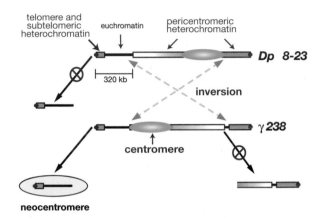

Figure 5. Neocentromere Formation in Humans and Flies

(*a*) Human chromosomes carrying neocentromeres, which exhibit centromere function/kinetochore formation in the absence of centromeric DNA, are usually associated with gross rearrangements (Amor and Choo 2002). In this classic example, a chromosome-10-derived neocentromere (mar(del)10), whose structure indicates formation via a large interstitial deletion that removed the endogenous centromere (*gray dotted lines*). Mar(del)10 was recovered in an individual whose karyotype also contained a ring chromosome (ring(del)10, not shown) that contains the DNA from the deleted region. The order of events for human neocentromeres is unclear; neocentromere formation could occur first, producing a dicentric chromosome that subsequently undergoes rearrangements, or neocentromere formation could occur after deletion of the endogenous centromere. (*b*) Neocentromeres can be generated experimentally in flies from a molecularly defined minichromosome. A 320-kb fragment of euchromatin and telomeric chromatin, which contains no centromeric DNA, can be separated from the rest of the minichromosome by irradiation. This fragment, which should be "acentric," can become a functional neocentromere that is propagated faithfully through mitosis and meiosis, and contains centromere and kinetochore proteins normally restricted to the endogenous centromere (Blower and Karpen 2001). However, neocentromere formation requires proximity to the endogenous centromere (420 kb), as in the inversion derivative *γ238*; furthermore, neocentromere formation does not occur on either side of the centromere when pericentric heterochromatin is present (Maggert and Karpen 2001). These results suggest that neocentromere formation occurs via epigenetic spreading of centromeric chromatin into adjacent euchromatin, followed by epigenetic propagation of centromere identity and function. The blocking of this process by heterochromatin is consistent with the observation that overexpressed CENP-A is incorporated ectopically into euchromatin but not heterochromatin (Heun et al. 2006) and suggests that the extent of centromeric chromatin is determined by two epigenetic processes: CENP-A loading and spreading, and heterochromatin formation/blocking.

sequences are associated with centromere function (Lam et al. 2006). Additionally, in functional chromosomes with two regions of centromeric satellites (dicentrics) observed in flies and humans, one of the regions loses the ability to form a kinetochore (Sullivan and Willard 1998). Second, centromeric sequences are not necessary for kinetochore formation, since non-centromeric DNA can acquire and faithfully propagate centromere function through a process known as "neocentromere formation" (Fig. 5a). Many functional neocentromeres have been identified in humans, and sequence analysis has shown that the new kinetochore-forming regions have not acquired satellite DNAs. The regions flanking the new kinetochore, however, have acquired epigenetic properties comparable to the corresponding regions in endogenous centromeres (i.e., pericentromeric heterochromatin), such as H3K9 methylation and HP1 binding (Lo et al. 2001).

Although the mechanism for neocentromere formation in humans is unknown, neocentromeres have been generated experimentally in a model system. In *Drosophila*, neocentromeres are produced from mini-chromosomes when non-centromeric DNA and an endogenous centromere are juxtaposed (Fig. 5b) (Maggert and Karpen 2001). Thus, proximity to a functional centromere is required for neocentromere activation in *Drosophila*, suggesting that one mechanism for centromere gain is spreading of centromeric proteins in *cis* onto adjacent, non-centromeric regions. Once this spreading has occurred, centromere function is then propagated epigenetically at this new site. Interestingly, neocentromere formation is inhibited when heterochromatin is present between the endogenous centromere and the neocentromere-forming region, suggesting that additional epigenetic mechanisms play a role in determining centromere size.

Finally, chromosome rearrangements are a hallmark of evolution and speciation. These changes are accompanied by centromere gains, losses, and movements with respect to genome sequences (Ferreri et al. 2005). Such plasticity is best explained if centromere identity is determined epigenetically, as described in Section 3.5.

3.3 The Unusual Composition of Centromeric Chromatin

The evidence for epigenetic regulation of centromere identity and propagation points to the likelihood that chromatin structure and composition are the key determinants, rather than primary DNA sequences. Here, we discuss the distinct components and structures found in CEN chromatin, and the surprising observation that these properties are conserved among distantly related eukaryotes.

The CENP-A family of centromere-specific histone H3-like proteins is present in centromeric nucleosomes in all eukaryotes (Fig. 6a). They serve as both the structural and functional foundations for the kinetochore and are excellent candidates for an epigenetic mark that establishes and propagates centromere identity (Cleveland et al. 2003). Unlike most kinetochore proteins that are assembled during mitosis, CENP-A is present at centromeres throughout the cell cycle, which is one of the first indications of its importance to centromere identity. CENP-A containing chromatin also provides the base that is essential for the recruitment of kinetochore proteins, the establishment of spindle attachments, and normal chromosome segregation in yeasts, worms, flies, and mammals (Carroll and Straight 2006). Reciprocal epistasis experiments have shown that CENP-A is the first protein in the kinetochore assembly pathway, consistent with its physical location in chromatin at the base of the kinetochore in mitotic chromosomes. Further evidence for

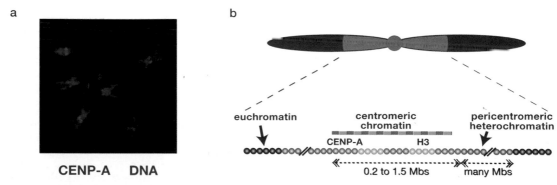

Figure 6. The Organization of Centromeric Chromatin

(*a*) CENP-A is a highly conserved, centromere-specific histone variant. Image shows localization exclusively to centromeres in *Drosophila* mitotic chromosomes. (*b*) CEN chromatin in flies and humans contains interspersed blocks of H3-containing and CENP-A–containing nucleosomes, and is flanked by pericentromeric heterochromatin (Blower et al. 2002).

the importance of CENP-A in kinetochore formation comes from overexpression studies in flies, in which CENP-A mislocalization to non-centromeric regions produces functional ectopic kinetochores (Heun et al. 2006). Therefore, because this histone variant is essential for centromere function, we specifically define CEN chromatin as the region of DNA and proteins associated with CENP-A.

The structure of CENP-A-containing nucleosomes is unusual compared to canonical histone cores containing H3, H2A, H2B, and H4. CENP-A nucleosomes can be assembled in vitro from purified CENP-A and histones H2A, H2B, and H4, consistent with previous observations indicating that they are homotypic in vivo (i.e., they contain two copies of CENP-A and not one copy of H3 and one of CENP-A) (Yoda et al. 2000). Detailed biophysical analysis showed that the interface between CENP-A and H4 is different from the H3–H4 interface, and the H4 interacting domain is sufficient to target CENP-A to centromeres in the presence of endogenous CENP-A (Black et al. 2004).

The replacement of H3 by CENP-A in centromeric nucleosomes initially suggested that CENP-A constituted all of the chromatin associated with the kinetochore. In S. pombe, CENP-A is uniformly distributed across the 5- to 7-kb central core regions. The large heterochromatic domains that contain H3K9 methylation and heterochromatin proteins flank these cores (Pidoux and Allshire 2004). However, detailed cytological and immunoprecipitation studies have revealed that centromeric chromatin has a more complex composition and organization in multicellular eukaryotes. Drosophila, human, and rice centromeres contain interspersed blocks of H3 and CENP-A-containing nucleosomes (collectively called CEN chromatin) flanked by even larger blocks (hundreds of kilobases to megabases) of pericentromeric heterochromatin (Fig. 6b) (Blower et al. 2002; Yan et al. 2005).

The interspersion of H3 and CENP-A domains raised key questions about the epigenetic nature of CEN chromatin. In particular, are H3 subdomains within the CEN chromatin of multicellular eukaryotes modified like heterochromatin or euchromatin? Or are they uniquely marked? Furthermore, is each interspersed CENP-A/H3 unit in larger eukaryotic centromeres equivalent to a single S. pombe centromere? These questions were addressed by examining the posttranslational modifications that characterize the interspersed blocks of H3 and CENP-A nucleosomes, which revealed even greater complexity. Surprisingly, the interspersed H3 domains in humans and flies contain H3K4me2, a mark usually associated with euchromatin and, moreover, lack

the H3K9me2 and me3 associated with flanking heterochromatin (Fig. 7a) (Sullivan and Karpen 2004; Lam et al. 2006). However, like heterochromatin, multiple forms of H3 and H4 acetylation were absent from the interspersed H3 nucleosomes, as was H3K4me3. Thus, the H3 nucleosomes within CEN chromatin display a pattern of modifications that are distinct from canonical euchromatin or heterochromatin (Fig. 7b). These results also suggest that fly and human centromeres are not composed simply of repeated, S. pombe-like centromeres. However, it is important to note that the overall organization of the centromere regions is conserved, such that the entire CENP-A chromatin domain is flanked by pericentromeric heterochromatin that contains H3K9 methylation in all multicellular eukaryotes and in S. pombe.

What are the possible functional roles of histone modifications in CEN and flanking chromatin? Distinct chromatin states in the CEN region are likely to contribute to the diverse properties of centromeric domains, such as differential replication timing of the CEN and flanking heterochromatin (Sullivan and Karpen 2001; Blower et al. 2002). Flanking pericentromeric heterochromatic modifications may also maintain centromere size by creating a barrier against expansion of CEN chromatin. In Drosophila, CEN chromatin readily spreads into neighboring sequences when flanking heterochromatin is removed, allowing neocentromere activation (Fig. 5b) (Maggert and Karpen 2001), and overexpression of CENP-A results in mislocalization to euchromatin, but not heterochromatin (Heun et al. 2006). Interestingly, overexpression of CENP-A in human cells results in spreading of CEN chromatin and alterations in H3K9 methylation in the flanking regions (Lam et al. 2006). Thus, centromere size appears to be determined by a balance between two epigenetic states: CEN chromatin and flanking pericentromeric heterochromatin.

Proteins required for sister chromatid cohesion, which is established in conjunction with DNA replication in S phase, are most highly concentrated in the heterochromatin that flanks the centromere. This distribution appears to contribute to proper bi-orientation of kinetochores in mitosis, as well as the maintenance of cohesion during metaphase despite spindle-mediated forces concentrated at kinetochores/centromeres (Watanabe 2005). Epigenetic regulation of cohesion involves the recruitment of cohesins by HP1 proteins (Swi6 in S. pombe), which is in turn mediated by the high concentrations of H3K9 methylation in the pericentromeric heterochromatin (Nonaka et al. 2002). Thus, CEN-spe-

a

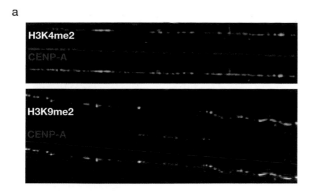

b

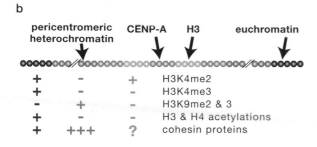

c

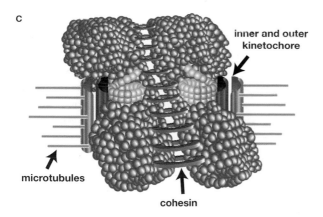

Figure 7. Distinct Patterns of Histone Modifications in Centromeric Chromatin

(a) Immunofluorescence using antibodies that recognize specific histone modifications on extended chromosome fibers showed that the interspersed H3-containing nucleosome blocks have a pattern of modifications that are distinct from canonical euchromatin and heterochromatin (Sullivan and Karpen 2004). For example, despite the fact that centromeres in most eukaryotes are embedded in large blocks of pericentric heterochromatin, the interspersed H3 blocks contain the H3K4me2 modification normally associated with "open" euchromatin (top), and lacks the heterochromatin marker H3K9me2 present in the pericentric flanking regions (bottom). (b) Summary of "2D" organization of centromeric chromatin in interphase based on extended chromatin fiber studies in flies and humans. + and – indicate the presence and absence of the indicated histone modification (respectively) in euchromatin, pericentromeric heterochromatin, and the interspersed blocks of H3 nucleosomes in centromeric chromatin (Sullivan and Karpen 2004; Lam et al. 2006). (c) Model for the 3D organization of chromatin in the centromere region of mitotic chromosomes. Associations between similarly modified nucleosomes are proposed to contribute to the formation of distinct 3D structures in centromeric and flanking chromatin. Interspersed CENP-A/CID and distinctly modified H3 and H4 may mediate formation of the "cylindrical" 3D structures observed in metaphase chromosomes (Blower et al. 2002; Sullivan and Karpen 2004). H3K9me2 chromatin, which recruits heterochromatin proteins such as HP1, and cohesion proteins such as RAD21/SCC1, is present in the inner kinetochore space between mitotic sister chromatids and in regions that flank centromeric chromatin. This arrangement may be necessary to "present" CENP-A toward the poleward face of the mitotic chromosome and facilitate recruitment of outer kinetochore proteins, and to promote HP1 self-interaction and proper chromosome condensation/cohesion. Cohesins are presented as ringed structures, in accord with recent models.

cific combinations of histone modifications could also be important for the recruitment of cohesion complexes to heterochromatin near sister kinetochores, while ensuring spatial separation of cohesion and kinetochore domains (Sullivan and Karpen 2004; see also Chapter 6).

In human and *Drosophila* mitotic chromosomes, CENP-A subdomains merge to form a 3D cylindrical structure that largely excludes H3 nucleosomes (Blower et al. 2002). Blocks of CENP-A nucleosomes are oriented on the poleward face of the chromosome, and blocks of H3 nucleosomes are located toward the inner chromatid region. Inner and outer kinetochore proteins are wrapped around the CENP-A cylinder; this 3D arrangement is consistent with CENP-A playing a central role in the recruitment of other kinetochore pro-

teins (Blower et al. 2002). In order to reconcile the 2D interspersion of CENP-A and H3 blocks (Figs. 6b and 7b) with separation in 3D mitotic chromosomes, it has been proposed that CEN DNA may spiral or loop through the cylindrical structure, leading to alignment or stacking of nucleosomes with the same composition (Fig. 7c). Thus, the distinctly modified interspersed H3 nucleosomes and flanking heterochromatin could be responsible for assembling the 3D structure of CEN chromatin in mitosis (Sullivan and Karpen 2004). This arrangement may be necessary to expose CENP-A chromatin to the outside of the chromosome, where it can recruit kinetochore proteins in a manner that establishes proper bi-orientation of sister kinetochores with respect to the spindle poles.

3.4 Models for Centromere Structure, Function, and Propagation

The key question under investigation at this time is how CENP-A, and other epigenetic marks, are specifically localized and propagated at centromeres. Another way to think about this question is to consider how CENP-A is assembled *only* at centromeric chromatin. One attractive model proposed that differential timing of both CEN DNA replication and CENP-A expression compared to bulk chromatin regulated CENP-A incorporation specifically at centromeres (Ahmad and Henikoff 2001). However, CEN DNA replication in humans and flies occurs throughout S phase, concurrent with bulk DNA replication (Shelby et al. 2000; Sullivan and Karpen 2001), and CENP-A incorporation occurs in the absence of DNA replication (Shelby et al. 2000; Ahmad and Henikoff 2001). These observations rule out a strict replication timing mechanism for propagation of CENP-A and centromere identity.

A more attractive mechanism is suggested by the intriguing observation that CENP-A is actively incorporated into nucleosomes in a replication-independent manner by a histone exchange complex (Shelby et al. 2000; Ahmad and Henikoff 2001). H3.3 is an H3 variant whose replication-independent assembly (Ahmad and Henikoff 2002) is mediated by a complex known as histone regulator A (HIRA), and not the chromatin assembly factor (CAF) complexes responsible for replication-dependent incorporation of canonical H3 nucleosomes (Nakatani et al. 2004). Depletion of HIRA components results in CENP-A mislocalization in *S. cerevisiae* (Sharp et al. 2002). However, it is currently unclear whether HIRA components affect centromeric chromatin in multicellular eukaryotes or *S. pombe*, where centromere identity is determined epigenetically. More importantly, these proteins play general roles in chromatin assembly and structure, such as H3.3 deposition; thus, the broad activity of the identified HIRA components does not explain the specificity of CENP-A incorporation at centromeres. It is possible that a subset of HIRA complexes contain factors that interact only with CENP-A, and recognize existing CENP-A nucleosomes in replicated CEN chromatin; however, no such specificity factors have been identified. One way to accommodate the involvement of nonspecific assembly factors is to imagine that specificity is provided by CENP-A or CENP-A nucleosomes. For example, the distinct structural relationship between CENP-A and H4 could provide specificity for assembly of new CENP-A at

centromeres, as suggested by the ability of the interacting domain to target CENP-A to centromeres (Black et al. 2004). However, it is unclear whether these domains are sufficient for targeting centromeres in the absence of endogenous CENP-A.

New ways of thinking about the epigenetic regulation of centromere identity and propagation are clearly required at this time. In *S. cerevisiae*, defects in CENP-A proteolysis result in misincorporation into normally non-centromeric regions, which is normally removed by an unknown "clearing" mechanism from everywhere except the endogenous centromere (Collins et al. 2004). This suggests that centromere identity may be regulated at a time subsequent to nucleosome assembly. However, mislocalization of CENP-A in flies results in ectopic kinetochore formation (Heun et al. 2006), suggesting that removal of misincorporated CENP-A may be specific to *S. cerevisiae*.

Nevertheless, variations of this kind of "negative specificity" model are worth considering. The key question for all centromere identity models is, What provides specificity? In this case, Why would proteins such as CENP-A be retained only at one site? One novel idea arises from the fact that stable association with the spindle is one property that is unique to functional centromeres/kinetochores (Mellone and Allshire 2003). Thus, centromere propagation and the site of CENP-A incorporation may be determined during mitosis, utilizing a mechanism that senses productive kinetochore-spindle attachments, or spindle-mediated tension. Another idea worth considering is that the modification pattern of interspersed H3 nucleosomes by histone modification proteins (e.g., acetyltransferases, methyltransferases, and kinases) may help propagate centromere identity, in lieu of (or in addition to) CENP-A-associated proteins (Sullivan and Karpen 2004). Distinctly modified interspersed H3 subdomains (Fig. 7) could create a "permissive" chromatin structure necessary for the assembly of new CENP-A.

Identification of factors required for CENP-A deposition at centromeres, without bias for a particular model, is a strategy that is likely to provide new insights. Biochemical and genetic studies have identified some as affecting CENP-A signals at centromeres, including previously known factors involved in replication-independent chromatin assembly. However, none of the factors identified to date interacts specifically with CENP-A or other centromeric chromatin proteins or modifications. Nevertheless, it is exciting that factors are being identified, and elucidating specific mechanisms should soon follow.

3.5 Epigenetics and Centromere Evolution

Given the importance of centromeres to cell and organismal viability, there should be no room for gain or loss of centromere function. Then why would centromeres utilize epigenetic mechanisms of regulation if there are significant advantages for the individual cell and organism to contain centromeres "hard-wired" into the primary DNA sequence?

A strong argument can be made that epigenetic regulation of centromere identity is necessary to accommodate changes occurring to chromosomes, sequences, and proteins during evolution. Studies in mammals (e.g., primates and marsupials), insects, and other taxa have shown that centromere gains and losses are a hallmark of chromosome evolution (Ferreri et al. 2005). Related species frequently differ in the arrangement and association of chromosome arms, even when the DNA sequences are nearly identical. These centromere gains and losses frequently accompany, and arguably are mandated by, translocations and other rearrangements. For example, the requirement for one and only one centromere would render many of the resulting

dicentric and acentric chromosome rearrangements useless, unless centromeres could be inactivated and neocentromeres activated (Fig. 8a). Thus, the ability to move the centromere from one DNA sequence to another, by spreading, hopping, or activation, may expedite chromosome evolution. In addition, expansions, contractions, and base changes occur very frequently in highly repetitive satellites during evolution and can be associated with changes in centromere positioning (Fig. 8b). Plant centromeres in particular display rapid and striking changes in DNA sequences and positioning during evolution (Hall et al. 2006). However, it is unclear whether DNA changes cause movement of centromeres, occur in response to centromere movement, or are completely independent. Nevertheless, there needs to be a mechanism for maintaining centromere function and propagation independent of sequence changes, which epigenetic regulation provides.

Cross-species comparisons of the amino acid sequences in centromere proteins led to the proposal that domains associated with CEN DNA "co-evolve" to accommodate changes in satellite sequences (Malik and Henikoff 2002; see also Chapter 13). An interesting com-

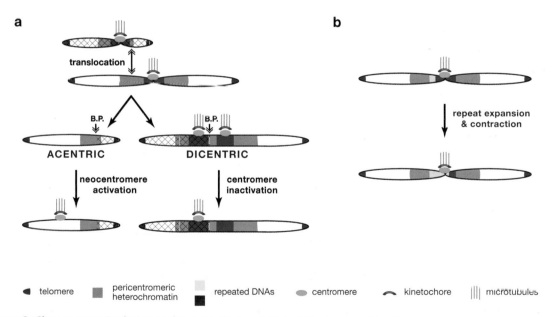

Figure 8. Chromosome Evolution and Epigenetic Regulation of Centromere Identity

Centromere plasticity with respect to association with specific DNA sequences may be necessary to accommodate the sequence changes and chromosome rearrangements that occur during evolution. (a) Translocations frequently observed during evolution can produce acentric and dicentric chromosomes, both of which are normally lost during mitotic or meiotic divisions. Epigenetic regulation and plasticity allow acentric fragments to acquire neocentromere function, as well as inactivation of one centromere on dicentric chromosomes, leading to normal inheritance of both translocation products. (B.P. = translocation breakpoints.) (b) Centromeres in most eukaryotes are embedded in heterochromatin and repeated DNAs, especially highly repeated satellite arrays. These sequences change at a dramatic rate during evolution and undergo frequent base changes and array expansions and contractions. Strict dependence of centromere identity on specific DNA sequences would result in loss of centromere and kinetochore functions, and detrimental chromosome loss. In contrast, epigenetic regulation of centromere identity and position provides a mechanism for maintaining centromeres despite sequence changes.

ponent of the model suggests that these changes are promoted by "meiotic drive," a phenomenon of allelic (and by default the whole chromosome) selection occurring during meiosis. Thus, a centromere that has the highest affinity for CEN chromatin proteins is the most successful at being incorporated into functional germ cells (i.e., the oocyte instead of the polar body in the case of female meiosis) (see Section 5 for other examples of meiotic drive). This in turn could force other centromeres to adopt the same sequences and protein variants in order to segregate efficiently.

Loss of an epigenetic mark, such as CENP-A, provides another mechanism for centromere inactivation without deletion of centromeric DNA, as demonstrated by the consequences of CENP-A depletion in numerous organisms (Cleveland et al. 2003). Identifying mechanisms for centromere gain is more challenging, as it requires acquisition of an epigenetic mark in the absence of DNA sequence changes. Studies of experimentally induced neocentromeres in flies suggest one molecular mechanism for centromere gain, by the *cis*-spreading of key centromere chromatin proteins such as CENP-A (Fig. 5b) (Maggert and Karpen 2001). However, this model cannot account for human neocentromere formation or most examples of centromere gain during evolution, which occur large distances away from the parental centromere. Perhaps a more appropriate mechanism for these cases of centromere acquisition involves CENP-A mislocalization in response to transient overexpression, resulting in the formation of ectopic kinetochores, as observed experimentally in flies (Heun et al. 2006). CENP-A overexpression has been observed in colon and breast tumors, suggesting a potential link with massive genome instability observed in cancer. Further investigations into the prevalence of CENP-A mislocalization in different types of human cancers, and its timing during cancer initiation and progression, are required to directly test this hypothesis, as well as its role in centromere evolution.

Finally, holocentric chromosomes could represent the first centromeres; kinetochore formation could have first evolved with random sequence specificity, followed by evolution of monocentric chromosomes that arose due to transposon invasion, satellite DNA expansion, and the formation of flanking heterochromatin. However, it is also possible that holocentric chromosomes evolved from monocentric chromosomes, due to the loss of heterochromatic boundary elements and *cis*-spreading of centromeric chromatin, in a manner that is analogous to the generation of *Drosophila* neocentromeres (Maggert and Karpen 2001).

4 Heterochromatin and Meiotic Pairing/Segregation

The meiotic process in most organisms comprises (1) pairing, which brings homologs into alignment; (2) synapsis, which connects homologs by a structure known as the synaptonemal complex (SC); (3) recombination, which physically links homologous chromosomes and exchanges genetic information; (4) segregation of homologs to opposite poles at meiosis I; and (5) separation of sister chromatids in meiosis II (Fig. 9). Heterochromatic elements such as centromeres and telomeres play important roles in controlling the position of recombination events within the euchromatin; and certainly, heterochromatic elements (most especially the centromere) are critical for segregation. Moreover, numerous recent studies argue strongly that histone modifications, and perhaps other epigenetic components and processes, play critical roles in facilitating the meiotic process. For example, the *C. elegans* HIM-17 protein, which is required for H3K9 methylation, is necessary for the formation of the DSBs that initiate meiotic recombination (Reddy and Villeneuve 2004). A histone kinase (Hsk1) is similarly required for the initiation of DSBs in *S. pombe* (Ogino et al. 2006). Histone methylations are also required to make the meiotic chromosomes competent to complete the meiotic divisions in *Drosophila* (Cullen et al. 2005; Ivanovska et al. 2005) and mammalian systems (De La Fuente 2006).

Although the use of recombination to ensure meiotic chromosome segregation is nearly universal, meiotic systems also exist in which homolog associations can be established without recombination (i.e., achiasmate meiotic systems). In such systems, heterochromatic pairings, and perhaps other substrates for epigenetic modification, appear to play critical roles. This is especially true in *Drosophila melanogaster* males and females, but may also be true in yeast as well. Finally, many meiotic drive systems—i.e., the favoring of one allele during meiotic segregation—are known in which the modification of specific heterochromatic elements causes or allows the subsequent loss or inactivation of a specific chromosome or of an entire chromosome set.

4.1 Discovery of a Heterochromatic Pairing Site in Drosophila Males

The first evidence for a heterochromatic element that mediated homolog pairing and segregation was the identification of structures in pericentromeric heterochromatin that mediate the pairing and segregation of achiasmate sex chromosome pairs in *Drosophila* male

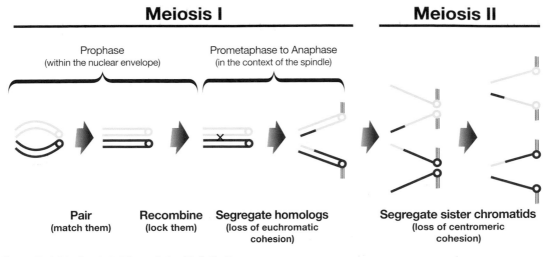

Figure 9. A Mechanistic View of the Meiotic Process

A pair of homologous chromosomes must do three things during the first meiotic division. First, the homologs must pair along their length. In virtually all organisms, this pairing culminates in an intimate association in which the homologs are connected along their entire length by the synaptonemal complex. This state is referred to as synapsis. Second, in virtually all organisms, homologous chromosomes are locked together by recombination, which is also called crossing-over. These exchange events (crossovers) form structures called chiasmata that physically interlock the homologous chromosomes by virtue of the sister chromatid present on each homolog on both sides of the crossover event. Both pairing and recombination occur during prophase (prior to nuclear envelope breakdown). The third major event, segregation, occurs on the MI spindle, which is created after nuclear envelope breakdown. During the early stages of spindle assembly (prometaphase), the chromosomes congress to create the metaphase plate. In most animals, males contain centriolar meiotic spindles, whereas in most animal females, the spindle is acentriolar. In this case, the chromosomes themselves form a mass at what will eventually become the metaphase plate, and organize a bipolar spindle around them. Once the chromosomes are properly co-oriented (i.e., balanced at the metaphase plate with homologous kinetochores attached to opposite poles of the spindle), a variety of mechanisms trigger the onset of anaphase. At anaphase, sister chromatid cohesion is released along the arms of the chromosomes (but not near the centromeres). This dissolves the connections, referred to as chiasmata, created by the crossovers, and thus allows the homologs to separate and proceed to opposite poles at anaphase I. Meiosis II is basically a haploid mitosis that occurs without either replication or recombination.

spermatogenesis. These structures, known as collochores, were shown to correspond to rDNA repeats; the integration of rDNA genes onto collochore-deleted X chromosomes was shown to restore X–Y pairing at the site of the rDNA insertion (McKee 1998). The crucial segment of the rDNA repeat, with respect to pairing, is a 240-bp repeat sequence in the intergenic spacer. When present in multiple copies, this 240-bp sequence facilitates the pairing and subsequent segregation of the X and Y chromosomes during meiosis in *Drosophila* males. Although the degree to which heterochromatic associations in *Drosophila* spermatocytes facilitate euchromatic pairings remains an open question, we discuss below a number of observations suggesting that the failure to mediate or maintain pairing appears to result in improper activation or inactivation of the heterochromatin. Such an event often results in spermatogenic failure (McKee 1998).

4.2 Heterochromatin Pairing Facilitates Segregation in *Drosophila* Females

There are two primary mechanisms for ensuring homolog segregation in *Drosophila* females: a chiasmate system that mediates the segregation of those homologs that have undergone crossing-over, and an achiasmate system that ensures the segregation of those pairs of homologs that fail to undergo crossing-over. For example, the small dot-like fourth chromosomes never undergo crossing-over, yet they segregate from each other with very high fidelity during meiosis. Cytological and genetic experiments have demonstrated that heterochromatin is required for achiasmate pairing and segregation. The pairing of homologous heterochromatic regions of achiasmate chromosomes is maintained from early prophase until the achiasmate homologs begin to separate from their partners at prometaphase I (Fig. 10) (Dernburg et al. 1996). Furthermore, deletion studies show that homologous achiasmate

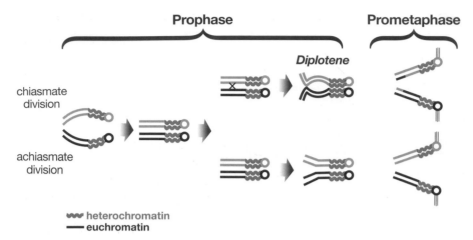

Figure 10. There Are Two Mechanisms for Ensuring the Segregation of Homologs in *Drosophila melanogaster* Females

The upper row of images reprises the canonical process as shown in Fig. 9. The lower row describes the meiotic process in *Drosophila* females which have an unusual diplotene. In a canonical meiosis, diplotene (sometimes referred to as diplotene-diakinesis) is defined as the last stage in prophase in which homologs repel, and are then held together only by chiasmata. However, in *Drosophila* females, only the euchromatic arms of chromosomes repel (or separate) at the end of prophase; the heterochromatic regions remain tightly paired even beyond the end of prophase (nuclear envelope breakdown) and into prometaphase (Dernburg et al. 1996). As discussed in the text, these heterochromatic pairings are both necessary and sufficient to ensure faithful segregation in the absence of crossing-over (Hawley et al. 1993; Karpen et al. 1996).

segregation in *Drosophila* females is entirely dependent on heterochromatic homologies (Hawley et al. 1993; Karpen et al. 1996). It is still not clear whether the ability to mediate these pairing and segregation events is a general property of heterochromatic sequences or chromatin proteins, or the result of a number of specific regions dispersed within the heterochromatin.

4.3 Role of the Centromere in Facilitating Achiasmate Segregation in Budding Yeast

Although flanking heterochromatic regions appear to be sufficient to mediate achiasmate segregation in *Drosophila* females, it seems likely that centromeric sequences alone may be able to play this role in budding yeast. Dean Dawson and his collaborators analyzed meiosis in yeast carrying a pair of homeologous (i.e., partially homologous) chromosomes, which are derived from different species of yeast (Kemp et al. 2004). Despite the fact that these chromosomes carry similar sets of genes, they have diverged sufficiently that they do not cross over; yet they segregate faithfully. Although the homeologous chromosome arms associate with each other no more frequently during meiotic prophase than do the arms of heterologous chromosomes, these homeologous chromosomes nonetheless experience a high degree of centromeric pairing (Kemp et al. 2004). Moreover, this pairing is diminished by the presence of a third

nonexchange chromosome, even if it carried a different centromere region, suggesting that centromere pairing is in fact sequence-independent.

If sequence homology is not required for centromere pairing, then why is yeast homeologous chromosome segregation not random? Kemp et al. (2004) suggest that "it is exchange and the synapsis that follows exchange, that juxtaposes homologous centromeres and blocks random centromere pairing." According to this model, centromeres might initially pair in a fully sequence-independent fashion. As the arms of homologous chromosomes undergo exchanges, their centromeres are "withdrawn" from potentially nonhomologous associations and forced into homologous couplings. Those centromeres left as "singlets" by such exchange-dependent re-sortings are then free to re-associate with other uncoupled centromeres, leading eventually to the pairings of centromeres of those chromosomes that for one reason or another did not exchange. While Kemp et al. draw an analogy of the progressive pairings of the more "attractive kids" at a high school dance leaving the least "attractive kids" to pair as dance partners by default, a more formal view of this process was described by Rhoda Grell decades ago. She suggested that once exchange pairings had occurred, the available chromosomes would be free to re-associate (Grell 1976). In fact, recent studies have shown that during meiosis, *all* yeast centromeres are nonhomologously associated and undergo a re-sorting process that couples homologous centromeres following recombination (Tsubouchi and Roeder 2005).

Taken together, these studies argue that sequence-independent recognition of centromeric regions plays an important role in creating early meiotic chromosome associations. Whereas these associations may often connect nonhomologous chromosomes, nonhomologous associations can be corrected by the physical linkages between homologous chromosomes, created by exchange. The mechanism by which such exchanges, occurring at great distances from the centromeres themselves, can facilitate such centromeric re-sorting remains unclear. However, it is tempting to at least suggest that "connecting" the chiasma (the physical manifestation of an exchange event) to the centromeres might be one function of the synaptonemal complex (SC). According to such a model, the SC would serve to communicate to the two homologous centromeres that a chiasma has indeed been formed, thus inducing their co-orientation, long before nuclear envelope breakdown.

4.4 *The Heterochromatin-associated* Ph1 *Locus in Maize and Its Role in Mediating Homologous Versus Homeologous Pairing*

In wheat, the correct pairing of homologous chromosomes, and the prevention of homeologous pairings, are mediated by the *Ph1* locus (Griffiths et al. 2006). In the absence of the *Ph1* gene, pairing and exchange between homeologous chromosomes becomes frequent (Luo et al. 1996) and reduces the fidelity of both meiotic and mitotic centromere pairing. The significance of centromere mis-pairing in terms of the more promiscuous pairing and exchange between the arms of homeologous chromosomes remains controversial (Dvorak and Lukaszewski 2000). However, it is clear that the effects of deleting *Ph1* can be suppressed by the addition of heterochromatic supernumerary B chromosomes to the genome (Bennett et al. 1974). Recently, the *Ph1* locus has been localized to a 2.5-megabase interval that contains subtelomeric heterochromatin inserted into a cluster of *cdc2*-related genes, which occurred after the polyploidization that characterized the evolution of modern wheat (Griffiths et al. 2006). Although these studies of the *Ph1* locus together suggest epigenetic regulation of partner choice in wheat meiosis, the role(s) of heterochromatin in mediating the prevention of homeologous pairing, and thus the promotion of homologous pairings, remains somewhat unclear.

5 Heterochromatin and Meiotic Drive

There are numerous cases in which heterochromatic regions or elements play roles in mediating a process called "meiotic drive." The term meiotic drive refers to the ability of one homolog to enhance its probability of transmission at the expense of its partner, such that in an *A/a* heterozygote, *A*-bearing gametes are produced or used more frequently than *a*-bearing gametes. We focus primarily on those cases where heterochromatic elements cause the loss or destruction of their homologs. Although we describe in detail only the *SD* system in *D. melanogaster*, and chromosome loss in *Sciara* and *Nasonia*, other systems of heterochromatin-related meiotic drive exist, such as heterochromatinization and subsequent loss of a germ line-restricted chromosome in the finch, and the accumulation of B chromosomes in maize.

5.1 Segregation Distorter *in* Drosophila *Males*

One of the more impressive examples of the role of heterochromatin in mediating meiotic drive is a haplotype in *D. melanogaster* isolated from a natural population, referred to as *Segregation Distorter* or just *SD* (Sandler and Hiraizumi 1959). A fruit fly second chromosome carrying the *SD* haplotype has the capacity to eliminate its wild-type homolog (denoted SD^+) during the process of sperm maturation by preventing that homolog from properly condensing. Only SD-bearing sperm complete maturation to become mature spermatids, and are therefore transmitted to progeny.

The *SD* chromosome carries several discrete genetic elements that contribute to its function. The first of these is the euchromatic *SD* mutant itself, which causes spermatid dysfunction by acting on a separate heterochromatic target element carried by the homolog called *Rsp* (Ganetzky 1977). *Rsp* itself comprises a repetitive element located in the pericentromeric heterochromatin of the second chromosome (Wu et al. 1988). The sensitivity of the homologous SD^+ chromosome to condensation failure depends on the number of copies of the *Rsp* repeat. There are also a number of other elements, usually carried by the *SD*-bearing chromosomes, that facilitate the ability of *SD* to cause distortion.

The SD mutation itself is the result of a small tandem duplication event involving the *RanGAP* gene. The rearrangement produces a mutant RanGAP protein, truncated by 234 amino acids at the carboxyl terminus, which causes a defect in nuclear transport (Kusano et al. 2003). Why such a defect might lead to a failure in chromatin condensation of chromosomes carrying the *Rsp* locus remains unclear (but see Kusano et al. 2003 for a discussion of the possible mechanisms).

5.2 Paternal Chromosome Loss in Sciara and Mapping of the Response Element

Sciarid flies undergo complex processes of paternal chromosome elimination in both the soma and the germ line. All embryos arise from zygotes of the genotype $A_mA_pX_mX_pX_p$, which is to say that they received one set of autosomes and one X chromosome from their mother, and one set of autosomes and two copies of the X chromosome from their father. The somatic cells of those embryos destined to become females will lose one of the two paternally derived X chromosomes, and the somatic cells of embryos destined to develop as males will lose two paternally derived X chromosomes. Critical insights into the actual mechanism of X-chromosome elimination, which involves incomplete sister chromatid separation, have been made by de Saint Phalle and Sullivan (1996).

Paternal X-chromosome loss also occurs in the cells that comprise the germ line. At later stages in germ-line development, a single X_p chromosome is discarded in both sexes. Following the loss of the paternal X chromosome, the entire paternal chromosome contribution appears to decondense, as revealed by lighter staining. This difference in condensation is maintained until the late first larval instar in both sexes, which is to say until the beginning of the gonial mitotic divisions. At that point, both maternal and paternal chromosomes appear to be fully condensed. Although the ensuing female meiosis is cytologically normal, male meiosis is unusual in two respects. First, the entire paternal chromosome set is eliminated, and second, the maternally derived X chromosome undergoes a directed nondisjunction at meiosis II to form a sperm with one set of autosomes and two X chromosomes.

Interestingly, histone H3 and H4 acetylation correlates with these patterns of chromosome elimination (Goday and Ruiz 2002). First, during early germ-line development, only half of the chromosomes, the paternal set, are highly acetylated at H3 and H4. As noted by Goday and Ruiz (2002), "the differential histone acetylation labeling of germline chromosomes demonstrated here is consistent with the early data on the existence of chromatin staining differences between both parental genomes in germ cells." The exception to this rule is the paternal X chromosome that will be eliminated during germ-line development, which is not highly acetylated. At later stages, prior to the onset of the gonial mitotic divi-

sions, all chromosomes, both maternal and paternal, are highly acetylated. Most interestingly, in male meiosis, in which the entire paternal set will be lost, the entire paternal set is under-acetylated for H3 and H4, when compared to the highly acetylated complement of maternally derived chromosomes. Although the relationship of these changes in histone modification to subsequent elimination events remains mechanistically unclear, it seems likely that such differences mark chromosomes for both meiotic and mitotic elimination.

5.3 Paternal Chromosome Loss in Nasonia

In the parasitic wasp *Nasonia vitripennis*, unfertilized (haploid) eggs develop into males, whereas fertilized (diploid) eggs usually develop into females (for review, see Beukeboom and Werren 1993). However, there exists a supernumerary or B chromosome, known as the *PSR* chromosome, that causes fertilized eggs to develop as males. The *PSR* chromosome accomplishes this bit of sex reversal by causing super-condensation and loss of all paternal chromosomes (except itself!) in fertilized eggs (Werren et al. 1987). The *PSR* chromosome is largely heterochromatic and contains tandemly repeated sequences that are not present on the autosomes. Efforts to deletion-map this chromosome did produce nonfunctional *PSR* chromosomes that can be transmitted to daughters (Beukeboom and Werren 1993). However, these studies failed to identify a specific domain or repeat class on this chromosome that was responsible for its ability to trigger the destruction of the paternal genome. Of note, however, was the recovery of one such deletion derivative which reverted from nonfunctionality to functionality within a single lineage. Beukeboom and Werren (1993) proposed that some set of repeats on the *PSR* chromosome may function as a "sink" for a product that is normally required for the proper processing of paternal chromosomes following fertilization. That there do indeed exist imprinted "state differences" between the paternal and maternal contributions on which the *PSR* chromosome might act is suggested by studies of the role of genomic imprinting in *Nasonia* sex determination (Dobson and Tanouye 1998).

5.4 Knob 10 in Corn—The Role of Heterochromatic "Knob" Sequences in Facilitating Chromosome Segregation at Meiosis I

Another example of heterochromatic elements associated with meiotic drive is observed in plants containing chromosomes with heterochromatic insertions, known as "knobs" or "pseudokinetochores"; they favor the transmission of the knob-bearing chromosome during female meiosis. The molecular organization of the knobs comprises two sets of tandem repeats, each of which appears to be present as a long uninterrupted array (Dawe and Hiatt 2004). The function of heterochromatic knobs as pseudokinetochores depends on a variant of chromosome 10, known as abnormal chromosome 10 (Ab10). When Ab10 is present, knobs form pseudokinetochores that move toward the poles of the first meiotic spindle ahead of endogenous centromeres, and the rate of movement is proportional to knob size. However, knobs only display pseudokinetochore activity when present in *cis* with an endogenous centromere, and they only exhibit lateral associations with microtubules, unlike the "end-on" associations displayed by fully functional endogenous kinetochores. Thus, despite nomenclature confusion, knobs are not "neocentromeres" (see Section 3.3) since they are unable to promote normal segregation on their own and do not recruit the proteins nor display the functions associated with endogenous centromeres and kinetochores.

In terms of meiotic drive, exchanges in meiocytes that are heterozygous for a given knob result in the formation of a crossover product that carries the knob on only one of its two chromatids (Rhoades and Dempsey 1966). The pseudokinetochore behavior of the knob directs the knob-bearing chromatid to one of the four products of meiosis, known as the basal megaspore, which is one of the two "outside" nuclei (of the four meiotic products), and the only one that is available for fertilization (Rhoades and Dempsey 1966; Yu et al. 1997). Thus, meiotic drive in this system is caused by heterochromatin-mediated directed segregation of one chromatid to the functional gamete, at the expense of its sister. How epigenetic mechanisms regulate knob behavior and pseudokinetochore activity is currently unknown.

6 The Silencing of Genes by Unpaired DNA during Meiosis

There are multiple examples of instances in which the failure of meiotic pairing results in the epigenetic silencing of the unpaired DNA. In the case of *Neurospora*, this silencing is generally limited to unpaired regions.

6.1 Meiotic Silencing of Unpaired DNA during Meiosis in **Neurospora**

An unusual example of gene silencing during meiosis, referred to as meiotic silencing by unpaired DNA (MSUD), was reported in *Neurospora crassa* (for review,

see Hynes and Todd 2003). This process allows an unpaired copy of a given gene to silence both itself and any other paired copies of that gene (Fig. 11). This process was identified by the characterization of heterozygous deletions in the *ascospore maturation 1* gene (*asm-1*). However, an unpaired sequence must be of sufficient size and have homology with the mRNA product of the gene to initiate silencing (Lee et al. 2004). Such sequences do not, however, need to contain the proper promoter element for the gene in question. Silencing is limited to the gene defined by the unpaired sequence and does not spread into neighboring, paired genes (Kutil et al. 2003). Mutations in the *sad-1* gene, which encodes a protein having substantial homology with the RNA-dependent RNA polymerases (RdRP) involved in RNAi pathways, block the process of MSUD (Shiu and Metzenberg 2002), suggesting that "the synthesis of a double-stranded RNA and probably RNA amplification is required for MSUD" (Hynes and Todd 2003). Although examples of MSUD have been observed for genes that are normally expressed in meiosis, MSUD has not been detected for genes expressed only in vegetative cells, strengthening the argument for RNAi involvement (Hynes and Todd 2003). Fur-

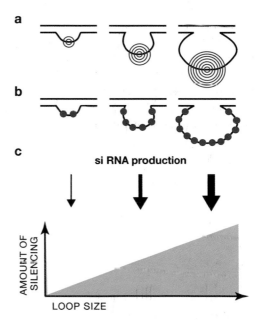

Figure 11. Models to Show How Transcription of Unpaired DNA Results in Different Amounts of Silencing during Meiosis

In model *a*, single transcription complexes are proposed to have more transcriptional activity in larger loops compared to smaller loops, as indicated by the number of red concentric circles. Model *b* suggests that larger loops may contain more transcription complexes (*red balls*) than smaller loops. (*c*) In both models, unpaired DNA in large loops would produce higher concentrations of siRNAs (indicated by size of arrows), and thus more silencing. (Adapted, with permission, from Lee et al. 2004 [© Genetics Society of America].)

ther support for this idea comes from observations that both an Argonaute-like and a Dicer-like protein are required for MSUD (Lee et al. 2003).

It has been suggested that MSUD is the culmination of a two-part process: trans-sensing and MSUD (Pratt et al. 2004). Trans-sensing, which appears to be dependent on DNA methylation, allows a given sequence of DNA to "sense" the identity or homology of paired regions in the context of meiotic prophase. Once unpaired sequences are identified in this fashion, the process of MSUD acts to suppress the expression of those sequences.

An RNA-dependent mechanism of meiotic silencing of unpaired DNA has also been documented in *C. elegans*. DNA lacking a pairing partner in meiosis is targeted for histone H3K9 methylation (Bean et al. 2004). This silencing mechanism requires an RNA-dependent RNA polymerase (EGO-1); however, it does not require Dicer (Maine et al. 2005).

6.2 Silencing of Unsynapsed Chromosomes in the Mouse

The meiotic silencing of unpaired DNA has also been demonstrated in both sexes of the mouse (Baarends et al. 2005; Turner et al. 2005). Specifically, unpaired autosomal regions in male meiosis, or unpaired X chromosomes in female meiosis, are both transcriptionally silent and marked by ubiquitination of histone H2A—both properties of the XY body that is normally present in male meiotic cells (Baarends et al. 2005). It was also reported that unpaired DNA arising from mouse translocation heterozygotes was silenced by the recruitment of the checkpoint kinase ATR by BRCA1, in a fashion similar to that observed for the silencing of the X and Y chromosomes during normal male meiosis (Turner et al. 2005).

6.3 Sex Chromosome Dysfunction in **Drosophila**

Euchromatic examples of MSUD are unknown in either sex in *D. melanogaster*. Moreover, the extensive study of euchromatic deletions and, more importantly, duplications, makes it highly unlikely that such genes exist. On the other hand, it is well documented that deletions of pericentromeric X heterochromatin interact with Y-autosome translocations to cause complete male sterility (for review, see McKee 1998). This sterility cannot be suppressed by adding back either a complete Y chromosome or a complete duplication of the X heterochromatin. McKee (1998) proposed that sterility of Y-autosome translocations is caused by the presence of a

meiotic checkpoint that detects unpaired or poorly paired sex chromosomes. On the basis of examples of MSUD in diverse eukaryotes described above, we suggest an alternative explanation, that the presence of unpaired sequences on the X or Y chromosome triggers the silencing of essential elements whose function is required for normal fertility.

7 Perspectives and Conclusions

Eukaryotic chromosomes were once viewed as trucks, semi-trailers filled with genes driven by "motors" at the kinetochore with telomeres as "bumpers." Early work in budding yeast supported a view in which the properties of the crucial elements of these vehicles devolved directly and only from their DNA sequence. However, the analysis of structures such as centromeres in higher organisms has led us to the view that inheritance elements are encoded more as a chromatin state than simply a DNA sequence. The concept of facultative heterochromatin developed to explain X inactivation, or the spreading of "inertness" to explain position-effect variegation, have become integral components of our understanding of genome duplication, repair, and mitotic and meiotic inheritance, and the activities of the relevant chromosomal regions.

In some ways, it is surprising that such essential functions are encoded epigenetically; "hard-wiring" in DNA sequences seems at first glance to be a more stable mechanism for ensuring successful chromosome inheritance. However, nature is filled with examples of functional "state changes" associated with chromosome duplication and segregation, such as neocentromeres in mammals and flies, "meiotic drive" in various systems, and a myriad of phenomena of directed chromosome loss. Many inheritance elements are associated with highly unstable repeated DNAs, for reasons that are currently unclear, and evolution is accompanied by large- and small-scale genome changes, such as mutations and chromosome rearrangements. Perhaps the surprising amount of plasticity in the regulation of inheritance elements is essential to accommodate changes in DNA sequences, and may be necessary for evolution to proceed.

To whatever extent we are permitted a quick glimpse into the future, our guess would be that the systems uncovered so far are but the tip of the epigenetic iceberg, that chromosome functions are more about chromatin than just DNA sequences. We expect many more examples to appear on the horizon.

References

Ahmad K. and Golic K.G. 1998. The transmission of fragmented chromosomes in *Drosophila melanogaster*. *Genetics* **148:** 775–792.

Ahmad K. and Henikoff S. 2001. Centromeres are specialized replication domains in heterochromatin. *J. Cell Biol.* **153:** 101–110.

———. 2002. The histone variant H3.3 marks active chromatin by replication-independent nucleosome assembly. *Mol. Cell* **9:** 1191–1200.

Aladjem M.I. and Fanning E. 2004. The replicon revisited: An old model learns new tricks in metazoan chromosomes. *EMBO Rep.* **5:** 686–691.

Amor D.J. and Choo K.H. 2002. Neocentromeres: Role in human disease, evolution, and centromere study. *Am. J. Hum. Genet.* **71:** 695–714.

Baarends W.M., Wassenaar E., van der Laan R., Hoogerbrugge J., Sleddens-Linkels E., Hoeijmakers J.H., de Boer P., and Grootegoed J.A. 2005. Silencing of unpaired chromatin and histone H2A ubiquitination in mammalian meiosis. *Mol. Cell. Biol.* **25:** 1041–1053.

Bassing C.H., Chua K.F., Sekiguchi J., Suh H., Whitlow S.R., Fleming J.C., Monroe B.C., Ciccone D.N., Yan C., et al. 2002. Increased ionizing radiation sensitivity and genomic instability in the absence of histone H2AX. *Proc. Natl. Acad. Sci.* **99:** 8173–8178.

Baum M., Ngan V., and Clarke L. 1994. The centromeric K-type repeat and the central core are together sufficient to establish a functional *Schizosaccharomyces pombe* centromere. *Mol. Biol. Cell* **5:** 747–761.

Bean C. J., Schaner C. E., and Kelly W. G. 2004. Meiotic pairing and imprinted X chromatin assembly in *Caenorhabditis elegans*. *Nat. Genet.* **36:** 100–105.

Bennett M. D., Dover G. A., and Riley R. 1974. Meiotic duration in wheat genotypes with or without homoeologous meiotic chromosome pairing. *Proc. R. Soc. Lond. B Biol. Sci.* **187:** 191–207.

Beukeboom L.W. and Werren J.H. 1993. Deletion analysis of the selfish B chromosome, *Paternal Sex Ratio* (PSR), in the parasitic wasp *Nasonia vitripennis*. *Genetics* **133:** 637–648.

Biessmann H. and Mason J.M. 2003. Telomerase-independent mechanisms of telomere elongation. *Cell. Mol. Life Sci.* **60:** 2325–2333.

Black B.E., Foltz D.R., Chakravarthy S., Luger K., Woods V.L., Jr., and Cleveland D.W. 2004. Structural determinants for generating centromeric chromatin. *Nature* **430:** 578–582.

Blasco M.A. 2005. Telomeres and human disease: Ageing, cancer and beyond. *Nat. Rev. Genet.* **6:** 611–622.

Bloom K., Hill A., Kenna M., and Saunders M. 1989. The structure of a primitive kinetochore. *Trends Biochem. Sci.* **14:** 223–227.

Blower M.D. and Karpen G.H. 2001. The role of *Drosophila* CID in kinetochore formation, cell-cycle progression and heterochromatin interactions. *Nat. Cell Biol.* **3:** 730–739.

Blower M.D., Sullivan B.A., and Karpen G.H. 2002. Conserved organization of centromeric chromatin in flies and humans. *Dev. Cell* **2:** 319–330.

Carroll C.W. and Straight A.F. 2006. Centromere formation: From epigenetics to self-assembly. *Trends Cell Biol.* **16:** 70–78.

Celeste A., Petersen S., Romanienko P.J., Fernandez-Capetillo O., Chen H.T., Sedelnikova O.A., Reina-San-Martin B., Coppola V., Meffre E., Difilippantonio M.J., et al. 2002. Genomic instability in mice lacking histone H2AX. *Science* **296:** 922–927.

Cenci G., Siriaco G., Raffa G.D., Kellum R., and Gatti M. 2003. The *Drosophila* HOAP protein is required for telomere capping. *Nat. Cell Biol.* **5:** 82–84.

Clarke L., Amstutz H., Fishel B., and Carbon J. 1986. Analysis of centromeric DNA in the fission yeast *Schizosaccharomyces pombe*. *Proc. Natl. Acad. Sci.* **83:** 8253–8257.

Cleveland D.W., Mao Y., and Sullivan K.F. 2003. Centromeres and kinetochores: From epigenetics to mitotic checkpoint signaling. *Cell* **112:** 407–421.

Collins K.A., Furuyama S., and Biggins S. 2004. Proteolysis contributes to the exclusive centromere localization of the yeast Cse4/CENP-A histone H3 variant. *Curr. Biol.* **14:** 1968–1972.

Cryderman D.E., Morris E., Biessmann H., Elgin S.C., and Wallrath L.L. 1999. Silencing at *Drosophila* telomeres: Nuclear organization and chromatin structure play critical roles. *EMBO J.* **18:** 3724–3735.

Cullen C.F., Brittle A.L., Ito T., and Ohkura H. 2005. The conserved kinase NHK-1 is essential for mitotic progression and unifying acentrosomal meiotic spindles in *Drosophila melanogaster*. *J. Cell Biol.* **171:** 593–602.

Dawe R.K. and Hiatt E.N. 2004. Plant neocentromeres: Fast, focused, and driven. *Chromosome Res.* **12:** 655–669.

De La Fuente R. 2006. Chromatin modifications in the germinal vesicle (GV) of mammalian oocytes. *Dev. Biol.* **292:** 1–12.

de Saint Phalle B. and Sullivan W. 1996. Incomplete sister chromatid separation is the mechanism of programmed chromosome elimination during early *Sciara coprophila* embryogenesis. *Development* **122:** 3775–3784.

Dernburg A.F. 2001. Here, there, and everywhere: Kinetochore function on holocentric chromosomes. *J. Cell Biol.* **153:** F33–38.

Dernburg A.F., Sedat J.W., and Hawley R.S. 1996. Direct evidence of a role for heterochromatin in meiotic chromosome segregation. *Cell* **86:** 135–146.

Dobson S.L. and Tanouye M.A. 1998. Evidence for a genomic imprinting sex determination mechanism in *Nasonia vitripennis* (Hymenoptera; Chalcidoidea). *Genetics* **149:** 233–242.

Donaldson K.M., Lui A., and Karpen G.H. 2002. Modifiers of terminal deficiency-associated position effect variegation in *Drosophila*. *Genetics* **160:** 995–1009.

Dvorak J. and Lukaszewski A.J. 2000. Centromere association is an unlikely mechanism by which the wheat *Ph1* locus regulates metaphase I chromosome pairing between homoeologous chromosomes. *Chromosoma* **109:** 410–414.

Fanti L., Giovinazzo G., Berloco M., and Pimpinelli S. 1998. The heterochromatin protein 1 prevents telomere fusions in *Drosophila*. *Mol. Cell* **2:** 527–538.

Ferguson B.M. and Fangman W.L. 1992. A position effect on the time of replication origin activation in yeast. *Cell* **68:** 333–339.

Fernandez-Capetillo O., Lee A., Nussenzweig M., and Nussenzweig A. 2004. H2AX: The histone guardian of the genome. *DNA Repair* **3:** 959–967.

Ferreri G.C., Liscinsky D.M., Mack J.A., Eldridge M.D., and O'Neill R.J. 2005. Retention of latent centromeres in the mammalian genome. *J. Hered.* **96:** 217–224.

Friedberg E., Walker G., and Siede W. 1995. *DNA repair and mutagenesis*. ASM Press, Washington D.C.

Ganetzky B. 1977. On the components of segregation distortion in *Drosophila melanogaster*. *Genetics* **86:** 321–355.

Garcia-Cao M., O'Sullivan R., Peters A.H., Jenuwein T., and Blasco M.A. 2004. Epigenetic regulation of telomere length in mammalian cells by the Suv39h1 and Suv39h2 histone methyltransferases. *Nat. Genet.* **36:** 94–99.

Goday C. and Ruiz M.F. 2002. Differential acetylation of histones H3 and H4 in paternal and maternal germline chromosomes during development of sciarid flies. *J. Cell Sci.* **115:** 4765–4775.

Grell R.F. 1976. Distributive pairing. In *The genetics and biology of Drosophila* (ed. E. Novitski and M. Ashburner), pp. 436–483. Academic Press, New York.

Griffiths S., Sharp R., Foote T.N., Bertin I., Wanous M., Reader S., Colas

I., and Moore G. 2006. Molecular characterization of *Ph1* as a major chromosome pairing locus in polyploid wheat. *Nature* **439:** 749–752.

Hall A.E., Kettler G.C., and Preuss D. 2006. Dynamic evolution at pericentromeres. *Genome Res.* **16:** 355–364.

Hassa P.O. and Hottiger M.O. 2005. An epigenetic code for DNA damage repair pathways? *Biochem. Cell Biol.* **83:** 270–285.

Hawley R.S., Irick H., Zitron A.E., Haddox D.A., Lohe A., New C., Whitley M.D., Arbel T., Jang J., McKim K., and Childs G. 1993. There are two mechanisms of achiasmate segregation in *Drosophila* females, one of which requires heterochromatic homology. *Dev. Genet.* **13:** 440–467.

Heun P., Erhardt S., Blower M.D., Weiss S., Skora A.D., and Karpen G.H. 2006. Mislocalization of the *Drosophila* centromere-specific histone CID promotes formation of functional ectopic kinetochores. *Dev. Cell* **10:** 303–315.

Huyen Y., Zgheib O., Ditullio R.A., Jr., Gorgoulis V.G., Zacharatos P., Petty T.J., Sheston E.A., Mellert H.S., Stavridi E.S., and Halazonetis T.D. 2004. Methylated lysine 79 of histone H3 targets 53BP1 to DNA double-strand breaks. *Nature* **432:** 406–411.

Hynes M.J. and Todd R.B. 2003. Detection of unpaired DNA at meiosis results in RNA-mediated silencing. *Bioessays* **25:** 99–103.

Ivanovska I., Khandan T., Ito T., and Orr-Weaver T.L. 2005. A histone code in meiosis: The histone kinase, NHK-1, is required for proper chromosomal architecture in *Drosophila* oocytes. *Genes Dev.* **19:** 2571–2582.

Joseph I., Jia D., and Lustig A.J. 2005. Ndj1p-dependent epigenetic resetting of telomere size in yeast meiosis. *Curr. Biol.* **15:** 231–237.

Karpen G.H. and Spradling A.C. 1992. Analysis of subtelomeric heterochromatin in the *Drosophila* minichromosome Dp1187 by single P element insertional mutagenesis. *Genetics* **132:** 737–753.

Karpen G.H., Le M.H., and Le H. 1996. Centric heterochromatin and the efficiency of achiasmate disjunction in *Drosophila* female meiosis. *Science* **273:** 118–122.

Kemp B., Boumil R.M., Stewart M.N., and Dawson D.S. 2004. A role for centromere pairing in meiotic chromosome segregation. *Genes Dev.* **18:** 1946–1951.

Kusano A., Staber C., Chan H.Y., and Ganetzky B. 2003. Closing the (Ran)GAP on segregation distortion in *Drosophila*. *Bioessays* **25:** 108–115.

Kutil B.L., Seong K.Y., and Aramayo R. 2003. Unpaired genes do not silence their paired neighbors. *Curr. Genet.* **43:** 425–432.

Lai S.R., Phipps S.M., Liu L., Andrews L.G., and Tollefsbol T.O. 2005. Epigenetic control of telomerase and modes of telomere maintenance in aging and abnormal systems. *Front. Biosci.* **10:** 1779–1796.

Lam A.L., Boivin C.D., Bonney C.F., Rudd M.K., and Sullivan B.A. 2006. Human centromeric chromatin is a dynamic chromosomal domain that can spread over noncentromeric DNA. *Proc. Natl. Acad. Sci.* **103:** 4186–4191.

Lee D.W., Pratt R.J., McLaughlin M., and Aramayo R. 2003. An argonaute-like protein is required for meiotic silencing. *Genetics* **164:** 821–828.

Lee D.W., Seong K.Y., Pratt R.J., Baker K., and Aramayo R. 2004. Properties of unpaired DNA required for efficient silencing in *Neurospora crassa*. *Genetics* **167:** 131–150.

Lo A.W., Magliano D.J., Sibson M.C., Kalitsis P., Craig J.M., and Choo K.H. 2001. A novel chromatin immunoprecipitation and array (CIA) analysis identifies a 460-kb CENP-A-binding neocentromere DNA. *Genome Res.* **11:** 448–457.

Louis E.J. and Vershinin A.V. 2005. Chromosome ends: Different sequences may provide conserved functions. *Bioessays* **27:** 685–697.

Lue N.F. 2004. Adding to the ends: What makes telomerase processive and how important is it? *Bioessays* **26:** 955–962.

Luo M.C., Dubcovsky J., and Dvorak J. 1996. Recognition of homeology by the wheat *Ph1* locus. *Genetics* **144:** 1195–1203.

Maggert K.A. and Karpen G.H. 2001. Neocentromere formation occurs by an activation mechanism that requires proximity to a functional centromere. *Genetics* **158:** 1615–1628.

Maine E.M., Hauth J., Ratliff T., Vought V.E., She X., and Kelly W.G. 2005. EGO-1, a putative RNA-dependent RNA polymerase, is required for heterochromatin assembly on unpaired dna during *C. elegans* meiosis. *Curr. Biol.* **15:** 1972–1978.

Malik H.S. and Henikoff S. 2002. Conflict begets complexity: The evolution of centromeres. *Curr. Opin. Genet. Dev.* **12:** 711–718.

McKee B.D. 1998. Pairing sites and the role of chromosome pairing in meiosis and spermatogenesis in male *Drosophila*. *Curr. Top. Dev. Biol.* **37:** 77–115.

Mellone B.G. and Allshire R.C. 2003. Stretching it: Putting the CEN(P-A) in centromere. *Curr. Opin. Genet. Dev.* **13:** 191–198.

Miller K.M. and Cooper J.P. 2003. The telomere protein Taz1 is required to prevent and repair genomic DNA breaks. *Mol. Cell* **11:** 303–313.

Nakaseko Y., Kinoshita N., and Yanagida M. 1987. A novel sequence common to the centromere regions of *Schizosaccharomyces pombe* chromosomes. *Nucleic Acids Res.* **15:** 4705–4715.

Nakatani Y., Ray-Gallet D., Quivy J.P., Tagami H., and Almouzni G. 2004. Two distinct nucleosome assembly pathways: Dependent or independent of DNA synthesis promoted by histone H3.1 and H3.3 complexes. *Cold Spring Harbor Symp. Quant. Biol.* **69:** 273–280.

Nimmo E.R., Pidoux A.L., Perry P.E., and Allshire R.C. 1998. Defective meiosis in telomere-silencing mutants of *Schizosaccharomyces pombe*. *Nature* **392:** 825–828.

Nonaka N., Kitajima T., Yokobayashi S., Xiao G., Yamamoto M., Grewal S.I., and Watanabe Y. 2002. Recruitment of cohesin to heterochromatic regions by Swi6/HP1 in fission yeast. *Nat. Cell Biol.* **4:** 89–93.

Ogino K., Hirota K., Matsumoto S., Takeda T., Ohta K., Arai K.I., and Masai H. 2006. Hsk1 kinase is required for induction of meiotic dsDNA breaks without involving checkpoint kinases in fission yeast. *Proc. Natl. Acad. Sci.* **103:** 8131–8136.

Pasero P., Bensimon A., and Schwob E. 2002. Single-molecule analysis reveals clustering and epigenetic regulation of replication origins at the yeast rDNA locus. *Genes Dev.* **16:** 2479–2484.

Perrini B., Piacentini L., Fanti L., Altieri F., Chichiarelli S., Berloco M., Turano C., Ferraro A., and Pimpinelli S. 2004. HP1 controls telomere capping, telomere elongation, and telomere silencing by two different mechanisms in *Drosophila*. *Mol. Cell* **15:** 467–476.

Pidoux A.L. and Allshire R.C. 2004. Kinetochore and heterochromatin domains of the fission yeast centromere. *Chromosome Res.* **12:** 521–534.

Prasanth S. G., Mendez J., Prasanth K. V., and Stillman B. 2004. Dynamics of pre-replication complex proteins during the cell division cycle. *Philos. Trans. R. Soc. Lond. B Biol. Sci.* **359:** 7–16.

Pratt R.J., Lee D.W., and Aramaya R. 2004. DNA methylation affects meiotic trans-sensing, not meiotic silencing, in *Neurospora*. *Genetics* **168:** 1925–1935.

Reddy K.C. and Villeneuve A.M. 2004. *C. elegans* HIM-17 links chromatin modification and competence for initiation of meiotic recombination. *Cell* **118:** 439–452.

Rhoades M.M. and Dempsey E. 1966. The effect of abnormal chromosome 10 on preferential segregation and crossing over in maize. *Genetics* **53:** 989–1020.

Rudd M.K., Mays R.W., Schwartz S., and Willard H.F. 2003. Human artificial chromosomes with alpha satellite-based de novo centromeres show increased frequency of nondisjunction and anaphase lag. *Mol. Cell Biol.* **23:** 7689–7697.

Sadaie M., Naito T., and Ishikawa F. 2003. Stable inheritance of telomere chromatin structure and function in the absence of telomeric repeats. *Genes Dev.* **17:** 2271–2282.

Sandler L. and Hiraizumi Y. 1959. Meiotic drive in natural populations of *Drosophila melanogaster*. II. Genetic variation at the *Segregation-distorter* locus. *Proc. Natl. Acad. Sci.* **45:** 1412–1422.

Sharp J.A., Franco A.A., Osley M.A., and Kaufman P.D. 2002. Chromatin assembly factor I and Hir proteins contribute to building functional kinetochores in *S. cerevisiae*. *Genes Dev.* **16:** 85–100.

Shelby R.D., Monier K., and Sullivan K.F. 2000. Chromatin assembly at kinetochores is uncoupled from DNA replication. *J. Cell Biol.* **151:** 1113–1118.

Shiu P.K. and Metzenberg R.L. 2002. Meiotic silencing by unpaired DNA: Properties, regulation and suppression. *Genetics* **161:** 1483–1495.

Steiner N.L. and Clarke L. 1994. A novel epigenetic effect can alter centromere function in fission yeast. *Cell* **79:** 865–874.

Sullivan B. and Karpen G. 2001. Centromere identity in *Drosophila* is not determined in vivo by replication timing. *J. Cell Biol.* **154:** 683–690.

Sullivan B.A. and Karpen G.H. 2004. Centromeric chromatin exhibits a histone modification pattern that is distinct from both euchromatin and heterochromatin. *Nat. Struct. Mol. Biol.* **11:** 1076–1083.

Sullivan B.A. and Willard H.F. 1998. Stable dicentric X chromosomes with two functional centromeres. *Nat. Genet.* **20:** 227–228.

Tsubouchi T. and Roeder G.S. 2005. A synaptonemal complex protein promotes homology-independent centromere coupling. *Science* **308:** 870–873.

Turner J.M., Mahadevaiah S.K., Fernandez-Capetillo O., Nussenzweig A., Xu X., Deng C.X., and Burgoyne P.S. 2005. Silencing of unsynapsed meiotic chromosomes in the mouse. *Nat. Genet.* **37:** 41–47.

Watanabe Y. 2005. Sister chromatid cohesion along arms and at centromeres. *Trends Genet.* **21:** 405–412.

Weinreich M., Palacios DeBeer M.A., and Fox C.A. 2004. The activities of eukaryotic replication origins in chromatin. *Biochim. Biophys. Acta* **1677:** 142–157.

Werren J.H., Nur U., and Eickbush D. 1987. An extrachromosomal factor causing loss of paternal chromosomes. *Nature* **327:** 75–76.

Wu H.Y. and Burgess S.M. 2006. Ndj1, a telomere-associated protein, promotes meiotic recombination in budding yeast. *Mol. Cell Biol.* **26:** 3683–3694.

Yan H., Jin W., Nagaki K., Tian S., Ouyang S., Buell C.R., Talbert P.B., Henikoff S., and Jiang J. 2005. Transcription and histone modifications in the recombination-free region spanning a rice centromere. *Plant Cell* **17:** 3227–3238.

Yoda K., Ando S., Morishita S., Houmura K., Hashimoto K., Takeyasu K., and Okazaki T. 2000. Human centromere protein A (CENP-A) can replace histone H3 in nucleosome reconstitution in vitro. *Proc. Natl. Acad. Sci.* **97:** 7266–7271.

Yu H.G., Hiatt E.N., Chan A., Sweeney M., and Dawe R.K. 1997. Neocentromere-mediated chromosome movement in maize. *J. Cell Biol.* **139:** 831–840.

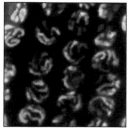

Epigenetic Regulation of the X Chromosomes in C. *elegans*

Susan Strome[1] and William G. Kelly[2]

[1]*Department of Biology, Indiana University, Bloomington, Indiana 47405*
[2]*Biology Department, Emory University, Atlanta, Georiga 30322*

CONTENTS

1. Sex Chromosome Imbalance in *C. elegans*, 293

2. The DCC Resembles the Condensin Complex, 293

3. Assessing the X:A Ratio, 295

4. Recruitment and Spread of the DCC, 296

5. Effects of the DCC: Down-Regulation of X-linked Genes and the Autosomal Gene *her-1*, 297

6. Regulation of X Chromosomes in the Germ Line, 297

7. Germ-line Development and X-Chromosome Silencing, 298

8. The Single X in Males Displays Marks of Heterochromatin, 299

9. Effects of MSUD on Gene Expression Patterns in the Germ Line, 300

10. Regulation of X-Chromosome Silencing by the MES Histone Modifiers, 301

11. The Sperm X Chromosome Is Imprinted, 302

12. Concluding Remarks, 304

References, 304

GENERAL SUMMARY

The mechanism of sex determination in the soil-dwelling nematode *Caenorhabditis elegans* relies on the ratio of X chromosomes to sets of autosomes (X:A ratio). Diploid animals with two X chromosomes (X:A ratio of 1) develop as hermaphrodites, whereas those with one X chromosome (X:A ratio of 0.5) develop as males. The difference in X-chromosome dosage between the sexes, if not modulated, would lead to different levels of expression of X-linked genes and to lethality in one sex. In this chapter, we discuss the different epigenetic strategies used in somatic tissues and in the germ line to modulate gene expression from different numbers of X chromosomes in the two worm sexes, and we compare the worm strategies to those used in other organisms to cope with similar X-chromosome dosage differences.

In the somatic tissues of worms, equalizing expression of X-linked genes between the sexes, termed "dosage compensation," occurs by down-regulating expression twofold from both X chromosomes in the XX sex. This down-regulation is accomplished by the dosage compensation complex (DCC), which resembles the condensin complex. The latter is involved in condensing all of a cell's chromosomes for their segregation during nuclear division to make two daughter cells. Thus, the DCC may achieve repression of X-linked genes by partially condensing the X chromosomes. Key issues in worm dosage compensation are how the X:A ratio is assessed, how the DCC is assembled uniquely in XX animals, how the DCC is targeted to the X chromosomes, and how it accomplishes precise twofold down-regulation of X-linked gene expression. Significant mechanistic advances have been made on each of these fronts.

The dosage compensation strategy used by worms differs from the strategies used by mammals and fruit flies (*Drosophila*). Mammals achieve dosage compensation by globally silencing one X in the XX sex. Fruit flies up-regulate gene expression from the single X in the XY sex. This diversity of strategies probably reflects the co-option of different preexisting systems for the specialized role of equalizing X-gene expression in animals with different numbers of X chromosomes.

In the germ-line tissue (i.e., the reproductive cells) of worms, a more extreme modulation of X-linked gene expression occurs: The single X in males and both Xs in hermaphrodites are globally silenced. In germ cells of both sexes, the X chromosomes lack histone modifications that are associated with actively expressed chromatin. This is regulated at least in part by the MES proteins. Furthermore, in males, the single X chromosome in each germ nucleus acquires histone modifications that are associated with heterochromatic silencing. This silencing depends on the unpaired status of the X in male meiosis. Genes expressed in the germ line are strikingly underrepresented on the X chromosome. A favored view is that the strong repression of the single X in male meiotic germ cells led to the paucity of germ-line-expressed genes on the X and to the need for X-chromosome silencing in the XX sex. Thus, the *C. elegans* germ line presents an interesting case study of how chromosome imbalances between the sexes have led to epigenetic regulation of chromosome states and have shaped the gene expression profile of the tissue.

1 Sex Chromosome Imbalance in *C. elegans*

Caenorhabditis elegans exists as two sexes that are genetically distinguished by their X-chromosome complement: XX worms are hermaphrodites and XO worms are males. There is no sex-specific chromosome, such as a Y chromosome. Hermaphrodites and males display numerous sex-specific anatomical features and have different germ-line programs (Fig. 1). These dramatic differences between the sexes are initiated in the early embryo, and result from counting and properly responding to the number of X chromosomes (Nigon 1951; for review, see Meyer 2000). How can a simple difference in sex chromosome number or ploidy translate into such dramatically different developmental programs? An important concept is that each *C. elegans* cell must assess not only its number of X chromosomes, but also the number of sets of autosomes. It is actually the ratio of these, the X:A ratio, that determines sex. Diploid animals with two X chromosomes (X:A ratio of 1) develop as hermaphrodites, whereas those with one X chromosome (X:A ratio of 0.5) develop as males. Many of the mechanistic details of appropriately responding to the X:A ratio have been elegantly dissected and are described below.

The difference in X-chromosome dosage between the sexes, if not modulated, would lead to different levels of expression of X-linked genes and to lethality in one sex. Intriguingly, somatic cells and germ cells have evolved different mechanisms to deal with this X-dosage challenge (Fig. 2). The germ line and somatic lineages are fully separated from each other by the 16- to 24-cell stage of embryogenesis. Starting at about the 30-cell stage, the somatic lineages initiate a process termed "dosage compensation." This constitutes a scheme whereby both X chromosomes in XX animals are repressed 2-fold. In contrast, as discussed in Chapters 16 and 17, mammals implement dosage compensation by globally silencing one X in the XX sex, and fruit flies implement dosage compensation by up-regulating expression from the single X in the XY sex. In *C. elegans* germ-line tissue, a more extreme modulation of X-linked gene expression occurs: The single X in males and both Xs in hermaphrodites are globally silenced. The epigenetic mechanisms that accomplish dosage compensation in the soma and X-chromosome silencing in the germ line are the subjects of this chapter.

2 The DCC Resembles the Condensin Complex

Understanding the assembly and composition of the dosage compensation complex (DCC) requires a brief introduction to the first few genes in the pathway that regulates both sex determination and dosage compensation (for review, see Meyer 1997). *xol-1* (*xol* for *XO-lethal*), the first gene in the pathway, is considered a master switch gene, because its activity is determined by the X:A ratio and it in turn dictates whether the pathway leads to male or hermaphrodite development. XOL-1 is a negative regulator of three *sdc* (*sdc* for sex determination

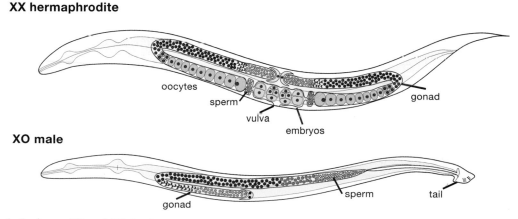

XX hermaphrodite

oocytes
sperm
vulva
embryos
gonad

XO male

gonad
sperm
tail

Figure 1. *C. elegans* XX and XO Anatomy

C. elegans naturally exists as two sexes, XX hermaphrodites and XO males. Hermaphrodites and males display several sex-specific anatomical features, most notably a male tail designed for mating, and a vulva on the ventral surface of hermaphrodites for reception of male sperm and for egg-laying. Their germ-line programs also differ. The two-armed gonad in hermaphrodites produces sperm initially and then oocytes throughout adulthood. The one-armed gonad in males produces sperm continuously. (Adapted with permission, from Hansen et al. 2004 [©Elsevier].)

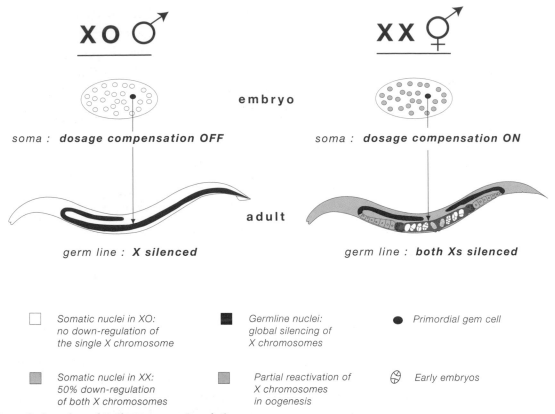

Figure 2. Overview of X-Chromosome Regulation

Dosage compensation occurs in somatic tissues uniquely in XX hermaphrodites. Silencing of the Xs in the germ line occurs in both XO males and XX hermaphrodites. Hermaphrodites display late and partial activation of X-linked genes during late pachytene of oogenesis. The arrows point out the single primordial germ cell in the embryo that generates the germ line in the adult gonad.

and dosage compensation defective) genes. The *sdc* genes encode components of the DCC and regulate the *her-1* (*her* for hermaphroditization of XO animals) sex determination gene. In XO embryos, an X:A ratio of 0.5 leads to high XOL-1 protein levels and low SDC protein levels; the DCC is not assembled, dosage compensation is not implemented, and the sex determination gene *her-1* is expressed, leading to male sexual development. In XX embryos, an X:A ratio of 1 results in low XOL-1 and high SDC levels; the DCC is assembled, dosage compensation is implemented, and *her-1* is repressed, leading to hermaphrodite sexual development. Of particular importance for thinking about assembly of the DCC is the fact that the SDC proteins are expressed specifically in XX embryos, in response to an X:A ratio of 1.

In addition to the SDC proteins (SDC-1, SDC-2, and SDC-3), the DCC also contains a set of DPY (DPY for dumpy) proteins (DPY-21, DPY-26, DPY-27, DPY-28, and DPY-30) and the MIX-1 (MIX for mitosis and X-associated) protein (Table 1) (for review, see Meyer 2005).

Significant insights into the mechanism of dosage compensation in worms came from the discovery that a portion of the *C. elegans* DCC resembles the 13S condensin complex (Table 1 and Fig. 3) (for review, see Meyer 2005). The condensin complex is conserved across species and is essential for proper chromosome compaction and segregation during mitosis and meiosis (for review, see Hirano 2002). MIX-1, DPY-27, DPY-28, and DPY-26 in particular are homologous to essential components of the condensin complex; MIX-1 is also an essential component of the canonical condensin complex that acts in mitosis and meiosis. The SDC proteins and DPY-30 do not resemble known condensin subunits. The current view is that the DCC complex has been adapted from an ancestral condensin complex for specific targeting to, and down-regulation of, genes on the X chromosome, likely through some degree of chromatin condensation.

Why is the DCC assembled only in XX embryos? Surprisingly, most of the DCC components are maternally supplied via the oocyte to both XX and XO embryos. The

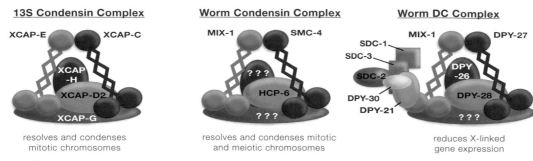

13S Condensin Complex
resolves and condenses mitotic chromosomes

Worm Condensin Complex
resolves and condenses mitotic and meiotic chromosomes

Worm DC Complex
reduces X-linked gene expression

Figure 3. The Dosage Compensation Complex (DCC) and Condensin Complex

The worm DCC resembles the condensin complex, which functions in condensing chromosomes as they go through nuclear division. In particular, the DCC contains several subunits that resemble the XCAP (XCAP for *Xenopus* chromosome-associated polypeptide) subunits of the 13S condensin complex originally characterized in *Xenopus*. The native worm condensin complex shows overall similarity to the homologous functional 13S complex in *Xenopus*. The SDC proteins and DPY-30 do not resemble known condensin subunits; they instead function in localizing the DCC complex to the X chromosome. (Adapted from Meyer 2005.)

key regulator of DCC assembly is SDC-2 (Dawes et al. 1999). SDC-2 is not maternally supplied and is produced only in XX embryos (Fig. 4) where it, along with SDC-3 and DPY-30, recruits the remaining DCC subunits to the X chromosomes. In fact, driving expression of SDC-2 in XO embryos is sufficient to cause assembly of the DCC on the single X chromosome and to trigger dosage compensation, which kills the embryos. SDC-2 thus directs the specific recruitment of other DCC components, most of which have other cellular roles, and co-opts their activities for dosage compensation and sex determination.

3 Assessing the X:A Ratio

How do worm cells count Xs and autosomes, and implement dosage compensation when the X:A ratio is 1? Four small regions of the X, termed X signal elements (XSEs), have been identified as contributing to the numerator portion of the X:A ratio. In two of the four regions, the genes responsible have been identified: *sex-1* and *fox-1* (*sex* for signal element on X, *fox* for feminizing gene on X) (Carmi et al. 1998; Skipper et al. 1999). Both gene products repress expression of *xol-1*, the most upstream gene in the sex determination and dosage compensation pathway, during a critical window of embryonic development (Fig. 4). Because XX embryos produce approximately twice as much SEX-1 and FOX-1 as XO embryos, *xol-1* expression is much lower in XX than in XO embryos. SEX-1 is a nuclear hormone receptor that represses transcription of *xol-1* (Carmi et al. 1998). FOX-1 is an RNA-binding protein that reduces the level of *xol-1* mRNA through a posttranscriptional mechanism (Skipper et al. 1999). X-chromosome dosage, therefore, is

equivalent to the dosage of X-linked factors expressed from XSEs that act to repress expression of *xol-1*. Thus far, only one autosomal signal element (ASE) has been identified as contributing to the denominator portion of the X:A ratio (Powell et al. 2005). The identified gene, *sea-1*, encodes a T-box transcription factor, which activates transcription of *xol-1*.

Table 1. Components that regulate gene expression from the X chromosome

Protein	Complex	Homolog and/or conserved domains
SDC-1	DCC	C_2H_2 zinc-finger domain
SDC-2	DCC	novel protein
SDC-3	DCC	C_2H_2 zinc-finger domain and myosin-like ATP-binding domain
DPY-21	DCC	conserved protein; no recognizable motifs
DPY-26	DCC	condensin subunit XCAP-H
DPY-27	DCC	condensin subunit SMC-4/XCAP-C
DPY-28	DCC	condensin subunit XCAP-D2/ Cnd1/Ycs4p
DPY-30	DCC	conserved protein; no recognizable motifs
MIX-1	DCC	condensin subunit SMC-2/XCAP-E
MES-2	MES-2/3/6	PRC2 subunit E(Z)/EZH2; SET domain
MES-3	MES-2/3/6	novel protein
MES-6	MES-2/3/6	PRC2 subunit ESC/EED; WD40 domains
MES-4	unknown	NSD1; PHD fingers and SET domain

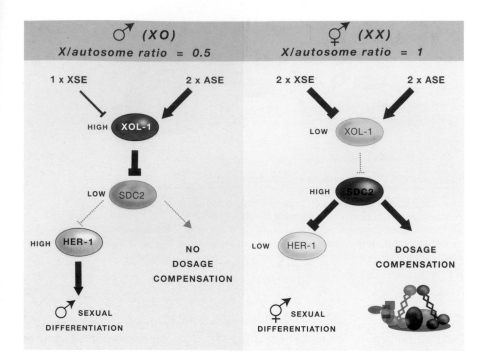

Figure 4. Pathway of Sex Determination and Dosage Compensation in *C. elegans*

This figure highlights the proposed roles of X signal elements (XSEs) and autosomal signal elements (ASEs) in regulating XOL-1 levels, and subsequent sexual differentiation and assembly of the dosage compensation complex (DCC), which represses X-linked gene expression about 2-fold in XX hermaphrodites.

Autosomal dosage (ASE dosage) therefore counterbalances X dosage (XSE dosage) effects through antagonistic action on the master switch gene, *xol-1* (Fig. 4). The working hypothesis for how this plays out in both sexes is as follows: Diploid XX embryos produce a double dose of XSE repressors of *xol-1*, which override the activating influence of ASEs. This keeps XOL-1 levels low and leads to hermaphrodite development and implementation of dosage compensation. Conversely, diploid XO embryos produce a single dose of XSE repressors, which are insufficient to counteract the activating influence of ASEs. High XOL-1 levels lead to male development and failure to implement dosage compensation. XSEs and ASEs translate the shift in X:A ratio from 0.5 to 1, into a switch in sex and the decision whether or not to implement dosage compensation.

4 Recruitment and Spread of the DCC

Research has focused on identifying features of the X chromosome involved in recruiting the DCC. It is interesting to note that in mammals and fruit flies, the DCC equivalent is targeted to the X chromosome by a combination of X-specific noncoding RNAs and X-chromosome sequence elements (see Chapters 16 and 17). There is no evidence to date that noncoding RNAs play a role in *C. elegans* dosage compensation. An elegant approach was used to investigate whether the worm DCC is recruited independently to many sites, or instead to only one or a few sites, after which complexes spread into adjoining

chromosomal regions (Csankovszki et al. 2004; Meyer 2005). Worm strains containing duplications of different regions of the X chromosome were stained for DCC components. Association of the DCC with a duplication was interpreted to mean that the duplication and the corresponding region of the intact X contains a DCC recruitment site. Lack of DCC association with the duplication, but association of DCC with the corresponding region of the intact X, was interpreted to mean that the region included in the X duplication lacks a DCC recruitment site and instead acquires DCC by spreading from adjoining regions. These experiments identified at least 13 regions that can independently recruit DCC, and provided evidence for DCC spreading along the X (Fig. 5) (Csankovszki et al. 2004). Some regions recruit strongly and others weakly, suggesting either the presence of varying numbers of recruitment sites, or of sites with varying capacity to recruit and/or promote spreading along the X. Testing progressively smaller regions narrowed the DCC recruitment site of one region down to 33 bp (Meyer 2005). This 33-bp sequence is unique in the genome. Defining other minimal recruitment sites is a high priority. Another critical issue is to elucidate how DCC spreading occurs from initial recruitment sites. Spreading may be mediated by cooperative interactions between DCCs, or by local modification of chromatin into a structure that facilitates more DCC binding in a self-reinforcing loop, as demonstrated for the spread of heterochromatin in *Schizosaccharomyces pombe* (see Chapter 8).

5 Effects of the DCC: Down-Regulation of X-linked Genes and the Autosomal Gene *her-1*

Regulating chromatin structure to achieve modest 2-fold effects on X-linked gene expression, as occurs in worms and fruit flies, seems mechanistically challenging. In fruit flies, modulation of gene expression is associated with particular histone modifications and with loss of linker histone from the X (see Chapter 16). In the somatic cells of worms, no striking differences have been reported in either the level or spectrum of histone modifications seen on the two dosage-compensated X chromosomes compared to the autosomes in hermaphrodites. Mammals, fruit flies, and worms appear to have adapted different preexisting systems to serve the specialized role of modulating gene expression from the X chromosome. Mammals have adapted heterochromatin-based silencing to inactivate one of the two Xs in the XX sex. Fruit flies have adapted the use of chromatin-modifying machinery to alter the state of the single X in XY animals, leading to up-regulation of gene expression. *C. elegans* has adapted a chromosome condensation mechanism used for mitosis and meiosis to down-regulate both Xs in the XX sex. The precise mechanism of down-regulation in worms and how it is limited to 2-fold are critical questions.

Given the similarity of the DCC to the 13S condensin complex, a likely mechanism for the down-regulation of gene expression is through DCC-mediated condensation of chromatin. This may restrict access of RNA polymerase and/or transcription factors to promoter regions, impede progression of RNA polymerase through transcription units, and/or slow reinitiation of transcription at each gene. Clues about the mechanism of repression may be found at the single autosomal *her-1* locus, which is the only known autosomal DCC target and which displays a more extreme 20-fold DCC-mediated down-regulation of gene expression (Fig. 6) (Dawes et al. 1999; Chu et al. 2002). Repression of *her-1* promotes hermaphrodite sexual development (Fig. 4). Therefore, the DCC is capable of achieving a 10-fold greater repression at *her-1* than it achieves on the X chromosomes. One known difference between the X-chromosome-associated DCC and the *her-1*-associated complex is DPY-21. DPY-21 is present on the X chromosomes but not at the *her-1* locus (Yonker and Meyer 2003).

The 13S condensin complex is needed to regulate chromosome structure on a genome-wide scale to facilitate chromosome segregation. Adaptation and modification of the condensin complex to form the DCC makes its effects subtler and narrows its regulation to (predominantly) a single chromosome. DCC with DPY-21 weakly

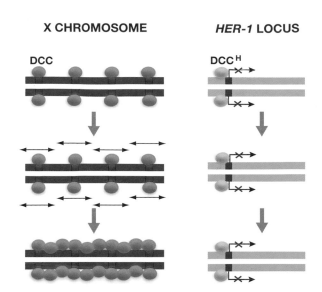

X CHROMOSOME   ***HER-1* LOCUS**

Figure 5. DCC-mediated Down-regulation of X Chromosomes and the *her-1* Locus

The DCC reduces X expression by half and represses *her-1* expression 20-fold in hermaphrodite somatic tissues. The DCC assembles in XX animals. It is recruited to recognition elements and spreads along both X chromosomes, reducing expression of numerous X-linked genes by 2-fold. It also binds to the upstream region of the autosomal gene *her-1*, and reduces expression by 20-fold (denoted DCC^H to show that the complex that binds to *her-1* lacks the DPY-21 protein). (Adapted, with permission, from Alekseyenko and Kuroda 2004; illustration by Katherine Sutliff [©AAAS].)

represses transcription of many genes on the X, whereas DCC without DPY-21 somehow strongly represses transcription at the *her-1* locus. It will be exciting to dissect in the future how the different condensin/DCC complexes serve their different functions.

6 Regulation of X Chromosomes in the Germ Line

In *C. elegans*, the germ line uses different strategies than somatic tissues to regulate expression from the X chromosomes, as the somatic mode of X-chromosome regulation (i.e., by the DCC) is inoperative in germ cells. Some essential components of the somatic DCC are not expressed in the germ line (e.g., SDC-2); others are expressed in germ cells but are found on *all* chromosomes (e.g., MIX-1, DPY-26, and DPY-28). MIX-1, as mentioned above, is a component of both the DCC and the canonical condensin complex in *C. elegans*. The current view is that a condensin-like complex has been adapted to serve a specialized X-chromosome-specific role, but that role appears to be served only in somatic tissues. In the germ line, those DCC proteins that are expressed are likely engaged in more general roles in mitotic and meiotic chromosome organization and segregation.

The absence of DCC members in the germ line raises the question of whether any mode of dosage compensation occurs in this tissue. Recall that the function of dosage compensation is to achieve equivalent expression of X-linked genes between the sexes for viability and normal development. Current evidence suggests that, in contrast to the 2-fold down-regulation of both X chromosomes in XX somatic cells, the X chromosomes are globally silenced during most stages of germ-line development in *both* XX and XO animals (Fig. 6). A lack of transcription from the X chromosomes in germ cells of both sexes would appear to dispense with a need for sex-specific regulation of this chromosome. There are nevertheless specific mechanisms that regulate the X chromosomes in germ cells, and as one would suspect, these bear little resemblance in either form or function to the somatic mechanisms already described in this chapter.

7 Germ-line Development and X-Chromosome Silencing

The adult gonads of both sexes in *C. elegans* contain an orderly progression of germ-cell stages. Germ cells proliferate in the distal region, enter meiosis in the middle region, and complete gametogenesis in the proximal region (Figs. 1 and 7) (for review, see Schedl 1997). In XO males, this progression occurs in a single tubular testis that continuously produces sperm. In XX hermaphrodites, two tubular gonad arms initially produce sperm during a late larval stage and then switch to the production of oocytes in adulthood. Mature oocytes are pushed into the spermathecae and fertilized by sperm residing there before further extrusion into the uterus. The production of some sperm by XX worms thus allows the hermaphroditic mode of reproduction in the adult. However, the ovary and certain other somatic tissues in adult XX animals can be considered essentially female in identity and function.

X chromosomes in the germ line are silenced both in XX hermaphrodites and in XO males. This silencing is characterized by repressive histone marks, such as H3K27 methylation; the X chromosomes in hermaphrodite germ cells are particularly enriched for H3K27me3 (Bender et al. 2004). There is also an absence of histone acetylation, commonly associated with active chromatin, and H3K4 methylation, which correlates with actively transcribed regions, as revealed by immunofluorescence analysis (Fig. 7) (Kelly et al. 2002). Silencing of the X chromosome in the *C. elegans* germ line is achieved predominantly by two

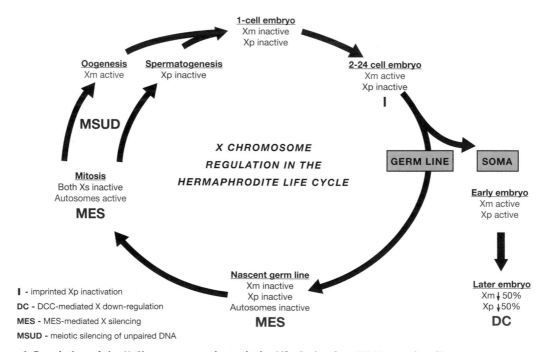

Figure 6. Regulation of the X Chromosomes through the Life Cycle of an XX Hermaphrodite

This figure illustrates that the X chromosomes are regulated by different mechanisms at different stages and in different tissues: imprinted X inactivation (I) of the paternal X in the early embryo, dosage compensation (DC) in the somatic tissues of 30-cell and later-stage embryos and worms, and MES-mediated silencing in the germ line. Meiotic silencing of unpaired DNA (MSUD) also occurs in the germ line; this silences the single X chromosome in XO males.

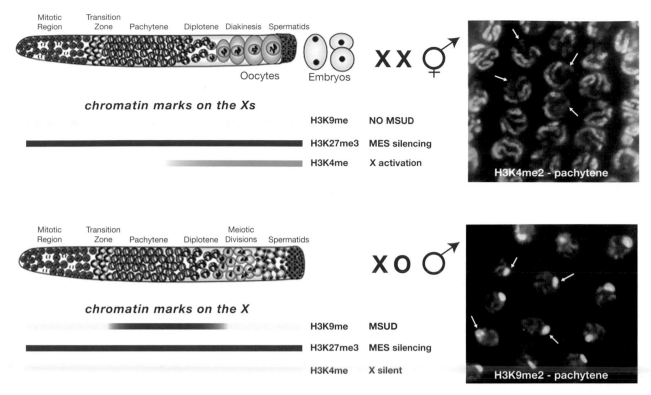

Figure 7. Epigenetic Regulation of the X Chromosomes during Germ-cell Development

In both sexes, germ cells progress through mitosis (*left*), enter meiosis in the transition zone, and progress through meiosis I prophase. Cells destined to form sperm in both sexes complete the meiotic divisions in the gonad. In hermaphrodites, cells destined to form oocytes progress through meiotic prophase in the gonad and complete the meiotic divisions after ovulation and fertilization. The presence of various histone modifications on the X chromosome(s) in germ cells is shown by red (for repressive modifications) and green bars (for activating modifications). As shown in the panels on the right, antibodies to particular histone modifications reveal that the X chromosomes in germ nuclei are marked differently from the autosomes and are silenced. H3K4me2 (*green*), a mark of actively expressed chromatin, is excluded from the Xs in XX pachytene nuclei. H3K9me2 (*green*), a mark of heterochromatin, is concentrated on the X in XO pachytene nuclei. DNA is stained red. Arrows indicate representative X chromosomes in each image.

mechanisms discussed below (Fig. 6). The first is through meiotic silencing of unpaired DNA (MSUD), which includes the single X chromosome in XO animals. The second is through histone modifications mediated by the MES proteins; this silencing mechanism has been most intensively studied in XX animals but likely operates in XO animals as well. In early embryos, silencing of the paternal X chromosome persists for a few cell cycles after fertilization and is termed imprinted X silencing (Fig. 6).

8 The Single X in Males Displays Marks of Heterochromatin

The earliest suggestion that the X chromosome differs from the autosomes in the germ line came from cytological observations. The single X in male worms assumes a characteristic structure during pachytene in meiotic

prophase—it hypercondenses, forming a ball-like structure reminiscent of the XY "sex body" seen during male meiosis in mammals (Goldstein and Slaton 1982; Handel 2004). The autosomes condense later in meiotic prophase, near the onset of spermiogenesis. Premature X condensation is also seen during sperm meiosis in XX hermaphrodites and in sexually transformed XX males, suggesting that premature X condensation is in response to germ cell sex and not to X chromosome ploidy or pairing status. Accordingly, in XX hermaphrodites, germ cells destined for oogenesis do not exhibit premature X condensation.

In addition to premature condensation, the single X chromosome in XO males transiently accumulates a striking enrichment of H3K9me2, which appears during pachytene and disappears in diplotene (Fig. 7) (Kelly et al. 2002). This enrichment for H3K9me2 does not occur during XX spermatogenesis, in either hermaphrodites or sexually transformed XX males. However, H3K9me2 enrichment on the X is observed in oogonia progressing through pachytene in sexually transformed XO hermaphrodites, with similar dynamics to those observed in XO males (Bean et al. 2004). The specific acquisition of a heterochromatin mark on the X in XO meiosis thus appears to be a consequence of its unpaired status, and not the sex of the germ line through which it is passing. Targeting of H3K9me2 to unpaired DNA is not limited to X-chromosome sequences; it is also found on unpaired autosomal fragments and translocations (Bean et al. 2004).

Such targeted repression of unpaired DNA in meiosis is not unique to *C. elegans*. Similar recognition and repression of unpaired DNA occurs in other organisms during meiosis, including *Neurospora* and mice. This is referred to as MSUD for meiotic silencing of unpaired DNA (Shiu et al. 2001; Baarends et al. 2005; Turner 2005; Turner et al. 2005). In mouse, for example, the poorly synapsed XY "sex body" during male mouse meiosis is similarly enriched in H3K9me2, and this is a consequence of its unpaired status (Cowell et al. 2002; Turner et al. 2005). Meiotic silencing in *Neurospora* requires the activity of proteins with conserved roles in RNA interference (RNAi). These include an RNA-directed RNA polymerase (RdRP), an Argonaute-related protein (conserved component of RNA-induced silencing complexes, RISCs), and the Dicer nuclease (for review, see Nakayashiki 2005; see Chapter 8). In *C. elegans*, the enrichment of H3K9me2 on unpaired DNA requires EGO-1, an RdRP that is restricted to the germ line, but does not require Dicer (Maine et al. 2005). This suggests that whereas there is conservation of meiotic silencing (repression of unpaired DNA), the mechanism by which this is achieved may have evolved differently in different organisms.

In contrast to what happens in XO male meiosis, the X chromosomes in either XX spermatogonia or XX oogonia do not appreciably accumulate H3K9me2. This difference is likely due to the complete synapsis of the Xs in hermaphrodite meiosis. Why would silencing of unpaired DNA be a conserved feature of sexual reproduction? In many organisms, homolog pairing is unique to meiosis, and during synapsis, novel insertions unique to one homolog would be exposed as regions of unpaired DNA. It has been proposed that the recognition and silencing of unpaired sequences during meiosis provides a mechanism for self-scanning of a diploid genome, and could provide protection against invasion by (or expansion of) transposable elements. As a further consequence, genes expressed during meiosis would encounter a strong selection against residence on an unpaired chromosome (such

as the male X). Such selection could lead to the unique genetic profile of the X that is observed in *C. elegans*. It is interesting to speculate that the genomic warfare between transposons and their hosts, and the epigenetic mechanisms that have evolved as weapons in this battle, have shaped genomes and led to genetic variability. This genetic variability, in turn, could create sufficient pairing defects to trigger meiotic silencing of essential loci and hence meiotic incompatibility. In this manner, epigenetic silencing during meiosis could, in theory, contribute to reproductive isolation and speciation.

9 Effects of MSUD on Gene Expression Patterns in the Germ Line

Genome-wide expression studies of germ-line transcription in *C. elegans* have shown that there is a remarkable difference in the profiles of mRNA accumulation from genes that map to the X chromosome compared to the autosomes (Reinke et al. 2000, 2004). Genes required for spermatogenesis, as well as general or "intrinsic" germ-line-expressed genes, are strikingly underrepresented on the X chromosome. Genes with enriched expression in oogenic germ lines are also underrepresented on the X, but not to the same degree as the spermatogenesis or "intrinsic" germ-line genes. In other words, genes that are required for the viability and function of germ cells are, for the most part, absent from the X. Indeed, most of the oogenesis-associated genes that reside on the X chromosome show late meiotic expression, which correlates with activation of the X chromosome (Figs. 6 and 7). Late meiosis activation is evidenced by the accumulation of "activating" histone modifications near the end of pachytene, including H3 and H4 acetylation and H3K4 methylation to levels similar to those of the autosomes (Kelly et al. 2002). This suggests that a bias against genes being located on the X is most stringent for those that function in the early germ-cell stages that are common to both sexes. In addition, a number of gene duplications with X/autosome paralogs have been identified in which the autosomal copy is uniquely required for germ-cell function. In these examples, the X-linked copy functions in somatic lineages but is not required in germ cells, where the autosomal copy is active (Maciejowski et al. 2005). Taken together, the above findings suggest that the X chromosome is an inhospitable environment for genes with essential early germ-cell functions in both sexes. Genes that are only required during late oogenesis (i.e., only required in late-stage female germ cells; Fig. 2) would be immune to the selective pressures caused by asynapsis in male germ cells, and thus their X linkage would be allowed.

C. elegans genes with essential early germ-line functions evidently have been excluded from the X, have been lost from the X, or have undergone gene duplication followed by germ-line reliance on the autosomal copy. The latter suggests a possible pathway for the evolution of the unique genetic profile of the X. Similar forces seem to be acting on X-linked genes in other species, including fruit flies and mammals (Wu and Xu 2003). At present, however, we do not understand the mechanisms that drive essential germ-line genes off the X.

10 Regulation of X-Chromosome Silencing by the MES Histone Modifiers

The previous section describes repression via heterochromatin formation that is specific to the single X in XO males and restricted to the pachytene stage of meiosis due to MSUD. How are the X chromosomes maintained in a repressed chromatin state during other stages of germ-line development and in XX hermaphrodites? Genetic screens for maternal-effect sterile (*mes*) mutants identified a set of four *mes* genes that participate in X-chromosome silencing in XX animals and likely in XO animals as well. A combination of genetic and molecular analyses have shown that the encoded MES proteins effect silencing through regulating part of the spectrum of histone modifications found on the X chromosomes in germ cells. Their functions are essential for the survival and development of germ cells.

Three of the MES proteins, MES-2, MES-3, and MES-6, function together in a complex that resembles the Polycomb Repressive Complex (PRC2) in fruit flies and vertebrates (see Xu et al. 2001; Bender et al. 2004; Ketel et al. 2005; Chapter 11). MES-2 and MES-6 are the worm orthologs of two PRC2 subunits, E(Z) (Enhancer of zeste) and ESC (Extra sex combs) (Table 1). E(Z) and ESC, in association with at least two other partners, catalyze methylation of histone H3K27, which, as mentioned above, is a repressive histone modification. Similarly, MES-2 and MES-6, in a complex with the novel MES-3 protein, mediate H3K27 methylation (Fig. 8). MES-2's SET domain is responsible for its histone lysine methyltransferase (HKMT) activity, whereas MES-6 and MES-3 appear to be required either for substrate binding or to boost catalytic activity.

MES-2, MES-3, and MES-6 are responsible for all detectable H3K27 di- and trimethylation (H3K27me2 and H3K27me3) in most regions of the germ line and in early embryos, but another HKMT catalyzes this methylation in somatic tissues. Importantly, within the germ line the fully repressive mark, H3K27me3, is enriched on the X chromosomes (Fig. 8) (Bender et al. 2004). One consequence of abolishing H3K27 methylation in the germ line is activation of genes on the X; in mutants of *mes-2*, *mes-3*, or *mes-6*, the X chromosomes in the hermaphrodite germ line lack H3K27me, acquire marks of active chromatin (e.g., H3K4me and H4K12ac), and become decorated with the transcriptionally active form of RNA polymerase (Fong et al. 2002; Bender et al. 2004). These findings suggest that the MES-2/3/6 complex participates, perhaps directly, in silencing of the X chromosomes in the germ line of worms. Indeed, de-silencing of the Xs is proposed to be the cause of germ-line degeneration observed in *mes* mutant hermaphrodites. Similar to the involvement of MES-2 and MES-6 in germ-line X silencing, the vertebrate homologs of MES-2 and MES-6 are involved in somatic X inactivation in XX mammals (see Chapters 11 and 17).

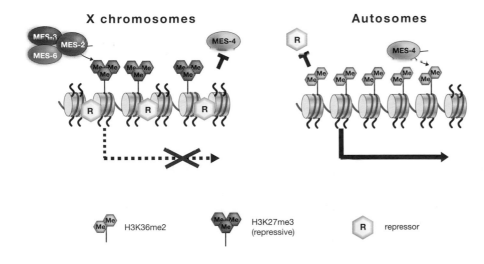

Figure 8. Model for How MES-2/3/6 and MES-4 Participate in X Silencing in the Germ Line

The MES-2/3/6 complex concentrates a repressive chromatin mark (H3K27me3) on the X chromosomes and repels MES-4 from the Xs. MES-4 concentrates a different chromatin mark (H3K36me2) on the autosomes and is hypothesized to repel an unknown repressor (*yellow hexagon*), thereby helping to focus that repressor's action on the Xs. The result is silencing of the Xs by the combined action of the "yellow hexagon" repressor and MES-2/3/6-catalyzed repressive histone methylation.

How is the MES-2/3/6 complex targeted to DNA regions destined to be repressed, and how does methylation of H3K27 repress gene expression? Answers to these questions are emerging in *Drosophila* (see Chapter 11). In flies, the E(Z)/ESC complex is recruited to particular sites called Polycomb Response Elements (PREs) by the DNA-binding protein PHO (Wang et al. 2004). After the complex locally methylates H3K27, gene repression is mediated, at least in part, by recruitment of the Polycomb Repressive Complex 1 (PRC1) to H3K27me (Wang et al. 2004). Surprisingly, the *C. elegans* genome does not contain obvious homologs of *pho* or of most PRC1-encoding genes. Targeting of the MES-2/3/6 complex and how it achieves repression are therefore likely to involve strategies distinct from those used in *Drosophila*.

The fourth MES protein involved in X repression in the germ line is MES-4. Its distribution is novel among chromosome-associated proteins and exactly opposite to what might be expected. In contrast to the other MES proteins, MES-4 associates with the five autosomes, in a banded pattern, and is strikingly absent from most of the length of the X chromosome (Fig. 8) (Fong et al. 2004). MES-4, like MES-2, contains a SET domain and also has HKMT activity (Bender et al. 2006). It is responsible for the majority of detectable H3K36 dimethylation (H3K36me2) in the germ line and early embryos. As predicted by the autosomal association of MES-4, H3K36me2 is dramatically concentrated on the autosomes. In somatic tissues and later-stage embryos, a different HKMT is responsible for H3K36 methylation. This latter activity appears to be associated with transcription elongation, as is the case in yeast. MES-4-mediated H3K36 methylation on the autosomes is not dependent on transcription and likely serves a different, germ-line-required, epigenetic role (Bender et al. 2006).

Key outstanding issues are determining how MES-4 is targeted to the autosomes and what role it serves there. Interestingly, the exclusion of MES-4 from the X chromosomes requires MES-2, MES-3, and MES-6 (Fig. 8); in the absence of any of these latter MES proteins, MES-4 spreads to the X (Fong et al. 2002). The normal concentration of MES-4 on the autosomes is thought to participate in silencing genes on the X. One model is that MES-4 and/or its methyl mark (H3K36me2) repels global chromatin repressors (the yellow hexagons in Fig. 8) from the autosomes and in that way focuses repressor action on the X chromosomes (Fong et al. 2002). Loss of MES-4 would titrate those repressors away from the X to autosomal regions, and this would result in de-silencing of the X. The MES-2/3/6 complex is the logical candidate for being the MES-4-repelled repressor. However, the distribution of MES-2/3/6-catalyzed methylation of H3K27 is not visibly altered in *mes-4* mutants. Consequently, the search continues for other repressors that may participate. This model for MES-4 action is similar to the "gaining specificity by preventing promiscuity" model suggested for *Saccharomyces cerevisiae* Dot1 (van Leeuwen and Gottschling 2002). Dot1-mediated methylation of H3K79 along chromosomes is thought to repel the Sir repressors and help focus their action on the telomeres. Loss of Dot1 allows spreading of the Sirs from telomeres and results in telomeric desilencing.

The *mes* phenotype is an excellent example of an epigenetic phenomenon with direct linkage to histone modifications. A maternal-effect mutation is defined as one whose mutant phenotype is not revealed in first-generation homozygous mutants, but instead shows up in their progeny. In the case of first-generation *mes/mes* mutants, the presence of some wild-type MES product produced by the *mes/+* mother and packaged into the oocyte is sufficient for the proper expansion of the two primordial germ cells into more than a thousand fully functional germ cells. The germ cells in these fertile *mes/mes* worms, however, cannot produce functional MES product for their offspring. As a result, in these offspring, the primordial germ cells undergo little proliferation and degenerate. The HKMT activities encoded by *mes-2* and *mes-4* must establish a heritable chromatin state that is properly maintained in the many descendants of the two initial primordial germ cells. The current model of MES protein function (Fig. 8) is that MES-2, MES-3, and MES-6 operate in a complex to concentrate a repressive chromatin modification (H3K27me) on the X chromosomes in the germ line, and directly participate in X silencing. Their activity also repels MES-4 from the X. As a result, MES-4 coats and modifies the autosomes with H3K36me—this activity is proposed to repel and help focus repressive activity on the X. The MES system is thought to act epigenetically in the mother's germ line and in early embryos to establish chromatin domains that are properly marked for subsequent expression (autosomal regions) or silencing (the X chromosomes) during germ-line development in larvae (see Fig. 6). Loss of the MES system leads to germ-line death and sterility, likely due at least in part to de-silencing of the X chromosomes.

11 The Sperm X Chromosome Is Imprinted

The genomes contributed by the different gametes arrive in the zygote with vastly different epigenetic histories (Fig. 6). Although the X chromosome is inactive in early germ-cell stages in both sexes, the X becomes transcriptionally active during oogenesis (Fong et al. 2002; Kelly et al. 2002). In contrast, during male meiosis the X is never activated, is subjected to premature condensation, and in XO meiosis is additionally enriched in H3K9me2 (Fig. 7) (Kelly et al. 2002; Bean et al. 2004). During spermatogenesis, histone modifications are progressively lost from all chromosomes as they progress into the meiotic divisions and become hypercondensed for packaging into spermatids (Fig. 7). The hypercondensation is presumably due to replacement of histones by specialized, highly basic proteins called protamines, as occurs in other organisms, although this has not yet been reported in worms. The sperm genome thus enters the egg mostly devoid of histone modifications (and likely depleted of histones themselves). This is in contrast to the oocyte chromatin, which continues to exhibit significant levels of most activating histone modifications even during chromosome condensation in diakinesis (Fig. 7). It is important to note that RNA polymerase is absent from the DNA during oocyte maturation, and therefore the histone modifications retained by the chromatin may reflect transcriptional "potential" rather than transcriptional activity. Thus, the zygote inherits two epigenetically different genomes, and in particular two X chromosomes with very different transcriptional histories—a recently active, oocyte-derived Xm and a sperm-derived Xp with little or no recent transcriptional activity. Note that due to the hermaphrodite mode of C. elegans reproduction and the ability of hermaphrodites to mate with males, an XX offspring can inherit an Xp from its XX parent or from an XO parent.

Shortly after entry into the oocyte, the sperm DNA simultaneously accumulates histone modifications and begins to decondense, forming the sperm pronucleus. In XX embryos, in striking contrast to the autosomes, the Xp chromatin does not accumulate histone H3 acetylation or H3K4me2 during decondensation (Fig. 6) (Bean et al. 2004). There is, however, no difference in histone H4 modifications between the Xp and the autosomes. In the oocyte pronucleus, all chromosomes, including the Xm, are similarly modified and remain so throughout embryogenesis. Amazingly, the Xp-specific absence of H3 modifications is maintained after DNA replication and survives multiple rounds of cell division, and thus has been termed an "epigenetic imprint." However, the imprint is gradually lost; i.e., H3-specific modifications become increasingly detectable on the Xp until there are no obvious differences in H3K4me2 levels between the Xp and other chromosomes (Fig. 6). The nature of the imprint is not yet known, but its gradual reversal during

several rounds of cell division is consistent with replication-coupled dilution of the mark. Since the absence of modifications is limited to histone H3-specific additions, it is conceivable that what is being replaced is an Xp-specific (or Xp-enriched) H3-like molecule that is either not modified or incapable of being modified, such as a histone H3 variant. Chromatin composition and structure during spermatogenesis are currently being investigated.

The Xp imprint in XX embryos is detected in both cross-progeny (from XO-derived sperm) and self-progeny (from XX-derived sperm). The pairing status of the X chromosome during spermatogenesis, and thus H3K9me2 targeting and enrichment on the X, therefore does not play an obvious role in establishment of the Xp imprint. However, the stability of the imprint—that is, the number of cell divisions after which it can still be readily observed—is significantly increased in offspring from XO-derived sperm relative to XX-derived sperm (Bean et al. 2004). Therefore, heterochromatin assembly on unpaired DNA during meiosis has effects that persist through early embryonic stages. Importantly, both pairing-based meiotic silencing and imprinted Xp inactivation are conserved in mammals, as discussed in Chapter 17.

The biochemical basis for the imprint and its effects are not known. In contrast to the situation in mammals, the consequences of inheriting both X chromosomes lacking or containing the imprint are not obvious in worms. Indeed, genetic imprinting per se (covered in detail in Chapter 19) has long been thought to be absent in C. elegans, since uniparental inheritance of any chromosome has no dramatic consequences on development and viability. It has been reported, however, that the Xp is preferentially and frequently lost during early development under stressful conditions (Prahlad et al. 2003), although there are no data regarding the mechanism that could cause such extreme chromosomal behavior. Numerous organisms exhibit preferential loss of paternally inherited chromosomes, and in extreme cases, the entire paternal genome is lost during embryonic development (Goday and Esteban 2001). The description of such dramatic differences in genome regulation, based solely on the parental sex from which the genetic contribution originates, gave rise to the original concept of genetic imprinting (for review, see Herrick and Seger 1999). Whether this is related to the epigenetic imprinting observed in C. elegans remains to be determined.

The Xp appears to be inactive in early XX embryos, yet there are no reported deleterious consequences of patriclinous inheritance of the X (i.e., XpXp animals; Haack and Hodgkin 1991). The unique genetic composition of the X chromosome may help to explain why uniparental inheritance is not obviously detrimental in *C. elegans*. In addition to a paucity of sperm and intrinsic germ-line-expressed loci on the X chromosome, genes encoding early zygotic transcripts and those required for early embryonic development are also strikingly underrepresented on the X chromosome (Piano et al. 2000; Baugh et al. 2003). Thus, most genes whose products are essential for the very early stages during which the Xp is inactive are unlikely to reside on the X, thereby rendering X-specific uniparental inheritance inconsequential in genetic tests.

Xp inactivation does, however, suggest a reason that somatic dosage compensation activation is not fully engaged in early embryos. The assembly of DCC components on the X chromosomes in XX embryos is not detectable by antibody staining until approximately the 30-cell stage. This is shortly after the Xp becomes fully decorated with H3 modifications, and is thus presumably fully activated. Activation of somatic dosage compensation is responsive to levels of X-linked products, called X signal elements (XSEs), that comprise the X portion of the X:A ratio. Full or partial repression of these elements on the Xp may render the early embryo functionally XO (i.e., the X:A is interpreted as 0.5). As the Xp reactivates with increased rounds of cell division, the level of X transcription may finally reach the critical threshold at which a 2 X chromosome dosage is sensed and X:A equals 1, triggering the dosage compensation cascade. One might consider this a switch from maternal/paternal control of dosage compensation to zygotic control.

12 Concluding Remarks

At this point, we have come full circle with respect to the specialized regulation of X-linked loci in *C. elegans*. The special dosage regulation of this chromosome in somatic cells is hypothesized to have arisen from the difference in X ploidy between the two sexes, as has happened in other species. Interestingly, the germ line and soma in worms have evolved distinct mechanisms to deal with this difference in X ploidy, which may reflect different requirements for X gene expression in those tissues. In the somatic lineages, a conserved complex that is normally used for chromatin condensation and segregation during mitosis and meiosis appears to have been co-opted and adapted for a remarkably specific 2-fold repression of both X chromosomes in XX animals. Although it is conceptually attractive to envision that an "inefficient" condensin complex might serve to decrease transcriptional efficiency by

2-fold, how the repression is *limited* to 2-fold is currently not understood.

In the germ line, the X chromosomes are globally silenced during early stages of germ-cell development in XX animals and during all stages of germ-cell development in XO animals. The MES system of chromatin modifiers participates in this silencing in XX animals and perhaps in XO animals as well. Such silencing is tolerated in part because the X chromosomes are underpopulated with germ-line-expressed genes. The paucity of X-linked genes essential for meiosis in both sexes is thought to have arisen from the lack of a pairing partner for the X during male meiosis and from repressive genome defense mechanisms that target unpaired chromosomal segments. A consequence of global X-chromosome silencing in the germ line is that the X ploidy difference between the sexes is of little consequence for this tissue. If some X-linked genes that operate in both sexes in fact escape silencing, then the question arises as to whether the germ line equalizes their expression in XX versus XO animals. The answer to this is not known. Equalization in the germ line would likely involve a mechanism other than the DCC that operates in somatic tissues.

The germ line and soma differ in many fundamental ways, and this chapter has highlighted their different regulation of the X chromosome. One interesting theme that has emerged is co-option of different preexisting mechanisms, such as utilization of a condensin-related complex to subtly down-regulate X expression in the soma versus utilization of a PRC2-related complex to silence the Xs in the germ line. The heterochromatinization of the single X in males is thought to have dramatically altered the representation of genes on the X, such that genes required for general germ-line functions and early embryonic development are significantly underrepresented on this chromosome. The X chromosome and its regulation in *C. elegans* thus represent an interesting intersection of epigenetic chromosome-wide regulation and its influences on, and responses to, genome evolution.

References

Alekseyenko A.A. and Kuroda M.I. 2004. Filling gaps in genome organization. *Science* **303:** 1148–1149.

Baarends W.M., Wassenaar E., van der Laan R., Hoogerbrugge J., Sleddens-Linkels E., Hoeijmakers J.H., de Boer P., and Grootegoed J.A. 2005. Silencing of unpaired chromatin and histone H2A ubiquitination in mammalian meiosis. *Mol. Cell. Biol.* **25:** 1041–1053.

Baugh L.R., Hill A.A., Slonim D.K., Brown E.L., and Hunter C.P. 2003. Composition and dynamics of the *Caenorhabditis elegans* early embryonic transcriptome. *Development* **130:** 889–900.

Bean C.J., Schaner C.E., and Kelly W.G. 2004. Meiotic pairing and imprinted X chromatin assembly in *Caenorhabditis elegans*. *Nat. Genet.* **36:** 100–105.

Bender L.B., Cao R., Zhang Y., and Strome S. 2004. The MES-2/MES-3/MES-6 complex and regulation of histone H3 methylation in *C. elegans*. *Curr. Biol.* **14:** 1639–1643.

Bender L.B., Suth J., Carroll C.R., Fong Y., Fingerman I.M., Cao R., Zhang Y., Briggs S.D., Reinke V., and Strome S. 2006. MES-4, an autosome-associated histone methyltransferase that participates in silencing the X chromosomes in the *C. elegans* germ line. *Development.* (In press.)

Carmi I., Kopczynski J.B., and Meyer B.J. 1998. The nuclear hormone receptor SEX-1 is an X-chromosome signal that determines nematode sex. *Nature* **396:** 168–173.

Chu D.S., Dawes H.E., Lieb J.D., Chan R.C., Kuo A.F., and Meyer B.J. 2002. A molecular link between gene-specific and chromosome-wide transcriptional repression. *Genes Dev.* **16:** 796–805.

Cowell I.G., Aucott R., Mahadevaiah S.K., Burgoyne P.S., Huskisson N., Bongiorni S., Prantera G., Fanti L., Pimpinelli S., Wu R., et al. 2002. Heterochromatin, HP1 and methylation at lysine 9 of histone H3 in animals. *Chromosoma* **111:** 22–36.

Csankovszki G., McDonel P., and Meyer B.J. 2004. Recruitment and spreading of the *C. elegans* dosage compensation complex along X chromosomes. *Science* **303:** 1182–1185.

Dawes H.E., Berlin D.S., Lapidus D.M., Nusbaum C., Davis T.L., and Meyer B.J. 1999. Dosage compensation proteins targeted to X chromosomes by a determinant of hermaphrodite fate. *Science* **284:** 1800–1804.

Fong Y., Bender L., Wang W., and Strome S. 2002. Regulation of the different chromatin states of autosomes and X chromosomes in the germline of *C. elegans*. *Science* **296:** 2235–2238.

Goday C. and Esteban M.R. 2001. Chromosome elimination in sciarid flies. *BioEssays* **23:** 242–250.

Goldstein P. and Slaton D.E. 1982. The synaptonemal complexes of *Caenorhabditis elegans*: Comparison of wild-type and mutant strains and pachytene karyotype analysis of wild-type. *Chromosoma* **84:** 585–597.

Haack H. and Hodgkin J. 1991. Tests for parental imprinting in the nematode *Caenorhabditis elegans*. *Mol. Gen. Genet.* **228:** 482–485.

Handel M.A. 2004. The XY body: a specialized meiotic chromatin domain. *Exp. Cell Res.* **296:** 57–63.

Hansen D., Hubbard E.J.A., and Schedl T. 2004. Multi-pathway control of the proliferation versus meiotic development decision in the *Caenorhabditis elegans* germline. *Dev. Biol.* **268:** 342–357.

Herrick G. and Seger J. 1999. Imprinting and paternal genome elimination in insects. *Results Probl. Cell Differ.* **25:** 41–71.

Hirano T. 2002. The ABCs of SMC proteins: Two-armed ATPases for chromosome condensation, cohesion, and repair. *Genes Dev.* **16:** 399–414.

Kelly W.G., Schaner C.E., Dernburg A.F., Ho-Lee M., Kim S.K., Villeneuve A.M., and Reinke V. 2002. X-chromosome silencing in the germline of *C. elegans*. *Development* **129:** 479–492.

Ketel C.S., Andersen E.F., Vargas M.L., Suh J., Strome S., and Simon J.A. 2005. Subunit contributions to histone methyltransferase activities of fly and worm polycomb group complexes. *Mol. Cell. Biol.* **25:** 6857–6868.

Maciejowski J., Ahn J.H., Cipriani P.G., Killian D.J., Chaudhary A.L., Lee J.I., Voutev R., Johnsen R.C., Baillie D.L., Gunsalus K.C., et al. 2005. Autosomal genes of autosomal/X-linked duplicated gene pairs and germ-line proliferation in *Caenorhabditis elegans*. *Genetics* **169:** 1997–2011.

Maine E.M., Hauth J., Ratliff T., Vought V.E., She X., and Kelly W.G. 2005. EGO-1, a putative RNA-dependent RNA polymerase, is required for heterochromatin assembly on unpaired DNA during *C. elegans* meiosis. *Curr. Biol.* **15:** 1972–1978.

Meyer B.J. 1997. Sex determination and X chromosome dosage compensation. In C. elegans *II* (ed. D.L. Riddle et al.), pp. 209–240. Cold Spring Harbor Laboratory Press, Cold Spring Harbor, New York.

———. 2000. Sex in the worm counting and compensating X-chromosome dose. *Trends Genet.* **16:** 247–53.

———. 2005. X-chromosome dosage compensation. In *WormBook*, http://www.wormbook.org/chapters/www_dosagecomp/dosage-comp.html.

Nakayashiki H. 2005. RNA silencing in fungi: Mechanisms and applications. *FEBS Lett.* **579:** 5950–5957.

Nigon V. 1951. Polyploidie experimentale chez un nematode libre, *Rhaditis elegans maupas*. *Bull. Biol. Fr. Belg.* **85:** 187–255.

Piano F., Schetter A.J., Mangone M., Stein L., and Kemphues K.J. 2000. RNAi analysis of genes expressed in the ovary of *Caenorhabditis elegans*. *Curr. Biol.* **10:** 1619–1622.

Powell J.R., Jow M.M., and Meyer B.J. 2005. The T-box transcription factor SEA-1 is an autosomal element of the X:A signal that determines *C. elegans* sex. *Dev. Cell* **9:** 339–349.

Prahlad V., Pilgrim D., and Goodwin E.B. 2003. Roles for mating and environment in *C. elegans* sex determination. *Science* **302:** 1046–1049.

Reinke V., Gil I.S., Ward S., and Kazmer K. 2004. Genome-wide germline-enriched and sex-biased expression profiles in *Caenorhabditis elegans*. *Development* **131:** 311–323.

Reinke V., Smith H.E., Nance J., Wang J., Van Doren C., Begley R., Jones S.J.M., Davis E.B., Scherer S., Ward S., and Kim S.K. 2000. A global profile of germ line gene expression in *C. elegans*. *Mol. Cell* **6:** 605–616.

Schedl T. 1997. Developmental genetics of the germ line. In C. elegans *II* (ed. D.L. Riddle et al.), pp. 241–269. Cold Spring Harbor Laboratory Press, Cold Spring Harbor, New York.

Shiu P.K., Raju N.B., Zickler D., and Metzenberg R.L. 2001. Meiotic silencing by unpaired DNA. *Cell* **107:** 905–916.

Skipper M., Milne C.A., and Hodgkin J. 1999. Genetic and molecular analysis of *fox-1*, a numerator element involved in *Caenorhabditis elegans* primary sex determination. *Genetics* **151:** 617–631.

Turner J.M. 2005. Sex chromosomes make their mark. *Chromosoma* **114:** 300–306.

Turner J.M., Mahadevaiah S.K., Fernandez-Capetillo O., Nussenzweig A., Xu X., Deng C.X., and Burgoyne P.S. 2005. Silencing of unsynapsed meiotic chromosomes in the mouse. *Nat. Genet.* **37:** 41–47.

van Leeuwen F. and Gottschling D.E. 2002. Genome-wide histone modifications: gaining specificity by preventing promiscuity. *Curr. Opin. Cell Biol.* **14:** 756–762.

Wang L., Brown J.L., Cao R., Zhang Y., Kassis J.A., and Jones R.S. 2004. Hierarchical recruitment of polycomb group silencing complexes. *Mol. Cell* **14:** 637–646.

Wu C.I. and Xu Y. 2003. Sexual antagonism and X inactivation—The SAXI hypothesis. *Trends Genet.* **19:** 243–247.

Xu L., Fong Y., and Strome S. 2001. The *Caenorhabditis elegans* maternal-effect sterile proteins, MES-2, MES-3, and MES-6, are associated in a complex in embryos. *Proc. Natl. Acad. Sci.* **98:** 5061–5066.

Yonker S.A., and Meyer B.J. 2003. Recruitment of *C. elegans* dosage compensation proteins for gene-specific versus chromosome-wide repression. *Development* **130:** 6519–6532.

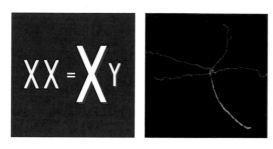

Dosage Compensation in *Drosophila*

John C. Lucchesi[1] and Mitzi I. Kuroda[2]

[1]*Department of Biology, Graduate Division of Biological and Biomedical Sciences, Emory University, Atlanta, Georgia 30322*
[2]*Harvard-Partners Center for Genetics & Genomics, Brigham & Women's Hospital, Department of Genetics,*
Harvard Medical School, Boston, Massachusetts 02446

CONTENTS

1. The Phenomenon of Dosage Compensation Was Discovered in *Drosophila*, 309

2. Dosage Compensation Involves Chromatin Modification, 309

3. Regulation of Dosage Compensation Starts with Measurement of the X:Autosome Ratio, 312

4. Noncoding roX RNAs Facilitate Assembly and Targeting of the MSL Complex on the X Chromosome, 313

5. A Balancing Act between Antagonistic Chromatin-remodeling Activities, 315

6. Outlook, 317

References, 317

GENERAL SUMMARY

The mechanism responsible for the transcription of genetic information, i.e., for the synthesis of messenger RNA, has been studied extensively. Factors involved in the activation of individual genes, in maintaining the active state, and in shutting down activity when it is no longer needed have been identified. As they were discovered, the interactions of these factors with the regulatory and coding portions of genes were studied and are still the subject of intensive investigations. In recent years, the long-standing dogma that cellular differentiation and development require the coordinate regulation of different sets of genes in time and space has led to the search for regulatory signals that would affect the activity of groups of functionally related genes. Long before these investigations were initiated, an example of coordinate regulation had been described in *Drosophila*. In this particular case, the group of genes whose activity was regulated in unison were not related by function, but rather shared a common location

in the genetic material: They were all present on one of the sex chromosomes, the X chromosome. The purpose of the regulation was to ensure that females with two X chromosomes and males with only one X would have equal levels of gene products; in other words, to compensate for differences in the doses of X-linked genes between the sexes. In studying this level of regulation, the question "How are groups of unrelated genes coordinately regulated?" became "What are the mechanisms that can regulate the activity of a whole chromosome?"

The study of dosage compensation in *Drosophila*—a mechanism that enhances the transcription of most of the genes on the single X chromosome in males—reveals the involvement of site-specific histone acetylation, X-specific noncoding RNAs (called roX1 and roX2), and chromosome-wide targeting of an evolutionarily conserved chromatin-modifying machine, shown localizing to the X chromosome in the title figure.

1 The Phenomenon of Dosage Compensation Was Discovered in *Drosophila*

The karyotypes (i.e., ensemble of chromosomes) of many organisms include a pair of sex chromosomes. In *Drosophila*, females have two sex chromosomes called the X chromosomes that are identical in shape and genetic content; both X chromosomes are active in all somatic cells. Males have one X and a chromosome that differs from the X in morphology and in the genetic information that it contains—the Y chromosome (Fig. 1). On the sex chromosomes there are genes that are responsible for sex determination and sexual differentiation. In addition, and this is particularly true of the X chromosome, there are many genes involved in basic cellular housekeeping functions or developmental pathways. Females with two X chromosomes have twice the number of these genes; males with a single X have only one dose. Yet, the level of the products of most of these genes is the same in the two sexes. In the early 1930s, this paradox was first noticed in *Drosophila* by H.J. Muller while he was studying the eye-pigment level of individuals carrying partial loss-of-function X-linked mutations (Fig. 2). Muller reasoned that there must be a regulatory mechanism that helps flies to compensate for the difference in dosage of X-linked genes in males and females by equalizing the level of X-linked gene products between the two sexes. He called this hypothetical regulatory mechanism "dosage compensation."

The first evidence that dosage compensation is achieved by regulating the transcription of X-linked genes was obtained more than 30 years later by A.S. Mukherjee and W. Beermann. Using transcription autoradiography of the giant polytenic chromosomes of larval salivary glands—a molecular technique that represented the state of the art at that time—these authors observed that the level of [³H]uridine incorporation by the single X in males and both Xs in females was equivalent. It appeared, therefore, that the rate of RNA synthesis by the single X chromosome in males was approximately twice the rate of each of the two Xs in females. The next experimental breakthrough consisted of the genetic identification by J. Belote and J. Lucchesi of four genes with loss-of-function mutations that were inconsequential in females but lethal in males; the latter exhibited approximately half of the normal level of [³H]uridine incorporation by their X chromosome. Furthermore, the X chromosome had lost its normal paler and somewhat puffed appearance that had been interpreted as an indication of an enhanced level of transcriptional activity in relation to each of the two X chromosomes in females. These results suggested that the equalization of X-linked gene products was achieved by doubling, on average, the transcriptional activity of the X chromosome in males rather than by halving the transcriptional activity of each X in females. Among the four genes, two were newly discovered (*male-specific lethal 1, msl1*; and *male-specific lethal 2, msl2*) whereas the other two (*maleless, mle*; and *male-specific lethal 3, msl3*) had been previously identified by other investigators in natural populations (specific references to this early phase of the study of dosage compensation can be found in Lucchesi and Manning 1987). For ease of reference, all of the gene products identified to date, on the basis of the male-specific lethal phenotype of their loss-of-function mutations, are called the MSLs.

The next phase in the study of dosage compensation was initiated with the cloning of *mle*, soon followed by the cloning of the three msl genes, in the laboratories of B. Baker, M. Kuroda, and J. Lucchesi. By cytoimmunofluorescence, the four gene products were found to associate in an identical pattern at numerous sites along the polytene X chromosome in males (Palmer et al. 1993; Bone et al. 1994; Gorman et al. 1995; Kelley et al. 1995; Gu et al. 1998; see title figure). This observation and the interdependence of the different gene products for X-chromosome binding suggested that they form a complex (Fig. 3).

2 Dosage Compensation Involves Chromatin Modification

The presence of the MSL complex on the X chromosome in males is correlated with the presence of a specific acetylated isoform of histone H4 at lysine 16 (Turner et al. 1992; Bone et al. 1994). In yeast, this particular covalent modification of histone H4 has been shown to play a key role in maintaining the boundary between silent chromatin and active chromatin: Loss of function of Sas2, the histone acetyltransferase (HAT) responsible for H4K16ac, allows the spreading of telomeric heterochromatin into adjacent subtelomeric chromatin (Suka et al. 2002; for

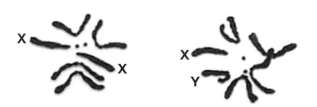

Figure 1. Camera Lucida Drawing of Female (*left*) and Male (*right*) Metaphase Plates from *Drosophila melanogaster*

Note that the X chromosome represents ~20% of the karyotypes. (Reprinted, with permission, from Morgan 1932.)

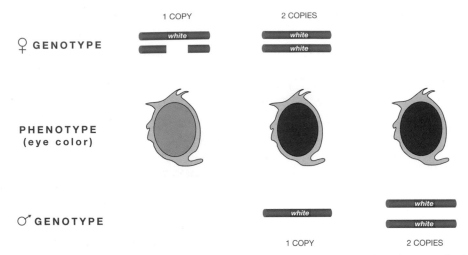

Figure 2. Diagrammatic Representation of the Results That Led H.J. Muller to Formulate the Hypothesis of Dosage Compensation

The mutant allele (w^a) of the X-linked white gene is a hypomorph and allows partial eye-pigment synthesis; its presence on the X chromosomes is indicated by the darker box. The level of pigmentation is directly proportional to the dosage of the w^a allele within each sex; yet, males with one dose and females with two doses have comparable amounts of pigment due to dosage compensation.

more detail, see Chapter 4). In *Drosophila*, H4K16ac is found highly enriched along the X chromosome in males. Recent structural studies have indicated that a key internucleosomal interaction may occur between an acidic patch of the histone H2A-H2B dimer on one nucleosome and a positively charged segment of the histone H4 tail (residues 16–26) extending from a neighboring nucleosome (Schalch et al. 2005). When lysine 16 is specifically acetylated, its positive charge becomes neutral, suggesting that weakening a repressive internucleosomal structure could play a key role in dosage compensation.

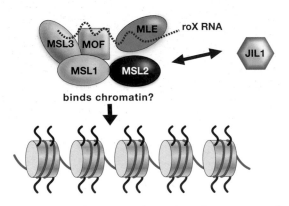

Figure 3. Diagram Illustrating the Various Components of the MSL Complex

Known functions of MSL components include the acetylation of histone H4K16 by MOF, MLE has ATPase and RNA/DNA helicase activity, and JIL-1 phosphorylates histone H3. It is not clear whether JIL-1 is a subunit of the complex.

None of the initially identified factors involved in dosage compensation exhibited HAT characteristics. Because all previous mutant searches had focused on screening the major autosomes, a new search for X-linked male-specific lethal mutations was carried out, and a new gene, *males absent on the first* (*mof*), was identified (Hilfiker et al. 1997). A recombinant protein encoded by mof is a HAT with specificity for acetylating histone H4 at lysine 16 (Akhtar and Becker 2000); it is an integral part of the MSL complex and is uniquely responsible for this isoform of H4 on the male X chromosome (Smith et al. 2000). MOF is a member of a subfamily of the MYST family of HATs. This subfamily, characterized by the presence of a chromodomain, can be further subdivided into enzymes that specifically acetylate H4K16 in vivo (MOF and human MOF; Smith et al. 2005) and those such as Esa1 (essential SAS-related acetyltransferase 1 protein) in yeast that acetylate all four terminal lysines of H4 (Smith et al. 1998). Another MYST family member, SAS2, specifically acetylates H4K16 in yeast but lacks a chromodomain.

At the same time that MOF was discovered and characterized, the catalytic properties of MLE, the other subunit with putative enzymatic activity, were established. MLE was found to exhibit RNA/DNA helicase, adenosine triphosphatase (ATPase), and single-stranded RNA/ssDNA-binding activities in vitro (Lee et al. 1997), foreshadowing a potential role for RNA in MSL function. Some multiprotein complexes that interact with chromatin to control the rate of transcription of euchromatic genes use

ATP hydrolysis to alter nucleosomal conformation. ATP hydrolysis is the responsibility of a family of helicase domain-containing proteins (Sif 2004); although it may not perform the same function in the MSL complex, this family is represented by MLE. Orthologs of MLE, which include human RNA helicase A (RHA), belong to the DEXH RNA helicase subfamily and are characterized by the acquisition of new domains (Kuroda et al. 1991) and of a conserved amino-terminal region of ~350 amino acids (Pannuti and Lucchesi 2000). Rather than assisting in remodeling chromatin, MLE may perform its function in dosage compensation by altering RNA structure.

The specific roles played by the remaining components of the MSL complex are not fully understood. MSL3, a subunit of the MSL complex characterized by the presence of a noncanonical chromodomain and a chromoshadow domain (Aasland and Stewart 1995; Koonin et al. 1995), is a member of a family of proteins that may have coevolved with the chromodomain-bearing HATs (Pannuti and Lucchesi 2000). In yeast, a member of this family, Eaf3 (Esa1-associated factor-3 protein), is found in the NuA4 complex with the MYST enzyme Esa1 (Eisen et al. 2001). In *Drosophila*, the Tip60 multiprotein complex includes MRG15, a paralog of MSL3 (Kusch et al. 2004). In humans, MRG15 is associated with MOF in the MAF2 complex (Pardo et al. 2002). Of particular interest is the existence of another human complex related to the MSL complex that includes human homologs of MOF, MSL1, MSL2, and MSL3. This complex, which specifically acetylates histone H4K16 and is responsible for the majority of this histone isoform in human cells, can include one of three different versions of a *Drosophila* MSL3 homolog encoded by two different genes (Smith et al. 2005).

In *Drosophila*, MSL1 and MSL2 appear to mediate the binding of the MSL complex to chromatin (Kelley et al. 1995; Copps et al. 1998; Gu et al. 1998). Neither of these proteins contains any identifiable DNA-binding domain and targeting to chromatin appears to require their association. MSL1 interacts with the RING finger domain of MSL2 via an amino-terminal coiled-coil domain (Copps et al. 1998; Scott et al. 2000) and with MSL3 and MOF via adjacent conserved carboxy-terminal domains (Morales et al. 2004). The association of MSL2 with the X chromosome appears to be remarkably stable (Straub et al. 2005c). Human homologs of MSL1 and MSL2 were reported by Marín (2003). Because in *Drosophila*, MSL1 and MSL2 are responsible for the association of the MSL complex with the X chromosome, it is reasonable to speculate that hMSL1 and hMSL2 may play a role in targeting the human MSL complex to its sites of action.

In addition to male-specific factors, it is likely that some general factors involved in chromatin organization and transcription in both sexes also participate in dosage compensation. JIL-1, a tandem kinase, is found along all chromosomes in both males and females, but is more highly concentrated on the male X chromosome. This enrichment is dependent on the MSL complex. JIL-1 mediates histone H3 phosphorylation and maintains open chromatin structure in transcriptionally active regions of the genome (Jin et al. 1999; Wang et al. 2001); yet, whether it plays a specific or a general role in dosage compensation is still to be determined. Recently, nuclear proteins including nucleoporins and subunits of the exosome complex have been found to co-immunoprecipitate with some of the MSLs (Mendjou et al. 2006). The role that these proteins may play in dosage compensation is not determined.

Activity of the complex must have evolved to control a preexisting group of genes, each with their own specific regulatory signals and intrinsic expression levels. That the MSL complex may be targeted to active genes is supported by the observation that ectopic MSL binding appears at regulated transgenes inserted on the X only upon induction of expression (Fig. 4) (Sass et al. 2003). At what step might general transcription be up-regulated twofold? In *Drosophila* males, H4K16ac is not limited to the promoter region of compensated X-linked genes;

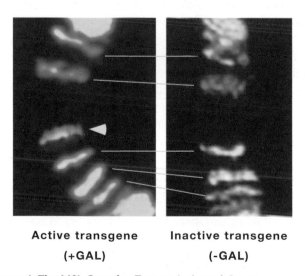

Active transgene   **Inactive transgene**

(+GAL)   **(-GAL)**

Figure 4. The MSL Complex Targets Activated Genes

A construct containing a promoter under the control of the *trans*-activator GAL4 has been inserted at a site that is normally devoid of the MSL complex in larval salivary gland chromosomes (*right panel*). When a gene expressing GAL4 is introduced in the genome (*left panel*), the *trans*-activator binds to the construct (*red color*) and recruits the MSL complex (*green color*). (Reprinted, with permission, from Sass et al. 2003 [© National Academy of Sciences].)

rather, it is found along the entire length of X-linked transcriptional units targeted by the MSL complex, with relatively modest levels of acetylation at the promoter regions and high levels in the middle and/or 3′ ends (Smith et al. 2001). This distribution of acetylation suggests that the MSL complex functions to increase the expression of X-linked genes by facilitating transcription elongation and perhaps RNA polymerase recycling, rather than by directly enhancing promoter accessibility.

An alternate hypothesis has been proposed to explain the function of the MSL complex (Birchler et al. 2003). This hypothesis is based on the notion that in females the activity of all chromosomes is set by, among other factors, the uniform distribution of MOF. In males, because of the absence of one X chromosome, a greater concentration of MOF and unknown factors is available to the autosomes and to the single X, driving their activity to higher levels; the MSL complex forms and is recruited to the X in order to sequester hypertranscription factors, including MOF, away from the autosomes. In this model, the principal role of the MSL complex is to down-regulate autosomal genes, rather than to up-regulate X-linked genes. However, two independent studies have contradicted this hypothesis by showing that loss of targeting of the MSL complex in male tissue-culture cells decreased transcription of genes on the X chromosome while the level of expression of autosomal genes was unaffected (Hamada et al. 2005; Straub et al. 2005b). Similar results have been reported recently for X-linked gene expression in the male germ line (Gupta et al. 2006).

Chromatin assembly has been recently subdivided into replication-coupled and replication-independent nucleosome deposition. The latter occurs in transcriptionally active regions of chromatin and involves the replacement of histone H3 with the variant H3.3 (see Chapter 13). Concordant with this observation, the rate of incorporation of histone H3.3 into the X chromosome in male cells is enhanced in relation to the autosomes (Mito et al. 2005).

3 Regulation of Dosage Compensation Starts with Measurement of the X:Autosome Ratio

Each embryo needs to count its X chromosomes in order to make the critical decision whether or not to implement dosage compensation. An incorrect decision, such as failure to up-regulate the single male X, or aberrant up-regulation of both female XXs, results in lethality. In *Drosophila*, the X-counting process is coordinated with the sex determination decision (for review, see Cline and Meyer 1996). Phenotypic sex is determined by the number of X chromosomes per nucleus, such that XX embryos are females and XY embryos are male. The Y chromosome is required for male fertility, but unlike in mammals, it plays no role in phenotypic sex. Formally, it is the X:autosome ratio that controls both sex and dosage compensation, as the X-counting mechanism is sensitive to the number of sets of autosomes. This becomes apparent in 2X:3A triploids, which have an intermediate X:A ratio between XY:2A males and XX:2A females. 2X:3A triploids differentiate as intersexes with a mixture of both male and female cells.

The X:A ratio controls both sex determination and dosage compensation by regulating a critical binary switch gene, Sex lethal (*Sxl*). *Sxl* encodes a female-specific RNA-binding protein that regulates splicing and translation of key mRNAs in the sex determination and dosage compensation pathways, respectively (Fig. 5). Sex lethal is encoded by the X chromosome and is positively regulated by transcription factors encoded by the X, such that embryos with two X chromosomes are able to initiate *Sxl* expression from an early, regulated promoter, P_e, whereas embryos with a single X per nucleus fail to express Sxl from P_e. This initial transient difference in activation of *Sxl* in early embryos is stabilized by an autoregulatory loop in which SXL protein positively regulates splicing of its own mRNA from a maintenance promoter that is expressed constitutively. *Sxl* initiates differentiation in the female mode by regulating the splicing of the *transformer* (*tra*) gene in a sex-specific manner. In turn, this gene product (together with the product of another gene, *transformer2* (*tra2*), present in both sexes) directs the splicing of the *doublesex* (*dsx*) primary transcript to yield a regulatory protein that acts to repress the male-specific realizer genes, thus achieving female sexual differentiation. In male embryos, an alternate mode of splicing of the *dsx* transcripts occurs by default and leads to a product that represses the female-specific realizer genes and results in male sexual differentiation.

The key target of SXL in the dosage compensation pathway is *msl2* mRNA (Bashaw and Baker 1997; Kelley et al. 1997). SXL-binding sites are located in both the 5′ and 3′ UTRs of *msl2* mRNA. SXL is normally present only in females, where it represses translation of the *msl2* mRNA through association with its UTRs (see Fig. 5). If SXL is absent in females, dosage compensation is aberrantly turned on, and these females die. Conversely, if SXL is ectopically expressed in males, dosage compensation is turned off, and males die. Ectopic expression of MSL2 in females is sufficient to assemble MSL complexes on both female X chromosomes, indicating that all other MSL components are either turned on or stabilized by expres-

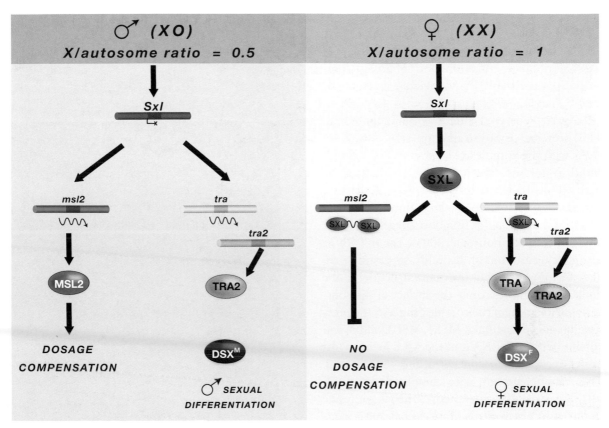

Figure 5. Diagram of the Control of Sex Determination and Dosage Compensation

If the X/A ratio equals 1, a regulatory cascade leads to female sexual development. In females, the presence of the Sxl gene product prevents the translation of the msl2 message and the assembly of the MSL complex. If the X/A ratio is only 0.5, absence of the cascade leads by default to male sexual development and to the formation of the MSL complex.

sion of MSL2. For example, MSL1 and MSL3 are normally expressed in females but are unstable in the absence of MSL2 protein. MLE and MOF are stable in females, but have no specific affinity for the X chromosome in the absence of MSL2.

In summary, dosage compensation must respond to the number of X chromosomes in the nucleus, and these are counted early in embryonic development. Males express MSL2 protein and thus the MSL complex by default, whereas females repress MSL2 translation, preventing inappropriate dosage compensation when two X chromosomes are present.

4 Noncoding roX RNAs Facilitate Assembly and Targeting of the MSL Complex on the X Chromosome

One of the most intriguing and mysterious aspects of dosage compensation in both mammals and *Drosophila* is the role of noncoding RNAs in targeting compensation to

the correct chromosome (for review, see Gilfillan et al. 2004; Kelley 2004; Straub et al. 2005a; also see Chapter 17). Two noncoding RNAs, called RNA on X (roX), are dissimilar in size and sequence, and yet function redundantly to target the MSL complex to the male X chromosome in *Drosophila* (Meller and Rattner 2002). Traditional mutant screens usually do not reveal the existence of genes that encode products with redundant functions such as the roX RNAs. Rather, roX RNAs were discovered by serendipity as male-specific RNAs in the adult brain (Amrein and Axel 1997; Meller et al 1997). Upon closer examination, both RNAs displayed a lack of significant open reading frames and colocalization with the MSL complex along the length of the X. roX RNA function was not revealed until an X chromosome mutant for both roX1 and roX2 was isolated. Most double mutant males die, with severely mis-localized MSL complexes, whereas single mutant males have no known phenotype (Meller and Rattner 2002). This is surprising in view of the fact that the two roX RNAs are very different in size (3.7 kb vs. 0.5–1.4 kb) and share little sequence similarity. Overexpression of MSL proteins can partially overcome the lack of roX RNAs, suggesting that the proteins possess all of the essential functions of dosage compensation but require the RNAs to facilitate assembly or localization (Oh et al. 2003).

roX RNAs are recovered after co-immunoprecipitation of MSL proteins, demonstrating physical association of the RNAs with the complex (Meller et al. 2000; Smith et al. 2000). It is not known whether the roX RNAs coexist, or form two distinct types of MSL species. A deletion series of roX1 revealed that no single 300-nucleotide segment is absolutely essential for function, suggesting internal redundancy (Stuckenholz et al. 2003). The roX RNAs display a surprising amount of flexibility, suggesting that, rather than forming a highly invariant structure in the MSL complex, they may decorate the surface of the complex. Partial purification of the complex suggests the presence of a tight core consisting of MSL1, MSL2, MSL3, and MOF proteins, with roX RNA and the MLE helicase lost except under very low salt concentrations (Smith et al. 2000). The minimal protein core complex lacking roX RNAs can still specifically acetylate histone H4 on lysine 16 within nucleosomes in vitro (Morales et al. 2004), consistent with the idea that roX RNAs participate in assembly or targeting of dosage compensation rather than playing a direct role in gene regulation.

How does the MSL complex bind the X chromosome, and what does it recognize? In the absence of both roX RNAs, or either MLE, MSL3, or MOF, partial MSL complexes bind a subset of ~35–70 sites (Fig. 6). These "high-affinity" sites have been compared in *mle*, *msl3*, and *mof* mutants and found to be largely the same at the cytological level, but their molecular identity is not yet known. A key defect in mle mutants may be the inability to incorporate roX RNAs into partial complexes, as the RNAs are colocalized with partial complexes in the absence of MSL3 or MOF but not in the absence of MLE (Meller et al. 2000). In the absence of either MSL1 or MSL2, none of the remaining MSL proteins or roX RNAs appear to retain any recognition of the X chromosome, leading to the hypothesis that MSL1 and MSL2 together provide at least the initial specificity for X targeting. Nevertheless, neither of these proteins carries a known DNA-binding motif, and in vitro binding of DNA has not been demonstrated.

What might be the specificity signal on the X chromosome that attracts the complex? One can imagine two very general models for regulation of a whole chromosome. A single site or a very limited number of sites might control the chromosome in *cis*, as is the case in mam-

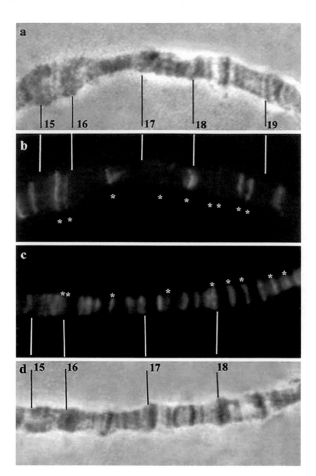

Figure 6. "High-affinity" Sites

Salivary gland chromosome preparations from larvae carrying a loss-of-function allele of msl3 (*a,b*) or from control wild-type larvae (*c,d*). *a* and *d* show images obtained with phase contrast whereas *b* and *c* show the same chromosomes immunostained for the presence of the complex. In *b* the complex (*green*) is found at a smaller number of sites than in *c*. (Reprinted, with permission, from Demakova et al. 2003 [© Springer].)

malian X inactivation (see Chapter 17). This mechanism requires either compartmentalization of the complex to a specific place in the nucleus or regulation over very long distances through the spreading of factors from the central control region to the rest of the chromosome. On the other extreme, a chromosome could have unique identifying sequences sprinkled along its entire length. In this case, any segment of the chromosome could be regulated autonomously. Positioned between these two models is a whole spectrum of possibilities, including a combination of central control regions with dispersed sequences that facilitate long-distance regulation.

roX RNAs are normally encoded by the X chromosome, with the *roX1* gene near the tip and the *roX2* gene around the middle of the euchromatic part of the X. Like the *Xist* gene in mammals, the *roX* genes may reside on the X in order to target MSL complex assembly to this chromosome. When roX genes are moved to the autosomes as transgenes, they potently attract MSL proteins to their novel insertion sites (Fig. 7), where the complex appears to spread in *cis*, variably into flanking sequences (Kelley et al. 1999; Kageyama et al. 2001). Under specific genetic conditions, for example when there are no competing endogenous roX genes on the X chromosome, extensive spreading from autosomal roX transgenes is consistently seen (Fig. 8) (Park et al. 2002). This extensive spreading is augmented by overexpression of MSL1 and MSL2, the key limiting MSL proteins, and diminished by overexpression of roX RNA from competing transgenes, suggesting that successful cotranscriptional assembly of MSL complexes may drive local spreading (Oh et al. 2003). Efficient assembly of functional MSL complexes within the nucleus may be the primary function of roX RNAs. Initial assembly at *roX* genes on the X chromosome is likely to enhance the efficiency of MSL targeting to the X, but apparently is not essential for ulti-mate targeting, because *roX* genes can function in *trans* (Meller and Rattner 2002).

In addition to the high-affinity sites mentioned above, it is clear that the complex exhibits different levels of affinity for a large number of additional sites along the X chromosome (Demakova et al. 2003). In stable translocation stocks, spreading of MSL complexes from the X into contiguous autosomal sequences is not evident (Fagegaltier and Baker 2004). Therefore, even if spreading of MSL complexes is a major mechanism for covering the X chromosome, there is very likely an additional characteristic of the X that causes the MSL complex to strongly favor X binding over autosomes. In fact, from examination of X:A translocations and of a limited set of X-derived transgenes, it appears that most ~30-kb segments of the X chromosome have the power to attract the MSL complex, with variable results for smaller segments (Oh et al. 2004).

How can these various observations be accommodated into a model for X-chromosome targeting of the MSL complex? A model for X-chromosome recognition that is the best fit for existing data is depicted in Figure 9. In this model, MSL complexes assemble at the sites of roX RNA transcription and subsequently access flanking and distant sites on the X based on their relative affinities for the MSL complex. Whether some or most targeting normally occurs solely in *trans*, or by some type of local spreading in *cis*, is unknown. However, it is clear that the MSL complex strongly prefers binding to X-chromosome segments rather than to autosomes. Although no simple sequence is known to uniquely define the X chromosome to date, perhaps a degenerate sequence will be revealed as whole genomes of multiple *Drosophila* species become available for comparison.

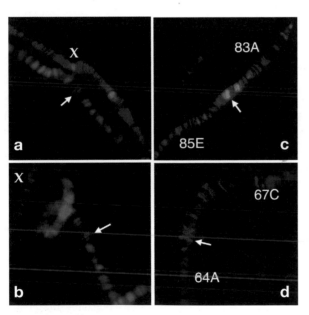

Figure 7. The Spread of the MSL Complex from an Autosomal Transgene Is Proportional to a Reduction in Function of the Endogenous, X-linked *roX* Genes

Maximal autosomal spreading is achieved when the transgene is the only source of roX RNA in the cell. (*a*) Chromosomes from a male with a wild-type X; the presence of the autosomal transgene is indicated by the narrow MSL band (*red*). (*b*) A male with only one active X-linked *roX* gene. The MSL complex spreads slightly more than in wild type. (*c,d*) Extensive spreading in males carrying two different transgenes and an X chromosome with both *roX* genes deleted. (Reprinted, with permission, from Park et al. 2002 [© AAAS].)

5 A Balancing Act between Antagonistic Chromatin-remodeling Activities

The male polytene X chromosome exhibits special sensitivity to changes in the dosage or activity of several chromatin regulators that are thought to be general, non-chromosome-specific factors. For example, a functional interaction between ISWI-bearing complexes and the MSL complex was brought to light by the observation that loss-of-function mutations of *Iswi* lead to a global structural effect on the X chromosome in males: In salivary gland preparations, this chromosome appears extremely short and fat (Fig. 10), with MSL complex and H4K16ac in an apparently continuous rather than finely banded pattern (Deuring et al. 2000).

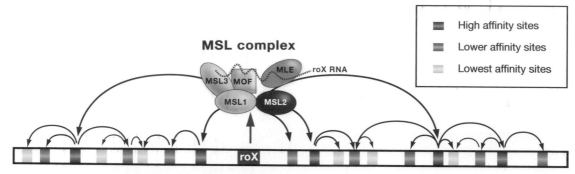

Figure 8. Model for the Targeting of the MSL Complex to the X Chromosome

The complex assembles at the site of transcription of the *roX* genes and accesses various sites along the X chromosome for which it has affinity ranging from very high to relatively low. Note that this model is applicable to the autosomal spreading of the complex from an ectopic *roX* gene.

The molecular basis of this cytological phenotype is not known, but it is compatible with a model in which the organization of the chromatin into distinct domains has been lost. Normal polytene chromosomes have a distinctive pattern of relatively condensed bands and decondensed interbands and puffs, and this organization appears to be largely absent on the *Iswi* mutant male X. The short length of the X may be due to a loss of structure and condensation, resulting in an abnormal increase in width at the expense of length.

ISWI (imitation switch protein) is an ATPase found in three chromatin-remodeling complexes of *Drosophila*: NURF (nucleosome remodeling factor), ACF (ATP-dependent chromatin assembly and remodeling factor), and CHRAC (chromatin accessibility complex) (Smith and Peterson 2005). In vitro studies have shown that ACF and CHRAC establish regularly ordered arrays of nucleosomes whereas NURF disrupts nucleosome periodicity. In vivo, ACF and CHRAC behave as chromatin assembly factors promoting the formation of chromatin in general and of repressive chromatin states in particular. In contrast, NURF alters the position of nucleosomes on a *hsp70* promoter reconstituted in vitro, and its remodeling activity is substantially enhanced if the nucleosomes are first hyperacetylated. Despite NURF's apparent biochemical activity in making chromatin more accessible, mutants in *nurf301* also exhibit the abnormally decondensed male X seen in *Iswi* mutants (Badenhorst et al 2002).

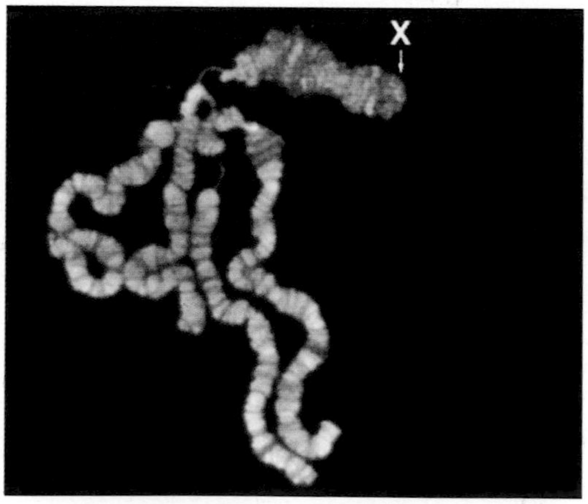

The X-chromosome defect visible in *Iswi* and *nurf* mutant salivary glands can be seen in transgenic females where the presence of the MSL complex has been induced; this defect does not occur in the presence of an inactive MSL complex in males; i.e., in the absence of H4K16ac. Furthermore, in vitro competition studies with purified ISWI suggest that its ability to interact with isolated nucleosomes is reduced by the acetylation of the histone H4 tail at lysine 16 (Corona et al. 2002). A speculation advanced by these authors would have MOF activity altering the expression of X-linked genes by reducing the ability of ISWI to promote the formation of regular nucleosomal arrays (Corona et al. 2002).

A very different chromosomal phenotype is seen when a heterochromatin protein, SU(VAR)3-7, is overexpressed (Delattre et al. 2004). The SU(VAR)3-7 protein is a structural component of heterochromatin that appears to colocalize with HP1 (heterochromatin protein 1) and with the histone lysine methyltransferase (HKMT)

Figure 9. Hypomorphic Mutations of the *Iswi* Gene Affect Preferentially the Structural Organization of the X Chromosome in Males

In wild-type males, the X chromosome is similar in general appearance to an autosomal arm. (Reprinted, with permission, from Deuring et al. 2000 [© Elsevier].)

SU(VAR)3-9 (see Chapter 5). Although overexpression of Su(var)3-7 causes morphological effects in the chromosomes of both males and females, the male X is most affected because it assumes a very small and highly compacted shape.

It is curious that the male polytene X is more susceptible than other chromosomes to changes in the balance of chromatin components with either positive or negative effects. An important mechanism for maintenance of heritable chromatin states may be competition for reassembly of factors after DNA synthesis. Perhaps the hyperactive male polytene X chromosome, because of its normally open state, acts as a leading indicator when the natural balance of key chromatin-modulating factors is altered (Lucchesi et al. 2005).

6 Outlook

With the increasing utility of microarray technologies, we are beginning to determine the precise genome-wide binding pattern of the MSL complex and the identities of the genes that it regulates (Alekseyenko et al. 2006; Gilfillan et al. 2006; Legube et al. 2006). This information will certainly lead to a better understanding of how the complex targets the X chromosome. Understanding targeting should in turn help us identify the mechanisms by which the MSL complex increases gene expression. Recent results reveal that coding regions of genes, rather than upstream regulatory regions, are bound by MSL complex, as seen for H4K16ac on selected genes. These results provide key support for the model that dosage compensation occurs by regulating the rate of transcriptional elongation and recycling of RNA (Smith et al. 2001).

A full elucidation of the molecular mechanism responsible for dosage compensation will require a precise determination of the impact of the MSL complex on the general transcriptional machinery. The MSL proteins and roX RNAs appear to be dedicated to dosage compensation, because none is required for female viability. However, it is likely that many additional factors facilitate dosage compensation, but are essential in both males and females because of their general nuclear, chromosomal, or transcriptional functions (Mendjan et al. 2006). In the upcoming years, the identification of these factors will be very important for understanding the biochemical basis of the twofold increase in X-linked gene expression seen in male *Drosophila*. Ultimately, understanding dosage compensation will contribute key information regarding the organization of chromatin into transcriptional domains in higher organisms and will also provide important insights into molecular mechanisms that fine-tune the heritable expression of genes within precise ranges.

References

Aasland R. and Stewart A.F. 1995. The chromo shadow domain, a second chromo domain in heterochromatin-binding protein 1, HP1. *Nucleic Acids Res.* **23:** 3168–3173.

Akhtar A. and Becker P.B. 2000. Activation of transcription through histone H4 acetylation by MOF, an acetyltransferase essential for dosage compensation in *Drosophila. Mol. Cell* **5:** 367–375.

Alekseyenko A.A., Larschan E., Lai W.R., Park P.J., and Kuroda M.I. 2006. High-resolution ChIP-chip analysis reveals that the *Drosophila* MSL complex selectively identifies active genes on the male X chromosome. *Genes Dev.* **20:** 848–857.

Amrein H. and Axel R. 1997. Genes expressed in neurons of adult male *Drosophila. Cell* **88:** 459–469.

Badenhorst P., Voas M., Rebay I., and Wu C. 2002. Biological functions of the ISWI chromatin remodeling complex NURF. *Genes Dev.* **16:** 3186–3198.

Bashaw G.J. and Baker B.S. 1997. The regulation of the *Drosophila* msl-2 gene reveals a function for sex-lethal in translational control. *Cell* **89:** 789–798.

Birchler J.A., Pal-Bhadra M., and Bhadra U. 2003. Dosage dependent gene regulation and the compensation of the X chromosome in *Drosophila* males. *Genetica* **117:** 179–190.

Bone J.R., Lavender J., Richman R., Palmer M.J., Turner B.M., and Kuroda M.I. 1994. Acetylated histone H4 on the male X chromosome is associated with dosage compensation in *Drosophila. Genes Dev.* **8:** 96–104.

Cline T.W. and Meyer B.J. 1996. Vive la difference: Males vs females in flies vs worms. *Annu. Rev. Genet.* **30:** 637–702.

Copps K., Richman R., Lyman L.M., Chang K.A., Rampersad-Ammons J., and Kuroda M.I. 1998. Complex formation by the *Drosophila* MSL proteins: Role of the MSL2 RING finger in protein complex assembly. *EMBO J.* **17:** 5409–5417.

Corona D.F., Clapier C.R., Becker P.B., and Tamkun J.W. 2002. Modulation of ISWI function by site-specific histone acetylation. *EMBO Rep.* **3:** 242–247.

Delattre M., Spierer A., Jaquet Y., and Spierer P. 2004. Increased expression of *Drosophila* Su(var)3-7 triggers Su(var)3-9-dependent heterochromatin formation. *J. Cell Sci.* **117:** 6239–6247.

Demakova O.V., Kotlikova I.V., Gordadze P.R., Alekseyenko A.A., Kuroda M.I., and Zhimulev I.F. 2003. The MSL complex levels are critical for its correct targeting to the chromosomes in *Drosophila melanogaster. Chromosoma* **112:** 103–115.

Deuring R., Fanti L., Armstrong J.A., Sarte M., Papoulas O., Prestel M., Daubresse G., Verardo M., Moseley S.L., Berloco M., et al. 2000. The ISWI chromatin-remodeling protein is required for gene expression and the maintenance of higher order chromatin structure in vivo. *Mol. Cell* **5:** 355–365.

Eisen A., Utley R.T., Nourani A., Allard S., Schmidt P., Lane W.S., Lucchesi J.C., and Cote J. 2001. The yeast NuA4 and *Drosophila* MSL complexes contain homologous subunits important for transcription regulation. *J. Biol. Chem.* **276:** 3484–3491.

Fagegaltier D. and Baker B.S. 2004. X chromosome sites autonomously recruit the dosage compensation complex in *Drosophila* males. *PLoS Biol.* **2:** e341.

Gilfillan G.D., Dahlsveen I.K., and Becker P.B. 2004. Lifting a chromosome: Dosage compensation in *Drosophila melanogaster*. *FEBS Lett.* **567:** 8–14.

Gilfillan G.D., Straub T., de Wit E., Greil F., Lamm R., van Steensel B., and Becker P.B. 2006. Chromosome-wide gene-specific targeting of the *Drosophila* dosage compensation complex. *Genes Dev.* **20:** 858–870.

Gorman M., Franke A., and Baker B.S. 1995. Molecular characterization of the male-specific lethal-3 gene and investigations of the regulation of dosage compensation in *Drosophila*. *Development* **121:** 463–475.

Gu W., Szauter P., and Lucchesi J.C. 1998. Targeting of MOF, a putative histone acetyl transferase, to the X chromosome of *Drosophila melanogaster*. *Dev. Genet.* **22:** 56–64.

Hamada F.N., Park P.J., Gordadze P.R., and Kuroda M.I. 2005. Global regulation of X chromosomal genes by the MSL complex in *Drosophila melanogaster*. *Genes Dev.* **19:** 2289–2294.

Hilfiker A., Hilfiker-Kleiner D., Pannuti A., and Lucchesi J.C. 1997. mof, a putative acetyl transferase gene related to the Tip60 and MOZ human genes and to the SAS genes of yeast, is required for dosage compensation in *Drosophila*. *EMBO J.* **16:** 2054–2060.

Jin Y., Wang Y., Walker D.L., Dong H., Conley C., Johansen J., and Johansen K.M. 1999. JIL-1: A novel chromosomal tandem kinase implicated in transcriptional regulation in *Drosophila*. *Mol. Cell* **4:** 129–135.

Kageyama Y., Mengus G., Gilfillan G., Kennedy H.G., Stuckenholz C., Kelley R.L., and Kuroda M.I. 2001. Association and spreading of the *Drosophila* dosage compensation complex from a discrete roX1 chromatin entry site. *EMBO J.* **20:** 2236–2245.

Kelley R.L. 2004. Path to equality strewn with roX. *Dev. Biol.* **269:** 18–25.

Kelley R.L., Wang J., Bell L., and Kuroda M.I. 1997. Sex lethal controls dosage compensation in *Drosophila* by a non-splicing mechanism. *Nature* **387:** 195–199.

Kelley R.L., Meller V.H., Gordadze P.R., Roman G., Davis R.L., and Kuroda M.I. 1999. Epigenetic spreading of the *Drosophila* dosage compensation complex from roX RNA genes into flanking chromatin. *Cell* **98:** 513–522.

Kelley R.L., Solovyeva I., Lyman L.M., Richman R., Solovyev V., and Kuroda M.I. 1995. Expression of msl-2 causes assembly of dosage compensation regulators on the X chromosomes and female lethality in *Drosophila*. *Cell* **81:** 867–877.

Koonin E.V., Zhou S., and Lucchesi J.C. 1995. The chromo superfamily: New members, duplication of the chromo domain and possible role in delivering transcription regulators to chromatin. *Nucleic Acids Res.* **23:** 4229–4233.

Kuroda M.I., Kernan M.J., Kreber R., Ganetzky B., and Baker B.S. 1991. The maleless protein associates with the X chromosome to regulate dosage compensation in *Drosophila*. *Cell* **66:** 935–947.

Kusch T., Florens L., Macdonald W.H., Swanson S.K., Glaser R.L., Yates J.R., III, Abmayr S.M., Washburn M.P., and Workman J.L. 2004. Acetylation by Tip60 is required for selective histone variant exchange at DNA lesions. *Science* **306:** 2084–2087.

Lee C.C., Chang K.A., Kurtoda M.I., and Hurwitz J. 1997. The NTPase/helicase activities of *Drosophila* maleless, an essential factor in dosage compensation. *EMBO J.* **16:** 2671–2681.

Legube G., McWeeney S.K., Lercher M.J., and Akhtar A. 2006. X-chromosome-wide profiling of MSL-1 distribution and dosage compensation in *Drosophila*. *Genes Dev.* **20:** 871–883.

Lucchesi J.C. and Manning J.E. 1987. Gene dosage compensation in *Drosophila melanogaster*. *Adv. Genet.* **24:** 371–429.

Lucchesi J.C., Kelly W.G., and Panning B. 2005. Chromatin remodeling in dosage compensation. *Annu. Rev. Genet.* **39:** 615–651.

Marin I. 2003. Evolution of chromatin-remodeling complexes: Comparative genomics reveals the ancient origin of "novel" compensasome genes. *J. Mol. Evol.* **56:** 527–539.

Meller V.H. and Rattner B.P. 2002. The roX genes encode redundant male-specific lethal transcripts required for targeting of the MSL complex. *EMBO J.* **21:** 1084–1091.

Meller V.H., Wu K.H., Roman G., Kuroda M.I., and Davis R.L. 1997. roX1 RNA paints the X chromosome of male *Drosophila* and is regulated by the dosage compensation system. *Cell* **88:** 445–457.

Meller V.H., Gordadze P.R., Park Y., Chu X., Stuckenholz C., Kelley R.L., and Kuroda M.I. 2000. Ordered assembly of roX RNAs into MSL complexes on the dosage-compensated X chromosome in *Drosophila*. *Curr. Biol.* **10:** 136–143.

Mendjan S., Taipale M., Kind J., Holz H., Gebhardt P., Schelder M., Vermeulen M., Buscaino A., Duncan K., Mueller J., et al. 2006. Nuclear pore components are involved in the transcriptional regulation of dosage compensation in *Drosophila*. *Mol. Cell* **21:** 811–823.

Mito Y., Henikoff J.G., and Henikoff S. 2005. Genome-scale profiling of histone H3.3 replacement patterns. *Nat. Genet.* **37:** 1090–1097.

Morales V., Straub T., Neumann M.F., Mengus G., Akhtar A., and Becker P.B. 2004. Functional integration of the histone acetyltransferase MOF into the dosage compensation complex. *EMBO J.* **23:** 2258–2268.

Morgan T.H. 1932. *The scientific basis of evolution*. Norton, New York, p. 80.

Oh H., Bone J.R., and Kuroda M.I. 2004. Multiple classes of MSL binding sites target dosage compensation to the X chromosome of *Drosophila*. *Curr. Biol.* **14:** 481–487.

Oh H., Park Y., and Kuroda M.I. 2003. Local spreading of MSL complexes from roX genes on the *Drosophila* X chromosome. *Genes Dev.* **17:** 1334–1339.

Palmer M.J., Mergner V.A., Richman R., Manning J.E., Kuroda M.I., and Lucchesi J.C. 1993. The male-specific lethal-one (msl-1) gene of *Drosophila melanogaster* encodes a novel protein that associates with the X chromosome in males. *Genetics* **134:** 545–557.

Pannuti A. and Lucchesi J.C. 2000. Recycling to remodel: Evolution of dosage-compensation complexes. *Curr. Opin. Genet. Dev.* **10:** 644–650.

Pardo P.S., Leung J.K., Lucchesi J.C., and Pereira-Smith O.M. 2002. MRG15, a novel chromodomain protein, is present in two distinct multiprotein complexes involved in transcriptional activation. *J. Biol. Chem.* **277:** 50860–50866.

Park Y., Kelley R.L., Oh H., Kuroda M.I., and Meller V.H. 2002. Extent of chromatin spreading determined by roX RNA recruitment of MSL proteins. *Science* **298:** 1620–1623.

Sass G.L., Pannuti A., and Lucchesi J.C. 2003. Male-specific lethal complex of *Drosophila* targets activated regions of the X chromosome for chromatin remodeling. *Proc. Natl. Acad. Sci.* **100:** 8287–8291.

Schalch T., Duda S., Sargent D.F., and Richmond T.J. 2005. X-ray structure of a tetranucleosome and its implications for the chromatin fibre. *Nature* **436:** 138–141.

Scott M.J., Pan L.L., Cleland S.B., Knox A.L., and Heinrich J. 2000. MSL1 plays a central role in assembly of the MSL complex, essential for dosage compensation in *Drosophila*. *EMBO J.* **19:** 144–155.

Sif S. 2004. ATP-dependent nucleosome remodeling complexes: Enzymes tailored to deal with chromatin. *J. Cell Biochem.* **91:** 1087–1098.

Smith C.L. and Peterson C.L. 2005. ATP-dependent chromatin remodeling. *Curr. Top. Dev. Biol.* **65:** 115–148.

Smith E.R., Allis C.D., and Lucchesi J.C. 2001. Linking global histone acetylation to the transcription enhancement of X-chromosomal genes in *Drosophila* males. *J. Biol.Chem.* **276:** 31483–31486.

Smith E.R., Cayrou C., Huang R., Lane W.S., Cote J., and Lucchesi J.C. 2005. A human protein complex homologous to the *Drosophila* MSL complex is responsible for the majority of histone H4 acetylation at lysine 16. *Mol. Cell. Biol.* **25:** 9175–9188.

Smith E.R., Pannuti A., Gu W., Steurnagel A., Cook R.G., Allis C.D., and Lucchesi J.C. 2000. The *Drosophila* MSL complex acetylates histone H4 at lysine 16, a chromatin modification linked to dosage compensation. *Mol. Cell. Biol.* **20:** 312–318.

Smith E.R., Eisen A., Gu W., Sattah M., Pannuti A., Zhou J., Cook R.G., Lucchesi J.C., and Allis C.D. 1998. ESA1 is a histone acetyltransferase that is essential for growth in yeast. *Proc. Natl. Acad. Sci.* **95:** 3561–3565.

Straub T., Dahlsveen I.K., and Becker P.B. 2005a. Dosage compensation in flies: Mechanism, models, mystery. *FEBS Lett.* **579:** 3258–3263.

Straub T., Gilfillan G.D., Maier V.K., and Becker P.B. 2005b. The *Drosophila* MSL complex activates the transcription of target genes. *Genes Dev.* **19:** 2284–2288.

Straub T., Neumann M.F., Prestel M., Kremmer E., Kaether C., Haass C., and Becker P.B. 2005c. Stable chromosomal association of MSL2 defines a dosage-compensated nuclear compartment. *Chromosoma* **114:** 352–364.

Stuckenholz C., Meller V.H., and Kuroda M.I. 2003. Functional redundancy within roX1, a noncoding RNA involved in dosage compensation in *Drosophila melanogaster*. *Genetics* **164:** 1003–1014.

Suka N., Luo K., and Grunstein M. 2002. Sir2p and Sas2p opposingly regulate acetylation of yeast histone H4 lysine 16 and spreading of heterochromatin. *Nat. Genet.* **32:** 378–383.

Turner B.M., Birley A.J., and Lavender J. 1992. Histone H4 isoforms acetylated at specific lysine residues define individual chromosomes and chromatin domains in *Drosophila* polytene nuclei. *Cell* **69:** 375–384.

Wang Y., Zhang W., Jin Y., Johansen J., and Johansen K.M. 2001. The JIL-1 tandem kinase mediates histone H3 phosphorylation and is required for maintenance of chromatin structure in *Drosophila*. *Cell* **105:** 433–443.

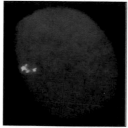

Dosage Compensation in Mammals

Neil Brockdorff[1] and Bryan M.Turner[2]

[1]X Inactivation Group (Developmental Epigenetics Group), MRC Clinical Sciences Centre, ISCM
Hammersmith Campus, London W12 ONN, United Kingdom
[2]Chromatin and Gene Expression Group, Institute of Biomedical Research, University of Birmingham
Medical School, Birmingham B15 2TT, United Kingdom

CONTENTS

1. Introduction, 323
   1.1 The Advantages of Sexual Reproduction, 323
   1.2 Methods of Sex Determination, 323
   1.3 Chromosomal Methods of Sex Determination Create a Need for Dosage Compensation, 324
   1.4 Identification of an Inactive X in Mammalian Females, 324

2. Key Concepts, 325
   2.1 X Inactivation is Developmentally Regulated, 325
   2.2 Chromosome Silencing Involves Multiple Levels of Epigenetic Modification, 325
   2.3 Some Genes Escape X Inactivation, 325
   2.4 X Inactivation Is Regulated by a Master Switch Locus, the X-Inactivation Center, 325

3. Initiation of X Inactivation, 325
   3.1 Imprinted Versus Random X Inactivation, 325
   3.2 Regulation of Imprinted X Inactivation, 326
   3.3 Regulation of Random X Inactivation— Counting, 327
   3.4 Regulation of Random X Inactivation— Choice, 328
   3.5 Switching Modes of Inactivation in Early Embryogenesis, 329

4. Propagation and Maintenance of the Inactive State, 330
   4.1 Xist RNA, Gene Silencing, and Heterochromatin Assembly, 330
   4.2 The Heterochromatic Structure of the Inactive X, 331
   4.3 The Enzymology of Histone Modifications on Xi, 333
   4.4 The Order of Events That Leads to X Inactivation; ES Cells as a Model System, 334
   4.5 Spreading of Silent Chromatin, 335
   4.6 Escape from X-Chromosome Inactivation, 336
   4.7 X Inactivation in Marsupial Mammals, 336

5. X-Chromosome Reactivation and Reprogramming, 336
   5.1 Stability of X Inactivation in Somatic Cells, 336
   5.2 X Reactivation in Normal Development, 337
   5.3 X Reactivation during Experimental Reprogramming, 337
   5.4 Lessons from Inducible Xist Transgenes, 337

6. Summary and Future Directions, 338

References, 338

GENERAL SUMMARY

Human DNA is packaged into 23 pairs of chromosomes of varying size. One of each pair is inherited from our fathers (the paternal homolog) and one from our mothers (the maternal homolog). Twenty-two pairs, collectively called the autosomes and numbered 1—22 in order of decreasing size, are the same in males and females, whereas one pair, the sex chromosomes, differ between the sexes. Females have two copies of a medium-sized chromosome designated the X chromosome, whereas males have one X and one copy of a smaller, gene-poor chromosome designated Y. In males, the X chromosome is always inherited from the mother and the Y from the father, whereas in females one X is maternal (Xm) and one paternal (Xp). This chromosomal difference between the sexes is common in mammals and many other organisms and is part of the biological mechanism by which sex is determined. However, it presents evolutionary problems for the organism in that the two sexes differ in the number of X-linked genes they carry, with females having twice as many as males. This can lead to an imbalance in the amount of gene products (RNAs and proteins), which would in turn require differences in metabolic control and other cellular processes. To avoid this, dosage compensation mechanisms have evolved that balance the level of X-linked gene products between the sexes.

In mammals, the mechanism of dosage compensation involves switching off (silencing) of most genes on just one of the two X chromosomes so that there is only one active X chromosome in females, as there is in males. This radical proposal, generally referred to as X inactivation, was first made in 1961 by Mary Lyon in order to explain the patterns of expression of X-linked coat-color genes in mice, similar to the coat-color pattern of the calico cat illustrated. More than 40 years of intensive research since then has been devoted to trying to resolve the intriguing and complex mechanisms by which the process operates. We know that X inactivation occurs early in development, but in a complex way. Very early on, when the embryo consists of only a few cells, the paternal X is selectively inactivated in all cells. Xp must somehow be marked, "imprinted," for inactivation. Later, at the blastocyst stage (just prior to implantation) when the embryo consists of 50–100 cells, in those cells that will go on to form the embryo itself (located in the inner cell mass, [ICM]), Xp is reactivated, so there are, briefly, two active Xs in females.

Then, either Xp or Xm is chosen at random for inactivation, and its genes are silenced. Intriguingly, in those cells of the blastocyst that will go on to form the extraembryonic tissues (placenta and yolk sac), Xp remains silent. How one X is "chosen" for inactivation in the ICM remains a fascinating unanswered question.

The X that is chosen for inactivation remains silent through all subsequent cell generations. This is one of the most stable forms of gene silencing that we know about, and attempts to reverse it experimentally have been consistently unsuccessful. However, oocytes, the female germ cells, are able to reverse the inactivation process such that they have two active Xs through meiosis, and the single X in the mature, haploid ovum is also active.

Studies of the X-inactivation process have revealed new molecular mechanisms for gene silencing. Initiation of silencing is driven by increased expression of a noncoding RNA transcribed from a gene designated *XIST*, from just one of the two female X chromosomes. This RNA coats the X chromosome containing the *XIST* gene that is switched on, shown as the region of green staining in the cell nucleus illustrated. This then initiates the silencing of genes all along that chromosome. *XIST* itself remains switched on. Following XIST coating, the inactive, silent X undergoes a series of changes. The major DNA-packaging proteins, the histones, undergo a series of chemical modifications at functionally important sites. For example, levels of acetylation at selected lysine residues fall dramatically, while methylation at other lysines increases. These changes are put in place by specific enzymes that are somehow targeted to the silent X. Furthermore, a histone variant, macro-H2A, replaces a proportion of the usual H2A on the inactive X. Following these changes, there is DNA methylation of selected regions on the inactive X, Xi, a process often associated with long-term gene silencing. All these changes, and others, give the inactive X a very characteristic structure, often described as condensed, such that it is visible in the cell nucleus as a distinctive patch of dense DNA, now known as the Barr body.

Over recent years, studies of X inactivation have provided crucial insights into fundamental epigenetic mechanisms of gene silencing and how patterns of gene expression are regulated through development. It can be confidently predicted that they will continue to do so.

1 Introduction

1.1 The Advantages of Sexual Reproduction

Sexual reproduction is common among eukaryotes. Even plants that can replicate themselves perfectly well by sending out shoots or runners often have an alternative sexual mode of reproduction. As is often the case, evolution provides a possible explanation for this in that sexual reproduction brings an enormous increase in genetic variability, upon which natural selection can operate, allowing evolutionary change. The reshuffling of alleles that occurs with every sexual generation could thus produce a population able to cope with environmental shifts more effectively than a relatively homogeneous population derived from asexual methods of reproduction. However, sex is complicated in higher eukaryotes, requiring developmental pathways that lead to male and female sexual organs, as well as physiological and biochemical apparatus required for meiosis, germ cell maturation, the attraction of partners, and mating (for further discussion of these issues, see Marshall Graves and Shetty 2001 and references therein). The crucial point, and definitive measure of evolutionary success, is that variable populations are better able to avoid the ultimate catastrophe of extinction.

1.2 Methods of Sex Determination

Genetic mechanisms for defining different sexes vary widely from one organism to another. The simplest systems involve a single locus that is homozygous in one sex (the *homogametic* sex) and heterozygous in the other (the *heterogametic* sex) (Fig. 1). This simple system has evolved in different ways to reach varying levels of complexity in different organisms. In some, mechanisms have been put in place that suppress meiotic recombination (crossing-over) of the sex-determining alleles in the heterogametic sex (Fig. 1), a step that helps prevent the generation of mixtures of alleles leading to intersex states. In many cases, the inability to recombine has spread to include part or all of one chromosome, with an accompanying loss of genetic information. The evolutionary pressures that have driven this chromosome degeneration are still not understood, but the end result in many species is that the two sexes show differences not just in alleles at one or a few loci, but in complete chromosomes. This is shown in the simple diagram in Figure 1. In most species, including our own, it is the males who carry the degenerate chromosome, although there are exceptions.

Sexual differentiation is usually triggered by the switching on or off of one or a small number of crucial genes during development. The products of these genes initiate a cascade of gene regulatory events that mediate progression down one or the other pathway of sex determination (see Chapter 15 for details in *Caenorhabditis elegans* and Chapter 16 for details in *Drosophila*). In humans, it is the protein product of the *SRY* gene on the Y chromosome that sends the early embryo down the male pathway of sexual development (for review, see McElreavey et al. 1995). A mechanism of this sort does not need major chromosome differences in order to operate successfully, so why have such differences arisen so often? It may be that

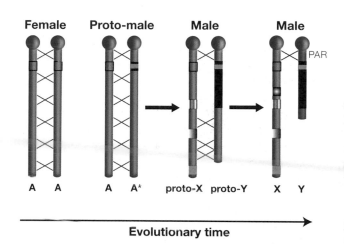

Figure 1. Evolution of the Y Chromosome

Early in evolution the two sexes may have differed at only a single, autosomal locus with one sex (designated proto-male) being heterozygous at this locus and the other sex (proto-female) being homozygous. The "male determining allele" is shown in yellow. If mating requires one member of each sex, then individuals homozygous for the male-determining allele cannot arise. At this early stage, physiological differences between the sexes will be subtle, comparable to those that distinguish the two mating types in yeast. To prevent the formation of intersex states, crossing-over will be suppressed within and around the male-determining locus (the suppressed area is shown as dark). Mutations will accumulate within this region because suppression of crossing-over will reduce their ability to spread through the population and, hence, the selection pressure against them. The degenerate region in which crossing-over is suppressed will gradually expand ("Muller's ratchet") until the chromosome has lost most of its active, functional genes. A small, active region must remain that is homologous to the X chromosome in order to allow pairing and crossing-over at meiosis. This is the pseudoautosomal region (PAR). The autosome originally homologous to the future Y (A in the diagram) will itself evolve, largely through translocations from other autosomes, eventually forming the distinctive X chromosome. The X, like other chromosomes, is a mosaic of DNA fragments put in place at different periods through evolution; some of these are ancient and some are relatively recent. This is illustrated by the differently patterned patches on the proto-X and X chromosomes. On the human X, the more recent arrivals are enriched in genes that escape X inactivation.

they have occurred as a by-product of the suppression of crossing-over required to prevent intersex states (Fig. 1). Mathematical analysis of the factors that influence the spread of alleles through populations shows that suppression of crossing-over will lead *inevitably* to the gradual accumulation of deleterious mutations (perhaps even of mutations that cause further suppression of crossing-over). This will lead to the progressive degeneration of one of the two originally homologous chromosomes (Fig. 1). The process has been termed "Muller's ratchet" in recognition of the geneticist who first proposed it. There is no *selection* for this eventuality, it just happens as a consequence of the initial step of adopting a two-sex strategy for reproduction (for discussion, see Charlesworth 1996 and references therein). Whatever the evolutionary drive behind chromosome degeneration, the fact that it has occurred (and is presumably continuing) has required the coevolution of mechanisms to cope with major chromosomal differences between members of the same species. This is addressed by mechanisms of dosage compensation.

1.3 Chromosomal Methods of Sex Determination Create a Need for Dosage Compensation

In both mammals and *Drosophila*, males have one copy of each sex chromosome, an X and a Y, whereas females have two copies of the X. In both groups of organisms, the Y is gene-poor and largely heterochromatic. It contains just a few genes needed for male development or fertility. In contrast, the X is a large, gene-rich chromosome. A twofold difference in copy number, if left uncorrected, would result in a twofold difference in the intracellular concentrations of several thousand gene products between the sexes. It is not surprising, then, that evolution cannot tolerate such a massive difference between members of the same species. It has instead developed ways of eliminating the difference through mechanisms of dosage compensation.

In general terms, there are just three ways in which levels of X-linked gene transcripts could be equalized between the heterogametic and homogametic sexes. These are (1) switching off genes on one of the two female Xs, (2) doubling the rate of transcription of genes on the single male X, and (3) halving the rate of transcription on each of the two female Xs. The end result in both (2) and (3) is the same in that genes on the single male X are transcribed twice as fast as the equivalent genes on the two female Xs, but the routes used to reach this state are fundamentally different. Examples of all three mechanisms have been identified. Mammals use the first, *Drosophila* the second (dealt with in Chapter 16), and the nematode worm *C. elegans*, the third (dealt with in Chapter 15). The fact that three very different mechanisms of dosage compensation have evolved, apparently independently, confirms the importance of the final objective, namely, equalizing the levels of X-linked gene products between the sexes (for review, see Marin et al. 2000). Consequently, mutations that disrupt dosage compensation in any of these three organisms are lethal in the affected sex.

1.4 Identification of an Inactive X in Mammalian Females

In 1949, Barr and Bertram described the sex chromatin body, a structure visible under the light microscope in the nuclei of female (but not male) cells of various mammalian species (see title page figure). The structure proved useful in studies of sexual abnormalities, but not until 1959 was it shown by Ohno and colleagues (see Ohno 1967) that this structure was derived from one of the two female X chromosomes. In 1961, Mary Lyon described genetic experiments for the expression of X-linked coat-color genes in female mice. To explain the patterns of inheritance for this variable patchwork (mosaic) of coat color in individual female mice, Lyon hypothesized that in each female cell one of the two female X chromosomes is stably inactivated early in development (Lyon 1961). The sex chromatin body, now known as the Barr body, is thus the cytological manifestation of the inactive X chromosome. Elegant experiments using skin fibroblasts from females heterozygous for a polymorphism of the X-linked enzyme glucose-6-phosphate dehydrogenase (G6PD) showed that only one of the two possible alleles was expressed in colonies grown from individual cells (clones), thereby demonstrating the heritability of the inactive state from one cell generation to the next (Davidson et al. 1963) and confirming the occurrence of X inactivation in human females (Beutler et al. 1962). Further studies of X inactivation in human females with multiple copies of the X (with karyotypes such as 47XXX or 48XXXX) showed that all X chromosomes in excess of one were inactivated. This has been generalized as the "*n*–1 rule," which states that if an individual has *n* X chromosomes, then *n*–1 will be inactivated (Ohno 1967). This rule explains the remarkably mild clinical symptoms associated with X-chromosome aneuploidies. The X-inactivation hypothesis has continued to provide an explanation for the peculiarities of X-linked gene expression in female cells and has remained essentially unchanged since first proposed. The past 40 years or so have been spent trying to work out the molecular mechanisms by which it operates.

2 Key Concepts

2.1 X Inactivation Is Developmentally Regulated

X inactivation in female mammals is developmentally regulated. Both X chromosomes are active in the early zygote (Epstein et al. 1978), and inactivation then proceeds coincident with cellular differentiation. Normally, there is an equal probability that cells will inactivate their maternally or paternally derived X chromosome (Xm and Xp, respectively). An exception to this is imprinted X inactivation, which occurs throughout marsupials and in early preimplantation mouse embryos, where it is always Xp that is inactivated. In the latter case, imprinted Xp inactivation is maintained in the first lineages to differentiate, namely the extraembryonic trophectoderm (TE) and primitive endoderm cells (PE), but the inactive X is reactivated in the inner cell mass (ICM) cells that give rise to the embryo. Reversal of X inactivation also occurs in developing primordial germ cells (PGCs), ensuring that the X chromosome is again active in the gamete. Figure 2 illustrates the cycle of X inactivation and reactivation in the female mouse.

2.2 Chromosome Silencing Involves Multiple Levels of Epigenetic Modification

Silencing of the X chromosome is achieved at the level of chromatin structure by modification of histone tails, incorporation or exclusion of variant histones, and DNA methylation of CpG-rich islands, all contributing to a stable heterochromatic structure. The layers of epigenetic modification are established progressively through ontogeny, as detailed in Section 4.4. Collectively, they ensure stable propagation of the inactive X through multiple rounds of cell division.

The contribution of individual modifications varies both temporally and in different lineages. For example, high levels of H3K27me3 (trimethylation of histone H3 at lysine 27) are required on the Xi during early development, but not later, in somatic cells. In contrast, CpG island methylation is only necessary in later development or not at all in trophectoderm cells.

2.3 Some Genes Escape X Inactivation

X inactivation affects most of the X chromosome, but some genes escape silencing. These include genes within a small region on the X chromosome that pairs with the Y chromosome during male meiosis, referred to as the pseudoautosomal (PAR) or XY pairing region (Fig. 1). Genes located in this region do not require dosage compensation, because two copies are present in both males and females.

Other escapees, both with and without Y-linked homologs, have also been characterized. A recent study demonstrated that approximately 15% of genes on the human X chromosome escape X inactivation (Carrel and Willard 2005). Interestingly, many of these genes lie on the short arm of the chromosome (referred to as Xp), which was acquired by the X chromosome relatively recently in evolutionary time. Studies in mouse indicate that escapees can begin by being inactivated in early ontogeny, with progressive reactivation occurring during development (Section 4.6). In marsupials, most genes studied have been found to escape X inactivation to some extent. This may reflect a failure to maintain silencing through ontogeny, possibly related to the lack of CpG island methylation on Xi in these species (see Section 4.7 for more detail).

2.4 X Inactivation Is Regulated by a Master Switch Locus, the X-Inactivation Center

Classic genetic studies demonstrated that X inactivation is mediated by a single *cis*-acting master switch locus, referred to as the X-inactivation center (Xic). The Xic was shown to be required for silencing the X chromosome in *cis*, and for ensuring correct and appropriate initiation of random X inactivation. More recent studies have characterized the Xic at the molecular level. The locus produces a large noncoding RNA termed *Xist* (*X inactive specific transcript*) that has the unique property of binding in *cis* and accumulating along the entire length of the chromosome from which it is transcribed (see Fig. 3) (Brown et al. 1991, 1992; Brockdorff et al. 1992). Coating of the chromosome with *Xist* RNA provides the trigger for X-chromosome silencing (Lee et al. 1996; Penny et al. 1996; Wutz and Jaenisch 2000). Studies to date indicate that this occurs, at least in part, through *Xist*-mediated recruitment of chromatin-modifying complexes (Fig. 3).

A second noncoding RNA, *Tsix*, is also located in the Xic region (Lee et al. 1999) and plays a key role in regulating *Xist* expression. *Tsix* overlaps with the *Xist* gene but is transcribed in the antisense direction, hence its name, which is *Xist* spelled backward!

3 Initiation of X Inactivation

3.1 Imprinted Versus Random X Inactivation

The decision to inactivate an X chromosome needs to be tightly regulated. Male cells must avoid silencing their single X chromosome, and female cells must avoid silenc-

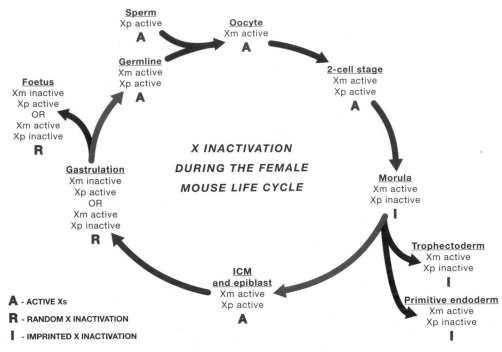

Figure 2. The Cycle of X Inactivation and Reactivation

The X chromosome undergoes a cycle of X inactivation and X reactivation during development. Red arrows indicate X-inactivation steps, and green arrows indicate X-reactivation steps. Inactivation first occurs in early preimplantation embryos (imprinted X inactivation) and subsequently in cells of the epiblast at the time of gastrulation (random X inactivation). The inactive X is reactivated in inner cell mass (ICM) cells when they are first allocated at the blastocyst stage, and also in the developing germ cells.

ing both X chromosomes or keeping both X chromosomes active. Two different modes of regulation have been shown to operate. The imprinted mode of X inactivation silences the paternally derived X chromosome. In the random mode, each cell has an equal probability of inactivating either the maternal or paternal X chromosome. Metatherian mammals (marsupials) use only the imprinted mode. Eutherian (placental) mammals, at least in some cases, use the imprinted mode in extraembryonic lineages, and the random mode in the embryo proper (Fig. 2). There are some indications that humans only have random X inactivation, but this remains unresolved.

3.2 Regulation of Imprinted X Inactivation

Paternally imprinted X inactivation was first observed in a marsupial (Sharman 1971). Subsequently, Takagi and Sasaki (1975) demonstrated imprinted X inactivation in the extraembryonic TE and PE lineages of mouse embryos. It is the parent of origin of the X that governs its status; i.e., paternal but not maternal X chromosomes are inactivated regardless of how many X chromosomes or chromosome sets are present. Note that the single X in XY

males is always maternally derived and therefore not inactivated in imprinted tissues.

What then is the nature of the imprint? Studies of *Xist* expression, which regulates in *cis* X inactivation, indicate that there is a repressive imprint on the Xm allele in mouse embryos (see Fig. 4). This imprint prevents *Xist* expression, keeping the X chromosome active. Nuclear transfer experiments demonstrated that the repressive *Xist* imprint is established during oocyte maturation (Tada et al. 2000). The molecular basis of the imprint is unknown, but DNA methylation is not required, contrasting with many other imprinted genes (see Chapter 19 for details on genomic imprinting).

One theory for the preferential inactivation of Xp in the zygote is that there is carryover of silencing of the XY bivalent that is established during the pachytene stage of male meiosis (*meiotic sex chromosome inactivation*, MSCI) (Huynh and Lee 2003). Recent studies argue against this. First, MSCI has been shown to be a distinct and *Xist*-independent mechanism that is triggered in pachytene by the presence of unpaired chromosomal regions, on both sex chromosomes and autosomes (Turner et al. 2005 and references therein). Second, expression analysis of a num-

a

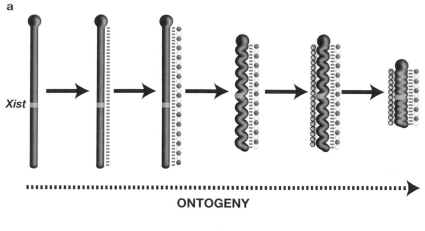

ONTOGENY

b

Xist at metaphase

Xist at interphase

hypoacetylated X

Figure 3. Progressive Chromosome-wide Heterochromatinization Induced by *Xist* RNA

When the *Xist* gene is expressed, the RNA binds to and coats the X chromosome from which it is transcribed (*green dashed line*). *Xist* RNA is thought to trigger silencing of the chromosome by recruiting chromatin-modifying activities (*red and yellow circles*). The initial wave of silencing in turn leads to recruitment of additional layers of epigenetic modification (*gray circles*), further stabilizing the heterochromatic structure. Establishment of these different levels of epigenetic silencing is achieved in a stepwise manner through development and ontogeny. Localization of *Xist* RNA along the X chromosomes is shown by in situ hybridization in both interphase and metaphase.

ber of X-linked genes in early zygotes has shown that Xp silencing occurs de novo in response to zygotic Xp *Xist* expression (Okamoto et al. 2005 and references therein).

The paternal *Xist* expression (and resultant Xp silencing) that begins at the onset of zygotic gene activation (at the 2- to 4-cell stage) indicates that the Xp *Xist* allele is poised to express (Fig. 4). However, in male somatic tissues, *Xist* is always repressed, so it follows that male germ cells must in some way remodel the *Xist* locus. Consistent with this, there is a region-specific demethylation of CpG sites in the *Xist* promoter during spermatogenesis (Norris et al. 1994).

In embryos with X chromosome aneuploidy, the parental imprints governing *Xist* expression result in inappropriate X-inactivation patterns during early development. This is also the case in androgenetic and gynogenetic (or parthenogenetic) embryos in which both chromosome sets originate from either the father or the mother, respectively. Although this generally disadvantages embryo viability, some embryos survive, apparently correcting inappropriate X-inactivation patterns to ensure that a single active X chromosome is present in all cells. This is thought to involve overriding of the imprint by the mechanism that normally regulates random X inactivation (see below).

The *Tsix* gene, an antisense regulator of *Xist*, is required for imprinted X inactivation, because deletion of the major promoter results in early embryo lethality when transmit-

ted by the maternal but not the paternal gamete (Lee 2000). Lethality appears to be attributable to inappropriate Xm *Xist* expression; i.e., a failure to retain an active X chromosome both in XmY and XmXp embryos. At present, it is not clear whether expression of *Tsix* RNA is the primary imprint or functions later to maintain the imprint.

3.3 Regulation of Random X Inactivation—Counting

In the random mode of X inactivation, cells utilize the $n-1$ rule, where all X chromosomes except one are inactivated per diploid chromosome set. The choice of which X is silenced (or kept active) is essentially random, although there are factors that can influence this (see Section 3.4). *Xist* regulation in random X inactivation is tightly controlled. Studies showed that single XX embryonic stem (ES) cells being differentiated only ever express a single *Xist* allele, whereas XY cells never initiate expression.

In X-chromosome aneuploid and polyploid mouse embryos, the outcome of random and imprinted X inactivation is different. A simple illustration of this is provided by X0 embryos. In imprinted X inactivation, Xm0 embryos are normal but Xp0 embryos are retarded because cells attempt to inactivate their single X chromosome, thereby compromising development of extraembryonic tissues. In random X inactivation, cells count their number of X chromosomes (obeying the $n-1$ rule) and

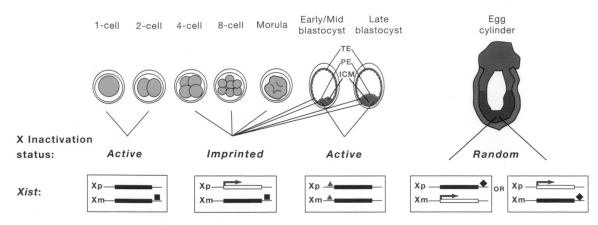

1-cell 2-cell 4-cell 8-cell Morula Early/Mid blastocyst Late blastocyst Egg cylinder

Figure 4. *Xist* **Gene Regulation in Early Development**

The figure illustrates current knowledge and models for imprinted and random *Xist* regulation in early XX mouse embryos. The Xm *Xist* allele arrives in the zygote with a repressive imprint, possibly mediated through the antisense Tsix locus (*black square*). The Xp allele is primed to be active and is expressed as soon as embryonic gene activation occurs at the 2-cell stage. From 2-cell up until morula stage, Xp *Xist* is expressed in all cells (expression indicated by *open rectangle* and *arrow* at 5′ end). This pattern is maintained at the early blastocyst stage and subsequently in TE and PE cells and their fully differentiated derivative tissues. In the late blastocyst ICM, *Xist* expression is extinguished, possibly by an ICM-specific repressor factor (*blue triangle*). *Xist* expression then commences subsequently at the time of gastrulation. Here the blocking factor (*black diamond*) ensures that *Xist* expression cannot occur on one of the two alleles (counting).

thus maintain the single X chromosome active, regardless of whether it is Xm or Xp, and are therefore unaffected.

The random mode of X inactivation requires that cells have a means of sensing the number of X chromosomes present, often referred to as "counting." A popular model for explaining counting is Rastan's blocking factor model (Rastan 1983). This proposes that a single Xic (*Xist* allele) is blocked in all cells, thereby designating the active X chromosome. Additional Xics (*Xist* alleles), where present, are expressed, thus inducing X inactivation (Fig. 5). Rastan's model implies that X inactivation is a default state, consisting of keeping one X chromosome active. Although this may seem counterintuitive, the model has the considerable merit of accounting for most experimental observations made to date. The nature of the blocking factor, however, remains elusive.

Our understanding of the mechanisms regulating *Xist* expression and initiation of X inactivation have benefited greatly from the use of ES cells cultured in vitro. ES cells are derived from mouse embryos at the blastocyst stage, specifically from the ICM (Fig. 4). They remain undifferentiated when cultivated in serum supplemented with leukemia inhibitory factor (LIF). Differentiation in vitro can be triggered by removing LIF from the culture medium and replating the cells onto dishes made of nonadherent plastic. Under these conditions, they round up and aggregate to form embryoid bodies, within which

they differentiate, over just a few days, into many different somatic cell types. In the undifferentiated state, XX ES cells have two active X chromosomes. When the cells are induced to differentiate, most cells of the embryoid body have undergone random X inactivation. Conversely, *Xist* expression and X inactivation do not occur in differentiating XY ES cells. ES cells thus provide a valuable model system, because they are amenable to genetic manipulation by gene targeting, and the different steps of the X-inactivation process can be studied. A key experiment was the demonstration that deletion of a 65-kb region located immediately downstream of *Xist* results in that allele expressing *Xist* (and therefore inactivating), even in differentiating XY ES cells (Clerc and Avner 1998). The implication of this finding is that the deleted sequences are required for binding of the putative blocking factor. Further delineation of the region responsible for this has been achieved using replacement targeting strategies deleting smaller regions within the 65 kb (see Fig. 6). Definition of the key elements and identification of the factors binding them represent key goals for future studies.

3.4 Regulation of Random X Inactivation—Choice

In certain circumstances, random X inactivation can be skewed. This occurs either as a result of a bias in the initial choice of which X to inactivate ("primary" nonrandom X

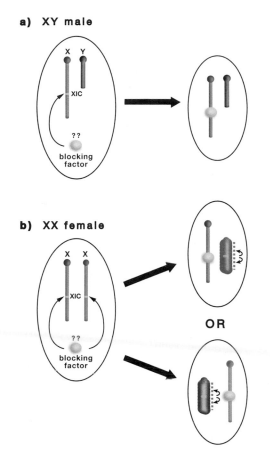

a) XY male

b) XX female

OR

Figure 5. The Blocking Factor Model for Random X Inactivation

The blocking factor model proposes that there is a factor (*yellow circle*) that blocks a single *Xist* allele (*green box*) in all cells such that at the onset of X inactivation that chromosome is protected from undergoing X inactivation. In male cells (*a*) there is only one allele present and the blocking factor always binds. In female cells (*b*) there is an equal probability that the blocking factor will bind either the Xm *Xist* allele (*red*) or the Xp *Xist* allele (*blue*). At the onset of X inactivation, only the unblocked allele will express *Xist* RNA (*green dashed line*). Different alleles of the X-inactivation center may have a greater or lesser affinity for the blocking factor such that in heterozygous females the factor preferentially binds one X chromosome. In some cases, this may underlie skewing in primary nonrandom X inactivation (see Fig. 7).

inactivation), or as a result of selection against cells that maintain a particular X active ("secondary" nonrandom X inactivation) (Fig. 7). In primary nonrandom X inactivation, choice is affected by *cis*-acting sequence variations or mutations that affect the probability of a given X being selected as the active/inactive X in heterozygous animals. According to Rastan's blocking factor model, these variations may manifest their effect by changing the probability of the blocking factor binding to a given allele.

An example of primary nonrandom X inactivation is provided by the X controlling element (*Xce*) in mouse. *Xce* is a classically defined locus, where different alleles

have been found to influence the probability of its X chromosome being the active X in XX heterozygotes (Cattanach and Isaacson 1967). Genetic mapping experiments position *Xce* immediately downstream of *Xist* (Fig. 6) and therefore in the correct location to affect blocking factor binding as defined by the 65-kb deletion. The underlying sequence variation remains to be identified.

A second element that can influence choice in the mouse is the *Tsix* antisense regulator (Lee and Lu 1999). This is thought to be mediated by *Tsix* antisense RNA, which is transcribed across the *Xist* locus prior to initiation of random X inactivation (Fig. 6). A chromosome bearing a *Tsix* promoter deletion is preferentially inactivated in XX cells undergoing random X inactivation. More subtle skewing effects are seen in cells with mutations in enhancer elements (Xite elements) that may govern *Tsix* expression levels. Although the *Tsix* promoter lies within the region defining the putative blocking-factor-binding site, targeted deletion experiments do not activate *Xist* by default in XY ES cells. This suggests that the loci are not synonymous, but that deletion of the *Tsix* promoter leaves binding sites for the blocking factor intact.

Tsix transcription is accompanied by low-level *Xist* transcription prior to the initiation of random X inactivation, suggesting that double-stranded RNA (i.e., *Tsix:Xist* hybrid strands) could mediate the effect on choice. Consistent with this, increasing the level of sense transcription across *Xist* promoters antagonizes the repressive effect of *Tsix* and results in an allele that is less likely to be the active X in a heterozygous XX cell (Nesterova et al. 2003). In summary, all of the different Xic elements known to influence the initiation of random and imprinted X inactivation, often described with regard to their ability to affect functions of counting and/or choice, are shown in Figure 6.

3.5 Switching Modes of Inactivation in Early Embryogenesis

How do early mouse embryos instigate the switch from the imprinted to the random mode of regulation? Until recently, it was thought that in both cases initiation of X inactivation is linked to cellular differentiation (Monk and Harper 1979). Thus, trophectoderm and primitive endoderm lineages were thought to inactivate Xp in response to the parental imprints on *Xist* when they first differentiate at the blastocyst stage, whereas ICM cells, which give rise to the embryo proper, were thought to first erase the *Xist* imprint and then to undergo random X inactivation when they differentiate into the three germ lineages at gastrula-

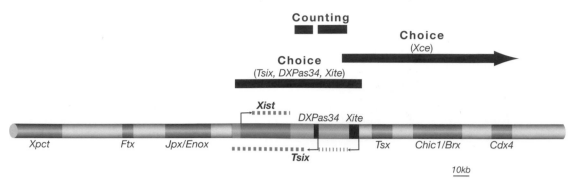

Figure 6. Genes and Regulatory Elements in the X-Inactivation Center Region

The key region regulating X inactivation is shown in green. Flanking genes are shown in gray. Arrows indicate promoters for the *Xist* (sense) and *Tsix* (antisense) genes. The extents of the respective noncoding RNAs are indicated with dashed green lines. Regulatory elements controlling *Tsix* expression, Xite and DXPas34, are indicated in black. Regions and loci involved in X-chromosome choice and X-chromosome counting are indicated above.

tion. More recent data, however, demonstrate that Xp inactivation occurs prior to the onset of cellular differentiation in cleavage-stage embryos, and that it occurs in all cells, including the precursors of the ICM (Mak et al. 2004; Okamoto et al. 2004). Thus, imprinted X inactivation in trophectoderm and primitive endoderm is a relic of the X-inactivation pattern established in early-cleavage embryos. ICM cells must thus instigate a program to reverse this initial wave of imprinted Xp inactivation, thereby setting the scene for subsequent random X inactivation (see Fig. 4). The basis for reversal of Xp inactivation is unknown but may involve an ICM-specific program that represses Xp *Xist* expression (see Section 5).

4 Propagation and Maintenance of the Inactive State

4.1 Xist RNA, Gene Silencing, and Heterochromatin Assembly

So what does the *Xist* gene do? There is strong evidence that the *Xist* gene and its RNA product provide both the switch that initiates X inactivation in *cis* and the means by which silencing spreads across the chromosome. Evidence indicates that (1) *Xist* is unique in being expressed only from Xi, (2) *Xist* RNA levels increase dramatically in pre-implantation embryos at the time of X inactivation, (3) *Xist* up-regulation precedes X inactivation and appears to be an absolute requirement for it to occur, (4) *Xist* RNA colocalizes with Xi in interphase nuclei and is distributed along one of the two metaphase X chromosomes (see Fig. 3), and (5) *Xist*-containing transgenes, when inserted into autosomes, can induce at least some of the properties of inactive chromatin. (Over)expression results in coating of

the autosome in *cis* and the parallel adoption of a heterochromatin-like, transcriptionally silent chromatin structure (Heard et al. 1999 and references therein). These findings suggest that *Xist* RNA is both necessary and sufficient to trigger heterochromatin formation and transcriptional silencing. However, continuing *Xist* expression is not required for the *maintenance* of X inactivation. For example, in human:rodent somatic cell hybrids where *Xist* expression is lost on the human Xi chromosome which is retained in a rodent background, silencing of X-linked genes is maintained (Brown and Willard 1994). This issue is discussed further in Section 5.

It is important to note that the association of *Xist* RNA with Xi is selective. It is not found along the PAR, which remains active and euchromatic, or at constitutive (centric) heterochromatin. Moreover, analysis of metaphase chromosomes demonstrates a banded localization that appears to correlate with gene-rich G-light bands (see Fig. 3) (Duthie et al 1999). These observations show that *Xist* RNA coats only certain chromatin regions (discussed further in Section 4.5).

The mechanism(s) by which *Xist* RNA brings about changes in chromatin structure, and its associated gene silencing, are still not understood in detail. We do know that different regions of the *Xist* RNA molecule are responsible for gene silencing and spreading along the X chromosome. Experiments with an inducible *Xist* expression system in mouse ES cells, in which the functions of *Xist* molecules carrying defined deletions could be tested, showed that silencing can be attributed to a conserved repeat sequence at the 5′ end of the molecule, whereas coating the X is mediated by sequences scattered throughout the rest of the molecule (Wutz et al. 2002).

a) Primary non-random

b) Secondary non-random

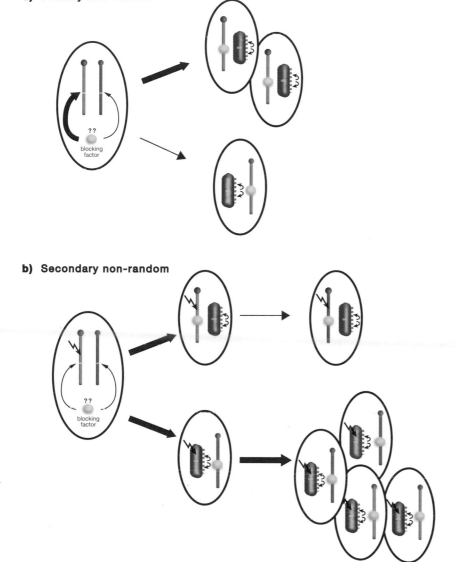

Figure 7. Models for Nonrandom X Inactivation

Primary nonrandom X inactivation refers to skewing of the initial choice of which X chromosome is inactivated. Theoretically, this could occur in heterozygous females where there is a bias in the probability of the two alleles binding the blocking factor. In secondary nonrandom X inactivation, the choice of which X is inactivated is random, but cell selection events result in progressive loss of cells inactivating one of the two X chromosomes. For example, where there is a deleterious mutation on one X chromosome, cells that inactivate the other, wild-type X chromosome will be preferentially lost.

4.2 The Heterochromatic Structure of the Inactive X

Since the very earliest light microscopy studies, it has been realized that Xi shares properties with heterochromatin. Like the constitutive heterochromatin found at and around centromeres, Xi remains visible, and apparently condensed throughout interphase (as the Barr body), and its DNA is usually replicated late in S phase. Xi is said to consist of *facultative* heterochromatin. However, it is important to remember that the DNA of constitutive heterochromatin is usually enriched in specific, repetitive satellite sequences that are responsible, at least in part, for its characteristic properties. The DNA of the X chromosome shows no such enrichment, although it does show more subtle differences in specific repeat elements which

may play a role in the inactivation process (see Section 4.5). In addition, although the Xi chromatin is often described as "condensed," careful microscopic analysis and 3D reconstruction of Xa and Xi chromosomes labeled with X-specific DNA probe suggest that the difference between them is more a matter of shape than of the amount of chromatin per unit volume (Eils et al. 1996).

Further parallels between Xi and constitutive heterochromatin have come from the use of indirect immunofluorescence microscopy to study the distribution of histone modifications and variants both along metaphase chromosomes and in interphase nuclei. The facultative heterochromatin of the inactive X chromosome in both human and mouse cells is depleted in acetylated histone H4 (Jeppesen and Turner 1993), and in this way resembles constitutive, centric heterochromatin. This was the first demonstration that the inactive X chromosome was marked by a specific type of histone modification. Subsequent experiments in several laboratories confirmed these observations and further showed that acetylated isoforms of all four core histones (H2A, H2B, H3, and H4) were depleted in both constitutive and facultative heterochromatin in interphase and metaphase cells (O'Neill et al. 2003 and references therein). In particular, both centric heterochromatin and Xi are depleted in di- and tri-methylated H3 at K4 (H3K4me2 and H3K4me3). These are generally thought, like acetylation, to be markers of transcriptionally active, or potentially active, chromatin.

The situation becomes more complex when one considers the appearance of marks associated with transcriptional silencing rather than the disappearance of those associated with transcriptional activity. For example, di- and tri-methylated H3 at K9 (H3K9me2/3) are consistent marks of transcriptionally silent genes and are frequently seen to be enriched in centric heterochromatin by immunofluorescence microscopy (Lachner et al. 2003). However, for a combination of technical and biological reasons, their enrichment on Xi has been uncertain. For this reason, the importance of antibody specificity in providing reliable immunofluorescence data should be highlighted. For example, lysines 9 and 27 of H3 are both part of the tetrapeptide ARKS, and antibodies raised against one can cross-react with the other. Such cross reactions must be rigorously excluded before results can be interpreted with confidence. Additionally, the immunization procedure used, and the immunogen used, can affect the detailed specificities of antisera. For example, antisera to H3K9me3 raised with cross-linked peptides bind more strongly to this modified histone when it is within a heterochromatic region than when it is within euchromatin (Maison et al. 2002). Antisera raised in this way are valuable reagents for studies on heterochromatin, but not ideal for quantitative comparisons of heterochromatin and euchromatin. Finally, it has been noted that enhanced immunofluorescence staining at interphase can result simply from a higher density of nucleosomes within heterochromatin (Perche et al. 2000).

A careful analysis of the distributions of histone modifications across Xi in human cultured cells has provided further insights into the complexity of the system (Chadwick and Willard 2004). H3K9me3 and H3K27me3 are both enriched at defined, and nonoverlapping, regions across Xi. Thus, unlike loss of histone

acetylation, enrichment in these modifications is a regional, not an overall, property of Xi. Intriguingly, those regions enriched in H3K27me3 are also found to be enriched in *Xist* RNA and the variant histone macroH2A1.2 (Costanzi and Pehrson 1998). How macroH2A1.2 might associate with Xi, and its possible role in the inactivation process, are discussed further in Section 4.4. Conversely, those regions of Xi that are enriched in H3K9me3 also show enhanced levels of heterochromatin protein HP1 (known to bind to methylated H3K9) and H4K20me3 (a mark also associated with constitutive, centric heterochromatin). Importantly, immunostaining of the Barr body in interphase cells showed the same co-staining patterns, suggesting that the different domains are retained through the cell cycle.

The picture that emerges is of combinations of histone modifications, histone variants, non-histone proteins, and *Xist* RNA, interacting to form chromatin with the distinctive properties of transcriptional silence, replication late in S phase, condensed appearance, and (possibly) nuclear localization that are characteristic of Xi. However, the exact functional significance of the distinctive chromatin domains on Xi remains to be established. An important and worrying observation is that the frequency of the observed domains varies widely from one human cell line to another (Chadwick and Willard 2004). This may be no more than confirmation of the known redundancy in the X-inactivation system (e.g., as noted earlier, silencing is maintained in mature cells even when *Xist* RNA is lost), but it also warns that cultured cell lines, particularly immortalized ones, are not always a completely accurate guide to what happens in primary tissues, and certainly not at the early stages of development.

The histone modifications associated with euchromatin and facultative and constitutive heterochromatin are summarized in Table 1. So far, only methylation of H3K27 and ubiquitylation of histone H2A at lysine 119 (H2AK119ub) have been found to be enriched on Xi but not on constitutive heterochromatin (Plath et al. 2003; Silva et al. 2003; de Napoles et al. 2004; Smith et al. 2004).

All the modifications listed in Table 1 are associated with the overall, heterochromatic conformation of the inactive X chromosome and have been identified by immunofluorescence analysis of either metaphase chromosomes or the Barr body in interphase cells. However, localized changes in histone modifications may also play important roles in the various stages of the X-inactivation process. Such changes can be identified by high-resolution microscopy, or by chromatin immunoprecipitation (ChIP) to map modifications explained in Chapter 10

Table 1. Histone modifications characteristic of constitutive and facultative heterochromatin

| Histone | Modification | Amount (relative to euchromatin) | |
		constitutive heterochromatin (centric)	facultative heterochromatin (Xi)
H2A, H2B, H3, H4	acetylation[a]	–	–
H2A	K119 ubiquitylation	≈	+
H3	K4 me2/me3	–	–
H3	K9 me2/me3	+	≈[b]
H3	K27 me2/me3	≈	+[c]
H4	K20 me3	+	≈

Histone modifications that are enriched (+) or depleted (–) in constitutive and facultative heterochromatin, relative to euchromatin, are indicated. The ≈ symbol indicates that the level of the modification is not detectably different from that in euchromatin. (me) methylation.

[a]Applies to all acetylatable lysines.

[b]Enriched in local "hot spots" but not overall.

[c]Enrichment is transient in some cell types.

within or adjacent to the crucial Xic region. For example, a large domain extending more than 340 kb 5′ of the *Xist* gene is characterized by H3K9 hypermethylation in undifferentiated ES cells. Hypermethylation diminishes as the cells differentiate and X inactivation proceeds (Heard et al. 2001). The same general chromatin region is enriched in methylated H3K27 in female ES cells (Rougeulle et al. 2004), and sites within it are enriched in acetylated H3 and H4 (O'Neill et al. 1999). Ongoing investigations will reveal to what extent these localized histone modifications in the Xic region are early causative events driving the X-inactivation process, or downstream events that are (possibly essential) components of an ongoing chromatin-remodeling process.

Constitutive centric heterochromatin is enriched in methylated DNA, primarily 5′-methylcytosine in CpG dimers (see Chapter 18). This is consistent with its low level of transcriptional activity. Perhaps surprisingly, the level of CpG methylation on Xi is not, overall, significantly higher than the rest of the genome. However, specific CpG islands associated with silenced genes are highly methylated, and experimental evidence suggests that DNA methylation plays an important role in the stabilization of the inactive state. Thus, mice lacking the enzymes that either methylate previously unmodified CpGs (the de novo DNA methyltransferases, Dnmta and Dnmtb) or maintain the modification of previously modified residues (the maintenance enzyme, Dnmt1) initiate and establish random X inactivation in the normal way (Sado et al. 2000, 2004). However, genes on the hypomethylated Xi in these mutant mice are more easily reactivated than an Xi in wild-type animals with normal levels of methylation (Sado et al. 2000).

4.3 The Enzymology of Histone Modifications on Xi

The enzymes responsible for the deacetylation of core histones (HDACs) during X inactivation, or for the demethylation of H3K4, are as yet unknown. Because the deacetylase inhibitor trichostatin A (TSA) can prevent, or at least delay, the appearance of a deacetylated X in differentiating female ES cells, it is reasonable to propose that HDACs are involved (O'Neill et al. 1999). However, we do not know which of the 11 class I and II HDACs are most likely to be responsible (class III enzymes are not inhibited by TSA). We also do not know the mechanism responsible for the removal of H3K4 methylation. Enzymes capable of removing methyl groups from H3K4 in the mono- or dimethylated state have only recently been identified, and enzymes capable of demethylating the H3K4 tri-methylated state remain to be discovered. At this stage, we cannot rule out that H3K4 methylation and/or histone deacetylation is lost from Xi by histone replacement or passively through DNA replication.

We currently know more about the enzymes responsible for putting histone modifications in place. Methylation of H3K27 is carried out by Ezh2/Enx1, the homolog of the *Drosophila* polycomb-group (PcG) protein, enhancer of zeste (E(Z)) (Silva et al. 2003). E(Z) is a histone methyltransferase (HKMT) that functions in the PRC2 PcG complex (see Chapter 11). PRC2 is recruited to Xi during ES cell differentiation with the same kinetics as H3K27 methylation. Interestingly, PcG proteins are also involved in ubiquitylation of H2A at lysine 119 on Xi. Specifically, the Ring1A/Ring1B protein, a core component of the PRC1 PcG complex, functions as the E3 ligase for H2A ubiquitylation. Deletion of both Ring1A and Ring1B results in loss of H2A ubiquitylation both on Xi and genome-wide (de Napoles et al. 2004).

4.4 The Order of Events That Leads to X Inactivation; ES Cells as a Model System

Mouse ES cells have provided an invaluable model system for studying the dynamics of X-chromosome inactivation. Increased levels of *Xist* RNA and its coating of one X chromosome are first detected in a high proportion of cells after 1–2 days of differentiation. There is evidence that both transcriptional and posttranscriptional mechanisms play a role in *Xist* up-regulation (Panning et al. 1997; Sheardown et al. 1997; Rougeulle et al. 2004). *Xist* is, however, transcribed from both X chromosomes in *undifferentiated* female ES cells, and from the single X in male ES cells, but the RNA product is rapidly degraded and only small amounts are detectable adjacent to the *Xist* locus. Evidence has been presented that as differentiation proceeds, there is stabilization of *Xist* RNA on one of the two X chromosomes in female cells (Panning et al. 1997; Sheardown et al. 1997). The mechanism that underlies this selective RNA stabilization step remains uncertain, as does its contribution to the overall increase in *Xist* RNA and coating of the chromosome in *cis*.

A number of X-inactivation steps have been found to occur coincident with the onset of *Xist* RNA accumulation in differentiating XX ES cells. These include recruitment of PcG proteins and associated methylation of H3K27, H2A monoubiquitylation, deacetylation of H3K9, and loss of methylation at H3K4 (Heard et al. 2001; Silva et al. 2003; de Napoles et al. 2004; Rougeulle et al. 2004). Global histone deacetylation is a relatively late event, occurring at days 3–5 in the majority of cells, and is therefore likely to be involved in maintenance and/or stabilization of the inactive state, rather than in its initiation (Keohane et al. 1996). This interpretation assumes that patterns of acetylation at the promoters of individual genes undergoing inactivation reflect those determined by immunofluorescence analysis of the whole chromosome, or of large domains. Initial studies by ChIP suggest that this is indeed the case, but further experimentation across larger numbers of genes is necessary (O'Neill et al. 2003).

Accumulation of the variant histone macroH2A1.2 on Xi occurs much later during XX ES cell differentiation (Mermoud et al. 1999). This variant histone has over 200 additional amino acids in its carboxy-terminal tail and several amino acid substitutions throughout the molecule. Interestingly, *Xist* RNA expression is required to retain macroH2A on Xi in somatic cells (Csankovszki et al. 1999)

but is not sufficient to recruit macroH2A in early differentiation stages (Mermoud et al. 1999; Wutz et al. 2002).

Selective DNA methylation on Xi is an even later event in ES cells. CpGs that are known to be highly methylated on Xi in adult cells do not become methylated in female ES cells until much later in differentiation, 14–21 days (Keohane et al. 1996). This is consistent with results in the developing embryo itself (Lock et al. 1987) and with the idea that DNA methylation is responsible for stabilization, or locking, of the inactive state rather than in initiation and spreading.

Thus, the picture that emerges is of a coordinated and carefully regulated sequence of events by which chromatin changes on the Xi are put in place as development proceeds (summarized in Fig. 8). It is remarkable that some of these changes, for example, histone deacetylation and DNA methylation, take place after the cells have started to progress down various different pathways of differentiation. It seems that the program responsible for the completion of X inactivation proceeds independently of other cell-differentiation programs. However, it is important to note that some aspects of random X inactivation can proceed only after differentiation has begun. For example, switching on expression of *Xist* transgenes in *undifferentiated* ES cells triggers various histone modifications associated with heterochromatinization, and also the transition to replication in late S phase (Wutz and Jaenisch 2000),

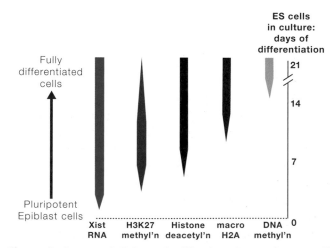

Figure 8. Layers of Epigenetic Silencing Accumulate on Xi through Differentiation

The diagram shows how five different epigenetic changes associated with transcriptional silencing are put in place on the inactive X chromosome at different stages of development and differentiation in both the developing embryo and ES cells in culture. In some cell types, methylation of H3K27 is transient, being seen only at early stages of differentiation and not in mature cells.

but there is no detectable incorporation of macroH2A; only after the cells have been induced to differentiate does macroH2A colocalize with *Xist* RNA on the chromosome containing the *Xist* transgene (Rasmussen et al. 2001). Association of macroH2A with *Xist*-coated chromatin is dependent on the continued presence of *Xist* RNA (Csankovszki et al. 1999) but does not require transcriptional silencing, because it is seen also in chromosomes coated with a mutant *Xist* RNA lacking regions necessary for silencing (Wutz et al. 2002). Thus, X inactivation can be seen as the end result of a series of parallel processes, only some of which are interdependent.

It is important to emphasize that particular enzyme complexes or histone modifications may have crucial roles at certain stages of the X-inactivation process but may become less important, or redundant, later on, perhaps after more permanent silencing mechanisms based on DNA methylation have been put in place. For example, the methylation of H3K27 catalyzed by PRC2 is essential for successful X inactivation early in ontogeny but seems to be dispensable later.

It should also be noted that a different order of events may occur during establishment of imprinted X inactivation in preimplantation embryos. Notably, enrichment of H3K27me3 is not detected until the 16-cell stage, considerably later than the onset of *Xist* expression (2- to 4-cell stage) (Mak et al. 2004; Okamoto et al. 2004). This may indicate a requirement for specific developmentally regulated cofactors in order to recruit the PRC2 PcG complex to Xi.

4.5 Spreading of Silent Chromatin

The XIC is essential for X inactivation, and it has been suggested that silencing spreads from the XIC, with proximal genes being silenced earlier than more distal ones. One explanation for this would be that *Xist* RNA spreads progressively along the chromosome from its site of synthesis. This would be consistent with the results of experiments in which *Xist* transgenes are inserted into autosomes and where coating of the autosome with *Xist* RNA leads to gene silencing and chromatin changes (see above).

Spreading of the inactive state from the XIC has been extensively studied in naturally occurring X;autosome translocations. Indeed, such translocations were crucial in demonstrating the existence of the XIC (Rastan 1983). Chromatin bearing the characteristic marks of facultative heterochromatin (e.g., loss of histone acetylation, late replication, and transcriptional silencing) has been shown to spread from the Xi into the autosomal part of the

hybrid chromosome (White et al. 1998; Duthie et al. 1999; Sharp et al. 2002). This establishes the important principle that facultative heterochromatin is not exclusive to the X chromosome, consistent with the results from *Xist* transgenes expressed on autosomes. However, the spreading of silent chromatin along the autosomal chromosome arm is variable between translocations and limited in extent.

There are two models to account for limited spreading of silencing into *cis*-linked autosomes. First, autosomes may resist the initial spreading of *Xist* RNA and associated gene silencing at the onset of X inactivation. Alternatively, initial spreading into autosomes could be efficient, but silencing may be poorly maintained through ontogeny, referred to as "spread and retreat." Evidence to date favors spread and retreat, but the two models are not mutually exclusive and further studies are required. It should be remarked that those cells in which extensive silencing of the autosome occurs may be selected against during early development because of loss of expression of crucial autosomal genes.

In some X;autosome translocations, the spread is discontinuous, appearing to skip over certain regions and leaving transcriptionally active, euchromatic regions surrounded by silent, heterochromatic regions (Sharp et al. 2002). The spread of constitutive heterochromatin into adjacent euchromatic regions that leads to position effect variegation in *Drosophila* (see Chapter 5) can show similar behavior. Such observations are easier to reconcile with an early cellular-selection mechanism based on the spread and retreat of silencing, than with the continuous spread of stable silencing.

It has been suggested that there are elements distributed along the X chromosome, called "way stations," that serve as assembly points for heterochromatin and thereby enhance the spread and/or maintenance of X inactivation (described in Gartler and Riggs 1983). These elements would have to be less common, or less regularly distributed, on autosomes. It has been proposed that a common dispersed repeat family, long interspersed repeats (LINES), are good candidates for the way-station elements (Lyon 1998, 2003). These repeat sequences are common in the human and mouse genomes but are particularly frequent along the X chromosome. Furthermore, LINE elements are most common in the more condensed, gene-poor, G-banded regions of the human and mouse genomes, suggesting that they may in some way favor a chromatin conformation associated with transcriptional silencing. The recent completion of the DNA sequence of the human X chromosome has revealed a distribution of LINE elements that is broadly consistent with a possible role as way stations, but the idea remains unproven.

4.6 Escape from X-Chromosome Inactivation

As discussed previously, a number of genes are known to escape inactivation. In mechanistic terms, escape from X inactivation has been attributed to the rarity of way stations adjacent to these genes, or to the presence of boundary elements that block the spread of *Xist* RNA and/or other silencing components. There is some evidence that genes which escape inactivation are silenced, at least to some extent, in early development. For example, in the case of the mouse *Smcx* gene, escape from inactivation varies with both developmental stage and tissue type (Sheardown et al. 1996). This is more consistent, therefore, with the spread and retreat idea discussed above in the context of silencing of autosomal genes in X;A translocations.

4.7 X Inactivation in Marsupial Mammals

The divergence of eutherian and marsupial mammals occurred about 130 million years ago. Marsupials, like eutherians, use an XY (male):XX (female) sex determination system and a dosage compensation mechanism in which one of the two female X chromosomes is inactivated. As mentioned previously, the inactive X in marsupials is always the paternally derived homolog (Xp). A further difference relative to eutherian mammals is that the extent of gene silencing of individual loci often varies between different tissues. There is also some evidence that the instability of silencing increases through development and ontogeny. This is again reminiscent of the spread and retreat model for silencing of autosomal loci in eutherians. Interestingly, enrichment of LINE elements on the X chromosome occurred after the eutherian–metatherian divergence, so it is possible that the instability of silencing on the marsupial X also connects with the idea of way stations. Furthermore, the lack of methylation at CpG islands associated with X-linked genes in marsupials could contribute additionally to the instability of Xi.

Relatively little is known about the molecular mechanism of X inactivation in marsupials and, as yet, no marsupial homolog of *XIST* has been identified. The absence of an *XIST* homolog does not, of course, preclude the existence of a nonhomologous RNA that carries out the same function as an initiator of X inactivation, but none has yet been found. The only two properties that are unequivocally shared by the inactive X in eutherian and marsupial mammals are that both replicate late in S phase and that both are marked by low levels of histone H4 acetylation (Wakefield et al. 1997).

In view of the male lethality for mutations that disrupt dosage compensation in other organisms, including other mammals, it is puzzling that incomplete dosage compensation is tolerated in marsupials. One possible explanation stems from the fact that the marsupial X carries fewer genes than its eutherian counterpart (Graves 1996; Marshall Graves and Shetty 2001). Only genes on the long arm of the eutherian X are present on the marsupial X. Genes on the short arm are distributed among the autosomes in marsupials. A large part of the marsupial X is thus gene-poor and constitutively heterochromatic. Perhaps this reduction in gene number has made it possible for some cell types to tolerate a relaxation of dosage compensation, while not being sufficiently great to allow the organism to dispense with dosage compensation altogether. In addition, marsupials may silence Xi more efficiently in early ontogeny, when dosage differences are likely to be most critical, as is similarly the case for imprinted loci. The two explanations may both have a role to play.

5 X-Chromosome Reactivation and Reprogramming

5.1 Stability of X Inactivation in Somatic Cells

Multiple layers of epigenetic modification contribute to the silencing of the inactive X chromosome, and as a result, the repressed state is generally highly stable. An illustration of this comes from early experiments that looked at the ability of 5-azacytidine (5azaC), an inhibitor of DNA methylation, to reverse X-chromosome silencing in XX cell lines (Mohandas et al. 1981). Sporadic reactivation of individual genes was seen to occur at a low frequency. However, analysis of cell lines in which a given gene had been reactivated revealed that other genes generally remained inactive, as did the entire chromosome as assessed at the cytogenetic level. Similar data have been obtained using TSA to inhibit type 1 histone deacetylases (Csankovszki et al. 2001).

Sporadic reactivation of individual X-linked genes has also been observed during aging in mice. This occurs even more so in marsupials, where X inactivation is not as stable and progressive reactivation occurs during ontogeny.

What about the role of *Xist* RNA? Data obtained using human cell lines demonstrated that loss of the region of the X chromosome encompassing *XIST* does not result in detectable X reactivation (Brown and Willard 1994). These results were confirmed in mouse fibroblast cell lines using a conditional knockout allele of *Xist* (Csankovszki et al. 1999). Here it was shown that loss of *Xist* results in delocalization of the variant histone macroH2A from Xi. More recent studies have demonstrated that the histone modifications catalyzed by PcG complexes, H3 lysine 27 trimethylation, and H2A ubiquitylation are also lost in

Xist conditional knockout fibroblast cell lines (Plath et al. 2004). Rare sporadic reactivation of individual genes was detected, and this was further increased by treatment with 5azaC or TSA. However, again there was no dramatic chromosome-wide reactivation. Thus, removal of multiple epigenetic silencing marks is still not sufficient to reverse chromosome-wide silencing.

The redundancy of Xi silencing mechanisms in somatic cells is often referred to as "belts and braces." By implication, it suggests that individual levels of epigenetic silencing are both self-maintaining and sustained through interrelated positive feedback mechanisms. Similar positive feedback mechanisms have been observed in other epigenetic systems; for example, linking maintenance of histone methylation and DNA methylation in pericentric heterochromatin (see Chapters 9 and 18).

5.2 X Reactivation in Normal Development

Whereas X inactivation in somatic cells is highly stable, there are circumstances in the course of normal development in which the entire X chromosome is reactivated. The best-studied example is reversal of X inactivation in developing primordial germ cells (PGCs). In mouse, PGCs are specified at about 7–8 days of development, shortly after gastrulation. At this time, cells of the embryo have already undergone random X inactivation. Subsequently, the developing PGCs migrate along the hindgut region of the embryo and arrive at the genital ridges, the structures that give rise to the adult gonads. It is at this time that XX PGCs reactivate their Xi (Monk and McLaren 1981). This event occurs coincident with a more general epigenetic reprogramming that includes erasure of parental imprints and genome-wide DNA demethylation (for more detail, see Chapter 20).

X reactivation in PGCs may indicate a specialized mechanism for reversing the multilayered heterochromatic structure. Extinction of *Xist* RNA expression has been seen to correlate with X reactivation, but given that silencing is *Xist*-independent in XX somatic cells, it is not certain that this is causative. It is possible that PGCs fail to establish all of the marks associated with silencing and are therefore more susceptible to reactivation. Consistent with this is the evidence that CpG island methylation does not occur on the Xi in developing PGCs in mouse (Grant et al. 1992).

A second example of X reactivation is the reversal of imprinted Xp inactivation during allocation of the ICM lineage of blastocyst-stage embryos, discussed in Section 3.5, which again is associated with wider genome reprogramming events. This reactivation also correlates with

extinction of *Xist* RNA and with a loss of many of the epigenetic marks associated with silencing.

5.3 X Reactivation during Experimental Reprogramming

X reactivation has also been observed under specific experimental circumstances. It occurs during nuclear transfer of somatic nuclei to unfertilized oocytes (Eggan et al. 2000) and following fusion of somatic cells with totipotent cell types such as ES, embryonic germ (EG), or embryonal carcinoma (EC) cells (see, e.g., Tada et al. 2001).

Nuclear transfer embryos provide a particularly interesting example. Experiments in mice (Eggan et al. 2000) demonstrated rapid reactivation of a marker gene on Xi in cleavage-stage nuclear transfer embryos. Despite this, the nucleus retained some memory of which X was inactive because in fetal-stage cloned embryos, the donor cell Xi was also the Xi in trophectoderm cells of the placenta. In contrast, cells of the embryo proper showed random X inactivation (see Fig. 9). Presumably, X reactivation and reprogramming that occur in the developing ICM give the embryo a second chance to reset epigenetic information from the donor nucleus.

X reactivation in fusions between XX somatic cells and pluripotent embryonic cells is less well studied. Presumably, this also occurs as a result of exposure of the somatic genome to factors present in the EC, ES, or EG cells. Reactivation has been demonstrated to occur relatively rapidly, within approximately 5 days following fusion, and high-level *Xist* expression is extinguished in fusion cells after long-term culture. Causal linkage of these events has not been demonstrated.

5.4 Lessons from Inducible Xist Transgenes

A series of experiments using inducible *Xist* transgenes in ES cells has greatly increased our understanding of stability versus reversibility of X inactivation. First, it was demonstrated that *Xist* RNA can establish X inactivation in undifferentiated ES cells and during very early stages of differentiation, but not subsequently (Wutz and Jaenisch 2000). This is referred to as the "window of opportunity." This ability of cells to respond to *Xist* RNA was found to broadly correlate with reversibility of X inactivation. Thus, silencing was reversed when the transgene was switched off in ES cells or during early differentiation stages but not in later differentiation or in somatic cells.

Returning to X reactivation and reprogramming, the inducible transgene data imply that in defined cellular environments, namely undifferentiated ES cells, X reacti-

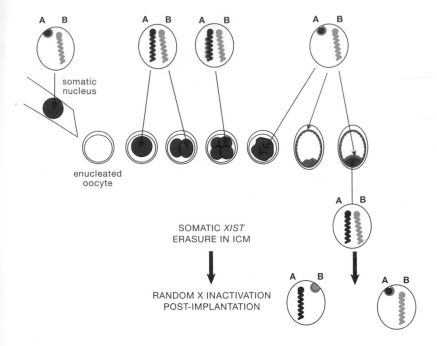

SOMATIC *XIST*
ERASURE IN ICM

↓

RANDOM X INACTIVATION
POST-IMPLANTATION

Figure 9. Regulation of X Inactivation in Cloned Mouse Embryos

The figure illustrates an XX donor cell with the inactive X chromosome (A) coated with *Xist* RNA (*green line*). In this model, transcription from the donor nucleus, including *Xist* RNA, is repressed by oocyte factors until the 2-cell stage, resulting in X reactivation. Recommencement of *Xist* expression then occurs at the 2-cell stage. *Xist* is then reexpressed, again from the inactive X allele from the donor cell. This would be attributable to retention of a mark such as DNA methylation at the *Xist* promoter. This pattern is maintained in cells allocated to the TE and PE lineages but not in pluripotent epiblast where *Xist* expression is again extinguished, leading to a second reactivation event. In the ICM, erasure of the epigenetic marks governing donor *Xist* expression allows subsequent random X inactivation in the embryo proper.

vation will occur when expression of *Xist* RNA is extinguished. If we consider that those cells in which X reactivation has been documented to occur (i.e., PGCs, ICM cells, EG and EC cells) are all similar to ES cells in terms of pluripotency and plasticity, then extinction of *Xist* expression may underlie X reactivation in all cases.

6 Summary and Future Directions

In recent years, there has been significant progress in our understanding of the molecular mechanism of X inactivation. To date, this progress has been fed by advances in related fields of epigenetic research and has, in turn, stimulated advances in other fields. An example of the latter is the growing evidence that some clusters of imprinted genes are regulated by *cis*-acting noncoding RNAs in much the same way that *Xist* regulates the X chromosome (see Chapter 19). There is every reason to think that this complementary progress will continue.

Many unanswered questions remain, however, and it is remarkable that despite over 40 years of research, we still do not understand, even in outline, the mechanisms involved in "counting" and "choice." The blocking factor hypothesis, now over 20 years old, provides an attractive conceptual guide, but the nature of the blocking factor itself, if it exists, remains elusive. Progress has been made in defining the *cis*-acting sequences and *trans*-acting factors that regulate counting, and their further elucidation provides an exciting challenge. Similarly, although we now know some of the chromatin-modifying complexes

involved in maintaining X inactivation, for example, the Polycomb-group complexes, the signal for establishing chromosome-wide silencing, triggered by *Xist* RNA, remains unknown. Possibly linked to this, we need to understand the mechanism of histone deacetylation and demethylation on the inactive X. Other key questions are to understand how silencing spreads across the chromosome and what role, if any, way stations (perhaps LINE elements) play in this process and in the stabilization/maintenance of the silent state. This may relate to the intriguing question of how X inactivation is reversed in some cell types and stages of development, but is essentially irreversible in others. This latter question relates to the wider and crucially important issue of understanding genome plasticity and reprogramming through development.

References

Beutler E., Yeh M. and Fairbanks V.F. 1962. The normal human female as a mosaic of X-chromosome activity: Studies using the gene for glucose-6-phosphate dehydrogenase deficiency as a marker. *Proc. Natl. Acad. Sci.* **48:** 9–16.

Brockdorff N., Ashworth A., Kay G.F., McCabe V.M., Norris D.P., Cooper P.J., Swift S., and Rastan S. 1992. The product of the mouse *Xist* gene is a 15 kb inactive X-specific transcript containing no conserved ORF and located in the nucleus. *Cell* **71:** 515–526.

Brown C.J. and Willard H.F. 1994. The human X-inactivation centre is not required for maintenance of X-chromosome inactivation. *Nature* **368:** 154–156.

Brown C.J., Ballabio A., Rupert J.L., Lafreniere R.G., Grompe M., Tonlorenzi R., and Willard H.F. 1991. A gene from the region of the human X inactivation centre is expressed exclusively from the inactive X chromosome. *Nature* **349:** 38–44.

Brown C.J., Hendrich B.D., Rupert J.L., Lafreniere R.G., Xing Y., Lawrence J., and Willard H.F. 1992. The human *XIST* gene: Analysis of a 17 kb inactive X-specific RNA that contains conserved repeats and is highly localized within the nucleus. *Cell* **71:** 527–542.

Carrel L. and Willard H.F. 2005. X-inactivation profile reveals extensive variability in X-linked gene expression in females. *Nature* **434:** 400–404.

Cattanach B.M., and Isaacson J.H. 1967. Controlling elements in the mouse X chromosome. *Genetics* **57:** 331–346.

Chadwick B.P. and Willard H.F. 2004. Multiple spatially distinct types of facultative heterochromatin on the human inactive X chromosome. *Proc. Natl. Acad. Sci.* **101:** 17450–17455.

Clerc P. and Avner P. 1998. Role of the region 3′ to *Xist* exon 6 in the counting process of X-chromosome inactivation. *Nat. Genet.* **19:** 249–253.

Charlesworth B. 1996. The evolution of chromosomal sex determination and dosage compensation. *Curr. Biol.* **6:** 149–162.

Costanzi C. and Pehrson J.R. 1998. Histone macroH2A is concentrated in the inactive X chromosome of female mammals. *Nature* **393:** 599–601.

Csankovszki G., Nagy A., and Jaenisch R. 2001. Synergism of Xist RNA, DNA methylation, and histone hypoacetylation in maintaining X chromosome inactivation. *J. Cell Biol.* **153:** 773–784.

Csankovszki G., Panning B., Bates B., Pehrson J.R., and Jaenisch R. 1999. Conditional deletion of *Xist* disrupts histone macroH2A localization but not maintenance of X inactivation. *Nat. Genet.* **22:** 323–324.

Davidson R.G., Nitowsky H.M., and Childs B. 1963. Demonstration of two populations of cells in the human female heterozygous for glucose-6-phosphate dehydrogenase variants. *Proc. Natl. Acad. Sci.* **50:** 481–485.

de Napoles M., Mermoud J.E., Wakao R., Tang Y.A., Endoh M., Appanah R. Nesterova T.B., Silva J., Otte A.P., Vidal M., et al. Polycomb group proteins Ring1A/B link ubiquitylation of histone H2A to heritable gene silencing and X inactivation. *Dev. Cell* **7:** 663–676.

Duthie S.M., Nesterova T.B., Formstone E.J., Keohane A.M., Turner B.M., Zakian S.M., and Brockdorff N. 1999. *Xist* RNA exhibits a banded localization on the inactive X chromosome and is excluded from autosomal material in *cis. Hum. Mol. Genet.* **8:** 195–204.

Eggan K., Akutsu H., Hochedlinger K., Rideout W., III, Yanagimachi R., and Jaenisch R. 2000. X-chromosome inactivation in cloned mouse embryos. *Science* **290:** 1578–1581.

Eils R., Dietzel S., Bertin E., Schrock E., and Speicher M.R. 1996. Three dimensional reconstruction of painted human interphase chromosomes: Active and inactive X chromosome territories have similar volumes but differ in shape and surface structure. *J. Cell. Biol.* **135:** 1427–1440.

Epstein C.J., Smith S., Travis B., and Tucker G. 1978. Both X chromosomes function before visible X-chromosome inactivation in female mouse embryos. *Nature* **274:** 500–502.

Gartler S.M. and Riggs A.D. 1983. Mammalian X-chromosome inactivation. *Annu. Rev. Genet.* **17:** 155–190.

Grant M., Zuccotti M., and Monk M. 1992. Methylation of CpG sites of two X-linked genes coincides with X-inactivation in the female mouse embryo but not in the germ line. *Nat. Genet.* **2:** 161–166.

Graves J.A. 1996. Mammals that break the rules: Genetics of marsupials and monotremes. *Annu. Rev. Genet.* **30:** 233–260.

Heard E., Mongelard F., Arnauld D., Chureau C., Vourch C., and Avner P. 1999. Human *XIST* yeast artificial chromosome transgenes show partial X inactivation center function in mouse embryonic stem cells. *Proc. Natl. Acad. Sci.* **96:** 6841–6846.

Heard E., Rougeulle C., Arnaud D., Avner P., Allis C.D., and Spector D.L. 2001. Methylation of histone H3 at lys-9 is an early mark on

the X chromosome during X inactivation. *Cell* **107:** 727–738.

Huynh K.D. and Lee J.T. 2003. Inheritance of a pre-inactivated paternal X chromosome in early mouse embryos. *Nature* **426:** 857–862.

Jeppesen P. and Turner B.M. 1993. The inactive X chromosome in female mammals is distinguished by a lack of histone H4 acetylation: A cytogenetic marker for gene expression. *Cell* **74:** 281–289.

Keohane A.M., Belyaev N.D., Lavender J.S., O'Neill L.P., and Turner B.M. 1996. X-inactivation and H4 acetylation in embryonic stem cells. *Dev. Biol.* **180:** 618–630.

Lachner M., O'Sullivan R.J., and Jenuwein T. 2003. An epigenetic road map for histone lysine methylation. *J. Cell Sci.* **116:** 2117–2124.

Lee J.T. 2000. Disruption of imprinted X inactivation by parent-of-origin effects at *Tsix. Cell* **103:** 17–27.

Lee J.T. and Lu N.F. 1999. Targeted mutagenesis of *Tsix* leads to nonrandom X inactivation. *Cell* **99:** 47–57.

Lee J.T., Davidow L.S., and Warshawsky D. 1999. *Tsix*, a gene antisense to *Xist* at the X-inactivation centre. *Nat. Genet.* **21:** 400–404.

Lee J.T., Strauss W.M., Dausman J.A., and Jaenisch R. 1996. A 450 kb transgene displays properties of the mammalian X-inactivation center. *Cell* **86:** 83–94.

Lock L.F., Takagi N., and Martin G.R. 1987. Methylation of the *Hprt* gene on the inactive X occurs after chromosome inactivation. *Cell* **48:** 39–46.

Lyon M.F. 1961. Gene action in the X-chromosome of the mouse (*Mus musculus* L.). *Nature* **190:** 372–373.

———. 1998. X-chromosome inactivation: A repeat hypothesis. *Cytogenet. Cell Genet.* **80:** 133–137.

———. 2003. The Lyon and the LINE hypothesis. *Semin. Cell Dev. Biol.* **14:** 313–318.

Maison C., Bailly D., Peters A.H., Quivy J.P., Roche D., Taddei A., Lachner M., Jenuwein T., and Almouzni G. 2002. Higher-order structure in pericentric heterochromatin involves a distinct pattern of histone modification and an RNA component. *Nat. Genet.* **30:** 329–334.

Mak W., Nesterova T.B., de Napoles M., Appanah R., Yamanaka S., Otte A.P., and Brockdorff N. 2004. Reactivation of the paternal X chromosome in early mouse embryos. *Science* **303:** 666–669.

Marin I., Siegal M.L., and Baker B.S. 2000. The evolution of dosage-compensation mechanisms. *BioEssays* **22:** 1106–1114.

Marshall Graves J.A. and Shetty S. 2001. Sex from W to Z: Evolution of vertebrate sex chromosomes and sex determining genes. *J. Exp. Zool.* **290:** 449–462.

McElreavey K., Barbaux S., Ion A., and Fellous M. 1995. The genetic basis of murine and human sex determination: A review. *Heredity* **75:** 599–611.

Mermoud J.E., Costanzi C., Pehrson J.R., and Brockdorff N. 1999. Histone MacroH2A1.2 relocates to the inactive X chromosome after initiation and propagation of X-inactivation. *J. Cell Biol.* **147:** 1399–1408.

Mohandas T., Sparkes R.S., and Shapiro L.J. 1981. Reactivation of an inactive human X chromosome: Evidence for X inactivation by DNA methylation. *Science* **211:** 393–396.

Monk M. and Harper M.I. 1979. Sequential X chromosome inactivation coupled with cellular differentiation in early mouse embryos. *Nature* **281:** 311–313.

Monk M. and McLaren A. 1981. X-chromosome activity in foetal germ cells of the mouse. *J. Embryol. Exp. Morphol.* **63:** 75–84.

Nesterova T.B., Johnston C.M., Appanah R., Newall A.E.T., Godwin J., Alexiou M., and Brockdorff N. 2003. Skewing X chromosome choice by modulating sense transcription across the *Xist* locus. *Genes Dev.* **17:** 2177–2190.

Norris D.P., Patel D., Kay G.F., Penny G.D., Brockdorff N., Sheardown

S.A., and Rastan S. 1994. Evidence that random and imprinted *Xist* expression is controlled by preemptive methylation. *Cell* **77:** 41–51.

Ohno S. 1967. *Sex chromosomes and sex-linked genes.* Springer-Verlag, Berlin pp. 1–140.

Okamoto I., Otte A.P., Allis C.D., Reinberg D., and Heard E. 2004. Epigenetic dynamics of imprinted X inactivation during early mouse development. *Science* **303:** 644–649.

Okamoto I., Arnaud D., Le Baccon P., Otte A.P., Disteche C.M., Avner P., and Heard E. 2005. Evidence for de novo imprinted X-chromosome inactivation independent of meiotic inactivation in mice. *Nature* **438:** 297–298.

O'Neill L.P., Randall T.E., Lavender J., Spotswood H.T., Lee J.T., and Turner B.M. 2003. X-linked genes in female embryonic stem cells carry an epigenetic mark prior to the onset of X inactivation. *Hum. Mol. Genet.* **12:** 1783–1790.

O'Neill L.P., Keohane A.M., Lavender J.S., McCabe V., Heard E., Avner P., Brockdorff N., and Turner B.M. 1999. A developmental switch in H4 acetylation upstream of *Xist* plays a role in X chromosome inactivation. *EMBO J.* **18:** 2897–2907.

Panning B., Dausman J. and Jaenisch R. 1997. X chromosome inactivation is mediated by *Xist* RNA stabilization. *Cell* **90:** 907–916.

Penny G.D., Kay G.F., Sheardown S.A., Rastan S., and Brockdorff N. 1996. Requirement for *Xist* in X chromosome inactivation. *Nature* **379:** 131–137.

Perche P.Y., Vourc'h C., Konecny L., Souchier C., Robert-Nicoud M., Dimitrov S., and Khochbin S. 2000. Higher concentrations of histone macroH2A in the Barr body are correlated with higher nucleosome density. *Curr Biol.* **10:** 1531-1534.

Plath K., Talbot D., Hamer K.M., Otte A.P., Yang T.P., Jaenisch R., and Panning B. 2004. Developmentally regulated alterations in Polycomb repressive complex 1 proteins on the inactive X chromosome. *J. Cell Biol.* **167:** 1025–1035.

Plath K., Fang J., Mlynarczyk-Evans S.K., Cao R., Worringer K.A., Wang H., de la Cruz C.C., Otte A.P., Panning B., and Zhang Y. 2003. Role of histone H3 lysine 27 methylation in X inactivation. *Science* **300:** 131–135.

Rasmussen T.P., Wutz A.P., Pehrson J.R., and Jaenisch R.R. 2001. Expression of *Xist* RNA is sufficient to initiate macrochromatin body formation. *Chromosoma* **110:** 411–420.

Rastan S. 1983. Non-random inactivation in mouse X-autosome translocation embryos—Location of the inactivation centre. *J. Embryol. Exp. Morphol.* **78:** 1–22.

Rougeulle C., Chaumeil J., Sarma K., Allis C.D., Reinberg D., Avner P., and Heard E. 2004. Differential histone H3 Lys-9 and Lys-27 methylation profiles on the X chromosome. *Mol. Cell. Biol.* **24:** 5475–5484.

Sado T., Okano M., Li E., and Sasaki H. 2004. De novo DNA methylation is dispensable for the initiation and propagation of X chromosome inactivation. *Development* **131:** 975–982.

Sado T., Fenner M.H., Tan S.S., Tam P., Shioda T., and Li E. 2000. X inactivation in the mouse embryo deficient for *Dnmt1*: Distinct effect of hypomethylation on imprinted and random X inactiva-

tion. *Dev. Biol.* **225:** 294–303.

Sharman G.B. 1971. Late DNA replication in the paternally derived X chromosome of female Kangaroos. *Nature* **230:** 231–232.

Sharp A.J., Spotswood H.T., Robinson D.O., Turner B.M., and Jacobs P.A. 2002. Molecular and cytogenetic analysis of the spreading of X inactivation in X;autosome translocations. *Hum. Mol. Genet.* **11:** 3145–3156.

Sheardown S., Norris D., Fisher A., and Brockdorff N. 1996. The mouse *Smcx* gene exhibits developmental and tissue specific variation in degreee of escape from X inactivation. *Hum. Mol. Genet.* **5:** 1355–1360.

Sheardown S.A., Duthie S.M., Johnston C.M., Newall A.E.T., Formstone E.J., Arkell R.M., Nesterova T.B., Alghisi G-C., Rastan S., and Brockdorff N. 1997. Stabilization of *Xist* RNA mediates initiation of X chromosome inactivation. *Cell* **91:** 99–107.

Silva J., Mak W., Zvetkova I., Appanah R., Nesterova T.B., Webster Z., Peters A.H.F.M., Jenuwein T., Otte A.P., and Brockdorff N. 2003. Establishment of histone H3 methylation on the inactive X chromosome requires transient recruitment of Eed-Enx1 polycomb group complexes. *Dev. Cell* **4:** 481–495.

Smith K.P., Byron M., Clemson C.M., and Lawrence J.B. 2004. Ubiquitinated proteins including uH2A on the human and mouse inactive X chromosome: Enrichment in gene rich bands. *Chromosoma* **113:** 324–335.

Tada M., Takahama Y., Abe K., Nakatsuji N., and Tada T. 2001. Nuclear reprogramming of somatic cells by in vitro hybridization with ES cells. *Curr. Biol.* **11:** 1553–1558.

Tada T., Obata Y., Tada M., Goto Y., Nakatsuji N., Tan S.S., Kono T., and Takagi N. 2000. Imprint switching for non-random X-chromosome inactivation during mouse oocyte growth. *Development* **127:** 3101–3105.

Takagi N. and Sasaki M. 1975. Preferential inactivation of the paternally derived X chromosome in the extraembryonic membranes of the mouse. *Nature* **256:** 640–642.

Turner J.M., Mahadevaiah S.K., Fernandez-Capetillo O., Nussenzweig A., Xu X., Deng C.X., and Burgoyne P.S. 2005. Silencing of unsynapsed meiotic chromosomes in the mouse. *Nat. Genet.* **37:** 41–47.

Wakefield M.J., Keohane A.M., Turner B.M., and Marshall Graves J.A. 1997. Histone underacetylation is an ancient component of mammalian X chromosome inactivation. *Proc. Natl. Acad. Sci.* **94:** 9665–9668.

White W.M., Willard H.F., Van Dyke D.L., and Wolff D.J. 1998. The spreading of X inactivation into autosomal material of an X; autosome translocation: Evidence for a difference between autosomal and X chromosomal DNA. *Am. J. Hum. Genet.* **63:** 20–28.

Wutz A. and Jaenisch R. 2000. A shift from reversible to irreversible X inactivation is triggered during ES cell differentiation. *Mol. Cell* **5:** 695–705.

Wutz A., Rasmussen T.P., and Jaenisch R. 2002 Chromosomal silencing and localization are mediated by different domains of *Xist* RNA. *Nat. Genet.* **30:** 167–174.

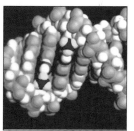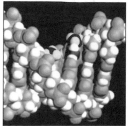

DNA Methylation in Mammals

En Li[1] and Adrian Bird[2]

[1]Novartis Institutes for BioMedical Research, Inc., Cambridge, Massachusetts 02139
[2]The Wellcome Trust Centre for Cell Biology, University of Edinburgh, Edinburgh, EH9 3JR, United Kingdom

CONTENTS

1. A Mechanism of Cell Memory, 343
   1.1 The Hypothesis, 343
   1.2 Evidence for Heritable Methylation Patterns, 343
   1.3 The Mammalian Maintenance DNA Methyltransferase, 343

2. The Origin of DNA Methylation Patterns, 344
   2.1 De Novo Methylation of DNA in Early Embryos, 344
   2.2 Discovery of De Novo Methyltransferases, 344
   2.3 CpG Islands and Patterns of DNA Methylution, 345
   2.4 Dynamic Changes in DNA Methylation Patterns during Development, 346
   2.5 Active Demethylation of the Zygotic Paternal Genome, 346
   2.6 What Protects CpG Islands from DNA Methylation?, 347
   2.7 DNA Methylation Triggered by Chromatin Structure?, 347
   2.8 The Role of SWI/SNF-like Chromatin-remodeling Proteins, 348

3. Regulation of Gene Expression by DNA Methylation, 348
   3.1 Early Evidence, 348
   3.2 Interference with Transcription Factor Binding, 348
   3.3 Attraction of Methyl-CpG-binding Proteins, 349
   3.4 MeCP2 and Rett Syndrome, 350
   3.5 MBD2 Mediates Methylation-dependent Transcriptional Repression, 351

4. DNA Methylation, Mutation, and Chromosomal Stability, 352
   4.1 DNA Methylation and Mutation, 352
   4.2 DNA Methylation and Chromosome Instability, 352

5. Future Directions, 352
   5.1 Environmental Factors That Induce Epigenetic Changes, 353
   5.2 Epigenetic Instability and Complex Diseases, 353
   5.3 Modulation of Reversible Epigenetic States, 353

References, 353

GENERAL SUMMARY

The DNA of vertebrate animals is covalently modified by methylation of the cytosine base in the dinucleotide sequence 5'CpG3'. CpG is an abbreviation for cytosine and guanine separated by a phosphate, which links the two nucleotides together in DNA. In mammals, DNA methylation patterns are established during embryonic development and maintained by a copying mechanism when cells divide. The heritability of DNA methylation patterns allows epigenetic marking of the genome to be stable through multiple cell divisions and therefore constitutes a form of cellular memory.

Molecular and genetics studies have shown that DNA cytosine methylation is associated with gene silencing and plays an important role in developmental processes such as X-chromosome inactivation and genomic imprinting. The methyl moiety of methyl cytosine resides in the major groove of the DNA helix, where many DNA-binding proteins make contact with DNA, and exerts its effect by attracting or repelling various DNA-binding proteins. A family of proteins that can bind to DNA containing methylated CpG dinucleotides, known as methyl-CpG-binding proteins, have been shown to recruit repressor complexes to methylated promoter regions and thereby contribute to transcriptional silencing. Certain transcription factors bind to CpG-containing DNA sequences only when they are nonmethylated. In these cases, CpG methylation can prevent protein binding and affect transcription.

Genetic ablation of genes encoding DNA methyltransferases or methyl-CpG-binding proteins has revealed diverse functions for DNA methylation in mammalian development. Establishment of normal methylation patterns of the genome is essential for embryonic development. DNA methylation is required for maintaining differential expression of the paternal and maternal copies of genes that are subjected to genomic imprinting and for stable silencing of genes on the inactive X chromosome. In addition, stable transcriptional repression of proviral genomes and endogenous retrotransposons depends on DNA methylation. We know examples of involvement of DNA methylation in the establishment and maintenance of tissue-specific gene expression patterns during development. There is also evidence that absence of DNA methylation compromises the faithful maintenance of chromosome number, leading to an increased frequency of chromosome loss.

The clinical relevance of DNA methylation first became apparent in relation to cancer. Reduced levels of DNA methylation, due to either genetic manipulation or treatment of DNA methyltransferase inhibitors, lead to suppression of some forms of tumors in mouse models of cancer. In contrast, formation of other tumor types is enhanced by low levels of DNA methylation. Several other human diseases have been linked to mutations of genes that encode critical components of the DNA methylation machinery. Mutations of the DNA methyltransferase Dnmt3b lead to immune deficiency, and mutations of the methyl-CpG-binding protein MeCP2 cause a severe neurological disorder known as Rett syndrome. It is apparent that the integrity of the DNA methylation system is of paramount importance for the health of mammals.

Although DNA methylation patterns can be transmitted from cell to cell, they are not permanent. In fact, changes in DNA methylation patterns can occur throughout the life of an individual. Some changes might be a physiological response to environmental changes, whereas others might be associated with a pathological process such as oncogenic transformation or cellular aging. The intrinsic and environmental factors that induce DNA methylation changes, however, remain largely unknown. The study of DNA methylation in human disease represents an important frontier in medicine and will contribute to our understanding of the impact of epigenetic modification on human life.

1 A Mechanism of Cell Memory

1.1 The Hypothesis

Cytosine methylation occurs predominantly in CpG dinucleotides in mammalian cells (Fig. 1). The idea that DNA methylation in animals could represent a mechanism of cell memory arose independently in two laboratories (Holliday and Pugh 1975; Riggs 1975). Recognizing that the CpG dinucleotide is self-complementary, both groups reasoned that patterns of methylated and nonmethylated CpG could be copied when cells divide. Immediately after replication of DNA, the parental DNA strand would maintain its pattern of modified cytosines, but the newly synthesized strand would be unmodified. To ensure copying of the parental pattern onto the progeny strand, they postulated a "maintenance methyltransferase" that would exclusively methylate CpGs base-paired with a methylated parental CpG. Nonmethylated CpGs would not be substrates for the maintenance methyltransferase (see Fig. 2). The consequence of this simple mechanism is that patterns of DNA methylation would be replicated semiconservatively, like the base sequence of DNA itself.

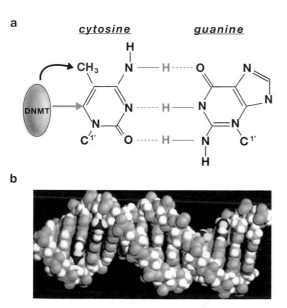

Figure 1. Cytosine Methylation in DNA

(*a*) Addition of a methyl group (*red*) at the 5 position of the cytosine pyrimidine ring (*black arrow*) does not sterically interfere with GC base-pairing (*blue lines*). DNA methyltransferases associate covalently with the carbon-6 position (*green arrow*) during methyl group transfer. (*b*) A model of B-form DNA methylated at cytosines in two self-complementary CpG sequences. The paired methyl moieties (*magenta* and *yellow*) lie in the major groove of the double helix.

1.2 Evidence for Heritable Methylation Patterns

The pattern of DNA methylation at a single genomic locus was initially established using DNA-methylation-sensitive restriction endonucleases. Many of these enzymes are prevented from cleaving their cognate recognition sequences in DNA if there is methylation of a specific base. A pattern of methylated and nonmethylated sites was mapped within the ribosomal RNA genes of *Xenopus laevis* using such enzymes which are known to cleave at sites containing CpG, but are blocked by cytosine methylation (Bird and Southern 1978). It was found that on a particular DNA strand, most CpGs were methylated, but that nonmethylated sites occurred at random. Significantly, these random sites were always symmetrical. In other words, either both CpGs in a complementary pair were methylated, or neither was methylated (Bird 1978). This finding fitted well with the predictions of the maintenance model (Holliday and Pugh 1975; Riggs 1975). A more direct test of the heritability of DNA methylation patterns was made possible by transfection of artificially methylated DNA into cultured cells (Wigler 1981). Nonmethylated plasmids usually did not become methylated, even after many cell generations. Plasmids that had been methylated at CCGG sites by the methyltransferase M.HpaII, however, retained their methylated status for many generations, although the fidelity of the process, in cultured cells at least, was less than 100%.

1.3 The Mammalian Maintenance DNA Methyltransferase

Progress in understanding any biological process at the molecular level depends on isolation of the key players. DNA methyltransferase activity was detected early in crude cellular extracts, but was finally purified as a 200-kD protein (Bestor and Ingram 1983). The enzyme Dnmt1 is specific for CpG and has significant activity against nonmethylated DNA. Its preferred DNA substrate, however, is DNA methylated at CpG on one strand only (so-called hemimethylated DNA). Because of this property, it seemed possible that this was the maintenance DNA methyltransferase. Subsequent studies have provided strong support for this view, because inactivation of Dnmt1 in mouse embryonic stem cells (Table 1) leads to genome-wide loss of CpG methylation (Li et al. 1992). The available evidence fits with the view that Dnmt1 maintains DNA methylation at CpG by completing hemimethylated sites, as postulated by Riggs and Holliday and Pugh (Fig. 2).

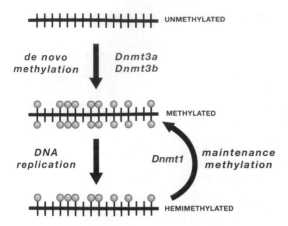

Figure 2. De novo Methylation and Maintenance Methylation of DNA

A stretch of genomic DNA is shown as a line with self-complementary CpG pairs marked as vertical strokes. Unmethylated DNA (*top*) becomes methylated "de novo" by Dnmt3a and Dnmt3b to give symmetrical methylation at certain CpG pairs. Upon semiconservative DNA replication, a progeny DNA strand is base-paired with one of the methylated parental strands (the other replication product is not shown). Symmetry is restored by the maintenance DNA methyltransferase Dnmt1, which completes half-methylated sites, but does not methylate unmodified CpGs.

2 The Origin of DNA Methylation Patterns

2.1 De Novo Methylation of DNA in Early Embryos

We need to understand not only how DNA methylation patterns are stably maintained, but also how they arise in the first place. Early cell-transfection experiments showed that nonmethylated DNA introduced into cultured somatic cells tended to remain in a nonmethylated state after many cell divisions. On the other hand, retroviral proviruses and other transgenes introduced into

mouse preimplantation embryos became stably methylated in cells of the animal (Jahner et al. 1982). This suggested that the process of de novo methylation of DNA is confined to totipotent stages of embryogenesis. This idea was tested by using mouse embryonal carcinoma (EC) cells, and subsequently, embryonic stem (ES) cells, as a model system. Retroviral DNA was noninfectious when introduced into these cells, in contrast to somatic cells, which supported the infectious cycle of the virus (Stewart et al. 1982). In addition, the proviral DNA became methylated at CpGs, and depletion of methylation by cloning their methylated genomes in bacteria (thereby erasing DNA methylation) restored the capacity for viral gene expression. These results showed that DNA methylation can silence expression of the viral genome in vivo, and also that embryonic cells do indeed have the capability to de novo methylate DNA. Initially, only one DNA methyltransferase, Dnmt1, was known, and it was therefore considered that this enzyme was able to de novo methylate DNA at these developmental stages. Deletion of the Dnmt1 gene, however, did not interfere with the de novo methylation of proviral DNA in ES cells (Lei et al. 1996), proving that other DNA methyltransferases are at work.

2.2 Discovery of De Novo Methyltransferases

All known prokaryotic cytosine DNA methyltransferases share a set of diagnostic protein motifs (Posfai et al. 1989). These features are also found in the maintenance DNA methyltransferase Dnmt1. Searches of expressed sequence tag (EST) databases showed three transcripts that could potentially encode additional DNA methyl-

Table 1. Functions of mammalian DNA methyltransferases

DNA methyltransferase	Species	Major activity	Major phenotypes of loss-of-function mutations
Dnmt1	mouse	maintenance methylation of CpG	genome-wide loss of DNA methylation, embryonic lethality at embryonic day 9.5 (E9.5), abnormal expression of imprinted genes, ectopic X-chromosome inactivation, activation of silent retrotransposon
Dnmt2	mouse	weak activity	no change in CpG methylation, no obvious developmental phenotypes
Dnmt3a	mouse	de novo methylation of CpG	postnatal lethality at 4–8 weeks, male sterility, and failure to establish methylation imprints in both male and female germ cells
Dnmt3b	mouse	de novo methylation of CpG	demethylation of minor satellite DNA, embryonic lethality around E14.5 days with vascular and liver defects (embryos lacking both Dnmt3a and Dnmt3b fail to initiate de novo methylation after implantation and die at E9.5)
DNMT3B	human	de novo methylation of CpG	ICF syndrome: immunodeficiency, centromeric instability, and facial anomalies; loss of methylation in repetitive elements and pericentromeric heterochromatin

transferases (Fig. 3). One candidate protein, Dnmt2, has minimal DNA methyltransferase activity in vitro, and its absence has no discernable effect on levels of DNA methylation. The other two candidates, Dnmt3a and Dnmt3b, encoded related catalytically active polypeptides that, unlike Dnmt1, showed no preference for methylating hemimethylated DNA in vitro (Okano et al. 1998). Disruption of the genes for Dnmt3a and Dnmt3b in mice confirmed that these constituted the missing de novo methyltransferases (Table 1), because ES cells and embryos lacking both proteins were unable to de novo methylate proviral genomes and repetitive elements (Okano et al. 1999). Furthermore, the male and female germ cells lacking Dnmt3a protein or an associated regulatory factor, Dnmt3L, fail to establish distinct methylation patterns at imprinted genes (Table 1) (Hata et al. 2002; Kaneda et al. 2004). Whereas inactivation of both Dnmt3a and Dnmt3b resulted in early embryonic lethality, loss of either gene caused severe defects that resulted in postnatal (Dnmt3a) or embryonic (Dnmt3b) lethality. Evidence for a related role in humans emerged with the discovery that ICF syndrome, a rare condition characterized by *i*mmunodeficiency, *c*entromeric instability, and *f*acial abnormalities, was associated with mutations in the gene for Dnmt3b (Ehrlich 2003). Analysis of genomic lymphoblastoid cell DNA from ICF patients showed reduced genomic methylation, specifically in repetitive DNA sequences that are associated with pericentromeric regions of the chromosomes.

2.3 CpG Islands and Patterns of DNA Methylation

DNA from mammalian somatic tissues is methylated at 70% of all CpG sites (Ehrlich 1982). Mapping studies (see Box on next page) indicate that highly methylated sequences include satellite DNAs, repetitive elements including transposons and their inert relics, nonrepetitive intergenic DNA, and exons of genes. Among these DNA sequence categories, there appears to be no reliable preference for methylating one type of sequence rather than another. CpGs in satellite DNAs are methylated to broadly the same degree as those in transposable elements or exons. Thus, most sequences are methylated according to their frequency of CpG dinucleotides, which usually reflects their base composition. Key exceptions to this global methylation of the mammalian genome are the CpG islands. CpG islands were detected as a fraction of vertebrate DNA that was cleaved unusually frequently by DNA methylation-sensitive restriction enzymes (Cooper et al. 1983). Cloning of the so-called "HpaII tiny fragments" showed that they were derived from GC-rich sequences of about 1 kb in length that are nonmethylated in germ cells, in the early embryo, and usually also in all somatic tissues (Bird et al. 1985). CpG islands (Fig. 4) are therefore exceptions to the "global" CpG methylation that prevails throughout most of the mammalian genome. Early mapping of individual gene promoters identified GC-rich regions near gene promoters (McKeon et al. 1982), and it is now evident that most (if not all) CpG islands mark the

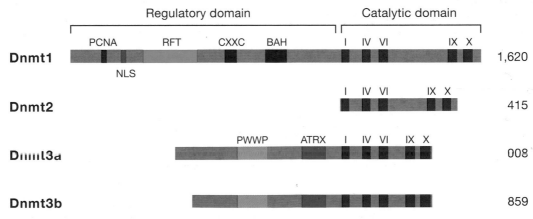

Figure 3. Mammalian DNA Methyltransferases

The catalytic domains of Dnmt1, Dnmt2, and the Dnmt3 family members are conserved (the signature motifs, I, IV, VI, IX, and X, are most conserved in all cytosine methyltransferases), but there is little similarity among their amino-terminal regulatory domains. (PCNA) PCNA-interacting domain; (NLS) nuclear localization signal; (RFT) replication foci-targeting domain; (CXXC) a cysteine-rich domain implicated in binding DNA sequences containing CpG dinucleotides; (BAH) bromo-adjacent homology domain implicated in protein–protein interactions; (PWWP) a domain containing a highly conserved "proline-tryptophan-tryptophan-proline" motif involved in heterochromatin association; (ATRX) an ATRX-related cysteine-rich region containing a C2-C2 zinc finger and an atypical PHD domain implicated in protein–protein interactions.

MAPPING DNA METHYLATION

To understand the functions of DNA methylation, it is first necessary to find out where it occurs in the genome. There are several methods of doing this, each with advantages and disadvantages.

- As methylation is mainly confined to CpG sequences, cleavage by restriction enzymes that recognize a CpG-containing DNA sequence has been extensively used for mapping (Bird and Southern 1978). This method has the advantage that large regions of the genome can be assayed, but it is limited to CpGs that are found within restriction enzyme sites.

- A reliable method for testing all cytosines within a region involves the **bisulfite modification** of single-stranded DNA (Frommer et al. 1992). This leads to deamination of unmodified cytosines, but 5-methylcytosine is protected. As a result, cytosines that survive bisulfite treatment are identified as methylated. Due to its high resolution and positive identification of methylated cytosine, this is the method of choice for analyzing DNA methylation patterns, although thorough analysis of large regions is time-consuming.

- Several **PCR-based methods** that depend on prior bisulfite treatment of DNA have been developed to accelerate the analysis of regions of interest (see, e.g., Herman et al. 1996). These methods are highly convenient, but by focusing on a few CpG sites within a region, they sacrifice the detailed information that would be revealed by bisulfite sequencing.

- Use of **microarrays** has recently been adapted for mapping DNA methylation. For example, DNA that is resistant to degradation by the 5-methylcytosine-specific nuclease McrBC can be probed against tiled arrays of genomic DNA sequences to give an overview of the methylation level across a specific region (Martienssen et al. 2005). Probes for tiled arrays can also be immunoprecipitated using 5-methylcytosine-specific antibodies, allowing a global survey of DNA methylation levels (Weber et al. 2005).

promoters and 5′ domains of genes. Approximately 60% of human genes have CpG island promoters.

2.4 Dynamic Changes in DNA Methylation Patterns during Development

DNA methylation patterns show apparent overall constancy when different somatic cell types are examined, but local changes are evident at specific DNA sequences. For example, CpG islands on one X chromosome become de novo methylated in large numbers during the embryonic process of X-chromosome inactivation in female placental mammals (Wolf et al. 1984). This process is essential for the leakproof silencing of genes on the inactivated chromosome (Fig. 4), because DNA methylation-deficient mice or cells show frequent transcriptional reactivation of X-linked genes. Programmed loss of DNA methylation is also implicated in the transcriptional activation of certain genes during differentiation. For example, the *interleukin-2* gene loses CpG methylation in its promoter region as the gene becomes expressed during T-cell differentiation (Bruniquel and Schwartz 2003). This demethylation event is an essential precondition for activation of the gene and is therefore a key part of the T-cell differentiation program.

2.5 Active Demethylation of the Zygotic Paternal Genome

In addition to these local changes in DNA methylation, there is a dramatic alteration in global DNA methylation levels in the fertilized egg. Analysis of chromosomal DNA methylation levels using an anti-5-methylcytosine antibody initially showed that one chromosome set was strikingly deficient in DNA methylation during the early embryonic cleavage stages (Rougier et al. 1998). The origin of this difference was discovered by examining the zygote prior to fusion of maternal and paternal pronuclei using 5-methylcytosine immunostaining (Mayer et al. 2000). At first, both maternal and paternal pronuclei showed equivalent staining, suggesting that the genomes of the egg and the sperm carried roughly equivalent levels of cytosine methylation. A few hours after fertilization, however, a dramatic loss of 5-methylcytosine exclusively from the paternal genome was observed. Bisulfite sequencing confirmed that one of the two genomes was indeed demethylated. The mechanism by which DNA methylation is lost is unknown, but it must involve "active" removal of the modification, because no DNA replication occurs during this period. Interestingly, the

CpG ISLAND PROMOTER

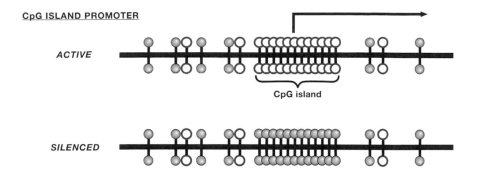

ACTIVE

CpG island

SILENCED

Figure 4. CpG Islands

CpG islands are regions of high CpG density that lack CpG methylation found at promoters of most human genes. Long-term silencing of the gene can be ensured by methylation of the CpG island region. For example, genes on the inactive X chromosome and certain imprinted genes are silenced in this way. Additionally, in cancer cells, certain genes are aberrantly silenced by CpG island methylation.

maternal genome also loses DNA methylation during early embryogenesis, but in this case the process is "passive," due to absence of maintenance DNA methylation during the early cleavage divisions. Altogether, it is thought that more than half of all genomic methylation is removed during this period. Re-methylation dependent on Dnmt3a and Dnmt3b then occurs at implantation.

Active demethylation of DNA has been reported in several other cases. For example, the demethylation of the *interleukin-2* promoter in differentiating T-helper cells (see above and Bruniquel and Schwartz 2003) occurs rapidly and in the absence of DNA replication. A comparable effect has been reproduced artificially in the frog oocyte system, where a silenced, methylated mammalian *Oct-4* gene can be transcriptionally reactivated via a process that depends on prior demethylation of the gene (Simonsson and Gurdon 2004). The stage appears to be set for isolation of the demethylase enzyme itself.

2.6 What Protects CpG Islands from DNA Methylation?

Studies of DNA methylation patterns have focused on the question of how mammalian CpG islands normally remain immune to otherwise global DNA methylation. The simplest possible explanation is (1) *CpG islands are intrinsically unmethylatable by the existing de novo DNA methyltransferases*, but this seems unlikely because they become densely methylated on the inactive X chromosome, whereas those on the active X chromosome in the same cell are resistant. Additionally, in cancer cells and cell lines, many normally nonmethylated CpG islands succumb to methylation. Several alternative (not necessarily mutually exclusive) explanations for the mammalian DNA methylation pattern have therefore been entertained: (2) *CpG islands are protected from methylation by the binding of factors which somehow exclude Dnmts*. There is evidence that bound factors do indeed exclude DNA methylation, but footprinting and nucle-

ase sensitivity assays show that CpG islands (Lin et al. 2000) are highly accessible in the nucleus. (3) *CpG islands are maintained in a methylation-free state with the aid of DNA demethylases that actively remove methyl-CpGs* (Frank et al. 1991). Despite several reports of demethylating activities, all current candidate DNA demethylases are unconfirmed. This scenario cannot, however, be ruled out. (4) *The atypical base composition and lack of methylation reflect abnormal DNA metabolism at these CpG islands*. For example, there is evidence that CpG islands are origins of DNA replication and may be affected by the structure of a replication initiation intermediate (Antequera and Bird 1999). Alternatively, recombination and/or repair may be concentrated at these sites, resulting in high levels of DNA turnover. How these putative activities might exclude Dnmts is unclear. (5) *Early embryonic transcription from a CpG island promoter is required to ensure that DNA methylation is excluded*. Promoter mutations provoke methylation of CpG islands in transgenic assays (Brandeis et al. 1994; MacLeod et al. 1994) and CpG island promoters of highly tissue-specific genes are usually expressed in early embryos, but formal proof that transcription excludes CpG methylation is lacking.

2.7 DNA Methylation Triggered by Chromatin Structure?

The above scenarios assume that the CpG methylation is the default state of the genome and that CpG islands therefore arise through exclusion of a global methylating activity. A somewhat different view has recently taken hold: (6) *DNA methylation patterns are determined by the modification state of the underlying chromatin*. DNA methylation, or lack of it, would in this case be triggered by an earlier decision to modify histone proteins in specific ways. In the fungus *Neurospora crassa* and the plant *Arabidopsis thaliana*, the evidence for this idea is strong (see Chapters 6 and 9). Methylation in these systems is not confined to CpG sites and has been shown to depend on the presence of methylation of histone H3 on lysine 9 (H3K9me) (Tamaru and Selker 2001; Jackson et al. 2002). Along related lines, it has emerged in plants that RNA interference (RNAi) can target chromatin modification, gene silencing, and DNA methylation (see Aufsatz et al. 2002; Chapter 8). Although the relationship has been less well established in mammals, where CpG methylation is predominant, absence of two histone lysine methyltransferases (HKMTs) specific for the H3K9 residues has been shown to reduce CpG methylation within heterochromatic repeat sequences (Lehnertz et al. 2003). In addition, depletion of the Polycomb group (PcG) protein EZH2, a HKMT specific for H3K27 residue, has been shown to cause loss of CpG methylation of EZH2-target promoters (Vire et al. 2006). There is also strong evidence that antisense transcription triggers DNA methylation in mammalian systems. Higgs and coworkers (Tufarelli et al. 2003) showed that driving an antisense transcript through the α-globin gene in differentiating mouse embryo cells ensured CpG island methylation. The mechanism behind this effect is unclear, but it has been speculated that RNAi plays a role, as it does in certain de novo methylation events in plants (Zilberman et al. 2003). De novo methylation triggered by RNAi has been reported in cultured mammalian cells (Kawasaki and Taira 2004), but current data suggest that the phenomenon is less clear-cut than in plants or fungi.

2.8 The Role of SWI/SNF-like Chromatin-remodeling Proteins

Evidence that chromatin accessory factors are also needed to ensure appropriate methylation came initially from plants, where the SNF2-like protein DDM1 was shown to be essential for full methylation of the *A. thaliana* genome (Jeddeloh et al. 1999). An equivalent dependence is seen in animals, because mutations in human *ATRX* (Gibbons et al. 2000) and mouse *Lsh2* genes (Dennis et al. 2001), both of which encode relatives of the chromatin-remodeling protein SNF2, have significant effects on global DNA methylation patterns. Loss of LSH2 protein, in particular, matches the phenotype of the *DDM1* mutation in *Arabidopsis*, because both mutants lose methylation of highly repetitive DNA sequences but retain some methylation elsewhere in the genome. Perhaps efficient global methylation of the genome requires perturbation of chromatin structure by these chromatin-remodeling proteins so that DNMTs can gain access to the DNA. Collaboration between DNMTs and chromatin factors that

allow them access to specialized chromosomal regions may be particularly important in regions that are "heterochromatic" and inaccessible.

3 Regulation of Gene Expression by DNA Methylation

3.1 Early Evidence

The effect of DNA methylation on gene expression has been tested in several ways. In an adenovirus reporter gene, artificial methylation of a subset of CpG sites using M.HpaII prevented expression of the gene when injected into frog oocyte nuclei (Vardimon et al. 1982). Similarly, the gene for adenine phosphoribosyltransferase (Stein et al. 1982) was silenced by CpG methylation when transfected into cultured mammalian cells. Studies of the effects of DNA methylation in natural genomic DNA became possible with the discovery that the drug 5-azacytidine could inhibit DNA methylation in living cells (Jones and Taylor 1980). This nucleoside analog is incorporated into DNA in place of cytidine and forms a covalent adduct with DNA methyltransferases, taking them out of circulation and preventing further DNA methylation. Silencing of several genes, including viral genomes (Harbers et al. 1981) and genes on the inactive X chromosome (Wolf et al. 1984), had previously been shown to correlate with their methylation. The ability of 5-azacytidine treatment to restore their expression (Mohandas et al. 1981) argued that this DNA methylation played a causal role in their repression. That the effect was genuinely due to changes in DNA methylation and not some other effect of the drug was demonstrated by testing purifed DNA extracted from the drug-treated cells. DNA from 5-azacytidine-treated cells, when transfected into cells, was capable of active expression of the silenced X-linked hypoxanthine phosphoribosyltransferase gene, whereas DNA from untreated control cells could not confer expression (Venolia et al. 1982).

3.2 Interference with Transcription Factor Binding

How does DNA methylation interfere with gene expression? One obvious possibility is that the presence of methyl groups in the major groove (see Fig. 1) interferes with the binding of transcription factors that activate transcription from a specific gene. A number of transcription factors recognize GC-rich sequence motifs that can contain CpG sequences. Several of these are unable to bind DNA when the CpG sequence is methylated (Watt and Molloy 1988). Evidence for involvement of this mechanism in gene regulation comes from studies of the role of the CTCF protein in imprinting at the *H19/Igf2* locus in mice (Bell and Felsenfeld 2000). CTCF is associated with transcriptional domain boundaries (Bell et al. 1999) and can insulate a promoter from the influence of remote enhancers. The maternally derived copy of the *Igf2* gene is silent because of the binding of CTCF between its promoter and a downstream enhancer. At the paternal locus, however, the CpG-rich CTCF-binding sites are methylated, preventing CTCF binding and thereby allowing the downstream enhancer to activate *Igf2* expression. Although there is evidence that *H19/Igf2* imprinting involves additional processes, the role of CTCF represents a clear example of transcriptional regulation by DNA methylation (for more details, see Chapter 19).

3.3 Attraction of Methyl-CpG-binding Proteins

The second mode of repression is opposite to the first, because it involves proteins that are attracted to, rather than repelled by, methyl-CpG (Fig. 5 and Table 2). This mode of repression was first detected in extracts from mammalian cells that were able to support transcription of added genes. Addition of trace amounts of DNA template allowed transcription from nonmethylated reporter genes, whereas methylated reporters were repressed (Boyes and Bird 1991). Increasing the amount of added DNA caused both methylated and unmethylated templates to be transcribed at equivalent levels, suggesting that limiting amounts of a DNA methylation-specific transcriptional inhibitor were being titrated out. In agreement with this interpretation, excess methylated nonspecific competitor DNA relieved the repression of methylated genes in these extracts. Evidence for indirect inhibition of transcription also came from experiments in which methylated DNA that was introduced into mammalian cells and frog oocytes only initially permitted gene transcription (Buschhausen et al. 1987; Kass et al. 1997). Silencing occurred several hours later, suggesting that silencing might depend on the assembly of chromatin. To find proteins that could associate specifically with methylated genes and bring about the observed repression, bandshift assays were performed using random methylated DNA sequences as probes. A DNA–protein complex that is specific for methylated DNA (MeCP1) was observed in a variety of mammalian cell types (Meehan et al. 1989). An individual methyl-CpG-binding protein, MeCP2, was, however, the first protein to be purified and cloned. Proteins with DNA-binding motifs related to that of MeCP2 were identified using database searches and designated the methyl-CpG-binding domain (MBD) family comprising MeCP2, MBD1, MBD2, MBD3, and MBD4 (Table 2) (Bird and Wolffe 1999). One of the resulting proteins (MBD2) is the DNA-binding component of MeCP1 complex (see above).

Three of the MBD proteins, MBD1, MBD2, and MeCP2, have been implicated in methylation-dependent repression of transcription (Bird and Wolffe 1999). An unrelated protein, Kaiso, has also been shown to bind methylated DNA and bring about methylation-dependent repression in model systems (Table 2) (Prokhortchouk et al. 2001; Yoon et al. 2003). An understanding of the mechanism of repression came from the realization that MeCP2 associates with the mSin3a corepressor complex and depends on histone deacetylation for its action (Jones et al. 1998; Nan et al. 1998). This finding showed that DNA

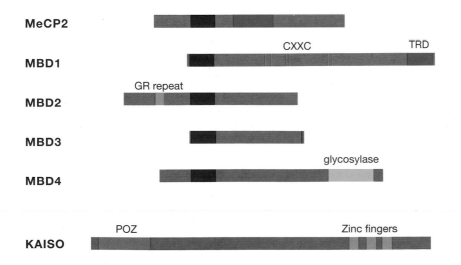

Figure 5. Proteins That Bind Methyl-CpG

Five members of the MBD protein family are aligned at their MBD domains (*purple*). Other domains are labeled and include transcriptional repression domains (TRD); CXXC domains, zinc fingers some of which are implicated in binding to non-methylated CpG; GR repeats of unknown function; a T:G mismatch glycosylase domain which is involved in repair of 5-methylcytosine deamination. Kaiso lacks the MBD domain but binds methylated DNA via zinc fingers (*orange*) and possesses a POB/BTB domain that is shared with other transcriptional repressors.

Table 2. Functions of methyl-CpG-binding proteins

MBP	Major activity	Species	Major phenotypes of loss-of-function mutations
MeCP2	binds mCpG with adjacent A/T run; transcriptional repressor	mouse	delayed onset neurological defects including inertia, hindlimb clasping, nonrhythmic breathing, and abnormal gait; postnatal survival ~10 weeks
MECP2	binds mCpG with adjacent AT run; transcriptional repressor	human	heterozygotes suffer from Rett syndrome, a profound neurological disorder characterized by apraxia, loss of purposeful hand use, breathing irregularities, and microcephaly
Mbd1	binds mCpG via MBD; a major splice form is also able to bind CpG via a CxxC domain	mouse	no overt phenotype, but subtle defects in neurogenesis detected
Mbd2	binds mCpG; transcriptional repressor	mouse	viable and fertile, but show reduced maternal nurturing behavior; defective gene regulation in T-helper-cell differentiation leading to altered response to infection; highly resistant to intestinal tumorigenesis
Mbd3	core component of NuRD co-repressor complex; does not show strong binding to mCpG	mouse	early embryonic lethal
Mbd4	DNA repair protein that binds mCpG and T:G mismatches at mCpG sites; thymine DNA glycosylase that excises T from T:G mismatches	mouse	viable and fertile; three- to fourfold increase in mutations at CpG sites; increased susceptibility to intestinal cancer correlates with C-to-T transitions within the Apc gene; Mbd4 functions to minimize the mutability of 5-methylcytosine
Kaiso	binds mCGmCG and CTGCNA; transcriptional repressor	mouse	no overt phenotype; small but significant delay in tumorigenesis on Min background

methylation could be read by MeCP2 and provide a signal to alter chromatin structure (Fig. 6). Each of the four methyl-CpG-binding proteins has since been shown to associate with a different corepressor complex. Of particular interest is MBD1, which associates with the histone lysine methyltransferase SETDB1 only during DNA replication (Sarraf and Stancheva 2004). The continued histone H3K9 methylation at chromosomal MBD1 target sequences, and stable silencing of the associated genes, depend on the periodic recruitment of this chromatin-modifying activity.

Knowledge of an MBD's protein target sites in the genome is a prerequisite for understanding its biological role. They could be expected to compete with one another for access to methylated sites due to their overlapping DNA sequence specificity. Surprisingly, MBD-binding sites were largely nonoverlapping in the genome when studied using cells from a primary human cell line, suggesting that each methyl-CpG-binding protein is targeted independently (Klose et al. 2005). This fits with emerging evidence showing that MBD binding shows DNA sequence specificity (see Table 2). For example, MeCP2 strongly prefers mCpG sites that are flanked by a run of AT-rich DNA (Klose et al. 2005). Furthermore, MBD1 has an additional DNA-binding domain that is specific for nonmethylated CpG, and Kaiso recognizes a pair of adjacent mCpG motifs (Prokhortchouk et

al. 2001). Only MBD2 so far appears to have an exclusive affinity for mCpG.

3.4 MeCP2 and Rett Syndrome

The existence of multiple methyl-CpG-binding proteins with repressive properties argues that these may be important mediators of the methylation signal. This is illustrated most strikingly by the finding that mutations in the human *MECP2* gene are responsible for a severe neurological disorder called Rett syndrome (RTT). RTT affects females that are heterozygous for new mutations in the X-linked *MECP2* gene (Table 2) (Amir et al. 1999). Due to random X-chromosome inactivation, the patients are mosaic for expression of either the mutant or the wild-type (*wt*) gene. Affected girls develop apparently normally for 6–18 months, at which time they enter a crisis that leaves them with greatly impaired motor skills, repetitive hand movements, abnormal breathing, microcephaly, and other symptoms (Table 2). Males who are hemizygous for comparable mutations do not survive. *Mecp2*-null mice are born and develop normally for several weeks, but they acquire a variety of neurological symptoms at about 6 weeks of age, which leads to death at approximately 10 weeks. Several features of this delayed-onset phenotype, which is fully penetrant, recall human Rett syndrome (Amir et al.

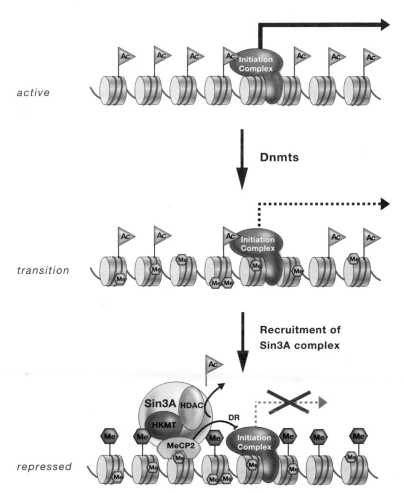

active

transition

Dnmts

Recruitment of
Sin3A complex

repressed

Figure 6. Recruitment of Corepressors by Methyl-CpG-binding Proteins

A hypothetical transition between an active, nonmethylated gene promoter and a repressed promoter whose silence is due to DNA methylation as mediated by MeCP2. The transition phase represents an intermediate step during which transcription is silenced and DNA methylation occurs. MeCP2 is envisaged to recruit the Sin3A histone deacetylase (HDAC) complex and histone lysine methyltransferase (HKMT) activity to the methylated sites. In addition, there is some evidence that MeCP2 can directly repress transcription by contact with the transcription initiation complex (DR). Other methyl-CpG-binding proteins interact with and potentially recruit distinct corepressors that include HKMT and/or HDAC activity.

1999). Conditional deletion of the *Mecp2* gene only in mouse brain causes the same symptoms as *Mecp2* deletion in the whole mouse (Table 2) (Chen et al. 2001; Guy et al. 2001). Therefore, although MeCP2 is ubiquitously expressed in cells of the mouse, the *Mecp2*-null phenotype appears to be entirely due to its absence in the brain. Consistent with this finding, biochemical and immunocytochemical studies have established that MeCP2 expression levels are highest in the brain—specifically in neurons. Significantly, expression of MeCP2 in neurons alone prevents onset of the mouse phenotype (Luikenhuis et al. 2004).

Given the role of MeCP2 as a transcriptional repressor, an attractive hypothesis to explain the disease is that genes in the brain needing to be silenced by MeCP2 escape repression in its absence, and this leads to aberrant neuronal function. The first mammalian MeCP2 target gene that has been identified encodes brain-derived neurotrophic factor (Bdnf) (Chen et al. 2003; Martinowich et al. 2003). Bdnf belongs to a set of pro-

teins synthesized in response to neuronal activity and is thought to be essential for converting transient stimuli into long-term changes in brain activity. Its mis-regulation may therefore be implicated in the pathology of RTT. Altered expression of several other genes in mice has been reported, but the magnitude of the effects is small (<threefold). Nevertheless, it is possible that one or several of these alterations also contribute to the RTT phenotype.

3.5 MBD2 Mediates Methylation-dependent Transcriptional Repression

MBD2 is the DNA-binding component of MeCP1 (see Table 2), which was initially implicated as a transcriptional repressor in cellular extracts (Meehan et al. 1989; Boyes and Bird 1991). MeCP1 is a large multiprotein complex that includes the NuRD (or Mi-2) corepressor complex and MBD2 (Wade et al. 1999). NuRD comprises histone deacetylases (HDAC) and a large chromatin-remodeling protein (Mi-2). NuRD can be recruited to DNA by several DNA-binding proteins besides MBD2. (Interestingly, MBD3, which resembles MBD2 but does not bind methylated DNA, is a component of the NuRD complex.) Cells that lack MBD2 are unable to effectively repress methylated reporter constructs, despite the presence of other methyl-CpG-binding proteins in these cells, arguing that it is an important component of the repression system (Hendrich et al. 2001). MBD2-deficient mice are viable and fertile, although they have a defect in maternal behavior (Hendrich et al. 2001), and careful examination has revealed aberrations in tissue-specific gene expression. Expression of the *interleukin-4* and *interferon-γ* genes during T-helper-cell differentiation is significantly disrupted (Hutchins et al. 2002). For example, a significant number of Th1 cells that should express only *interferon-γ* also express *interleukin-4* (Table 2). Because MBD2 is found bound to the *interleukin-4* gene, it is likely that its absence weakens repression of the gene in Th1 cells.

4 DNA Methylation, Mutation, and Chromosomal Stability

4.1 DNA Methylation and Mutation

Set against the advantages of DNA methylation as an epigenetic system of cellular memory is the disadvantage of 5-methylcytosine mutability. Cytosine (C) deaminates spontaneously to give uracil (U), which is then mispaired with guanine (Lindahl 1974). This potential mutation is recognized by uracil DNA glycosylases, which efficiently remove the inappropriate base and initiate repair to restore C in place of U. When 5-methylcytosine deaminates, however, thymine (T) is formed. This also results in a mismatch, but the fact that T, unlike U, is a natural DNA base appears to interfere with the efficient repair of the lesion. As a result, the mutant thymine base can persist through DNA replication and is passed on to progeny cells as a C-to-T transition mutation. Mutations of this kind appear to be one of the most frequent single causes of genetic disease in humans, because approximately one-third of all point mutations are C-to-T transitions at CpG sequences (Cooper and Youssoufian 1988). The instability of CpG over evolutionary time is further demonstrated by the four- to fivefold underrepresentation of CpG in the mammalian genome (Bird 1980). The only exceptions are CpG islands, within which CpGs are nonmethylated and therefore stable.

MBD4 is so far unique among methyl-CpG-binding proteins in that it has enzyme activity. The MBD4 car-

boxy-terminal domain is a thymine DNA glycosylase that can selectively remove T from a T-G mismatch in vitro (Hendrich et al. 1999). This activity would be expected of a DNA repair system that corrects 5-methylcytosine deamination. Confirming this hypothesis, mice lacking MBD4 show enhanced mutability of methylated cytosine residues at a chromosomal reporter sequence (Millar et al. 2002). In addition, *Mbd4*-null mice acquire C-to-T transition mutations within the adenomatous polyposis coli (APC) gene and have an increased frequency of intestinal tumorigenesis (Table 2). It is noteworthy that, despite the existence of a dedicated repair system, sites of cytosine methylation persist as hot spots for mutation.

4.2 DNA Methylation and Chromosome Instability

Although DNA methylation is clearly mutagenic, there is evidence that its presence is beneficial with respect to chromosomal stability. Mice possessing about 10% of normal levels of DNA methylation due to a hypomorphic mutation of Dnmt1 acquire aggressive T-cell lymphomas that often display trisomy of chromosome 15 (Gaudet et al. 2003). Mutations of DNMT3B in patients with ICF syndrome, or inactivation of Dnmt3b in mice, lead to various chromosomal aberrations, including chromosome fusion, breakage, and aneuploidy (Ehrlich 2003; Dodge et al. 2005). These results are of interest because cancers often display reduced levels of DNA methylation, which may contribute to tumor initiation or progression. One possible explanation for the result is that DNA methylation contributes to accurate chromosome segregation and, in its absence, it is more frequent to have nondisjunction leading to chromosome aberrations. Alternatively, DNA methylation may suppress the expression and recombination of retrotransposons in the mammalian genome, thereby protecting chromosomes from deleterious recombination. Indeed, DNA methylation has been shown to play a critical role in silencing the transcription of retrotransposons during embryonic development and spermatogenesis (Walsh et al. 1998; Bourc'his and Bestor 2004).

5 Future Directions

Our understanding of the biological functions of DNA methylation in mammals has been growing steadily but is far from complete. For instance, unlike genetic mutations, we know very little about the rate of changes in CpG methylation in mammals and the intrinsic and environmental factors that induce changes in DNA methylation patterns. Accumulating evidence is indicating that changes in DNA methylation and histone modification may contribute to the pathogenesis of many complex diseases. Modulation of epigenetic states of the genome thus has the potential to become a new therapeutic approach for the treatment of these diseases. In the future, we expect to see advances in these exciting areas of research.

5.1 Environmental Factors That Induce Epigenetic Changes

It is well established that DNA methylation patterns of the mammalian genome are highly regulated during development. How environmental factors may affect DNA methylation and gene expression is less well understood. Some recent studies are beginning to shed light on how environmental factors may induce epigenetic changes that can have long-lasting biological effects. One such example is the observation that rat maternal behavior produces stable alterations in DNA methylation in the offspring. Weaver and coworkers have reported that baby rats receiving different levels of maternal care have differences in DNA methylation in the promoter region of the glucocorticoid receptor (GR) gene, which are inversely correlated with GR expression, and these differences persist into adulthood (Weaver et al. 2004). Another example is the report that the diet of an adult female mouse can alter DNA methylation in the offspring. In mice, *agouti* is a dominant trait that gives the coat the brownish (agouti) color. Several *agouti viable yellow* alleles arise spontaneously through insertion of a transposable retroviral element into the gene. In mice with such an allele, the expression of *agouti* is controlled by the long terminal repeat (LTR) of the retroviral element. The coat color of these mice, which varies from yellow to mottled to wild-type agouti, is determined by the methylation states of the LTR promoter. Thus, in this system, coat color can serve as a readout for DNA methylation. When pregnant females are fed diets supplemented with methyl donors such as folate, choline, and betain, their offspring show a shift of coat color toward agouti, which is correlated with increased methylation of the LTR promoter (Cooney et al. 2002; Waterland and Jirtle 2003). These results suggest that environmental factors can induce stable alterations of epigenetic states, providing a mechanism by which environmental factors may bring about long-term biological effects. Further studies are required to determine to what extent epigenetic mechanisms are involved in gene–environment interactions in mammals and how environmental factors may transduce into epigenetic states.

5.2 Epigenetic Instability and Complex Diseases

Many complex diseases, such as type II diabetes, schizophrenia, autoimmune diseases, and cancer, exhibit a heritable component but do not demonstrate a clear Mendelian pattern of inheritance. The dynamic epigenetic mechanism provides an alternative explanation for some of the features of complex diseases, which include late onset, gender effects, parent-of-origin effects, discordance of monozygotic twins, and fluctuation of symptoms in complex diseases (Petronis 2001). Although growing evidence has linked aberrant DNA methylation and histone modifications to cancer, the role of epigenetic mechanisms in the etiology of many other complex diseases is largely unknown. Comparative studies of genome-wide DNA methylation patterns between normal and disease populations may provide insights into the epigenetic basis for various complex diseases in which genetic mutations are difficult to detect.

5.3 Modulation of Reversible Epigenetic States

Most, if not all, epigenetic modifications are reversible, which makes modulation of epigenetic states a potential new therapeutic option for cancer and other diseases. A number of agents that alter patterns of DNA methylation or inhibit HDACs are currently being tested in clinical trials (Egger et al. 2004). Some of them have shown promising antitumor effects. In fact, the demethylating agent 5-azacytidine has been recently approved by the U.S. Food and Drug Administration for the treatment of myelodysplastic syndrome, a heterogeneous disease characterized by morphologic dysplasia of hematopoietic cells. The clinical use of 5-azacytidine and other nucleoside analogs is limited by their toxicity, partly because these compounds are being incorporated into DNA. This has encouraged the search for agents that can inhibit DNA methyltransferases directly or target other epigenetic regulators. Because DNA methylation is just one component of the complex epigenetic regulatory network, one approach to maximize therapeutic effects and minimize toxicity is combination therapy using DNA methyltransferase or HDAC inhibitors in combination with other anticancer therapeutics (for more detail, see Chapter 24).

References

Amir R.E., Van den Veyver I.B., Wan M., Tran C.Q., Francke U., and Zoghbi H.Y. 1999. Rett syndrome is caused by mutations in X-linked MECP2, encoding methyl-CpG-binding protein 2. *Nat. Genet.* **23:** 185–188.

Antequera F. and Bird A. 1999. CpG islands as genomic footprints of promoters that are associated with replication origins. *Curr. Biol.* **9:** R661–667.

Aufsatz W., Mette M.F., van der Winden J., Matzke A.J., and Matzke M. 2002. RNA-directed DNA methylation in *Arabidopsis. Proc. Natl. Acad. Sci.* **99:** 16499–16506.

Bell A.C. and Felsenfeld G. 2000. Methylation of a CTCF-dependent boundary controls imprinted expression of the Igf2 gene. *Nature* **405:** 482–485.

Bell A.C., West A.G., and Felsenfeld G. 1999. The protein CTCF is required for the enhancer blocking activity of vertebrate insulators. *Cell* **98:** 387–396.

Bestor T.H. and Ingram V.M. 1983. Two DNA methyltransferases from murine erythroleukemia cells: Purification, sequence specificity, and mode of interaction with DNA. *Proc. Natl. Acad. Sci.* **80:** 5559–5563.

Bird A.P. 1978. Use of restriction enzymes to study eukaryotic DNA methylation. II. The symmetry of methylated sites supports semi-conservative copying of the methylation pattern. *J. Mol. Biol.* **118:** 48–60.

———. 1980. DNA methylation and the frequency of CpG in animal DNA. *Nucleic Acids Res.* **8:** 1499–1594.

Bird A.P. and Southern E.M. 1978. Use of restriction enzymes to study eukaryotic DNA methylation. I. The methylation pattern in ribosomal DNA from *Xenopus laevis. J. Mol. Biol.* **118:** 27–47.

Bird A.P and Wolffe A.P. 1999. Methylation-induced repression—Belts, braces and chromatin. *Cell* **99:** 451–454.

Bird A., Taggart M., Frommer M., Miller O.J., and Macleod D. 1985. A fraction of the mouse genome that is derived from islands of non-methylated, CpG-rich DNA. *Cell* **40:** 91–99.

Bourc'his D. and Bestor T.H. 2004. Meiotic catastrophe and retrotransposon reactivation in male germ cells lacking Dnmt3L. *Nature* **431:** 96–99.

Boyes J. and Bird A. 1991. DNA methylation inhibits transcription indirectly via a methyl-CpG-binding protein. *Cell* **64:** 1123–1134.

Brandeis M., Frank D., Keshet I., Siegried Z., Mendelsohn M., Nemes A., Temper V., Razin A., and Cedar H. 1994. Sp1 elements protect a CpG island from de novo methylation. *Nature* **371:** 435–438.

Bruniquel D. and Schwartz R.H. 2003. Selective, stable demethylation of the interleukin-2 gene enhances transcription by an active process. *Nat. Immunol.* **4:** 235–240.

Buschhausen G., Wittig B., Graessmann M., and Graessmann A. 1987. Chromatin structure is required to block transcription of the methylated herpes simplex virus thymidine kinase gene. *Proc. Natl. Acad. Sci.* **84:** 1177–1181.

Chen R.Z., Akbarian S., Tudor M., and Jaenisch R. 2001. Deficiency of methyl-CpG-binding protein-2 in CNS neurons results in a Rett-like phenotype in mice. *Nat. Genet.* **27:** 327–331.

Chen W.G., Chang Q., Lin Y., Meissner A., West A.E., Griffith E.C., Jaenisch R., and Greenberg M.E. 2003. Derepression of BDNF transcription involves calcium-dependent phosphorylation of MeCP2. *Science* **302:** 885–889.

Cooney C.A., Dave A.A., and Wolff G.L. 2002. Maternal methyl supplements in mice affect epigenetic variation and DNA methylation of offspring. *J. Nutr.* **132:** 2393S–2400S.

Cooper D.N. and Youssoufian H. 1988. The CpG dinucleotide and human genetic disease. *Hum. Genet.* **78:** 151–155.

Cooper D.N., Taggart M.H., and Bird A.P. 1983. Unmethylated domains in vertebrate DNA. *Nucleic Acids Res.* **11:** 647–658.

Dennis K., Fan T., Geiman T., Yan Q., and Muegge K. 2001. Lsh, a member of the SNF2 family, is required for genome-wide methylation. *Genes Dev.* **15:** 2940–2944.

Dodge J.E., Okano M., Dick F., Tsujimoto N., Chen T., Wang S., Ueda Y., Dyson N., and Li E. 2005. Inactivation of Dnmt3b in mouse embryonic fibroblasts results in DNA hypomethylation, chromosomal instability, and spontaneous immortalization. *J. Biol. Chem.* **280:** 17986–17991.

Egger G., Liang G., Aparicio A., and Jones P.A. 2004. Epigenetics in human disease and prospects for epigenetic therapy. *Nature* **429:** 457–463.

Ehrlich M. 1982. Amount and distribution of 5-methycytosine in human DNA from different types of tissues or cells. *Nucleic Acids Res.* **10:** 2709–2721.

———. 2003. The ICF syndrome, a DNA methyltransferase 3B deficiency and immunodeficiency disease. *Clin. Immunol.* **109:** 17–28.

Frank D., Keshet I., Shani M., Levine A., Razin A., and Cedar H. 1991. Demethylation of CpG islands in embryonic cells. *Nature* **351:** 239–241.

Frommer M., McDonald L.E., Millar D.S., Collis C.M., Watt J., Grigg G.W., Molloy P.L., and Paul C.L. 1992. A genomic sequencing protocol that yields a positive display of 5-methylcytosine residues in individual DNA strands. *Proc. Natl. Acad. Sci.* **89:** 1827–1831.

Gaudet F., Hodgson J.G., Eden A., Jackson-Grusby L., Dausman J., Gray J.W., Leonhardt H., and Jaenisch R. 2003. Induction of tumors in mice by genomic hypomethylation. *Science* **300:** 489–492.

Gibbons R.J., McDowell T.L., Raman S., O'Rourke D.M., Garrick D., Ayyub H., and Higgs D.R. 2000. Mutations in ATRX, encoding a SWI/SNF-like protein, cause diverse changes in the pattern of DNA methylation. *Nat. Genet.* **24:** 368–371.

Guy J., Hendrich B., Holmes M., Martin J.E., and Bird A. 2001. A mouse *Mecp2*-null mutation causes neurological symptoms that mimic Rett syndrome. *Nat. Genet.* **27:** 322–326.

Harbers K., Schnieke H., Stuhlmann H., Jahner D., and Jaenisch B. 1981. DNA methylation and gene expression; endogenous retroviral genome becomes infectious after molecular cloning. *Proc. Natl. Acad. Sci.* **78:** 7609–7613.

Hata K., Okano M., Lei H., and Li E. 2002. Dnmt3L cooperates with the Dnmt3 family of de novo DNA methyltransferases to establish maternal imprints in mice. *Development* **129:** 1983–1993.

Hendrich B., Guy J., Ramsahoye B., Wilson V.A., and Bird A. 2001. Closely related proteins Mbd2 and Mbd3 play distinctive but interacting roles in mouse development. *Genes Dev.* **15:** 710–723.

Hendrich B., Hardeland U., Ng H.-H., Jiricny J., and Bird A. 1999. The thymine glycosylase MBD4 can bind to the product of deamination at methylated CpG sites. *Nature* **401:** 301–304.

Herman J.G., Graff J.R., Myohanen S., Nelkin B.D., and Baylin S.B. 1996. Methylation-specific PCR: A novel PCR assay for methylation status of CpG islands. *Proc. Natl. Acad. Sci.* **93:** 9821–9826.

Holliday R. and Pugh J.E. 1975. DNA modification mechanisms and gene activity during development. *Science* **186:** 226–232.

Hutchins A., Mullen A., Lee H., Barner K., High F., Hendrich B., Bird A., and Reiner S. 2002. Gene silencing quantitatively controls the function of a developmental *trans*-activator. *Mol. Cell* **10:** 81–91.

Jackson J.P., Lindroth A.M., Cao X., and Jacobsen S.E. 2002. Control of CpNpG DNA methylation by the KRYPTONITE histone H3 methyltransferase. *Nature* **416:** 556–560.

Jahner D., Stuhlmann H., Stewart C.L., Harbers K., Lohler J., Simon I., and Jaenisch R. 1982. De novo methylation and expression of retroviral genomes during mouse embryogenesis. *Nature* **298:** 623–628.

Jeddeloh J.A., Stokes T.L., and Richards E.J. 1999. Maintenance of genomic methylation requires a SWI2/SNF2-like protein. *Nat. Genet.* **22:** 94–97.

Jones P.A. and Taylor S.M. 1980. Cellular differentiation, cytidine analogues and DNA methylation. *Cell* **20:** 85–93.

Jones P.L., Veenstra G.J., Wade P.A., Vermaak D., Kass S.U., Landsberger N., Strouboulis J., and Wolffe A.P. 1998. Methylated DNA and MeCP2 recruit histone deacetylase to repress transcription. *Nat. Genet.* **19:** 187–191.

Kaneda M., Sado T., Hata K., Okano M., Tsujimoto N., Li E., and Sasaki H. 2004. Role of de novo DNA methyltransferases in initiation of genomic imprinting and X-chromosome inactivation. *Cold Spring Harbor Symp. Quant. Biol.* **69:** 125–129.

Kass S.U., Landsberger N., and Wolffe A.P. 1997. DNA methylation directs a time-dependent repression of transcription initiation. *Curr. Biol.* **7:** 157–165.

Kawasaki H. and Taira K. 2004. Induction of DNA methylation and gene silencing by short interfering RNAs in human cells. *Nature* **431:** 211–217.

Klose R.J., Sarraf S.A., Schmiedeberg L., McDermott S.M., Stancheva I., and Bird A.P. 2005. DNA binding specificity of MeCP2 due to a requirement for A/T sequences adjacent to methyl-CpG. *Mol. Cell* **19:** 667–678.

Lehnertz B., Ueda Y., Derijck A.A., Braunschweig U., Perez-Burgos L., Kubicek S., Chen T., Li E., Jenuwein T., and Peters A.H. 2003. Suv39h-mediated histone H3 lysine 9 methylation directs DNA methylation to major satellite repeats at pericentric heterochromatin. *Curr. Biol.* **13:** 1192–1200.

Lei H., Oh S.P., Okano M., Juttermann R., Gos K.A., Jaenisch R., and Li E. 1996. De novo DNA cytosine methyltransferase activities in mouse embryonic stem cells. *Development* **122:** 3195–3205.

Li E., Bestor T.H., and Jaenisch R. 1992. Targeted mutation of the DNA methyltransferase gene results in embryonic lethality. *Cell* **69:** 915–926.

Lin I.G., Tomzynski T.J., Ou Q., and Hsieh C.L. 2000. Modulation of DNA binding protein affinity directly affects target site demethylation. *Mol. Cell. Biol.* **20:** 2343–2349.

Lindahl T. 1974. An *N*-glycosidase from *Escherichia coli* that releases free uracil from DNA containing deaminated cytosine residues. *Proc. Natl. Acad. Sci.* **71:** 3649–3653.

Luikenhuis S., Giacometti E., Beard C.F., and Jaenisch R. 2004. Expression of MeCP2 in postmitotic neurons rescues Rett syndrome in mice. *Proc. Natl. Acad. Sci.* **101:** 6033–6038.

MacLeod D., Charlton J., Mullins J., and Bird A.P. 1994. Sp1 sites in the mouse aprt gene promoter are required to prevent methylation of the CpG island. *Genes Dev.* **8:** 2282–2292.

Martienssen R.A., Doerge R.W., and Colot V. 2005. Epigenomic mapping in *Arabidopsis* using tiling microarrays. *Chromosome Res.* **13:** 299–308.

Martinowich K., Hattori D., Wu H., Fouse S., He F., Hu Y., Fan G., and Sun Y.E. 2003. DNA methylation-related chromatin remodeling in activity-dependent BDNF gene regulation. *Science* **302:** 890–893.

Mayer W., Niveleau A., Walter J., Fundele R., and Haaf T. 2000. Demethylation of the zygotic paternal genome. *Nature* **403:** 501–502.

McKeon C., Ohkubo H., Pastan I., and de Crombrugghe B. 1982. Unusual methylation pattern of the alpha 2(1) collagen gene. *Cell* **29:** 203–210.

Meehan R.R., Lewis J.D., McKay S., Kleiner E.L., and Bird A.P. 1989. Identification of a mammalian protein that binds specifically to DNA containing methylated CpGs. *Cell* **58:** 499–507.

Millar C.B., Guy J., Sansom O.J., Selfridge J., MacDougall E., Hendrich B., Keightley P. D., Bishop S.M., Clarke A.R., and Bird A. 2002. Enhanced CpG mutability and tumorigenesis in MBD4-deficient mice. *Science* **297:** 403–405.

Mohandas T., Sparkes R.S., and Shapiro L.J. 1981. Reactivation of an inactive human X-chromosome: Evidence for X-inactivation by DNA methylation. *Science* **211:** 393–396.

Nan X., Ng H.-H., Johnson C.A., Laherty C.D., Turner B.M., Eisenman R.N., and Bird A. 1998. Transcriptional repression by the methyl-CpG-binding protein MeCP2 involves a histone deacetylase complex. *Nature* **393:** 386–389.

Okano M., Xie S., and Li E. 1998. Cloning and characterization of a family of novel mammalian DNA (cytosine-5) methyltransferases. *Nat. Genet.* **19:** 219–220.

Okano M., Bell D.W., Haber D.A., and Li E. 1999. DNA methyltransferases Dnmt3a and Dnmt3b are essential for de novo methylation and mammalian development. *Cell* **99:** 247–257.

Petronis A. 2001. Human morbid genetics revisited: Relevance of epigenetics. *Trends Genet.* **17:** 142–146.

Posfai J., Bhagwat A.S., Posfai G., and Roberts R.J. 1989. Predictive motifs derived from cytosine methyltransferases. *Nucleic Acids Res.* **17:** 2421–2435.

Prokhortchouk A., Hendrich B., Jorgensen H., Ruzov A., Wilm M., Georgiev G., Bird A., and Prokhortchouk E. 2001. The p120 catenin partner Kaiso is a DNA methylation-dependent transcriptional repressor. *Genes Dev.* **15:** 1613–1618.

Riggs A.D. 1975. X-inactivation, differentiation and DNA methylation. *Cytogenet. Cell Genet.* **14:** 9–25.

Rougier N., Bourc'his D., Gomes D.M., Niveleau A., Plachot M., Paldi A., and Viegas-Pequignot E. 1998. Chromosome methylation patterns during mammalian preimplantation development. *Genes Dev.* **12:** 2108–2113.

Sarraf S.A. and Stancheva I. 2004. Methyl-CpG-binding protein MBD1 couples histone H3 methylation at lysine 9 by SETDB1 to DNA replication and chromatin assembly. *Mol. Cell* **15:** 595–605.

Simonsson S. and Gurdon J. 2004. DNA demethylation is necessary for the epigenetic reprogramming of somatic cell nuclei. *Nat. Cell Biol.* **6:** 984–990.

Stein R., Razin A., and Cedar H. 1982. In vitro methylation of the hamster adenine phosphoribosyl transferase gene inhibits its expression in mouse L cells. *Proc. Natl. Acad. Sci.* **79:** 4418–3422.

Stewart C.L., Stuhlmann H., Jahner D., and Jaenisch R. 1982. De novo methylation and infectivity of retroviral genomes introduced into embryonal carcinoma cells. *Proc. Natl. Acad. Sci.* **79:** 4098–4102.

Tamaru H. and Selker E.U. 2001. A histone H3 methyltransferase controls DNA methylation in *Neurospora crassa*. *Nature* **414:** 277–283.

Tufarelli C., Stanley J.A., Garrick D., Sharpe J.A., Ayyub H., Wood W.G., and Higgs D.R. 2003. Transcription of antisense RNA leading to gene silencing and methylation as a novel cause of human genetic disease. *Nat. Genet.* **34:** 157–165.

Vardimon L., Kressmann A., Cedar H., Maechler M., and Doerfler W. 1982. Expression of a cloned adenovirus gene is inhibited by in vitro methylation. *Proc. Natl. Acad. Sci.* **79:** 1073–1077.

Venolia L., Gartler S.M., Wasserman E.R., Yen P., Mohandas T., and Shapiro L.J. 1982. Transformation with DNA from 5 azacytidine-reactivated X chromosomes. *Proc. Natl. Acad. Sci.* **79:** 2352–2354.

Vire E., Brenner C., Deplus R., Blanchon L., Fraga M., Didelot C., Morey L., Van Eynde A., Bernard D., Vanderwinden J.M., et al. 2006. The Polycomb group protein EZH2 directly controls DNA methylation. *Nature* **439:** 871–874.

Wade P.A., Gegonne A., Jones P.L., Ballestar E., Aubry F., and Wolffe A.P. 1999. Mi-2 complex couples DNA methylation to chromatin remodeling and histone deacetylation. *Nat. Genet.* **23:** 62–66.

Walsh C.P., Chaillet J.R., and Bestor T.H. 1998. Transcription of IAP endogenous retroviruses is constrained by cytosine methylation. *Nat. Genet.* **20:** 116–117.

Waterland R.A. and Jirtle R.L. 2003. Transposable elements: Targets for early nutritional effects on epigenetic gene regulation. *Mol. Cell. Biol.* **23:** 5293–5300.

Watt F. and Molloy P.L. 1988. Cytosine methylation prevents binding to DNA of a HeLa cell transcription factor required for optimal expression of the adenovirus late promoter. *Genes Dev.* **2:** 1136–1143.

Weaver I.C., Cervoni N., Champagne F.A., D'Alessio A.C., Sharma S., Seckl J.R., Dymov S., Szyf M., and Meaney M.J. 2004. Epigenetic programming by maternal behavior. *Nat. Neurosci.* **7:** 847–854.

Weber M., Davies J.J., Wittig D., Oakeley E.J., Haase M., Lam W.L., and Schubeler D. 2005. Chromosome-wide and promoter-specific analyses identify sites of differential DNA methylation in normal and transformed human cells. *Nat. Genet.* **37:** 853–862.

Wigler M.H. 1981. The inheritance of methylation patterns in vertebrates. *Cell* **24:** 285–286.

Wolf S.F., Jolly D.J., Lunnen K.D., Friedman T., and Migeon B.R. 1984. Methylation of the hypoxanthine phosphoribosyltransferase locus on the human X-chromosome: Implications for X-chromosome inactivation. *Proc. Natl. Acad. Sci.* **81:** 2806–2810.

Yoon H.G., Chan D.W., Reynolds A.B., Qin J., and Wong J. 2003. N-CoR mediates DNA methylation-dependent repression through a methyl CpG-binding protein Kaiso. *Mol. Cell* **12:** 723–734.

Zilberman D., Cao X., and Jacobsen S.E. 2003. ARGONAUTE4 control of locus-specific siRNA accumulation and DNA and histone methylation. *Science* **299:** 716–719.

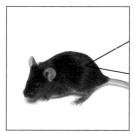
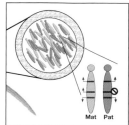

Genomic Imprinting in Mammals

Denise P. Barlow[1] and Marisa S. Bartolomei[2]

[1]CeMM Research Center for Molecular Medicine of the Austrian Academy of Sciences, Institute of Genetics,
Max Perutz Laboratories, A-1030 Vienna, Austria
[2]Department of Cell & Developmental Biology, University of Pennsylvania School of Medicine,
Philadelphia, Pennsylvania 19104-6148

CONTENTS

1. Historical Overview, 359

2. Genomic Imprinting—An Epigenetic Gene Regulatory System, 361

3. Key Discoveries in Genomic Imprinting, 363

   3.1 Imprinted Genes Control Embryonic and Neonatal Growth, 363

   3.2 The Function of Genomic Imprinting in Mammals, 364

   3.3 Imprinted Genes Are Clustered and Controlled by Imprint Control Elements, 365

   3.4 Imprinted Gene Clusters Contain at Least One ncRNA, 368

   3.5 The Role of DNA Methylation in Genomic Imprinting, 369

   3.6 Two Types of cis-Acting Silencing Identified in Imprinted Gene Clusters, 371

4. Genomic Imprinting—A Model for Mammalian Epigenetic Regulation, 372

5. Future Directions, 373

Acknowledgments, 373

References, 374

GENERAL SUMMARY

Mammals are diploid organisms whose cells possess two matched sets of chromosomes, one inherited from the mother and one from the father. Thus, mammals have two copies of every gene. Normally, both the maternal and paternal copies of each gene have the same potential to be active in any cell. Genomic imprinting is an epigenetic mechanism that changes this potential because it restricts the expression of a gene to one of the two parental chromosomes. It is a phenomenon displayed by only a few hundred of the approximately 25,000 genes in our genome, the majority being expressed equally when inherited from either parent. Genomic imprinting affects both male and female offspring and is therefore a consequence of parental inheritance, not of sex. As an example of what is meant by this, an imprinted gene that is active on a maternally inherited chromosome will be active on the maternal chromosome and silent on the paternal chromosome in all males and females.

The definition of genomic imprinting is restricted here to "parental-specific gene expression in diploid cells." Thus, diploid cells that contain two parental copies of all genes will express only one parental copy of an imprinted gene and silence the other parental copy. In contrast, non-imprinted genes will be expressed by both parental gene copies in a diploid cell. In understanding the concept of genomic imprinting, it is important to distinguish between imprinted genes and those showing apparent parental-specific expression because of unequal parental genetic contribution to the embryo. Examples of unequal parental genetic contribution include Y-chromosome-linked genes present only in males, genes that escape X-inactivation in females, mitochondrial genes contributed mainly by the maternal parent, and mRNAs and proteins present only in the sperm or egg cytoplasm.

Many features of genomic imprinting in mammals make it a fascinating biological problem in post-genomic times. For instance, it is providing clues as to a possible evolutionary response to parental conflict, to the adaptation of the maternal parent to an internal reproduction system, and, perhaps, providing just a glimpse of the way the mammalian genome protects itself against invading DNA sequences. Genomic imprinting is an intellectually challenging phenomenon, not least because it raises the question of why a diploid organism would evolve a silencing system that forsakes the advantages of the diploid state. Perhaps most intriguing is that the subset of genes subject to genomic imprinting largely code for factors regulating embryonic and neonatal growth. Thus, it is likely that genomic imprinting evolved to play a specific role in mammalian reproduction.

At this stage of our knowledge, genomic imprinting does not appear to be widespread among the four eukaryotic kingdoms that include protista, fungi, plants, and animals. However, it does exist in a possibly related form in two invertebrate arthropods—Coccidae and Sciaridae—and in the endosperm of some seed-bearing plants, such as maize and *Arabidopsis*. This distribution indicates that genomic imprinting arose independently at least three times during the evolution of life. Surprisingly, despite this predicted independent evolution of genomic imprinting, some similarities among the imprinting mechanism are emerging. It is likely that this reflects conservation of basic epigenetic regulatory mechanisms that underlie both genomic imprinting and normal gene regulation.

1 Historical Overview

The presence of genomic imprinting in mammals has considerable medical, societal, and intellectual implications in terms of (1) the clinical management of genetic traits and diseases, (2) the capacity to control human and animal breeding by assisted reproductive technologies, and (3) the progress of biotechnology and post-genomic medical research. Any modern-day discussion of genetic problems, whether in research or medicine, necessarily needs to consider whether a gene shows a biparental (i.e., diploid) mode of expression or is subject to genomic imprinting and shows parental-specific (i.e., haploid) expression. Despite the importance of genomic imprinting to human health and well-being, surprisingly, widespread acceptance of its existence and significance did not happen until the early 1990s after three genes were unequivocally shown to display parental-specific expression in mice.

Parental-specific behavior of whole chromosomes had, however, already been observed in cytogenetic studies of chromosomes in arthropods as early as the 1930s (Chandra and Nanjundiah 1990). Interestingly, the term "chromosome imprinting" was first coined to describe paternal-specific chromosome elimination that plays a role in sex determination in some arthropod species (Crouse et al. 1971). Chromosomal imprinting of the mammalian X chromosome was also noted, which leads to paternal-specific inactivation of one of the two X chromosomes in all cells of female marsupials and in the extraembryonic tissues of the mouse (Cooper et al. 1971). During the same period, classic geneticists were generating mouse mutants carrying chromosomal translocations that laid the foundation for the observation of imprinted gene expression. Some of these "translocation" mice, initially used to map the positions of genes on chromosomes, demonstrated a parental-specific phenotype when certain chromosomal regions were inherited as duplications of one parental chromosome in the absence of the other parental chromosome (known as uniparental disomy or UPD, Fig. 1). These results indicated the possibility "that haploid expression of particular maternal or paternal genes is important for normal mouse development" (Searle and Beechey 1978). At the same time, other geneticists used an unusual mouse mutant, known as the "Hairpin-tail" mouse that carried a large deletion of chromosome 17, to unequivocally set aside a basic tenet of genetics "that organisms heterozygous at a given locus are phenotypically identical irrespective of which gamete contributes which allele to the genotype" (Johnson 1974). Instead, offspring who received the Hairpin-tail deletion from a maternal parent were increased in

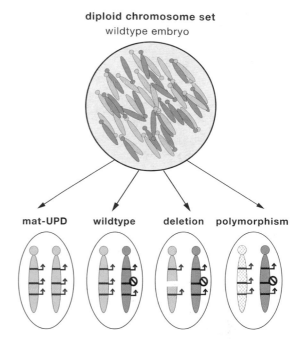

Figure 1. Mouse Models to Study Genomic Imprinting That Allow the Maternal and Paternal Chromosome to Be Distinguished

Mammals are diploid and inherit a compete chromosome set from the maternal and paternal parents. However, mice can be generated that (1) inherit two copies of a chromosome pair from one parent and no copy from the other parent (known as uniparental disomy or UPD); (2) inherit a partial chromosomal deletion from one parent and a wild-type chromosome from the other parent; (3) inherit chromosomes carrying single-nucleotide polymorphisms (known as SNPs) from one parent and a wild-type chromosome from the other parent. Offspring with UPDs or deletions are likely to display lethal phenotypes, whereas SNPs will allow the production of viable offspring. (mat) Maternal, (stop sign) the imprint.

size and died midway through embryonic development, whereas paternal transmission of the genetically identical chromosome produced viable and fertile mice (Fig. 1). It is notable with hindsight that despite the existence of imprinted X-chromosome inactivation in mammals, the favored interpretation of these genetic translocation and deletion experiments was that genes on these autosomes primarily acted in the haploid egg or sperm to modify proteins used later in embryonic development. Despite this, the concept of differential functioning of the maternal and paternal genome was gaining ground, and a suggestion was made that "the maternal genome might be normally active at the Hairpin-tail chromosomal region while its paternal counterpart is preferentially inactivated" (McLaren 1979).

A major step forward in establishing the existence of genomic imprinting in mammals came several years later with the development of an improved nuclear transfer technology being used to test the possibility of generating diploid uniparental embryos solely from mouse egg nuclei. The nuclear transfer technique took a donor male or female pronucleus from a newly fertilized egg and used a fine micropipette to place it inside a host fertilized egg from which either the maternal or paternal pronucleus had been removed. This recreated diploid embryos, but with the difference that they had two maternal or two paternal genomes (known respectively as gynogenetic and androgenetic embryos) (Fig. 2). The technique was first used to show that nuclei from fertilized Hairpin-tail mutant embryos could not be rescued when transferred into a wild-type host egg. This provided proof that the embryonic genome, and not the oocyte cytoplasm, carried the Hairpin-tail defect. It also confirmed the suggestion that genes on the maternal and paternal copies of chromosome 17 functioned differently during embryonic

development (McGrath and Solter 1984b). Subsequently, nuclear transfer was used to demonstrate that embryos reconstructed from two maternal pronuclei (known as gynogenetic embryos) or two paternal pronuclei (androgenetic embryos) failed to survive; only embryos reconstructed from one maternal and one paternal pronucleus produced viable and fertile offspring (McGrath and Solter 1984a; Surani et al. 1984). This work overturned a previous claim that uniparental mice could develop to adulthood (Hoppe and Illmensee 1982). Gynogenetic embryos at the time of death were defective in extraembryonic tissues that contribute to the placenta, and androgenetic embryos were defective in embryonic tissue. This led to the hypothesis that embryonic development required imprinted genes expressed from the maternal genome whereas the paternal genome expressed imprinted genes required for extraembryonic development (Barton et al. 1984). Subsequent identification of imprinted genes in the mouse did not confirm a bias in the function of imprinted genes, indicating that the observed differences between gynogenetic and androgenetic embryos may be explained by a dominant effect of one or a few imprinted genes.

The nuclear transfer experiments, combined with supporting data from mouse genetics, provided convincing evidence that both parental genomes were required for embryogenesis in mice and laid a strong foundation for the existence of genomic imprinting in mammals (Fig. 2). An extensive survey of parental chromosome contribution to embryonic development, using "translocation" mice to create UPD chromosomes (Fig. 1), identified two regions on mouse chromosomes 2 and 11 that showed opposite phenotypes when present either as two maternal or two paternal copies. This further strengthened the argument for parental-specific gene expression in mammals (Cattanach and Kirk 1985). In addition, data from human genetics clinics were being collected which strongly indicated that some genetic conditions, most notably the Prader-Willi syndrome that appeared to arise exclusively by paternal transmission, could best be explained by parental-specific gene expression (Reik 1989). Further clues came from applying the newly developed technology to make transgenic mice by microinjecting gene sequences into a fertilized mouse egg. This was often beset by the problem of DNA methylation, unexpectedly inducing silencing of the transgene in somatic tissues. This "problem," however, added weight to the argument that parental chromosomes behave differently, when it was demonstrated that some transgenes showed parental-specific differences in their

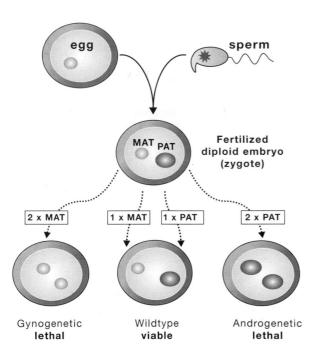

Figure 2. A Maternal Genome and a Paternal Genome are Needed for Mammalian Reproduction

The nuclear transfer technique used micropipettes and high-powered microscopes to remove the male or female nuclei from a newly fertilized egg and place them in various combinations into a second "host" fertilized egg that had already been enucleated, thereby generating anew, diploid embryos with two maternal (Gynogenetic) or two paternal (Androgenetic) genomes or a biparental genome (Wildtype). Gynogenetic and androgenetic embryos were lethal at early embryonic stages. Only reconstituted embryos that received both a maternal and paternal nucleus (Wildtype) survived to produce living young. These experiments show the necessity for both the maternal and paternal genome in mammalian reproduction and indicate that the two parental genomes express different sets of genes needed for complete embryonic development.

ability to acquire DNA methylation. This normally followed the pattern that maternal-transmitted transgenes were methylated whereas paternal-transmitted transgenes were not. However, only in a few cases did DNA methylation differences correlate with parental-specific expression. Although many similarities were later found to exist between "transgene" methylation imprinting and genomic imprinting of endogenous mouse genes, several features distinguish them (Reik et al. 1990). This includes a high susceptibility to background effects that in most cases required a mixed genetic background to reveal imprinted behavior, an inability to maintain imprinted expression at different chromosomal integration sites, and a requirement for foreign DNA sequences to produce the imprinted effect (Chaillet et al. 1995).

Despite the wealth of supportive data, final proof of the existence of genomic imprinting in mammals depended on the identification of genes showing imprinted parental-specific expression. This occurred in 1991 when three imprinted mouse genes were described. The first of these, *Igf2r* (Insulin-like growth factor type 2 receptor that is a "scavenger" receptor for the growth hormone *Igf2*), was identified as a maternally expressed imprinted gene. This gene was later shown to explain the overgrowth phenotype of the Hairpin-tail mutant mouse (Barlow et al. 1991). A few months later, the *Igf2* gene (Insulin-like growth factor type 2), which was known to function as a growth hormone, was identified as a paternally expressed imprinted gene (DeChiara et al. 1991; Ferguson-Smith et al. 1991). Finally, the *H19* gene (cDNA clone No. *19* isolated from a fetal *H*epatic library), an unusual noncoding RNA (ncRNA), was subsequently shown to be a maternally expressed imprinted gene (Bartolomei et al. 1991). Diverse strategies were used to identify these three imprinted genes, each of which depended on emerging technologies in mouse genetics. For *Igf2r*, positional cloning was used to identify genes that mapped

to the Hairpin-tail deletion on chromosome 17, and mice inheriting the deletion from one parent were used to identify those genes showing maternal-specific expression (Fig. 1). For *Igf2*, the physiological role of this growth factor in embryonic development was being tested by insertional mutagenesis. Surprisingly, mice carrying the mutant nonfunctional allele showed a phenotype following paternal transmission, but no phenotype on maternal transmission. In view of previous findings that foreign DNA sequences can induce imprinted expression of mouse genes, paternal-specific expression of *Igf2* from wild-type unmodified chromosomes was also confirmed using mice carrying reciprocal parental duplications and deficiencies of chromosome 7 (Fig. 1). The *H19* ncRNA was identified as an imprinted gene by testing the hypothesis that imprinted genes could be clustered together, after this gene was mapped close to the *Igf2* locus on chromosome 7. Although all these strategies were to prove useful in subsequent attempts to identify imprinted genes, the demonstration that imprinted genes were closely clustered has proven to be a pivotal discovery in understanding the mechanism controlling genomic imprinting in mammals.

2 Genomic Imprinting—An Epigenetic Gene Regulatory System

The defining characteristic of genomic imprinting is that it is *cis*-acting (see box below). Thus, the imprinting mechanism acts only on one chromosome. This contrasts with *trans*-acting gene regulatory mechanisms that are free to act on any chromosome in the nucleus. The two parental chromosomes will normally contain many single base pair differences (known as single-nucleotide polymorphisms, SNPs) if the population is outbred, but they can be genetically identical if inbred mouse strains are used. Because genomic imprinting is seen in inbred mice that have genetically identical parental chromosomes, the process must

---

KEY FEATURES OF GENOMIC IMPRINTING IN MAMMALS

- *cis*-Acting mechanism
- A consequence of inheritance, not sex
- Imprints are epigenetic modifications acquired by one parental gamete
- Imprinted genes are mostly clustered together with a noncoding RNA
- Imprints can modify long-range regulatory elements that act on multiple genes
- Imprinted genes play a role in mammalian development

use an epigenetic mechanism to modify the information carried by the DNA sequence and create an expression difference between the two parental gene copies. These observations also indicate that a *cis*-acting silencing mechanism is operating which is restricted to one chromosome so that the silencing factors cannot freely diffuse through the nucleus to reach the active gene copy. Although imprinted genes are repressed on one parental chromosome and active on the other, we do not know a priori that genomic imprinting is only a silencing mechanism. We must also consider that it could, conversely, be an activating mechanism directed toward a gene that is silenced by default in the mammalian genome.

The starting point for genomic imprinting must therefore depend on an epigenetic system that modifies or "imprints" one of the two parental chromosomes (Fig. 3). We can reason that this imprint is subsequently used to attract or repel transcriptional factors and so change expression of the imprinted gene on one parental chromosome. Because we know that inbred mice with genetically identical chromosomes also show genomic imprinting, we can reason that parental imprints cannot be acquired after the embryo becomes diploid because there would be no way for the cell's epigenetic machinery to distinguish between identical parental gene copies. Thus, parental imprints must be acquired when the two parental chromosome sets are separate, and this only occurs during gamete formation and for about 12 hours postfertilization (Fig. 3). The most likely scenario is that gametic imprints are placed on paternally imprinted genes during sperm production and on maternally imprinted genes during egg formation. A key feature about the "imprinted" DNA sequence is that it would only be modified in one of the two parental gametes; thus, two types of recognition systems are required, one sperm-specific and one oocyte-specific, each directed toward a different DNA sequence. Several other features are required of the imprint. First, once established, it must remain on the same parental chromosome after fertilization when the embryo is diploid. Second, the imprint must be inherited by the same parental chromosome following each cell division of the embryo and adult animal. Last, it must be erasable. The latter is necessary because the embryo will follow either a male or female developmental path midway through development, and its gonads will need to produce only one type of imprinted haploid parental gamete. Because germ cells have arisen from embryonic diploid cells (Fig. 3), they must first lose their inherited maternal and paternal imprints before they gain that of the gamete.

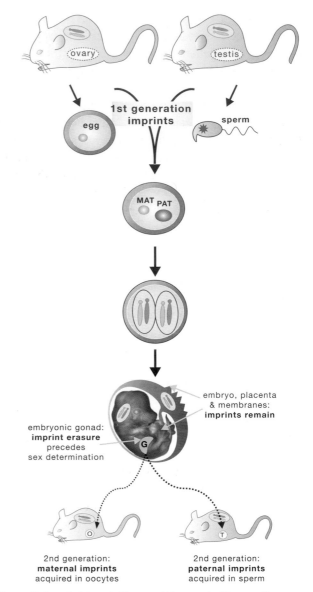

Figure 3. Imprint Acquisition and Erasure in Mammalian Development

Imprints are acquired by the gametes; thus, oocytes and sperm already carry imprinted chromosomes (1st generation imprints). After fertilization when the embryo is diploid, the imprint is maintained on the same parental chromosome after each cell division in cells of the embryo, membranes, placenta, and also in the adult. The germ cells are formed in the embryonic gonad, and the imprints are erased only in these cells prior to sex determination. As the embryo develops into a male, the gonads differentiate to testes that produce haploid sperm which acquire a paternal imprint on their chromosomes. Similarly, in developing females, chromosomes in the ovaries acquire maternal imprints (2nd generation imprints).

How are gametic imprints identified? Without being too caught up in semantics, an imprint can be defined as the epigenetic modification that distinguishes the maternal gene copy from the paternal gene copy. The imprint, once formed, must also allow the transcription machinery to treat the maternal and paternal gene copy differently, within the same nucleus. A gametic imprint is predicted to be continuously present at all developmental stages (Fig. 3); thus, imprints can be found by comparing epigenetic modifications on maternal and paternal chromosomes in embryonic or adult tissues (using strategies outlined in Fig. 1) and tracing them back in development to one of the two gametes. Gametic imprints could be modifications of DNA or of histone proteins that pack DNA into chromosomes. Although there is only one type of epigenetic DNA modification known in mammals, which is DNA methylation (see Chapter 18), histones can bear multiple types of modifications, including methylation, acetylation, phosphorylation, sumoylation, and ubiquitylation (see Chapter 10 for more detail). They can also be replaced by variant histones with specific functions (see Chapters 3 and 13). Any of these epigenetic modifications could theoretically qualify as an imprint. We can reason that enzymes responsible for these epigenetic modifications would be exclusively expressed in one of the two gametes and specifically associate with one parental chromosome to copy the modification when the cell divides. However, as described in Section 3, only DNA methylation has been clearly demonstrated to function as the gametic imprint for imprinted genes in mammals and, to date, is the only known heritable modification.

How does a gametic imprint operate to control imprinted expression? As reasoned above, we need to "keep an open mind" as to whether the imprint leads to activation or repression of one parental copy of an imprinted gene. To understand how the imprint operates, we need three pieces of information: which parental chromosome carries the imprint, which parental chromosome carries the expressed allele of the imprinted gene, and the position of the imprinted sequence relative to the expressed or silenced allele of the imprinted gene. Using this type of approach, we now know that gametic imprints can act on whole clusters of genes at once. These imprinted clusters contain between 3 and 10 imprinted genes and span from 100 kb to 3000 kb of genomic DNA (for more details, see http://www.mgu.har.mrc.ac.uk/research/imprinting/). The majority of genes in any one cluster are imprinted protein-coding mRNA genes; however, at least one is always an imprinted ncRNA (noncoding RNA).

Because of the arrangement of imprinted genes in clusters, with some genes expressed from one parental chromosome and some from the other, it is not always a simple matter to determine how the imprint operates. It is possible to study the effect of the imprint on single genes in the cluster, but it may prove more informative to study the effects of the imprint on the entire cluster. This is described in more detail in Section 3. One thing, however, is clear. Nature has not chosen the simplest model, whereby the imprint is directed toward a promoter to preemptively silence an imprinted gene in one gamete. Instead, imprints appear, in general, to be directed toward long-range *cis*-acting repressors that influence the expression of multiple genes, located a long distance away on the same chromosome.

3 Key Discoveries in Genomic Imprinting

3.1 Imprinted Genes Control Embryonic and Neonatal Growth

What is the function of genomic imprinting in mammals? One way to answer this question would be to determine the function of known imprinted genes in vivo. Modern technology now allows the function of mouse genes to be determined by mutating the gene sequence to impair its function. Using this "homologous recombination" technique, the functions of 26 of the 78 known imprinted genes have been determined (for original references, see http://www.mgu.har.mrc.ac.uk/research/imprinting/function.html). Table 1 lists these genes according to their function in mouse development and their expression from the maternal or paternal allele. The largest category to date comprises imprinted genes that affect growth of the embryo, or placenta, or the neonate fully dependent on its mother's milk. In this category, approximately half are paternally expressed imprinted genes that function as growth promoters (as demonstrated by a growth retardation in embryos deficient for the gene). The other half are maternally expressed imprinted genes that function as growth repressors (as demonstrated by a growth enhancement in embryos deficient for the gene). The next-largest category includes genes with no obvious defects in embryonic development, followed by the category with behavioral or neurological defects. The remaining 3 tested genes have various, apparently unlinked, defects. These results are at one level disappointing, since they do not identify one function for all imprinted genes. However, there may be light at the end of the tunnel, because these results do tell us that more than 50% of imprinted genes function as embryonic or neonatal growth regulators. More interestingly, the ability to regulate growth appears to be neatly divided, with maternally expressed growth-regulating genes acting to repress growth of the offspring, whereas paternally expressed

Table 1. The function of imprinted genes as determined by gene inactivation

Maternal	Gene function	Paternal
–Igf2r	growth defects in	+Igf2
–Gnas	embryo, placenta,	+Gnasxl
–Tssc3/Ipl	or postnatal stage	+Peg1/Mest
–Mash2		+Peg3/Pw1
–Grb10/Meg1		+Rasgrf1
–/+ Cdkn1c		+Dlk1
Nesp	behavioral or	+Peg1/Mest
Ube3a	neurological	+Peg3/Pw1
Kcnq1*	defects	+Rasgrf1
Asb1[spermatogenesis]	other defects	Ndn[strain-specific lethality]
Dcn[Tumor suppressor]		
H19 ncRNA	no obvious defects	Snrpn/Snurf
Slc22a2	in embryo or	Frat3
Slc22a3	neonate	Ins2

(Maternal) Maternally expressed imprinted gene, (Paternal) paternally expressed imprinted genes, (+) growth promoting effect, (–) growth suppressing effect, (–/+) defect in differentiation but growth regulatory status unclear, (*) additional differentiation defect. (Reference to the primary data can be found at: http://www.mgu.har.mrc.ac.uk/research/imprinting/function.html).

genes in this category act to increase growth. 20% of tested imprinted genes are active in neurological processes, some of which affect neonatal growth rate by altering maternal behavior. The most puzzling category, in view of attempts to identify a selective force driving the acquisition of imprinted gene expression in all extant mammals, is that of imprinted genes with no obvious biological function in embryonic development, which contains 25% of tested imprinted genes.

3.2 The Function of Genomic Imprinting in Mammals

Can analyses of gene function help us understand why genes are imprinted in mammals? A look at genomic imprinting in different types of mammals has shed some light. Placental mammals such as mice and humans, and marsupials such as opossum and wallaby, have genomic imprinting. Egg-laying mammals, such as platypus and echidna, appear to lack imprinted genes, although extensive studies have not yet been performed. Placental mammals and marsupials are distinguished by a reproductive strategy that allows the embryo to directly influence the amount of maternal resources used for its own growth. In contrast, embryos that develop within eggs are unable to directly influence maternal resources. Most invertebrates

and vertebrates use an egg-laying reproductive strategy. Notably, they can also undergo parthenogenesis—a form of reproduction in which the female gamete develops into a new diploid individual without fertilization by a male gamete (note that parthenogenetic embryos arise from the duplication of the same maternal genome, whereas the gynogenetic embryos described in Fig. 3 arise from two different maternal genomes). The ability of organisms to undergo parthenogenesis most likely indicates a complete absence of genomic imprinting, as it shows that the paternal genome is dispensable. In mammals, however, a direct consequence of imprinted gene expression controlling fetal growth is that parthenogenesis is not possible. Both the maternal and paternal parents are necessary to produce viable offspring, making mammals completely reliant on sexual reproduction to produce viable offspring (Fig. 4). Parthenogenesis has not yet been observed in mammals despite claims to the contrary, although some rare mice with a diploid maternal genome were recently created by manipulating expression of the *Igf2* imprinted cluster (Kono et al. 2004).

Why should genomic imprinting have evolved only in some mammals but not in vertebrates in general? Three features of genomic imprinting—the growth regulatory

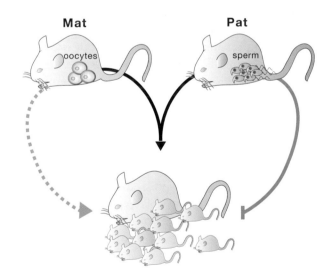

Figure 4. Imprinted Genes Play a Role in Mammalian Reproduction

Mammals are diploid, and reproduction requires fertilization of a haploid female egg by a haploid male sperm to recreate a diploid embryo. Only females are anatomically equipped for reproduction, but they cannot use parthenogenesis to reproduce because essential imprinted genes needed for fetal growth are imprinted and silenced on maternal chromosomes. These genes are expressed only from paternal chromosomes; thus, both parental genomes are needed for reproduction in mammals. Parthenogenesis is the production of diploid offspring from two copies of the same maternal genome.

function of many imprinted genes, the restriction of imprinted genes to placental and marsupial mammals, and last, the necessity of the paternal genome for fetal development—provide evidence that can fit two equally attractive hypotheses.

The first hypothesis proposes that genomic imprinting evolved in response to a "parental conflict" situation (Moore and Haig 1991). This arises from the opposing interests of the maternal and paternal genomes: Embryonic growth is dependent on one parent but influenced by an embryo whose genome comes from two parents. Paternally expressed imprinted genes are proposed to increase embryonic growth, thereby maximizing the competitiveness of individual offspring bearing a particular paternal genome. Maternally expressed imprinted genes are proposed to suppress fetal growth. This would allow a more equal distribution of maternal resources to all offspring and increase transmission of the maternal genome to multiple offspring, which may have different paternal genomes.

The second hypothesis is named "trophoblast defense" (Varmuza and Mann 1994). This proposes that the maternal genome is at risk from the consequences of being anatomically equipped for internal reproduction, should spontaneous oocyte activation lead to full embryonic development. Because males lack the necessary anatomical equipment for internal reproduction, they do not share the same risks should spontaneous activation of spermatozoa occur. Imprinting is thus proposed either to silence genes on the maternal chromosome that promote placental development, or to activate genes that limit this process. The genes necessary for placental formation would consequently only be expressed from a paternal genome after fertilization has occurred.

Which, if either, of these hypotheses is the right explanation for the evolution of genomic imprinting in mammals? Both hypotheses indicate a role for imprinted genes in regulating the development and function of the placenta; however, neither the parental conflict nor the trophoblast defense models can provide a full explanation for all the data (Wilkins and Haig 2003). It is interesting to note that imprinted genes have also been identified in the plant endosperm, a tissue that has been compared to the placenta of mammalian embryos because it transfers nutrient resources from the parent plant to the embryo (for more information on genomic imprinting in plants, see Chapter 11). This finding strengthens arguments that genomic imprinting evolved as a means to regulate nutrient transfer between the parent and offspring, but it does not tell us why. Fuller or alternative explanations of the function of genomic imprinting in mammals could come from two sources. The first would be to examine the function of "imprinting" across a complete gene cluster per se, in contrast to examining the phenotype of mice lacking a single imprinted gene product. This would require an ability to reverse an imprint and generate biparental gene expression across the whole imprinted cluster. The second approach is to learn exactly how genes are imprinted. It is possible that not all genes in a cluster are a deliberate target of the imprinting mechanism and that some may just be "innocent bystanders" of the process so their function would not be informative about the role of genomic imprinting. The existence of innocent-bystander genes affected by the imprinting mechanism may satisfactorily explain the curious abundance of imprinted genes with no obvious biological function in development (Table 1).

3.3 Imprinted Genes Are Clustered and Controlled by Imprint Control Elements

To date, about 80 imprinted genes have been mapped to ten mouse chromosomes, the majority of which are found in clusters (Verona et al. 2003). Eleven clusters of imprinted genes have been assigned to eight chromosomes (numbers 2, 6, 7, 9, 11, 12, 15, and 17), whereas solo imprinted genes have only been identified on three chromosomes (numbers 2, 14, and 18). The existence of clusters of imprinted genes was a strong indication that a common DNA element may regulate imprinted expression of multiple genes in *cis*. To date, only six of the imprinted clusters have been well characterized, and these are listed in Table 2 by the name of the principal imprinted mRNA gene in the cluster (i.e., the *Igf2r*, *Igf2*, *Kcnq1*, *Gnas*, *Dlk1* imprinted gene clusters) or after a disease association (the *Pws* cluster, Prader-Willi syndrome, discussed in more detail in Chapter 23). These six clusters contain from three to ten imprinted genes and are spread over 100–3000 kb of DNA.

Figure 5 shows the parental-specific expression pattern in a typical imprinted gene cluster. A common feature of these six clusters is the presence of a DNA sequence carrying a gametic methylation imprint that is known as a gametic DMR (Differentially DNA-Methylated Region). A gametic methylation imprint is defined as a methylation imprint established in one gamete and maintained only on one parental chromosome in diploid cells of the embryo. In four clusters (*Igf2r*, *Kcnq1*, *Gnas*, and *Pws*), the gametic DMR has a maternal methylation

Table 2. Features of imprinted gene clusters in the mouse genome

Cluster type	Cluster name	Chromosome mouse/human	Gametic methylation imprint	Cluster size (kb)	Gene number in cluster	mRNAs and expression	ncRNA and expression	ncRNA orientation
Type I	Igf2r	17 / 6	M	400	4	Igf2r (M) Slc22a2 (M) Slc22a3 (M)	Air (P)	antisense to Igf2r
	Kcnq1	7 / 11	M	700	10	Mash2 (M) Kcnq1 (M) Cd81 (M) Cdkn1c (M) Msuit (M) Slc22l1 (M) Ipl (M) Tssc4 (M) Obph1 (M)	Kcnq1ot1 (P)	antisense to Kcnq1
	Pws	7 / 15	M	3000	~7	Ube3a (M) Atp10c (M) Frat3 (P) Mkrn3 (P) Ndn (P) Magel2 (P) Snrpn (P)	*Ube3aas (P) *IPW (P) *Mkrn3as (P) *PEC2 (P) *PEC3 (P) *Pwcr1 (P) *may be one long ncRNA	antisense to Ube3a (also overlaps Snrpn in sense orientation)
	Gnas	2 / 20	M (x 2)	100	5	Nesp (M) Gnas (M) Gnasxl (P)	[1]Nespas(P) [2]Exon1A (P)	[1]antisense to Nesp [2]sense to Gnas
Type II	Igf2	7 / 11	P	100	3	Igf2 (P) Ins2 (P)	H19 (M)	sense no overlaps
	Dlk1	9 / 14	P	1000	7	Dlk1 (P) Dio3 (P) Rtl1 (P)	Gtl2 (M)* Rian (M)* Rtl1as (M)* Mirg (M)* *region may contain longer ncRNAs	sense to Dlk1 and also antisense to Rtl1

(M) Maternal, (P) paternal, (DMR) differentially methylated region. Details are given in the text. (Modified from Beechey et al. 2005 [http://www.mgu.har.mrc.ac.uk/research/imprinting].)

imprint acquired in oogenesis, whereas in two clusters (*Igf2* and *Dlk1*), it has a paternal methylation imprint acquired during spermatogenesis. In all six examples, the gametic DMR has been shown to control imprinted expression of the whole or part of the cluster and is therefore designated as the imprint control element (ICE) for the cluster (Spahn and Barlow 2003).

Table 2 shows that each imprinted gene cluster contains multiple mRNAs and at least one ncRNA. Four clusters (*Igf2r*, *Kcnq1*, *Igf2*, and *Dlk1*) have a simple pattern in which the chromosome carrying the methylated gametic DMR expresses multiple mRNAs but does not express the ncRNA (as illustrated in Fig. 5 for a maternal gametic DMR). The chromosome carrying the unmethylated gametic DMR shows the reciprocal expression pattern: repression of the multiple mRNAs and expression

of the ncRNA. The remaining two clusters (*Gnas* and *Pws*) have a complex pattern where imprinted mRNAs are expressed from both chromosomes while the imprinted ncRNA is expressed only from the chromosome carrying the unmethylated gametic DMR. Table 2 shows that in three clusters (*Igf2r*, *Kcnq1*, and *Gnas*) the ncRNA promoter sits in an intron of one of the imprinted mRNAs, whereas in the remaining clusters, the ncRNA promoter is separated but lies close to the imprinted mRNA genes. This close intermingling of active and silent genes in an imprinted cluster indicates that the silencing and activating mechanisms affecting imprinted genes do not spread and may be restricted to the affected gene. In particular, the fact that the promoter of a silent ncRNA can reside in the intron of an actively transcribed gene indicates that silencing mechanisms

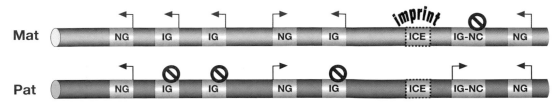

Mat

Pat

Figure 5. Imprinted Genes Are Expressed from One Parental Allele and Often Clustered

Most imprinted genes are found in clusters that include multiple protein-coding mRNAs and at least one noncoding RNA (ncRNA). Non-imprinted genes can also be present. The imprinting mechanism is *cis*-acting, and imprinted expression is controlled by an imprint control element that carries an epigenetic imprint inherited from one parental gamete. One pair of diploid chromosomes is shown pink (maternally expressed imprinted gene) and blue (paternally expressed imprinted gene). (IG) Imprinted mRNA gene, (IG-nc) imprinted ncRNA gene, (NG) non-imprinted gene, (ICE) imprint control element, (*arrow*) expressed gene, (*filled circle*) repressed gene.

may not even spread throughout the length of a gene, but may just be restricted to regulatory elements.

What is the role of the gametic DMR? Despite the fact that the gametic DMRs can be maternally or paternally methylated, experiments that deleted these elements have produced broadly similar results, albeit with a few interesting exceptions (Fig. 6). For three clusters (*Igf2r*, *Kcnq1*, *Dlk1*), experimental deletion of the methylated gametic DMR produced no effect. However, deletion of the unmethylated gametic DMR completely reversed the parental-specific expression pattern such that ncRNA expression was lost and biallelic mRNA expression was obtained (Lin et al. 1995; Zwart et al. 2001; Fitzpatrick et al. 2002). Two clusters (*Gnas* and *Pws*) appear to contain

more than one gametic DMR and show a more complex behavior that still shares some similarities with this pattern (Williamson et al. 2006). The *Igf2* cluster, however, behaves differently: Deletion of both the methylated and unmethylated gametic DMR causes changes in mRNA and ncRNA expression in *cis* (Thorvaldsen et al. 1998).

The results from the above gametic DMR deletion experiments do not at first glance indicate a common function for gametic DMRs. However, an understanding of their exact function depends on knowing the position of the DMR with respect to the imprinted genes in each cluster. In the three clusters with the simplest pattern (*Igf2r*, *Kcnq1*, and *Dlk1*), the gametic DMR either contains or controls expression of the ncRNA; thus, deletion

IMPRINTED CHROMOSOME

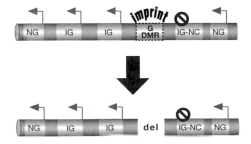

no changes in gene expression

NON-IMPRINTED CHROMOSOME

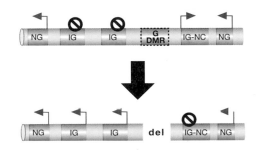

expression resembles 'imprinted' chromosome

Figure 6. Imprinted Expression Is Regulated by Gametic DMRs

Left panel shows the effect of deleting the gametic DMR from the imprinted chromosome (*green*). Right panel shows the effect of deleting the gametic DMR from the non-imprinted chromosome (*yellow*). In many imprinted clusters (e.g., *Igf2r*, *Kcnq1*, and *Dlk1*), experimental deletion of the G-DMR only affects the chromosome carrying the non-imprinted G-DMR. This results in a loss of repression of the imprinted protein-coding mRNA genes (IG) and a gain of repression of the imprinted ncRNA gene (IG-NC). Note that in some imprinted clusters (Igf2 and Pws) that are not illustrated here, the methylated G-DMR appears also to be required for expression of some of the imprinted mRNAs in *cis*. (del) Deleted DNA, (G-DMR) gametic differentially DNA-methylated region, (NG) non-imprinted gene, (*arrow*) expressed allele, (*black stop sign*) repressed allele, (IMPRINT) epigenetic modification leading to a change in gene expression in *cis*.

of this element will clearly lead to loss of ncRNA expression. The gametic DMR in the *Igf2* cluster, however, does not directly control the *H19* promoter, but changes the interaction between *Igf2* and *H19* and their shared enhancers, and in this way regulates their expression. Despite these differences, in general, the unmethylated gametic DMR is implicated in all six clusters as a positive regulator of ncRNA expression, and the presence of the DNA methylation imprint is associated with repression of the ncRNA. The data obtained from deletion of gametic DMRs clearly identify these regions as the major ICE, whose activity is regulated by DNA methylation.

Although the data are still in an early stage, attempts have already been made to define two types of imprinted clusters (Table 2). The type I and type II clusters both show the same general behavior of reciprocal parental-specific expression between the ncRNA and multiple mRNA genes but can be differentiated in three ways. First, type I clusters carry a maternal DNA methylation imprint (*Igf2r*, *Kcnq1*, *Pws*, and *Gnas*), whereas type II clusters carry a paternal DNA methylation imprint (*Igf2* and *Dlk1*). Second, type I clusters express a ncRNA from the paternal chromosome, whereas the type II express ncRNA from the maternal chromosome. Third, in type I clusters the ncRNA promoter lies in an intron and generates a transcript in an antisense orientation with respect to one of the multiple imprinted mRNA genes, whereas in type II clusters the promoter lies downstream and has a sense orientation with respect to at least one imprinted mRNA gene (Regha et al. 2006).

3.4 Imprinted Gene Clusters Contain at Least One ncRNA

Of the 12 known imprinted gene clusters, 8 are associated with a ncRNA (O'Neill 2005). ncRNAs, with the exception of those involved in RNA processing and translation, such as splicing, transfer, and ribosomal RNAs, were earlier thought to be a rarity in the mammalian genome. Now, thanks to the availability of the mouse and human genome sequences, transcriptome analyses can be performed that list all RNA transcripts in a given cell population. This has already shown that the majority of the mammalian transcriptome (not counting those ncRNAs associated with processing and translation) is, surprisingly, largely composed of ncRNAs instead of protein-coding mRNAs. The sheer abundance of ncRNA transcription in the mammalian genome, and the fact that a large number of ncRNAs overlap a known protein-coding gene, indicate that this cannot be viewed as "transcription noise" but is

likely to constitute a new, hitherto unknown, gene regulatory system (Mattick 2005). A Web site named NONCODE has been formed to collect data from all organisms on functional ncRNAs (http://noncode.bioinfo.org.cn/). There are several types of mammalian ncRNAs that have been shown to possess gene regulatory functions including "short" ncRNAs from 21 bp to 30 bp that participate in RNA interference pathways (Chapter 8), "intergenic" unprocessed transcripts that regulate local chromatin activity (Haussecker and Proudfoot 2005), and "long" processed ncRNAs such as *Xist*, which is involved in X-chromosome inactivation (Chapter 17).

What types of ncRNAs are associated with imprinted gene clusters? The analysis of the ncRNAs associated with the six well-characterized imprinted clusters shown in Table 2 is still incomplete, highlighting some similarities but also some differences. Three imprinted ncRNAs are unusually long mature RNAs: *Air* is 108 kb (Lyle et al. 2000), *Kcnq1ot1* is at least 60 kb, but the final size has not yet been determined (Mitsuya et al. 1999), and *Ube3aas* may be in excess of 1000 kb (Landers et al. 2004). The *H19* ncRNA, in contrast, is only 2.3 kb (Brannan et al. 1990). The *Gtl2* ncRNA contains multiple alternatively spliced transcripts; however, downstream intergenic transcription has also been noted, suggesting that longer transcription units are likely (Tierling et al. 2005). *Nespas* ncRNA is larger than can be resolved on RNA blots, and the full size is unknown (Wroe et al. 2000). All these imprinted ncRNAs appear to be intron-poor with a low intron–exon ratio or are unspliced as mature transcripts. It was earlier suggested that all imprinted genes were intron-poor; however, this may only be true for imprinted ncRNAs (Hurst et al. 1996). One further feature is that two imprinted ncRNAs (*Ube3aas* and the *Gtl2 downstream transcripts*) act as host transcripts for snoRNAs (small nucleolar RNAs that direct modifications to rRNA, snRNAs, and possibly mRNAs, thereby acting as posttranscription regulators) and miRNAs (microRNAs involved in transcriptional and posttranscriptional mRNA regulation). The snoRNAs are not directed toward the imprinted mRNA genes in the cluster, and it is likely that they play no role in the imprinting mechanism itself (Seitz et al. 2004). Similarly, the miRNAs in the *Dlk1* cluster are involved in posttranscriptional repression of one of the mRNA genes in the cluster, but their role in regulating imprinted expression of the cluster has not yet been tested (Davis et al. 2005).

Two features of imprinted ncRNAs indicate that they may play a role in the silencing of the imprinted mRNA (i.e., protein-coding) genes in the cluster. The first is that the ncRNA generally shows reciprocal parental-specific expression compared to the imprinted mRNA genes (Table 2). Second, the DMR that carries the gametic methylation imprint, which controls imprinted expression of the whole cluster, overlaps the ncRNA promoter in three instances (*Air*, *Kcnq1ot1*, *Gnas Exon1a*). This could indicate that imprints evolved to regulate the ncRNA in each imprinted cluster. This interpretation is supported by experiments that deleted the unmethylated sequence carrying the gametic DMR, causing a loss of ncRNA expression concomitant with a gain of expression of imprinted mRNA genes (Fig. 6), as tested at the *Igf2r*, *Kcnq1*, *Gnas*, *Pws*, and *Dlk1* clusters (Wutz et al. 1997; Bielinska et al. 2000; Fitzpatrick et al. 2002; Lin et al. 2003; Williamson et al. 2006). The *Igf2* cluster is an exception, because the unmethylated gametic DMR appears not to be a direct regulator of the *H19* ncRNA (Thorvaldsen et al. 1998).

Experiments that directly test the role of the ncRNA itself have now been performed for two ncRNAs from type I imprinted gene clusters (*Air* and *Kcnq1ot1*) and one ncRNA (*H19*) from a type II cluster. Truncation of the 108-kb *Air* ncRNA to 3 kb showed that the ncRNA itself is necessary to silence all three mRNA genes in the *Igf2r* cluster, indicating a clear regulatory role for this ncRNA (Sleutels et al. 2002). In addition, truncation of the 60-kb *Kcnq1ot1* ncRNA to 1.5 kb also showed that this ncRNA was directly needed to silence all ten mRNA genes in the larger *Kcnq1* cluster (Mancini-DiNardo 2006). In contrast, precise deletion of the *H19* ncRNA and promoter had no effect on imprinting in the *Igf2* cluster in endoderm tissues, although some loss of imprinting was seen in mesoderm tissue (Schmidt et al. 1999). Thus, two of the type I maternally imprinted clusters share a common ncRNA-dependent silencing mechanism, whereas the single type II paternally imprinted cluster so far examined uses a different, insulator-dependent model. The results from other imprinted clusters are eagerly awaited to see whether this indicates that there are only two types of basic imprinting mechanisms in mammals, one for paternally imprinted clusters and the other for maternally imprinted clusters.

3.5 The Role of DNA Methylation in Genomic Imprinting

The identification of the first three endogenous imprinted genes in 1991 enabled investigators to study how the cell's epigenetic machinery marked an imprinted gene with its parental identity. The first and most easily testable candidate was DNA methylation, a modification in mammals that covalently adds a methyl group to the cytosine residue in any CpG dinucleotide. DNA methylation is acquired through the action of de novo methyltransferases and maintained in situ each time the cell divides by the action of maintenance methyltransferases (described in Chapter 18). Hence, this modification fulfils the criteria outlined in Figure 3 for a parental identity mark or "imprint" because (1) it can be established in either the sperm or oocyte by de novo methyltransferases that act only in one gamete, (2) it can be stably propagated at each embryonic cell division by a maintenance methyltransferase, and (3) it can be erased in the germ line to reset the imprint in the next generation, either by passive demethylation or possibly through the action of a demethylase.

DNA methylation could potentially perform two different functions in genomic imprinting. It could act as the imprinting mark by being acquired de novo only by the chromosomes of one gamete. It could also serve to silence one of the parental alleles, since DNA methylation is associated with gene repression. To determine which function it has, it is first necessary to show that DNA methylation is present only on one parental chromosome (i.e., that it is a DMR). Second, it is necessary to identify which imprinted gene in the cluster and which part of the gene's regulatory apparatus are marked by DNA methylation. The location of methylation marks on a promoter, or on distant positive or negative regulatory elements, will have different consequences for gene expression. Finally, it is necessary to identify when the DMR forms during development. If it forms during gametogenesis and is maintained in place in somatic cells (known as a gametic DMR), it may serve as the imprinting mark. If, however, it is placed on the gene after the embryo has become diploid when both parental chromosomes are in the same cell (known as a somatic DMR), it is unlikely to serve as the identity mark, but it may serve to maintain parental-specific silencing.

Parental allele-specific DNA methylation has been found at most imprinted clusters that have been examined. For example, the *Igf2* cluster has a gametic DMR located 2 kb upstream of the *H19* ncRNA promoter, which is methylated only in the paternal gamete and is maintained thereafter in all somatic tissues (Bartolomei et al. 1993; Ferguson-Smith et al. 1993). A similar gametic DMR was identified covering the promoter of the *Air* ncRNA, present only on the silent maternal gene copy and acquired in the female gamete (Stoger et al. 1993). Surprisingly, gametic DMRs were not identified at the promoters of the principal imprinted protein-coding genes in these clusters (respectively, *Igf2* and

Igf2r). Instead, the silenced *Igf2* promoter is free of DNA methylation, whereas the silenced *Igf2r* promoter lies within a somatic DMR (Sasaki et al. 1992; Stoger et al. 1993). Similar findings of gametic DMRs methylated on the chromosome carrying the silent copy of the imprinted ncRNA have been made for the four other well-studied imprinted gene clusters, *Pws*, *Kcnq1*, *Gnas*, and *Dlk1* (Shemer et al. 1997; Liu et al. 2000; Takada et al. 2002; Yatsuki et al. 2002).

Somatic DMRs occur more rarely and have been reported only on a few imprinted protein-coding genes in each cluster, indicating that DNA methylation may play only a limited role in maintaining imprinted gene expression (Stoger et al. 1993; Moore et al. 1997; Yatsuki et al. 2002). Deletions of gametic DMRs in mice result in complete loss of imprinting for multiple genes, thereby proving that this class of DMRs also serves as a major ICE for the whole cluster (Fig. 6) (Wutz et al. 1997; Thorvaldsen et al. 1998; Bielinska et al. 2000; Fitzpatrick et al. 2002; Lin et al. 2003; Williamson et al. 2006). In contrast, deletion of the somatic DMRs affects expression of the associated imprinted protein-coding gene, but imprinted expression is maintained by other genes in the cluster (Constancia et al. 2000; Sleutels et al. 2003).

A genome-wide deficiency in DNA methylation, caused by mutations in the *Dnmt* gene family, underscores its essential role in regulating imprinted gene expression. Mutations in the de novo methylase *Dnmt3a*, the methylase stimulatory factor *Dmnt3L*, or the *Dnmt1* maintenance methylase generate DNA methylation-deficient embryos that all exhibit alterations in imprinted gene expression (see Chapter 18). The type of perturbation shown for four imprinted clusters (*Igf2*, *Igf2r*, *Kcnq1*, and *Dlk1*) indicates that DNA methylation is generally acting to suppress the action of the gametic DMR on the same parental chromosome that expresses the clustered mRNA genes. Thus, in the absence of DNA methylation, the gametic DMR cannot function appropriately. As a consequence, several imprinted protein-coding genes, including *Igf2*, *Igf2r*, *Kcnq1*, and *Dlk1*, become repressed on both parental chromosomes. This indicates that these mRNA genes are silenced by default in the mammalian genome and require epigenetic activation to be expressed. Notably, the *H19* ncRNA that is normally only expressed on the chromosome carrying the unmethylated gametic DMR becomes expressed on both parental chromosomes (Chapter 18). Some exceptions to this general pattern have been reported for genes that show imprinted expression only in the placenta (Lewis et al. 2004).

Are other types of epigenetic modifications used as gametic imprints? Given the sheer abundance of epigenetic mechanisms acting to modify genetic information in the mammalian genome (described in Chapter 3), DNA methylation is unlikely to be the only imprinting mechanism. Histone modifications that affect chromatin activity states are also likely candidates for parental imprints since they could fulfil many of the prerequisites shown in Figure 3. To date, however, only a Polycomb Group component protein known as *Eed* (that facilitates histone H3 lysine 27 methylation) has been shown to affect a few paternally repressed genes. The effects of *Eed* on genomic imprinting, however, are relatively minor compared to that of DNA methylation, and it may only serve a maintenance function (Mager et al. 2003).

How are DMRs selected by the gametic methylation machinery? A sequence comparison of known gametic DMRs reveals no striking sequence conservation, although some have been reported to contain a series of direct repeats that may adopt a secondary structure which attracts DNA methylation (Neumann et al. 1995). The sequence of DMRs is markedly CpG-rich compared to the remainder of the genome and resembles that of CpG islands associated with the promoters of more than half of the genes in the mammalian genome (Chapter 18). Strikingly, a key feature of CpG island promoters is that they normally lack DNA methylation and that mechanisms exist in early embryonic cells to keep promoter CpG islands methylation-free (Antequera 2003). However, CpG islands can become methylated in tumors and during aging.

Two observations have shed light on how DNA methylation could target DMR sequences. The first observation is that the paternal specificity of methylation at the *H19* gametic DMR depends on prevention of default methylation of this region in the maternal gamete by the CTCF protein (Fedoriw et al. 2004, and see Section 3.6). This may indicate a lack of sequence specificity of the DNA methylation system. The second observation is that the methylation accessory protein *Dnmt3L* plays different roles in male and female gametes. In male gametes, *Dnmt3L* plays a major role in the methylation and silencing of retrotransposons and a minor role in DMR methylation. In contrast, in oocytes, *Dnmt3L* plays a major role in DMR methylation and plays no role in retrotransposon methylation (Bourc'his and Bestor 2006). Retrotransposons are mobile genetic elements that are present in very high numbers in the mammalian genome and can be copied via RNA intermediates and inserted at new genomic sites (Kazazian 2004). The finding that the same protein which directs methylation to silence retrotransposons can also direct DMR methylation indicates that mechanisms needed for genome defense against invading DNA have been co-opted to make imprints in the maternal germ line (Barlow 1993).

3.6 Two Types of cis-Acting Silencing Identified in Imprinted Gene Clusters

Currently, two classes of *cis*-acting silencing mechanisms are hypothesized to govern imprinting at various clusters: the insulator model applicable to the *Igf2* cluster and the ncRNA-mediated silencing model applicable to the *Igf2r* and *Kcnq1* clusters. Although not yet completely defined, most of the clusters in Table 2 incorporate aspects of one of the two models. The breakthrough that led to the definition of the insulator model at the *Igf2* locus was the deletion of the gametic DMR that is located 2 kb upstream of the start of *H19* transcription and 80 kb downstream of *Igf2* (Fig. 7) (Thorvaldsen et al. 1998). When deleted, *H19* and *Igf2* exhibited a loss of imprinting regardless of whether the deletion was inher-

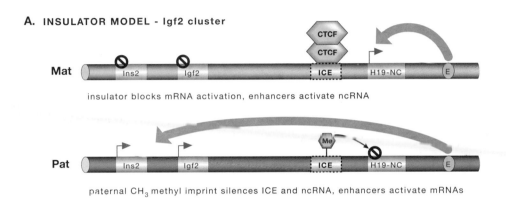

A. INSULATOR MODEL - Igf2 cluster

insulator blocks mRNA activation, enhancers activate ncRNA

paternal CH$_3$ methyl imprint silences ICE and ncRNA, enhancers activate mRNAs

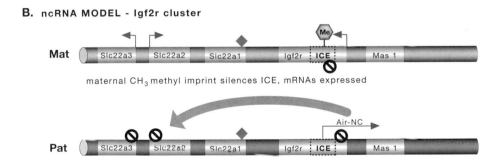

B. ncRNA MODEL - Igf2r cluster

maternal CH$_3$ methyl imprint silences ICE, mRNAs expressed

Air ncRNA silences 3 genes *in cis*

Figure 7. Two *cis*-Acting Silencing Mechanisms at Imprinted Gene Clusters

(*a*) Insulator model for the Igf2 cluster. The expression pattern for endoderm is shown. On the maternal chromosome the unmethylated ICE binds the CTCF protein and forms an insulator that prevents the common endoderm enhancers (E) from activating Igf2 and Ins2. Instead, the enhancers activate the nearby H19 ncRNA promoter. On the paternal chromosome, the methylated ICE cannot bind CTCF and an insulator does not form, hence the Igf2 and Ins2 mRNA genes are expressed only on this chromosome. The H19 ncRNA is methylated most likely because of spreading from the 2-kb distant methylated ICE, and silenced. (*b*) ncRNA model for the *Igf2r* cluster. The expression pattern for placenta is shown. On the maternal chromosome, the methylated ICE contains the Air ncRNA promoter that is directly silenced by the DNA methylation imprint. The *Igf2r*, *Slc22a2*, and *Slc22a3* mRNA genes are expressed only on this chromosome. *Mas1* and *Slc22a1* are not expressed in placenta. On the paternal chromosome, the Air ncRNA promoter lying in the unmethylated ICE is expressed, and silences *Igf2r*, *Slc22a2*, and *Slc22a3* in *cis*. Note that in both models, the DNA methylation imprint silences the ncRNA and permits mRNA expression. (ICE) Imprint control element, (*gray box*) imprinted mRNA gene, (gene-NC) imprinted ncRNA gene, (*arrow*) expressed allele of an imprinted gene, (*black stop sign*) repressed allele of an imprinted gene, (*filled diamond*) tissue-specific gene silenced on both parental chromosomes, (*gray arrows*) long-distance effect in *cis*.

ited maternally or paternally, identifying this DMR as an ICE. It was subsequently shown that this ICE bound CTCF, a protein shown to mediate insulator activity at the *beta-globin* locus, and that the ICE itself functioned as an insulator (Bell and Felsenfeld 2000; Hark et al. 2000). In this context, an insulator is defined as an element that blocks enhancer and promoter interactions when placed between them. Thus, the model for imprinted gene expression at this locus is as follows: On the maternal allele, CTCF binds to the ICE and blocks the access of *Igf2* and *Ins2* to enhancers shared with the *H19* ncRNA, which are located downstream of the three genes. This thereby allows *H19* exclusive access to the enhancers (Fig. 7). On the paternal allele, the ICE acquires DNA methylation in the male germ line, preventing CTCF from binding to it. Thus, on the paternal chromosome, *Igf2* and *Ins2* interact with the enhancers and are expressed from this chromosome. The presence of DNA methylation on the paternal ICE leads to secondary methylation of the *H19* promoter by an unknown mechanism, and it becomes silenced on the paternal chromosome. The involvement of CTCF in the insulator model has led to the identification of CTCF-binding sites at other imprinted genes such as *Rasgrf1*, *Grb10*, and *Kcnq1ot1*, indicating that the insulator model may operate in other imprinted clusters.

The ncRNA class of imprinting model may, however, be more common. The breakthrough that led to the identification of functional ncRNAs in imprinted clusters was an experiment that truncated the 108-kb *Air* ncRNA to 3 kb (Sleutels et al. 2002). This shortened ncRNA retained imprinted expression and the *Air* promoter retained imprinted DNA methylation—yet silencing of all three mRNA genes in the *Igf2r* cluster was lost (Fig. 7). ncRNA-mediated silencing has also now been shown to operate at the *Kcnq1* cluster (Mancini-DiNardo 2006). At this time, it is not known how the *Air* or *Kcnq1ot1* ncRNAs silence genes in their respective imprinted clusters. Many models are possible that can be applied to both clusters. Two possibilities arise from the sense–antisense overlap between a mRNA and the ncRNA that occurs in each cluster. The first possibility is that double-stranded RNA can form between the mRNA and ncRNA and induce RNA interference (RNAi) (see Chapter 8). A second possibility is that this sense–antisense overlap causes a form of transcriptional interference between the two promoters, which only affects transcription from the mRNA promoter. In both these cases, the first event would be posttranscriptional silencing of the overlapped mRNA followed by accumulation of repressive chromatin that

can spread and induce transcriptional gene silencing throughout the whole cluster.

It is, however, also possible that imprinted ncRNAs act by coating the local chromosomal region and directly recruiting repressive chromatin proteins to the imprinted cluster, in a manner similar to that described for the action of the *Xist* ncRNA in X-chromosome inactivation (Chapter 17). There are many similarities between silencing mediated by imprinted ncRNAs and that mediated by the *Xist* ncRNA. Most significantly, both are *cis*-acting epigenetic silencing mechanisms, and both show a positive correlation between expression of a ncRNA and silencing of multiple mRNA genes. It has also been suggested that genomic imprinting and X inactivation evolved from a common epigenetic mechanism (Reik and Lewis 2005). If this were the case, this would predict that imprinted ncRNAs could silence genes by targeting repressive chromatin to the imprinted gene cluster. However, we have no clear information yet about how imprinted ncRNAs perform their silencing function. We also do not yet know how many other imprinted ncRNAs also play a functional role in gene silencing.

4 Genomic Imprinting—A Model for Mammalian Epigenetic Regulation

Genomic imprinting has an advantage over other mammalian epigenetic gene regulation models, because both the active and inactive parental alleles reside in the same nucleus and are exposed to the same transcriptional environment. As a result, any epigenetic difference between the two parental alleles is more likely to correlate to their transcriptional state in contrast to "before and after" epigenetic systems, where epigenetic changes may also reflect the altered differentiation state of the cell. The presence of both the active and silent parental alleles in the same nucleus makes genomic imprinting an ideal system to study epigenetic gene regulation, but also imposes a difficulty since it is necessary first to distinguish one of the parental alleles to identify specific features associated with gene activity and silencing. This difficulty has been largely overcome in the mouse by the development of model systems that allow the maternal and paternal chromosomes to be distinguished (Fig. 1).

Despite the fact that epigenetic gene regulatory pathways are highly conserved in evolution, there are likely to be differences that relate to the type of genome organization for each organism. The mammalian genome shows an unusual organization that intersperses genes with high-copy-number repeats (also known as transposable elements). This greatly increases the length of most genes, as well as the distance between neighboring genes (Kazazian 2004). This contrasts with other model organisms such as yeast, ciliates, fungi, nematodes, plants, and *Drosophila*, whose genomes show a tendency toward remaining repeat-free or at least to separate repeats from genes (Rabinowicz et al. 2003).

It has been noted that high-copy-number repeats attract DNA methylation and repressive histone modifications in many organisms. This is thought to be largely a defensive adaptation against invading foreign DNA sequences (e.g., retrotransposons, transposons, and viruses). A consideration of how epigenetic mechanisms operate in mammals must, therefore, take into consideration the interspersed nature of repeats and genes (Goll and Bestor 2005). Notably, the fact that mammalian introns are repeat-rich and yet genes are able to be highly transcribed makes it less likely that the mammalian genome is organized into large blocks of silent heterochromatin or active euchromatin. This viewpoint is receiving some support from genome-wide analyses of human chromatin patterns, showing that active histone modifications are generally restricted to promoters or short regions presumed to be regulatory elements (Kim et al. 2005). The arrangement of imprinted genes into clusters that contain reciprocally expressed overlapping genes, as well as genes that escape imprinting altogether, supports the emerging picture that chromatin modifications may not spread far in the mammalian genome.

How can genomic imprinting contribute to an understanding of mammalian epigenetics? Although the characterization of imprinted gene clusters is far from complete, they clearly have the potential to provide information about how genes are controlled in local regions or domains. To date, imprinted gene clusters have already provided examples of *cis*-acting DNA sequences that are regulated by DNA methylation; of genes that are silenced by default in the mammalian genome and require epigenetic activation to be expressed; of long-range regulatory elements that can act as insulators; and of unusual ncRNAs that silence large domains of genes in *cis*. Time will tell whether these types of epigenetic regulatory mechanisms are unique to imprinted clusters or whether they can also be found regulating expression of non-imprinted genes in the mammalian genome.

5 Future Directions

Genomic imprinting has been the focus of intense interest since the discovery of the first imprinted genes in mammals in 1991. Some questions still await conclusive answers, particularly those concerning why mammals alone among the vertebrates use imprinted genes to regulate embryonic and neonatal growth. This contrasts with the extensive progress during the intervening 15 years on elucidating the epigenetic mechanisms controlling imprinted expression in mammals. From this information, we think we understand the general principles of how the imprinting mechanism operates at imprinted gene clusters, although all the details are still not clear.

At this stage, we know that genomic imprinting uses the cell's normal epigenetic machinery to regulate parental-specific expression, and that everything is set in motion by restricting this machinery in the gamete, to just one parental allele. We know there are general similarities in the mechanism controlling imprinted expression of different gene clusters, but we do not yet understand how many variants of this mechanism exist in the mammalian genome.

In the future, we would very much like to know to what degree non-imprinted genes are controlled by the epigenetic mechanisms described for imprinted gene clusters. Ultimately, we want to know whether we can transfer this knowledge for therapeutic use in humans; for example, by inducing re-expression of the silent mRNA genes in patients with the Prader-Willi and Angelman syndromes that show behavioral and growth defects, due to a deletion of the chromosome carrying the expressed alleles of imprinted mRNA genes (Jiang et al. 2004). An understanding of the way the cell controls epigenetic information is of increasing importance with the realization that epigenetic regulation can also be disturbed in cancers (see Chapter 24), in assisted reproductive technologies (see Chapter 22), and also in the aging process (Egger et al. 2004). We anticipate that an improved understanding of genomic imprinting will continue to provide an important model to discover how the mammalian genome uses epigenetic mechanisms to regulate gene expression.

Acknowledgments

We are grateful to the past and present members of the Barlow and Bartolomei laboratories for discussions and debates on the ideas presented here. We apologize that limitations on the number of references prevented citation of all original data. Finally, we are very grateful to Marie-Laure Caparros, Anne Ferguson-Smith, Thomas Jenuwein, Davor Solter, and Shirley Tilghman for comments on the manuscript.

References

Antequera F. 2003. Structure, function and evolution of CpG island promoters. *Cell. Mol. Life Sci.* **60:** 1647–1658.

Barlow D.P. 1993. Methylation and imprinting: From host defense to gene regulation? *Science* **260:** 309–310.

Barlow D.P., Stoger R., Herrmann B.G., Saito K., and Schweifer N. 1991. The mouse insulin-like growth factor type-2 receptor is imprinted and closely linked to the Tme locus. *Nature* **349:** 84–87.

Bartolomei M.S., Zemel S., and Tilghman S.M. 1991. Parental imprinting of the mouse H19 gene. *Nature* **351:** 153–155.

Bartolomei M.S., Webber A.L., Brunkow M.E., and Tilghman S.M. 1993. Epigenetic mechanisms underlying the imprinting of the mouse H19 gene. *Genes Dev.* **7:** 1663–1673.

Barton S.C., Surani M.A., and Norris M.L. 1984. Role of paternal and maternal genomes in mouse development. *Nature* **311:** 374–376.

Bell A.C. and Felsenfeld G. 2000. Methylation of a CTCF-dependent boundary controls imprinted expression of the Igf2 gene. *Nature* **405:** 482–485.

Bielinska B., Blaydes S.M., Buiting K., Yang T., Krajewska-Walasek M., Horsthemke B., and Brannan C.I. 2000. De novo deletions of SNRPN exon 1 in early human and mouse embryos result in a paternal to maternal imprint switch. *Nat. Genet.* **25:** 74–78.

Bourc'his D. and Bestor T.H. 2006. Origins of extreme sexual dimorphism in genomic imprinting. *Cytogenet. Genome Res.* **113:** 36–40.

Brannan C.I., Dees E.C., Ingram R.S., and Tilghman S.M. 1990. The product of the H19 gene may function as an RNA. *Mol. Cell Biol.* **10:** 28–36.

Cattanach B.M. and Kirk M. 1985. Differential activity of maternally and paternally derived chromosome regions in mice. *Nature* **315:** 496–498.

Chaillet J.R., Bader D.S., and Leder P. 1995. Regulation of genomic imprinting by gametic and embryonic processes. *Genes Dev.* **9:** 1177–1187.

Chandra H.S. and Nanjundiah V. 1990. The evolution of genomic imprinting. *Dev. Suppl.* **1990:** 47–53.

Constancia M., Dean W., Lopes S., Moore T., Kelsey G., and Reik W. 2000. Deletion of a silencer element in Igf2 results in loss of imprinting independent of H19. *Nat. Genet.* **26:** 203–206.

Cooper D.W., VandeBerg J.L., Sharman G.B., and Poole W.E. 1971. Phosphoglycerate kinase polymorphism in kangaroos provides further evidence for paternal X inactivation. *Nat. New Biol.* **230:** 155–157.

Crouse H.V., Brown A., and Mumford B.C. 1971. Chromosome inheritance and the problem of chromosome "imprinting" in Sciara (Sciaridae, Diptera). *Chromosoma* **34:** 324–398.

Davis E., Caiment F., Tordoir X., Cavaille J., Ferguson-Smith A., Cockett N., Georges M., and Charlier C. 2005. RNAi-mediated allelic *trans*-interaction at the imprinted *Rtl1/Peg11* locus. *Curr. Biol.* **15:** 743–739.

DeChiara T.M., Robertson E.J., and Efstratiadis A. 1991. Parental imprinting of the mouse insulin-like growth factor II gene. *Cell* **64:** 849–859.

Egger G., Liang G., Aparicio A., and Jones P.A. 2004. Epigenetics in human disease and prospects for epigenetic therapy. *Nature* **429:** 457–463.

Fedoriw A.M., Stein P., Svoboda P., Schultz R.M., and Bartolomei M.S. 2004. Transgenic RNAi reveals essential function for CTCF in H19 gene imprinting. *Science* **303:** 238–240.

Ferguson-Smith A.C., Sasaki H., Cattanach B.M., and Surani M.A. 1993. Parental-origin-specific epigenetic modification of the mouse H19 gene. *Nature* **362:** 751–755.

Ferguson-Smith A.C., Cattanach B.M., Barton S.C., Beechey C.V., and Surani M.A. 1991. Embryological and molecular investigations of parental imprinting on mouse chromosome 7. *Nature* **351:** 667–670.

Fitzpatrick G.V., Soloway P.D., and Higgins M.J. 2002. Regional loss of imprinting and growth deficiency in mice with a targeted deletion of KvDMR1. *Nat. Genet.* **32:** 426–431.

Goll M.G. and Bestor T.H. 2005. Eukaryotic cytosine methyltransferases. *Annu. Rev. Biochem.* **74:** 481–514.

Hark A.T., Schoenherr C.J., Katz D.J., Ingram R.S., Levorse J.M., and Tilghman S.M. 2000. CTCF mediates methylation-sensitive enhancer-blocking activity at the *H19/Igf2* locus. *Nature* **405:** 486–489.

Haussecker D. and Proudfoot N.J. 2005. Dicer-dependent turnover of intergenic transcripts from the human β-globin gene cluster. *Mol. Cell Biol.* **25:** 9724–9733.

Hoppe P.C. and Illmensee K. 1982. Full-term development after transplantation of parthenogenetic embryonic nuclei into fertilized mouse eggs. *Proc. Natl. Acad. Sci.* **79:** 1912–1916.

Hurst L.D., McVean G., and Moore T. 1996. Imprinted genes have few and small introns. *Nat. Genet.* **12:** 234–237.

Jiang Y.H., Bressler J., and Beaudet A.L. 2004. Epigenetics and human disease. *Annu. Rev. Genomics Hum. Genet.* **5:** 479–510.

Johnson D.R. 1974. Hairpin-tail: A case of post-reductional gene action in the mouse egg. *Genetics* **76:** 795–805.

Kazazian H.H., Jr. 2004. Mobile elements: Drivers of genome evolution. *Science* **303:** 1626–1632.

Kim T.H., Barrera L.O., Zheng M., Qu C., Singer M.A., Richmond T.A., Wu Y., Green R.D., and Ren B. 2005. A high-resolution map of active promoters in the human genome. *Nature* **436:** 876–880.

Kono T., Obata Y., Wu Q., Niwa K., Ono Y., Yamamoto Y., Park E.S., Seo J.S., and Ogawa H. 2004. Birth of parthenogenetic mice that can develop to adulthood. *Nature* **428:** 860–864.

Landers M., Bancescu D.L., Le Meur E., Rougeulle C., Glatt-Deeley H., Brannan C., Muscatelli F., and Lalande M. 2004. Regulation of the large (~1000 kb) imprinted murine *Ube3a* antisense transcript by alternative exons upstream of *Snurf/Snrpn*. *Nucleic Acids Res.* **32:** 3480–3492.

Lewis A., Mitsuya K., Umlauf D., Smith P., Dean W., Walter J., Higgins M., Feil R., and Reik W. 2004. Imprinting on distal chromosome 7 in the placenta involves repressive histone methylation independent of DNA methylation. *Nat. Genet.* **36:** 1291–1295.

Lin M.S., Zhang A., and Fujimoto A. 1995. Asynchronous DNA replication between 15q11.2q12 homologs: Cytogenetic evidence for maternal imprinting and delayed replication. *Hum. Genet.* **96:** 572–576.

Lin S.P., Youngson N., Takada S., Seitz H., Reik W., Paulsen M., Cavaille J., and Ferguson-Smith A.C. 2003. Asymmetric regulation of imprinting on the maternal and paternal chromosomes at the *Dlk1-Gtl2* imprinted cluster on mouse chromosome 12. *Nat. Genet.* **35:** 97–102.

Liu J., Litman D., Rosenberg M.J., Yu S., Biesecker L.G., and Weinstein L.S. 2000. A GNAS1 imprinting defect in pseudohypoparathyroidism type IB. *J. Clin. Invest.* **106:** 1167–1174.

Lyle R., Watanabe D., te Vruchte D., Lerchner W., Smrzka O.W., Wutz A., Schageman J., Hahner L., Davies C., and Barlow D.P. 2000. The imprinted antisense RNA at the *Igf2r* locus overlaps but does not imprint *Mas1*. *Nat. Genet.* **25:** 19–21.

Mager J., Montgomery N.D., de Villena F.P., and Magnuson T. 2003. Genome imprinting regulated by the mouse Polycomb group protein Eed. *Nat. Genet.* **33:** 502–507.

Mancini-DiNardo D.S., Levorse J.M., Ingram R.S., and Tilghman S.M.

2006. Elongation of the Kcnqot1 transcript is required for genomic imprinting of neighboring genes. *Genes Dev.* **20:** 1268–1282.

Mattick J.S. 2005. The functional genomics of noncoding RNA. *Science* **309:** 1527–1528.

McGrath J. and Solter D. 1984a. Completion of mouse embryogenesis requires both the maternal and paternal genomes. *Cell* **37:** 179–183.
———. 1984b. Maternal Thp lethality in the mouse is a nuclear, not cytoplasmic, defect. *Nature* **308:** 550–551.

McLaren A. 1979. *Maternal effects in development: The fourth symposium of the British Society for Developmental Biology* (ed. D.R. Newth and M. Balls). Cambridge University Press, Cambridge, United Kingdom.

Mitsuya K., Meguro M., Lee M.P., Katoh M., Schulz T.C., Kugoh H., Yoshida M.A., Niikawa N., Feinberg A.P., and Oshimura M. 1999. *LIT1*, an imprinted antisense RNA in the human *KvLQT1* locus identified by screening for differentially expressed transcripts using monochromosomal hybrids. *Hum. Mol. Genet.* **8:** 1209–1217.

Moore T. and Haig D. 1991. Genomic imprinting in mammalian development: A parental tug-of-war. *Trends Genet.* **7:** 45–49.

Moore T., Constancia M., Zubair M., Bailleul B., Feil R., Sasaki H., and Reik W. 1997. Multiple imprinted sense and antisense transcripts, differential methylation and tandem repeats in a putative imprinting control region upstream of mouse *Igf2. Proc Natl. Acad. Sci.* **94:** 12509–12514.

Neumann B., Kubicka P., and Barlow D.P. 1995. Characteristics of imprinted genes. *Nat. Genet.* **9:** 12–13.

O'Neill M.J. 2005. The influence of non-coding RNAs on allele-specific gene expression in mammals. *Hum. Mol. Genet.* **14:** R113–120.

Rabinowicz P.D., Palmer L.E., May B.P., Hemann M.T., Lowe S.W., McCombie W.R., and Martienssen R.A. 2003. Genes and transposons are differentially methylated in plants, but not in mammals. *Genome Res.* **13:** 2658–2664.

Regha K., Latos P.A., and Spahn L. 2006. The imprinted mouse *Igf2r/Air* cluster—A model maternal imprinting system. *Cytogenet. Genome Res.* **113:** 165–177.

Reik W. 1989. Genomic imprinting and genetic disorders in man. *Trends Genet* **5:** 331–336.

Reik W. and Lewis A. 2005. Co-evolution of X-chromosome inactivation and imprinting in mammals. *Nat. Rev. Genet.* **6:** 403–410.

Reik W., Howlett S.K., and Surani M.A. 1990. Imprinting by DNA methylation: From transgenes to endogenous gene sequences. *Dev. Suppl.* **1990:** 99–106.

Sasaki H., Jones P.A., Chaillet J.R., Ferguson-Smith A.C., Barton S.C., Reik W., and Surani M.A. 1992. Parental imprinting: Potentially active chromatin of the repressed maternal allele of the mouse insulin-like growth factor II (Igf2) gene. *Genes Dev.* **6:** 1843–1856.

Schmidt J.V., Levorse J.M., and Tilghman S.M. 1999. Enhancer competition between *H19* and *Igf2* does not mediate their imprinting. *Proc. Natl. Acad. Sci.* **96:** 9733–9738.

Searle A.G. and Beechey C.V. 1978. Complementation studies with mouse translocations. *Cytogenet. Cell Genet.* **20:** 282–303.

Seitz H., Royo H., Lin S.P., Youngson N., Ferguson-Smith A.C., and Cavaille J. 2004. Imprinted small RNA genes. *Biol. Chem.* **385:** 905–911.

Shemer R., Birger Y., Riggs A.D., and Razin A. 1997. Structure of the imprinted mouse *Snrpn* gene and establishment of its parental-specific methylation pattern. *Proc. Natl. Acad. Sci.* **94:** 10267–10272.

Sleutels F., Zwart R., and Barlow D.P. 2002. The non-coding Air RNA is required for silencing autosomal imprinted genes. *Nature* **415:** 810–813.

Sleutels F., Tjon G., Ludwig T., and Barlow D.P. 2003. Imprinted silencing of *Slc22a2* and *Slc22a3* does not need transcriptional overlap between *Igf2r* and *Air. EMBO J.* **22:** 3696–3704.

Spahn L. and Barlow D.P. 2003. An ICE pattern crystallizes. *Nat. Genet.* **35:** 11–12.

Stoger R., Kubicka P., Liu C.G., Kafri T., Razin A., Cedar H., and Barlow D.P. 1993. Maternal-specific methylation of the imprinted mouse *Igf2r* locus identifies the expressed locus as carrying the imprinting signal. *Cell* **73:** 61–71.

Surani M.A., Barton S.C., and Norris M.L. 1984. Development of reconstituted mouse eggs suggests imprinting of the genome during gametogenesis. *Nature* **308:** 548–550.

Takada S., Paulsen M., Tevendale M., Tsai C.E., Kelsey G., Cattanach B.M., and Ferguson-Smith A.C. 2002. Epigenetic analysis of the *Dlk1-Gtl2* imprinted domain on mouse chromosome 12: Implications for imprinting control from comparison with *Igf2-H19. Hum. Mol. Genet.* **11:** 77–86.

Thorvaldsen J.L., Duran K.L., and Bartolomei M.S. 1998. Deletion of the *H19* differentially methylated domain results in loss of imprinted expression of *H19* and *Igf2. Genes Dev.* **12:** 3693–3702.

Tierling S., Dalbert S., Schoppenhorst S., Tsai C.E., Oliger S., Ferguson-Smith A.C., Paulsen M., and Walter J. 2005. High-resolution map and imprinting analysis of the *Gtl2-Dnchc1* domain on mouse chromosome 12. *Genomics.* **87:** 225–235.

Varmuza S. and Mann M. 1994. Genomic imprinting—Defusing the ovarian time bomb. *Trends Genet.* **10:** 118–123.

Verona R.I., Mann M. R., and Bartolomei M.S. 2003. Genomic imprinting: Intricacies of epigenetic regulation in clusters. *Annu. Rev. Cell Dev. Biol.* **19:** 237–259.

Wilkins J.F. and Haig D. 2003. What good is genomic imprinting: The function of parent-specific gene expression. *Nat. Rev. Genet.* **4:** 359–368.

Williamson C.M., Turner M.D., Ball S.T., Nottingham W.T., Glenister P., Fray M., Tymowska-Lalanne Z., Plagge A., Powles-Glover N., Kelsey G., et al. 2006. Identification of an imprinting control region affecting the expression of all transcripts in the *Gnas* cluster. *Nat. Genet.* **38:** 350–355.

Wroe S.F., Kelsey G., Skinner J.A., Bodle D., Ball S.T., Beechey C.V., Peters J., and Williamson C.M. 2000. An imprinted transcript, antisense to *Nesp*, adds complexity to the cluster of imprinted genes at the mouse *Gnas* locus. *Proc. Natl. Acad. Sci.* **97:** 3342–3346.

Wutz A., Smrzka O. W., Schweifer N., Schellander K., Wagner E. F., and Barlow D. P. 1997. Imprinted expression of the *Igf2r* gene depends on an intronic CpG island. *Nature* **389:** 745–749.

Yatsuki H., Joh K., Higashimoto K., Soejima H., Arai Y., Wang Y., Hatada I., Obata Y., Morisaki H., Zhang Z., et al. 2002. Domain regulation of imprinting cluster in Kip2/Lit1 subdomain on mouse chromosome 7F4/F5: Large-scale DNA methylation analysis reveals that DMR-Lit1 is a putative imprinting control region. *Genome Res.* **12:** 1860–1870.

Zwart R., Sleutels F., Wutz A., Schinkel A. H., and Barlow D. P. 2001. Bidirectional action of the *Igf2r* imprint control element on upstream and downstream imprinted genes. *Genes Dev.* **15:** 2361–2366.

WWW Resources

Beechey C.V., Cattanach B.M., Blake A., and Peters J. 2005. MRC Mammalian Genetics Unit, Harwell, Oxfordshire. World Wide Web Site—Mouse Imprinting data and references (http://www.mgu.har.mrc.ac.uk/research/imprinting/). http://nocode.bioinfo.org.cn/ NONCODE—the database of noncoding RNA

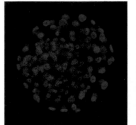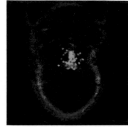

Germ Line and Pluripotent Stem Cells

M. Azim Surani[1] and Wolf Reik[2]

[1]Wellcome Trust Cancer Research UR Gurdon Institute, University of Cambridge,
Cambridge CB2 1QN, United Kingdom
[2]The Babraham Institute, Babraham Research Campus, Cambridge CB2 4AT, United Kingdom

CONTENTS

1. Introduction, Life of Mammals: A Genetic and Epigenetic Continuum, 379

2. Genetic and Epigenetic Mechanisms Regulating Germ-Cell Specification (From the Early Embryo to Germ Cells), 381

   2.1 Principles of Germ-line Development in Different Animal Groups, 381

   2.2 Early Germ-line Development in Mammals, 381

   2.3 The Role of Blimp1 in Specification of PGCs, 383

   2.4 Repression of the Somatic Program in Germ Cells—An Evolutionarily Conserved Phenomenon, 383

   2.5 Regulation of Epigenetic Programming after PGC Specification in Mice, 384

   2.6 Germ Line and Stem Cells—A Reversible Phenotype, 385

   2.7 Development of Germ Cells from Pluripotent ES Cells, 385

   2.8 From Primordial Germ Cells to Gametes, 385

3. From the Oocyte to the Early Embryo, 386

   3.1 Maternal Inheritance and Potential Asymmetric Distribution?, 386

   3.2 Epigenetic Events at Fertilization, 387

   3.3 From the Zygote to the Blastocyst, 388

4. From Pluripotent Stem Cells to Somatic Cells and Back to Germ Cells, 389

   4.1 Derivation of Pluripotent Stem Cells, 389

   4.2 Epigenetic Properties of Pluripotential Cell Lines, 390

   4.3 Reprogramming Capacity of Stem Cells, 391

5. Perspective, 392

Acknowledgments, 392

References, 392

GENERAL SUMMARY

An egg or oocyte is a most remarkable cell because it is the only cell in the body that is potentially capable of developing into a whole organism. William Harvey was the first to recognize this in 1651 when he remarked "Ex Ovo Omni" or "everything comes from an egg." He recognized that an egg probably develops progressively into an organism, and this insight was important for the concept of *epigenesis* or progressive development. This eventually led to the demise of the *preformationist* view of development, a theory proposing that individuals develop from the enlargement of tiny fully formed organisms (the so-called homunculus) contained in the germ cells. Conrad Waddington later depicted this concept in his famous illustration, as an "epigenetic landscape," a symbolic representation of sequential development from an egg. Development of an entire organism from an egg is possible in some organisms without any contribution from a male, which is called *parthenogenesis*, but this cannot occur in mammals due to the phenomenon of *genomic imprinting*, where fertilization of an egg by sperm is obligatory for development to adulthood.

In most organisms, development commences following fusion between sperm and eggs to generate a zygote, which gives rise not only to a new individual but, theoretically at least, to an endless series of generations. In this way, germ cells provide the enduring link between all generations. The newly fertilized egg or zygote is therefore unique, because no other cell has the potential to develop into an entirely new organism. This property is referred to as *totipotency*. As transmitters of both genetic and epigenetic information to subsequent generations, germ cells are unique, and they exhibit many exceptional properties that are required to fulfil this potential. The oocyte also has the striking property of conferring totipotency on cell nuclei from *somatic cells* such as nerve cells when it is transplanted into the egg, a process referred to as cloning or nuclear reprogramming.

During development from a zygote onward, there is a progressive decline in totipotency of the newly dividing cells. In mammals, only the products of very early cell divisions retain totipotency where each of the cells is, in principle, separately capable of generating a new organism.

Early development of the mammalian embryo gives rise to a blastocyst (left title image), a structure with an outer group of trophectoderm cells destined to form the placenta, and an inner group of cells that will give rise to the entire fetus and eventually a new organism. These inner cells will therefore differentiate into all the known 200 or so specialized somatic cells found in adults, and they are therefore referred to as *pluripotent*. Under certain culture conditions, these pluripotent cells can be "rescued" from early embryos and made to grow indefinitely in vitro while still retaining the ability to differentiate into any specific cell type found in embryos and adults, including sperm and eggs themselves. Such cells have been derived from human and mouse embryos and are called *pluripotent embryonic stem cells* or *ES cells*. The capacity to generate pluripotent stem cells is lost quite rapidly when the embryo implants and commences the program of embryonic development.

Among the earliest cell types to emerge during embryonic development are the precursors of sperm and eggs called the *primordial germ cells* (PGCs) highlighted as the green cells in the right-hand title image. This developmental event ensures that the cells that will eventually give rise to subsequent generations of the developing embryo are established first. Primordial germ cells are highly specialized cells that will eventually develop into mature sperm or eggs in the adult organism, thus repeating the cycle, while the rest of the body cells will eventually perish. PGCs are therefore very special cells. They can also be used to derive pluripotent stem cells called *embryonic germ cells* or *EG cells*.

Stem cells are also present in adults. For example, *adult stem cells* generate billions of different blood cells that arise from blood stem cells in the bone marrow. Similarly, our skin cells or the cells in the gut are continually replaced through differentiation of appropriate stem cells. These adult stem cells normally only have the potential to generate cells of specific tissues and not the diverse cell types that can be made from pluripotent stem cells. One of the key objectives is to understand the similarities and differences between the pluripotent ES and adult stem cells, including the underlying epigenetic mechanisms that regulate their properties. Understanding the unique epigenetic properties of germ cells and pluripotent stem cells will eventually enable us to develop new concepts for therapies, particularly in regenerative medicine.

1 Introduction, Life of Mammals: A Genetic and Epigenetic Continuum

The genetic information encoded in an individual's genome is established at fertilization and does not change during development, with the exception of mutations and some directed sequence changes occurring, for example, in the immune system. The egg or oocyte—the female contribution to a fertilized zygote—contributes three types of heritable information to the developing embryo (Fig. 1a). It contributes half of the genetic information, complemented by the male haploid genome derived from the sperm. But also, as host to the developing embryo, certain early developmental events are governed by maternal inheritance through the contribution of maternal RNAs and proteins. Finally, there is epigenetic information contained in the oocyte that affects the developmental regulation of the genome.

Epigenetic information is encoded by DNA methylation, histone modifications, histone variants, and non-histone chromatin proteins, and undergoes major changes during development and differentiation. The key feature of these epigenetic marks is that they can be heritable from one cell generation to the next, and they can regulate gene expression. Epigenetic information is thus thought to be of critical importance for the determination and maintenance of defined and stable gene expression programs that underlie cell fate decisions during development. In totipotent and pluripotent cells, it is imagined that epigenetic marks are less stable and more plastic, but as development

proceeds and the potency of cells becomes more and more restricted, epigenetic marks become more rigid and restrictive. Totipotent and pluripotent cells, such as germ cells or embryonic stem cells (Fig. 1b), also have the unique property of being able to reprogram the genome and erase existing epigenetic marks, and this ability may underlie their developmental plasticity.

The interdependence of developmental decisions and epigenetic gene regulation sets up a continuum of genetic and epigenetic events in mammals. This is because developmental events, for example, through signaling between cells, result in specific programs of gene expression that can be epigenetically fixed. Developmental events, in addition, can set up new epigenetic events (for example, the methylation or demethylation of imprinted genes in germ cells). The setting or erasure of epigenetic marks in turn determines new gene expression programs and, hence, influences the way individual cells respond to developmental cues. The resulting developmental and epigenetic continuum is particularly fascinating when it includes the germ line, because this extends to the next generation and possibly beyond that into the future.

Does life therefore really begin at fertilization? It is true that genetically a new individual can be identified from the time of fertilization of an egg by sperm. This is when a haploid set of chromosomes from the mother comes together with a haploid set of chromosomes from the father, and the diploid genome of the offspring is formed. But epigenetic information that will be transmitted to the offspring is also present in the gametes. For

a

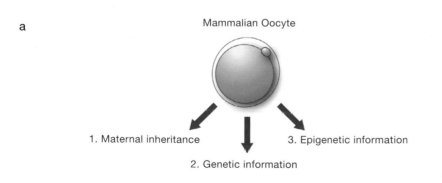

Mammalian Oocyte

1. Maternal inheritance 　 2. Genetic information 　 3. Epigenetic information

b

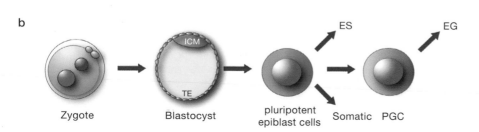

Zygote 　 Blastocyst 　 pluripotent epiblast cells 　 Somatic 　 PGC 　 ES 　 EG 　 ICM 　 TE

Figure 1. The Mammalian Oocyte and Zygote

(*a*) The mammalian oocyte contains maternal RNAs and maternal proteins (maternal inheritance), which can determine early developmental events, genetic information (maternal chromosomes), and epigenetic information (DNA methylation and chromatin marks). (*b*) The zygote gives rise to the blastocyst with its inner cell mass cells, which give rise to ES cells (in culture), all somatic cells, and primordial germ cells (PGC), which can give rise to EG cells in culture.

example, imprinted genes carry DNA methylation marks that differ between the male and female germ cells, and these preexisting patterns are inherited by offspring (see Chapter 19). These epigenetic marks are introduced into the germ cell genomes during fetal or early postnatal development of parents.

This program of epigenetic modifications depends on genetic determination of germ cell allocation and development, in the early postimplantation embryo. The earliest precursors of the germ cells arise from a small group of cells formed in the early postimplantation embryo, as a result of receiving signaling molecules that come from another part of the conceptus, the extraembryonic lineages, which include the placenta. Germ cell development, subsequent to their allocation, is genetically regulated, affecting and determining their epigenetic program, such that gene products that are needed in somatic cells are repressed in germ cells. On the other hand, other special genetically governed functions are needed in germ cells, such as, for example, epigenetic reprogramming, establishment of imprinting, chromosome recombination during meiosis, and reduction divisions to form haploid gametes.

Before germ cell development begins, the earliest lineage decisions are made in the embryo. The pluripotent epiblast cells in early postimplantation embryos are the source of all the somatic tissues as well as primordial germ cells. Differentiation of somatic tissues into the three primary cell lineages (i.e., ectoderm, mesoderm, and endoderm)

commences in response to signals from the extraembryonic tissues. At the same time, specification of primordial germ cells also arises from a specific set of cells located in the proximal epiblast. The trophectoderm (TE) of the blastocyst arises earlier, which is the result of the first lineage allocation event in the mammalian embryo that generates the placenta (see Fig. 2). The ICM gives rise to the whole of the adult organism, and pluripotent embryonic stem (ES) cells are derived from the inner cell mass (ICM) cells. This group of cells arises at the morula stage of development in the embryo before it has implanted, where the ICM cells, and the outer trophectoderm (TE) layer, are set aside to form the blastocyst, which then implants into the uterus. Epigenetic regulation differs considerably between the extraembryonic lineages (TE) and embryonic lineages (ICM). For example, the overall levels of DNA methylation are lower in the extraembryonic tissues, and maintenance of imprinting and of imprinted X inactivation can be different. ICM and TE cells, like primordial germ cells (PGCs), are determined by a genetic program involving transcription factors and, where appropriate, pluripotency genes. How these different genetic programs first arise in the early embryo, and the relative contributions of cell–cell interactions, epigenetic modifications, or egg cytoplasmic factors in setting them up, is unclear.

In addition to epigenetic information that is carried in imprinted genes from the gametes into the embryo, other dramatic epigenetic events occur around the time of fer-

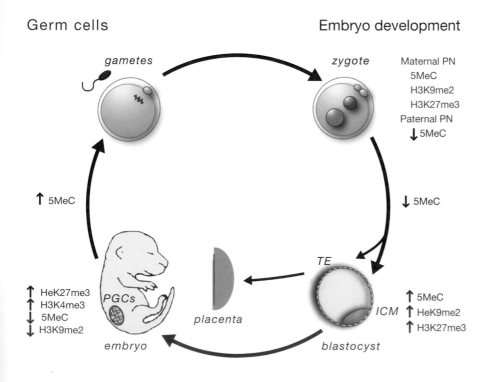

Germ cells

Embryo development

gametes

zygote

Maternal PN
5MeC
H3K9me2
H3K27me3
Paternal PN
↓5MeC

↑ 5MeC

↓ 5MeC

↑ HeK27me3
↑ H3K4me3
↓ 5MeC
↓ H3K9me2

PGCs

placenta

TE

embryo

blastocyst

ICM

↑ 5MeC
↑ HeK9me2
↑ H3K27me3

Figure 2. Epigenetic Reprogramming Cycle in Mammalian Development

Immediately after fertilization in the zygote, the paternal pronucleus (PN) is packaged with histones that lack H3K9me2 and H3K27me3, whereas the maternal chromatin contains these marks. The paternal PN also rapidly loses 5-methylcytosine (5MeC) on a genome-wide scale, while the maternal does not. Passive loss of 5MeC occurs during preimplantation development until the blastocyst stage, when the inner cell mass (ICM) cells begin to acquire high levels of 5MeC, H3K9me2, and H3K27me3. The placenta, which is largely derived from the trophectoderm (TE) of the blastocyst, remains relatively hypomethylated. Primordial germ cells (PGCs) undergo demethylation of 5MeC and H3K9me2 before and after entry into the gonads. De novo DNA methylation, including parent-specific imprinting, takes place at later stages of germ-cell development.

tilization. The sperm genome loses DNA methylation rapidly in the fertilized zygote, and then regains DNA and histone modifications over the subsequent cell divisions (Fig. 2). The maternal genome resists zygotic DNA demethylation, but then becomes demethylated in a more protracted fashion during cleavage divisions of the early embryo. Although these somewhat opposing and distinct epigenetic programs lead to an overall loss of gametic epigenetic information, it is likely that this dynamic reprogramming event interacts with the cellular and genetic processes that determine the earliest processes of cell allocation to ICM and TE lineages, respectively. In this way, the epigenetic life cycle never ends and never begins, but constantly interacts with genetic programs that determine the development of the ICM and TE, pluripotent stem cells, germ cells, and perhaps other lineages.

2 Genetic and Epigenetic Mechanisms Regulating Germ-Cell Specification (From the Early Embryo to Germ Cells)

2.1 Principles of Germ-line Development in Different Animal Groups

In animals, the specification of germ cells is one of the earliest events during development that segregates germ cells from somatic cells (Surani et al. 2004). Germ cells eventually generate the totipotent state. There are two key modes of initiation of the germ-cell lineage that are referred to as preformation (this is distinct from the old usage of the word as in preformationism) and epigenesis (Extavour and Akam 2003). The first involves inheritance of preformed germ cell determinants by specific cells as occurs in *Caenorhabditis elegans* and *Drosophila melanogaster* (Leatherman and Jongens 2003; Blackwell 2004). In contrast, the epigenesis mode of germ cell specification is a process where a group of potentially equivalent pluripotent cells acquire a germ cell fate in response to inductive signals, while the remaining cells acquire the somatic fate (Lawson and Hage 1994; McLaren 2003). This mechanism for germ cell specification operates in mice and probably in other mammals such as humans.

2.2 Early Germ-line Development in Mammals

Primordial germ cells in mice are first detected at E7.5 (at the early bud, EB, stage), as a cluster of approximately 40 cells that constitutes the founder population of the germ-cell lineage (Lawson and Hage 1994; McLaren 2003). They are positive for alkaline phosphatases and located within the extraembryonic mesoderm at the base of the allantois

(see below Fig. 4, right panel). Clonal analysis reveals that the proximal epiblast cells at E6.0–E6.5 (prestreak [PS] and early streak [ES] stage embryos) give rise to PGCs as well as the tissues of the extraembryonic mesoderm (Lawson and Hage 1994). PGCs are formed in response to signaling molecules that are produced by extraembryonic ectoderm and primary endoderm (Fig. 3). Bmp4 and Bmp8b are among the key signaling molecules that confer germ cell competence and specification (Lawson et al. 1999).

To gain detailed insights into the genetic program of PGC specification, single-cell cDNAs were generated from the founder PGCs and their neighboring somatic cells (Saitou et al. 2002). A variety of markers were used to distinguish between PGCs and somatic cells. This screen initially identified *fragilis*, a novel member of the interferon-inducible transmembrane protein family implicated in cell aggregation, and *stella*, a nucleo-cytoplasmic protein. Further investigations showed that *fragilis* is expressed in the proximal epiblast cells at E6.25 (Fig. 3) when they gain competence to give rise to both PGCs and the neighboring extraembryonic mesoderm cells. The *fragilis*-positive cells move to the posterior proximal region during gastrulation. The founder population of PGCs are subsequently detected among this population of cells where they show expression of *stella*. At the same time, founder PGCs show expression of pluripotency genes, including *Sox2* and *Oct4*, suggesting that PGCs

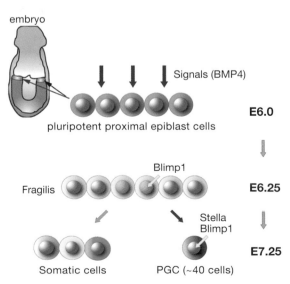

Figure 3. Early Germ-Cell Determination in the Mouse

Pluripotent proximal epiblast cells (*blue*) which are derived from the ICM (see Figs. 1 and 2) receive an extracellular signal from BMP4 which confers germ-line competence (*pink*). Activation of Blimp1 commits these cells to the primordial germ cell lineage (*red*), whereas other cells become somatic cells.

exhibit an underlying pluripotency, which is lost in the neighboring somatic cells (Fig. 3). In contrast, the founder PGCs show repression of some genes, including *Hoxb1* and *Hoxa1*, which are at this time significantly up-regulated in somatic neighbors. The repression of *Hox* genes is part of an important mechanism that underlies repression of the somatic cell fate in founder PGCs (Saitou et al. 2002).

Based on the analysis of the emergence of founder PGCs, it is evident that, as in other organisms, repression of the somatic program is likely a key feature of PGC specification in mice (Seydoux and Strome 1999; Blackwell 2004; Surani et al. 2004). During an analysis of candidate genes for gene repression, histone lysine methyltransferases (HKMTs) were analyzed to determine whether any of them showed differential expression between PGCs and neighboring somatic cells. Expression of some of these genes, such as *G9a*, *Pfm1*, *Set1* and *Ezh2*, was detected in both the founder PGCs and the somatic cells. However, one of these genes, *Blimp1* (*B-lymphocyte maturation-induced protein-1*) showed expression exclusively in the founder PGCs and not in the neighboring somatic cells at E7.5 (Ohinata et al. 2005). *Blimp1* is a transcriptional repressor with a SET/PR domain, a proline-rich region that can recruit Groucho and HDAC2, five C_2H_2 zinc fingers that can form a complex with G9a, and an acidic tail (Shaffer et al. 2002; Shapiro-Shelef et al. 2003; Gyori et al. 2004; Sciammas and Davis 2004). *Blimp1* was first identified for its role during specification of plasma cells following repression of the B-cell program in the precursor cells (Turner et al. 1994). *Blimp1* is indeed widely expressed during mouse development (Chang et al. 2002).

Detailed analysis of *Blimp1* in early mouse embryos led to some unexpected findings. Among these was the discovery that *Blimp1* expression commences in the proximal epiblast cells at E6.25 at the onset of gastrulation, initially in only 4–6 cells that are in direct contact with the extraembryonic ectoderm cells (Fig. 3) (Ohinata et al. 2005). *Blimp1* expression is detected at one end of the short axis in a region that is destined to form the posterior proximal region. The number of *Blimp1*-positive cells increases progressively so that there are approximately 20 cells at the mid-streak (MS) stage that are seen to form a tight cluster in the posterior proximal region at E6.75. At E7.5 early bud stage (EB), the number of *Blimp1*-positive cells increases to approximately 40 (see Fig. 4). These cells constitute the founder population of PGCs and show expression of the classic alkaline phosphatase PGC marker and commence expression of *stella* (Fig. 3). A genetic lineage-tracing experiment confirmed that all the *Blimp1*-positive cells originating in the epiblast from E6.25 onward are indeed lineage-restricted PGC precursor cells. These data contrast with the previous hypothesis, based on clonal analysis, which suggested that the proximal epiblast cells at E6.0–E6.5 are not lineage restricted to give rise exclusively to PGCs, since clonal descendants of individual cells could give rise to both a somatic and a germ cell (Lawson and Hage 1994; McLaren and Lawson 2005). A likely explanation for this discrepancy is that in the clonal analysis, the marked cells may initially have been negative for *Blimp1* and they subsequently divided to generate a positive cell that gave rise to PGCs, while the daughter cell produced a somatic descendant. The mechanism that regulates the accretion of *Blimp1*-positive cells is currently unknown.

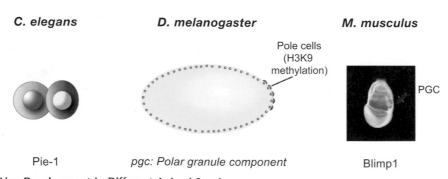

Figure 4. Germ-Line Development in Different Animal Species

In *C. elegans*, the germ-line lineage (*red*) is specified after the first division of the zygote by expression of Pie1, which confers transcriptional quiescence. The other cell (*blue*) gives rise to somatic tissues. In *D. melanogaster*, the precursors of the germ cells are the so-called pole cells contained on one side of the zygote syncytium (i.e., multinucleated); transcriptional quiescence in these cells depends on localized RNA from the gene *Pgc*. In *M. musculus*, the earliest precursors of the germ cells are visible by expression of Blimp1 at the base of the allantois. Blimp1 initiates transcriptional quiescence in these cells.

2.3 The Role of Blimp1 in Specification of PGCs

Further analysis of the role of *Blimp1* in PGC specification has generated insights into the underlying mechanism of germ-cell specification in mice. Loss of function of *Blimp1* showed that this is a key determinant of PGC specification in mice (Ohinata et al. 2005; Vincent et al. 2005). At E7.5, *Blimp1* mutant embryos contain an aberrant cluster of approximately 20 PGC-like cells, unlike control embryos, where the PGCs continue to proliferate and commence migration out of the cluster. Furthermore, the number of aberrant PGC-like cells fails to increase when examined at E8.5 (Ohinata et al. 2005).

Single-cell analysis of mutant PGC cells revealed a lack of consistent repression of *Hox* genes. Therefore, it is likely that *Blimp1* has a role in the repression of the somatic program in founder PGCs. There was also inconsistency in the up-regulation of PGC-specific genes such as *stella* and *Nanos3*, and of some pluripotency-specific genes such as *Sox2* in mutants. These findings stress that *Blimp1* has a critical role as a transcriptional regulator during PGC specification and in the prevention of these cells from acquiring a somatic cell fate.

Studies on B cells have revealed that Blimp1 is sufficient to induce differentiation into plasma cells, through repression of key molecules that maintain B-cell identity (Turner et al. 1994; Shaffer et al. 2002; Shapiro-Shelef et al. 2003; Sciammas and Davis 2004; for details, see Chapter 21). It does this through the formation of a Groucho and HDAC2 repressor complex (Ren et al. 1999). Its zinc fingers also seem important for the formation of a complex with G9a (Gyory et al. 2004), an HKMT that is required for H3K9me2. However, it is still unclear whether Blimp1 functions during PGC specification by forming a novel complex, as no HKMT activity has been ascribed to the Blimp1 SET/PR domain.

Blimp1 is an evolutionarily conserved gene in both vertebrates and invertebrates, and it has a variety of functions. For example, it has a role in the development of several lineages in vertebrates such as the zebrafish and *Xenopus* (de Souza et al. 1999; Roy and Ng 2004; Hernandez-Lagunas et al. 2005), although not specifically in germ-cell specification. This implies that the gene has acquired a new role in PGC specification in mice and perhaps in all mammals. For this highly conserved gene, it suggests that additional control elements must have evolved to drive its expression in PGC precursors and founder cells.

2.4 Repression of the Somatic Program in Germ Cells— An Evolutionarily Conserved Phenomenon

The mechanism of germ-cell specification is not evolutionarily conserved, which is evident when comparing the mechanism in mice with the events in two other well-studied model organisms, *D. melanogaster* and *C. elegans* (Seydoux and Strome 1999). The differences in the mechanism of germ-cell specification are primarily accounted for by the differences in the mode of early development in these different organisms, as well as by the additional complexities imposed by the phenomenon of genomic imprinting in mammals. Importantly, however, the repression of the somatic gene expression program during specification of germ cells is a shared phenomenon in diverse organisms, even though the molecular mechanisms may differ (Seydoux and Strome 1999; Leatherman and Jongens 2003; Saitou et al. 2003; Blackwell 2004).

In *C. elegans*, the first cell division of the zygote is asymmetric: It establishes a somatic cell (AB), whereas the second cell (P1) is set aside to establish the germ-cell lineage (Fig. 4). Indeed, each of the P1, P2, and P3 cells produces a somatic cell when it divides, and the latter commence transcription and differentiation. The P1–P3 cells destined for the germ-cell lineage remain transcriptionally quiescent. This transcriptional quiescence is maintained by a zinc finger protein, PIE-1, which inhibits transcriptional elongation by competing with RNA polymerase II (pol II) carboxy-terminal domain (CTD) phosphorylation at Ser-2 (Van Doren et al. 1998; Seydoux and Strome 1999; Zhang et al. 2003; for more detail, see Chapter 10). However, both the somatic and germ-cell blastomeres exhibit transcriptionally permissive chromatin states, as shown by high levels of genome-wide H3K4me. Later, when the P4 blastomere divides to form two germ-line cells, Z2 and Z3, a repressive chromatin state becomes evident, with loss of H3K4me and acquisition of high levels of repressive H3K9me (Schaner et al. 2003). Thus, during the establishment of the germ-cell lineage in *C. elegans*, the chromatin changes from a transcriptionally permissive state to an inactive state.

The establishment of the germ-cell lineage in *D. melanogaster* is again distinct from what is observed in the mouse and worm. The germ-line precursors, called the pole cells, are detected before the onset of embryonic development, in the fertilized syncytial (multinucleated) egg (Fig. 4), and these are again transcriptionally quiescent, due to a lack of phosphorylation of CTD of pol II, as observed in *C. elegans* (Seydoux and Dunn 1997; Van Doren et al. 1998; Schaner et al. 2003). Although this transcriptional silencing is also associated with the repressive chromatin modification H3K9me, pole cells are destined to form only germ cells. These cells are thus equivalent to the Z2 and Z3 cells of *C. elegans* that are destined for the germ-cell lineage only. Furthermore, transcriptional silencing in pole cells is apparently regulated by the *polar granule component (pgc)* gene, since loss of *pgc* causes a loss of repression, although the pole cells are still detected. The mutant pole cells exhibit pol II CTD phosphorylation of Ser-2 (Deshpande et al. 2004; Martinho et al. 2004). It has been suggested that *pgc* might sequester critical components needed for phosphorylation of pol II CTD, thus inhibiting the transition from preinitiation complex to the elongation complex.

The analysis of specification of germ cells during development in mice, flies, and worms clearly illustrates the fact that transcriptional repression that is presumably essential to repress the somatic cell fate is found in all three organisms, although the precise mechanisms by which this is achieved differ markedly. This is evidently due to the differences in events associated with early development of the different species.

2.5 Regulation of Epigenetic Programming after PGC Specification in Mice

Extensive epigenetic programming and reprogramming continue in the germ-cell lineage following the specification of PGCs (Seki et al. 2005). This period of development is marked by the erasure of some of the repressive epigenetic modifications that allows the germ-cell lineage to acquire an underlying pluripotent characteristic, which may be a prerequisite for subsequent totipotency.

Among the key changes observed is the erasure of H3K9me2 at E8.0, together with a decrease in the levels of HP1α by E9.0 within the euchromatic and the pericentric hetrochromatic regions (Seki et al. 2005). At the same time, there is also a decline in the overall levels of DNA methylation in PGCs from E8.0 onward. While there is a decline in H3K9me2 and DNA methylation, there is also a progressive increase in H3K27me3, a repressive modification mediated by the polycomb group protein, Ezh2 (Fig. 5). The loss of DNA methylation is accompanied by repression of de novo DNA methyltransferases Dnmt3a and Dnmt3b, as well as by a transient decline in the maintenance methyltransferase, Dnmt1. It is noteworthy that the loss of H3K9me2 DNA methylation also coincides with the re-expression of a key pluripotency-associated gene, *Nanog* (Yamaguchi et al. 2005). *Nanog* is first expressed in the inner cells of the late morula and in the inner cell mass cells of blastocysts. However, expression of this gene is rapidly down-regulated after implantation, and the gene is only reexpressed following specification of PGCs. Together with the expression of other pluripotency genes, including *Oct4*, *Sox2*, and *Esg1*, this shows that germ cells acquire characteristics of pluripotency (Fig. 5).

Additional extensive epigenetic programming events ensue when PGCs enter into the developing gonads (Surani et al. 2004). First, there are increases in H3K4 methylation and in H3K9 acetylation, which are characteristic of permissive chromatin states, excluding H3K9me. In addition, there is very extensive genome-wide DNA demethylation (Fig. 5) that includes erasure of parental imprints and of methylation in single-copy genes. In female embryos, the inactive X chromosome is also reactivated at this time.

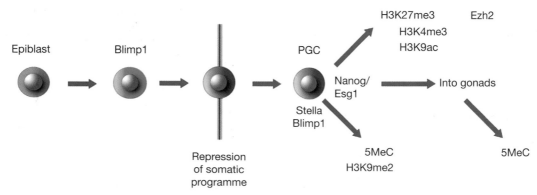

Figure 5. Early Epigenetic Events during Germ-Cell Specification

Expression of Blimp1 in descendants of epiblast cells leads to repression of the somatic gene expression program (*red*). This is followed by expression of Stella, Nanog, and Esg1; increase in H3K27me3, H3K4me3, and H3K9ac; and loss of H3K9me2 and 5MeC. The primordial germ cells then enter the gonads, after which time further epigenetic reprogramming takes place.

Although there is therefore an effective mechanism that erases "acquired" epigenetic modifications, not all epigenetic marks are completely removed during germ-cell development. For example, DNA methylation of the IAP retrotransposon family is only partially reprogrammed (Lane et al. 2003). Incomplete removal of some epigenetic marks during gametogenesis can apparently lead to epigenetic inheritance through the germ line, of which there are a number of examples now in mammals (Chong and Whitelaw 2004). How widespread this phenomenon is and how many gene loci it could involve needs to be established.

2.6 Germ Line and Stem Cells—A Reversible Phenotype

Germ cells and ICM cells are the cells from which it has proved possible to derive pluripotent stem cells into culture. PGCs, when cultured in the presence of FGF2, can undergo changes to form pluripotent embryonic germ (EG) cells (Matsui et al. 1992; Resnick et al. 1992), which in many respects are similar to pluripotent ES cells, except that EG cells exhibit extensive erasure of parental imprints during their derivation (Tada et al. 1998).

As PGCs exhibit some characteristics of pluripotency, it is probable that mechanisms exist to allow PGCs to retain their distinct lineage-specific characteristics. How this may be achieved is as yet unclear, but it is possible that Blimp1 may have a continuing role following specification of PGCs to ensure that this is the case. During derivation of EG cells, it is assumed that this restriction is relieved and PGCs acquire an overtly pluripotent character with the ability to differentiate into many distinct cell types, which seldom occurs with germ cells in vivo. It is noteworthy that the derivation of EG cells becomes progressively less efficient when PGCs from E11.5 and E12.5 are used, further suggesting a change in the characteristics of these cells from E11.5, when they begin their differentiation pathway toward definitive male and female germ cells.

2.7 Development of Germ Cells from Pluripotent ES Cells

Pluripotent stem cells can differentiate into all types of somatic tissues when introduced into blastocysts, including germ cells (see below, Fig. 7). Increasing efforts are being made to generate different tissues from ES cells more efficiently in culture. Recently, it has been shown that it may be possible to generate PGCs, and possibly sperm- and egg-like structures from ES cells in culture (Hubner et al. 2003; Toyooka et al. 2003; Geijsen et al.

2004). It is therefore possible that with increasing knowledge of the genetic program of specification of PGCs and gametes, the mechanism of germ-cell specification could be examined in vitro. This may also provide a model system to examine regulation of epigenetic reprogramming in this lineage. Such an approach could ultimately advance our understanding of the human germ-cell lineage. Furthermore, if it becomes possible to direct differentiation toward human oocytes from cultured ES cells, it may be possible to use them for "therapeutic" cloning, circumventing the need for donor oocytes that are difficult to obtain. These oocytes could then be used for somatic nuclear transplantation to generate blastocysts, and subsequently, ES cells. Somatic nuclei undergo reprogramming to totipotency when transplanted into oocytes (for detail on nuclear transplantation and reprogramming, see Chapter 22). This procedure is likely to have a great impact on biomedicine, because it could allow the development of "personalized" ES cells from patients. In addition, it could allow generation of a very large repertoire of human ES cells from a variety of patients with specific diseases. These human ES cells may in turn open up many opportunities to study the underlying causes of human diseases; they could be used, for example, to test a variety of therapeutic compounds to alleviate diseases.

The use of human embryos and hES cells in research and therapy does raise many ethical issues. A variety of guidelines and regulations exist in different countries to monitor research in this area. If generation of viable gametes from ES cells becomes possible, and they are capable of fertilization and development to term, this will raise substantial ethical issues on how and whether this approach should be used in human medicine.

2.8 From Primordial Germ Cells to Gametes

The next stage in the development of the germ-cell lineage is the initiation of gametogenesis and the entry of germ cells into meiosis. The gonadal somatic environment regulates the timing of this event. In females, germ cells arrest in meiotic prophase, whereas male germ cells enter into mitotic arrest. A number of environmental signals dictate whether germ cells enter meiosis or not. Recently, a novel gene, *Meisetz*, was identified and shown to play a crucial role in initiating meiosis (Hayashi et al. 2005). This gene also contains the SET/PR domain and multiple zinc fingers, which have been demonstrated to have catalytic activity for H3K4me3. Expression of *Meisetz* is specific to germ cells, and it is detected at the time of their entry into meiotic prophase in females at

E13.5 and in postnatal testis. Mutation in *Meisetz* results in sterility in both males and females, which demonstrates its essential role in germ cells. Mutant germ cells show marked deficiency in the DNA double-strand-break repair pathway, and deficiency in pairing of homologous chromosomes during meiosis. This study demonstrates the significant role of epigenetic mechanisms in germ cells during meiosis.

Extensive epigenetic modifications continue during spermatogenesis, and eventually, the somatic linker histones are replaced by testis-specific variants (Kimmins and Sassone-Corsi 2005), which is followed by replacement of most histones with protamines. Studies have shown that Suv39, an H3K9 histone methyltransferase, is involved in gene repression and chromosome pairing. Furthermore, two SET-domain proteins, Suv39h1 and Suv39h2, have roles in male germ cells, the latter being expressed preferentially in the testis, and accumulating in the chromatin of the sex body. Mutations in both Suv39h1 and Suv39h2 result in infertility due to the arrest of spermatogenic cells (Peters et al. 2001). In addition, there is also a chromatoid body, a cloud-like cytoplasmic structure that is present in male germ cells. It is an RNA-processing body consisting of Dicer and Argonaute proteins and microRNAs, a germ-cell-specific cytoplasmic organelle that interacts with the nucleus and contains compacted mRNA.

Germ-line stem-cell self-renewal also implicates a role for noncoding RNAs in spermatogenesis, which may be mediated through the RNA interference (RNAi) machinery and HMTases. The members of the Piwi/Argonaute (called *Miwi* in mice) family have been reported to play a role in RNAi phenomena. Loss of Miwi-Like proteins (Mili) results in sterility in males (Kuramochi-Miyagawa et al. 2004), causing elevated expression of retrotransposon transcripts, IAP and Line 1. The involvement of Miwi-like proteins in their repression, however, has not been directly demonstrated.

Recent studies have succeeded in deriving pluripotent stem cells from spermatogonial stem cells recovered from neonatal and even from adult testis (Kanatsu-Shinohara et al. 2004; Guan et al. 2006). These cells may be maintained in culture indefinitely, but unlike ES cells, they have a paternal (androgenetic) imprint. They can differentiate into a variety of somatic cell types in vitro and in vivo, and are viable germ cells in vivo. These cells should provide an important tool to study many aspects of spermatogenesis and the role of epigenetic mechanisms that regulate their property as stem cells and differentiation into male gametes.

Erasure of imprints in early germ cells leads to epigenetically equivalent parental chromosomes for the first and only time in the life of mammals. Transplantation of such "imprint-free" nuclei directly into oocytes leads to development of embryos that are aberrant and die at early embryonic stages, presumably because without the appropriate epigenetic modifications, there is misexpression of genes that normally undergo imprinting. The experiment also shows that imprints cannot be acquired by imprint-free nuclei if they are transplanted directly to the oocyte. The initiation of DNA methylation imprints begins after birth, during the growth of oocytes in female germ cells. It occurs at later fetal stages in male germ cells. The de novo methyltransferase Dnmt3a and its cofactor Dnmt3L play a critical role in this process (for details, see Chapter 19). Imprinting is a major barrier to parthenogenetic development in mammals. Attempts to manipulate the epigenotype of female gametes may make it possible to allow development of mammalian embryos that are of maternal origin only.

3 From the Oocyte to the Early Embryo

We have seen in the previous sections how mature sperm and oocytes acquire very specific and different epigenetic marks during gametogenesis. Some of these differences, such as parental imprints, are maintained faithfully in the embryo after fertilization. Many others become dramatically reprogrammed as the embryonic genome attains totipotency. It is important to ask to what extent epigenetic marks inherited from the gametes play a role in the earliest differentiation events in the embryo. If you start life with one cell (the fertilized zygote) with a complete genome, which then divides, how do you ever get differentiation of gene expression and developmental programs in daughter cells?

3.1 Maternal Inheritance and Potential Asymmetric Distribution?

In organisms with a large egg cell (such as *Drosophila*, *Xenopus*, or chicken), some maternally made proteins or RNAs are located asymmetrically in the egg. They are then only inherited by some of the descendant cells, which subsequently develop a particular fate, while others that do not inherit these determinants develop differently (Huynh and St Johnston 2004). Such a strategy is possible with relatively large eggs (such as those of *Drosophila*), but becomes more difficult with smaller mammalian eggs. However, the developmental program may not be dictated simply by the size of the egg but, more importantly, by the necessity to generate a blastocyst in mammals that has to implant and generate a placenta to sustain the embryo. The ICM could be considered as developmentally equivalent to a *Drosophila* oocyte, since it undergoes patterning in response to signals from the extraembryonic somatic tissues during early development. Although there have been some recent suggestions that link the symmetry of the fertilized zygote to the symmetry of the blastocyst and even the postimplantation embryo (Gardner 1997; Weber et al. 1999), no asymmetrically localized determinants of differentiation have been found so far in mammalian eggs. Furthermore, mammalian embryos show a remarkable ability to "regulate" development; that is, when cells are removed or perturbed, compensatory growth or cell movements will often be able to keep the embryo developing normally (Kelly 1977). Nevertheless, there are probably slightly different propensities of individual cells (blastomeres) to develop along the ICM and TE lineages as early as the 4-cell stage (Fujimori et al. 2003, Piotrowska-Nitsche et al. 2005).

3.2 Epigenetic Events at Fertilization

During development and differentiation, somatic-cell lineages acquire very specific and specialized DNA methylation and histone modification patterns. These patterns are apparently difficult to erase or reverse when a somatic nucleus is transferred to an oocyte (see Chapter 22). The epigenetic marks of the oocyte and the sperm are specialized, too, but these are reprogrammed efficiently at fertilization, so that the embryonic genome can take up its new function, namely, to become totipotent (Reik et al. 2001; Surani 2001). A number of features of the epigenetic makeup of the gametes, and the epigenetic reprogramming after fertilization, are now known (Fig. 2). Both oocyte and sperm genomes have considerable levels of DNA methylation; as an example of a particular sequence class, the retrotransposon family, intracisternal A particles (IAP), which occur in a copy number of approximately 1000 in the mouse genome, are highly methylated both in the oocyte and the sperm genome (Lane et al. 2003). There are, in contrast, certain sequences, particularly differentially methylated regions (DMRs) in imprinted genes, that are methylated only in the oocyte, or in the sperm (see Chapter 19).

The oocyte genome also has high levels of histone modifications, both active ones (e.g., H3K9 acetylation, H3K4 methylation), and repressive ones (H3K9 methylation, H3K27 methylation) (Morgan et al. 2005). At this point before fertilization, the oocyte genome is transcriptionally inactive, but contains maternally inherited transcripts and proteins which are needed during the first few cleavage divisions, including those required for important reprogramming events (Fig. 1a). The sperm genome is highly specialized, in that the majority of the histones have been replaced during spermatogenesis by highly basic protamines, which may facilitate packaging of the DNA into the compacted sperm head (McLay and Clarke 2003). It is currently not known what modifications the remaining histones have, and where they are located in the genome, given the difficulty in studying these low-abundance chromatin regions.

Shortly after fertilization, a highly regulated sequence of reprogramming events occur to the sperm genome. Protamines are rapidly removed and replaced by histones. It is likely that being DNA replication independent (RI), this involves incorporation of the histone variant H3.3 by the histone chaperone HIRA (van der Heijden et al. 2005; see Chapter 13). At the same time, there is genome-wide demethylation of DNA in the male pronucleus, involving single-copy and repetitive sequences, but not paternally methylated imprinted genes (Olek and Walter 1997; Mayer et al. 2000; Oswald et al. 2000; Dean et al. 2001; Santos et al. 2002; Lane et al. 2003).

Prior to DNA replication, the histones in the paternal pronucleus are acetylated (H3 and H4), H3K4 methylated, and rapidly acquire H3K9me1 and H3K27me1 (Arney et al. 2002; Santos et al. 2002, 2005; Lepikhov and Walter 2004). H3K9me2/3 and H3K27me2/3, however, only occur subsequent to DNA replication, likely in conjunction with the incorporation of core histone H3.1, instead of H3.3 (Santos et al. 2005). At the first mitosis, most histone marks analyzed so far begin to be quite similar on the maternal and paternal chromosomes, at least as determined by the low level of resolution using immunofluorescence staining (Santos et al. 2005).

The enzyme activities that are responsible for these early reprogramming steps are all likely to be present in the oocyte, either as protein or as RNA molecules that can be rapidly translated. We already mentioned HIRA, but after DNA replication it is CAF-1 which is needed for replication dependent (RD) incorporation of histone H3.1. The su(var) enzymes methylate H3K9, and Ezh2, together with its cofactor Eed, methylates H3K27 (Erhardt et al. 2003; Santos et al. 2005). It is likely that the dramatic DNA demethylation of the paternal genome is caused by a process of "active demethylation," for which various mechanisms have been suggested but no definitive enzyme(s) isolated (Morgan et al. 2005). Possible candidates for a demethylation mechanism in mammals are the Aid/Apobec enzyme family, which can deaminate 5-methylcytosine in DNA leading to thymine, and the resulting T:G mismatch can be repaired by base excision repair (Morgan et al. 2004). A demethylation pathway in plants also involves base excision repair which is initiated by the DNA glycosylase Demeter (Gehring et al. 2006). Similar demethylases may also be present in the oocyte cytoplasm at fertilization. This raises the question of why the maternal genome is not demethylated at the same time as the paternal one. The maternal chromatin or pronucleus must possess some specific protection mechanism; the colocalization of methylated DNA in the maternal pronucleus, together with H3K9me2, may suggest this histone modification as a candidate for protection (Arney et al. 2002; Santos et al. 2002, 2005).

Although the evidence mainly suggests that histone modifications are acquired, rather than lost at the global level during this period, it is possible that histone arginine methylation is more dynamic. Indeed, a candidate for erasing histone arginine methylation by "deimination," Padi4, is present in the oocyte (Sarmento et al. 2004).

The main result of the rapid chromatin changes which occur at fertilization seems to be that at the 2-cell stage, the paternal genome is similar to the maternal one. This excludes DNA methylation, which differs considerably between the two genomes largely as a result of the demethylation of the sperm genome. In addition, the level of analysis so far has not excluded that there are gene-specific differences in histone modifications that are established at this stage.

3.3 From the Zygote to the Blastocyst

The general theme of further reprogramming, particularly of genome-wide DNA methylation patterns, continues from the 2-cell stage through to the cleavage stages of preimplantation development, until the embryo reaches the blastocyst stage (Monk et al. 1987; Howlett and Reik 1991; Rougier et al. 1998). The precise dynamics of histone modifications are not fully described yet in the mouse, but DNA methylation is reduced stepwise with each nuclear division until the 16-cell morula stage. The reason for this is that Dnmt1, the methyltransferase which maintains methylation at CpG dinucleotides in a semiconservative fashion during DNA replication, is excluded from the nucleus (Carlson et al. 1992). Therefore, at each division, 50% of all genomic DNA methylation is lost. The only known, well-documented exception to this are DMRs in imprinted genes. It is not clear whether their methylation is maintained over this period by an unknown Dnmt, or by the action of a small amount of Dnmt1 that is able to enter the nucleus and be specifically targeted to DMRs. Remarkably, at the 8-cell stage, Dnmt1 protein appears to enter the nucleus for one replication cycle. If this Dnmt1 protein is removed (by its genetic ablation in the oocyte, which provides most if not all of the protein during the cleavage divisions), methylation in DMRs is indeed reduced by 50%, consistent with its being needed for maintenance of methylation for one round of replication only (Howell et al. 2001).

At the 8- to 16-cell stage, the outer cells of the morula flatten and become epithelial; this is called compaction. This is the first outward sign of differentiation in the mammalian embryo. Over the next 2–3 divisions, the morula then cavitates (i.e., a cavity forms) and the blastocyst is distinguished by its inner cell mass (ICM) and outer trophectoderm (TE) cells. The ICM cells go on to form all lineages of the embryo and fetus, whereas the TE cells form most (but not all) lineages of the placenta (extraembryonic lineages). Shortly after this stage, another epithelial layer of cells forms on the surface of the ICM; these are primitive endoderm cells which again contribute to the placenta and the yolk sac, but not the embryo. A few genetic determinants of these very early allocation events are known: *Oct4*, *Nanog*, and *Sox2* are important for the determination or maintenance of ICM cells, whereas *Cdx2* is required for the early maintenance of the TE cell fate (Nichols et al. 1998; Avilion et al. 2003; Chambers et al. 2003; Mitsui et al. 2003; Niwa et al. 2005). To what extent maternal protein (present in the oocyte), or epigenetic regulation of these genes in the early embryo, could contribute to the early cell-fate decisions, or their maintenance, is currently unknown (Dean and Ferguson-Smith 2001).

Major epigenetic programming events occur right at this developmental stage, however. The ICM cells acquire a high level of DNA methylation, at least as judged by immunofluorescence (as illustrated by the red inner cells of the blastocyst in the left image of the title page), which is probably brought about by the de novo DNA methyltransferase Dnmt3b (Santos et al. 2002). This is accompanied by increases in histone H3K9 and H3K27 methylation, transduced by G9a and Eset, and Ezh2, respectively (Erhardt et al. 2003). Although de novo methylation of DNA is not critical for the initial establishment of ICM cells, histone methylation by Ezh2 and Eset is: In gene knock-outs of either gene, the ICM cells do not develop properly (O'Carroll et al. 2001; Dodge et al. 2004).

In contrast to the increase in epigenetic modifications in the ICM, the TE remains largely DNA hypomethylated, as do most of the cell lineages in the later placenta (Chapman et al. 1984; Santos et al. 2002). It is thought that placental cell types need less epigenetic stability, because, after all, their lifetime is much more restricted than that of the fetus, which develops into the adult organism.

In addition to these large-scale and genome-wide epigenetic events, more locus-specific reprogramming also takes place at these stages. In XX female embryos, the paternally inherited X chromosome is inactivated during the cleavage stages and remains so in the extraembryonic tissues (i.e., the TE and the placenta) (Huynh and Lee 2003; Okamoto et al. 2005). In the ICM, however, the inactive X is reactivated, and this is followed by random inactivation of one X chromosome after differentiation in the ICM-derived lineages (Mak et al. 2004; for more detail, see Fig. 4 of Chapter 17). Mechanistically, imprinted X inactivation in the preimplantation embryo involves expression of the noncoding RNA *Xist* from the paternal X chromosome, whose "coating" of the chromosome is thought to lead to gene silencing and the establishment of repressive epigenetic modifications (Heard 2004). In the newly formed ICM cells, *Xist* transcription is down-regulated, repressive histone modifications are subsequently lost, and the chromosome becomes reactivated (Mak et al. 2004; Okamoto et al. 2004). This is followed, shortly afterward, by the initiation of random X inactivation in epiblast cells. We show in the next section that ES cells are "frozen" at the stage after reactivation of the X chromosome, such that female ES cells contain two active X chromosomes.

4 From Pluripotent Stem Cells to Somatic Cells and Back to Germ Cells

4.1 Derivation of Pluripotent Stem Cells

In the previous chapter, we learned that there are dramatic epigenetic reprogramming events in the zygote, and others in the cleavage-stage embryo and the blastocyst, resulting in different epigenetic patterns in the ICM and the TE. We now consider the genetic and epigenetic properties of early stem cells derived into culture from the blastocyst and later lineages, such as embryonic stem cells (Smith 2001), trophoblast stem (TS) cells (Rossant 2001), endoderm stem (XEN) cells (Kunath et al. 2005), and embryonic germ cell (EG) stem cells (Matsui et al. 1992).

What is common to these cell types is that they can be isolated or established from intact embryos into culture under certain culturing conditions. Once established, they can be cultured for extended periods of time and show no signs of senescence. They can also be genetically manipulated during culture, and then can be reintroduced into living embryos to participate in the development of the appropriate lineages.

One of the most important discoveries in mammalian embryology during the 1980s was development of methods to generate pluripotent embryonic stem (ES) cells from ICMs. The ES cells, explanted from mouse blastocysts into culture, can be maintained for extended culture periods, and when microinjected back into blastocysts, they colonize all embryonic lineages, thus forming chimeras (Fig. 6; Evans and Kaufman 1981; Martin 1981). What was particularly striking was that descendants of the ES cells could colonize the germ cells and give rise to normal offspring, which were derived wholly from the ES cell genotype. This, together with the ability to genetically manipulate the ES cell genome by homologous recombination techniques (leading to gene knock-outs), has revolutionized mouse genetics and has made the mouse the mammalian genetic model organism of choice.

ES cells share properties with ICM/epiblast cells, but there are also substantial differences, making it likely that they are a "synthetic" cell type which does not exist in the normal embryo (the same is likely to apply to the other pluripotent cell lines) (Smith 2001). For example, whereas self-renewal of mouse ES cells requires the

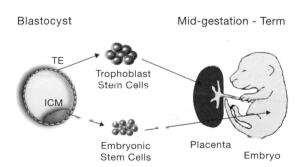

Figure 6. Derivation of ES and TS Cells from the Blastocyst

ES cells are derived from inner cell mass (ICM) cells and can be kept in culture without differentiating. They can be genetically manipulated while in culture. ES cells can be reintroduced into blastocysts and can then colonize all tissues in the embryo, including the germ line, but excluding the trophoblast cells of the placenta. TS cells can be established similarly into culture from the trophectoderm cells of the blastocyst, and when reintroduced into blastocysts, contribute to placental cell types.

Lif/gp130/Stat3 signaling pathway, embryos with muta-tions in this pathway still develop a normal ICM (Smith 2001). It is thus likely that epigenetic changes occur, and may be necessary, for the derivation and maintenance of ES cells from ICM cells. Outgrowths of ICM cells into culture rapidly lose expression of Oct4, and only a mouse strain from which it is relatively easy to derive ES cells, called 129Sv, retains some Oct4-expressing cells upon culture (Buehr et al. 2003). Epigenetic changes have also been reported in imprinted genes in mouse and rhesus monkey ES cells, and in the mouse this can result in aber-rant development of the cells when they are reintroduced into chimeras (Dean et al. 1998; Humpherys et al. 2001; Fujimoto et al. 2005).

4.2 Epigenetic Properties of Pluripotential Cell Lines

ES cells can be differentiated in vitro into a number of different cell lineages (Fig. 7). To what extent do epige-netic mechanisms maintain cells in an undifferentiated or differentiated state? Clearly, there are epigenomic dif-ferences between undifferentiated ES cells, differentiated ES cells, and somatic cells, where pluripotential cells are particularly characterized by hyperdynamic plasticity of the chromatin (Meshorer et al. 2006). Deletion in ES cells of Parp (poly-ADP ribosylase), which is involved in controlling the alteration of histone marks, leads to a low frequency of transdifferentiation into trophoblast cells, suggesting that epigenetic marks in ES cells are needed to maintain their identity (Hemberger et al. 2003). DNA methylation may also contribute to this

because severe depletion of DNA methylation results in a partial block in an ES cell's ability to differentiate, and those that do differentiate express markers of trophecto-derm tissues (Jackson et al. 2004).

The maintenance of pluripotency of ES cells depends on the transcription factors Oct4, Nanog, and Sox2. These bind alone, or in combination, to many gene loci in ES cells, which need to be either expressed or silenced for pluripotency to be maintained (Boyer et al. 2005; Loh et al. 2006). Differentiation of ES cells in vitro is charac-terized by the transcriptional silencing of pluripotency genes, which then remain repressed in all somatic tissues. Epigenetic mechanisms are indeed important for their silencing: The Oct4 promoter, for example, accumulates repressive histone modifications and DNA methylation during differentiation (Feldmann et al. 2006). Loss of DNA methylation, for example, using Dnmt1 knock-out embryos, leads to reexpression of Oct4 when differenti-ated (Feldman et al. 2006).

ES cells and their differentiated derivatives have also served as a model for the epigenetic regulation of X-chro-mosome inactivation. Female ICM and ES cells have a down-regulated Xist gene and two active X chromosomes. Upon differentiation in vitro, Xist becomes up-regulated on one of the X chromosomes, the Xist RNA begins to "coat" this chromosome in cis, and silencing of genes on the X concomitant with the accumulation of repressive histone modifications and DNA methylation ensues (Heard 2004).

Other pluripotent cell types can similarly be estab-lished in culture, but their epigenetic properties are less well characterized than those of ES cells. It is, however, known from epigenetic studies of X inactivation that in female TS cells there is a paternally inactivated X chromo-some containing repressive histone marks (Huynh and Lee 2003). Female XEN cells also have a paternal X chro-mosome which is inactive (Kunath et al. 2005).

Pluripotent cell lines can also be established from pri-mordial germ cells (PGCs) either during their period of migration in the embryo (E8–E10.5) or once they have reached the embryonic gonads (E11.5–E13.5) (Fig. 8) (Matsui et al. 1992). PGCs, during the E8.5–12.5 stages of development, undergo extensive epigenetic reprogram-ming including DNA demethylation of imprinted genes and other sequences in the genome (Hajkova et al. 2002; Lee et al. 2002). Indeed, most EG cells have undergone DNA demethylation erasure of imprinted genes and other sequences, and this alters their developmental potential, as shown when introduced into chimeras (Tada et al. 1998). It is not clear yet whether there are epigenetic

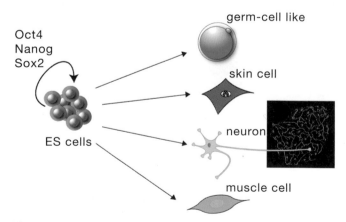

Figure 7. Differentiation of ES Cells into Different Cell Types In Vitro

ES cells can be differentiated in vitro under suitable culture condi-tions into many different cell types, such as neurons, muscle cells, and even germ cells (oocytes). The inset image shows neuroglial cells derived from ES cells in culture.

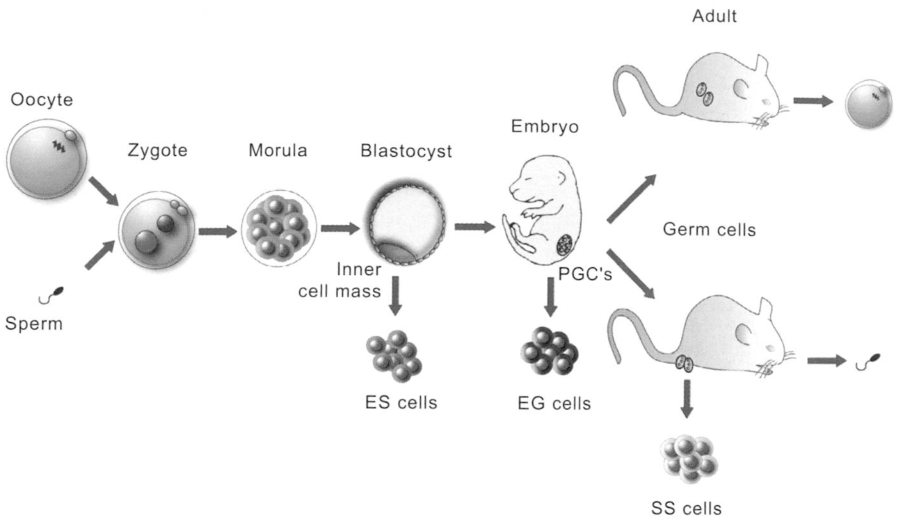

Adult

Oocyte

Zygote Morula Blastocyst Embryo

Sperm

Inner cell mass

Germ cells

PGC's

ES cells EG cells

SS cells

Figure 8. Link between Germ-Line Development and Pluripotency

The oocyte and zygote are totipotent (the oocyte has restrictions imposed by parental imprinting) and so are morula cells up to a certain stage. Pluripotent cells can be derived into culture from the inner cell mass (ICM) cells, primordial germ cells (PGCs), and spermatogonia stem (SS) cells in the neonatal and adult testis. There is thus an intimate link between germ-line cells throughout their development, and pluripotency.

differences between endogenous PGCs and in vitro cultured EG cells, similar to those suspected to exist between ICM and ES cells.

4.3 Reprogramming Capacity of Stem Cells

The continued state of pluripotency in culture without cell senescence may well require continued epigenetic reprogramming of stem cells. That these cells indeed have reprogramming activities has been shown in fusion experiments in which EG or ES cells were fused to differentiated somatic cells (Tada et al. 1997, 2001; Cowan et al. 2005). In the tetraploid cell lines resulting from fusion, the somatic epigenotype has been shown to be reprogrammed (Fig. 9). In EG–somatic cell fusions, the somatic genome loses DNA methylation of imprinted

genes as well as other sequences in the genome (Tada et al. 1997). In ES–somatic cell fusions, in contrast, imprinted gene DNA methylation is retained, but the inactive X chromosome (in female cells) is reactivated, and the promoter of the Oct4 gene becomes DNA demethylated, resulting in Oct4 reexpression (Tada et al. 2001; Cowan et al. 2005; Surani 2005).

The identities of any reprogramming factors affecting DNA methylation or histone modifications in these cell hybrids are currently unknown. It is not known, for example, whether demethylation in EG or ES cell hybrids requires DNA demethylase activities, or whether DNA replication occurs in the absence of Dnmt1, resulting in passive demethylation. Nevertheless, ES and EG cells should be seen as an important resource for the isolation and characterization of epigenetic reprogramming factors.

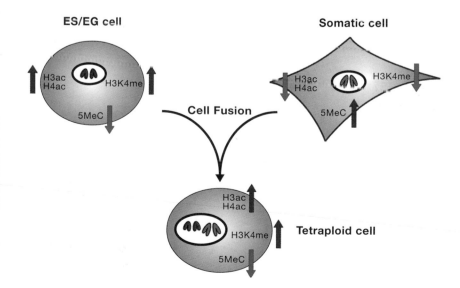

ES/EG cell

H3ac
H4ac   H3K4me

5MeC

Somatic cell

H3ac
H4ac   H3K4me

5MeC

Cell Fusion

H3ac
H4ac

H3K4me

5MeC

Tetraploid cell

Figure 9. Pluripotent Cells Have the Capacity to Reprogram Somatic Cells

ES or EG cells can be fused with somatic cells, resulting in tetraploid hybrids. This leads to epigenetic reprogramming of the somatic nucleus, with changes, for example, in 5meC, H3ac, H4ac, and H3K4me. The tetraploid cells resulting from this fusion and reprogramming also have a pluripotent phenotype: When injected into blastocysts they can contribute to many different cell types in the embryo.

5 Perspective

The next few years will see decisive and exciting advances in our understanding of the genetic and epigenetic factors that are critical for totipotency and pluripotency of germ cells and stem cells. High throughput and sensitive methods for determining the various layers of epigenetic information of the genome are becoming available. Factors that regulate epigenetic information, and particularly those that are needed to reprogram epigenetic marks in somatic cells to those found in pluripotent cells, are being identified. Better methods to selectively and safely manipulate epigenetic states in vivo also need to be developed.

Pluripotent stem cells present many exciting opportunities for fundamental studies as well as for their potential applications in biomedicine. As for fundamental research, the uniqueness of the pluripotent state may provide insights into the mechanisms that regulate cell-fate decisions. The ability to differentiate into diverse cell types also provides a potential to generate replacement cells in the quest to repair diseased tissues. They could also be used to develop disease models to explore how various human diseases originate from the very beginning as a result of specific mutations and epimutations (i.e., mutations caused by alterations of DNA methylation or the chromatin template not involving changes to the DNA sequence), which in turn may allow development of new drugs to cure or even prevent diseases.

The use of human pluripotent stem cells raises sensitive and critical ethical issues, which are being debated by the wider public. It is imperative that appropriate ethical and regulatory frameworks are established for the use of stem cells for research and biomedical applications, because the generation of pluripotent stem cells involves the use of early embryos.

Deeper insights into the epigenetic mechanisms that regulate pluripotency and reprogramming of the genome may one day allow generation of pluripotent stem cells directly from adult stem cells or even differentiated cells without the necessity to use embryos. This is one aspect where advances in our understanding of epigenetic mechanisms may prove decisive for advances in medicine in the future.

Acknowledgments

M.A.S. thanks all the present and past members of the lab for their original work, which has been funded by the Wellcome Trust, BBSRC, EU, DTI, and CellCentric. W.R. thanks all his colleagues in the lab, especially Wendy Dean, for their contributions to work and ideas discussed in this chapter. Work in W.R.'s lab is funded by BBSRC, MRC, EU, DTI, and CellCentric.

References

Arney K.L., Bao S., Bannister A.J., Kouzarides T., and Surani M.A. 2002. Histone methylation defines epigenetic asymmetry in the mouse zygote. *Int. J. Dev. Biol.* **46:** 317–320.

Avilion A.A., Nicolis S.K., Pevny L.H., Perez L., Vivian N., and Lovell-Badge R. 2003. Multipotent cell lineages in early mouse development depend on SOX2 function. *Genes Dev.* **17:** 126–140.

Blackwell T.K. 2004. Germ cells: Finding programs of mass repression. *Curr. Biol.* **14:** R229–230.

Boyer L.A., Lee T.I., Cole M.F., Johnstone S.E., Levine S.S., Zucker J.P., Guenther M.G., Kumar R.M., Murray H.L., Jenner R.G., et al. 2005. Core transcriptional regulatory circuitry in human embryonic stem cells. *Cell* **122:** 947–956.

Buehr M., Nichols J., Stenhouse F., Mountford P., Greenhalgh C.J., Kantachuvesiri S., Brooker G., Mullins J., and Smith A.G. 2003. Rapid loss of Oct-4 and pluripotency in cultured rodent blastocysts and derivative cell lines. *Biol. Reprod.* **68:** 222–229.

Carlson L.L., Page A.W., and Bestor T.H. 1992. Properties and localization of DNA methyltransferase in preimplantation mouse embryos: Implications for genomic imprinting. *Genes Dev.* **6:** 2536–2541.

Chambers I., Colby D., Robertson M., Nichols J., Lee S., Tweedie S., and Smith A. 2003. Functional expression cloning of Nanog, a pluripotency sustaining factor in embryonic stem cells. *Cell* **113:** 643–655.

Chang D.H., Cattoretti G., and Calame K.L. 2002. The dynamic expression pattern of B lymphocyte induced maturation protein-1 (Blimp-1) during mouse embryonic development. *Mech. Dev.* **117:** 305–309.

Chapman V., Forrester L., Sanford J., Hastie N., and Rossant J. 1984. Cell lineage-specific undermethylation of mouse repetitive DNA. *Nature* **307:** 284–286.

Chong S. and Whitelaw E. 2004. Epigenetic germline inheritance. *Curr. Opin. Genet. Dev.* **14:** 692–696.

Cowan C.A., Atienza J., Melton D.A., and Eggan K. 2005. Nuclear reprogramming of somatic cells after fusion with human embryonic stem cells. *Science* **309:** 1369–1373.

Dean W. and Ferguson-Smith A. 2001. Genomic imprinting: Mother maintains methylation marks. *Curr. Biol.* **11:** R527–530.

Dean W., Santos F., Stojkovic M., Zakhartchenko V., Walter J., Wolf E., and Reik W. 2001. Conservation of methylation reprogramming in mammalian development: Aberrant reprogramming in cloned embryos. *Proc. Natl. Acad. Sci.* **98:** 13734–13738.

Dean W., Bowden L., Aitchison A., Klose J., Moore T., Meneses J.J., Reik W., and Feil R. 1998. Altered imprinted gene methylation and expression in completely ES cell-derived mouse fetuses: Association with aberrant phenotypes. *Development* **125:** 2273–2282.

Deshpande G., Calhoun G., and Schedl P. 2004. Overlapping mechanisms function to establish transcriptional quiescence in the embryonic *Drosophila* germline. *Development* **131:** 1247–1257.

de Souza F.S., Gawantka V., Gomez A.P., Delius H., Ang S.L., and Niehrs C. 1999. The zinc finger gene Xblimp1 controls anterior endomesodermal cell fate in Spemann's organizer. *EMBO J.* **18:** 6062–6072.

Dodge J.E., Kang Y.K., Beppu H., Lei H., and Li E. 2004. Histone H3-K9 methyltransferase ESET is essential for early development. *Mol. Cell. Biol.* **24:** 2478–2486.

Erhardt S., Su I.H., Schneider R., Barton S., Bannister A.J., Perez-Burgos L., Jenuwein T., Kouzarides T., Tarakhovsky A., and Surani M.A. 2003. Consequences of the depletion of zygotic and embryonic enhancer of zeste 2 during preimplantation mouse development. *Development* **130:** 4235–4248.

Evans M.J. and Kaufman M.H. 1981. Establishment in culture of pluripotential cells from mouse embryos. *Nature* **292:** 154–156.

Extavour C.G. and Akam M. 2003. Mechanisms of germ cell specification across the metazoans: Epigenesis and preformation. *Development* **130:** 5869–5884.

Feldman N., Gerson A., Fang J., Li E., Zhang Y., Shinkai Y., Cedar H., and Bergman Y. 2006. G9a-mediated irreversible epigenetic inactivation of Oct-3/4 during early embryogenesis. *Nat. Cell. Biol.* **8:** 188–194.

Fujimori T., Kurotaki Y., Miyazaki J., and Nabeshima Y. 2003. Analysis of cell lineage in two- and four-cell mouse embryos. *Development* **130:** 5113–5122.

Fujimoto A., Mitalipov S.M., Kuo H.C., and Wolf D.P. 2005. Aberrant genomic imprinting in rhesus monkey ES cells. *Stem Cells* **24:** 595–603.

Gardner R.L. 1985. Clonal analysis of early mammalian development. *Philos. Trans. R. Soc. Lond. B Biol. Sci.* **312:** 163–178.

———. 1997. The early blastocyst is bilaterally symmetrical and its axis of symmetry is aligned with the animal-vegetal axis of the zygote in the mouse. *Development* **124:** 289–301.

Gehring M., Huh J.H., Hsieh T.F., Penterman J., Choi Y., Harada J.J., Goldberg R.B., and Fischer R.L. 2006. DEMETER DNA glycosylase establishes MEDEA polycomb gene self-imprinting by allele-specific demethylation. *Cell* **124:** 495–506.

Geijsen N., Horoschak M., Kim K., Gribnau J., Eggan K., and Daley G.Q. 2004. Derivation of embryonic germ cells and male gametes from embryonic stem cells. *Nature* **427:** 148–154.

Guan K., Nayernia K., Maier L.S., Wagner S., Dressel R., Lee J.H., Nolte J., Wolf F., Li M., Engel W., and Hasenfuss G. 2006. Pluripotency of spermatogonial stem cells from adult mouse testis. *Nature* **440:** 1199–1203.

Gyory I., Wu J., Fejer G., Seto E., and Wright K.L. 2004. PRDI-BF1 recruits the histone H3 methyltransferase G9a in transcriptional silencing. *Nat. Immunol.* **5:** 299–308.

Hajkova P., Erhardt S., Lane N., Haaf T., El-Maarri O., Reik W., Walter J., and Surani M.A. 2002. Epigenetic reprogramming in mouse primordial germ cells. *Mech. Dev.* **117:** 15–23.

Hayashi K., Yoshida K., and Matsui Y. 2005. A histone H3 methyltransferase controls epigenetic events required for meiotic prophase. *Nature* **438:** 374–378.

Heard E. 2004. Recent advances in X-chromosome inactivation. *Curr. Opin. Cell Biol.* **16:** 247–255.

Hemberger M., Nozaki T., Winterhager E., Yamamoto H., Nakagama H., Kamada N., Suzuki H., Ohta T., Ohki M., Masutani M., and Cross J.C. 2003. Parp1-deficiency induces differentiation of ES cells into trophoblast derivatives. *Dev. Biol.* **257:** 371–381.

Hernandez-Lagunas L., Choi I.F., Kaji T., Simpson P., Hershey C., Zhou Y., Zon L., Mercola M., and Artinger K.B. 2005. Zebrafish *narrowminded* disrupts the transcription factor *prdm1* and is required for neural crest and sensory neuron specification. *Dev. Biol.* **278:** 347–357.

Howell C.Y., Bestor T.H., Ding F., Latham K.E., Mertineit C., Trasler J.M., and Chaillet J.R. 2001. Genomic imprinting disrupted by a maternal effect mutation in the Dnmt1 gene. *Cell* **104:** 829–838.

Howlett S.K. and Reik W. 1991. Methylation levels of maternal and paternal genomes during preimplantation development. *Development* **113:** 119–127.

Hubner K., Fuhrmann G., Christenson L.K., Kehler J., Reinbold R., De La Fuente R., Wood J., Strauss J.F., III, Boiani M., and Scholer H.R. 2003. Derivation of oocytes from mouse embryonic stem cells. *Science* **300:** 1251–1256.

Humpherys D., Eggan K., Akutsu H., Hochedlinger K., Rideout W.M., III, Biniszkiewicz D., Yanagimachi R., and Jaenisch R. 2001. Epigenetic instability in ES cells and cloned mice. *Science* **293:** 95–97.

Huynh J.R. and St Johnston D. 2004. The origin of asymmetry: Early polarisation of the *Drosophila* germline cyst and oocyte. *Curr. Biol.* **14:** R438–449.

Huynh K.D. and Lee J.T. 2003. Inheritance of a pre-inactivated paternal X chromosome in early mouse embryos. *Nature* **426:** 857–862.

Jackson M., Krassowska A., Gilbert N., Chevassut T., Forrester L., Ansell J., and Ramsahoye B. 2004. Severe global DNA hypomethylation blocks differentiation and induces histone hyperacetylation in embryonic stem cells. *Mol. Cell. Biol.* **24:** 8862–8871.

Kanatsu-Shinohara M., Inoue K., Lee J., Yoshimoto M., Ogonuki N., Miki H., Baba S., Kato T., Kazuki Y., Toyokuni S., et al. 2004. Generation of pluripotent stem cells from neonatal mouse testis. *Cell* **119:** 1001–1012.

Kelly S.J. 1977. Studies of the developmental potential of 4- and 8-cell stage mouse blastomeres. *J. Exp. Zool.* **200:** 365–376.

Kimmins S. and Sassone-Corsi P. 2005. Chromatin remodelling and epigenetic features of germ cells. *Nature* **434:** 583–589.

Kunath T., Arnaud D., Uy G.D., Okamoto I., Chureau C., Yamanaka Y., Heard E., Gardner R.L., Avner P., and Rossant J. 2005. Imprinted X-inactivation in extra-embryonic endoderm cell lines from mouse blastocysts. *Development* **132:** 1649–1661.

Kuramochi-Miyagawa S., Kimura T., Ijiri T.W., Isobe T., Asada N., Fujita Y., Ikawa M., Iwai N., Okabe M., Deng W., et al. 2004. *Mili,* a mammalian member of *piwi* family gene, is essential for spermatogenesis. *Development* **131:** 839–849.

Lane N., Dean W., Erhardt S., Hajkova P., Surani A., Walter J., and Reik W. 2003. Resistance of IAPs to methylation reprogramming may provide a mechanism for epigenetic inheritance in the mouse. *Genesis* **35:** 88–93.

Lawson K.A. and Hage W.J. 1994. Clonal analysis of the origin of primordial germ cells in the mouse. *CIBA Found. Symp.* **182:** 68–84.

Lawson K.A., Dunn N.R., Roelen B.A., Zeinstra L.M., Davis A.M., Wright C.V., Korving J.P., and Hogan B.L. 1999. *Bmp4* is required for the generation of primordial germ cells in the mouse embryo. *Genes Dev.* **13:** 424–436.

Leatherman J.L. and Jongens T.A. 2003. Transcriptional silencing and translational control: Key features of early germline development. *Bioessays* **25:** 326–335.

Lee J., Inoue K., Ono R., Ogonuki N., Kohda T., Kaneko-Ishino T., Ogura A., and Ishino F. 2002. Erasing genomic imprinting memory in mouse clone embryos produced from day 11.5 primordial germ cells. *Development* **129:** 1807–1817.

Lepikhov K. and Walter J. 2004. Differential dynamics of histone H3 methylation at positions K4 and K9 in the mouse zygote. *BMC Dev. Biol.* **4:** 12.

Loh Y.H., Wu Q., Chew J.L., Vega V.B., Zhang W., Chen X., Bourque G., George J., Leong B., Liu J., et al. 2006. The Oct4 and Nanog transcription network regulates pluripotency in mouse embryonic stem cells. *Nat. Genet.* **38:** 431–440.

Mak W., Nesterova T.B., de Napoles M., Appanah R., Yamanaka S., Otte A.P., and Brockdorff N. 2004. Reactivation of the paternal X chromosome in early mouse embryos. *Science* **303:** 666–669.

Martin G.R. 1981. Isolation of a pluripotent cell line from early mouse embryos cultured in medium conditioned by teratocarcinoma stem cells. *Proc. Natl. Acad. Sci.* **78:** 7634–7638.

Martinho R.G., Kunwar P.S., Casanova J., and Lehmann R. 2004. A noncoding RNA is required for the repression of RNApolII-dependent transcription in primordial germ cells. *Curr. Biol.* **14:** 159–165.

Matsui Y., Zsebo K., and Hogan B.L. 1992. Derivation of pluripotential embryonic stem cells from murine primordial germ cells in culture. *Cell* 70, 841–847.

Mayer W., Niveleau A., Walter J., Fundele R., and Haaf T. 2000. Demethylation of the zygotic paternal genome. *Nature* 403: 501–502.

McLaren A. 2003. Primordial germ cells in the mouse. *Dev. Biol.* 262: 1–15.

McLaren A. and Lawson K.A. 2005. How is the mouse germ-cell lineage established? *Differentiation* 73: 435–437.

McLay D.W. and Clarke H.J. 2003. Remodelling the paternal chromatin at fertilization in mammals. *Reproduction* 125: 625–633.

Meshorer E., Yellajoshula D., George E., Scambler P.J., Brown D.T., and Misteli T. 2006. Hyperdynamic plasticity of chromatin proteins in pluripotent embryonic stem cells. *Dev. Cell.* 10: 105–116.

Mitsui K., Tokuzawa Y., Itoh H., Segawa K., Murakami M., Takahashi K., Maruyama M., Maeda M., and Yamanaka S. 2003. The homeoprotein Nanog is required for maintenance of pluripotency in mouse epiblast and ES cells. *Cell* 113: 631–642.

Monk M., Boubelik M., and Lehnert S. 1987. Temporal and regional changes in DNA methylation in the embryonic, extraembryonic and germ cell lineages during mouse embryo development. *Development* 99: 371–382.

Morgan H.D., Dean W., Coker H.A., Reik W., and Petersen-Mahrt S.K. 2004. Activation-induced cytidine deaminase deaminates 5-methylcytosine in DNA and is expressed in pluripotent tissues: Implications for epigenetic reprogramming. *J. Biol. Chem.* 279: 52353–52360.

Morgan H.D., Santos F., Green K., Dean W., and Reik W. 2005. Epigenetic reprogramming in mammals. *Hum. Mol. Genet.* 14: R47–58.

Nichols J., Zevnik B., Anastassiadis K., Niwa H., Klewe-Nebenius D., Chambers I., Scholer H., and Smith A. 1998. Formation of pluripotent stem cells in the mammalian embryo depends on the POU transcription factor Oct4. *Cell* 95: 379–391.

Niwa H., Toyooka Y., Shimosato D., Strumpf D., Takahashi K., Yagi R., and Rossant J. 2005. Interaction between Oct3/4 and Cdx2 determines trophectoderm differentiation. *Cell* 123: 917–929.

O'Carroll D., Erhardt S., Pagani M., Barton S.C., Surani M.A., and Jenuwein T. 2001. The polycomb-group gene Ezh2 is required for early mouse development. *Mol. Cell. Biol.* 21: 4330–4336.

Ohinata Y., Payer B., O'Carroll D., Ancelin K., Ono Y., Sano M., Barton S.C., Obukhanych T., Nussenzweig M., Tarakhovsky A., et al. 2005. Blimp1 is a critical determinant of the germ cell lineage in mice. *Nature* 436: 207–213.

Okamoto I., Otte A.P., Allis C.D., Reinberg D., and Heard E. 2004. Epigenetic dynamics of imprinted X inactivation during early mouse development. *Science* 303: 644–649.

Okamoto I., Arnaud D., Le Baccon P., Otte A.P., Disteche C.M., Avner P., and Heard E. 2005. Evidence for de novo imprinted X-chromosome inactivation independent of meiotic inactivation in mice. *Nature* 438: 369–373.

Olek A. and Walter J. 1997. The pre-implantation ontogeny of the H19 methylation imprint. *Nat. Genet.* 17: 275–276.

Oswald J., Engemann S., Lane N., Mayer W., Olek A., Fundele R., Dean W., Reik W., and Walter J. 2000. Active demethylation of the paternal genome in the mouse zygote. *Curr. Biol.* 10: 475–478.

Peters A.H., O'Carroll D., Scherthan H., Mechtler K., Sauer S., Schofer C., Weipoltshammer K., Pagani M., Lachner M., Kohlmaier A., et al. 2001. Loss of the *Suv39h* histone methyltransferases impairs mammalian heterochromatin and genome stability. *Cell* 107: 323–337.

Piotrowska-Nitsche K., Perea-Gomez A., Haraguchi S., and Zernicka-Goetz M. 2005. Four-cell stage mouse blastomeres have different developmental properties. *Development* 132: 479–490.

Reik W., Dean W., and Walter J. 2001. Epigenetic reprogramming in mammalian development. *Science* 293: 1089–1093.

Ren B., Chee K.J., Kim T.H., and Maniatis T. 1999. PRDI-BF1/Blimp-1 repression is mediated by corepressors of the Groucho family of proteins. *Genes Dev.* 13: 125–137.

Resnick J.L., Bixler L.S., Cheng L., and Donovan P.J. 1992. Long-term proliferation of mouse primordial germ cells in culture. *Nature* 359: 550–551.

Rossant J. 2001. Stem cells from the mammalian blastocyst. *Stem Cells* 19: 477–482.

Rougier N., Bourc'his D., Gomes D.M., Niveleau A., Plachot M., Paldi A., and Viegas-Pequignot E. 1998. Chromosome methylation patterns during mammalian preimplantation development. *Genes Dev.* 12: 2108–2113.

Roy S. and Ng T. 2004. Blimp-1 specifies neural crest and sensory neuron progenitors in the zebrafish embryo. *Curr. Biol.* 14: 1772–1777.

Saitou M., Barton S.C., and Surani M.A. 2002. A molecular program for the specification of germ cell fate in mice. *Nature* 418: 293–300.

Saitou M., Payer B., Lange U.C., Erhardt S., Barton S.C., and Surani M.A. 2003. Specification of germ cell fate in mice. *Philos. Trans. R. Soc. Lond. B Biol. Sci.* 358: 1363–1370.

Santos F., Hendrich B., Reik W., and Dean W. 2002. Dynamic reprogramming of DNA methylation in the early mouse embryo. *Dev. Biol.* 241: 172–182.

Santos F., Peters A.H., Otte A.P., Reik W., and Dean W. 2005. Dynamic chromatin modifications characterise the first cell cycle in mouse embryos. *Dev. Biol.* 280: 225–236.

Sarmento O.F., Digilio L.C., Wang Y., Perlin J., Herr J.C., Allis C.D., and Coonrod S.A. 2004. Dynamic alterations of specific histone modifications during early murine development. *J. Cell Sci.* 117: 4449–4459.

Schaner C.E., Deshpande G., Schedl P.D., and Kelly W.G. 2003. A conserved chromatin architecture marks and maintains the restricted germ cell lineage in worms and flies. *Dev. Cell* 5: 747–757.

Sciammas R. and Davis M.M. 2004. Modular nature of Blimp-1 in the regulation of gene expression during B cell maturation. *J. Immunol.* 172: 5427–5440.

Seki Y., Hayashi K., Itoh K., Mizugaki M., Saitou M., and Matsui Y. 2005. Extensive and orderly reprogramming of genome-wide chromatin modifications associated with specification and early development of germ cells in mice. *Dev. Biol.* 278: 440–458.

Seydoux G. and Dunn M.A. 1997. Transcriptionally repressed germ cells lack a subpopulation of phosphorylated RNA polymerase II in early embryos of *Caenorhabditis elegans* and *Drosophila melanogaster*. *Development* 124: 2191–2201.

Seydoux G. and Strome S. 1999. Launching the germline in *Caenorhabditis elegans*: Regulation of gene expression in early germ cells. *Development* 126: 3275–3283.

Shaffer A.L., Lin K.I., Kuo T.C., Yu X., Hurt E.M., Rosenwald A., Giltnane J.M., Yang L., Zhao H., Calame K., and Staudt L.M. 2002. Blimp-1 orchestrates plasma cell differentiation by extinguishing the mature B cell gene expression program. *Immunity* 17: 51–62.

Shapiro-Shelef M., Lin K.I., McHeyzer-Williams L.J., Liao J., McHeyzer-Williams M.G., and Calame K. 2003. Blimp-1 is required for the formation of immunoglobulin secreting plasma cells and pre-plasma memory B cells. *Immunity* 19: 607–620.

Short R.V. 2000. Where do babies come from? *Nature* 403: 705.

Smith A.G. 2001. Embryo-derived stem cells: Of mice and men. *Annu. Rev. Cell Dev. Biol.* 17: 435–462.

Surani M.A. 2001. Reprogramming of genome function through epigenetic inheritance. *Nature* 414: 122–128.

————. 2005. Nuclear reprogramming by human embryonic stem cells. *Cell* **122:** 653–654.

Surani M.A., Ancelin K., Hajkova P., Lange U.C., Payer B., Western P., and Saitou M. 2004. Mechanism of mouse germ cell specification: A genetic program regulating epigenetic reprogramming. *Cold Spring Harbor Symp. Quant. Biol.* **69:** 1–9.

Tada M., Tada T., Lefebvre L., Barton S.C., and Surani M.A. 1997. Embryonic germ cells induce epigenetic reprogramming of somatic nucleus in hybrid cells. *EMBO J.* **16:** 6510–6520.

Tada M., Takahama Y., Abe K., Nakatsuji N., and Tada T. 2001. Nuclear reprogramming of somatic cells by in vitro hybridization with ES cells. *Curr. Biol.* **11:** 1553–1558.

Tada T., Tada M., Hilton K., Barton S.C., Sado T., Takagi N., and Surani M.A. 1998. Epigenotype switching of imprintable loci in embryonic germ cells. *Dev. Genes Evol.* **207:** 551–561.

Toyooka Y., Tsunekawa N., Akasu R., and Noce T. 2003. Embryonic stem cells can form germ cells in vitro. *Proc. Natl. Acad. Sci.* **100:** 11457–11462.

Turner C.A., Jr., Mack D.H., and Davis M.M. 1994. Blimp-1, a novel zinc finger-containing protein that can drive the maturation of B lymphocytes into immunoglobulin-secreting cells. *Cell* **77:** 297–306.

van der Heijden G.W., Dieker J.W., Derijck A.A., Muller S., Berden J.H., Braat D.D., van der Vlag J., and de Boer P. 2005. Asymmetry in histone H3 variants and lysine methylation between paternal and maternal chromatin of the early mouse zygote. *Mech. Dev.* **122:** 1008–1022.

Van Doren M., Williamson A.L., and Lehmann, R. 1998. Regulation of zygotic gene expression in *Drosophila* primordial germ cells. *Curr. Biol.* **8** 243–246.

Vincent S.D., Dunn N.R., Sciammas R., Shapiro-Shalef M., Davis M.M., Calame K., Bikoff E.K., and Robertson E.J. 2005. The zinc finger transcriptional repressor Blimp1/Prdm1 is dispensable for early axis formation but is required for specification of primordial germ cells in the mouse. *Development* **132:** 1315–1325.

Waddington C. 1956. *Principles of embryology*. Allen & Unwin, London, United Kingdom.

Weber R.J., Pedersen R.A., Wianny F., Evans M.J., and Zernicka-Goetz M. 1999. Polarity of the mouse embryo is anticipated before implantation. *Development* **126:** 5591–5598.

Yamaguchi S., Kimura H., Tada M., Nakatsuji N., and Tada T. 2005. Nanog expression in mouse germ cell development. *Gene Expr. Patterns* **5:** 639–646.

Zhang F., Barboric M., Blackwell T.K., and Peterlin B.M. 2003. A model of repression: CTD analogs and PIE-1 inhibit transcriptional elongation by P-TEFb. *Genes Dev.* **17:** 748–758.

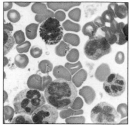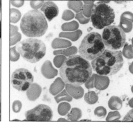

Epigenetic Control of Lymphopoiesis

Meinrad Busslinger[1] and Alexander Tarakhovsky[2]

[1] Research Institute of Molecular Pathology, Vienna Biocenter, A-1030 Vienna, Austria
[2] Laboratory of Lymphocyte Signaling, The Rockefeller University, New York, New York 10021

CONTENTS

1. Introduction, 399

2. Lineage Commitment in Early
   Lymphopoiesis, 400
   2.1 Extracellular Signals, 400
   2.2 Transcription Factors, 401
   2.3 Epigenetic Control of Gene Expression, 402

3. Epigenetic Control of Antigen Receptor
   Diversity, 403
   3.1 Developmental Regulation of Antigen
       Receptor Gene Rearrangements, 403
   3.2 Subnuclear Relocation of Immunoglobulin
       Genes, 405

3.3 Locus Contraction of Immunoglobulin
    Genes, 405
3.4 Control of Allelic Exclusion at the IgH and
    Igκ Loci, 406

4. Terminal Differentiation of Mature B Cells, 408
   4.1 Plasma Cell Differentiation, 408
   4.2 Developmental Plasticity of Mature
       B Cells, 410

5. Concluding Remarks, 411

References, 411

GENERAL SUMMARY

Every immunologist carries a simple wisdom—there are no better cells than lymphocytes for studying any puzzling biological process. The epigenetic mechanisms employed for cell regulation follow this rule. Early during development, the common lymphocyte progenitor decides whether to become "boring" (B cell) or "terrific" (T cell). For those who are interested in the physiology of epigenetics, developing lymphocytes offer a lavish opportunity to dissect discrete developmental steps in the most precise way. Due to the high efficacy of cell-sorting methods, even minute subpopulations of developing and mature lymphocytes can be isolated in quantities that will satisfy the most scrupulous technique freaks. Therefore, lymphocytes provide an opportunity to identify the genes and their regulators that contribute to the reprogramming of stem cells into mature cells. Transcription factors play a critical role in this decision, but the "writers" and "readers" of the "histone modification code" are quickly catching up.

In the course of B- or T-cell development, every developing lymphocyte must declare its "social" identity; i.e., its antigenic specificity. The expression of a unique antigen receptor by a developing B or T cell is essential for its subsequent selection against harmful autoreactive cells and facilitates the development of cells with distinct receptors that recognize a limitless number of environmental antigens. To achieve such a remarkable specificity, each B or T cell undergoes the process of immunoglobulin or T-cell receptor gene recombination. To make this process possible, the immunoglobulin or T-cell receptor loci, which are guarded by the chromatin, become accessible to the gene recombination machinery. Gene segments, separated by as much as 3 Mb of DNA, are brought into proximity and ligated to generate a unique antigen receptor gene. The immunoglobulin gene recombination process contains all the possible flavors that appeal to veterans of epigenetics. The fact that the immunoglobulin loci have to be transcribed before recombination is obviously reminiscent of Xist transcription, which induces epigenetic changes leading to X-chromosome inactivation. To support this notion, the Ezh2 histone methyltransferase, which is involved in X inactivation, also has an important role in immunoglobulin gene recombination. For those who are excited about small RNAs, developing B cells show bidirectional transcription within the immunoglobulin locus, thus offering a chance to generate small RNA duplexes that may act as guide RNAs to tag chromatin with marks permissive for recombination.

Once developed, B and T cells must stay alive long enough to have their Warhol's 15 minutes of fame. In the world of immune cells, this means encountering the antigen and responding to it. To achieve this noble aim, B and T cells use all possible tools, from transcription factors to histone-modifying enzymes. After activation by the antigen, B and T cells follow very different paths. Activated T cells differentiate into effector cells that produce a distinct set of cytokines, which either help other immune cells to exert their function or facilitate direct killing of the targets. Unlike T cells, which retain their lineage identity until death, the ultimate aim of B cells is to lose their identity and become a source of highly specific soluble antibodies that neutralize pathogens. To reach this goal, activated B cells enter the path of antibody perfection and mutate their immunoglobulin genes to improve the affinity of their antigen receptors. This process is carried out by the activation-induced deaminase (AID), but the mechanism that targets this powerful "mutator" to a tiny segment within the 2 meters of genomic DNA is unknown and is likely to be epigenetically controlled. B cells that express high-affinity antigen receptors enter another round of AID-dependent DNA "bashing" that results in immunoglobulin class switching leading to the generation of antibodies of distinct isotypes, which have different potentials to protect the organism. Curious epigeneticists will definitely find this so-called immunoglobulin class switch recombination to be enriched with phenomena reminiscent of those that occur during DNA elimination in Tetrahymena, telomeric silencing in yeast, and X inactivation in mammals. Sooner or later, B cells that have exhausted their zest for antibody improvement will meet the antigen. This event leads to major genome reprogramming that strips B cells of their identity and leads to the generation of antibody-producing plasma cells. Although transcriptional control of this vital process is well established, the mechanism of massive B-cell dedifferentiation is of great interest for those who look for ways to change the cell's identity at will.

The practical significance of analyzing epigenetic mechanisms underlying lymphocyte differentiation and function is hard to overestimate. Many lymphoid cancers overexpress various histone methyltransferases, and their targeting by specific inhibitors may help to stop cancer cells from growing and spreading. Given the wealth of knowledge about factors that control lymphocyte differentiation from stem cells, it is likely that lymphocytes will be among the first cells that could be reprogrammed from stem cells for therapeutic purposes. To conclude, excitement about science is good for immunity, and this important wisdom makes epigenetics the best immunostimulator.

1 Introduction

Tissue-specific stem cells are responsible for the development and regeneration of entire organs such as the skin, gut, and blood system throughout life. For this, the stem cells are equipped with two unique properties. First, they have an extensive self-renewal potential and can therefore be propagated in their uncommitted state. Second, they are pluripotent and thus give rise to all cell types of an organ by differentiating to multipotent progenitors with gradually restricted developmental potential (Fig. 1). These progenitors subsequently undergo commitment to one of several lineages and then differentiate along the selected pathway into a functionally specialized cell type of the organ (Fig. 1).

How lineage commitment of multipotent progenitors is controlled at the molecular level is an important question in developmental biology. Transcription factors play an essential role in this process, as they are able to reprogram the expression of large gene sets. To this end, they use multiple mechanisms to activate or repress gene transcription in response to extracellular signals (Fig. 2) (Fisher 2002). They can indirectly affect gene expression programs by antagonizing other transcription factors through protein–protein interaction. More commonly, transcription factors directly control gene transcription by recruiting coactivators or corepressors with histone-modifying or chromatin-remodeling activities to regulatory DNA elements (Fig. 2). Gene expression is determined not only by the availability of a combination of transcription factors, but also by the chromatin context, which reflects the ontogenetic his-

tory of a cell. In particular, the activity of a gene is influenced by the local DNA methylation pattern, the histone modification state of the chromatin, the nuclear position of the gene relative to repressive heterochromatin domains, and the architecture of the gene locus (Fig. 2) (Fisher 2002).

The hematopoietic system, and in particular lymphocyte development, are well suited to study the epigenetic mechanisms controlling lineage commitment and differentiation for the following reasons. The lineage diagram of the hematopoietic system has been largely elucidated as the different developmental stages between the hematopoietic stem cell (HSC) and the various mature blood cell types have been well characterized (Fig. 3) (Akashi et al. 2000; Busslinger 2004). As a consequence, cells at different developmental stages can be isolated by FACS sorting from hematopoietic organs, cultured in vitro, analyzed at the molecular level, modified by viral expression systems, and reinjected into recipient mice to study the functional consequences of genetic manipulations in vivo. Although the epigenetic mechanisms underlying hematopoietic development are just beginning to be elucidated, most of our knowledge about epigenetic control is currently available for B-lymphopoiesis.

B-cell development can be roughly divided into three phases characterized by the entry of progenitors into the lineage at B-cell commitment, the generation of antigen receptor diversity by V(D)J recombination, and the terminal differentiation of mature B cells into immunoglobulin-secreting plasma cells (Fig. 3). In this chapter, we discuss the epigenetic mechanisms control-

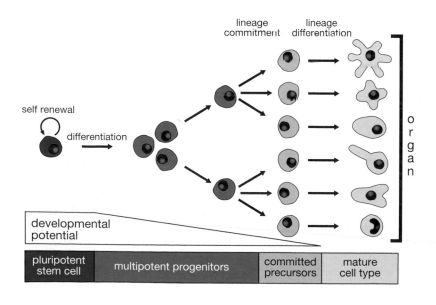

Figure 1. Schematic Diagram of Organ Development from a Single Stem Cell

The development of an organ is initiated by tissue-specific stem cells, which have extensive self-renewal capacity and are pluripotent, thus giving rise to all cell types of the organ. The stem cells first differentiate into multipotent progenitors (also known as transit-amplifying cells) with increasingly restricted developmental potential. These progenitors subsequently undergo commitment to one of several lineages and then differentiate along the selected pathway into a functionally specialized cell type of the organ.

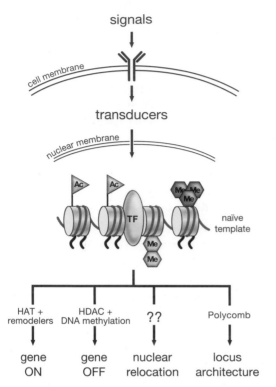

signals

cell membrane

transducers

nuclear membrane

naïve
template

HAT +
remodelers

HDAC +
DNA methylation

??

Polycomb

gene
ON

gene
OFF

nuclear
relocation

locus
architecture

Figure 2. Central Role of Transcription Factors in Epigenetic Gene Control

Transcription factors (TF), which are frequently regulated in response to extracellular signals, are responsible for gene activation, transcriptional repression, gene relocation within nuclear compartments, and chromatin architecture of gene loci. The transcription factors fulfill these diverse functions by interacting with coactivators (including histone acetyltransferases [HATs]), corepressors (including histone deacetylases [HDACs]), chromatin-remodeling machines, and Polycomb protein complexes.

ling these three developmental aspects, which have revealed fundamental principles with implications also for other developmental systems.

2 Lineage Commitment in Early Lymphopoiesis

2.1 Extracellular Signals

Hematopoietic development is initiated in the bone marrow of adult mice by the decision of HSCs to differentiate to common myeloid progenitors (CMPs) (Akashi et al. 2000) or lymphoid-primed multipotent progenitors (LMPPs) (Adolfsson et al. 2005). Erythroid cell types arise exclusively from CMPs, whereas myeloid cells can develop from either CMPs or LMPPs. Activation of the lymphoid *RAG1* and *RAG2* genes in LMPP cells characterizes the emergence of the earliest lymphocyte progenitors (ELPs), which differentiate to common lymphoid progenitors (CLPs) with their characteristic B-, T-, and NK-cell developmental potential (Fig. 3) (Busslinger 2004).

Extracellular signals and transcription factors are both essential for the differentiation of HSCs to CLPs and early B-cell progenitors (pro-B cells) in the bone marrow. The generation of lymphoid progenitors and B lymphocytes depends on signaling through the c-Kit, Flt3, and IL-7 receptors (Waskow et al. 2002; Vosshenrich et al. 2003). In particular, mice double-deficient for Flt3 and IL-7Rα entirely fail to generate pro-B cells (Vosshenrich et al. 2003). Flt3 and IL-7R signaling activates the closely related transcription factors Stat5a and Stat5b, which in turn facilitate differentiation of lymphoid progenitors to

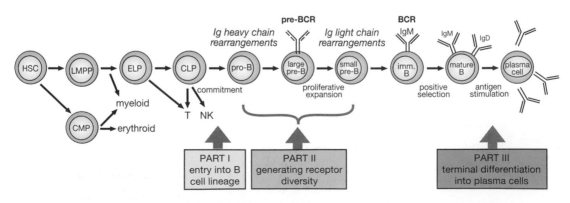

Figure 3. Schematic Diagram of B-Cell Development

Hematopoietic stem cells (HSC) differentiate via the indicated developmental stages to immunoglobulin-secreting plasma cells. (LMPP) Lymphoid-primed multipotent progenitors; (CMP) common myeloid progenitor; (ELP) earliest lymphocyte progenitor; (CLP) common lymphoid progenitor. Parts I–III indicated below refer to the stages of lymphopoietic development in which epigenetic mechanisms have been studied and which are discussed in this chapter.

the pro-B-cell stage (Fig. 4) (Zhang et al. 2000; Kovanen and Leonard 2004). B-cell development, furthermore, depends on the zinc-finger transcription factor Ikaros and the Ets-domain protein PU.1, because CLPs with their characteristic IL-7R expression are absent in the bone marrow of *Ikaros* and *PU.1* mutant mice (Fig. 4) (Allman et al. 2003; Dakic et al. 2005).

2.2 Transcription Factors

The entry of lymphoid progenitors into the B-cell pathway depends on the helix-loop-helix transcription factor E2A (Tcfe2a), the early B-cell factor EBF1, and the paired domain protein Pax5 (BSAP) (Fig. 4). E2A and EBF1 coordinately activate the expression of B-cell-specific genes, which are essential for the generation of early pro-B cells (Busslinger 2004). Activation of the B-cell-specific transcription program is, however, not sufficient to commit early progenitors to the B-lymphoid lineage in the absence of Pax5. Homozygous *Pax5* mutation arrests B-cell development at an early pro-B-cell stage in the bone marrow (Nutt et al. 1997). *Pax5⁻/⁻* pro-B cells retain an extensive self-renewal and broad developmental potential characteristic of uncommitted lymphoid progenitors (Nutt et al. 1999). Interestingly, ectopic expression of myeloid transcription factors efficiently induces a lymphoid-to-myeloid lineage switch in *Pax5⁻/⁻* progenitors, demonstrating that the *Pax5⁻/⁻* pro-B cells are in vitro clonable lymphoid progenitors with a latent myeloid differentiation potential (Busslinger 2004). Importantly, the restoration of *Pax5* expression suppresses the multilineage potential of *Pax5⁻/⁻* pro-B cells, while simultaneously promoting their development to mature B cells (Nutt et al. 1999). These experiments identified Pax5 as the critical B-lineage commitment factor that restricts

the developmental options of lymphoid progenitors to the B-cell pathway.

Uncommitted hematopoietic progenitors promiscuously express genes of different lineage programs by a process known as "lineage priming" (Miyamoto et al. 2002). In agreement with this finding, the *Pax5⁻/⁻* pro-B cells express not only genes characteristic of the pro-B cell, but also genes of other lineage-specific programs (Nutt et al. 1999). At lineage commitment, Pax5 fulfills a dual role by activating B-cell-specific genes and simultaneously repressing lineage-inappropriate genes (Fig. 5) (Nutt et al. 1999). Pax5 activates several target genes coding for essential components of the (pre)B-cell receptor (BCR) signaling pathway, such as the signal-transducing chain Igα (mb-1, Cd79a), the stimulatory coreceptors CD19 and CD21, as well as the central adapter protein BLNK (Fig. 5) (Busslinger 2004). Hence, the transactivation function of Pax5 facilitates signal transduction from the pre-BCR and BCR, which constitute important checkpoints in B-cell development. On the other hand, the Pax5-repressed genes code for a plethora of secreted proteins, cell adhesion molecules, signal transducers, and nuclear proteins, which are expressed in erythroid, myeloid, and/or T-lymphoid cells (Fig. 5) (Delogu et al. 2006). Among them are the *Csf1r* (*M-CSFR*) and *Notch1* genes, which nicely exemplify how their Pax5-dependent down-regulation renders committed B cells no longer responsive to the myeloid cytokine M-CSF (Nutt et al. 1999) or to T-cell-inducing Notch ligands (Busslinger 2004). Hence, the repression function of Pax5 contributes to B-cell commitment by shutting down inappropriate signaling systems. By interacting with corepressors of the Groucho protein family, Pax5 is able to recruit histone deacetylases (HDACs) to regulatory elements, thus resulting in gene repression (Linderson et al. 2004).

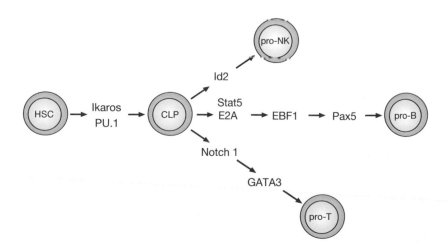

Figure 4. Transcriptional Control of Early Lymphopoiesis

Progenitors of the natural killer (NK), B-, and T-cell lineages develop from the hematopoietic stem cell (HSC) under the control of the indicated transcription factors. For detailed description, see Busslinger (2004).

B-lineage-inappropriate genes

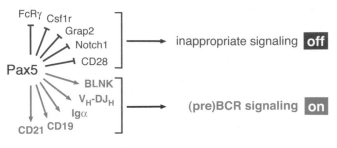

Figure 5. Dual Role of Pax5 in B-Cell Development

Pax5 activates B-cell-specific genes involved in (pre)BCR signaling and represses lineage-inappropriate genes which are essential for signal transduction in myeloid (FcRγ, Csf1r) or T-lymphoid (Notch1, CD28, Grap2) cells.

E2A, EBF1, and their target genes are normally expressed in *Pax5*$^{-/-}$ pro-B cells, indicating that Pax5 functions downstream of E2A and EBF1 in B-cell development (Nutt et al. 1997; Busslinger 2004). Consistent with this observation, *Pax5* transcripts are absent in *E2A*$^{-/-}$ or *EBF1*$^{-/-}$ progenitors (Ikawa et al. 2004; Medina et al. 2004). *E2A*$^{-/-}$ progenitors express *EBF1* at a low level (Ikawa et al. 2004), whereas *E2A* is normally transcribed in *EBF1*$^{-/-}$ progenitors (Medina et al. 2004). Moreover, retroviral restoration of *EBF1* expression in *E2A*$^{-/-}$ progenitors activates *Pax5* transcription, thereby initiating pro-B-cell differentiation (Seet et al. 2004). Hence, the three transcription factors promote early B-cell development in the genetic hierarchy E2A–EBF1–Pax5. Within the B-lymphoid lineage, *EBF1* expression is, however,

maintained through feedback regulation by Pax5 (see Fig. 4) (Horcher et al. 2001; Fuxa et al. 2004).

2.3 Epigenetic Control of Gene Expression

Most genes are transcriptionally repressed in any given cell of a multicellular eukaryotic organism and are selectively activated only along specific developmental pathways. Gene repression is mediated by multiple layers of epigenetic mechanisms including DNA methylation, histone modifications, nucleosome positioning, and higher-order chromatin structure (Smale 2003). First insight into how gene repression is relieved at the onset of B-cell development has been provided by analysis of the B-cell-specific gene *Cd79a* (*mb-1*, *Igα*; Maier et al. 2004). The promoter of the *Cd79a* gene contains functional recognition sequences for the three transcription factors E2A, EBF1, and Pax5, in addition to Ets- and Runx-binding sites (Fig. 6) (Maier et al. 2004). This transcriptional control region is methylated at CpG dinucleotides in HSCs, partially demethylated in CLPs, and completely unmethylated in committed pro-B cells (Maier et al. 2004). The *Cd79a* promoter is also fully methylated in *E2A*$^{-/-}$ or *EBF1*$^{-/-}$ progenitors but can be demethylated by the synergistic action of EBF1 and Runx1, which leads, in cooperation with Pax5, to transcriptional activation of the *mb-1* gene in committed pro-B cells (Fig. 6) (Maier et al. 2004). Hence, EBF1 and possibly E2A function as "pioneer" factors in early B-cell development by initiating epigenetic changes required for gene activation during the differentiation of early progenitors to committed pro-B cells. Although changes in histone modification and chromatin structure have not yet been studied, it is likely that

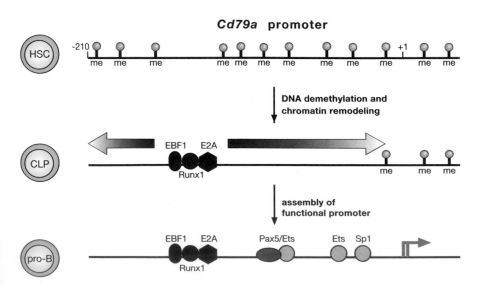

Figure 6. Epigenetic Activation of the *Cd79a* Gene in Early Lymphopoiesis

A schematic diagram of the *Cd79a* (*mb-1*, *Igα*) promoter is shown together with the CpG methylation (me) pattern and sequential binding of the different transcription factors during the transition of HSCs to committed pro-B cells (Maier et al. 2003).

E2A and Pax5 contribute to the local formation of open acetylated chromatin through their interaction with SAGA and p300/CBP histone acetyltransferase (HAT) complexes (Massari et al. 1999; Barlev et al. 2003).

3 Epigenetic Control of Antigen Receptor Diversity

3.1 Developmental Regulation of Antigen Receptor Gene Rearrangements

The guiding principle of the acquired immune system is that every newly generated lymphocyte recognizes a unique antigen and that the overall diversity of lymphocytes is great enough to counteract any possible antigen. To this end, B and T cells express lineage-specific antigen receptors that mediate antibody-dependent humoral or cellular immunity, respectively. The BCR consists of the immunoglobulin heavy chain (IgH) and an Igκ or Igλ light chain (IgL). T cells of the αβ lineage, which comprise the majority of T lymphocytes in mouse and man, express the T-cell receptor (TCR) β polypeptide in association with TCRα, while the functionally distinct γδ T cells contain TCRγ paired with TCRδ on their cell surface. These antigen receptor proteins are encoded by large gene loci containing discontinuous variable (V), diversity (D), and joining (J) gene segments, which are assembled by V(D)J recombination into a functional gene during lymphocyte development. The multiplicity of D, J, and especially V gene segments, combined with the randomness of their recombination, is responsible for the virtually unlimited diversity of the immune repertoire (Bassing et al. 2002).

The mechanics of V(D)J recombination at the DNA level is rather simple. All V, D, and J gene segments are flanked by recombination signal sequences (RSSs), which consist of relatively conserved heptamer and nonamer elements separated by a spacer of either 12 or 23 bp. The lymphoid-specific recombinase proteins RAG1 and RAG2, assisted by high-mobility group proteins, assemble 12-bp and 23-bp RSSs into a synaptic complex and then generate double-strand DNA breaks between the RSSs and coding segments. These DNA breaks are subsequently processed and religated by ubiquitous repair factors of the nonhomologous end-joining machinery to form coding and signal joints (Bassing et al. 2002).

The simplicity of the V(D)J recombination process at the DNA template level poses logistic problems for the assembly of the different antigen receptors, because the RAG proteins are expressed in all immature B and T lymphocytes. Hence, stringent regulation must be in place to restrict the access of RAG proteins to only specific subsets of all the recombination substrates (Yancopoulos and Alt 1985; Stanhope-Baker et al. 1996). V(D)J recombination is tightly controlled in a lineage- and stage-specific manner. Within the B-lymphoid lineage, the *IgH* locus is rearranged in pro-B cells prior to recombination of *Igκ* and *Igλ* genes in pre-B cells, whereas the *TCRβ* and *TCRα* genes are rearranged in pro-T and pre-T cells, respectively. Moreover, V(D)J recombination of the *IgH* gene occurs in a defined temporal order with D_H-J_H rearrangements preceding V_H-DJ_H recombination. Rearrangements of the TCRβ locus also proceed in the same order ($D_β$-$J_β$ before $V_β$-$DJ_β$) during pro-T-cell development. Control mechanisms must therefore exist to shield all V genes from RAG-mediated cleavage during D-J recombination and to facilitate rearrangement of only one out of a hundred V genes during V-DJ recombination. Consequently, the process of antigen receptor generation entirely depends on accurate regulation of the accessibility of RSSs for the RAG1/2 recombinase.

Successful V-DJ recombination of the *IgH* or *TCRβ* gene leads to expression of the Igμ or TCRβ protein as part of the pre-BCR or pre-TCR complex, which acts as an important checkpoint to inhibit V-DJ recombination of the second DJ-rearranged allele and to promote development to pre-B or pre-T cells that initiate *IgL* or *TCRα* gene rearrangements, respectively. Finally, the expression of a signaling-competent BCR or TCR completely arrests V(D)J recombination by transcriptional repression of the *RAG1/2* genes in immature B or T cells (Jankovic et al. 2004). Signaling of an autoreactive BCR can, however, restart immunoglobulin light-chain gene rearrangement, which results in the generation of a BCR with a novel antigen specificity (receptor editing; Jankovic et al. 2004). Moreover, signaling of the cytokine IL-7 is essential for promoting recombination of the TCRγ gene in pro-T cells (Schlissel et al. 2000). Hence, V(D)J recombination is controlled not only intrinsically by developmental and lineage-specific nuclear mechanisms, but also extrinsically by signals generated at the cell surface.

The developmental and locus-specific constraints on V(D)J recombination are largely imposed at the epigenetic level (Krangel 2003). In non-lymphoid cells, the *Ig* and *TCR* genes are present in inaccessible chromatin, as exogenously expressed RAG proteins readily cleave transfected episomal recombination substrates but not endogenous antigen receptor genes in kidney cells (Romanow et al. 2000). Moreover, recombinant RAG proteins added to isolated lymphocyte nuclei can only cleave the *Ig* or *TCR* gene that is actively undergoing V(D)J recombination at the developmental stage used for nucleus preparation (Stanhope-Baker et al. 1996). Hence, the lineage specificity and temporal ordering of gene rearrangements is caused by the sequential opening of local chromatin that renders specific RSSs accessible to the V(D)J recombinase. The ability of chromatin to both protect RSSs and to direct their cleavage suggests the existence of a "chromatin code" that marks the sites of recombination and/or facilitates RAG-mediated cleavage.

Acetylation of histones on lysine residues is not only a characteristic feature of open chromatin, but also plays an important role in determining the chromatin accessibility of *Ig* and *TCR* loci, as it demarcates domains of recombination-competent gene segments (McMurry and Krangel 2000; Chowdhury and Sen 2001). Analysis of the histone acetylation state has revealed a stepwise activation of discrete chromatin domains in the *IgH* locus (Chowdhury and Sen 2001). A 120-kb genomic region encompassing the D_H, J_H, and C_μ gene segments is first hyperacetylated prior to V(D)J recombination. D_H-J_H rearrangements subsequently induce histone acetylation and rearrangements of the D_H-proximal V_H genes (Chowdhury and Sen 2001). Finally, the distal 2-Mb domain containing the majority of V_H genes appears to be activated by IL-7 signaling (Chowdhury and Sen 2001). Detailed analysis of the *TCRα/δ* locus in developing T cells also revealed a complete overlap between regions displaying histone H3 hyperacetylation and accessibility to the V(D)J recombinase (McMurry and Krangel 2000). Hence, histone acetylation appears to be an essential part of the chromatin modification pattern that controls the initiation and/or progression of recombination (Krangel 2003).

Acetylation per se is, however, insufficient to facilitate recombination, as inhibitors of histone deacetylases have little impact on V(D)J recombination in vivo (McBlane and Boyes 2000). Furthermore, a striking dichotomy between high levels of histone H3 acetylation and poor V(D)J recombination has been observed in pro-B cells (Hesslein et al. 2003; Su et al. 2003). Normal levels of histone acetylation in the D_H-distal V_H gene cluster fail to support distal V_H-DJ_H recombination in pro-B cells lacking the histone lysine methyltransferase (HKMT) Ezh2 that trimethylates histone H3 at lysine 27 (H3K27me3) (Su et al. 2003). The observation that higher levels of H3K27me3 are associated with distal compared to proximal V_H genes suggests a domain-specific role of H3K27 methylation in V_H gene recombination (Su et al. 2003). The selectivity of Ezh2-mediated regulation for the *IgH* locus is underscored by the equal efficiency of *TCRβ* gene recombination in wild-type and Ezh2-deficient pro-T cells (Su et al. 2005). Hence, additional chromatin modifications are likely to be involved in controlling V(D)J recombination of proximal V_H genes in pro-B cells and *TCRβ* rearrangements in pro-T cells. Dimethylation of histone H3 on lysine 4 (H3K4me2) is an active histone mark, which also correlates with the accessible state of *IgH* and *TCRβ* gene segments (Morshead et al. 2003). In contrast, dimethylation of H3 on lysine 9 (H3K9me2) is a repressive chromatin mark that inversely correlates with V(D)J recombination of *IgH* and *TCRβ* gene segments (Morshead et al. 2003). An essential role for H3K9me2 in suppressing recombination was recently demonstrated by targeting the H3K9 HKMT G9a to the PDβ1 germ-line promoter of a *TCRβ* minilocus, which prevented V(D)J rearrangements by rendering the local chromatin inaccessible (Osipovich et al. 2004).

The histone modification pattern facilitating V(D)J recombinase access must be established by processes that occur within the antigen receptor loci prior to rearrangement. Before the mapping of histone modifications became experimentally feasible, it was already known that germ-line transcription of short sense RNA from unrearranged gene segment precedes V(D)J recombination (Yancopoulos and Alt 1985). A possible role of transcription in controlling locus accessibility was furthermore supported by findings demonstrating that enhancers and promoters located within the antigen receptor loci are essential for V(D)J recombination to occur. Deletion of endogenous enhancers and promoters reduces or abolishes V(D)J recombination of antigen receptor loci, whereas the insertion of additional lineage-specific enhancers leads to a novel V(D)J recombination pattern (Bassing et al. 2002; Krangel 2003). Numerous promoters, associated with V, D, and J segments, control rearrangements of promoter-proximal sequences within relatively short distances, whereas enhancers exert long-range control of V(D)J recombination (Bassing et al. 2002; Krangel 2003). The assembly of a pre-initiation complex at a promoter may locally disrupt nucleosomes and thereby facilitate access to recombination enzymes, even in the absence of histone modification changes. More likely, however, promoters actively contribute to the establishment of a recombination-permissive chromatin structure, as the elongating RNA polymerase II complex carries its own histone acetyltransferase that may help to spread histone acetylation along transcribed regions (Orphanides and Reinberg 2000). Gene transcription also results in local exchange of the replication-dependent histone H3 by the replacement variant H3.3, which has been implicated in maintaining the accessible chromatin state of transcribed regions (Chow et al. 2005; Mito et al. 2005).

As mentioned above, every antigen receptor locus contains hundreds of RSSs, although only a few of them will be cleaved in an individual lymphocyte at a defined developmental stage. It is thus conceivable that DNA sequence variations of individual RSS sites may also contribute to their cleavage efficiency. The analysis of artificial V(D)J recombination substrates indeed demonstrated that naturally occurring differences in RSS heptamer and nonamer elements, as well as in the less well conserved spacer and flanking coding sequences, influence the recombination frequency and thus contribute to the differential usage of particular V, D, and J gene segments in the primary antigen receptor repertoire (Lee et al. 2003). In the framework of a "histone code"-centric model, the cleavage selectivity should be determined by a process that would translate the unique features of individual RSSs or adjacent sequences into a specific histone modification pattern marking the site for RAG-mediated cleavage. This code may be established with the help of antisense transcripts. Indeed, antisense intergenic transcription throughout the entire V_H gene cluster precedes V_H-DJ_H recombination of the IgH locus in pro-B cells (Bolland et al. 2004). These long antisense transcripts may direct chromatin remodeling to open up the large V_H gene domain prior to recombination. Alternatively, these antisense transcripts could form double-stranded RNA hybrids with the short sense germ-line V_H transcripts and then be processed by the RNA interference machinery to generate microRNAs that recruit HKMTs to the recombination sites (Bolland et al. 2004; see Chapter 8 for detail on the RNAi machinery). As an extension of this hypothesis, we speculate that specific sense germ-line transcription of a defined RSS site may generate double-stranded RNA, which could target histone-modifying enzymes to this but not other RSS sequences. If experimentally verified, this hypothetical mechanism could account for the precision and selectivity of RAG-mediated cleavage of individual RSS sites. Interestingly, the RAG2 protein was recently shown to directly interact with histones and could thus play an important role in reading the specific histone modification pattern at individual RSS sequences (West et al. 2005).

3.2 Subnuclear Relocation of Immunoglobulin Genes

The nuclear periphery and pericentromeric heterochromatin are two major repressive compartments in the nucleus that are important for propagating the inactive state of genes in hematopoietic cells (Brown et al. 1997; Baxter et al. 2002). Depending on their activity state, genes are repositioned between these repressive compartments and central nuclear positions that facilitate gene transcription (Brown et al. 1997; Baxter et al. 2002). Interestingly, the IgH and Igκ loci are located in their default state at the nuclear periphery in all non-B cells, including uncommitted lymphoid progenitors (Kosak et al. 2002). The IgH locus is thereby anchored via the distal V_H genes at the nuclear periphery and is oriented with the proximal IgH domain toward the center of the nucleus, which facilitates D_H-J_H rearrangements in lymphoid progenitors (Fuxa et al. 2004). An initial step of IgH locus activation consists of relocation of the IgH and Igκ loci from the nuclear periphery to more central positions within the nucleus at the onset of B-cell development (Kosak et al. 2002). This subnuclear repositioning likely facilitates chromatin opening and germ-line transcription, leading to proximal V_H-DJ_H rearrangements. Circumstantial evidence suggests a role for EBF1 and Pax5 in the central relocation of IgH and Igκ loci, respectively (Fuxa et al. 2004; Sato et al. 2004).

Although both alleles of the IgH and Igκ loci are repositioned together to central nuclear positions in pro-B cells (Kosak et al. 2002; Fuxa et al. 2004), the two alleles behave differently following successful V(D)J recombination in mature B cells (Skok et al. 2001). Following B-cell activation, the productively rearranged Ig alleles remain positioned away from centromeric clusters, thus reinforcing their expression (Skok et al. 2001). At the same time, the nonfunctional IgH and Igκ alleles are relocated to, and thus silenced at, centromeric heterochromatin following B-cell activation. Interestingly, the centromeric recruitment of nonfunctional IgH and Igκ loci occurs via their distal V gene region, suggesting that the same DNA sequences are involved in the recruitment of silent Ig loci to either the nuclear periphery or centromeric heterochromatin (Roldán et al. 2005).

3.3 Locus Contraction of Immunoglobulin Genes

The approximately 200 V_H genes of the IgH locus are spread over a 2.4-Mb region and can be divided into 15 distal, central, or proximal V_H gene families according to their sequence similarity and position relative to the proximal D_H segments. In non-B-lymphoid cells and lymphoid progenitors, the two IgH alleles are present in an extended conformation at the nuclear periphery (Kosak et al. 2002). In contrast, the IgH locus undergoes long-range contraction in committed pro-B cells, which juxtaposes distal V_H genes next to the rearranged proximal DJ_H domain, thus facilitating V_H-DJ_H rearrangements (Fig. 7) (Kosak et al. 2002; Fuxa et al. 2004). The Igκ locus with its approximately 140 $V_κ$ genes also extends over a 3-Mb region and is thus as large as the *IgH* locus. Similar to the *IgH* gene in pro-B cells, the *Igκ* locus undergoes contraction in small pre-B and immature B cells, demonstrating that both *Ig* loci are in a contracted state in rearranging cells (Roldán et al. 2005). Fluorescent in situ hybridization (FISH) analysis with distal, central, and proximal gene probes, furthermore, demonstrated that looping of individual *Ig* subdomains is responsible for long-range contraction of the *IgH* and *Igκ* loci (Fig. 7) (Roldán et al. 2005).

Distal V_H-DJ_H rearrangements do not take place in *Pax5*$^{-/-}$ pro-B cells (Nutt et al. 1997) despite the fact that the V_H genes are accessible in a hyperacetylated chromatin state along the entire V_H gene cluster including the most distal V_HJ558 family (Hesslein et al. 2003). The failure of distal V_H-DJ_H rearrangements correlates with the absence of *IgH* locus contraction (Fig. 7), which can, however, be restored by retroviral Pax5 expression in *Pax5*$^{-/-}$ pro-B cells (Fuxa et al. 2004). Hence, Pax5 is a key regulator of *IgH* locus contraction in pro-B cells. The histone methyltransferase Ezh2 has also been implicated in *IgH* locus contraction, as conditional *Ezh2* inactivation in HSCs results in a reduction of distal V_H-DJ_H rearrangements despite full chromatin accessibility of distal V_H genes in Ezh2-deficient pro-B cells (Su et al. 2003). Interestingly, there is no genetic relationship between *Pax5* and *Ezh2* despite the similar *IgH* rearrangement phenotype of the respective mutant pro-B cells (Fuxa et al. 2004). It is therefore possible that Pax5 functions as a sequence-specific targeting factor to recruit the Ezh2-containing Polycomb repressive complex 2 (PRC2) to selected regions in the *IgH* locus. The resulting methylation of local chromatin at histone H3 on lysine 27 (H3K27me3) may attract the PRC1 complex to induce chromatin compaction of the targeted regions (Francis et al. 2004; discussed in Chapter 11), thus leading to looping and contraction of the *IgH* locus. Alternatively, locus contraction may not depend on histone modifications in the nucleus, but rather requires lysine methylation of signaling proteins by the cytoplasmic Ezh2-containing methyltransferase complex, which is known to regulate actin polymerization by binding to the GTP/GDP exchange factor Vav1 (Su et al. 2005).

3.4 Control of Allelic Exclusion at the IgH and Igκ Loci

Allelic exclusion ensures the productive rearrangement of only one of the two *Ig* alleles, which leads to the expression of a single antibody molecule with a unique antigen specificity in B cells. The process of allelic exclusion can be divided into two distinct steps. During the initiation phase, one of the two *Ig* alleles is selected by differential epigenetic marking to rearrange first, which precludes simultaneous recombination of the two alleles. Expression of the productively rearranged allele subsequently prevents recombination of the second allele by feedback inhibition, thus maintaining allelic exclusion. The process of allelic exclusion is already initiated in early development at the time of implantation, when the two alleles of the antigen receptor genes start to replicate asynchronously in each cell (Mostoslavsky et al. 2001). The paternal or maternal *Ig* gene, which is stochastically selected for early replication by a so-far-unknown chromosomal mark, is almost invariably the first allele to undergo rearrangements in immature B lymphocytes (Mostoslavsky et al. 2001). The second V_H-DJ_H rearrangement of the *IgH* locus is thereby the regulated step, as D_H-J_H recombination occurs on both *IgH* alleles during pro-B-cell development (Bassing et al. 2002). However, nothing is yet known about how the allele-specific epigenetic mark (established in the early embryo) is translated into sequential activation of V_H-DJ_H recombination at the two *IgH* alleles. Successful

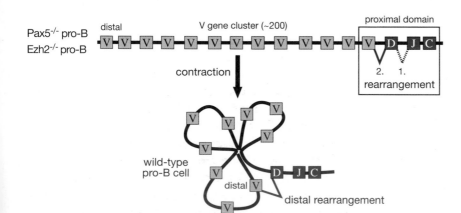

Figure 7. Contraction of the Immunoglobulin Heavy-Chain Locus in pro-B Cells

The *IgH* locus consists of a proximal domain containing diversity (D), joining (J), and constant (C) gene segments, and a large variable (V) gene cluster with ~200 V genes spread over a 2.4-Mb region. The *IgH* locus is in an extended configuration in *Pax5*$^{-/-}$ or *Ezh2*$^{-/-}$ pro-B cells, which allows V(D)J recombination to take place only in the proximal domain. In wild-type pro-B cells, all V_H genes participate in V_H-DJ_H rearrangements due to contraction of the *IgH* locus by looping.

rearrangement of one *IgH* allele leads to cell-surface expression of the Igμ protein as part of the pre-B-cell receptor (pre-BCR). This receptor functions as an important checkpoint to signal proliferative expansion of large pre-B cells, to induce subsequent differentiation to small pre-B cells, and to maintain allelic exclusion at the DJ_H-rearranged *IgH* allele (Kitamura and Rajewsky 1992; Bassing et al. 2002). RAG protein expression is rapidly lost upon pre-BCR signaling (Fig. 8), which halts all further V(D)J recombination and prepares the ground for establishing allelic exclusion in large pre-B cells (Grawunder et al. 1995). Pre-BCR signaling also leads to histone deacetylation and thus reduced accessibility of the V_H genes in small pre-B cells, which may be a possible feedback mechanism underlying allelic exclusion (Chowdhury and Sen 2003). A more plausible mechanism is, however, provided

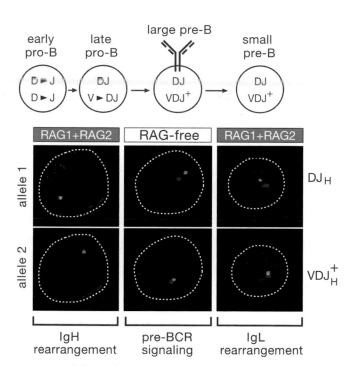

Figure 8. Allelic Exclusion by Decontraction of the *IgH* Locus in pre-B Cells

In early pro-B cells, D_H-J_H rearrangements occur simultaneously on both *IgH* alleles, whereas only one allele undergoes V_H-DJ_H recombination at a time in late pro-B cells. The nuclei of sorted pro-B and pre-B cells were analyzed by three-dimensional DNA-FISH with fluorescent probes from the distal (*red*) and proximal (*green*) regions of the *IgH* locus. The two *IgH* alleles of the same cell are shown on two representative confocal sections. Pre-BCR signaling results not only in rapid loss of the RAG protein, but also in decontraction of the *IgH* locus. Although both alleles are decontracted, the *IgH* locus is fully extended only in the case of the incompletely DJ_H-rearranged allele. The two signals of the functionally rearranged allele (VDJ⁺) are separated by a shorter distance due to the deletion of intervening DNA sequences. The FISH data are taken from Roldán et al. (2005).

by the rapid reversal of *IgH* locus contraction in response to pre-BCR signaling, which physically separates the V_H genes from the proximal *IgH* domain (Fig. 8), thus preventing V_H-DJ_H rearrangement on the second DJ_H-rearranged IgH allele (Roldán et al. 2005). Pre-BCR signaling, furthermore, leads to rapid repositioning of the nonfunctional *IgH* allele to repressive centromeric domains (Roldán et al. 2005). Hence, locus decontraction and centromeric recruitment alter the DJ_H-rearranged *IgH* allele during the RAG-free window of pre-BCR signaling in such a way that it can no longer undergo V_H-DJ_H rearrangement after subsequent re-expression of RAG proteins in small pre-B cells (Fig. 8).

The initiation of allelic exclusion at the *Igκ* locus has been extensively studied by investigating the DNA methylation pattern with methyl-sensitive restriction enzymes (Mostoslavsky et al. 1998; Goldmit et al. 2002; 2005), as well as by analyzing heterozygous κ°-GFP reporter mice that contain a *GFP* gene insertion in the $J_κ1$ element of the endogenous *Igκ* locus (Liang et al. 2004). The *Igκ* locus is heavily methylated at CpG dinucleotides in all non-B and pro-B cells, but becomes specifically demethylated on only one allele in pre-B cells (Fig. 9) (Mostoslavsky et al. 1998; Liang et al. 2004). This monoallelic demethylation precedes rearrangement and is dependent on the activity of both the intronic and 3′ κ enhancers (Mostoslavsky et al. 1998). The demethylated allele is present in accessible chromatin, as it is DNase-I-sensitive, hyperacetylated at histones H3 and H4, and positioned away from centromeric heterochromatin in pre-B cells (Fig. 9) (Goldmit et al. 2002, 2005). As a consequence, only the unmethylated *Igκ* allele initiates germ-line transcription and $V_κ$-$J_κ$ rearrangements (Goldmit et al. 2002; Liang et al. 2004), whereas both alleles undergo locus contraction in small pre-B cells (Fig. 9) (Roldán et al. 2005). Surprisingly, the second *Igκ* allele is relocated to centromeric heterochromatin in pre-B cells (Goldmit et al. 2005) similar to the *IgH* locus (Roldán et al. 2005). This monoallelic centromeric recruitment (Fig. 9) may explain why the DNA-methylated allele is depleted in histone acetylation and is associated with the proteins HP1γ and Ikaros, which are enriched together with histone deacetylase complexes at centromeric heterochromatin (Goldmit et al. 2005). Interestingly, it is the late-replicating *Igκ* allele which is repositioned to the centromeric clusters (Goldmit et al. 2005) in agreement with the finding that the asynchronous replication pattern established already in the early embryo correlates with monoallelic initiation of *Igκ* rearrangements in pre-B cells (Mostoslavsky et al. 2001). Surprisingly, only a very small fraction (5%) of all pre-B cells undergo *Igκ* locus activation in the κ°-GFP

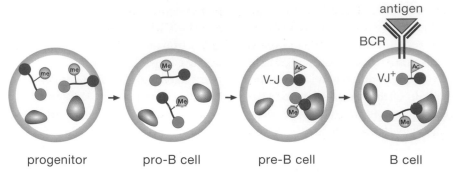

Figure 9. Mechanisms Controlling Allelic Exclusion at the *Igκ* Locus

Subnuclear relocation, DNA demethylation, and histone acetylation contribute to the selection of one *Igκ* allele for V_κ-J_κ recombination in pre-B cells. See text for detailed explanation. The distal V_κ region (*red*) and proximal J_κ-C_κ domain (*green*) of the *Igκ* locus are indicated, together with their location relative to the repressive compartments at the nuclear periphery (*gray*) and centromeric heterochromatin (*blue*). The locus contraction, DNA methylation (me), and histone acetylation (ac) states of the two *Igκ* alleles are schematically shown for different developmental stages including activated mature B cells.

reporter mice (Liang et al. 2004). On the basis of this result, it was hypothesized that certain transcription factors binding to *Igκ* *cis*-regulatory elements are present in limiting amounts in pre-B cells and that the cooperative binding of such factors to *Igκ* enhancers is a rare event, occurring stochastically at only one allele. Hence, stochastic enhancer activation by allelic competition for limiting transcription factors may contribute to allelic exclusion at the *Igκ* locus (Liang et al. 2004). Successful rearrangement of one *Igκ* allele leads to cell-surface expression of the BCR, which subsequently maintains allelic exclusion at the second *Igκ* allele by repressing *RAG1/2* recombinase expression (Jankovic et al. 2004).

4 Terminal Differentiation of Mature B Cells

4.1 *Plasma Cell Differentiation*

Completion of V(D)J recombination and expression of the immunoglobulin (Ig) receptor on the B-cell surface mark the end of the antigen-independent phase of B-lymphopoiesis. From this point on, the fate of B cells becomes dependent on antigen-induced receptor signaling (Rajewsky 1996). In the absence of antigen, peripheral B cells are maintained in a resting state where their survival is supported by tonic signals from the cell-surface BCR (Kraus et al. 2004). A comparison of the chromatin organization in resting and activated B cells shows that B-cell quiescence is characterized by low levels of global histone methylation (Baxter et al. 2004). The relatively high levels of histone acetylation in quiescent B cells remain stable during cell activation (Baxter et al. 2004). The global reduction in histone lysine methylation, including the virtual absence of histone H3K9 methylation, corre-

lates with the lack of other hallmarks of constitutive heterochromatin such as Ikaros association and HP1 binding in quiescent B cells (Baxter et al. 2004). The activation of B cells reinstates the methylation of histones, which leads to an increase of the active H3K4me3 mark on genes required for B-cell function (Pax5) and to a simultaneous increase of the repressive H3K9me2 modification on silent genes (Dntt) (Baxter et al. 2004).

Activation-induced chromatin reorganization may maintain B-cell identity during the immune response. However, the B-cell genome must remain amenable to antigen-induced reprogramming, since activated B cells, following antigen encounter, are able to differentiate directly into antibody-producing plasma cells (Calame et al. 2003). Alternatively, antigen stimulation can initiate the germinal center reaction. During this reaction, mature IgM^low IgD^high B cells switch their immunoglobulin isotypes and mutate their *Ig* genes with the help of activation-induced cytidine deaminase (AID; Honjo et al. 2004). The Ig proteins generated by these processes are perfectly suited for the differentiation and maintenance of memory B cells or the development of plasma cells, which produce antibodies with high affinity for a particular antigen (Honjo et al. 2004).

The timing of germinal center reactions and the conversion of B cells into plasma cells are regulated by two mutually exclusive transcriptional repressors, Bcl6 and Blimp1 (Fig. 10) (Turner et al. 1994; Ye et al. 1997). Bcl6 is expressed at low levels in mature naïve B cells but is rapidly up-regulated in some B cells after antigenic stimulation (Fukuda et al. 1997). Cells that do not up-regulate Bcl6 upon antigen encounter differentiate into plasma cells that serve as an initial source of low-affinity antibodies (Fukuda et al. 1997). In contrast, B cells that up-regulate Bcl6 enter the germinal center reaction (Fukuda et al. 1997) and are maintained as B cells by Bcl6-mediated repression of genes that control plasma cell differentiation (Shaffer et al. 2000). One key target of Bcl6 is the *Prdm1* gene, which encodes the transcription factor Blimp1 (Fig. 10). Interestingly, Pax5 appears to assist Bcl6 in repressing the *Blimp1* (*Prdm1*) gene (Fig. 10) (Delogu et al. 2006). However, once Blimp1 is expressed, it extinguishes the B-cell transcriptional program, including *Bcl6* and *Pax5* expression, and simultaneously induces the transcription of plasma-cell-specific genes (Shaffer et al. 2002; Calame et al. 2003).

Bcl6 and Blimp1 use a wide arsenal of repressive mechanisms to inactivate gene transcription. One peculiar aspect of Bcl6 is the utilization of lysine acetylation beyond histone modification to control gene repression (Bereshchenko et al. 2002). Bcl6 interacts with MTA3, a subunit of the corepressor complex Mi-2/NuRD, which is highly expressed in germinal center B cells (Fujita et al. 2004). Association of Bcl6 with the MTA3-containing Mi-2/NuRD complex is essential for gene repression, as RNAi-mediated depletion of MTA3 leads to the reactivation of Bcl6-repressed target genes in B cells (Fujita et al. 2004). The repression function of the Bcl6/MTA3/Mi-2/NuRD complex depends on the acetylation status of lysine residues in both Bcl6 and the histones associated with the repressed gene locus. The central domain of Bcl6 needs to be deacetylated to promote its interaction with MTA3, whereas gene repression by the MTA3/Mi-2/NuRD complex depends on the class I histone deacetylases HDAC1 and HDAC2 (Fujita et al. 2004). Bcl6,

furthermore, associates via its amino-terminal POZ domain with the three corepressors SMRT, NCoR, and BCoR in a mutually exclusive manner (Huynh and Bardwell 1998; Huynh et al. 2000). These three corepressors, which additionally interact with the class II enzyme HDAC3 (Huynh et al. 2000), may enhance MTA3-mediated repression of the same Bcl6 target genes or silence a different gene set in germinal center B cells.

Antigen stimulation of the high-affinity Ig receptors on germinal center B cells is accompanied by a reduction of the Bcl6 protein level. Receptor activation thereby leads to MAP kinase-induced Bcl6 phosphorylation, which triggers rapid degradation of the Bcl6 protein by the ubiquitin/proteasome pathway (Niu et al. 1998). The drop in Bcl6 levels alleviates *Prdm1* gene repression, resulting in increased Blimp1 protein expression and subsequent development to plasma cells (Fig. 10) (Shaffer et al. 2000; Calame et al. 2003). Blimp1 controls multiple aspects of plasma cell differentiation. First, Blimp1 targets the transcriptional core program of B-cell differentiation by repressing Pax5 (Fig. 10) (Shaffer et al. 2002), which is essential for the maintenance of B cell function and identity (Horcher et al. 2001; Mikkola et al. 2002). Second, by down-regulating the expression of other transcription factors (such as Spi-B, EBF1, CIITA, Id3, Oct2, and OBF1), Blimp1 indirectly terminates the transcription of genes that code for essential proteins in antigen receptor signaling and antigen presentation (Shaffer et al. 2002). Third, to ensure the resting state of plasma cells, Blimp1 directly represses *c-myc* transcription (Shaffer et al. 2002). Fourth, the lineage-inappropriate genes, which are repressed by Pax5 in B cells, are reactivated upon Blimp1-mediated down-regulation of Pax5 expression in plasma cells (Delogu et al. 2006). Hence, by repressing other repressors, Blimp1 may indirectly activate the expression of additional genes with essential plasma cell functions. Fifth, Blimp1 is essential for the expression of secreted immunoglobulins (Calame et al. 2003), which accumulate in the endoplasmic reticulum, thereby activating *XBP1* expression as part of the unfolded protein response pathway. The transcription factor XBP1 regulates antibody secretion and is thus indispensable for plasma cell differentiation (Reimold et al. 2001; Shaffer et al. 2004).

Interestingly, Blimp1 is also a key determinant of primordial germ cell specification in early embryogenesis, which is discussed in Chapter 20. The mechanism of Blimp1-mediated repression in developing primordial germ cells and plasma cells is largely unknown. It is, however, likely that Blimp1 uses the same repression principles in plasma cells as in fibroblasts where PRDI-BF1, the

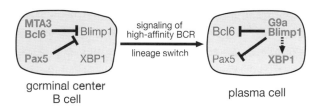

Figure 10. Transcriptional Repression Determines the Germinal Center B Cell and Plasma Cell Fates

Pax5 and Bcl6 (together with MTA3) regulate the B-cell gene expression program and maintain the GC (germinal center) B-cell fate by transcriptional repression of the plasma cell regulator Blimp1. Strong BCR signaling at the end of GC B-cell development leads to degradation of the Bcl6 protein and concomitant expression of Blimp1, which subsequently represses *Bcl6* and *Pax5* and, together with XBP1, induces the plasma cell transcription program. Blimp1 most likely activates *XBP1* expression indirectly as part of the unfolded protein response, by inducing the expression of secreted immunoglobulins. For detailed description, see text.

human ortholog of mouse Blimp1, is involved in postin-duction repression of the interferon-β (*IFNB1*) gene during viral infection (Keller and Maniatis 1991). Several mechanisms account for PRDI-BF1-mediated repression of the *IFNB1* gene. Binding of PRDI-BF1 is able to displace transcriptional activators from the *IFNB1* promoter (Keller and Maniatis 1991). In addition, PRDI-BF1 interacts with corepressors of the Groucho protein family that employ histone deacetylases as a part of their repression mechanism (Ren et al. 1999). Further silencing is obtained through the association of PRDI-BF1 with the G9a protein (Györy et al. 2004), which belongs to the subfamily of histone methyltransferases with specificity for H3 lysine 9 (Tachibana et al. 2002). In contrast to Suv39h1, which uses H3K9me3 to build a transcriptionally repressive environment at centromeric heterochromatin (Peters et al. 2001), G9a contributes to H3K9me2 and gene silencing in euchromatic regions (Tachibana et al. 2002). The catalytic activity of G9a is required for the repression function of PRDI-BF1, since a catalytically inactive G9a protein reverses the inhibitory effect of PRDI-BF1 on *IFNB1* transcription (Györy et al. 2004). Furthermore, deletion of the G9a interaction domain prevents H3K9 methylation and transcriptional silencing by PRDI-BF1 (Györy et al. 2004). In view of these data, it is likely that G9a-mediated histone methylation is an essential mechanism by which Blimp1 generates a stable gene expression pattern in plasma cells. Interestingly, the Blimp1 protein also contains a SET domain of the PR (RIZ) subfamily. The SET domain of the related RIZ1 (Prdm2) protein has been implicated in tumor suppression and methylation of histone H3 on lysine 9 (Kim et al. 2003). Hence, it remains to be seen whether the SET domain of Blimp1 also contributes to gene repression during plasma cell differentiation.

4.2 Developmental Plasticity of Mature B Cells

The generation of plasma cells is usually considered to be the terminal process of B-cell development, as the expression of immunoglobulin genes is an essential function of both B cells and plasma cells. Interestingly, the immunoglobulin genes are expressed under the combinatorial control of ubiquitous rather than B-lymphoid transcription factors in the two cell types. Apart from immunoglobulin genes, B cells and plasma cells differ radically, however, in their gene expression pattern (Shaffer et al. 2002, 2004). With regard to Pax5 function, the plasma cells even seem to go into reverse gear, as the plasma-cell-specific silencing of *Pax5* expression leads to the reactivation of B-lineage-inappropriate genes that are

normally repressed by Pax5 at the onset of B-cell development (Delogu et al. 2006). These gene expression data therefore support the alternative view that the differentiation of antigen-stimulated B cells to plasma cells is a true "lineage" switch. This hypothesis predicts that the developmental potential of mature B cells should be plastic rather than being restricted to the B-lymphocyte fate, which is supported by the following evidence.

Ectopic expression of the B-cell transcription factor Bcl6 and its corepressor MTA3 in established plasma cell lines leads to the repression of plasma-cell-specific genes, including the regulatory genes *Blimp1* and *XBP1* (Fujita et al. 2004). At the same time, multiple B-cell-specific genes are reactivated, including the Pax5 target genes *CIITA* and *BLNK*, and by inference, *Pax5* itself (Fujita et al. 2004). Hence, Bcl6 and its partner protein MTA3 are sufficient to reprogram plasma cells to a B-cell fate, at least under the in vitro culture conditions analyzed (Fig. 11).

The transcription factor C/EBPα, which is essential for granulocyte development, is exclusively expressed in myeloid progenitors and their differentiated progeny (Akashi et al. 2000). Forced expression of C/EBPα in B lymphocytes from the bone marrow or spleen leads to efficient transdifferentiation of the infected B cells into functional macrophages within 5 days (Xie et al. 2004). C/EBPα thereby activates the myeloid gene program and concomitantly represses B-cell-specific genes by interfering with the transcriptional activity of Pax5 (Xie et al. 2004). Hence, the loss of Pax5 function is likely to facilitate the myeloid lineage conversion of B cells in response to ectopic C/EBPα expression (Fig. 11).

Conditional gene inactivation unequivocally identified a critical role for Pax5 in controlling the identity of B cells throughout B lymphopoiesis. Cre-mediated gene deletion in committed pro-B cells demonstrated that Pax5 is required not only to initiate its B-lymphoid transcription program, but also to maintain it in early B-cell development (Mikkola et al. 2002). As a consequence of *Pax5* inactivation, previously committed pro-B cells regain the capacity to differentiate into macrophages in vitro and to reconstitute T-cell development in vivo (Fig. 11) (Mikkola et al. 2002). Conditional *Pax5* deletion in mature B cells also leads to loss of the Pax5-dependent gene expression program (Horcher et al. 2001; Delogu et al. 2006). More surprisingly, however, the mature *Pax5*-deleted B cells retrodifferentiate all the way to *Pax5* mutant progenitors in vivo following injection into RAG2-deficient mice. These *Pax5* mutant progenitors home to the bone marrow, from where they seed the thymus and fully reconstitute T-cell development in the RAG2-deficient host (Fig. 11). The

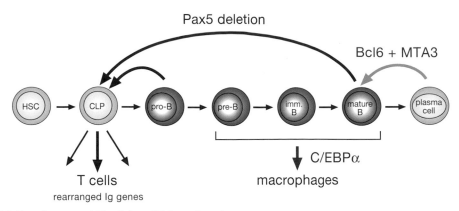

Figure 11. Developmental Plasticity of B Lymphocytes

Ectopic expression of Bcl6 and MTA3 in established plasma cell lines silences the transcription of plasma-cell-specific genes and simultaneously reactivates the expression program of B cells (*orange arrow*) (Fujita et al. 2004). CD19+ B lymphocytes, which were not further characterized with regard to their developmental stage, undergo rapid transdifferentiation in vitro to macrophages in response to forced C/EBPα expression (*red arrow*) (Xie et al. 2004). Conditional *Pax5* deletion allows committed pro-B cells and even mature B cells to retrodifferentiate in vivo to uncommitted progenitors, which then develop into other hematopoietic cell types in the bone marrow or T cells in the thymus (*black arrows*) (Mikkola et al. 2002; C. Cobaleda and M. Busslinger, unpublished data). The blue color denotes Pax5 expression during B-cell development.

fact that the corresponding CD4+CD8+ double positive thymocytes carry *IgH* and *Igκ* as well as *TCRα* and *TCRβ* rearrangements unambiguously demonstrates that mature B cells, following Pax5 loss, can be converted into T cells via retrodifferentiation to an uncommitted progenitor cell stage (C. Cobaleda and M. Busslinger, unpubl.). Hence, Pax5 expression is continuously required to maintain the identity of B lymphocytes from the pro-B-cell to the mature B-cell stage. Based on the analyses of other developmental systems in flies and vertebrates, transcription factors are thought to initiate cell-fate decision by altering gene expression patterns, while the transcriptional state of committed cells is subsequently maintained by epigenetic factors encoded by the *Polycomb* and *Trithorax* group genes (Ringrose and Paro 2004; discussed further in Chapters 11 and 12). The permanent requirement of the transcription factor Pax5 could argue against an important role of these epigenetic memory systems in B-cell development. More likely, however, Pax5 may maintain B-cell identity by acting as a crucial recruitment factor to target Polycomb or Trithorax protein complexes to gene regulatory elements.

5 Concluding Remarks

In summary, various epigenetic mechanisms are involved in regulating and guiding lymphocyte development. Of all the different epigenetic regulators, we currently know most about the role of transcription factors, which control entire gene expression patterns by recruiting chromatin-modifying activities (such as histone acetyltransferases or deacetylases) to gene regulatory elements. Less is known about the control of gene expression by histone methyltransferases, by Trithorax and Polycomb group proteins, or by microRNA and siRNA pathways. Unraveling the role of these regulatory systems will require experimentally engineered conditional gene inactivation, because histone methyltransferases, Trithorax and Polycomb proteins, as well as components of the RNAi machinery, are of fundamental importance not only for lymphopoiesis, but also for embryonic development. Moreover, the development and availability of global ChIP-on-chip technologies will allow high-resolution mapping of epigenetic modifications along entire chromosomes and complex loci (such as the antigen receptor loci) at different stages of lymphopoiesis. These recent advances are likely to provide important novel insight into the epigenetic control mechanisms underlying lymphocyte development.

References

Adolfsson J., Månsson R., Buza-Vidas N., Hultquist A., Liuba K., Jensen C.T., Bryder D., Yang L., Borge O.J., Thoren L.A.M., et al. 2005. Identification of Flt3+ lympho-myeloid stem cells lacking erythro-megakaryocytic potential: A revised road map for adult blood lineage commitment. *Cell* **121:** 295–306.

Akashi K., Traver D., Miyamoto T., and Weissman I.L. 2000. A clonogenic common myeloid progenitor that gives rise to all myeloid lineages. *Nature* **404:** 193–197.

Allman D., Sambandam A., Kim S., Miller J.P., Pagan A., Well D., Meraz A., and Bhandoola A. 2003. Thymopoiesis independent of common lymphoid progenitors. *Nature Immunol.* **4:** 168–174.

Barlev N.A., Emelyanov A.V., Castagnino P., Zegerman P., Bannister A.J., Sepulveda M.A., Robert F., Tora L., Kouzarides T., Birshtein B.K., and Berger S.L. 2003. A novel human Ada2 homologue functions with Gcn5 or Brg1 to coactivate transcription. *Mol. Cell. Biol.* **23:** 6944–6957.

Bassing C.H., Swat W., and Alt F.W. 2002. The mechanism and regulation of chromosomal V(D)J recombination. *Cell* (suppl.) **109:** S45–S55.

Baxter J., Merkenschlager M., and Fisher A.G. 2002. Nuclear organisation and gene expression. *Curr. Opin. Cell Biol.* **14:** 372–376.

Baxter J., Sauer S., Peters A., John R., Williams R., Caparros M.L., Arney K., Otte A., Jenuwein T., Merkenschlager M., and Fisher A.G. 2004. Histone hypomethylation is an indicator of epigenetic plasticity in quiescent lymphocytes. *EMBO J.* **23:** 4462–4472.

Bereshchenko O.R., Gu W., and Dalla-Favera R. 2002. Acetylation inactivates the transcriptional repressor BCL6. *Nat. Genet.* **32:** 606–613.

Bolland D.J., Wood A.L., Johnston C.M., Bunting S.F., Morgan G., Chakalova L., Fraser P.J. and Corcoran A.E. 2004. Antisense intergenic transcription in V(D)J recombination. *Nat. Immunol.* **5:** 630–637.

Brown K.E., Guest S.S., Smale S.T., Hahm K., Merkenschlager M., and Fisher A.G. 1997. Association of transcriptionally silent genes with Ikaros complexes at centromeric heterochromatin. *Cell* **91:** 845–854.

Busslinger M. 2004. Transcriptional control of early B cell development. *Annu. Rev. Immunol.* **22:** 55–79.

Calame K.L., Lin K.I., and Tunyaplin C. 2003. Regulatory mechanisms that determine the development and function of plasma cells. *Annu. Rev. Immunol.* **21:** 205–230.

Chow C.M., Georgiou A., Szutorisz H., Maia e Silva A., Pombo A., Barahona I., Dargelos E., Canzonetta C., and Dillon N. 2005. Variant histone H3.3 marks promoters of transcriptionally active genes during mammalian cell division. *EMBO Rep.* **6:** 354–360.

Chowdhury D. and Sen R. 2001. Stepwise activation of the immunoglobulin μ heavy chain gene locus. *EMBO J.* **20:** 6394–6403.

———. 2003. Transient IL-7/IL-7R signaling provides a mechanism for feedback inhibition of immunoglobulin heavy chain gene rearrangements. *Immunity* **18:** 229–241.

Dakic A., Metcalf D., Di Rago L., Mifsud S., Wu L., and Nutt S.L. 2005. PU.1 regulates the commitment of adult hematopoietic progenitors and restricts granulopoiesis. *J. Exp. Med.* **201:** 1487–1502.

Delogu A., Schebesta A., Sun Q., Aschenbrenner K., Perlot T., and Busslinger M. 2006. Gene repression by Pax5 in B cells is essential for blood cell homeostasis and is reversed in plasma cells. *Immunity* **24:** 269–281.

Fisher A.G. 2002. Cellular identity and lineage choice. *Nat. Rev. Immunol.* **2:** 977–982.

Francis N.J., Kingston R.E., and Woodcock C.L. 2004. Chromatin compaction by a polycomb group protein complex. *Science* **306:** 1574–1577.

Fujita N., Jaye D.L., Geigerman C., Akyildiz A., Mooney M.R., Boss J.M., and Wade P.A. 2004. MTA3 and the Mi-2/NuRD complex regulate cell fate during B lymphocyte differentiation. *Cell* **119:** 75–86.

Fukuda T., Yoshida T., Okada S., Hatano M., Miki T., Ishibashi K., Okabe S., Koseki H., Hirosawa S., Taniguchi M., et al. 1997. Disruption of the *Bcl6* gene results in an impaired germinal center formation. *J. Exp. Med.* **186:** 439–448.

Fuxa M., Skok J., Souabni A., Salvagiotto G., Roldán E., and Busslinger M. 2004. Pax5 induces *V*-to-*DJ* rearrangements and locus contraction of the *immunoglobulin heavy-chain* gene. *Genes Dev.* **18:** 411–422.

Goldmit M., Ji Y., Skok J., Roldan E., Jung S., Cedar H., and Bergman Y. 2005. Epigenetic ontogeny of the *Igk* locus during B cell development. *Nat. Immunol.* **6:** 198–203.

Goldmit M., Schlissel M., Cedar H., and Bergman Y. 2002. Differential accessibility at the κ chain locus plays a role in allelic exclusion. *EMBO J.* **21:** 5255–5261.

Grawunder U., Leu T.M.J., Schatz D.G., Werner A., Rolink A.G., Melchers F., and Winkler T.H. 1995. Down-regulation of *RAG1* and *RAG2* gene expression in preB cells after functional immunoglobulin heavy chain rearrangement. *Immunity* **3:** 601–608.

Györy I., Wu J., Fejér G., Seto E., and Wright K.L. 2004. PRDI-BF1 recruits the histone H3 methyltransferase G9a in transcriptional silencing. *Nat. Immunol.* **5:** 299–308.

Hesslein D.G.T., Pflugh D.L., Chowdhury D., Bothwell A.L.M., Sen R., and Schatz D.G. 2003. *Pax5* is required for recombination of transcribed, acetylated, 5′ IgH V gene segments. *Genes Dev.* **17:** 37–42.

Honjo T., Muramatsu M., and Fagarasan S. 2004. AID: How does it aid antibody diversity? *Immunity* **20:** 659–668.

Horcher M., Souabni A., and Busslinger M. 2001. Pax5/BSAP maintains the identity of B cells in late B lymphopoiesis. *Immunity* **14:** 779–790.

Huynh K.D. and Bardwell V.J. 1998. The BCL-6 POZ domain and other POZ domains interact with the co-repressors N-CoR and SMRT. *Oncogene* **17:** 2473–2484.

Huynh K.D., Fischle W., Verdin E., and Bardwell V.J. 2000. BCoR, a novel corepressor involved in BCL-6 repression. *Genes Dev.* **14:** 1810–1823.

Ikawa T., Kawamoto H., Wright L.Y.T., and Murre C. 2004. Long-term cultured E2A-deficient hematopoietic progenitor cells are pluripotent. *Immunity* **20:** 349–360.

Jankovic M., Casellas R., Yannoutsos N., Wardemann H., and Nussenzweig M.C. 2004. RAGs and regulation of autoantibodies. *Annu. Rev. Immunol.* **22:** 485–501.

Keller A.D. and Maniatis T. 1991. Identification and characterization of a novel repressor of β-interferon gene expression. *Genes Dev.* **5:** 868–879.

Kim K.C., Geng L., and Huang S. 2003. Inactivation of a histone methyltransferase by mutations in human cancers. *Cancer Res.* **63:** 7619–7623.

Kitamura D. and Rajewsky K. 1992. Targeted disruption of μ chain membrane exon causes loss of heavy-chain allelic exclusion. *Nature* **356:** 154–156.

Kosak S.T., Skok J.A., Medina K.L., Riblet R., Le Beau M.M., Fisher A.G., and Singh H. 2002. Subnuclear compartmentalization of immunoglobulin loci during lymphocyte development. *Science* **296:** 158–162.

Kovanen P.E. and Leonard W.J. 2004. Cytokines and immunodeficiency diseases: Critical roles of the γ_c-dependent cytokines interleukins 2, 4, 7, 9, 15, and 21, and their signaling pathways. *Immunol. Rev.* **202:** 67–83.

Krangel M.S. 2003. Gene segment selection in V(D)J recombination: Accessibility and beyond. *Nat. Immunol.* **4:** 624–630.

Kraus M., Alimzhanov M.B., Rajewsky N., and Rajewsky K. 2004. Survival of resting mature B lymphocytes depends on BCR signaling via the Igα/β heterodimer. *Cell* **117:** 787–800.

Lee A.I., Fugmann S.D., Cowell L.G., Ptaszek L.M., Kelsoe G., and Schatz D.G. 2003. A functional analysis of the spacer of V(D)J recombination signal sequences. *PLoS Biol.* **1:** 56–69.

Liang H.E., Hsu L.Y., Cado D., and Schlissel M.S. 2004. Variegated transcriptional activation of the immunoglobulin κ locus in pre-B cells contributes to the allelic exclusion of light-chain expression. *Cell* **118:** 19–29.

Linderson Y., Eberhard D., Malin S., Johansson A., Busslinger M., and Pettersson S. 2004. Corecruitment of the Grg4 repressor by PU.1 is critical for Pax5-mediated repression of B-cell-specific genes. *EMBO Rep.* **5:** 291–296.

Maier H., Colbert J., Fitzsimmons D., Clark D.R., and Hagman J. 2003. Activation of the early B-cell-specific *mb-1* (Ig-α) gene by Pax-5 is dependent on an unmethylated Ets binding site. *Mol. Cell. Biol.* **23:** 1946–1960.

Maier H., Ostraat R., Gao H., Fields S., Shinton S.A., Medina K.L., Ikawa T., Murre C., Singh H., Hardy R.R., and Hagman J. 2004. Early B cell factor cooperates with Runx1 and mediates epigenetic changes associated with *mb-1* transcription. *Nat. Immunol.* **5:** 1069–1077.

Massari M.E., Grant P.A., Pray-Grant M.G., Berger S.L., Workman J.L., and Murre C. 1999. A conserved motif present in a class of helix-loop-helix proteins activates transcription by direct recruitment of the SAGA complex. *Mol. Cell.* **4:** 63–73.

McBlane F. and Boyes J. 2000. Stimulation of V(D)J recombination by histone acetylation. *Curr. Biol.* **10:** 483–486.

McMurry M.T. and Krangel M.S. 2000. A role for histone acetylation in the developmental regulation of V(D)J recombination. *Science* **287:** 495–498.

Medina K.L., Pongubala J.M., Reddy K.L., Lancki D.W., DeKoter R., Kieslinger M., Grosschedl R., and Singh H. 2004. Assembling a gene regulatory network for specification of the B cell fate. *Dev. Cell* **7:** 607–617.

Mikkola I., Heavey B., Horcher M., and Busslinger M. 2002. Reversion of B cell commitment upon loss of *Pax5* expression. *Science* **297:** 110–113.

Mito Y., Henikoff J.G., and Henikoff S. 2005. Genome-scale profiling of histone H3.3 replacement patterns. *Nat. Genet.* **37:** 1090–1097.

Miyamoto T., Iwasaki H., Reizis B., Ye M., Graf T., Weissman I.L., and Akashi K. 2002. Myeloid or lymphoid promiscuity as a critical step in hematopoietic lineage commitment. *Dev. Cell* **3:** 137–147.

Morshead K.B., Ciccone D.N., Taverna S.D., Allis C.D., and Oettinger M.A. 2003. Antigen receptor loci poised for V(D)J rearrangement are broadly associated with BRG1 and flanked by peaks of histone H3 dimethylated at lysine 4. *Proc. Natl. Acad. Sci. USA* **100:** 11577–11582.

Mostoslavsky R., Singh N., Kirillov A., Pelanda R., Cedar H., Chess A., and Bergman Y. 1998. κ chain monoallelic demethylation and the establishment of allelic exclusion. *Genes Dev.* **12:** 1801–1811.

Mostoslavsky R., Singh N., Tenzen T., Goldmit M., Gabay C., Elizur S., Qi P., Reubinoff B.E., Chess A., Cedar H., and Bergman Y. 2001. Asynchronous replication and allelic exclusion in the immune system. *Nature* **414:** 221–225.

Niu H., Ye B.H., and Dalla-Favera R. 1998. Antigen receptor signaling induces MAP kinase-mediated phosphorylation and degradation of the BCL-6 transcription factor. *Genes Dev.* **12:** 1953–1961.

Nutt S.L., Heavey B., Rolink A.G., and Busslinger M. 1999. Commitment to the B-lymphoid lineage depends on the transcription factor Pax5. *Nature* **401:** 556–562.

Nutt S.L., Urbánek P., Rolink A., and Busslinger M. 1997. Essential functions of Pax5 (BSAP) in pro-B cell development: Difference between fetal and adult B lymphopoiesis and reduced *V*-to-*DJ* recombination at the *IgH* locus. *Genes Dev.* **11:** 476–491.

Orphanides G. and Reinberg D. 2000. RNA polymerase II elongation through chromatin. *Nature* **407:** 471–475.

Osipovich O., Milley R., Meade A., Tachibana M., Shinkai Y., Krangel M.S., and Oltz E.M. 2004. Targeted inhibition of V(D)J recombination by a histone methyltransferase. *Nat. Immunol.* **5:** 309–316.

Peters A.H., O'Carroll D., Scherthan H., Mechtler K., Sauer S., Schöfer C., Weipoltshammer K., Pagani M., Lachner M., Kohlmaier A., et al. 2001. Loss of the *Suv39h* histone methyltransferases impairs mammalian heterochromatin and genome stability. *Cell* **107:** 323–337.

Rajewsky K. 1996. Clonal selection and learning in the antibody system. *Nature* **381:** 751–758.

Reimold A.M., Iwakoshi N.N., Manis J., Vallabhajosyula P., Szomolanyi-Tsuda E., Gravallese E.M., Friend D., Grusby M.J., Alt F., and Glimcher L.H. 2001. Plasma cell differentiation requires the transcription factor XBP-1. *Nature* **412:** 300–307.

Ren B., Chee K.J., Kim T.H., and Maniatis T. 1999. PRDI-BF1/Blimp-1 repression is mediated by corepressors of the Groucho family of proteins. *Genes Dev.* **13:** 125–137.

Ringrose L. and Paro R. 2004. Epigenetic regulation of cellular memory by the Polycomb and Trithorax group proteins. *Annu. Rev. Genet.* **38:** 413–443.

Roldán E., Fuxa M., Chong W., Martinez D., Novatchkova M., Busslinger M., and Skok J.A. 2005. Locus 'decontraction' and centromeric recruitment contribute to allelic exclusion of the immunoglobulin heavy-chain gene. *Nat. Immunol.* **6:** 31–41.

Romanow W.J., Langerak A.W., Goebel P., Wolvers-Tettero I.L.M., van Dongen J.J.M., Feeney A.J., and Murre C. 2000. E2A and EBF act in synergy with the V(D)J recombinase to generate a diverse immunoglobulin repertoire in nonlymphoid cells. *Mol. Cell* **5:** 343–353.

Sato H., Saito-Ohara F., Inazawa J., and Kudo A. 2004. Pax-5 is essential for κ sterile transcription during Igκ chain gene rearrangement. *J. Immunol.* **172:** 4858–4865.

Schlissel M.S., Durum S.D., and Muegge K. 2000. The interleukin 7 receptor is required for T cell receptor γ locus accessibility to the V(D)J recombinase. *J. Exp. Med.* **191:** 1045–1050.

Seet C.S., Brumbaugh R.L., and Kee B.L. 2004. Early B cell factor promotes B lymphopoiesis with reduced interleukin 7 responsiveness in the absence of E2A. *J. Exp. Med.* **199:** 1689–1700.

Shaffer A.L., Yu X., He Y., Boldrick J., Chan E.P., and Staudt L.M. 2000. BCL-6 represses genes that function in lymphocyte differentiation, inflammation and cell cycle control. *Immunity* **13:** 199–212.

Shaffer A.L., Lin K.I. Kuo T.C., Yu X., Hurt E.M., Rosenwald A., Giltnane J.M., Yang L., Zhao H., Calame K., and Staudt L.M. 2002. Blimp-1 orchestrates plasma cell differentiation by extinguishing the mature B cell gene expression program. *Immunity* **17:** 51–62.

Shaffer A.L., Shapiro-Shelef M., Iwakoshi N.N., Lee A.-H., Qian S.B., Zhao H., Yu X., Yang L., Tan B.K., Rosenwald A., et al. 2004. XBP1, downstream of Blimp-1, expands the secretory apparatus and other organelles, and increases protein synthesis in plasma cell differentiation. *Immunity* **21:** 81–93.

Skok J.A., Brown K.E., Azuara V., Caparros M.L., Baxter J., Takacs K., Dillon N., Gray D., Perry R.P., Merkenschlager M., and Fisher A.G. 2001. Nonequivalent nuclear location of immunoglobulin alleles in B lymphocytes. *Nat. Immunol.* **2:** 848–854.

Smale S.T. 2003. The establishment and maintenance of lymphocyte identity through gene silencing. *Nat. Immunol.* **4:** 607–615.

Stanhope-Baker P., Hudson K.M., Shaffer A.L., Constantinescu A., and Schlissel M.S. 1996. Cell type-specific chromatin structure determines the targeting of V(D)J recombinase activity in vitro. *Cell* **85:** 887–897.

Su I.H., Basavaraj A., Krutchinsky A.N., Hobert O., Ullrich A., Chait B.T., and Tarakhovsky A. 2003. Ezh2 controls B cell development through histone H3 methylation and *Igh* rearrangement. *Nat. Immunol.* **4:** 124–131.

Su I.H., Dobenecker M.W., Dickinson E., Osler M., Basavaraj A., Marqueron R., Viale A., Reinberg D., Wülfing C., and Tarakhovsky A. 2005. Polycomb group protein Ezh2 controls actin polymerization and cell signaling. *Cell* **121:** 425–436.

Tachibana M., Sugimoto K., Nozaki M., Ueda J., Ohta T., Ohki M., Fukuda M., Takeda N., Niida H., Kato H., and Shinkai Y. 2002. G9a histone methyltransferase plays a dominant role in euchromatic histone H3 lysine 9 methylation and is essential for early embryogenesis. *Genes Dev.* **16:** 1779–1791.

Turner C.A.J., Mack D.H., and Davis M.M. 1994. Blimp-1, a novel zinc finger-containing protein that can drive the maturation of B lymphocytes into immunoglobulin-secreting cells. *Cell* **77:** 297–306.

Vosshenrich C.A.J., Cumano A., Müller W., Di Santo J.P., and Vieira P. 2003. Thymic stroma-derived lymphopoietin distinguishes fetal from adult B cell development. *Nat. Immunol.* **4:** 773–779.

Waskow C., Paul S., Haller C., Gassmann M., and Rodewald H. 2002.

Viable c-Kit$^{w/w}$ mutants reveal pivotal role for c-Kit in the maintenance of lymphopoiesis. *Immunity* **17:** 277–288.

West K.L., Singha N.C., De Ioannes P., Lacomis L., Erdjument-Bromage H., Tempst P., and Cortes P. 2005. A direct interaction between the RAG2 C terminus and the core histones is required for efficient V(D)J recombination. *Immunity* **23:** 203–212.

Xie H., Ye M., Feng R., and Graf T. 2004. Stepwise reprogramming of B cells into macrophages. *Cell* **117:** 663–676.

Yancopoulos G.D. and Alt F.W. 1985. Developmentally controlled and tissue-specific expression of unrearranged V$_H$ gene segments. *Cell* **40:** 271–281.

Ye B.H., Cattoretti G., Shen Q., Zhang J., Hawe N., de Waard R., Leung C., Nouri-Shirazi M., Orazi A., Chaganti R.S.K., et al. 1997. The *BCL*-6 proto-oncogene controls germinal-centre formation and Th2-type inflammation. *Nature Genet.* **16:** 161–170.

Zhang S., Fukuda S., Lee Y., Hangoc G., Cooper S., Spolski R., Leonard W.J., and Broxmeyer H.E. 2000. Essential role of signal transducer and activator of transcription (Stat)5a but not Stat5b for Flt3-dependent signaling. *J. Exp. Med.* **192:** 719–728.

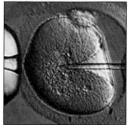

Nuclear Transplantation and the Reprogramming of the Genome

Rudolf Jaenisch[1] and John Gurdon[2]

[1]Whitehead Institute, and Department of Biology, Massachusetts Institute of Technology,
Cambridge, Massachusetts 02142-1479
[2]The Wellcome Trust/Cancer Research UK Gurdon Institute, The Henry Wellcome Building of Cancer and Developmental
Biology, University of Cambridge, Cambridge CB2 1QN, United Kingdom

CONTENTS

1. History, 417

2. Nuclear Transfer Procedures, 417
   2.1   Amphibians, 417
   2.2   Mammals, 417

3. Phenotype of Cloned Animals, 419
   3.1   Amphibians, 419
   3.2   Mammals, 420
   3.3   Derivation of Cloned Mammals from Terminally Differentiated Cells, 421

4. Changes Associated with Nuclear Reprogramming, 425

4.1   Amphibians, 425
4.2   Mammals, 427

5. Epigenetic Memory, 428

6. Medical Implications of Nuclear Transplantation, 429
   6.1   Reproductive Cloning, 430
   6.2   Therapeutic Application of Nuclear Transplantation, 431
   6.3   Reproductive Versus Therapeutic Cloning: What Is the Difference? , 432

References, 432

GENERAL SUMMARY

The body plan of animals is constructed from hundreds of different cell types that perform the various physiological functions of the organism. A key question posed early on was the mechanism of differential gene expression which assures that the appropriate genes are active or silent, respectively, for the normal function of a given differentiated cell. In the early days, before the molecular basis of gene expression was appreciated, it had been hypothesized that the basis for tissue-specific gene expression might be the genetic elimination or permanent inactivation of silent genes from those tissues that do not express silent genes and retention of those that are expressed. Indeed, in some organisms, such as the insect *Sciara*, genetic material is eliminated from somatic tissues with the full genetic complement being retained only in the germ-line cells. This raised the question of "nuclear equivalence," i.e., whether the genome of somatic cells retains the full complement of genetic material. The most unbiased approach to this question is nuclear cloning, where the potency of a somatic donor cell nucleus to direct the development of a new organism is tested by transplantation into an enucleated egg. Indeed, the generation of cloned animals from somatic cell nuclei proved beyond doubt that major genetic changes which would prevent a somatic nucleus from generating all tissue types are not part of the normal developmental process.

1 History

The first success in transplanting the nucleus of one cell into another in a multicellular organism was obtained by Briggs and King in 1952 (Briggs and King 1952), although nuclear transfer had been achieved before in single-celled organisms, including amoeba, ciliates, and *Acetabularia* (Gurdon 1964). Briggs and King (1952) obtained normal swimming tadpoles by transplanting the nuclei of a blastula cell into an enucleated egg of *Rana pipiens*. In this and in subsequent work with *R. pipiens*, up to 30% of blastula nuclear transfers developed to morphologically normal postneurula stages. The importance of this early success was that it opened the way to testing whether the nuclei of differentiating cells could also support normal development of recipient eggs; that is, substitute for the zygote nucleus of normally fertilized eggs. The next important paper by Briggs and King (1957) reported that very soon after the blastula stage, somatic cell nuclei (in this case of the endoderm) lose their ability to support normal development. Indeed, by the tail-bud stage, endoderm nuclei no longer gave any normal development.

Soon after this, successful nuclear transfer was reported in *Xenopus* (Fischberg et al. 1958). In this species, a genetic marker was used to prove that nuclear transplant embryos were derived entirely by activity of the transplanted nucleus with no contribution of the egg nucleus, which had been killed by ultraviolet irradiation. Nuclear transplant embryos developed more normally in *Xenopus* than in *Rana*, and developed quite rapidly to the adult stage. The first sexually mature adult cloned animals were obtained by the transfer of endoderm nuclei in *Xenopus* (Gurdon et al. 1958), and most appeared normal in all respects.

For nuclear cloning to succeed, an egg must "reprogram" the somatic donor nucleus to an "embryonic" epigenetic state so that the genetic program required for embryonic development can be activated. It is a major focus of current research to understand the molecular basis of reprogramming in nuclear transplantation experiments. Because embryonic development in amphibians and mammals is very different, each system can provide different insights into the problem of reprogramming in nuclear transplantation experiments. For example, because a great number of the large eggs can be easily obtained in amphibians, this system is particularly useful for biochemical analyses. In contrast, mammals, in particular the mouse, allow the application of tissue culture and genetic approaches as methods of choice. For these reasons, observations in the amphibian and the mammalian systems are complementary and are discussed

side by side to emphasize the similarities and the differences in the two systems. First, we describe the phenotype of cloned amphibians and mammals and how it is influenced by the differentiation state of the donor nucleus. This is followed by a description of molecular mechanisms that have been recognized to be important for reprogramming. Finally, the potential implications of the nuclear transplantation technology for therapy are discussed.

2 Nuclear Transfer Procedures

2.1 Amphibians

The ready availability of *Xenopus* eggs throughout the year, and the easy maintenance of an aquatic animal in the laboratory, have resulted in most amphibian nuclear transfer work, after the initial success with *Rana*, being undertaken with *Xenopus*. Two kinds of experiments need to be distinguished according to whether nuclei are injected into enucleated eggs or non-enucleated oocytes (Fig. 1). Nuclear transfer requires the injection of a complete donor cell whose plasma membrane has been made permeable, either physically, by sucking the cell into a pipette too small for the size of the cell, or chemically, by a short exposure to a membrane-integrating substance. The amount of donor cell cytoplasm introduced with a nucleus is about 10^{-5} of the egg volume and has no effect. Eggs that have received a transplanted nucleus are cultured in a simple non-nutrient saline solution equivalent to pond water, and therefore develop independently of the culture solution.

Nuclear transfer to oocytes is entirely different. Full-grown oocytes taken from the ovary of a female are in the prophase of first meiosis (Fig. 1). About 50 somatic cell nuclei injected into the nucleus (germinal vesicle) of these growing oocytes do not replicate DNA or divide but become increasingly active in transcription over the course of a few days. Oocytes that have received transplanted nuclei are cultured, like eggs, in a non-nutrient saline medium and undergo no morphological change for the whole of the culture period of 2 weeks or more.

2.2 Mammals

The earliest attempts to clone mammals were performed with rabbit embryos. In these experiments, oocytes were fused with cells from morula-stage embryos, and the resulting triploid clones were seen to undergo a few cleavage divisions (Bromhall 1975). In 1984, McGrath and Solter introduced Sendai-virus-mediated fusion to efficiently transfer donor nuclei into enucleated zygotes, a

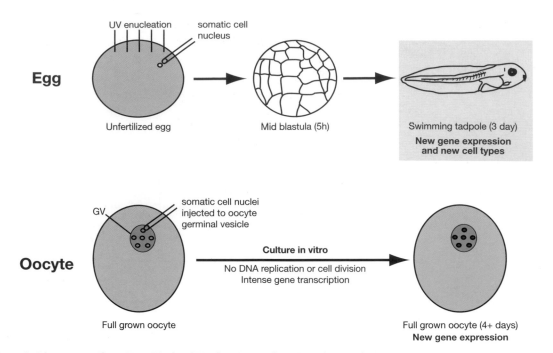

Figure 1. Diagram to Show Two Kinds of Nuclear Transplant Experiments in *Xenopus*

The upper figure illustrates single nuclear transfers to eggs enucleated by ultraviolet irradiation (UV). A swimming tadpole is formed 3 days after nuclear transfer. The lower figure shows multiple nuclear transfers to growing oocytes in first meiotic prophase. (GV) Germinal vesicle or nucleus of the oocyte. Injected oocytes do not divide, and can be cultured for many days. Injected somatic cell nuclei undergo changes in gene expression during the first 4 days of culture.

method that was used for two types of experiments: (1) Uniparental embryos were generated by replacing the male with a female pronucleus (giving rise to a gynogenetic embryo) or by replacing the female with a male pronucleus (giving rise to an androgenetic embryo) (McGrath and Solter 1984a,b; Surani et al. 1984). Uniparental embryos invariably failed to develop, providing the first evidence that normal development of mammals crucially depends on parent-specific genomic imprinting. (2) Cloned embryos were produced by transferring nuclei of cleavage-stage donor embryos into enucleated zygotes. None of the reconstructed embryos developed beyond late-cleavage stages, leading to the conclusion that totipotency of the genome is rapidly lost during early development and that cloning of mammals, in contrast to amphibians, may not be possible (McGrath and Solter 1984b). However, results obtained in other mammalian species soon challenged this conclusion.

In 1986, Willadsen succeeded in cloning live lambs from 4- to 16-cell donor embryo nuclei, and this was shortly followed by the generation of cloned cattle and pigs (Robi et al. 1987; Prather et al. 1989). Why did cloning of farm animals succeed in contrast to the well-controlled mouse cloning experiments (McGrath and Solter 1984b)? A major developmental difference between these animal species is the timing of the transition from maternal control of development (relying on maternally stored RNA) to zygotic control of development (relying on zygotically produced RNA). The major transition appears to occur at the 8- to 16-cell stage for sheep and bovine embryos (Calarco and McLaren 1976; Camous et al. 1986) but already at the 2-cell stage for mouse embryos (Bolton et al. 1984). Thus, the time constraints for activating the donor genome may be more relaxed in cloned sheep or bovine embryos than in cloned mouse embryos, where the donor nucleus must be activated soon after nuclear transfer in order for cleavage to proceed.

The first successful derivation of cloned animals by somatic cell nuclear transfer, as opposed to nuclear transfer from donor cells of the cleavage embryo, was the generation of sheep from cultured fibroblast donor cells (Campbell et al. 1996b). This was soon followed by the creation of "Dolly" from a mammary gland donor nucleus (Wilmut et al. 1997), which constituted the first mammal cloned from an adult donor cell. Since then, a total of 15 mammalian species, including mice, goats, pigs, cows, rabbits, rats, cats, and dogs, have been cloned (for review, see Campbell et al. 2005).

The procedure of mammalian cloning, like amphibian cloning, involves two steps. In most, if not all, successful somatic cell nuclear transfer experiments, the donor nuclei were transferred into enucleated oocytes (in contrast to the early mouse cloning experiments where the cleavage embryo donor nucleus was introduced into enucleated zygotes; McGrath and Solter 1984). In a first step, the metaphase spindle of the recipient egg is removed with a pipette, followed by the transfer of the donor nucleus into the enucleated egg. In most mammals, the donor nucleus is introduced into the egg by electrofusion with the enucleated egg. Mice have particularly fragile eggs, and nuclear transfer in this species is most efficient by physical transfer of the donor nucleus into the egg (Wakayama et al. 1998). As depicted in Figure 2, the donor nucleus is introduced into the enucleated egg by the use of a piezo element that facilitates the penetration of the zona pellucida and the cytoplasmic membrane (Wakayama et al. 1998). Cloned blastocysts either are implanted into the uterus of a pseudopregnant foster mother to generate cloned mice or are explanted in tissue culture to generate nuclear transfer embryonic stem (NT-ES) cells. NT-ES

cells are genetically identical to the donor and can, therefore, be used for "customized" cell therapy, also designated as "therapeutic cloning" (see below).

3 Phenotype of Cloned Animals

The derivation of animals by nuclear transfer from somatic donor nuclei is inefficient, and those rare animals that survive to adulthood often display multiple abnormalities. In contrast, when embryonic cells are used as nuclear donors, the survival of both amphibian and mammalian clones is much higher. The following chapter focuses on the relationship between the age of the donor nucleus and its effect on clone survival and juxtaposes the phenotype of amphibian and mammalian clones derived from adult and embryonic donor cells.

3.1 Amphibians

When blastula nuclei are transplanted to good quality recipient eggs of *Xenopus*, over 30% of all such eggs develop into normal tadpoles, and most of these can be reared to fertile adults. As donor cells differentiate, the

a

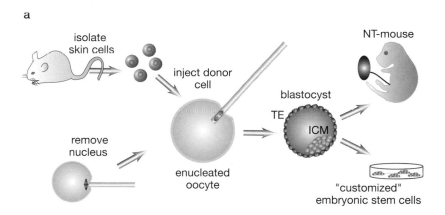

b

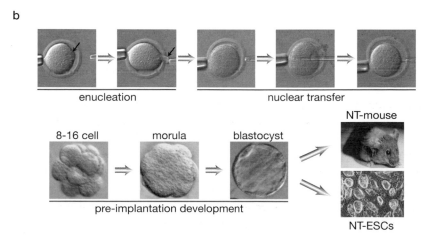

Figure 2. Murine Nuclear Transfer by Micro-injection

(*a*) Schematic drawing of the nuclear transfer procedure. The inner cell mass (ICM) gives rise to ES cells, when explanting the blastocyst onto irradiated feeder cells. Alternatively, when transferred to a synchronized pseudopregnant female, the blastocyst can generate a mouse. The outer cells of the blastocyst, the trophectoderm (TE), give rise to the extraembryonic tissues (placenta), and the ICM cells generate the embryo. (*b*) The same steps as in *a* shown by light microscopy. (Reprinted, with permission, from Meissner 2006.)

normality of development of nuclear transplant embryos decreases (Fig. 3), less rapidly with endoderm donor nuclei than with others. In amphibians, it has been informative to carry out serial nuclear transfers in which donor nuclei are taken from a blastula that has itself resulted from the transfer of a somatic cell nucleus (Fig. 4). This is done because the cells of a first-transfer embryo are often a mosaic of chromosomally normal and abnormal cells due to the difference in DNA replication rate between a somatic cell and an activated egg; serial nuclear transplantation shows the developmental potential of nuclei of a first-transfer embryo that are least damaged by nuclear transfer. Even with the intestinal epithelium of feeding larvae, some normal sexually mature, genetically marked, male and female frogs were obtained (Gurdon and Uehlinger 1966). This result showed that the process of differentiation does not necessarily involve the loss of ability to promote normal development, and hence, the principle of the conservation of the genome as cells differentiate.

In nuclear transfer experiments with amphibians, it was not possible to obtain a normal adult animal from the nucleus of another adult. However, morphologically normal tadpoles were obtained from the nuclei of cells from many different adult tissues (Laskey and Gurdon 1970), and these tadpoles contained the normal range of functional specialized cell types. Therefore, cells that are committed to one pathway of differentiation nevertheless contain the genetic potential to promote, in combination with egg cytoplasm, most kinds of unrelated cell differentiation.

The ability of nuclei of one cell type to promote other kinds of cell differentiation may be quantitated by asking to what extent functional muscle and nerve differentiation, as judged by embryos that make swimming movements after stimulation, can be generated from the nuclei of feeding larval intestinal epithelium. Such embryos can often be obtained by the serial transfer of nuclei from first-transfer embryos that are not completely normal (see Fig. 4) (Gurdon 1962). In addition, blastula cells from morphologically defective first-transfer embryos can form muscle when grafted to normal hosts grown from fertilized eggs (Fig. 4) (Byrne et al. 2003). The result shows that up to 30% of the cells of intestinal epithelium can lead, after nuclear transfer, to functional axial muscle cells (Table 1). The range of abnormalities resulting from nuclear transfer in amphibians does not show any consistent pattern that can be related at the morphological level to donor cell origin. Irrespective of the cell type and developmental stage of donor nuclei, nuclear transplant embryos die with a similar range of defects, including incomplete cleavages, failure to gastrulate, defective axis formation, and lack of head structures. The apparently haphazard nature of these defects is not surprising, since it is known that large chromosomal abnormalities are often seen in cells of first-transfer embryos (see Section 4.1, Reprogramming in clones).

3.2 Mammals

COMMON ABNORMALITIES IN CLONED ANIMALS

The majority of cloned mammalian embryos fail to develop soon after implantation. Those that live to birth often display common abnormalities irrespective of the donor cell type (see below; Table 2). For instance, newborn clones are frequently overgrown and show an enlarged placenta, symptoms referred to as Large Offspring Syndrome (Young et al. 1998; Hill et al. 2000; Tanaka et al. 2001). Moreover, neonate clones often suffer from respiratory distress, and kidney, liver, heart, and brain defects. Even long-term survivors can show abnormalities later in life. For example, aging cloned mice frequently become obese, develop severe immune problems, or die prematurely (Ogonuki et al. 2002; Tamashiro et al. 2002). As schematically shown in Figure 5, the two stages when the majority of clones fail are immediately after implantation and at birth. These are two critical stages of development that may be particularly vulnerable to faulty gene expression (see below). However, the generation of adult and seemingly

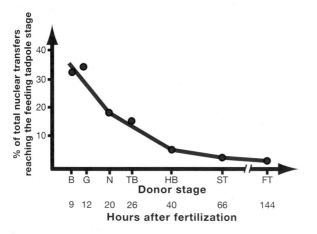

Figure 3. The Survival of *Xenopus* Nuclear Transfer Embryos Decreases as Donor Nuclei Are Taken from More Specialized Donor Cells

Donor stage abbreviations: (B) blastula; (G) gastrula; (N) neurula; (TB) tail bud; (HB) heart beat; (ST) swimming tadpole; (FT) feeding tadpole. (Reprinted, with permission, from Gurdon 1960.)

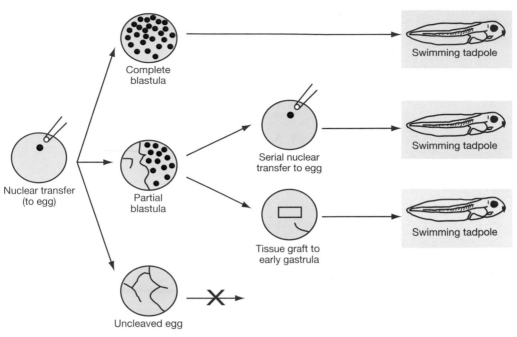

Figure 4. Serial Nuclear Transfers and Grafts in *Xenopus*

For standard first transfers, a single somatic cell nucleus is transplanted to an enucleated egg, which is grown directly to a larva. For serial nuclear transfers, a first-transfer embryo at an early stage is used to provide donor nuclei for a further set of nuclear transfers to eggs. These serial nuclear transfer embryos are grown to the tadpole stage. Last, a first-transfer embryo that is partly defective is used to provide a piece of tissue for grafting to a host embryo reared from a fertilized egg. The graft contributes to part of the resulting tadpole.

healthy adult cloned animals has been taken as evidence that nuclear transfer can generate normal cloned animals, albeit with low efficiency. Importantly, serious abnormalities in cloned animals may often become manifest only when the animals age (Ogonuki et al. 2002; Tamashiro et al. 2002). The stochastic occurrence of disease and other defects at a later age in many or most adult clones implies that compensatory mechanisms which allow the survival of cloned animals do not guarantee their "normalcy." Rather, the phenotype of surviving cloned animals appears to be distributed over a wide spectrum, including abnormalities causing sudden demise at early postnatal age or more subtle abnormalities allowing survival to advanced age (Fig. 5). These considerations illustrate the complexity of defining subtle gene expression defects and emphasize the need for more sophisticated test criteria such as environmental stress or behavior tests.

EPIGENETIC VERSUS GENETIC CAUSES

The abnormalities that are characteristic for cloned animals are not inherited by offspring from the clones, indicating that "epigenetic" rather than genetic aberrations are the cause. This is because epigenetic changes, in con-

trast to genetic changes, are reversible modifications of DNA or chromatin that are erased when the genome is passed through the germ line (Ogonuki et al. 2002; Tamashiro et al. 2002). Thus, the problems associated with cloning are due to faulty "epigenetic/genomic reprogramming" of the transplanted donor nucleus rather than to somatic mutations acquired in the somatic donor cells.

3.3 Derivation of Cloned Mammals from Terminally Differentiated Cells

STATE OF DONOR CELL DIFFERENTIATION AND THE EFFICIENCY OF NUCLEAR REPROGRAMMING

A question already raised in the seminal cloning experiments with amphibians suggested an inverse relationship between cellular differentiation state of the donor

Table 1. Efficiency of nuclear reprogramming: *Xenopus* larval endoderm cells

First transfers only	muscle and nerve of embryo	15%
First + serial transfers	muscle and nerve of embryo	22%
First + serial transfers with grafts to hosts	muscle and nerve of embryo	30%

Table 2. Inverse relation between differentiation state of donor cells and reprogramming efficiency

Donor cell	Blastocysts (% of oocytes)	Cloned mice (% of implanted blastocysts)	ES cell derivation (% of explanted blastocysts)	Reference
Fertilized egg		60–80	25–65	1
ES cells	6–15	10–26	50	2
EC cells	6–15	N.D.	50	3
Sertoli	10–50	6	25	4
Cumulus	10–50	1–3	13–33	5
Fibroblast	10–50	1	13–33	5
NKT	70	4	N.D.	6
B, T cells	4	N.D.	7	7
Neurons	15	N.D.	6–28	8
Melanoma	1.5	N.D.	25	9

N.D. indicates not determined.

References: [1] Wakayama et al. 2005. [2] Wakayama et al. 1999; Rideout et al. 2000; Eggan et al. 2001; Humpherys et al. 2001; [3] Blelloch et al. 2004. [4] Ogura et al. 2000. [5] Wakayama et al. 1998, 2005; Wakayama and Yanagimachi 1999. [6] Inoue et al. 2005. [7] Hochedlinger and Jaenisch 2002. [8] Eggan et al. 2004; Li et al. 2004. [9] Hochedlinger et al. 2004.

nucleus and its potency to direct development after transfer into the egg (see above, and Fig. 3). An important issue has been whether the state of donor cell differentiation affects the efficiency of reprogramming also in mammals. As summarized in Table 2, reprogramming can be measured functionally by evaluating clone development at several different levels, including (1) the rate of blastocyst formation following nuclear transfer into the egg, (2) the fraction of cloned embryos surviving to birth or adulthood after implantation into the uterus, and (3) the frequency with which pluripotent embryonic stem (ES) cells can be derived from cloned blastocysts explanted into culture.

The efficiency of preimplantation development of reconstructed oocytes into blastocysts is particularly sensitive to experimental parameters such as the cell cycle stage and physical condition of the transferred nucleus. For example, cloning of nondividing donor cells is more efficient than cloning of actively proliferating cells (Campbell et al. 1996a; Cibelli et al. 1998). Thus, as expected from this relationship, eggs reconstructed with donor nuclei from fibroblasts, Sertoli, cumulus, or NKT cells that are in G_1 or G_0 of the cell cycle reach the blastocyst stage with relatively high efficiency, in contrast to ES or embryonal carcinoma (EC) cells that are actively dividing with a major fraction of the cells in S phase (Table 2). Due to this experimental variability during cleavage, measuring the fraction of blastocysts derived from reconstructed oocytes is not a reliable criterion to quantify "reprogram-ability." However, once a cloned embryo has reached the blastocyst stage, the development to birth after implantation into the uterus depends on the differentiation state of the donor nucleus. Cloned embryos derived from embryonic donors such as ES or EC cells develop to term at a 10- to 20-fold higher efficiency than embryos derived from cumulus or fibroblast donor cells

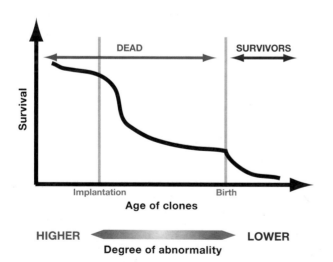

Figure 5. The Survival of Mammalian Clones

The phenotypes of clones are distributed over a wide range of abnormalities. Most clones fail at two defined developmental stages, implantation and birth. More subtle gene expression abnormalities result in disease and death at later ages.

(Eggan et al. 2001; Wakayama and Yanagimachi 2001), presumably because the nucleus of an undifferentiated embryonic cell is more amenable to, or requires less, reprogramming than the nucleus of a differentiated somatic cell. This indicates an inverse relationship between the stage of donor cell differentiation and the efficiency of reprogramming. Finally, once an embryo has reached the blastocyst stage, it has a rather consistent probability of giving rise to ES cells, indicating that the derivation of ES cells from explanted blastocysts is much less dependent on the state of differentiation of the donor nucleus (Table 2).

CAN NUCLEI OF TERMINALLY DIFFERENTIATED CELLS BE REPROGRAMMED TO TOTIPOTENCY?

In the early amphibian and mammalian cloning experiments, the donor cell populations used for nuclear transplantation were heterogeneous, and it could not be excluded that rare adult stem cells present in the donor population instead of nuclei of the differentiated cells gave rise to the rare surviving clones. For example, the epigenetic state of somatic stem cells may resemble that of embryonic stem cells and may be easier to reprogram and thus may preferentially have generated the surviving clones. To resolve the question whether the nucleus of a terminally differentiated cell could be sufficiently reprogrammed to yield an adult animal, genetic markers were required that would retrospectively identify the donor nucleus of a surviving clone. Such markers were used to demonstrate unambiguously that nuclei from mature immune cells, from terminally differentiated neurons, and from malignant cancer cells can be reprogrammed and generate adult cloned mice.

MONOCLONAL MICE FROM MATURE IMMUNE CELLS

The monoclonal mice were generated from nuclei of peripheral lymphocytes where the genetic rearrangements of the immunoglobulin (Ig) and T-cell receptor (TCR) genes could be used as stable markers revealing the identity and differentiation state of the donor nucleus of a given clone. Because previous attempts to generate monoclonal mice had been unsuccessful, two-step cloning was used to produce first ES cells from cloned blastocysts, and in a second step, monoclonal mice from the cloned ES cells (Fig. 6). Animals generated from a B- or T-cell donor nucleus were viable and carried fully rearranged immunoglobulin or TCR genes in all tissues (Hochedlinger and Jaenisch 2002). As expected, the immune cells of the monoclonal mice expressed only those alleles of the Ig and TCR locus that had been productively rearranged in the respective donor cells used for nuclear transfer, and the rearrangement of other Ig or TCR genes was inhibited. These results unequivocally demonstrated that nuclei from terminally differentiated donor cells can be reprogrammed to pluripotency by nuclear cloning. The frequency of directly deriving cloned embryos from mature B and T cells (instead of the two-step procedure used in our experiments), although difficult to estimate, is likely significantly lower than that

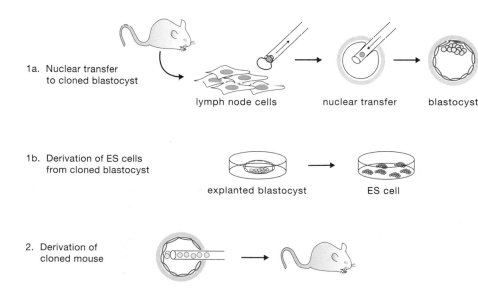

1a. Nuclear transfer to cloned blastocyst

      lymph node cells   nuclear transfer   blastocyst

1b. Derivation of ES cells from cloned blastocyst

      explanted blastocyst   ES cell

2. Derivation of cloned mouse

Figure 6. Two-step Procedure for the Derivation of Monoclonal Mice from Mature Lymphoid Donor Cells

(1) Nuclei from peripheral lymph node cells were transferred into enucleated eggs, and cloned blastocysts were derived. The blastocysts were explanted in vitro, and cloned ES cells were derived. (2) In a second step, monoclonal mice were derived by tetraploid complementation (Eggan et al. 2001; Hochedlinger and Jaenisch 2002).

of deriving clones from fibroblasts or cumulus cells (possibly less than 1 in 2000 operated embryos; Table 2). More recently, terminally differentiated NKT cells were directly cloned (Inoue et al. 2005). NKT cells, like B and T cells, have genetic rearrangements that allow retrospective identification of the differentiation state of the cells. However, although T cells and NKT cells are part of the same cell lineage, their respective nuclear transfer efficiency was significantly different (Hochedlinger and Jaenisch 2002; Inoue et al. 2005).

CLONED MICE FROM MATURE OLFACTORY NEURONS

In contrast to B or T cells, nuclei of postmitotic neurons have irreversibly exited the cell cycle as part of their program of differentiation. To assess whether the nucleus of a mature neuron could be reprogrammed to totipotency, fertile cloned adult mice were generated from postmitotic olfactory neurons using a similar approach as used for the generation of the monoclonal mice (cf. Fig. 6) (Eggan et al. 2004; Li et al. 2004). As summarized in Table 2, the efficiency of deriving cloned ES cells from olfactory neurons was in the same range as that for nuclei from immune cells. These observations indicate that a postmitotic neuronal nucleus can reenter the cell cycle and can be reprogrammed to pluripotency.

In the mouse, each of the two million cells in the olfactory epithelium expresses only one of approximately 1500 odorant receptor (OR) genes, such that the functional identity of a neuron is defined by the nature of the receptor it expresses (this is analogous to monoallelic expression of immune globulin or TCR genes in B or T cells discussed in Chapter 21). One mechanism to permit the stochastic choice of a single olfactory receptor could involve DNA rearrangements. The generation of mice cloned from a mature olfactory neuron made it possible to investigate whether olfactory receptor choice involves irreversible DNA rearrangements. If olfactory receptor choice involved DNA rearrangements, the prediction would be, in analogy with monoclonal mice described above, that a mouse cloned from a P2-expressing neuron would express this receptor in all olfactory neurons and the repertoire of receptor expression might be altered (these would be monosmic mice that can detect only one odorant) (Fig. 7). Alternatively, if OR choice involved a reversible epigenetic mechanism, the cloned animals should have an identical P2 expression pattern to the donor mouse and a normal repertoire of receptor expression. The analysis of olfactory receptor expression showed that the mechanism of receptor choice is fully reversible and does not involve genetic alterations as seen in the maturation of B and T cells (Eggan et al. 2004).

CANCER AND THE REVERSION OF THE MALIGNANT STATE BY NUCLEAR TRANSPLANTATION

The cloning of mice from terminally differentiated lymphocytes and postmitotic neurons demonstrated that nuclear transfer provides a tool to selectively reprogram the epigenetic state of a cellular genome without altering its genetic constitution. Cancer is caused by genetic as well as epigenetic alterations, but the impact of epigenetics on the malignant phenotype of a cancer cell has not been defined. Nuclear transplantation of cancer donor cells was used as an unbiased approach to assess the reversibility of the transformed state. Indeed, previous

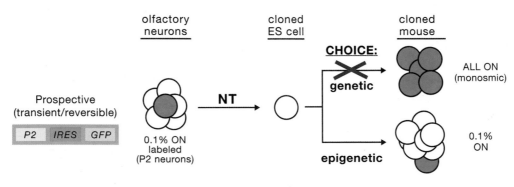

Figure 7. Nuclear Cloning of Mature Olfactory Neurons

Mice cloned from mature olfactory neurons had the normal repertoire of olfactory receptor (OR) expression with only 0.1% of neurons expressing the P2 receptor, as in the donor mice. P2 receptor expression was determined by using donor mice that had a GFP marker gene inserted in the P2 receptor gene. Results demonstrated that the choice of receptor expression is not determined by genetic alteration, but by a reversible epigenetic mechanism.

experiments with amphibians showed that nuclei from a kidney carcinoma cell could be reprogrammed to support early development to the tadpole stage (McKinnell 1962). A similar result was obtained in mice where nuclei from a medulloblastoma cell line were able to direct early development, albeit with low efficiency, resulting in arrested embryos (Li et al. 2003). However, these experiments did not unequivocally demonstrate that the clones were derived from cancer cells as opposed to contaminating nontransformed cells. When the nuclei of a variety of tumor cells, including leukemia, lymphoma, breast cancer, and melanoma cells, were transferred into enucleated mouse eggs, most were able to support preimplantation development into normal-appearing blastocysts (Hochedlinger et al. 2004). Therefore, the malignant phenotype of these tumor types could be suppressed by the oocyte environment and permitted apparently normal early development. However, only the genome from a RAS-induced melanoma model gave rise to a cloned ES cell line that was able to differentiate into most, if not all, somatic cell lineages in chimeric mice. However, because of genetic alterations present in the donor cells, all chimeras developed cancer. These findings demonstrated that the cancer nucleus after exposure to the egg cytoplasm directed differentiation of all lineages, indicating that the malignant phenotype of this cancer was largely determined by epigenetic alterations. A different conclusion was derived from the cloning of EC cell donor nuclei. In contrast to the somatic cancer nucleus, the malignant phenotype of the embryonal tumors was caused by genetic alterations, because it was not reversible by exposure to the egg cytoplasm (Blelloch et al. 2004).

4 Changes Associated with Nuclear Reprogramming

The strategies used for early development are very different in amphibians and mammals. For example, cleavage of the frog embryo is rapid, with about 30 minutes per cell cycle, in contrast to the mammalian embryo that has only cleaved once within 24 hours after fertilization. Additionally, the zygotic genome of the frog becomes expressed only after 12 mitotic cycles at the mid-blastula transition, in contrast to the genome of the mouse embryo that is activated at the 2-cell stage. Thus, it may not be surprising that the different developmental strategies used in amphibians and mammals affect reprogramming of the somatic donor nucleus. This chapter contrasts epigenetic reprogramming that takes place in normal development with reprogramming in

frog and mammalian clones. An interesting question is whether the epigenetic state of the somatic donor nucleus influences gene expression patterns in cloned embryos. This is designated as "epigenetic memory," as discussed later in the chapter.

4.1 Amphibians

REPROGRAMMING IN NORMAL DEVELOPMENT

In amphibians, the nuclei and chromosomes of oocytes and eggs are in a state entirely different from those of somatic cells. The germinal vesicle of an oocyte contains the maximally expanded lampbrush chromosomes that are intensely active in transcription (Callan and Lloyd 1960), apparently reflecting not only the high proportion of genes being transcribed, but also the dense packing of RNA polymerases on the DNA of most genes. This exceptional state of transcription is reached during early oogenesis and probably continues in the ovary throughout the life of an adult female. Mature sperm, conversely, are maximally condensed and entirely inactive in transcription. The usual chromosomal histones are replaced in sperm by protamines, which are exchanged in sperm nuclei that have entered an egg at fertilization, and sperm nuclei undergo immensely rapid decondensation within about 20 minutes. In amphibians, there are no equivalent processes to X-chromosome inactivation and imprinting that take place in mammals. A decrease in DNA methylation takes place from fertilization to the mid-blastula transition (5 hours), after which it gradually increases as development proceeds (Meehan 2003). In summary, substantial nuclear reprogramming events take place in normal development during gametogenesis and for a few hours immediately after fertilization.

The most obvious change undergone by transplanted nuclei in amphibians is a volume increase and dispersion of chromatin. This takes place more rapidly in the nuclei of embryonic cells compared to those of differentiated or adult cells. In each respect, the transplanted nuclei come to adopt the condition of nuclei normally resident in eggs or oocytes. Changes in nucleic acid synthesis also follow nuclear transplantation. DNA synthesis is rapidly induced by eggs in the nuclei of nondividing cells such as those of adult brain. Nuclear transplant embryos, derived from single nuclear transfers to eggs, synthesize ribosomal RNA and tRNA to the same extent as endogenous nuclei of embryos grown from fertilized eggs. The pattern of gene transcription is changed from that characteristic of donor cells to that of early embryos; for example, all gene transcription is switched off during cleavage of nuclear-transplant embryos, and is then reactivated in surviving nuclear-transplant embryos according to cell type. Muscle genes are expressed in the muscle of nuclear-transplant embryos, even when they were derived from intestine nuclei (Gurdon et al. 1984).

In the case of nuclear transfers to oocytes, extensive changes in transcription take place by transplanted nuclei in the absence of any DNA replication. For example, *Xenopus* kidney-derived nuclei extinguish kidney-specific genes and activate oocyte-specific genes. Some of the newly activated genes are embryo-specific, as is the case for mouse thymus nuclei, which express the stem-cell marker gene *Oct-4* but extinguish the thymus-specific gene *Thy-1* (Byrne et al. 2003). In conclusion, amphibian nuclear transfers to eggs or oocytes show an extensive reprogramming of gene transcription so that somatic cell nuclei (and in the case of eggs, their mitotic progeny) change their transcription to accord with that of the recipient cells.

REPROGRAMMING IN CLONES

It has long been thought that the most likely explanation for the increasing proportion of developmental abnormalities that are seen in amphibian nuclear transfer experiments with more differentiated donor cells relates to incomplete DNA replication. In normal *Xenopus* development, egg and sperm pronuclei commence chromosome duplication 20 minutes after fertilization, and it is complete 20 minutes later. In contrast, the nuclei of dividing cultured cells take about 6 hours to complete one round of DNA replication. It is not surprising, therefore, that transplanted somatic cell nuclei have often been seen to continue DNA synthesis for much longer than 40 minutes after nuclear injection, and to do so right up to the time when chromosomes condense for the first mitosis. As a result, chromosome replication can be incomplete, and incompletely replicated chromosomes torn apart, as transplanted nuclei are forced into their first mitosis. Broken chromosome fragments have been seen in nuclear transplant embryos (Di Berardino and Hoffner 1970), and this incompatibility between the rate of DNA replication and cell division in zygote as compared to somatic nuclei, resulting in aneuploidy, seems likely to account for many abnormalities of nuclear transplant embryo development, and especially for the high proportion of eggs that fail to undergo any regular cleavage at all; these can constitute up to 75% of all eggs receiving nuclei from nondividing differentiated cells. It has been noticed that the serial transfer of nuclei

from partially cleaved first-transfer embryos often yields normal tadpole development (see above). A good explanation for this is that the incubation of somatic cell nuclei in a mitosis-phase extract of eggs greatly increases the abundance of sites of the origin of DNA replication, thereby enabling such nuclei to complete chromosome replication more rapidly than can nuclei from terminally differentiated cells such as erythrocytes (Lemaitre et al. 2005).

Two other explanations may help to account for nuclear transplant abnormalities that arise after zygotic transcription starts at the mid-blastula transition. One is the quantitative irregularity of early zygotic gene activation (Byrne et al. 2003), and the other is the persistence of donor-specific gene expression in the incorrect germ line of nuclear transplant embryos (see below). However, it has not been demonstrated that these differences from normal gene expression are directly responsible for the observed developmental abnormalities.

MECHANISMS OF REPROGRAMMING

The abundance and large size of amphibian eggs and oocytes encourage attempts to understand the molecular basis of reprogramming. A preferred route is to obtain cell-free extracts that can reproduce in vitro the events that follow nuclear transfer to living eggs and oocytes. Depletion of extracts could identify necessary components. This approach has been particularly successful in identifying egg components that initiate DNA synthesis. Notable is the identification of nucleoplasmin (Laskey et al. 1978; Philpott et al. 1991), an abundant component of *Xenopus* eggs that can decondense sperm and promote histone protein exchange. These same processes take place when somatic nuclei are added to egg extracts (Dimitrov and Wolffe 1996; Tamada et al. 2006). Other egg extract components that may contribute to the nuclear reprogramming process include the remodeling complex ISWI (Kikyo et al. 2000) and the germ-cell proteins FRGY2 that function to reversibly disassemble nucleoli (Gonda et al. 2003). It has been suggested that by permeabilizing and resealing nuclei in extracts, the remodeling complex BRG-1 may have a role in eggs and early embryos (Hansis et al. 2004). These experiments are not easy to interpret because cell-free extracts are not yet known to be able to initiate transcription of nuclei. Therefore, the treatment of nuclei in vitro, followed by transfer to the living oocyte to test transcription (Byrne et al. 2003; Tamada et al. 2006), is the best that can be done.

At present, it seems that three steps are necessary for successful nuclear reprogramming: (1) the removal of epigenetic marks on DNA or protein that characterize the differentiated state; (2) the provision of necessary transcription factors for those genes that need to be newly expressed; and (3) the decondensation of chromatin, to give transcription factors access to the genes on which they act.

4.2 Mammals

Successive epigenetic reprogramming is an important aspect of normal development (Rideout et al. 2001). Changes of DNA methylation as well as of histones are imposed on the two parental genomes successively during gametogenesis. Following fertilization, the embryo's genome is further modified during cleavage and after implantation. Table 3 summarizes some of the epigenetic differences that distinguish cloned from normal animals as a result of faulty reprogramming. For the following discussion, we highlight the epigenetic differences between fertilized and cloned embryos at different stages of development. The stages of development that are depicted in Table 3 and that are discussed in sequence are (1) gametogenesis, (2) cleavage, (3) postimplantation, and (4) postnatal development.

GAMETOGENESIS

The most important epigenetic reprogramming in normal development occurs during gametogenesis, a process that renders both sperm and oocyte genomes "epigenetically competent" for subsequent fertilization and for faithful activation of the genes that are crucial for early development (Latham 1999). In cloning, this process is cut short, and most problems affecting the "normalcy" of cloned animals may be due to the inadequate reprogramming of the somatic nucleus following transplantation into the egg. Because the placenta is derived from the trophectoderm lineage that constitutes the first differentiated cell type of the embryo, one might speculate that reprogramming and differentiation into this early lineage are compromised in most cloned animals. Indeed, as summarized below, the fraction of abnormally expressed genes in cloned newborns is substantially higher in the placenta as compared to somatic tissues.

CLEAVAGE

During cleavage, a wave of genome-wide demethylation removes the epigenetic modification present in the zygote so that the DNA of the blastocyst is largely devoid of methylation. Between implantation and gastrulation, a wave of global de novo methylation reestablishes the overall methylation pattern, which is then maintained throughout life in the somatic cells of the animal. In cloned embryos, methylation of repetitive sequences is abnormal (Bourc'his et al. 2001; Dean et al. 2001; Kang et al. 2003, Mann et al. 2003). To investigate gene expression, the activity of "pluripotency genes" such as Oct-4 that are silent in somatic cells but active in embryonic cells was examined in cloned embryos. Strikingly, the reactivation of Oct-4 and of "Oct-4-like" genes was shown to be faulty and random in a large fraction of somatic clones (Boiani et al. 2002; Bortvin et al. 2003). Because embryos lacking Oct-4 arrest early in development, incomplete reactivation of Oct-4-like genes in clones might be causal to the frequent failure of the great majority of nuclear transfer embryos to survive the postimplantation period. Moreover, a number of studies have detected abnormal DNA methylation in cloned

Table 3. Normal versus cloned embryos

Stage	Normal embryos	Cloned embryos
Gametogenesis	genome "competent" for activation of "early" genes, establishment of imprints	none
Cleavage	global demethylation of DNA	abnormal methylation of DNA
	activation of embryonic ("Oct4-like") genes	stochastic / faulty activation of "Oct4-like" genes
Postimplantation	telomere length adjustment	normal
	global de novo DNA methylation, X inactivation	abnormal in some cloned animals
	normal imprinting and gene expression	abnormal imprinting, global gene dysregulation
Postnatal	normal animal	large offspring syndrome, premature death, etc.

embryos. Although it is still an unresolved question to what extent the epigenetic modification of chromatin structure and DNA methylation, which occurs in normal development, needs to be mimicked for nuclear cloning to succeed, the available evidence is entirely consistent with faulty epigenetic reprogramming causing abnormal gene expression in cloned animals.

POSTIMPLANTATION DEVELOPMENT

Following implantation and prior to gastrulation, three key events shape the epigenetic state of the embryo's genome: (1) A wave of global de novo methylation reestablishes the overall methylation pattern that is characteristic of the adult and that is then maintained throughout life in the somatic cells (Dean et al. 2003); (2) dosage compensation in female embryos is accomplished by the random inactivation of one of the two X chromosomes; (3) the telomeres are adjusted to a length that is characteristic of the somatic cells. Because all these events are only initiated in the postzygotic embryo, little disturbance in the regulation of these epigenetic events might be expected in cloned animals. However, lower global methylation levels were seen in cloned bovine fetuses but not in postnatal cows (Cezar et al. 2003). X inactivation was random and undisturbed in healthy but not in abnormal cloned mouse fetuses (Eggan et al. 2000; Senda et al. 2004; Nolen et al. 2005). However, it is not clear whether these disturbances are causally involved in abnormal clone development rather than being a consequence of abnormal reprogramming during preimplantation development. In contrast, telomere length adjustment is faithfully accomplished in cloned cows and mice (Lanza et al. 2000; Tian et al. 2000; Wakayama et al. 2000; Betts et al. 2001) and thus would not be expected to impair survival of cloned animals.

POSTNATAL DEVELOPMENT

The most extensive analysis of gene expression has been performed in newborn cloned mice. Expression profiling showed that 4–5% of the genome and 30–50% of imprinted genes are abnormally expressed in placentas of newborn cloned mice (Humpherys et al. 2002; Kohda et al. 2005). This argues that mammalian development is surprisingly tolerant to widespread gene dysregulation and that compensatory mechanisms assure survival of some clones to birth. However, the results suggest that even surviving clones may have subtle defects that, although not severe enough to jeopardize immediate survival, will cause an abnormal phenotype at a later age.

5 Epigenetic Memory

Two kinds of epigenetic modifications to the genome are known to take place in vertebrate development. These include a methylated cytosine in many regions of DNA where a CpG is present, and various modifications of histone tails. These changes are acquired during gametogenesis and early development and are closely associated with the activity or inactivity of genes. It would therefore be expected that these epigenetic modifications would be reversed by nuclear transfer, or if not, that they may help to account for some of the failures of nuclear transplant embryo development.

In amphibians, some insight into DNA demethylation has been achieved in experiments where mammalian somatic cell nuclei were injected into *Xenopus* oocytes (Simonsson and Gurdon 2004). The mouse Oct-4 promoter is methylated in adult thymus cells where *Oct-4* is not expressed. However, the promoter region but not the enhancer region of the regulatory part of this gene was demethylated when thymus nuclei were injected into oocytes, a result that shows the selectivity of the demethylation process. When complete nuclei were injected, the demethylation of DNA seemed to precede induced *Oct-4* transcription. It is likely that a DNA demethylase activity is a special property of oocytes (see above), and that the demethylation of the promoter DNA of developmentally repressed genes may also be an important and necessary step when somatic cell nuclei are reprogrammed in egg nuclear transfer experiments. Changes in histone modifications have not yet been examined in amphibian nuclear transfer experiments.

Another design of amphibian nuclear transfer experiment has shown that the epigenetic state of somatic cells is by no means always reversed. In view of the failure of nuclear transplants, even from early tail-bud endoderm donors of *R. pipiens* (above), Briggs and King (1957) asked whether the morphology of abnormal embryos reflected their origin; they described a preferential survival of endoderm tissues in embryos of endodermal nuclear origin and called this an "endoderm syndrome." It was pointed out, however, that the endoderm differentiates later than other germ layers, and that this might account for its better survival (Gurdon 1963). Indeed, Simnett (1964) reported that nuclear transplant embryos of neural origin also showed the same preferential survival of their endoderm. The same question has recently been approached again, using cell-type-specific gene markers. In these experiments (Ng and Gurdon 2005), nuclei from the neurectoderm or endoderm, already expressing the cell-

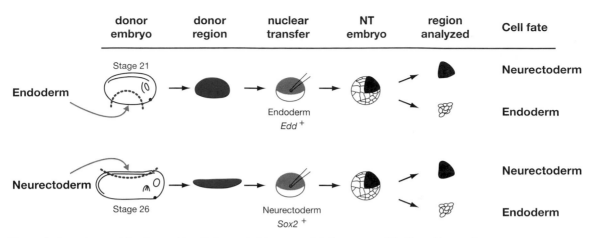

Figure 8. Experimental Design to Test Epigenetic Memory of Cell-type-specific Gene Expression

Donor nuclei are taken from stage-21 endoderm or stage-26 neurectoderm cells. The resulting nuclear transfer embryos usually form partial blastulae, the normally cleaved parts of which are divided into the future neurectodermal or endodermal regions, and analyzed for gene expression. See Table 4. (Reprinted, with permission, from Ng and Gurdon 2005.)

type-specific markers Sox2 or endodermin, respectively, were transplanted to enucleated eggs; the resulting nuclear transplant embryos were divided into neurectodermal or endodermal parts, and these parts were tested for the same Sox2 or endodermin markers (Fig. 8 and Table 4). It was found that both genes were preferentially expressed in the inappropriate cell type. For example, over half of the embryos of neurectodermal origin overexpressed the neural marker Sox2 in their *endo*derm cells. In some cloned embryos, there appeared to have been no reduction in the level of Sox2 gene expression compared to that of the donor cells. Transplanted nuclei, like those of normal embryos, are wholly inactive in transcription until the blastula stage. Therefore, remarkably, the active state of gene transcription established in the course of cell differentiation can be maintained in cloned embryos, in the complete absence of the conditions that induced that gene for more than 12 mitotic cell divisions (from egg to blastula). This striking example of epigenetic memory is seen in some nuclear transplant embryos but is wholly absent in others, in which gene expression has been successfully reprogrammed.

6 Medical Implications of Nuclear Transplantation

It is important to distinguish between "reproductive cloning" and "nuclear transplantation therapy" (also referred to as SCNT or therapeutic cloning). In reproductive cloning, an embryo is generated by transfer of a somatic nucleus into an enucleated egg with the goal to create a cloned individual. In contrast, the purpose of nuclear transplantation therapy is to generate an embry-

Table 4. Epigenetic memory of cell-type-specific gene expression

			Gene expression (%)	
			Edd	Sox2
Endoderm nuclei (*edd*⁺) ⟶ NT embryo		neurectoderm	45	5
		endoderm	12	0
Neurectoderm nuclei (*Sox2*⁺) ⟶ NT embryo		neurectoderm	6	22
		endoderm	0	81

The percent values represent the proportion of nuclear transplant embryos (assayed individually by RT-PCR) that express the genes *Edd* or *Sox2* at two or more times greater than the normal (or background) level. (Data from Ng and Gurdon 2005.)

onic stem cell line (referred to as ntES cells) that is "tailored" to the needs of a patient who served as the nuclear donor (Hochedlinger and Jaenisch 2003). The ntES cells could be used as a source of functional cells that would be suitable for treating an underlying disease by transplantation. Figure 9 juxtaposes normal development from a fertilized embryo, reproductive cloning, and therapeutic cloning.

6.1 Reproductive Cloning

As outlined above, all evidence obtained from the cloning of eight different mammalian species indicates that the production of normal individuals by nuclear transfer faces major hurdles. It is a key question in the public debate whether it would ever be possible to produce a normal individual by nuclear cloning. The available evidence suggests that it may be difficult if not impossible to produce normal clones for the following reasons: (1) As summarized above, all analyzed clones at birth showed dysregulation of hundreds of genes. Nevertheless, the development of clones to birth and beyond despite widespread epigenetic abnormalities suggests that mammalian development can tolerate dysregulation of many genes. (2) Some clones survive to adulthood by compensating for gene dysregulation. Although this "compensation" assures *survival*, it may not prevent maladies that become manifest at later ages. Therefore, most if not all clones are expected to have at least subtle abnormalities that may not be severe enough to result in an obvious phenotype at birth but will cause serious problems later, as seen in aged mice. Different clones may just differ in the extent of abnormal gene

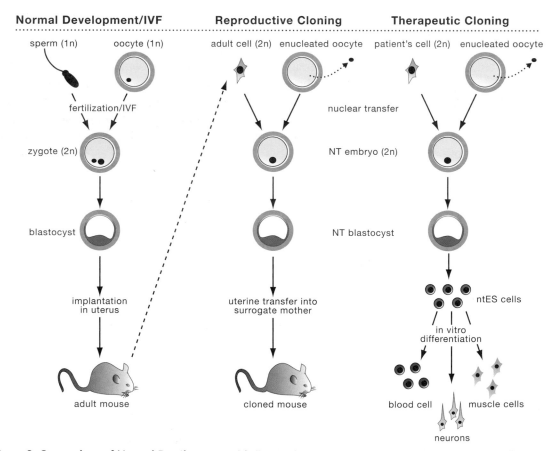

Figure 9. Comparison of Normal Development with "Reproductive Cloning" and "Therapeutic Cloning"

During normal development (*left*), a haploid (1n) sperm cell fertilizes a haploid oocyte to form a diploid (2n) zygote that undergoes cleavage to become a blastocyst embryo. Blastocysts implant in the uterus and ultimately give rise to a newborn animal. During "reproductive cloning" (*center*), the diploid nucleus of an adult donor cell is introduced into an enucleated oocyte recipient which, after artificial activation, divides into a cloned blastocyst. Upon transfer into surrogate mothers, a few of the cloned blastocysts will give rise to a newborn clone. In contrast, the derivation of ntES cells by nuclear transfer (*right*) requires the explantation of cloned blastocysts in culture to derive an ES cell line that can be differentiated in vitro into potentially any cell type of the body for research or therapeutic purposes. (Reprinted, with permission, from Hochedlinger and Jaenisch 2003 [© Massachusetts Medical Society].)

expression: If the key "*Oct-4 like*" genes are not activated, clones die immediately after implantation. If those genes are activated, the clone may survive to birth and beyond. These considerations argue that cloned animals, even if appearing "normal" at superficial inspection, may not be so but may harbor subtle abnormalities that become phenotypically manifest only at later ages (Jaenisch 2004). These considerations preclude the application of this approach as a potential human reproductive technology.

6.2 Therapeutic Application of Nuclear Transplantation

Immune rejection is a frequent complication of allogeneic organ transplantation due to immunological incompatibility. To treat this "host versus graft" disease, immunosuppressive drugs are routinely given to transplant recipients, a treatment that has serious side effects. Embryonic stem cells derived by nuclear transplantation are genetically identical to the patient's cells, thus eliminating the risk of immune rejection and the requirement for immunosuppression. Most importantly, protocols are being developed that allow the generation of functional cells such as neurons, muscle cells, and islet cells that can be used for therapy of patients afflicted with serious disorders such as Parkinson's, heart failure, or diabetes. Moreover, embryonic stem cells provide a renewable source of replacement tissue, allowing repeated therapy whenever needed.

Indeed, the feasibility of therapeutic cloning has been demonstrated in an animal model of disease. For this, a mouse strain carrying a deletion of the *Rag2* gene was used as a "patient" (Rideout et al. 2002). *Rag2* mutant mice suffer from severe combined immune deficiency (SCID) due to a mutation in the gene catalyzing immune receptor rearrangements in lymphocytes. These mice are devoid of mature B and T cells, a condition resembling a human disorder ("bubble babies"). Figure 10 summarizes the steps involved in this experiment. In a first step, nuclei of somatic (fibroblast) donor cells from the tails of *Rag2*-deficient mice were injected into enucleated eggs. The resultant embryos were cultured to the blastocyst stage, and autologous ES cells were isolated. Subsequently, one of the mutant *Rag2* alleles was targeted by homologous recombination in ES cells to restore normal gene structure. To obtain somatic cells for treatment, these ES cells were differentiated into embryoid bodies (embryo-like structures that contain various somatic cell types) and further into hematopoietic precursors by expressing *HoxB4*. Resulting hematopoietic precursors were transplanted into irradiated *Rag2*-deficient animals to treat the disease. The cells generated functional B and T cells which had undergone proper rearrangements of their immunoglobulin and T-cell-receptor alleles as well as serum immunoglobulins in the transplanted *Rag2* mice. This experiment demonstrated that nuclear transfer, in combination with gene therapy, can be used to treat a genetic disorder. Consequently, therapeutic cloning should be applicable to other diseases where the genetic lesion is known, such as sickle cell anemia or

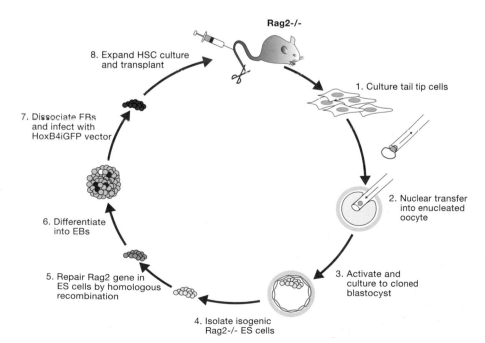

Rag2-/-

8. Expand HSC culture and transplant

7. Dissociate EBs and infect with HoxB4iGFP vector

6. Differentiate into EBs

5. Repair Rag2 gene in ES cells by homologous recombination

4. Isolate isogenic Rag2-/- ES cells

3. Activate and culture to cloned blastocyst

2. Nuclear transfer into enucleated oocyte

1. Culture tail tip cells

Figure 10. Scheme for Therapeutic Cloning Combined with Gene and Cell Therapy

A piece of tail from a mouse homozygous for the recombination activating gene 2 (*Rag2*) mutation was removed and cultured. After fibroblast-like cells grew out, they were used as donors for nuclear transfer by direct injection into enucleated MII oocytes using a Piezoelectric driven micromanipulator. Embryonic stem (ES) cells isolated from the nuclear transfer-derived blastocysts were genetically repaired by homologous recombination. After repair, the ntES cells were differentiated in vitro into embryoid bodies (EBs), infected with the HoxB4iGFP retrovirus, expanded, and injected into the tail vein of irradiated, *Rag2*-deficient mice. (Reprinted, with permission, from Hochedlinger and Jaenisch 2003 [© Massachusetts Medical Society].)

beta-thalassemia. It is an unresolved issue whether nuclear transplantation in humans will be as efficient as bovine, or as inefficient as murine, cloning.

The use of nuclear cloning to generate "customized" ES cells for tissue repair is controversial because the very generation of the ES cells would, so goes the argument, necessarily involve the destruction of potential human life. As a possible solution to this ethical dilemma, altered nuclear transfer (ANT) was suggested as a modification of the nuclear transfer procedure (Hurlbut 2005). ANT involves the disabling of a gene in the somatic donor cells that is essential for placental development and thus prevents the formation of a fetus because the ANT blastocyst would be unable to implant (and thus no potential human life would be destroyed) but would still allow the generation of "customized" ES cells. Using Cdx2 as a target gene, a proof-of-principle experiment in mice verified the ANT approach (Meissner and Jaenisch 2006). It remains to be seen whether the ANT modification will satisfy those who are opposed to the generation of ES cells from cloned human blastocysts.

6.3 Reproductive Versus Therapeutic Cloning: What Is the Difference?

Why is faulty reprogramming problematic for reproductive cloning but not for therapeutic applications? The most important reason for this seeming paradox is that, in contrast to reproductive cloning, the therapeutic use of nuclear transfer does not require the formation of a fetus, relying instead on the direct differentiation of functional cells in culture. Because there is no requirement for the development of a fetus, the functionality of the differentiated cells that result from this process would not be expected to be affected by the disturbed imprinting that contributes substantially to the developmental failure of clones (Jaenisch 2004). Because ES cells derived from fertilized embryos are able to participate in the generation of all normal embryonic tissues, ES cells generated through nuclear transfer should have a similar potential to generate the full range of normal tissues. Indeed, all the available evidence is consistent with the conclusion that ES cells derived from cloned embryos are biologically and molecularly indistinguishable from ES cells derived from fertilized embryos (Brambrink et al. 2006). Thus, if human ES cells derived from IVF embryos are appropriate to treat diseases, so are "customized" ES cells derived by nuclear cloning from the cells of a patient.

References

Betts D., Bordignon V., Hill J., Winger Q., Westhusin M., Smith L., and King W. 2001. Reprogramming of telomerase activity and rebuilding of telomere length in cloned cattle. *Proc. Natl. Acad. Sci.* **98:** 1077–1082.

Blelloch R.H., Hochedlinger K., Yamada Y., Brennan C., Kim M., Mintz B., Chin L., and Jaenisch R. 2004. Nuclear cloning of embryonal carcinoma cells. *Proc. Natl. Acad. Sci.* **101:** 13985–13990.

Boiani M., Eckardt S., Scholer H.R., and McLaughlin K.J. 2002. Oct4 distribution and level in mouse clones: Consequences for pluripotency. *Genes Dev.* **16:** 1209–1219.

Bolton V.N., Oades P.J., and Johnson M.H. 1984. The relationship between cleavage, DNA replication, and gene expression in the mouse 2-cell embryo. *J. Embryol. Exp. Morphol.* **79:** 139–163.

Bortvin A., Eggan K., Skaletsky H., Akutsu H., Berry D.L., Yanagimachi R., Page D.C., and Jaenisch R. 2003. Incomplete reactivation of *Oct4*-related genes in mouse embryos cloned from somatic nuclei. *Development* **130:** 1673–1680.

Bourc'his D., Le Bourhis D., Patin D., Niveleau A., Comizzoli P., Renard J.P., and Viegas-Pequignot E. 2001. Delayed and incomplete reprogramming of chromosome methylation patterns in bovine cloned embryos. *Curr. Biol.* **11:** 1542–1546.

Brambrink T., Hochedlinger K., Bell G., and Jaenisch R. 2006. ES cells derived from cloned and fertilized blastocysts are transcriptionally and functionally indistinguishable. *Proc. Natl. Acad. Sci.* **103:** 933–938.

Briggs R. and King T.J. 1952. Transplantation of living nuclei from blastula cells into enucleated frogs' eggs. *Proc. Natl. Acad. Sci.* **38:** 455–463.

———. 1957. Changes in the nuclei of differentiating endoderm cells as revealed by nuclear transplantation. *J. Morphol.* **100:** 269–312.

Bromhall J.D. 1975. Nuclear transplantation in the rabbit egg. *Nature* **258:** 719–722.

Byrne J.A., Simonsson S., Western P.S., and Gurdon J.B. 2003. Nuclei of adult mammalian somatic cells are directly reprogrammed to *oct-4* stem cell gene expression by amphibian oocytes. *Curr. Biol.* **13:** 1206–1213.

Calarco P.G. and McLaren A. 1976. Ultrastructural observations of preimplantation stages of the sheep. *J. Embryol. Exp. Morphol.* **36:** 609–622.

Callan H.G. and Lloyd L. 1960. Lampbrush chromosomes of crested newts *Triturus cristatus* (Laurenti). *Philos. Trans. R. Soc. Lond. B Biol. Sci.* **243:** 135–219.

Camous S., Kopecny V., and Flechon J.E. 1986. Autoradiographic detection of the earliest stage of [3H]-uridine incorporation into the cow embryo. *Biol. Cell* **58:** 195–200.

Campbell K.H., Loi P., Otaegui P.J., and Wilmut I. 1996a. Cell cycle coordination in embryo cloning by nuclear transfer. *Rev. Reprod.* **1:** 40–46.

Campbell K.H., McWhir J., Ritchie W.A., and Wilmut I. 1996b. Sheep cloned by nuclear transfer from a cultured cell line. *Nature* **380:** 64–66.

Campbell K.H., Alberio R., Choi I., Fisher P., Kelly R.D., Lee J.H., and Maalouf W. 2005. Cloning: Eight years after Dolly. *Reprod. Domest. Anim.* **40:** 256–268.

Cezar G.G., Bartolomei M.S., Forsberg E.J., First N.L., Bishop M.D., and Eilertsen K.J. 2003. Genome-wide epigenetic alterations in cloned bovine fetuses. *Biol. Reprod.* **68:** 1009–1014.

Cibelli J.B., Stice S.L., Golueke P.J., Kane J.J., Jerry J., Blackwell C., Ponce de Leon F. A., and Robl J.M. 1998. Cloned transgenic calves produced from nonquiescent fetal fibroblasts. *Science* **280:** 1256–1258.

Dean W., Santos F., and Reik W. 2003. Epigenetic reprogramming in early mammalian development and following somatic nuclear transfer. *Semin. Cell Dev. Biol.* **14:** 93–100.

Dean W., Santos F., Stojkovic M., Zakhartchenko V., Walter J., Wolf E., and Reik W. 2001. Conservation of methylation reprogramming in mammalian development: Aberrant reprogramming in cloned embryos. *Proc. Natl. Acad. Sci.* **98:** 13734–13738.

Di Berardino M.A. and Hoffner N. 1970. Origin of chromosomal abnormalities in nuclear transplants—A reevaluation of nuclear differentiation and nuclear equivalence in amphibians. *Dev. Biol.* **23:** 185–209.

Dimitrov S. and Wolffe A.P. 1996. Remodeling somatic nuclei in *Xenopus laevis* egg extracts: Molecular mechanisms for the selective release of histones H1 and H1⁰ from chromatin and the acquisition of transcriptional competence. *EMBO J.* **15:** 5897–5906.

Eggan K., Akutsu H., Hochedlinger K., Rideout W., Yanagimachi R., and Jaenisch R. 2000. X-Chromosome inactivation in cloned mouse embryos. *Science* **290:** 1578–1581.

Eggan K., Akutsu H., Loring J., Jackson-Grusby L., Klemm M., Rideout W.M., III, Yanagimachi R., and Jaenisch R. 2001. Hybrid vigor, fetal overgrowth, and viability of mice derived by nuclear cloning and tetraploid embryo complementation. *Proc. Natl. Acad. Sci.* **98:** 6209–6214.

Eggan K., Baldwin K., Tackett M., Osborne J., Gogos J., Chess A., Axel R., and Jaenisch R. 2004. Mice cloned from olfactory sensory neurons. *Nature* **428:** 44–49.

Fischberg M., Gurdon J.B., and Elsdale T.R. 1958. Nuclear transplantation in *Xenopus laevis. Nature* **181:** 424.

Gonda K., Fowler J., Katoku-Kikyo N., Haroldson J., Wudel J., and Kikyo N. 2003. Reversible disassembly of somatic nucleoli by the germ cell proteins FRGY2a and FRGY2b. *Nat. Cell Biol.* **5:** 205–210.

Gurdon J.B. 1960. The developmental capacity of nuclei taken from differentiating endoderm cells of *Xenopus laevis. J. Embryol. Exp. Morphol.* **8:** 505–526.

———. 1962. The developmental capacity of nuclei taken from intestinal epithelium cells of feeding tadpoles. *J. Embryol. Exp. Morphol.* **10:** 622–640.

———. 1963. Nuclear transplantation in Amphibia and the importance of stable nuclear changes in cellular differentiation. *Q. Rev. Biol.* **38:** 54–78.

———. 1964. The transplantation of living cell nuclei. *Adv. Morphog.* **4:** 1–43.

Gurdon J.B. and Uehlinger V. 1966. "Fertile" intestine nuclei. *Nature* **210:** 1240–1241.

Gurdon J.B., Elsdale T.R., and Fischberg M. 1958. Sexually mature individuals of *Xenopus laevis* from the transplantation of single somatic nuclei. *Nature* **182:** 64–65.

Gurdon J.B., Brennan S., Fairman S., and Mohun T.J. 1984. Transcription of muscle-specific actin genes in early *Xenopus* development: Nuclear transplantation and cell dissociation. *Cell* **38:** 691–700.

Hansis C., Barreto G., Maltry N., and Niehrs C. 2004. Nuclear reprogramming of human somatic cells by *Xenopus* egg extract requires BRG1. *Curr. Biol.* **14:** 1475–1480.

Hill J.R., Burghardt R.C., Jones K., Long C.R., Looney C.R., Shin T., Spencer T.E., Thompson J.A., Winger Q.A., and Westhusin M.E. 2000. Evidence for placental abnormality as the major cause of mortality in first-trimester somatic cell cloned bovine fetuses. *Biol. Reprod.* **63:** 1787–1794.

Hochedlinger K. and Jaenisch R. 2002. Monoclonal mice generated by nuclear transfer from mature B and T donor cells. *Nature* **415:** 1035–1038.

———. 2003. Nuclear transplantation, embryonic stem cells, and the potential for cell therapy. *N. Engl. J. Med.* **349:** 275–286.

Hochedlinger K., Blelloch R., Brennan C., Yamada Y., Kim M., Chin L., and Jaenisch R. 2004. Reprogramming of a melanoma genome by nuclear transplantation. *Genes Dev.* **18:** 1875–1885.

Humpherys D., Eggan K., Akutsu H., Friedman A., Hochedlinger K., Yanagimachi R., Lander E., Golub T.R., and Jaenisch R. 2002. Abnormal gene expression in cloned mice derived from embryonic stem cell and cumulus cell nuclei. *Proc. Natl. Acad. Sci.* **99:** 12889–12894.

Humpherys D., Eggan K., Akutsu H., Hochedlinger K., Rideout W., Biniszkiewicz D., Yanagimachi R., and Jaenisch R. 2001. Epigenetic instability in ES cells and cloned mice. *Science* **293:** 95–97.

Hurlbut W.B. 2005. Altered nuclear transfer as a morally acceptable means for the procurement of human embryonic stem cells. *Perspect. Biol. Med.* **48:** 211–228.

Inoue K., Wakao H., Ogonuki N., Miki H., Seino K., Nambu-Wakao R., Noda S., Miyoshi H., Koseki H., Taniguchi M., and Ogura A. 2005. Generation of cloned mice by direct nuclear transfer from natural killer T cells. *Curr. Biol.* **15:** 1114–1118.

Jaenisch R. 2004. Human cloning—The science and ethics of nuclear transplantation. *N. Engl. J. Med.* **351:** 2787–2791.

Kang Y.K., Lee K.K., and Han Y.M. 2003. Reprogramming DNA methylation in the preimplantation stage: Peeping with Dolly's eyes. *Curr. Opin. Cell Biol.* **15:** 290–295.

Kikyo N., Wade P.A., Guschin D., Ge H., and Wolffe A.P. 2000. Active remodelling of somatic nuclei in egg cytoplasm by the nucleosomal ATPase ISWI. *Science* **289:** 2360–2362.

Kohda T., Inoue K., Ogonuki N., Miki H., Naruse M., Kaneko-Ishino T., Ogura A., and Ishino F. 2005. Variation in gene expression and aberrantly regulated chromosome regions in cloned mice. *Biol. Reprod.* **73:** 1302–1311.

Lanza R., Cibelli J., Blackwell C., Cristofalo V., Francis M., Baerlocher G., Mak J., Schertzer M., Chavez E., Sawyer N., et al. 2000. Extension of cell life-span and telomere length in animals cloned from senescent somatic cells. *Science* **288:** 665–669.

Laskey R.A. and Gurdon J.B. 1970. Genetic content of adult somatic cells tested by nuclear transplantation from cultured cells. *Nature* **228:** 1332–1334.

Laskey R.A., Honda B.M., Mills A.D., and Finch J.T. 1978. Nucleosomes are assembled by an acidic protein which binds histones and transfers them to DNA. *Nature* **275:** 416–420.

Latham K.E. 1999. Mechanisms and control of embryonic genome activation in mammalian embryos. *Int. Rev. Cytol.* **193:** 71–124.

Lemaitre J.M., Danis E., Pasero P., Vassetzky Y., and Mechali M. 2005. Mitotic remodeling of the replicon and chromosome structure. *Cell* **123:** 787–801.

Li J., Ishii T., Feinstein P., and Mombaerts P. 2004. Odorant receptor gene choice is reset by nuclear transfer from mouse olfactory sensory neurons. *Nature* **428:** 393–399.

Li L., Connelly M.C., Wetmore C., Curran T., and Morgan J.I. 2003. Mouse embryos cloned from brain tumors. *Cancer Res.* **63:** 2733–2736.

Mann M.R., Chung Y.G., Nolen L.D., Verona R.I., Latham K.E., and Bartolomei M.S. 2003. Disruption of imprinted gene methylation and expression in cloned preimplantation stage mouse embryos. *Biol. Reprod.* **69:** 902–914.

McGrath J. and Solter D. 1984a. Completion of mouse embryogenesis requires both the maternal and paternal genomes. *Cell* **37:** 179–187.

McGrath J. and Solter D. 1984b. Inability of mouse blastomere nuclei transferred to enucleated zygotes to support development in vitro. *Science* **226:** 1317–1319.

McKinnell R.G. 1962. Development of *Rana pipiens* eggs transplanted with Lucke tumor cells. *Am. Zool.* **2:** 430–431.

Meehan R.R. 2003. DNA methylation in animal development. *Semin. Cell Dev. Biol.* **14:** 53–65.

Meissner A. 2006. "Conditional RNA interference, altered nuclear transfer and genome-wide DNA methylation analysis." Ph.D. thesis. University Saarland, Saarbrücken, Germany.

Meissner A. and Jaenisch R. 2006. Generation of nuclear transfer-derived pluripotent ES cells from cloned *Cdx2*-deficient blastocysts. *Nature* **439:** 212–215.

Ng R.K. and Gurdon J.B. 2005. Epigenetic memory of active gene transcription is inherited through somatic cell nuclear transfer. *Proc. Natl. Acad. Sci.* **102:** 1957–1962.

Nolen L.D., Gao S., Han Z., Mann M.R., Gie Chung Y., Otte A.P., Bartolomei M.S., and Latham K.E. 2005. X chromosome reactivation and regulation in cloned embryos. *Dev. Biol.* **279:** 525–540.

Ogonuki N., Inoue K., Yamamoto Y., Noguchi Y., Tanemura K., Suzuki O., Nakayama H., Doi K., Ohtomo Y., Satoh M., et al. 2002. Early death of mice cloned from somatic cells. *Nat. Genet.* **30:** 253–254.

Ogura A., Inoue K., Ogonuki N., Noguchi A., Takano K., Nagano R., Suzuki O., Lee J., Ishino F., and Matsuda J. 2000. Production of male cloned mice from fresh, cultured, and cryopreserved immature Sertoli cells. *Biol. Reprod.* **62:** 1579–1584.

Philpott A., Leno G.H., and Laskey R.A. 1991. Sperm decondensation in *Xenopus* egg cytoplasm is mediated by nucleoplasmin. *Cell* **65:** 569–578.

Prather R.S., Sims M.M., and First N.L. 1989. Nuclear transplantation in early pig embryos. *Biol. Reprod.* **41:** 414–418.

Rideout W.M., III, Eggan K., and Jaenisch R. 2001. Nuclear cloning and epigenetic reprogramming of the genome. *Science* **293:** 1093–1098.

Rideout W.M., III, Hochedlinger K., Kyba M., Daley G.Q., and Jaenisch R. 2002. Correction of a genetic defect by nuclear transplantation and combined cell and gene therapy. *Cell* **109:** 17–27.

Rideout W.M., III, Wakayama T., Wutz A., Eggan K., Jackson-Grusby L., Dausman J., Yanagimachi R., and Jaenisch R. 2000. Generation of mice from wild-type and targeted ES cells by nuclear cloning. *Nat. Genet.* **24:** 109–110.

Robi J.M., Prather R., Barnes F., Eyestone W., Northey D., Gilligan B., and First N.L. 1987. Nuclear transplantation in bovine embryos. *J. Anim. Sci.* **64:** 642–647.

Senda S., Wakayama T., Yamazaki Y., Ohgane J., Hattori N., Tanaka S., Yanagimachi R., and Shiota K. 2004. Skewed X-inactivation in cloned mice. *Biochem. Biophys. Res. Commun.* **321:** 38–44.

Simnett J.D. 1964. The development of embryos derived from the transplantation of neural ectoderm cell nuclei in *Xenopus laevis*. *Dev. Biol.* **10:** 467–486.

Simonsson S. and Gurdon J. 2004. DNA demethylation is necessary for the epigenetic reprogramming of somatic cell nuclei. *Nat. Cell Biol.* **6:** 984–990.

Surani M.A., Barton S.C., and Norris M.L. 1984. Development of reconstituted mouse eggs suggests imprinting of the genome during gametogenesis. *Nature* **308:** 548–550.

Tamada H., Van Thuan N., Reed P., Nelson D., Katoku-Kikyo N., Wudel J., Wakayama T., and Kikyo N. 2006. Chromatin decondensation and nuclear reprogramming by nucleoplasmin. *Mol. Cell. Biol.* **26:** 1259–1271.

Tamashiro K.L., Wakayama T., Akutsu H., Yamazaki Y., Lachey J.L., Wortman M.D., Seeley R.J., D'Alessio D.A., Woods S.C., Yanagimachi R., and Sakai R.R. 2002. Cloned mice have an obese phenotype not transmitted to their offspring. *Nat. Med.* **8:** 262–267.

Tanaka S., Oda M., Toyoshima Y., Wakayama T., Tanaka M., Yoshida N., Hattori N., Ohgane J., Yanagimachi R., and Shiota K. 2001. Placentomegaly in cloned mouse concepti caused by expansion of the spongiotrophoblast layer. *Biol. Reprod.* **65:** 1813–1821.

Tian X.C., Xu J., and Yang X. 2000. Normal telomere lengths found in cloned cattle. *Nat. Genet.* **26:** 272–273.

Wakayama S., Ohta H., Kishigami S., Van Thuan N., Hikichi T., Mizutani E., Miyake M., and Wakayama T. 2005. Establishment of male and female nuclear transfer embryonic stem cell lines from different mouse strains and tissues. *Biol. Reprod.* **72:** 932–936.

Wakayama T. and Yanagimachi R. 1999. Cloning of male mice from adult tail-tip cells. *Nat. Genet.* **22:** 127–128.

———. 2001. Mouse cloning with nucleus donor cells of different age and type. *Mol. Reprod. Dev.* **58:** 376–383.

Wakayama T., Perry A.C., Zuccotti M., Johnson K.R., and Yanagimachi R. 1998. Full-term development of mice from enucleated oocytes injected with cumulus cell nuclei. *Nature* **394:** 369–374.

Wakayama T., Rodriguez I., Perry A.C., Yanagimachi R., and Mombaerts P. 1999. Mice cloned from embryonic stem cells. *Proc. Natl. Acad. Sci.* **96:** 14984–14989.

Wakayama T., Shinkai Y., Tamashiro K.L., Niida H., Blanchard D.C., Blanchard R.J., Ogura A., Tanemura K., Tachibana M., Perry A.C., et al. 2000. Cloning of mice to six generations. *Nature* **407:** 318–319.

Wilmut I., Schnieke A.E., McWhir J., Kind A.J., and Campbell K.H. 1997. Viable offspring derived from fetal and adult mammalian cells. *Nature* **385:** 810–813.

Young L.E., Sinclair K.D., and Wilmut I. 1998. Large offspring syndrome in cattle and sheep. *Rev. Reprod.* **3:** 155–163.

Epigenetics and Human Disease

Huda Y. Zoghbi[1] and Arthur L. Beaudet[2]

[1]Howard Hughes Medical Institute, and [2]Department of Molecular and Human Genetics,
Baylor College of Medicine, Houston, Texas 77030

CONTENTS

1. Introduction, 437

2. Studies of Human Cases Uncover the Role of Epigenetics in Biology, 438

3. Human Diseases, 439

    3.1 Disorders of Genomic Imprinting, 439

    3.2 Disorders Affecting Chromatin Structure in trans, 443

    3.3 Disorders Affecting Chromatin Structure in cis, 447

    3.4 Epigenetics–Environment Interactions, 449

4. Looking into the Future, 451

Acknowledgments, 451

References, 451

GENERAL SUMMARY

The last two decades have witnessed unparalleled success in identifying the genetic bases for hundreds of human disorders. Studies of genotype–phenotype relationships challenged clinicians and researchers, because some observations could not be easily explained. For example, monozygotic twins carrying the same disease mutation can be quite different clinically. A mutation passed on in a multigeneration family can cause vastly different diseases depending on the sex of the transmitting parent. The study of such unusual cases uncovered the role of the epigenome (altered genetic information without change in DNA sequence) in health and disease. These studies showed that some regions of the mammalian genome are not functionally equivalent on the maternal and paternal alleles. Patients who inherit both homologous chromosomes (or segments thereof) from the same parent—uniparental disomy (UPD)—have loss of expression of some genes that are only expressed on maternal alleles (in case of paternal UPD) and increased levels for paternally expressed genes. UPD as well as altered DNA modifications (epigenetic mutations that might alter DNA methylation) quickly became recognized as the molecular bases for a variety of developmental and neurological disorders. It is interesting that for many of these disorders, either epigenetic or genetic mutations can lead to the same phenotype. This is often because the genetic mutations disrupt the function of a gene that is typically misregulated when epigenetic defects affect the locus.

In another class of diseases where genetic mutations cause loss of function of proteins involved in DNA methylation or chromatin remodeling, the phenotypes result from altered epigenetic states at one or more loci. The relationships between the genome and epigenome have broadened the types of molecular events that cause human diseases. These could be de novo or inherited, genetic or epigenetic, and most interestingly, some might be influenced by environmental factors. The finding that environmental factors such as diet and experience alter the epigenome (specifically DNA methylation) is likely to provide mechanistic insight about disorders with genetic predisposition and which are highly influenced by the environment. Such disorders include neural tube defects and psychiatric illnesses. Identifying environmental factors that can affect the epigenome provides hope for developing interventions that might decrease the risk or the burden of developmental abnormalities, cancer, and neuropsychiatric disorders.

1 Introduction

Two genetically identical male monozygotic twins, raised in the same environment, manifested very different neurological functions. Both twins carried the same mutation in the X-linked adrenoleukodystrophy (ALD) gene, yet one developed blindness, balance problems, and loss of myelin in the brain—features typical of the progressive and lethal neurological disease—while the other remained healthy. The conclusion of the investigators reporting the unusual occurrence was "some nongenetic factors may be important for different adrenoleukodystrophy phenotypes" (Korenke et al. 1996). That indeed was a valid conclusion in 1996, given the focus of medical genetics on DNA sequence. If the DNA sequence could not explain a phenotypic variation, then environmental factors did. Similar to the case of the ALD-discordant monozygotic twins, many monozygotic twins have been found to be discordant for schizophrenia despite similar environmental rearing conditions (Petronis 2004). Thankfully, research during the past decade has finally focused attention on epigenetic changes, which are modifications of the genetic information that do not alter DNA sequence, as a potential explanation for discordant phenotypes in monozygotic twins and in individuals who otherwise share similar DNA sequence alterations (Dennis 2003; Fraga et al. 2005).

Epigenetic modifications control gene expression patterns in a cell. These modifications are stable and heritable such that a mother liver cell will indeed give rise to more liver cells after it divides. In the case of nondividing cells such as neurons, adaptation of chromosomal regions through chromatin modifications offers a mechanism for maintaining epigenetic information and possibly mediating the reproducible response of neurons to specific stimuli. An epigenotype (the epigenetic state of a genomic locus) is established based on the methylation state of the DNA, chromatin modifications, and the yet-to-be elucidated various activities of noncoding RNAs.

In mammals, DNA methylation, which is the best-studied epigenetic signal, occurs predominantly at the carbon-5 position of symmetrical CpG dinucleotides. The state of DNA methylation is maintained after cell division through the activity of DNA-methyltransferase 1, which methylates hemimethylated CpG dinucleotides in daughter cells. Chromatin modifications involve covalent posttranslational modifications of the protruding amino-terminal histone tails by the addition of acetyl, methyl, phosphate, ubiquitin, or other groups. Methyl modifications can be mono-, di-, or tri-methylation. These modifications constitute the potential "histone code" that underlies a specific chromatin structure, which in turn affects the expression of adjacent genes. Because chromatin consists of densely packed DNA strands wrapped around histones, the folding pattern of DNA into chromatin is clearly at the root of gene activity changes. Although histone codes and chromatin structures can be stably transmitted from a parent cell to daughter cells, the mechanisms underlying the replication of such structures are not fully understood. The epigenotype shows plasticity during development and postnatally, depending on environmental factors and experiences (see Section 3.4); thus, it is not surprising that epigenotypes could contribute not only to developmental human disorders, but also to postnatal and even adult diseases. The most recent class of molecules contributing to the epigenetic signal is that of noncoding RNAs. For years the class of non-protein-coding RNA (ncRNA) included tRNA, rRNA, and spliceosomal RNA. More recently, because of the availability of genome sequence from multiple organisms, together with cross-species molecular genetic studies (from *Escherichia coli* to humans), the list of ncRNAs has expanded and resulted in the identification of hundreds of small ncRNAs, including small nucleolar RNA (snoRNA), microRNA (miRNA), short-interfering RNA (siRNA), and small double-stranded RNA. Some of these small RNA molecules regulate chromatin modifications, imprinting, DNA methylation, and transcriptional silencing, as discussed in detail in Chapter 8.

The first definitive evidence of a role for epigenetics in human disease came about after the understanding of genomic imprinting and the finding that several genes are subject to regulation by this mechanism (Reik 1989). Genomic imprinting is a form of epigenetic regulation in which the expression of a gene depends on whether it is inherited from the mother or the father. Thus, at an imprinted diploid locus, there is unequal expression of the maternal and paternal alleles. In each generation, the parent-specific imprinting marks have to be erased, reset, and maintained, thus rendering imprinted loci vulnerable to any errors that may occur during this process. Such errors, as well as mutations in genes encoding proteins involved in DNA methylation, binding to methylated DNA, and histone modifications, all contribute to the fast-growing class of human disorders affecting the epigenome (Fig. 1).

CHROMATIN RELATED DISEASES

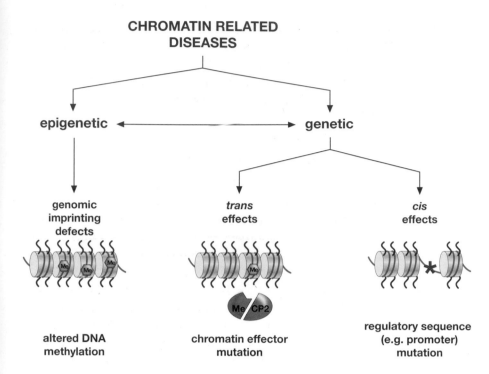

Figure 1. Genetic and Epigenetic Mechanisms Underlying Chromatin-related Disorders

Epigenetic mechanisms typically involve the alteration of DNA methylation or chromatin at imprinted loci, so disrupting monoallelic expression. Genetic mechanisms can be categorized into two classes. *trans* effects include the loss or dysfunction of chromatin-associated factors which can in turn alter chromatin structure and gene expression at certain genomic regions. *cis* effects represent mutations in noncoding regions that may be necessary for regulation. These mutations, which may include the expansion of DNA repeats, can lead to chromatin alterations which affect genome stability and gene expression.

2 Studies of Human Cases Uncover the Role of Epigenetics in Biology

There is no doubt that the study of model organisms has been crucial for understanding many biological principles, especially in the fields of genetics, development, and neuroscience. It is often forgotten, however, that humans represent one of the most important model organisms when it comes to all aspects of biology. The characterization of thousands of human diseases represents the largest mutant screen for any species, and if carefully and systematically studied, these phenotypes are likely to reveal biological insights in addition to the medical benefits. It is therefore not surprising that the genotype–phenotype relationships that challenged Mendelian inheritance in the case of "dynamic mutations" were revealed through the study of patients with fragile X syndrome (Pieretti et al. 1991). Patients with unique features and the observant physicians who study them often break open a new field in biology, revealing novel genetic and molecular mechanisms. This indeed proved to be the case in revealing the role of epigenetics in human development and disease.

A female patient made medical history for being reported twice by the physicians who saw her over the span of ten years. At the age of 7 years, she was reported in the medical literature because she suffered from cystic fibrosis (CF) and growth hormone deficiency, and was very short (Hubbard et al. 1980). During the race to find

the CF gene, Beaudet and colleagues sought unusual patients who had CF plus additional features in hope of identifying small deletions or chromosomal rearrangements that might facilitate the mapping and identification of the CF gene. Hence, this patient was brought to their attention. She was 16 years of age, measured 130 cm, had normal intelligence, but clearly had some body asymmetry (see right panel of title page figure). Analysis of her DNA revealed that she is homozygous for multiple polymorphic DNA markers on chromosome 7, including the centromeric alphoid repeats (Spence et al. 1988). After excluding non-paternity and hemizygosity, and after analyzing grand-maternal DNA (mother was deceased), Spence and colleagues concluded that this patient inherited two identical copies of the centromeric region of chromosome 7 from her maternal grandmother (Spence et al. 1988). Given Engel's theoretical proposal that uniparental disomy (UPD) is a possibility in humans (Engel 1980), Beaudet and colleagues immediately recognized that maternal UPD for chromosome 7 uncovered a recessive mutation in the CF gene and accounted for the additional somatic features. The constellation of clinical features in the patient, together with the laboratory evaluations, not only resulted in the identification of the first human case of UPD, but also illustrated that the maternal and paternal genomes are not equivalent for at least some portion of chromosome 7. This provided a novel mechanism of non-Mendelian inheritance to explain disease and developmental abnormalities (Fig. 2).

Although in 1988 it was thought by some that UPD of a chromosome was a rare event, today we know that UPD has been reported thus far for all human chromosomes except chromosomes 3 and 19.

The study of unusual patients not only identified cases of UPD for additional chromosomes, but in 1989 also led to the proposal that UPD causes disease due to changes in epigenotype and disruption of genomic imprinting (Nicholls et al. 1989). Nicholls et al. studied a patient with Prader-Willi syndrome (PWS) who had a balanced Robertsonian translocation t(13;15) that was also present in his asymptomatic mother and maternal relatives. The fact that the proband inherited his second free chromosome 15 from his mother (while all asymptomatic individuals inherited it from their fathers) led the authors to conclude that maternal UPD had led to the PWS phenotype. After confirming maternal UPD15 in a second PWS patient with an apparently normal karyotype, the authors proposed a role for genomic imprinting in the etiology of

PWS. Furthermore, they concluded that either paternal deletions or maternal UPD from 15q11-13 will lead to PWS, and they predicted that paternal UPD15 would lead to Angelman syndrome, just as maternal deletions of this region do. All of these predictions proved true (Fig. 3).

3 Human Diseases

3.1 Disorders of Genomic Imprinting

The discovery of UPD was the clinical entry point into disorders of genomic imprinting in humans. Whereas PWS and Angelman syndrome were the first genomic imprinting disorders to be studied, Beckwith-Wiedemann syndrome, pseudohypoparathyroidism, and Silver-Russell syndrome expanded the list and introduced many intriguing questions about how epigenetic defects lead to the disease phenotype. In the following section, we give a brief review of the clinical features of each disorder, the various mechanisms leading to epigenotypic defects, and the phenotypes and biological insight gained from the study of this class of disorders (see Table 1).

SISTER SYNDROMES: PRADER-WILLI AND ANGELMAN

Prader-Willi syndrome (PWS; OMIM 176270) and Angelman syndrome (AS; OMIM 105830) are caused in the majority of cases by the same 5- to 6-Mb deletion in 15q11-q13, but their phenotypes are vastly different. Genomic imprinting in the region of 15q11-q13 accounts for the phenotypic differences, given that PWS is caused by paternally inherited deletions whereas in AS, the deletion is of maternal origin (Ledbetter et al. 1981; Magenis et al. 1987; Nicholls et al. 1989). PWS, which occurs in approximately 1/10,000 births, was described almost 50 years ago and is characterized by infantile hypotonia, developmental delay, failure to thrive due to poor feeding, and lethargy, followed by hyperphagia, severe obesity, short stature, secondary hypogonadism with genital hypoplasia, and mild cognitive impairment. PWS patients also have distinct physical characteristics such as small hands and feet, almond-shaped eyes, and thin upper lip. Most PWS patients have mild to moderate mental retardation, and the vast majority display a variety of obsessive-compulsive behaviors, anxiety, and sometimes a withdrawn, unhappy disposition (Fig. 4a). In contrast, patients with AS have a "happy disposition," smile frequently, and have unexplained bouts of laughter. AS patients suffer from severe developmental delay, very minimal (if any) verbal skills, balance problems (ataxia), abnormal hand-

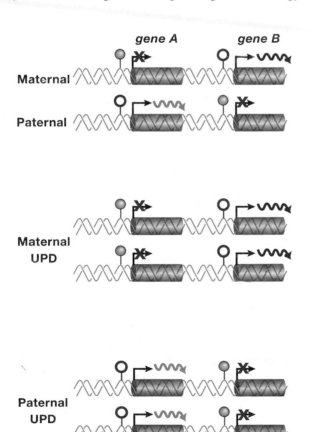

Figure 2. Consequences of Uniparental Disomy (UPD)

The DNA methylation states of upstream CpG islands are indicated by pink circles when methylated and open circles when unmethylated. The DNA methylation state affects the expression of its downstream gene. Maternally inherited alleles are doubled (gene *B*); whereas those that are on the paternal alleles are lost (gene *A*).

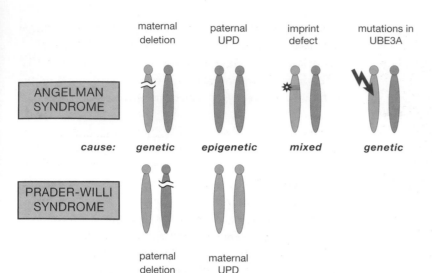

Figure 3. Prader-Willi Syndrome and Angelman Syndrome

Both syndromes can be caused by genetic, epigenetic, or mixed defects.

flapping movements, microcephaly, seizures, and some dysmorphic features such as prominent mandible and wide mouth (Fig. 4b).

Hypotonia, hypopigmentation of the skin and irides, and strabismus can be seen in both disorders. The majority of PWS and AS (~70%) are caused by paternal and maternal deletions of 15q11-q13, respectively. About 25% of PWS cases are caused by maternal UPD of 15q11-q13, whereas paternal UPD of this region accounts for 2–5% of AS patients (Fig. 3). The difference in frequency of UPD between PWS and AS is usually initiated by maternal nondisjunction, as influenced by maternal age leading to a conception with trisomy or monosomy 15. These are then "rescued," leading to maternal UPD and PWS or paternal UPD and AS, respectively. The difference in frequency of the two UPDs is presumably related to the frequency of the two abnormal eggs and the probability of rescue for the two circumstances. Translocations within the PWS/AS critical

region account for less than 10% of the cases, but it is of note that such translocations are associated with a high recurrence risk (up to 50%) depending on the sex of the transmitting parent. In fact, PWS and AS co-occurred in some families due to translocations or other structural abnormalities of 15q11-q13, and the phenotype was determined by the sex of the transmitting parent (Hasegawa et al. 1984; Smeets et al. 1992).

Imprinting defects represent another class of mutations leading to PWS or AS phenotypes. These defects, which involve a bipartite imprinting center (IC) within 15q11-q13 (Ohta et al. 1999), cause a chromosome of one parental origin to have an altered epigenotype, typically that of the chromosome of an opposite parental origin. Imprinting defects often involve deletion of the IC, but there are instances when such defects appear to be due to an epigenetic mutation that does not involve the DNA sequence. The outcome of such diverse imprinting defects is the same and includes alterations in DNA methylation,

Table 1. Selected disorders of genomic imprinting

Disorder	Gene	Comments	Gene(s) involved
Prader-Willi syndrome	deletion, UPD, imprint defect	15q11-q13	snoRNAs and other (?)
Angelman syndrome	deletion, UPD, imprint defect, point mutation, duplication[a]	15q11-q13	*UBE3A*
Beckwith-Wiedemann syndrome	imprint defect, UPD, 11p15.5 duplication, translocation point mutation	11p15.5	*IGF2, CDKN1C*
Silver-Russell syndrome	UPD, duplication translocation, inversion	7p11.2	several candidates in the region
	epimutation	11p15.5	biallelic expression of H19 and decrease of IGF2
Pseudohypoparathyroidism	point mutation, imprint defect, UPD	20q13.2	*GNAS1*

[a]Maternal duplications, trisomy, and tetrasomy for this region cause autism and other developmental abnormalities.

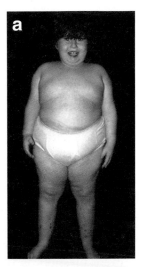

Figure 4. Images of a Prader-Willi Syndrome Patient (*a*) and Angelman Syndrome Patient (*b*)

These pictures illustrate the dramatic differences in the clinical features of the disorders resulting from defects in an imprinted region. Images kindly provided by Drs. Daniel J. Driscoll and Carlos A. Bacino, respectively.

chromatin structure, and, ultimately, gene expression patterns. Imprinting defects account for 2–5% of PWS and AS cases, and the IC deletions are typically associated with 50% recurrence risk, depending on the sex of the transmitting parent, whereas the recurrence risk is low for families without IC deletions. The identification of imprinting defects in a handful of AS patients who were conceived after intracytoplasmic sperm injection (ICSI) raised the possibility that this approach of in vitro fertilization might cause imprinting defects (Cox et al. 2002; Orstavik et al. 2003). The finding of imprinting defects among AS cases born to sub-fertile couples who did not receive ICSI (but did receive hormonal stimulation) raises further questions about whether there are common mechanisms for infertility and imprinting defects or whether indeed assisted reproductive technology (hormones and/or ICSI) has epigenetic consequences (Ludwig et al. 2005).

Exactly which gene(s) is affected by genomic imprinting in 15q11-q13 is known for AS but not for PWS. About 10–15% of AS cases are caused by loss-of-function mutations in the ubiquitin E3 ligase gene (*UBE3A*) encoding the E6-associated protein (E6-AP) (Kishino et al. 1997; Matsuura et al. 1997). Expression studies demonstrated that *Ube3a* is expressed exclusively from the maternal allele in cerebellar Purkinje cells and hippocampal neurons. Furthermore, *Ube3a*[+/-] mice lacking the maternal allele reproduce features of AS (Jiang et al. 1998). These results, together with human data, pinpoint the *UBE3A* gene as the causative gene in AS. Paternal UPD or maternal deletions of 15q11-q13 lead to loss of expression of *UBE3A* in Purkinje cells. In the case of IC imprinting defects, it appears that loss of silencing of an antisense transcript leads to suppression of *UBE3A* expression (Rougeulle et al. 1998). It is intriguing that about 10% of AS cases remain without a molecular diagnosis. A subset of these patients appear to have mutations in a chromatin-remodeling protein, methyl-CpG-binding protein 2, as discussed below.

In the case of PWS, there are several candidate imprinting genes that are only expressed from the paternal allele; however, it is not clear which of these genes is contributing to the PWS phenotype. The best candidate genes thus far are in a cluster of noncoding snoRNAs. The best protein-coding candidate genes are *SNURF-SNRPN* and *Necdin* (*NDN*). *SNURF-SNRPN* has its major transcriptional start site at the IC, and it encodes a small nuclear ribonucleoprotein (SNRPN) that functions in the regulation of splicing. Another gene, a "SNRPN upstream reading frame" or *SNURF*, along with upstream noncoding exons, is thought to be the major site of imprinting defects, because disruption of this gene leads to altered imprinting of *SNRPN* and other 15q11-q13 imprinted genes. Mice lacking *Snrpn* appear normal, but mice with deletions spanning *Snrpn* and other genes homologous to those in 15q11-q13 are hypotonic, develop growth retardation, and die before weaning (Tsai et al. 1999). Several small nucleolar RNA (snoRNA) genes are expressed from the paternal allele and are suspected to contribute to the PWS phenotype (Meguro et al. 2001). A recent study showed that loss of the paternal allele from one cluster of these genes (HBII-52) does not cause PWS (Runte et al. 2005). However, a study in mice suggests loss of Pwcr1/MBII-85 snoRNA is likely responsible for the neonatal lethality in PWS mouse models (Ding et al. 2005). Therefore, PWS may be caused by loss of one or more snoRNA genes, possibly in combination with loss of other paternally expressed genes in 15q11-q13. Careful studies of rare translocation and deletion families support the interpretation that deficiency of *PWCR1/HBII-85* snoRNAs causes PWS (Schule et al. 2005).

BECKWITH-WIEDEMANN SYNDROME

The story of Beckwith-Wiedemann syndrome (BWS; OMIM 130650) represents an excellent example of how a human disorder uncovered the importance of epigenetics not only in normal development, but in the regulation of cell growth and tumorigenesis. BWS is characterized by somatic overgrowth, congenital abnormalities, and a predisposition to childhood embryonal malignancies (Weksberg et al. 2003). BWS patients typically manifest gigantism, macroglossia (large tongue), hemihypertrophy, variable degrees of ear and other organ anomalies, and omphalocele (protrusion of abdominal organs through the navel). In addition, many patients suffer from increased size of internal organs; embryonic tumors such as Wilms' tumor, hepatoblastoma, or rhabdomyosarcoma; and hyperplasia and hypertrophy of pancreatic islets, often leading to neonatal hypoglycemia.

The majority of BWS cases are sporadic, but a small number of families with an autosomal dominant inheritance pattern (in retrospect, modified by genomic imprinting) suggested genetic etiology and linked the syndrome to 11p15 (Ping et al. 1989). Preferential loss of maternal alleles in BWS-related tumors, an excess of transmitting females in the dominant form of the disease, and paternal UPD of 11p15.5 in some cases of BWS provided evidence that epigenetics and imprinting must play an important role in the etiology of BWS, and that the disease might result from a mixture of genetic and epigenetic abnormalities either de novo or inherited. The cluster of imprinted genes implicated in BWS maps to an approximately 1-Mb region in 11p15.5 and includes at least 12 imprinted genes. These genes are thought to be regulated by two imprinting centers separated by a nonimprinted region (Weksberg et al. 2003). The reciprocally imprinted *H19* and insulin-like growth factor (*IGF2*) and a differentially methylated region are thought to represent one imprinting control region (ICR1) (Joyce et al. 1997; Weksberg et al. 2003). *H19* encodes a maternally expressed noncoding pol II RNA, and *IGF2* encodes a paternally expressed growth factor. These two genes share a common set of enhancers, access to which is affected by the methylation state of ICR1 and binding of CTCF, a zinc finger protein (Hark et al. 2000). The second imprinting control region (ICR2) contains several maternally expressed genes, including the cyclin-dependent kinase inhibitor (*CDKN1C* encoding p57^{kip2}), a component of the potassium channel (*KCNQ1*), and a putative cation transporter (*SLC22A1L*). The differentially methylated region in ICR2 maps to an intron of *KCNQ1* and is unmethylated on paternal alleles, leading to expression of *KCNQ1OT1* in an antisense direction of *KCNQ1*. Methylation of ICR2 on the maternal allele is believed to silence maternal expression of *KCNQ1OT1*, allowing expression of the maternally expressed *KCNQ1* and *CDKN1C* (Lee et al. 1999; Smilinich et al. 1999).

Various epigenetic as well as genetic molecular defects provided some insight about which genes contribute to the BWS phenotype. On unmethylated maternal alleles, CTCF binds ICR1 and establishes a chromatin boundary whereby the *IGF2* promoter is insulated from enhancers. These enhancers can then access the *H19* promoter (proximal to the boundary), permitting transcription of *H19*. Methylation of ICR1 on paternal alleles abrogates the binding of CTCF, permitting expression of *IGF2* and silencing of *H19*. The findings that either duplications in 11p15.5 that span the *IGF2* locus or paternal UPD of this region (expected to lead to overexpression of *IGF2*), coupled with data showing that transgenic mice overexpressing *IGF2* develop overgrowth and large tongues, implicated *IGF2* overexpression as one potential cause of the overgrowth phenotype in BWS (Henry et al. 1991; Weksberg et al. 1993; Sun et al. 1997). It is quite intriguing that loss-of-function mutations in *CDKN1C* give rise to BWS, similar to those caused by overexpression of *IGF2*. Mice lacking *Cdkn1c* develop omphaloceles but not overgrowth. However, when loss of *Cdkn1c* is coupled with increased expression of *Igf2*, the animals reproduce many features of BWS (Caspary et al. 1999). To date, the molecular lesions that cause BWS include (1) paternal duplications encompassing *IGF2*, (2) paternal UPD for 11p15.5, (3) loss-of-function mutations in the maternal allele of *CDKN1C*, (4) translocations on the maternal chromosome disrupting *KCNQ1* which affect imprinting of *IGF2* but curiously not ICR2, and (5) most commonly, loss of imprinting for ICR2/*KCNQ1OT1* which again alters imprinting of *IGF2* and suggests some regulatory interactions between ICR1 and ICR2 (Cooper et al. 2005). Some of the epigenetic changes identified in BWS, such as methylation defects at the *H19* ICR1, have also been confirmed in individuals who develop Wilms' tumor but not BWS, suggesting that the timing of the epigenetic defect might dictate whether abnormal growth regulation will affect the whole organism or a specific organ. The fact that aberrant methylation at IRC1 often leads to Wilms' tumor, and at ICR2 often leads to rhabdomyosarcoma and hepatoblastoma in BWS, suggests that there is more than one locus in 11p15.5 predisposing to tumorigenesis (Weksberg et al. 2001; DeBaun et al. 2003; Prawitt et al. 2005).

SILVER-RUSSELL SYNDROME

Silver-Russell syndrome (SRS; OMIM 180860) is a developmental disorder characterized by growth retardation, short stature often with asymmetry, and some dysmorphic facial and cranial features as well as digit abnormalities. The most prominent feature is the somatic growth abnormality, with other features being highly variable. SRS is genetically heterogeneous, but it is estimated that about 10% of the cases result from maternal UPD for chromosome 7 (Eggermann et al. 1997). It is proposed that loss of function of a paternally expressed gene, possibly one that promotes growth, causes SRS, but an alternate model of overexpression of a maternally expressed growth-suppressing gene cannot be excluded. It is interesting that an epigenetic mutation causing demethylation of the ICR1 on chromosome 11p15 has been identified in several individuals with SRS. This epigenetic defect causes biallelic expression of H19 and decreased expression of IGF2 (Gicquel et al. 2005).

PSEUDOHYPOPARATHYROIDISM

Pseudohypoparathroidism (PHP) represents a group of phenotypes that result from functional hypoparathyroidism despite normal parathyroid hormone (PTH) levels. These patients are resistant to PTH. There are several clinical subtypes—Ia, Ib, Ic, II, and Albright hereditary osteodystrophy (OMIM 103580). In addition to the functional hypoparathyroidism and osteodystrophy, these clinical variants may exhibit a variety of developmental and somatic defects. The clinically heterogeneous phenotypes result from mutations in the *GNAS1* gene encoding the α-stimulating activity polypeptide 1 (G$_s$α), a guanine nucleotide-binding protein. *GNAS1* maps to chromosome 20q13.2. The *GNAS1* locus has three upstream alternative first exons (exons 1A, XL, and NESP55) that are spliced to exons 2–13 to produce different transcripts and, in the case of NESP55 and XL, this alternative splicing produces unique proteins. There are differentially methylated regions near these exons, causing NESP55 to be expressed exclusively from maternal alleles, whereas XL, exon 1A, and an antisense transcript for NESP55 are paternally expressed. Although the transcript encoding the G$_s$α protein is biallelically expressed, the maternal allele is preferentially expressed in some tissues such as the proximal renal tubule. The combination of genomic and tissue-specific imprinting accounts for the variable phenotypes and parent-of-origin effect even for mutations that have a clear autosomal dominant inheritance pattern (Hayward et al. 1998). Of note is the finding that one patient with paternal uniparental disomy of the *GNAS1* region developed PHP type Ib disease (Bastepe et al. 2003).

The genotype–phenotype studies of these clinical disorders demonstrate that with the exception of SRS, all the other genomic imprinting disorders (PWS, AS, BWS, and PHP) can be caused by a mixture of genetic or epigenetic abnormalities, either de novo or inherited. It is hard to believe that such a mixed genetic model for disease would remain unique for this small subset of disorders. A little over a decade ago, UPD was only a theoretical possibility, but now it is established to occur in many chromosomal regions and to result in diverse diseases and developmental phenotypes. One challenge in human genetics research is to uncover which genes are responsible for which UPD-associated phenotypes in order to establish a list of diseases that are likely to result from mixed genetic/epigenetic mechanisms.

3.2 Disorders Affecting Chromatin Structure in trans

The importance of finely tuned chromatin structure for human health has been highlighted through the rapidly growing list of human diseases caused by mutations in genes encoding proteins essential for chromatin structure and remodeling. These disorders themselves do not have epigenetic mutations but alter chromatin states that are critical components of the epigenotype. The vast differences in phenotypes, as well as the fact that subtle changes in protein levels or even conserved amino acid substitution can lead to human disease, are beginning to provide clues about the tightly controlled regulation and interactions of chromatin-remodeling proteins. Disorders that affect chromatin in *trans* result either from disruption of function of proteins directly involved in chromatin remodeling, such as CREB-binding protein (CBP), EP300, or methyl-CpG-binding protein (MeCP2), or from loss of function of proteins involved in DNA methylation such as de novo DNA methyltransferase 3B (DNMT3B) or methylene tetrahydrofolate reductase (MTHFR) (see Table 2). Disruption of the function of any of these genes causes complex multisystem phenotypes or neoplasia owing to the downstream effects of misregulation of expression of a large number of target genes. Although yet to be discovered, there is an ample opportunity for diseases caused by mutation in noncoding RNAs acting in *trans*.

RUBINSTEIN-TAYBI SYNDROME

Rubinstein-Taybi syndrome (RSTS; OMIM 180849) is characterized by mental retardation, broad thumbs and toes, facial abnormalities, congenital heart defects, and increased risk of tumor formation. The high concordance rate in monozygotic twins, together with a few cases of

Table 2. Selected genetic disorders affecting chromatin structure in *trans*

Disorder	Gene	Comments
Rubinstein-Taybi syndrome	*CREBBP, EP300*	
Rett syndrome	*MECP2*	loss of function as well as duplication causes a broad spectrum of phenotypes
α-Thalassemia and X-linked mental retardation	*ATRX*	somatic mutations cause α-thalassemia and myelodysplastic syndrome
ICF Syndrome	*DNMT3B*	
Schimke immuno-osseous dysplasia	*SMARCAL1*	
Mental retardation	*MTHFR*	

mother-to-child transmission, suggested that this disease has a genetic basis and that an autosomal dominant inheritance was most likely. Cytogenetic abnormalities involving 16p13.3 were identified in several RSTS patients (Tommerup et al. 1992) and found to map to the region that contains the CREB-binding protein gene (CREBBP or CBP). Heterozygous mutations in *CREBBP* demonstrated that haploinsufficiency of CBP causes RSTS (Petrij et al. 1995). CBP was first described as a coactivator of the cAMP-responsive binding protein CREB. When cellular levels of cAMP increase, protein kinase A (PKA) translocates to the nucleus and phosphorylates CREB, which leads to its activation and binding to cAMP-response elements (CREs) (Mayr and Montminy 2001). CBP is a large protein (~250 kD) with a bromodomain that has been shown to bind PKA-phosphorylated CREB (Chrivia et al. 1993). CBP in turn activates transcription from a CRE-containing promoter through the acetylation of all four core histones in the adjacent nucleosomes (Ogryzko et al. 1996). CBP also interacts through a region in its carboxyl terminus directly with the basal transcription factor TFIIIB (Arias et al. 1994; Kwok et al. 1994). in vitro functional analysis of one of the CBP missense mutations (Arg-1378 to proline) that cause RSTS revealed that this mutation abolishes the histone acetyltransferase (HAT) activity of CBP (Murata et al. 2001). These data, together with the finding that mice haploinsufficient for CBP have impaired learning and memory, altered synaptic plasticity, and abnormal chromatin acetylation, support the conclusion that decreased HAT activity of CBP is a key contributor to the RSTS phenotype (Alarcon et al. 2004). Consistent with the role of decreased HAT activity in disease is the recent discovery that mutations in a second gene, *p300*, encoding a potent HAT and transcriptional coactivator cause some cases of RSTS (Roelfsema et al. 2005). The finding that some of the synaptic plasticity defects, as well as learning and memory deficits of the CBP$^{+/-}$ mice, can be reversed by using histone deacetylase (HDAC) inhibitors (Alarcon et al. 2004) raises the question whether pharmacologic therapy using such reagents can ameliorate some of the mental deficits in RSTS.

RETT SYNDROME

Rett syndrome (RTT, OMIM 312750) is a dominant X-linked postnatal neurodevelopmental disorder characterized by motor abnormalities, ataxia, seizures, replacement of hand use by purposeless hand-wringing, and language regression (Hagberg et al. 1983). RTT is classified as one of the autistic spectrum disorders (ASD) in DSMIV and shares three main features with ASD: Both manifest postnatally, often after a period of apparent normal development; both disrupt social and language development, and both are accompanied by unusual stereotyped hand or arm movements (Fig. 5a). Although RTT is a sporadic disorder in the vast majority of cases (>99%), the discovery of a handful of families in whom the gene was transmitted through maternal lines suggested a genetic basis for this disorder. Such families, together with findings that RTT was typically observed in females and that obligate carrier females can be asymptomatic, led to the hypothesis that RTT is an X-linked dominant disorder. An exclusion mapping strategy localized the RTT gene to Xq27-qter, and candidate gene analysis pinpointed the gene encoding methyl-CpG-binding protein 2 (*MECP2*) as the causative gene (Amir et al. 1999).

The discovery of mutations in *MECP2* as the major cause of RTT provided molecular evidence for a relationship between RTT and autism. Mutations in *MECP2* are now known to cause a broad spectrum of phenotypes in females, including learning disabilities, isolated mental retardation, Angelman-like syndrome, and ASD. X-chromosome inactivation (XCI) patterns are the major molecular determinants for this clinical variability. Females with *MECP2* mutations and balanced XCI patterns typically have classic RTT with the exception of a few hypomorphic

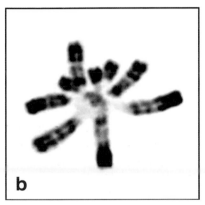

Figure 5. Genetic Disorders Affecting Chromatin in *cis*

(*a*) This photo of a Rett syndrome patient illustrates the unusual stereotyped hand movements, teeth grinding, and abnormal posture. Photo kindly provided by Dr. Daniel G. Glaze. (*b*) Micrograph of chromosomes from an ICF patient, courtesy of Drs. Timothy H. Bestor, Robert A. Rollins, and Deborah Bourc'his.

alleles. Females with unbalanced XCI patterns favoring the wild-type allele typically have the milder phenotypes (Wan et al. 1999; Carney et al. 2003). Males with *MECP2* mutations display a broader phenotype than females, due to their hemizygosity for the locus. RTT-causing mutations typically cause neonatal lethality unless the male is mosaic for the mutations or has XXY karyotype, in which case, all the phenotypes seen in females are also seen in these males (Zeev et al. 2002; Neul and Zoghbi 2004). On the other hand, males that have hypomorphic alleles which barely cause a phenotype in females develop any combination of features including mental retardation, seizures, tremors, enlarged testes, bipolar disease, or schizophrenia (Meloni et al. 2000; Couvert et al. 2001).

MeCP2 was identified on the basis of its ability to bind symmetrically methylated CpG dinucleotides (Lewis et al. 1992). It localizes to heterochromatin and acts as a transcriptional repressor in a methylation-dependent manner (Nan et al. 1997). MeCP2 binds methylated DNA

through its methyl-CpG-binding domain and interacts with corepressors Sin3A and HDACs through its transcription repression domain. MeCP2 also associates with Brahma, a component of the SWI-SNF chromatin-remodeling complex (Harikrishnan et al. 2005).

An intriguing feature of RTT is the delayed postnatal onset of phenotypes in the absence of neurodegeneration. Studies on the distribution and abundance of MeCP2 revealed that it is detected in mature neurons, probably after synapse formation (Shahbazian et al. 2002a; Kishi and Macklis 2004; Mullaney et al. 2004). Such a distribution suggests that MeCP2's neuronal function is essential after neuronal maturation and activity have been established and that it plays a role in regulating neuronal activity. Some targets of MeCP2 are beginning to be identified, but exactly which of these targets mediate the diverse RTT phenotypes remains to be determined (Chen et al. 2003; Martinowich et al. 2003; Horike et al. 2005; Nuber et al. 2005). Studies using cell extracts from RTT patients or brain extracts from mouse models that lack functional MeCP2 have revealed altered histone acetylation (Wan et al. 2001; Shahbazian et al. 2002b; Kaufmann et al. 2005), consistent with a proposed role for this protein in deacetylation of histones based on its interactions with HDACs. It is interesting that doubling the dose of MeCP2 in mice and humans leads to progressive postnatal phenotypes that are in fact more severe than some of the loss-of-function phenotypes (Collins et al. 2004; Meins et al. 2005; Van Esch et al. 2005). Whether increasing MeCP2 levels results in titration of key interactors and/or aberrant expression of its targets remains to be seen. In pursuit of revealing potentially novel functions for MeCP2, Young and colleagues discovered that MeCP2 interacts with Y box-binding protein 1 (YB-1), an RNA-binding protein that affects splicing (Young et al. 2005). MeCP2 regulates RNA splicing of reporter minigenes, but most importantly, seems to affect RNA splicing in vivo based on altered RNA splicing patterns in brain tissue from a mouse model for RTT (Young et al. 2005). The importance of MeCP2 in epigenetic regulation of neuronal gene expression and its effects on RNA splicing are likely to be at the root of loss of developmental milestones and abnormal neurological function in RTT and related disorders.

α-THALASSEMIA X-LINKED MENTAL RETARDATION

Males with α-thalassemia X-linked mental retardation syndrome (ATRX; OMIM 301040) display α-thalassemia, moderate to severe mental retardation, dysmorphic facial features, microcephaly, skeletal and genital abnormalities, and, usually, inability to walk. Heterozygous females are typically asymptomatic. Mutations in the *ATRX* gene, which maps to Xq13, cause this syndrome, as well as a host of additional phenotypes, including variable degrees of X-linked mental retardation (XLMR), severe MR with spastic paraplegia, and acquired α-thalassemia in myelodysplastic syndrome (ATMDS) owing to somatic mutations (Gibbons et al. 1995, 2003; Villard et al. 1996; Yntema et al. 2002). The ATRX protein contains a plant homeodomain (PHD-like) zinc finger motif as well as a DNA-dependent ATPase of the SNF2 family. This, together with its localization to pericentromeric heterochromatic domains and association with heterochromatin1α (HP1α) (McDowell et al. 1999), suggests a role as a chromatin-remodeling protein. Mutations in ATRX cause down-regulation of the α-globin locus and abnormal methylation of several highly repeated sequences, including subtelomeric repeats, Y-specific satellite, and ribosomal DNA arrays. A recent study demonstrated that ATRX is essential for the survival of cortical neurons, hinting that increased neuronal loss might contribute to the severe mental retardation and spasticity seen in patients with ATRX mutations (Berube et al. 2005).

It is interesting that levels of ATRX are tightly regulated and that either decreases or increases cause major neurodevelopmental problems. For example, human patients with mutations that result in 10–30% of normal ATRX levels display the full ATRX phenotype despite having significant amounts of the normal ATRX protein (Picketts et al. 1996). Too much of ATRX seems to be equally devastating. Transgenic mice that overexpress ATRX develop neural tube defects, have growth retardation, and die during embryogenesis. Those that survive develop craniofacial abnormalities, compulsive facial scratching, and seizures. The features are reminiscent of clinical features of patients with loss-of-function mutations of ATRX, raising the possibility that levels of ATRX are tightly regulated for the functional integrity of the protein complex within which it resides.

IMMUNODEFICIENCY, CENTROMERIC REGION INSTABILITY, AND FACIAL ANOMALIES SYNDROME

The immunodeficiency, centromeric region instability, and facial anomalies syndrome (ICF, OMIM 242860) is a rare autosomal recessive chromosome breakage disorder. ICF patients display two invariant phenotypes, immunodeficiency and cytogenetic abnormalities. Highly variable and less penetrant phenotypes include craniofacial defects

such as a broad and flat nasal bridge, epicanthal folds, high forehead and low-set ears, psychomotor retardation, and intestinal dysfunction (Smeets et al. 1994). The immunodeficiency is typically severe and is often the cause of premature death during childhood due to respiratory or gastrointestinal infections. A decrease in serum IgG levels is the most common immunological defect, but decreased numbers of B or T cells are also observed (Ehrlich 2003). Cytogenetic abnormalities primarily affecting chromosomes 1 and 16, and to a lesser degree 9, are seen on routine karyotype analysis of blood and in cultured cells of ICF patients (Fig. 5b) (Tuck-Muller et al. 2000).

Hypomethylation of juxtacentromeric repeat sequences on chromosomes 1, 9, and 16 had been discovered well before the identification of the ICF gene (Jeanpierre et al. 1993). These chromosomes contain the largest blocks of classic satellite (satellites 2 and 3) tandem repeats near their centromeres. The finding that ICF is caused by loss-of-function mutations in the de novo DNA methyltransferase gene (*DNMT3B*) provided insight into the decrease in methylation at centromeric satellites 2 and 3 (Hansen et al. 1999; Okano et al. 1999; Xu et al. 1999). However, it remains unclear why loss of function of a widely expressed de novo methyltransferase selectively affects specific repetitive sequences. One possible explanation entails the subcellular distribution and/or context-specific protein interaction of DNMT3B (Bachman et al. 2001). Another possibility is that the catalytic activity of DNMT3B is more essential for methylating sequences that have a high density of CpGs over large genomic regions, as in the case of satellite 2 (Gowher and Jeltsch 2002) or the D4Z4 repetitive sequence, implicated in facioscapulohumeral muscular dystrophy (Kondo et al. 2000). Whether additional specific sequences are hypomethylated remains to be determined, but it is predicted that DNA hypomethylation leads to altered expression of genes that play an important role in craniofacial, nervous system, and immunological development.

Gene expression studies using RNA from lymphoblastoid cell lines of ICF patients and healthy controls revealed several alterations in genes involved in maturation, migration, activation, and homing of lymphocytes (Ehrlich et al. 2001). It is not clear, however, whether loss of DNMT3B causes dysregulation of such genes, because the methylation patterns at their promoter did not seem to be altered. Given that the only hypomethylation detected so far in ICF is at satellite DNA, it is hypothesized that some of the genes altered in ICF might associate with satellite DNA. Such sequences typically behave as heterochromatin when methylated; thus, in ICF there is dysregulated gene expression due to trans-effects of heterochromatic regions rich in satellite 2 and 3 domains (Bickmore and van der Maarel 2003).

SCHIMKE IMMUNO-OSSEOUS DYSPLASIA

Schimke immuno-osseous dysplasia (SIOD, OMIM 242900) is an autosomal recessive multisystem disorder characterized by dysplasia of the spine and ends of long bones, growth deficiency, renal function abnormalities due to focal and segmental glomerulosclerosis, hypothyroidism, and defective T-cell-mediated immunity (Schimke et al. 1971; Spranger et al. 1991). SIOD is caused by mutations in *SMARCAL1* (SW1/SNF2-related, matrix associated; actin-dependent regulator of chromatin, subfamily a-like1), which encodes a protein proposed to regulate transcriptional activity through chromatin remodeling (Boerkoel et al. 2002). Nonsense and frameshift mutations cause severe phenotypes, whereas some of the missense mutations cause milder or partial phenotypes (Boerkoel et al. 2002). Recently, a patient with B-cell lymphoma and SIOD was found to have mutations in *SMARCAL1*, suggesting that loss of function of this protein can cause a fatal lymphoproliferative disorder (Taha et al. 2004). The exact mechanism by which loss of SMARCAL1 causes the phenotypes of SIOD remains to be elucidated.

METHYLENE TETRAHYDROFOLATE REDUCTASE DEFICIENCY

Methylene tetrahydrofolate reductase (MTHFR) is involved in the conversion of 5,10-methylene tetrahydrofolate (5,10-MTHF) to 5-methyl tetrahydrofolate (5MTHF). A methyl group is then acquired from 5MTHF during the conversion of homocysteine to methionine by methionine synthase. Methionine is further converted to S-adenosyl methionine (SAM), the major methyl donor for all methyl transferases. Deficiency of MTHFR causes a rare autosomal recessive disorder characterized by mental retardation (Rozen 1996). A common thermolabile polymorphism (677C>T, which changes alanine to valine) causes reduced activity of MTHFR and has been associated, especially in homozygotes whose diets are low in folate, with hyperhomocysteinemia (Goyette et al. 1994). This polymorphism has been investigated as a risk factor of atherosclerosis, neural tube defects, and cancer (Ma et al. 1997; Brattstrom et al. 1998; Chen et al. 1999; Botto and Yang 2000; Schwahn and Rozen 2001). Mice heterozygous or homozygous for a null allele of MTHFR have decreased levels of SAM and decreased global DNA methylation. Furthermore, the null mutants have aortic lipid deposition and neuronal degeneration (Chen et al. 2001). The global alteration in DNA methylation associated with partial or complete loss of MTHFR suggests that some of the phenotypes associated with its dysfunction might result from disturbances of chromatin due to decreased DNA (and possibly histone) methylation. There is one report of MTHFR deficiency causing an Angelman syndrome phenotype (Arn et al. 1998), and there is considerable phenotypic overlap of severe MTHFR deficiency with AS and RTT (Fattal-Valevski et al. 2000).

3.3 Disorders Affecting Chromatin Structure in cis

The genes for most Mendelian disorders are usually identified by finding mutations in either exons or splice sites, whereby the gene products, RNA or protein, are altered or not produced. For many of these disorders, however, there is frequently a small group of patients in whom mutations cannot be identified after sequencing of coding and noncoding regions of the gene despite linkage to the specific locus. It is becoming increasingly clear that epigenetic or genetic abnormalities which affect gene expression in *cis* underlie some Mendelian disorders and cases lacking exonic mutations. The following three examples demonstrate how *cis*-linked alterations in chromatin structure can result in human disease (see Table 3).

αδβ- AND δβ-THALASSEMIA

The thalassemias are the most common single-gene disorders in the world. They are a heterogeneous group of hemoglobin synthesis disorders caused by reduced levels of one or more of the globin chains of hemoglobin. The

Table 3. Selected genetic disorders affecting chromatin structure in *cis*

Disorder	Gene	Comments
αδβ- and δβ-thalassemia	deletion of LCR causes decreased globin expression	
Fragile X syndrome	expansion of CCG repeat leads to abnormal methylation and silencing of *FMR1*	premutation alleles (60–200) cause a neurodegenerative disorder
FSH dystrophy	contraction of D4Z4 repeats causes less repressive chromatin	
Multiple cancers	germ-line epimutation of *MLH1*	

imbalance in synthesis of various globin chains leads to abnormal erythropoiesis and profound anemia (Weatherall et al. 2001). Hundreds of coding and splicing mutations have been identified, but it was the deletions of the regulatory sequences that pinpointed how changes in chromatin structure can explain some subtypes of thalassemia. The discovery that deletions of approximately 100 kb which removed the upstream part of the β-globin gene (while leaving the gene intact) caused αδβ-thalassemia helped identify the locus control region (LCR) that regulates β-globin expression (Kioussis et al. 1983; Forrester et al. 1990). Smaller deletions involving part of the LCR caused δβ-thalassemia (Curtin et al. 1985; Driscoll et al. 1989). These deletions resulted in an altered chromatin state at the β-globin locus despite being tens of kilobases upstream of the coding region (Grosveld 1999).

FRAGILE X SYNDROME

Fragile X mental retardation (OMIM 309550) is one of the most common causes of inherited mental retardation. Over 60 years ago, Martin and Bell described a family which showed that mental retardation segregated as an X-linked disorder (Martin and Bell 1943). In 1969, Lubs reported on the constriction on the long arm of the X chromosome in some mentally retarded males and one asymptomatic female (Lubs 1969). This chromosomal variant was mapped to Xq27.3 and dubbed the fragile X chromosome (Harrison et al. 1983). Cytogenetic studies, especially those using culture media deficient in folic acid and thymidine, revealed the fragile site in families with X-linked mental retardation, and they were then diagnosed as having fragile X syndrome (Sutherland 1977; Richards et al. 1981). Affected males have moderate to severe mental retardation, macroorchidism, connective tissue abnormalities such as hyperextensibility of joints, and large ears (Fig. 6) (Hagerman et al. 1984). The gene responsible for fragile X syndrome is *FMR1*, which encodes FMRP protein. The most common mutational mechanism is an expansion of an unstable noncoding CGG repeat (Warren and Sherman 2001). Normal alleles contain 6–60 repeats, premutation alleles have 60–200, and the full mutation contains >200 repeats. The repeat expansion at the 5′UTR of the *FMR1* gene provides an excellent example of a genetic disorder that is mediated through altered chromatin structure in *cis*. A CpG island in the 5′ regulatory region of *FMR1* becomes aberrantly methylated upon repeat expansion in the case of the full mutation (Verkerk et al. 1991). Decreased histone acetylation at the 5′ end is documented in cells from fragile X patients com-

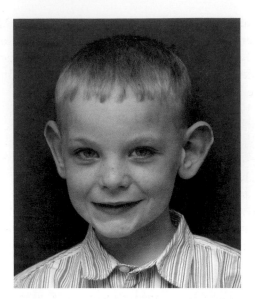

Figure 6. Example of a Genetic Disorder Affecting Chromatin in *trans*

The photograph is of a patient with fragile X syndrome who, in addition to mental retardation, has the typical features of prominent forehead and large ears. Photograph kindly provided by Dr. Stephen T. Warren.

pared to healthy controls (Coffee et al. 1999). In turn, the altered DNA methylation and histone acetylation patterns lead to loss of expression of *FMR1* and, therefore, loss of FMRP function in patients with fragile X syndrome. Thus, these patients have a primary genetic mutation and a secondary epigenetic mutation.

An interesting epigenetic mechanism has been proposed to explain how the CGG FMR1 repeat gets methylated and subsequently silenced. The finding that a premutation CGG repeat forms a single and stable hairpin structure (Handa et al. 2003), together with findings that rCGG repeats can be cleaved by Dicer, raised the possibility that expanded CGG repeats (which are unmethylated during early development) can be transcribed and that the resulting RNA forms a hairpin structure that can be cleaved by Dicer to produce small noncoding RNAs. These small RNA molecules associate with RNA-induced initiator of transcriptional gene silencing (RITS) and recruit DNA de novo methyltransferases and/or histone methyltransferases to the 5′UTR of *FMR1*, leading to full methylation of the CGG repeat and transcriptional repression of *FMR1* as development progresses (Jin et al. 2004a).

FMRP is a selective RNA-binding protein that contains 2 KH domains and an RGG box. It associates with polysomes in an RNA-dependent manner through messenger ribonucleoprotein particles and has been implicated in suppressing translation both in vitro and in vivo (Laggerbauer et al. 2001; Li et al. 2001). The localization of FMRP with mRNA and polyribosomes in dendritic spines provided evidence for its role in regulating local protein synthesis in response to synaptic stimulation (Feng et al. 1997; Weiler and Greenough 1999; Brown et al. 2001; Darnell et al. 2001, 2005). Putative targets of FMRP have been identified that play a role in synaptic development and that could explain partially the neurodevelopmental phenotypes (Brown et al. 2001; Darnell et al. 2001).

Several studies suggest that the RNA interference (RNAi) pathway is a major mechanism by which FMRP regulates translation. *Drosophila* fragile X homolog (Dfmr1) associates with Argonaute (ARGO2) and the RNA-induced silencing complex (RISC), and mammalian FMRP interacts with EIF$_2$C2 and associates with Dicer activity (Caudy et al. 2002; Ishizuka et al. 2002; Jin et al. 2004b). The favored proposed mechanism for FMRP's role as a translational suppressor is that FMRP binds to specific mRNA ligands, recruits RISC along with miRNAs, and facilitates the recognition between the miRNAs and the mRNA ligands (Jin et al. 2004a).

Carriers of the fragile X premutation (60–200 repeats) develop a distinct neurodegenerative syndrome characterized by tremor and ataxia (Hagerman and Hagerman 2004). Interestingly, these premutations may induce pathogenesis at the RNA level because the *FMR1* RNA and protein are present. Studies in animal models suggest that the RNA encoded by CGG repeats binds to and alters the function of some cellular proteins, causing them to accumulate (Jin et al. 2003; Willemsen et al. 2003).

FACIOSCAPULOHUMERAL DYSTROPHY

Facioscapulohumeral dystrophy (FSHD; OMIM 158900) is an autosomal dominant muscular dystrophy characterized by progressive wasting of the muscles of the face, upper arm, and shoulder. The more severe cases have hearing loss, and a very small subset of severely affected children are mentally retarded and have seizures (Mathews 2003). The major locus for FSHD (FSHD1) maps to the subtelomeric region of chromosome 4q35 near D4Z4, a low-copy repeat that contains an array of 3.3-kb GC-rich units. This repeated array is polymorphic and contains 11–150 units on normal chromosomes whereas it is in the 1–10-unit range on FSHD chromosomes (Wijmenga et al. 1992; van Deutekom et al. 1993). A second variable satellite repeat sequence (β-68bp *Sau*3A) distal to D4Z4 appears to play a role in developing FSHD. The 4qA variant at the β-satellite repeat, along with the contraction of the D4Z4 repeat, is necessary for the manifestation of FSHD (Lemmers et al. 2002). Exactly how contractions of D4Z4 together with 4qA β-satellite variant cause disease is not quite understood. The 4q35 region containing D4Z4 shares similarities with other subtelomeres and displays features typical of heterochromatic regions (Flint et al. 1997; Tupler and Gabellini 2004). In vitro and in vivo studies identified a 27-bp sequence in D4Z4 that binds to a complex termed the D4Z4-repressing complex (DRC) which comprises the transcriptional repressor Ying Yang1 (YY1), high mobility group box 2 (HMGB2), and nucleolin (Gabellini et al. 2002). Bickmore and van der Maarel, and Gabellini and colleagues, proposed that contraction of the repeats causes a less repressive chromatin state leading to increased transcription of 4q35-qter genes (Bickmore and van der Maarel 2003; Tupler and Gabellini 2004). The finding of increased expression of three genes—*FRG1* and *2* (FSHD region genes 1 and 2), and adenine nucleotide transporter 1 (ANT1)—in FSHD muscle compared to normal muscle is consistent with this hypothesis (Gabellini et al. 2002). Whether these gene expression changes are a direct or indirect consequence of D4Z4 contractions and whether misregulation of additional genes contributes to the disease phenotype remains to be seen.

EPIMUTATIONS AND HUMAN DISEASE

Epimutations in the DNA mismatch repair gene *MLH1* have been identified in two individuals who have had multiple cancers (Suter et al. 2004). Abnormal methylation of the promoter region of the *MLH1* gene was detected in all available normal tissues from these two individuals. Deletion or loss of heterozygosity of *MLH1* in tumor tissue led to the complete loss of MLH1 protein. These patients both suffered from colorectal cancer; one of them had duodenal cancer and the other developed endometrial and breast cancer as well as melanoma. The extent of the role of epimutations in human disease will only become apparent when investigators begin to search for such mutations systemically.

3.4 Epigenetics–Environment Interactions

Data from human studies as well as animal models are providing evidence that the environment can affect epigenetic marks and, as a result, gene function. The finding that monozygotic twins have similar epigenotypes during early years of life, but exhibit remarkable differences in the content and distribution of 5-methylcytosines and acetylated histones, provides strong evidence that the epigenotype is metastable and displays temporal variability (Fraga et al. 2005). It is likely that many environmental factors and stochastic events contribute to the variations in the epigenome (Fig. 7) (Anway et al. 2005), but diet and early experiences are emerging as potential key players.

DIET AND EPIGENOTYPES IN AGING AND DISEASE

Several reports indicate that there is an age-dependent decrease of global DNA methylation while concurrently there might be site-specific hypermethylation (Hoal-van Helden and van Helden 1989; Cooney 1993; Rampersaud et al. 2000). Given the large body of data linking altered DNA methylation to cancer risk or progression (Mays-Hoopes 1989; Issa et al. 1994), such epigenetic changes might contribute to the age-related increase in cancer risk. The role of diet as a contributing factor in controlling global methylation status has been best illustrated in adult males suffering from uremia and undergoing hemodialysis. The presence of hyperhomocystinemia in these patients suggests low methionine content, presumably due to folate depletion. These males had reduced global and locus-specific DNA methylation that was reversed after the administration of high doses of folic acid (Ingrosso et al. 2003).

Because several of the neuropsychiatric features resulting from folate and B12 deficiencies overlap with those seen with sporadic neuropsychiatric disorders, it was proposed that the latter might be caused by alterations in methylation patterns in the central nervous system (Reynolds et al. 1984). Low levels of SAM were found in folate-responsive depression; furthermore, SAM supplementation is helpful as an adjunct therapeutic in some forms of depression (Bottiglieri et al. 1994). Last, although it is unclear how increased folic acid intake by childbearing women reduces the risk of neural tube defect, it is tempting to propose some epigenetic-mediated effects on DNA or histone methylation. The finding that supplementing maternal diets with extra folic acid, B12, and betaine alters the epigenotype and phenotype of the offspring of *agouti viable yellow* mice is likely to be the first of many examples yet to be discovered in humans and other mammals (Wolff et al. 1998; Waterland and Jirtle 2003).

EARLY EXPERIENCES AND EPIGENOTYPES

The best example of how early experiences and maternal behavior might alter the mammalian epigenotype has so far been described only in rats. Frequent licking and grooming by rat mothers altered the DNA methylation status in the promoter region of the glucocorticoid receptor (GR) gene in the hippocampus of their pups. The highly licked and groomed pups have decreased DNA methylation and increased histone acetylation at the GR promoter compared to pups that were raised by low-licking and grooming mothers (Weaver et al. 2004). The increased levels of GR, secondary to the epigenotype change, affect the regulation of stress hormone levels and the lifelong response to stress in the rat pups (Liu et al. 1997; Weaver et al. 2004). Although such data are not

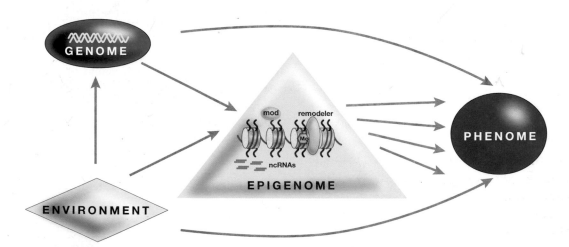

Figure 7. The Epigenotype Plays a Critical Role along with the Genotype and Environmental Factors in Determining Phenotypes

Known epigenetic factors affecting gene expression and genome stability include DNA methylation, chromatin-remodeling complexes, covalent histone modifications, the presence of histone variants, or noncoding regulatory RNAs (ncRNA).

available for humans yet, they certainly raise questions about the role of early experiences in modulating epigenotypes and risk for psychiatric disorders in humans.

4 Looking into the Future

During the next decade, we anticipate that mutations which alter the epigenotype will become increasingly recognized as mutational mechanisms that cause a variety of human disorders. Traditionally, the identification of disease-causing genes has focused on disorders where familial cases or patients with chromosomal abnormalities facilitated the positional cloning of the responsible gene. At this time, we are challenged as we attempt to discover the mutational bases for some of the most common and devastating disorders such as schizophrenia, autism, and mood disorders. Familial cases are not very common; genetic heterogeneity is very likely; and last but not least, genetic data—especially the rate of discordances in monozygotic twins—do not always support a straightforward Mendelian inheritance model. These findings, coupled with the strong environmental effects on the penetrance of some of these disorders, underscore the importance of investigating the epigenomes in such diseases. Even single-gene disorders such as Angelman syndrome, Beckwith-Wiedemann syndrome, and Silver-Russell syndrome can be caused either by genomic mutations or by mutations that affect the epigenotype, and can be either inherited or de novo. Such molecular variations will undoubtedly be unearthed for other human disorders. Furthermore, data demonstrating that the levels of several proteins involved in epigenetic regulation are tightly regulated and that perturbations of such levels either through loss-of-function mutations or duplications cause human disorders, suggest that epigenetic mutations that will affect transcription, RNA splicing, or protein modifications are also likely to cause disease.

Acknowledgments

We thank Drs. Timothy H. Bestor, Robert A. Rollins, and Deborah Bourc'his for the image of chromosomes from an ICF syndrome patient; Dr. Daniel J. Driscoll for the image of a Prader-Willi syndrome patient, Dr. Carlos A. Bacino for the image of an Angelman syndrome patient, Dr. Daniel G. Glaze for the image of a Rett syndrome patient, and Dr. Stephen T. Warren for the image of a fragile X syndrome patient. We gratefully acknowledge colleagues in the Zoghbi laboratory for reading the chapter and providing excellent suggestions. We also thank past and current laboratory members who have contributed to our work on Rett syndrome, Prader-Willi syndrome, and Angelman syndrome. Last, but not least, we are grateful to our patients with Rett syndrome, UPD of chromosome 7, Prader-Willi syndrome, Angelman syndrome, and autism, and to their families for enlightening us about the role of epigenetics in human diseases. Our work has been supported by grants from the National Institutes of Health (5 P01 HD040301-05; 5 P30 HD024064-17; 5 P01 HD37283); International Rett Syndrome Association; Rett Syndrome Research Foundation; Cure Autism Now; The Simons Foundation; March of Dimes (12-FY03-43); and the Blue Bird Clinic Rett Center. H.Y.Z. is an investigator with the Howard Hughes Medical Institute. We regret that due to space constraints, we had to eliminate many important and relevant citations.

References

Alarcon J.M., Malleret G., Touzani K., Vronskaya S., Ishii S., Kandel E.R., and Barco A. 2004. Chromatin acetylation, memory, and LTP are impaired in CBP$^{+/-}$ mice: A model for the cognitive deficit in Rubinstein-Tayhi syndrome and its amelioration. *Neuron* **42:** 947–959.

Amir R.E., Van den Veyver I.B., Wan M., Tran C.Q., Francke U., and Zoghbi H.Y. 1999. Rett syndrome is caused by mutations in X-linked *MECP2*, encoding methyl-CpG-binding protein 2. *Nat. Genet.* **23:** 185–188.

Anway M.D., Cupp A.S., Uzumcu M., and Skinner M.K. 2005. Epigenetic transgenerational actions of endocrine disruptors and male fertility. *Science* **308:** 1466–1469.

Arias J., Alberts A.S., Brindle P., Claret F.X., Smeal T., Karin M., Feramisco J., and Montminy M. 1994. Activation of cAMP and mitogen responsive genes relies on a common nuclear factor. *Nature* **370:** 226–229.

Arn P.H., Williams C.A., Zori R.T., Driscoll D.J., and Rosenblatt D.S. 1998. Methylenetetrahydrofolate reductase deficiency in a patient with phenotypic findings of Angelman syndrome. *Am. J. Med. Genet.* **77:** 198–200.

Bachman K.E., Rountree M.R., and Baylin S.B. 2001. Dnmt3a and Dnmt3b are transcriptional repressors that exhibit unique localization properties to heterochromatin. *J. Biol. Chem.* **276:** 32282–32287.

Bastepe M., Frohlich L.F., Hendy G.N., Indridason O.S., Josse R.G., Koshiyama H., Korkko J., Nakamoto J.M., Rosenbloom A.L., Slyper A.H., et al. 2003. Autosomal dominant pseudohypoparathyroidism type Ib is associated with a heterozygous microdeletion that likely disrupts a putative imprinting control element of *GNAS*. *J. Clin. Invest.* **112:** 1255–1263.

Berube N.G., Mangelsdorf M., Jagla M., Vanderluit J., Garrick D., Gibbons R.J., Higgs D.R., Slack R.S., and Picketts D.J. 2005. The chromatin-remodeling protein ATRX is critical for neuronal survival during corticogenesis. *J. Clin. Invest.* **115:** 258–267.

Bickmore W.A. and van der Maarel S.M. 2003. Perturbations of chromatin structure in human genetic disease: Recent advances. *Hum. Mol. Genet.* **12:** R207–R213.

Boerkoel C.F., Takashima H., John J., Yan J., Stankiewicz P., Rosenbarker L., Andre J.L., Bogdanovic R., Burguet A., Cockfield S., et al. 2002.

Mutant chromatin remodeling protein SMARCAL1 causes Schimke immuno-osseous dysplasia. *Nat. Genet.* **30:** 215–220.

Bottiglieri T., Hyland K., and Reynolds E.H. 1994. The clinical potential of ademethionine (S-adenosylmethionine) in neurological disorders. *Drugs* **48:** 137–152.

Botto L.D. and Yang Q. 2000. 5,10-Methylenetetrahydrofolate reductase gene variants and congenital anomalies: A HuGE review. *Am. J. Epidemiol.* **151:** 862–877.

Brattstrom L., Wilcken D.E., Ohrvik J., and Brudin L. 1998. Common methylenetetrahydrofolate reductase gene mutation leads to hyperhomocysteinemia but not to vascular disease: The result of a meta-analysis. *Circulation* **98:** 2520–2526.

Brown V., Jin P., Ceman S., Darnell J.C., O'Donnell W.T., Tenenbaum S.A., Jin X., Feng Y., Wilkinson K.D., Keene J.D., et al. 2001. Microarray identification of FMRP-associated brain mRNAs and altered mRNA translational profiles in fragile X syndrome. *Cell* **107:** 477–487.

Carney R.M., Wolpert C.M., Ravan S.A., Shahbazian M., Ashley-Koch A., Cuccaro M.L., Vance J.M., and Pericak-Vance M.A. 2003. Identification of MeCP2 mutations in a series of females with autistic disorder. *Pediatr. Neurol.* **28:** 205–211.

Caspary T., Cleary M.A., Perlman E.J., Zhang P., Elledge S.J., and Tilghman S.M. 1999. Oppositely imprinted genes p57^Kip2 and Igf2 interact in a mouse model for Beckwith-Wiedemann syndrome. *Genes Dev.* **13:** 3115–3124.

Caudy A.A., Myers M., Hannon G.J., and Hammond S.M. 2002. Fragile X-related protein and VIG associate with the RNA interference machinery. *Genes Dev.* **16:** 2491–2496.

Chen J., Giovannucci E.L., and Hunter D.J. 1999. MTHFR polymorphism, methyl-replete diets and the risk of colorectal carcinoma and adenoma among U.S. men and women: An example of gene-environment interactions in colorectal tumorigenesis. *J. Nutr.* **129:** S560–S564.

Chen W.G., Chang Q., Lin Y., Meissner A., West A.E., Griffith E.C., Jaenisch R., and Greenberg M.E. 2003. Derepression of BDNF transcription involves calcium-dependent phosphorylation of MeCP2. *Science* **302:** 885–889.

Chen Z., Karaplis A.C., Ackerman S.L., Pogribny I.P., Melnyk S., Lussier-Cacan S., Chen M.F., Pai A., John S.W., Smith R.S., et al. 2001. Mice deficient in methylenetetrahydrofolate reductase exhibit hyperhomocysteinemia and decreased methylation capacity, with neuropathology and aortic lipid deposition. *Hum. Mol. Genet.* **10:** 433–443.

Chrivia J.C., Kwok R.P., Lamb N., Hagiwara M., Montminy M.R., and Goodman R.H. 1993. Phosphorylated CREB binds specifically to the nuclear protein CBP. *Nature* **365:** 855–859.

Coffee B., Zhang F., Warren S.T., and Reines D. 1999. Acetylated histones are associated with FMR1 in normal but not fragile X-syndrome cells (erratum *Nat. Genet.* **22:** 209 [1999]). *Nat. Genet.* **22:** 98–101.

Collins A.L., Levenson J.M., Vilaythong A.P., Richman R.D., Armstrong L., Noebels J.L., Sweatt J.D., and Zoghbi H.Y. 2004. Mild overexpression of MeCP2 causes a progressive neurological disorder in mice. *Hum. Mol. Genet.* **13:** 2679–2689.

Cooney C.A. 1993. Are somatic cells inherently deficient in methylation metabolism? A proposed mechanism for DNA methylation loss, senescence and aging. *Growth Dev. Aging* **57:** 261–273.

Cooper W.N., Luharia A., Evans G.A., Raza H., Haire A.C., Grundy R., Bowdin S.C., Riccio A., Sebastio G., Bliek J., et al. 2005. Molecular subtypes and phenotypic expression of Beckwith-Wiedemann syndrome. *Eur. J. Hum. Genet.* **13:** 1025–1032.

Couvert P., Bienvenu T., Aquaviva C., Poirier K., Moraine C., Gendrot

C., Verloes A., Andres C., Le Fevre A.C., Souville I., et al. 2001. *MECP2* is highly mutated in X-linked mental retardation. *Hum. Mol. Genet.* **10:** 941–946.

Cox G.F., Burger J., Lip V., Mau U.A., Sperling K., Wu B.L., and Horsthemke B. 2002. Intracytoplasmic sperm injection may increase the risk of imprinting defects. *Am. J. Hum. Genet.* **71:** 162–164.

Curtin P., Pirastu M., Kan Y.W., Gobert-Jones J.A., Stephens A.D., and Lehmann H. 1985. A distant gene deletion affects β-globin gene function in an atypical γζβ-thalassemia. *J. Clin. Invest.* **76:** 1554–1558.

Darnell J.C., Jensen K.B., Jin P., Brown V., Warren S.T., and Darnell R.B. 2001. Fragile X mental retardation protein targets G quartet mRNAs important for neuronal function. *Cell* **107:** 489–499.

Darnell J.C., Fraser C.E., Mostovetsky O., Stefani G., Jones T.A., Eddy S.R., and Darnell R.B. 2005. Kissing complex RNAs mediate interaction between the Fragile-X mental retardation protein KH2 domain and brain polyribosomes. *Genes Dev.* **19:** 903–918.

DeBaun M.R., Niemitz E.L., and Feinberg A.P. 2003. Association of in vitro fertilization with Beckwith-Wiedemann syndrome and epigenetic alterations of *LIT1* and *H19*. *Am. J. Hum. Genet.* **72:** 156–160.

Dennis C. 2003. Epigenetics and disease: Altered states. *Nature* **421:** 686–688.

Ding F., Prints Y., Dhar M.S., Johnson D.K., Garnacho-Montero C., Nicholls R.D., and Francke U. 2005. Lack of *Pwcr1/MBII-85* snoRNA is critical for neonatal lethality in Prader-Willi syndrome mouse models. *Mamm. Genome* **16:** 424–431.

Driscoll M.C., Dobkin C.S., and Alter B.P. 1989. γζβ-thalassemia due to a de novo mutation deleting the 5′ β-globin gene activation-region hypersensitive sites. *Proc. Natl. Acad. Sci.* **86:** 7470–7474.

Eggermann T., Wollmann H.A., Kuner R., Eggermann K., Enders H., Kaiser P., and Ranke M.B. 1997. Molecular studies in 37 Silver-Russell syndrome patients: Frequency and etiology of uniparental disomy. *Hum. Genet.* **100:** 415–419.

Ehrlich M. 2003. The ICF syndrome, a DNA methyltransferase 3B deficiency and immunodeficiency disease. *Clin. Immunol.* **109:** 17–28.

Ehrlich M., Buchanan K.L., Tsien F., Jiang G., Sun B., Uicker W., Weemaes C.M., Smeets D., Sperling K., Belohradsky B.H., et al. 2001. DNA methyltransferase 3B mutations linked to the ICF syndrome cause dysregulation of lymphogenesis genes. *Hum. Mol. Genet.* **10:** 2917–2931.

Engel E. 1980. A new genetic concept: Uniparental disomy and its potential effect, isodisomy. *Am. J. Med. Genet.* **6:** 137–143.

Fattal-Valevski A., Bassan H., Korman S.H., Lerman-Sagie T., Gutman A., and Harel S. 2000. Methylenetetrahydrofolate reductase deficiency: Importance of early diagnosis. *J. Child Neurol.* **15:** 539–543.

Feng Y., Absher D., Eberhart D.E., Brown V., Malter H.E., and Warren S.T. 1997. FMRP associates with polyribosomes as an mRNP, and the I304N mutation of severe fragile X syndrome abolishes this association. *Mol Cell* **1:** 109–118.

Flint J., Thomas K., Micklem G., Raynham H., Clark K., Doggett N.A., King A., and Higgs D.R. 1997. The relationship between chromosome structure and function at a human telomeric region. *Nat. Genet.* **15:** 252–257.

Forrester W.C., Epner E., Driscoll M.C., Enver T., Brice M., Papayannopoulou T., and Groudine M. 1990. A deletion of the human beta-globin locus activation region causes a major alteration in chromatin structure and replication across the entire beta-globin locus. *Genes Dev.* **4:** 1637–1649.

Fraga M.F., Ballestar E., Paz M.F., Ropero S., Setien F., Ballestar M.L., Heine-Suner D., Cigudosa J.C., Urioste M., Benitez J., et al. 2005. Epigenetic differences arise during the lifetime of monozygotic twins. *Proc. Natl. Acad. Sci.* **102:** 10604–10609.

Gabellini D., Green M.R., and Tupler R. 2002. Inappropriate gene activation in FSHD: A repressor complex binds a chromosomal repeat deleted in dystrophic muscle. *Cell* **110:** 339–348.

Gibbons R.J., Picketts D.J., Villard L., and Higgs D.R. 1995. Mutations in a putative global transcriptional regulator cause X-linked mental retardation with alpha-thalassemia (ATR-X syndrome). *Cell* **80:** 837–845.

Gibbons R.J., Pellagatti A., Garrick D., Wood W.G., Malik N., Ayyub H., Langford C., Boultwood J., Wainscoat J.S., and Higgs D.R. 2003. Identification of acquired somatic mutations in the gene encoding chromatin-remodeling factor ATRX in the α-thalassemia myelodysplasia syndrome (ATMDS). *Nat. Genet.* **34:** 446–449.

Gicquel C., Rossignol S., Cabrol S., Houang M., Steunou V., Barbu V., Danton F., Thibaud N., Le Merrer M., Burglen L., et al. 2005. Epimutation of the telomeric imprinting center region on chromosome 11p15 in Silver-Russell syndrome. *Nat. Genet.* **37:** 1003–1007.

Gowher H. and Jeltsch A. 2002. Molecular enzymology of the catalytic domains of the Dnmt3a and Dnmt3b DNA methyltransferases. *J. Biol. Chem.* **277:** 20409–20414.

Goyette P., Sumner J.S., Milos R., Duncan A.M., Rosenblatt D.S., Matthews R.G., and Rozen R. 1994. Human methylenetetrahydrofolate reductase: Isolation of cDNA, mapping and mutation identification. *Nat. Genet.* **7:** 195–200.

Grosveld F. 1999. Activation by locus control regions? *Curr. Opin. Genet. Dev.* **9:** 152–157.

Hagberg B., Aicardi J., Dias K., and Ramos O. 1983. A progressive syndrome of autism, dementia, ataxia, and loss of purposeful hand use in girls: Rett's syndrome: Report of 35 cases. *Ann. Neurol.* **14:** 471–479.

Hagerman P.J. and Hagerman R.J. 2004. The fragile-X premutation: A maturing perspective. *Am. J. Hum. Genet.* **74:** 805–816.

Hagerman R.J., Van Housen K., Smith A.C., and McGavran L. 1984. Consideration of connective tissue dysfunction in the fragile X syndrome. *Am. J. Med. Genet.* **17:** 111–121.

Handa V., Saha T., and Usdin K. 2003. The fragile X syndrome repeats form RNA hairpins that do not activate the interferon-inducible protein kinase, PKR, but are cut by Dicer. *Nucleic Acids Res.* **31:** 6243–6248.

Hansen R.S., Wijmenga C., Luo P., Stanek A.M., Canfield T.K., Weemaes C.M., and Gartler S.M. 1999. The *DNMT3B* DNA methyltransferase gene is mutated in the ICF immunodeficiency syndrome. *Proc. Natl. Acad. Sci.* **96:** 14412–14417.

Harikrishnan K.N., Chow M.Z., Baker E.K., Pal S., Bassal S., Brasacchio D., Wang L., Craig J.M., Jones P.L., Sif S., and El-Osta A. 2005. Brahma links the SWI/SNF chromatin-remodeling complex with MeCP2-dependent transcriptional silencing. *Nat. Genet.* **37:** 254–264.

Hark A.T., Schoenherr C.J., Katz D.J., Ingram R.S., Levorse J.M., and Tilghman S.M. 2000. CTCF mediates methylation-sensitive enhancer-blocking activity at the *H19/Igf2* locus. *Nature* **405:** 486–489.

Harrison C.J., Jack E.M., Allen T.D., and Harris R. 1983. The fragile X: A scanning electron microscope study. *J. Med. Genet.* **20:** 280–285.

Hasegawa T., Hara M., Ando M., Osawa M., Fukuyama Y., Takahashi M., and Yamada K. 1984. Cytogenetic studies of familial Prader-Willi syndrome. *Hum. Genet.* **65:** 325–330.

Hayward B.E., Kamiya M., Strain L., Moran V., Campbell R., Hayashizaki Y., and Bonthron D.T. 1998. The human *GNAS1* gene is imprinted and encodes distinct paternally and biallelically expressed G proteins. *Proc. Natl. Acad. Sci.* **95:** 10038–10043.

Henry I., Bonaiti-Pellie C., Chehensse V., Beldjord C., Schwartz C., Utermann G., and Junien C. 1991. Uniparental paternal disomy in a genetic cancer-predisposing syndrome. *Nature* **351:** 665–667.

Hoal-van Helden E.G. and van Helden P.D. 1989. Age-related methylation changes in DNA may reflect the proliferative potential of organs. *Mutat. Res.* **219:** 263–266.

Horike S., Cai S., Miyano M., Cheng J.F., and Kohwi-Shigematsu T. 2005. Loss of silent-chromatin looping and impaired imprinting of *DLX5* in Rett syndrome. *Nat. Genet.* **37:** 31–40.

Hubbard V.S., Davis P.B., di Sant'Agnese P.A., Gorden P., and Schwartz R.H. 1980. Isolated growth hormone deficiency and cystic fibrosis: A report of two cases. *Am. J. Dis. Child* **134:** 317–319.

Ingrosso D., Cimmino A., Perna A.F., Masella L., De Santo N.G., De Bonis M.L., Vacca M., D'Esposito M., D'Urso M., Galletti P., and Zappia V. 2003. Folate treatment and unbalanced methylation and changes of allelic expression induced by hyperhomocysteinaemia in patients with uraemia. *Lancet* **361:** 1693–1699.

Ishizuka A., Siomi M.C., and Siomi H. 2002. A *Drosophila* fragile X protein interacts with components of RNAi and ribosomal proteins. *Genes Dev.* **16:** 2497–2508.

Issa J.P., Ottaviano Y.L., Celano P., Hamilton S.R., Davidson N.E., and Baylin S.B. 1994. Methylation of the oestrogen receptor CpG island links ageing and neoplasia in human colon. *Nat. Genet.* **7:** 536–540.

Jeanpierre M., Turleau C., Aurias A., Prieur M., Ledeist F., Fischer A., and Viegas-Pequignot E. 1993. An embryonic-like methylation pattern of classical satellite DNA is observed in ICF syndrome. *Hum. Mol. Genet.* **2:** 731–735.

Jiang Y.H., Armstrong D., Albrecht U., Atkins C.M., Noebels J.L., Eichele G., Sweatt J.D., and Beaudet A.L. 1998. Mutation of the Angelman ubiquitin ligase in mice causes increased cytoplasmic p53 and deficits of contextual learning and long-term potentiation (see comments). *Neuron* **21:** 799–811.

Jin P., Alisch R.S., and Warren S.T. 2004a. RNA and microRNAs in fragile X mental retardation. *Nat. Cell Biol.* **6:** 1048–1053.

Jin P., Zarnescu D.C., Zhang F., Pearson C.E., Lucchesi J.C., Moses K., and Warren S.T. 2003. RNA-mediated neurodegeneration caused by the fragile X premutation rCGG repeats in *Drosophila*. *Neuron* **39:** 739–747.

Jin P., Zarnescu D.C., Ceman S., Nakamoto M., Mowrey J., Jongens T.A., Nelson D.L., Moses K., and Warren S.T. 2004b. Biochemical and genetic interaction between the fragile X mental retardation protein and the microRNA pathway. *Nat. Neurosci.* **7:** 113–117.

Joyce J.A., Lam W.K., Catchpoole D.J., Jenks P., Reik W., Maher E.R., and Schofield P.N. 1997. Imprinting of IGF2 and H19: Lack of reciprocity in sporadic Beckwith-Wiedemann syndrome. *Hum. Mol. Genet.* **6:** 1543–1548.

Kaufmann W.E., Jarrar M.H., Wang J.S., Lee Y.J., Reddy S., Bibat G., and Naidu S. 2005. Histone modifications in Rett syndrome lymphocytes: A preliminary evaluation. *Brain Dev.* **27:** 331–339.

Kioussis D., Vanin E., deLange T., Flavell R.A., and Grosveld F.G. 1983. Beta globin gene inactivation by DNA translocation in gamma beta-thalassaemia. *Nature* **306:** 662–666.

Kishi N. and Macklis J.D. 2004. MECP2 is progressively expressed in post-migratory neurons and is involved in neuronal maturation rather than cell fate decisions. *Mol. Cell. Neurosci.* **27:** 306–321.

Kishino T., Lalande M., and Wagstaff J. 1997. *UBE3A*/E6-AP mutations cause Angelman syndrome. *Nat. Genet.* **15:** 70–73.

Kondo T., Bobek M.P., Kuick R., Lamb B., Zhu X., Narayan A., Bourc'his D., Viegas-Pequignot E., Ehrlich M., and Hanash S.M. 2000. Whole-genome methylation scan in ICF syndrome: Hypomethylation of non-satellite DNA repeats *D4Z4* and *NBL2*. *Hum. Mol. Genet.* **9:** 597–604.

Korenke G.C., Fuchs S., Krasemann E., Doerr H.G., Wilichowski E., Hunneman D.H., and Hanefeld F. 1996. Cerebral adrenoleukodystrophy (ALD) in only one of monozygotic twins with an identical ALD genotype. *Ann. Neurol.* **40:** 254–257.

Kwok R.P., Lundblad J.R., Chrivia J.C., Richards J.P., Bachinger H.P., Brennan R.G., Roberts S.G., Green M.R., and Goodman R.H. 1994. Nuclear protein CBP is a coactivator for the transcription factor CREB. *Nature* **370:** 223–226.

Laggerbauer B., Ostareck D., Keidel E.M., Ostareck-Lederer A., and Fischer U. 2001. Evidence that fragile X mental retardation protein is a negative regulator of translation. *Hum. Mol. Genet.* **10:** 329–338.

Ledbetter D.H., Riccardi V.M., Airhart S.D., Strobel R.J., Keenan B.S., and Crawford J.D. 1981. Deletions of chromosome 15 as a cause of the Prader-Willi syndrome. *N. Engl. J. Med.* **304:** 325–329.

Lee M.P., DeBaun M.R., Mitsuya K., Galonek H.L., Brandenburg S., Oshimura M., and Feinberg A.P. 1999. Loss of imprinting of a paternally expressed transcript, with antisense orientation to K_VLQT1, occurs frequently in Beckwith-Wiedemann syndrome and is independent of insulin-like growth factor II imprinting. *Proc. Natl. Acad. Sci.* **96:** 5203–5208.

Lemmers R.J., de Kievit P., Sandkuijl L., Padberg G.W., van Ommen G.J., Frants R.R., and van der Maarel S.M. 2002. Facioscapulohumeral muscular dystrophy is uniquely associated with one of the two variants of the 4q subtelomere. *Nat. Genet.* **32:** 235–236.

Lewis J.D., Meehan R.R., Henzel W.J., Maurer-Fogy I., Jeppesen P., Klein F., and Bird A. 1992. Purification, sequence, and cellular localization of a novel chromosomal protein that binds to methylated DNA. *Cell* **69:** 905–914.

Li Z., Zhang Y., Ku L., Wilkinson K.D., Warren S.T., and Feng Y. 2001. The fragile X mental retardation protein inhibits translation via interacting with mRNA. *Nucleic Acids Res.* **29:** 2276–2283.

Liu D., Diorio J., Tannenbaum B., Caldji C., Francis D., Freedman A., Sharma S., Pearson D., Plotsky P.M., and Meaney M.J. 1997. Maternal care, hippocampal glucocorticoid receptors, and hypothalamic-pituitary-adrenal responses to stress. *Science* **277:** 1659–1662.

Lubs H.A. 1969. A marker X chromosome. *Am. J. Hum. Genet.* **21:** 231–244.

Ludwig M., Katalinic A., Gross S., Sutcliffe A., Varon R., and Horsthemke B. 2005. Increased prevalence of imprinting defects in patients with Angelman syndrome born to subfertile couples. *J. Med. Genet.* **42:** 289–291.

Ma J., Stampfer M.J., Giovannucci E., Artigas C., Hunter D.J., Fuchs C., Willett W.C., Selhub J., Hennekens C.H., and Rozen R. 1997. Methylenetetrahydrofolate reductase polymorphism, dietary interactions, and risk of colorectal cancer. *Cancer Res.* **57:** 1098–1102.

Magenis R.E., Brown M.G., Lacy D.A., Budden S., and LaFranchi S. 1987. Is Angelman syndrome an alternate result of del(15)(q11q13)? *Am. J. Med. Genet.* **28:** 829–838.

Martin J. and Bell J. 1943. A pedigree of mental defect showing sex-linkage. *Arch. Neurol. Psychiat.* **6:** 154–157.

Martinowich K., Hattori D., Wu H., Fouse S., He F., Hu Y., Fan G., and Sun Y.E. 2003. DNA methylation-related chromatin remodeling in activity-dependent *Bdnf* gene regulation. *Science* **302:** 890–893.

Mathews K.D. 2003. Muscular dystrophy overview: Genetics and diagnosis. *Neurol. Clin.* **21:** 795–816.

Matsuura T., Sutcliffe J.S., Fang P., Galjaard R.J., Jiang Y.H., Benton C.S., Rommens J.M., and Beaudet A.L. 1997. De novo truncating mutations in E6-AP ubiquitin-protein ligase gene (*UBE3A*) in Angelman syndrome. *Nat. Genet.* **15:** 74–77.

Mayr B. and Montminy M. 2001. Transcriptional regulation by the phosphorylation-dependent factor CREB. *Nat. Rev. Mol. Cell. Biol.* **2:** 599–609.

Mays-Hoopes L.L. 1989. Age-related changes in DNA methylation: Do they represent continued developmental changes? *Int. Rev. Cytol.* **114:** 181–220.

McDowell T.L., Gibbons R.J., Sutherland H., O'Rourke D.M., Bickmore W.A., Pombo A., Turley H., Gatter K., Picketts D.J., Buckle V.J., et al. 1999. Localization of a putative transcriptional regulator (ATRX) at pericentromeric heterochromatin and the short arms of acrocentric chromosomes. *Proc. Natl. Acad. Sci.* **96:** 13983–13988.

Meguro M., Mitsuya K., Nomura N., Kohda M., Kashiwagi A., Nishigaki R., Yoshioka H., Nakao M., Oishi M., and Oshimura M. 2001. Large-scale evaluation of imprinting status in the Prader-Willi syndrome region: An imprinted direct repeat cluster resembling small nucleolar RNA genes. *Hum. Mol. Genet.* **10:** 383–394.

Meins M., Lehmann J., Gerresheim F., Herchenbach J., Hagedorn M., Hameister K., and Epplen J.T. 2005. Submicroscopic duplication in Xq28 causes increased expression of the *MECP2* gene in a boy with severe mental retardation and features of Rett syndrome. *J. Med. Genet.* **42:** e12.

Meloni I., Bruttini M., Longo I., Mari F., Rizzolio F., D'Adamo P., Denvriendt K., Fryns J.P., Toniolo D., and Renieri A. 2000. A mutation in the Rett syndrome gene, *MECP2*, causes X-linked mental retardation and progressive spasticity in males. *Am. J. Hum. Genet.* **67:** 982–985.

Mullaney B.C., Johnston M.V., and Blue M.E. 2004. Developmental expression of methyl-CpG binding protein 2 is dynamically regulated in the rodent brain. *Neuroscience* **123:** 939–949.

Murata T., Kurokawa R., Krones A., Tatsumi K., Ishii M., Taki T., Masuno M., Ohashi H., Yanagisawa M., Rosenfeld M.G., et al. 2001. Defect of histone acetyltransferase activity of the nuclear transcriptional coactivator CBP in Rubinstein-Taybi syndrome. *Hum. Mol. Genet.* **10:** 1071–1076.

Nan X., Campoy F.J., and Bird A. 1997. MeCP2 is a transcriptional repressor with abundant binding sites in genomic chromatin. *Cell* **88:** 471–481.

Neul J.L. and Zoghbi H.Y. 2004. Rett syndrome: A prototypical neurodevelopmental disorder. *Neuroscientist* **10:** 118–128.

Nicholls R.D., Knoll J.H.M., Butler M.G., Karam S., and Lalande M. 1989. Genetic imprinting suggested by maternal heterodisomy in nondeletion Prader-Willi syndrome. *Nature* **342:** 281–285.

Nuber U.A., Kriaucionis S., Roloff T.C., Guy J., Selfridge J., Steinhoff C., Schulz R., Lipkowitz B., Ropers H.H., Holmes M.C., and Bird A. 2005. Up-regulation of glucocorticoid-regulated genes in a mouse model of Rett syndrome. *Hum. Mol. Genet.* **14:** 2247–2256.

Ogryzko V.V., Schiltz R.L., Russanova V., Howard B.H., and Nakatani Y. 1996. The transcriptional coactivators p300 and CBP are histone acetyltransferases. *Cell* **87:** 953–959.

Ohta T., Gray T.A., Rogan P.K., Buiting K., Gabriel J.M., Saitoh S., Muralidhar B., Bilienska B., Krajewska-Walasek M., Driscoll D.J., et al. 1999. Imprinting-mutation mechanisms in Prader-Willi syndrome. *Am. J. Hum. Genet.* **64:** 397–413.

Okano M., Bell D.W., Haber D.A., and Li E. 1999. DNA methyltransferases Dnmt3a and Dnmt3b are essential for de novo methylation and mammalian development. *Cell* **99:** 247–257.

Orstavik K.H., Eiklid K., van der Hagen C.B., Spetalen S., Kierulf K., Skjeldal O., and Buiting K. 2003. Another case of imprinting defect in a girl with Angelman syndrome who was conceived by intracytoplasmic semen injection. *Am. J. Hum. Genet.* **72:** 218–219.

Petrij F., Giles R.H., Dauwerse H.G., Saris J.J., Hennekam R.C., Masuno M., Tommerup N., van Ommen G.J., Goodman R.H., Peters D.J., et al. 1995. Rubinstein-Taybi syndrome caused by mutations in the transcriptional co-activator CBP. *Nature* **376:** 348–351.

Petronis A. 2004. The origin of schizophrenia: Genetic thesis, epigenetic antithesis, and resolving synthesis. *Biol. Psychiatry* **55:** 965–970.

Picketts D.J., Higgs D.R., Bachoo S., Blake D.J., Quarrell O.W., and Gibbons R.J. 1996. ATRX encodes a novel member of the SNF2 family of proteins: Mutations point to a common mechanism underlying the ATR-X syndrome. *Hum. Mol. Genet.* **5:** 1899–1907.

Pieretti M., Zhang F., Fu Y.-H., Warren S.T., Oostra B.A., Caskey C.T., and Nelson D.L. 1991. Absence of expression of the *FMR-1* gene in fragile X syndrome. *Cell* **66:** 817–822.

Ping A.J., Reeve A.E., Law D.J., Young M.R., Boehnke M., and Feinberg A.P. 1989. Genetic linkage of Beckwith-Wiedemann syndrome to 11p15. *Am. J. Hum. Genet.* **44:** 720–773.

Prawitt D., Enklaar T., Gartner-Rupprecht B., Spangenberg C., Oswald M., Lausch E., Schmidtke P., Reutzel D., Fees S., Lucito R., et al. 2005. Microdeletion of target sites for insulator protein CTCF in a chromosome 11p15 imprinting center in Beckwith-Wiedemann syndrome and Wilms' tumor. *Proc. Natl. Acad. Sci.* **102:** 4085–4090.

Rampersaud G.C., Kauwell G.P., Hutson A.D., Cerda J.J., and Bailey L.B. 2000. Genomic DNA methylation decreases in response to moderate folate depletion in elderly women. *Am. J. Clin. Nutr.* **72:** 998–1003.

Reik W. 1989. Genomic imprinting and genetic disorders in man. *Trends Genet.* **5:** 331–336.

Reynolds E.H., Carney M.W., and Toone B.K. 1984. Methylation and mood. *Lancet* **2:** 196–198.

Richards B.W., Sylvester P.E., and Brooker C. 1981. Fragile X-linked mental retardation: The Martin-Bell syndrome. *J. Ment. Defic. Res.* **25:** 253–256.

Roelfsema J.H., White S.J., Ariyurek Y., Bartholdi D., Niedrist D., Papadia F., Bacino C.A., den Dunnen J.T., van Ommen G.J., Breuning M.H., et al. 2005. Genetic heterogeneity in Rubinstein-Taybi syndrome: Mutations in both the *CBP* and *EP300* genes cause disease. *Am. J. Hum. Genet.* **76:** 572–580.

Rougeulle C., Cardoso C., Fontes M., Colleaux L., and Lalande M. 1998. An imprinted antisense RNA overlaps UBE3A and a second maternally expressed transcript. *Nat. Genet.* **19:** 15–16.

Rozen R. 1996. Molecular genetics of methylenetetrahydrofolate reductase deficiency. *J. Inherit. Metab. Dis.* **19:** 589–594.

Runte M., Varon R., Horn D., Horsthemke B., and Buiting K. 2005. Exclusion of the C/D box snoRNA gene cluster *HBII-52* from a major role in Prader-Willi syndrome. *Hum. Genet.* **116:** 228–230.

Schimke R.N., Horton W.A., and King C.R. 1971. Chondroitin-6-sulphaturia, defective cellular immunity, and nephrotic syndrome. *Lancet* **2:** 1088–1089.

Schule B., Albalwi M., Northrop E., Francis D.I., Rowell M., Slater H.R., Gardner R.J., and Francke U. 2005. Molecular breakpoint cloning and gene expression studies of a novel translocation t(4;15)(q27;q11.2) associated with Prader-Willi syndrome. *BMC Med. Genet.* **6:** 18.

Schwahn B. and Rozen R. 2001. Polymorphisms in the methylenetetrahydrofolate reductase gene: Clinical consequences. *Am. J. Pharmacogenomics* **1:** 189–201.

Shahbazian M.D., Antalffy B., Armstrong D.L., and Zoghbi H.Y. 2002a. Insight into Rett syndrome: MeCP2 levels display tissue- and cell-specific differences and correlate with neuronal maturation. *Hum. Mol. Genet.* **11:** 115–124.

Shahbazian M., Young J., Yuva-Paylor L., Spencer C., Antalffy B., Noebels J., Armstrong D., Paylor R., and Zoghbi H. 2002b. Mice with truncated MeCP2 recapitulate many Rett syndrome features and display hyperacetylation of histone H3. *Neuron* **35:** 243–254.

Smeets D.F., Moog U., Weemaes C.M., Vaes-Peeters G., Merkx G.F., Niehof J.P., and Hamers G. 1994. ICF syndrome: A new case and review of the literature. *Hum. Genet.* **94:** 240–246.

Smeets D.F., Hamel B.C., Nelen M.R., Smeets H.J., Bollen J.H., Smits

A.P., Ropers H.H., and van Oost B.A. 1992. Prader-Willi syndrome and Angelman syndrome in cousins from a family with a translocation between chromosomes 6 and 15. *N. Engl. J. Med.* **326:** 807–811.

Smilinich N.J., Day C.D., Fitzpatrick G.V., Caldwell G.M., Lossie A.C., Cooper P.R., Smallwood A.C., Joyce J.A., Schofield P.N., Reik W., et al. 1999. A maternally methylated CpG island in *KvLQT1* is associated with an antisense paternal transcript and loss of imprinting in Beckwith-Wiedemann syndrome. *Proc. Natl. Acad. Sci.* **96:** 8064–8069.

Spence J.E., Perciaccante R.G., Greig G.M., Willard H.F., Ledbetter D.H., Hejtmancik J.F., Pollack M.S., O'Brien W.E., and Beaudet A.L. 1988. Uniparental disomy as a mechanism for human genetic disease. *Am. J. Hum. Genet.* **42:** 217–226.

Spranger J., Hinkel G.K., Stoss H., Thoenes W., Wargowski D., and Zepp F. 1991. Schimke immuno-osseous dysplasia: A newly recognized multisystem disease. *J. Pediatr.* **119:** 64–72.

Sun F.L., Dean W.L., Kelsey G., Allen N.D., and Reik W. 1997. Transactivation of *Igf2* in a mouse model of Beckwith-Wiedemann syndrome. *Nature* **389:** 809–815.

Suter C.M., Martin D.I., and Ward R.L. 2004. Germline epimutation of MLH1 in individuals with multiple cancers. *Nat. Genet.* **36:** 497–501.

Sutherland G.R. 1977. Fragile sites on human chromosomes: Demonstration of their dependence on the type of tissue culture medium. *Science* **197:** 265–266.

Taha D., Boerkoel C.F., Balfe J.W., Khalifah M., Sloan E.A., Barbar M., Haider A., and Kanaan H. 2004. Fatal lymphoproliferative disorder in a child with Schimke immuno-osseous dysplasia. *Am. J. Med. Genet. A* **131:** 194–199.

Tommerup N., van der Hagen C.B., and Heiberg A. 1992. Tentative assignment of a locus for Rubinstein-Taybi syndrome to 16p13.3 by a de novo reciprocal translocation, t(7;16)(q34;p13.3). *Am. J. Med. Genet.* **44:** 237–241.

Tsai T.F., Jiang Y.H., Bressler J., Armstrong D., and Beaudet A.L. 1999. Paternal deletion from Snrpn to Ube3a in the mouse causes hypotonia, growth retardation and partial lethality and provides evidence for a gene contributing to Prader-Willi syndrome. *Hum. Mol. Genet.* **8:** 1357–1364.

Tuck-Muller C.M., Narayan A., Tsien F., Smeets D.F., Sawyer J., Fiala E.S., Sohn O.S., and Ehrlich M. 2000. DNA hypomethylation and unusual chromosome instability in cell lines from ICF syndrome patients. *Cytogenet. Cell Genet.* **89:** 121–128.

Tupler R. and Gabellini D. 2004. Molecular basis of facioscapulohumeral muscular dystrophy. *Cell. Mol. Life Sci.* **61:** 557–566.

van Deutekom J.C., Wijmenga C., van Tienhoven E.A., Gruter A.M., Hewitt J.E., Padberg G.W., van Ommen G.J., Hofker M.H., and Frants R.R. 1993. FSHD associated DNA rearrangements are due to deletions of integral copies of a 3.2 kb tandemly repeated unit. *Hum. Mol. Genet.* **2:** 2037–2042.

Van Esch H., Bauters M., Ignatius J., Jansen M., Raynaud M., Hollanders K., Lugtenborg D., Bienvenu T., Jensen L.R., Gecz J., et al. 2005. Duplication of the *MECP2* region is a frequent cause of severe mental retardation and progressive neurological symptoms in males. *Am. J. Hum. Genet.* **77:** 442–453.

Verkerk A.J.M.H., Pieretti M., Sutcliffe J.S., Fu Y.-H., Kuhl D.P.A., Pizzuti A., Reiner O., Richards S., Victoria M.F., Zhang R., et al. 1991. Identification of a gene (FMR-1) containing a CGG repeat coincident with a breakpoint cluster region exhibiting length variation in fragile X syndrome. *Cell* **65:** 905–914.

Villard L., Gecz J., Mattei J.F., Fontes M., Saugier-Veber P., Munnich A., and Lyonnet S. 1996. XNP mutation in a large family with Juberg-Marsidi syndrome. *Nat. Genet.* **12:** 359–360.

Wan M., Zhao K., Lee S.S., and Francke U. 2001. *MECP2* truncating mutations cause histone H4 hyperacetylation in Rett syndrome. *Hum. Mol. Genet.* **10:** 1085–1092.

Wan M., Lee S.S., Zhang X., Houwink-Manville I., Song H.R., Amir R.E., Budden S., Naidu S., Pereira J.L., Lo I.F., et al. 1999. Rett syndrome and beyond: Recurrent spontaneous and familial *MECP2* mutations at CpG hotspots. *Am. J. Hum. Genet.* **65:** 1520–1529.

Warren S.T. and Sherman S.L. 2001. The fragile X syndrome. In *The metabolic and molecular bases of inherited disease*, 8th edition (ed. C.R. Scriver et al.), vol. 1, pp. 1257–1289. McGraw-Hill, New York.

Waterland R.A. and Jirtle R.L. 2003. Transposable elements: Targets for early nutritional effects on epigenetic gene regulation. *Mol. Cell. Biol.* **23:** 5293–5300.

Weatherall D.J., Clegg M.B., Higgs D.R., and Wood W.G. 2001. The hemoglobinopathies. In *The metabolic & molecular bases of inherited disease*, 8th edition (ed. C.R. Scriver et al.), pp. 4571–4636. McGraw-Hill, New York.

Weaver I.C., Cervoni N., Champagne F.A., D'Alessio A.C., Sharma S., Seckl J.R., Dymov S., Szyf M., and Meaney M.J. 2004. Epigenetic programming by maternal behavior. *Nat. Neurosci.* **7:** 847–854.

Weiler I.J. and Greenough W.T. 1999. Synaptic synthesis of the Fragile X protein: Possible involvement in synapse maturation and elimination. *Am. J. Med. Genet.* **83:** 248–252.

Weksberg R., Smith A.C., Squire J., and Sadowski P. 2003. Beckwith-Wiedemann syndrome demonstrates a role for epigenetic control of normal development. *Hum. Mol. Genet.* **12:** R61–R68.

Weksberg R., Nishikawa J., Caluseriu O., Fei Y.L., Shuman C., Wei C., Steele L., Cameron J., Smith A., Ambus I., et al. 2001. Tumor development in the Beckwith-Wiedemann syndrome is associated with a variety of constitutional molecular 11p15 alterations including imprinting defects of *KCNQ1OT1*. *Hum. Mol. Genet.* **10:** 2989–3000.

Weksberg R., Teshima I., Williams B.R., Greenberg C.R., Pueschel S.M., Chernos J.E., Fowlow S.B., Hoyme E., Anderson I.J., Whiteman D.A., et al. 1993. Molecular characterization of cytogenetic alterations associated with the Beckwith-Wiedemann syndrome (BWS) phenotype refines the localization and suggests the gene for BWS is imprinted. *Hum. Mol. Genet.* **2:** 549–556.

Wijmenga C., Hewitt J.E., Sandkuijl L.A., Clark L.N., Wright T.J., Dauwerse H.G., Gruter A.M., Hofker M.H., Moerer P., Williamson R., et al. 1992. Chromosome 4q DNA rearrangements associated with facioscapulohumeral muscular dystrophy. *Nat. Genet.* **2:** 26–30.

Willemsen R., Hoogeveen-Westerveld M., Reis S., Holstege J., Severijnen L.A., Nieuwenhuizen I.M., Schrier M., Van Unen L., Tassone F., Hoogeveen A.T., et al. 2003. The *FMR1* CGG repeat mouse displays ubiquitin-positive intranuclear neuronal inclusions; implications for the cerebellar tremor/ataxia syndrome. *Hum. Mol. Genet.* **12:** 949–959.

Wolff G.L., Kodell R.L., Moore S.R., and Cooney C.A. 1998. Maternal epigenetics and methyl supplements affect *agouti* gene expression in A^{vy}/a mice. *Faseb J.* **12:** 949–957.

Xu G.L., Bestor T.H., Bourc'his D., Hsieh C.L., Tommerup N., Bugge M., Hulten M., Qu X., Russo J.J., and Viegas-Pequignot E. 1999. Chromosome instability and immunodeficiency syndrome caused by mutations in a DNA methyltransferase gene. *Nature* **402:** 187–191.

Yntema H.G., Poppelaars F.A., Derksen E., Oudakker A.R., van Roosmalen T., Jacobs A., Obbema H., Brunner H.G., Hamel B.C., and van Bokhoven H. 2002. Expanding phenotype of XNP mutations: Mild to moderate mental retardation. *Am. J. Med. Genet.* **110:** 243–247.

Young J.I., Hong E.P., Castle J., Crespo-Barreto J., Bowman A.B., Rose M.F., Kang D., Richman R., Johnson J.M., Berget S., and Zoghbi H.Y. 2005. Inaugural article: Regulation of RNA splicing by the methylation-dependent transcriptional repressor methyl-CpG binding protein 2. *Proc. Natl. Acad. Sci.* **102:** 17551–17558.

Zeev B.B., Yaron Y., Schanen N.C., Wolf H., Brandt N., Ginot N., Shomrat R., and Orr-Urtreger A. 2002. Rett syndrome: Clinical manifestations in males with *MECP2* mutations. *J. Child Neurol.* **17:** 20–24.

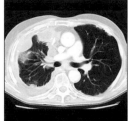

Epigenetic Determinants of Cancer

Stephen B. Baylin[1] and Peter A. Jones[2]

[1]Cancer Biology Program, The Sidney Kimmel Cancer Center, Johns Hopkins Medical Institutions, Baltimore, Maryland 21231
[2]Department of Urology, Biochemistry and Molecular Biology, USC/Norris Comprehensive Cancer Center, Keck School of Medicine, University of Southern California, Los Angeles, California 90089-9181

CONTENTS

1. The Biological Basis of Cancer, 459

2. The Importance of Chromatin to Cancer, 459

3. The Role of DNA Methylation in Cancer, 461

4. Hypermethylated Gene Promoters in Cancer, 462
    4.1 The Genes Involved, 462
    4.2 Searching for New Genes Epigenetically Silenced in Cancer, 463
    4.3 Determining the Functional Importance of Genes Hypermethylated in Cancer, 464

5. Epigenetic Gene Silencing and Its Role in the Evolution of Cancer—Importance for Early Tumor Progression Stages, 465

6. The Molecular Anatomy of Epigenetically Silenced Cancer Genes, 467

7. Summary of Major Research Issues for Understanding Epigenetic Gene Silencing in Cancer, 470

8. Detection of Cancer by DNA Methylation, 471

9. Epigenetic Therapy, 471

References, 473

GENERAL SUMMARY

Cancer is caused by the heritable deregulation of genes, which control when cells divide, die, and move from one part of the body to another. During the process of carcinogenesis, genes can become activated in ways that enhance division or prevent cell death, or alternatively, they can become inactivated so that they no longer are available to apply the brakes to these processes. The first class of genes is called "oncogenes" and the second "tumor suppressor genes." It is the interplay between these two gene classes that results in the formation of cancer.

Genes can become inactivated by at least three pathways, including (1) a gene can be mutated so that its function becomes disabled; (2) a gene can be completely lost and thus not be available to work appropriately; and (3) a gene, which has not been mutated or lost, can be switched off in a heritable fashion by epigenetic changes. This epigenetic silencing can involve histone modifications, the binding of repressive proteins, and inappropriate methylation of cytosine (C) residues in CpG sequence motifs that reside within control regions which govern gene expression.

This chapter focuses on this third pathway. The basic molecular mechanisms responsible for maintaining the silenced state are quite well understood, as outlined in this book. Consequently, we also know that epigenetic silencing has profound implications for cancer prevention, detection, and therapies. We now have drugs available approved by the American FDA which can reverse epigenetic changes and restore gene activity to cancer cells. Additionally, because the changes in DNA methylation can be analyzed with a high degree of sensitivity, many strategies to detect cancer early rely on finding DNA methylation changes. The translational opportunities for epigenetics in human cancer research, detection, prevention, and treatment are therefore quite extraordinary.

1 The Biological Basis of Cancer

Cancer is ultimately a disease of gene expression in which the complex networks governing homeostasis in multicellular organisms become deranged, allowing cells to grow without reference to the needs of the organism as a whole. Great advances have been made in the delineation of the subset of cellular control pathways subject to derangement in human cancer (Table 1). That this limited number of cellular control pathways are affected and heritably disabled in almost all cancers is a key concept that has advanced the field (Hanahan and Weinberg 2000). The focus, until the last several years, has largely been on the genetic basis of cancer, particularly on the mutational activation of oncogenes or inactivation of tumor suppressor genes. However, a growing body of data has appeared since the mid-1990s to indicate that heritable changes, regulated by epigenetic alterations, may also be critical for the evolution of all human cancer types (Fig. 1) (Jones and Laird 1999; Jones and Baylin 2002; Herman and Baylin 2003). These data, particularly DNA and chromatin methylation patterns that are fundamentally altered in cancers, have led to new opportunities for the understanding, detection, treatment, and prevention of cancer.

Genetic and epigenetic abnormalities can cause heritable disruptions to homeostatic pathways by two different mechanisms. Either the activation of an oncogene can occur, generally through activating point mutations, or tumor suppressor genes can be inactivated (Jones and Laird 1999; Hanahan and Weinberg 2000; Jones and Baylin 2002; Herman and Baylin 2003). For example, mutations in a signaling gene (oncogene) such as *RAS*, which enhance the activity of the gene product to stimulate growth, are often found in human cancers. These mutations are often dominant and drive the formation of cancers. Genetic mutations and epigenetic silencing of tumor suppressor genes, on the other hand, are often recessive, requiring disruptive events in both allelic copies of a gene for the full expression of the transformed phenotype. The idea that both copies of a tumor suppressor gene had to be incapacitated in a malignant cell line was proposed by Knudson (2001) and has found wide acceptance. It is now realized that three classes of "hits" can participate in different combinations to cause a complete loss of activity of tumor suppressor genes. Direct mutations in the coding sequence, loss of parts or entire copies of genes, or epigenetic silencing can cooperate with each other to result in the disablement of key control genes (Fig. 2).

2 The Importance of Chromatin to Cancer

Despite the major advances in understanding the key molecular lesions in cellular control pathways that contribute to cancer, it remains true that microscopic examination of nuclear structure by a pathologist is a gold standard in cancer diagnosis. The human eye can accurately discern changes in nuclear architecture, which largely involve the state of chromatin configuration, and definitively diagnose the cancer phenotype in a single cell. Foremost in the cues used by pathologists are the size of the nucleus, nuclear outline, a condensed nuclear membrane, prominent nucleoli, dense "hyperchromatic" chromatin, and a high nuclear/cytoplasmic ratio. These structural features, visible under a microscope (Fig. 3), likely correlate with profound alterations in chromatin function and resultant changes in gene expression states and/or chromosome stability. Linking changes observable at a microscopic level with the molecular marks discussed throughout this book remains one of the great challenges in cancer research. In this chapter, we review epigenetic marks, typified by changes in DNA cytosine methylation

Table 1. Key cellular pathways disrupted in human cancers by genetic and epigenetic mechanisms

Pathway	Example of genetic alteration	Example of epigenetic alteration
Self-sufficiency in growth signals	mutations in *RAS* gene	methylation of *RASSFIA* gene
Insensitivity to antigrowth signals	mutation in *TGFβ* receptor genes	down-regulation of TGFβ receptors
Tissue invasion and metastasis	mutation in *E-Cadherin* gene	methylation of *E-Cadherin* promoter
Limitless replicative potential	mutations in *p16* and *Rb* genes	silencing of *p16* or *Rb* genes by promoter methylation
Sustained angiogenesis		silencing of thrombospondin-1
Evading apoptosis	mutation in *p53*	methylation of *DAP-kinase*, *ASC/TMS1*, and *HIC1*
DNA repair capacity	mutations in *MLH1*, *MSH2*	methylation of *GST Pi*, *O6-MGMT*, *MLH1*
Monitoring genomic stability	mutations in *Chfr*	methylation of *Chfr*
Protein ubiquination functions	mutations in *Chfr*	methylation of *Chfr*

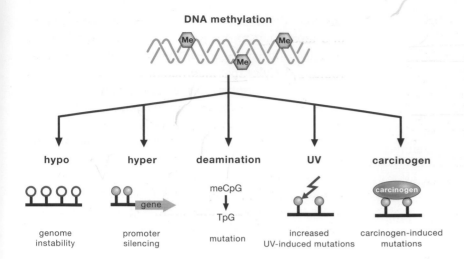

Figure 1. Epigenetic Alterations Involving DNA Methylation Can Lead to Cancer by Various Mechanisms

Loss of DNA cytosine methylation (hypo) results in genome instability. Focal hypermethylation in gene promoters (hyper) causes heritable silencing and therefore inactivation of tumor suppressor genes. Additionally, methylated CpG sites are hotspots for C→T transition mutations caused by spontaneous hydrolytic deamination. Methylation of CpG sites also increases the binding of some chemical carcinogens to DNA and increases the rate of UV-induced mutations.

at CpG dinucleotides and histone modifications, which are abnormally distributed in cancer cells. They are increasingly being linked to heritable events that affect the stability and function of the genome and, thus, contribute very significantly to the cancer phenotype.

Several examples of the roles of chromatin-modifying activities in human cancer are known (Wolffe 2001). For example, acute myeloid leukemia (AML) and acute promyelocytic leukemia (PML) are both caused by chromosomal translocations that alter the use of histone deacetylases (HDACs). In PML, the PML gene is fused to the retinoic acid receptor (RAR). This receptor recruits HDAC activity and DNA methylation, and causes a state of transcriptional silencing, as shown with experimental promoter constructs. The data suggest that this targeting of chromatin change can potentially lead to tumor suppressor gene silencing, which participates in a cellular differentiation block (Di Croce et al. 2002). In AML, the DNA-binding domain of the transcription factor AML-1 is fused to a protein called ETO, which interacts with a HDAC. Repression of cellular differentiation by the mistargeted HDAC contributes to aberrant gene repression and, ultimately, leukemia (Amann et al. 2001). These are just two examples of the direct involvement of chromatin modifications in the oncogenic phenotype. It has, however, become clear that chromatin modifications can

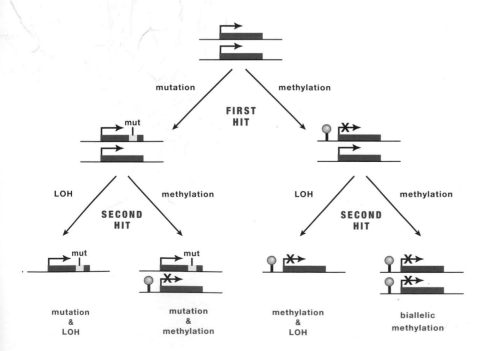

Figure 2. How DNA Methylation Can Contribute to the Inactivation of Tumor Suppressor Genes

Two active alleles of a tumor suppressor gene are shown as the two blue boxes at the top. The first step of gene inactivation is shown as a localized mutation (*left*) or gene silencing by DNA methylation (*right*). The second hit is shown as either a loss of heterozygosity (LOH) or transcriptional silencing by additional epigenetic events. In this way, DNA methylation can contribute as one of the pathways to satisfy Knudson's hypothesis.

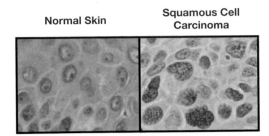

Figure 3. Chromatin Structural Changes in Cancer Cells

These two photomicrographs were taken from a patient with a squamous cell carcinoma of the skin. The left panel shows normal epidermal cells within one millimeter of the contiguous tumor, shown at the same magnification on the right. The chromatin, which stains purple due to its affinity to hematoxylin, appears much more coarse and granular in the cancer cells than in normal epidermis. Such changes in the staining characteristics of chromatin are used by pathologists as diagnostic criteria for cancer.

directly and indirectly alter the patterns of cytosine methylation, an epigenetic change of the DNA which can either initiate or "lock in" silencing of key genes leading to heritable perturbations in key cellular pathways.

3 The Role of DNA Methylation in Cancer

The initial discovery that DNA contained 5-methylcytosine, in addition to the four bases directly incorporated into DNA, soon led to the proposal that alterations in DNA methylation may contribute to oncogenesis (Table 2). Over the last 40 years, there have been many studies

that have shown alterations in the patterns of distribution of 5-methylcytosine between cancer and normal cells in human DNA. Among these, there are at least three major routes by which CpG methylation can contribute to the oncogenic phenotype. These include hypomethylation of the cancer genome, focal hypermethylation of the promoters of tumor suppressor genes, and direct mutagenesis (Fig. 1) (Jones and Laird 1999; Jones and Baylin 2002; Herman and Baylin 2003). Although each of these alterations individually could contribute to cancer causation in humans, it is, perhaps, most significant that all three occur simultaneously, thus indicating that alterations in the homeostasis of epigenetic mechanisms are central contributors to human cancer.

The most prominent, and the earliest recognized, change in DNA methylation patterns in cancer cells is an overall decrease in this modification, which could contribute to genomic instability (for further discussion, see Chapter 18). This is well known to be a hallmark of human cancer (Feinberg and Vogelstein 1983; Feinberg et al. 1988; Jones and Laird 1999; Jones and Baylin 2002; Herman and Baylin 2003). More recently, a mounting body of data has illustrated that the abnormal methylation of CpG islands in the 5′ regions of cancer-related genes is integral to their transcriptional silencing, providing an alternative mechanism to mutation for the inactivation of genes with tumor suppressor function (Jones and Laird 1999; Jones and Baylin 2002; Herman and Baylin 2003). Finally, in addition to the above roles of cytosine methyla-

Table 2. Time line for elucidating the role of DNA methylation in cancer

Observation	Reference
Hypothesis of "methylases as oncogenic agents"	Srinivasan and Borek (1964)
Decreased levels of 5-methylcytosine in animal tumors	Lapeyre and Becker (1979)
5-Azacytidine and 5-aza-2′-deoxycytidine inhibit methylation and activate genes	Jones and Taylor (1980)
Decreased genomic and gene-specific methylation in human tumors	Ehrlich et al. (1982); Feinberg and Vogelstein (1983); Flatau et al. (1983)
Inhibitors of DNA methylation alter tumorigenic phenotype	Frost et al. (1984)
Methylation of a CpG island in cancer	Baylin et al. (1987)
Hot spots for p53 mutations are methylated CpG sites	Rideout et al. (1990)
Allele-specific methylation of the retinoblastoma tumor suppressor gene	Sakai et al. (1991)
Loss of imprinting in cancer	Rainier et al. (1993)
Hypermethylation of CpG islands is associated with aging	Issa et al. (1994)
Mice with decreased methylation develop fewer tumors	Laird et al. (1995)
Coupling DNA methylation and HDAC inhibitors leads to rapid isolation of tumor suppressor genes	Suzuki et al. (2002); Yamashita et al. (2002)
DNA repair gene (*MLH1*) is methylated in somatic cells	Gazzoli et al. (2002)
5-Azacytidine is FDA-approved for treatment of myelodysplastic syndrome	Kaminskas et al. (2005)

instability and gene silencing, 5-methyl-
a highly unstable base, and hence muta-
gens. This can contribute directly to cancer by causing
transition mutations in which meCpG is converted to
TpG (Rideout et al. 1990). The fact that these modifications are so prevalent in cancers and are now known to
contribute directly to carcinogenesis has also led to new
possibilities in which epigenetic changes are targeted for
therapeutic reversal (Egger et al. 2004).

DNA cytosine methylation is therefore now acknowl-
edged to play a critical role in human carcinogenesis.
Almost all human genes contain methylated cytosine
residues in their coding regions, known for some time to
contribute disproportionately to the formation of disease-causing mutations. The methylation of the carbon 5
of the cytosine ring increases the rate of hydrolytic deamination of the base in double-stranded DNA. However,
the deamination product of 5-methylcytosine is thymine
rather than uracil (see Fig. 12 in Chapter 3). DNA repair
mechanisms are subsequently less efficient at repairing
deamination-induced mismatches in DNA. Methylated
CpG sites are known to contribute to more than 1/3 of all
transition mutations in the human germ line (Rideout et
al. 1990). This is also true for cancer-causing genes such
as *p53* (Rideout et al. 1990). More surprising is the observation that this mechanism also contributes significantly
to the formation of inactivating mutations in tumor suppressor genes in somatic tissues. For example, more than
50% of all of the *p53* mutations which are acquired in
sporadic colorectal cancers occur at sites of cytosine
methylation (Greenblatt et al. 1994). Thus, the modification of DNA by the DNA methyltransferases (DNMTs)
substantially increases the risk of getting cancer by this
endogenous mechanism.

Methylation of cytosine residues has also been
shown to favor the formation of carcinogenic adducts
between DNA and carcinogens such as benzo(a)pyrene
in cigarette smoke. In this case, methylation of the cytosine residue increases the formation of carcinogenic
adducts between an adjacent guanine residue and
benzo(a)pyrene diol epoxide, resulting in increased
mutations at CpG sites in the lungs of cigarette smokers
(Greenblatt et al. 1994; Pfeifer et al. 2000). Interestingly,
methylation can also alter the rate of mutations in the
p53 gene in sunlight-exposed skin (Greenblatt et al.
1994; Pfeifer et al. 2000). This is because the methyl
group changes the absorption spectrum for cytosine into
the range of incident sunlight, thereby increasing the formation of pyrimidine dimers in the DNA of skin cells.
Thus, the epigenetic modification of DNA not only
increases spontaneous mutagenesis, but also can influence the way DNA interacts with carcinogens and ultraviolet light (Pfeifer et al. 2000).

Hypomethylation of DNA, which has long been
known to occur in animal and human tumors (Table 2),
affects chromosomal stability and increases aneuploidy.
Genomic instability is a hallmark of cancer, and the
increased chromosomal fragility caused by hypomethylation of satellite and other sequences could conceivably
contribute to cancer formation by decreasing the stability
of the genome (Narayan et al. 1998; Gaudet et al. 2003).
The exact mechanisms by which this instability is mediated are not yet fully understood but could easily be the
result of altered DNA–protein interactions caused by
hypomethylation.

4 Hypermethylated Gene Promoters in Cancer

4.1 *The Genes Involved*

The best-understood mechanism by which DNA methylation contributes to cancer is through the focal hypermethylation of promoters of tumor suppressor genes. Exact
mechanisms by which this hypermethylation occurs are
discussed in detail below. However, this clearly is a significant pathway resulting in the heritable silencing of genes
that suppress cancer development (Jones and Laird 1999;
Jones and Baylin 2002; Herman and Baylin 2003). Usually, DNA hypermethylation occurs at CpG-rich regions,
or CpG islands, which are located in and around the transcriptional start site of abnormally silenced genes in cancer. It is important to recognize that cytosine methylation
in CpG islands in the vicinity of the gene start-site position is most critical since this same DNA modification
occurring within bodies of genes generally bears no correlation to transcription status (Jones 1999).

The list of cancer-related genes affected by the above
transcription disruption is growing steadily. As previously
reviewed, this involves genes in all chromosome locations
(Jones and Baylin 2002). Indeed, this epigenetic change
may now outnumber those genes which are frequently
mutated in human tumors. As mentioned earlier (Table
1), loss of tumor suppressor gene function through CpG
methylation, which causes gene silencing, affects virtually
every pathway known. To understand the significance of
the genes for the process of tumorigenesis, and the challenges for the future in this field, the genes may, perhaps,
be divided into three groups.

The first group of genes comprises those which were
instrumental in defining promoter hypermethylation and
gene silencing as an important mechanism for loss of

tumor suppressor gene function in cancer (Table 3). These were already recognized as classic tumor suppressor genes which, when mutated in the germ line of families, cause inherited forms of cancer (Jones and Laird 1999; Jones and Baylin 2002; Herman and Baylin 2003). They are also often mutated in sporadic forms of cancers but can frequently be hypermethylated on one or both alleles in such tumors (Jones and Laird 1999; Jones and Baylin 2002; Herman and Baylin 2003). In addition, for these genes, promoter hypermethylation can sometimes constitute the "second hit" in Knudson's hypothesis by being associated with loss of function of the second copy of the gene in familial tumors where the first hit is a germ-line mutation (Grady et al. 2000; Esteller et al. 2001b). In some instances, 5-azacytidine-induced reactivation of these genes in cultured tumor cells has been shown to restore the key tumor suppressor gene function lost during tumor progression. An example of such is mismatch repair function in colon cancer cells where the *MLH1* gene is silenced (Herman et al. 1998).

The second group of epigenetically silenced genes are those previously identified as candidate tumor suppressor genes by virtue of their function, but they have not been found to have an appreciable frequency of mutational inactivation. These genes may be those emerging as candidate suppressors because they reside in chromosome positions that frequently suffer deletions in cancers (Table 3). Examples include *RASFF1A* and *FHIT* on chromosome 3p in lung and other types of tumors (Dammann et al. 2000; Burbee et al. 2001). Others are those known to encode proteins which subserve functions critical to prevention of tumor progression, such as the pro-apoptotic gene, *DAP-kinase* (Katzenellenbogen et al. 1999). These genes present an important challenge for the field of cancer epigenetics in that despite their having been identified as having frequent promoter hypermethylation in tumors, it must be proven, since many of these genes are not frequently, or at all, mutated, how the genes actually

contribute to tumorigenesis. We return to this issue, and the steps being taken to address it, in a later section.

The third group of genes is being identified through strategies employed to randomly identify aberrantly silenced genes associated with promoter hypermethylation (Suzuki et al. 2002; Yamashita et al. 2002; Ushijima 2005). As compared to those genes in the second group, it is a challenge to place these genes into a functional context for cancer progression because their functions may be totally unknown.

4.2 Searching for New Genes Epigenetically Silenced in Cancer

The most commonly used approach for identifying new genes that are epigenetically silenced in cancer is to consider any potential tumor suppressor gene a candidate if mutations are not found, or if gene expression is low or absent in tumors of interest. Another approach being utilized is to employ techniques that randomly screen cancer genomes for hypermethylated genes (Toyota et al. 1999; Suzuki et al. 2002; Yamashita et al. 2002; Ushijima 2005). This avenue presents great opportunities for enriching our knowledge of cancer biology but also generates many challenges. As recently reviewed (Ushijima 2005), each approach has strengths and weaknesses.

Several approaches rely on an initial step in which DNA is digested with one or more restriction enzymes that differentially cleave CpG sites according to whether they are methylated. To identify potentially hypermethylated genes, products are then analyzed on two-dimensional gels (Restriction Landmark Genomic Sequencing, RLGS), randomly amplified using arbitrary primers or subtraction techniques to differentiate the methylated sequences between normal and tumor DNA. These analyses have the power to identify large numbers of hypermethylated genes, and each of them has contributed to new knowledge about important candidate tumor suppressor genes. However, the sequences identified may or may not be associated with CpG islands, which are strategically located to participate in gene silencing. Repetitive sequences, which are often highly methylated, are sometimes included in the final products of the procedures. Consequently, the efficiency of identifying hypermethylated tumor suppressor genes is often not high, and genome-wide coverage may thus be difficult.

Other approaches are relying on spotting CpG island sequences contained in the genome onto a microarray (C.M. Chen et al. 2003), often taking into account their relationships to gene start sites, and probing these arrays

---

Table 3. Discovery classes of hypermethylated genes

1. Classic tumor suppressor genes known to be mutated in the germ line of families with hereditary cancer syndromes:

    Some examples = *VHL, E-cadherin, p16Ink4a, MLH1, APC, Stk4, Rb*

2. Candidate tumor suppressor genes:

    Some examples = *FHIT, RASSF1A, O6-MGMT, Gst-Pi, GATAs 4 and 5, DAP-kinase*

3. Genes discovered through random screens for hypermethylated genes:

    Some examples = *HIC-1, SFRPs 1,2,4,5, BMP-3, SLC5A8, SSI1*

with either genomic DNA which has been digested with methylation-sensitive enzymes, or cDNAs to take into account gene expression status. This approach has a powerful potential for identifying hypermethylated tumor suppressors but is limited by the number of candidate CpG islands that can be arrayed.

Another approach is to manipulate cultured tumor cells with agents that cause DNA demethylation, such as 5-azacytidine or 5-aza-2′-deoxycytidine, and hybridizing RNA from before and after drug treatment to gene microarrays to detect up-regulated genes (Suzuki et al. 2002; Yamashita et al. 2002). This approach has the potential to identify all genes hypermethylated in cell cultures of all human cancer types. However, gene changes caused by effects other than demethylation activities of the treatment agents can decrease the efficiency for hypermethylated gene identification. What is less recognized is that the very low expression levels of the genes which are being sought, both before and after drug treatment, severely challenge the sensitivity of most gene microarray platforms and markedly reduce the efficiency of this approach (Suzuki et al. 2002). Use of subtraction techniques after drug treatment to enrich for gene transcripts which are increased can improve the sensitivity of the gene microarray approach (Suzuki et al. 2002) but must be adapted to fit fluorescent probe-labeling procedures for microarrays that readily provide full-genome coverage. Simultaneously employing drugs that alter chromatin changes which collaborate with promoter DNA methylation, such as HDAC inhibitors, can help with gene microarray approaches. This maneuver helps to more specifically identify the genes being sought by taking into account the roles chromatin changes have in silencing hypermethylated genes, as discussed in detail in a later section. This latter approach has recently identified important genes silenced in colon cancer (Suzuki et al. 2002).

4.3 Determining the Functional Importance of Genes Hypermethylated in Cancer

The rapidity with which hypermethylated genes are being discovered in cancer has presented a formidable research challenge. Frequent promoter hypermethylation in a given gene does not in and of itself guarantee functional significance for the attendant gene silencing as would loss of function due to a genetic mutation. This is especially the case when the hypermethylated gene is not a known classic tumor suppressor and when there is no evidence that the gene may also be frequently mutated in cancers. Thus, it is obligatory that the gene in question be studied in such a way that the significance of loss of function is determined in terms of both the processes controlled by the encoded protein and the implications for tumor progression. There are several stages for such investigations, each of increasing importance for firmly documenting the role in cancer formation, which are outlined in Table 4.

First, of course, is the documentation of the hypermethylation and its consequences for the expression state of the gene, including the ability of the gene to undergo reexpression with promoter demethylation. Second, the incidence for hypermethylation and silencing of the gene must be well established in primary as well as cultured tumor samples. Third, as further explained below, it is often essential to know at what point the silencing of the gene occurs in tumor progression (Fig. 4). Fourth, the contribution of loss of function of the gene to tumorigenicity must be directly assessed. This can begin with routine studies of cultured cells through assessment of gene reinsertion effects on cellular properties such as induction of apoptosis, effects on soft agar cloning, and effects on tumorigenicity of the cells when grown as heterotransplants in athymic mice. Fifth, the function of the encoded protein must be established either through having previous knowledge of the type of protein involved, through recognition of suggested functions by nature of the protein structure, or through studies of the biology of the protein in cell culture models. Ultimately, however, step six must be taken, which may often involve trans-

Table 4. Steps in documenting the importance of a hypermethylated gene for tumorigenesis

1. Document CpG island promoter methylation and correlate with transcriptional silencing of the gene and ability to reverse the silencing with demethylating drugs in cell culture.

2. Document correlation of promoter hypermethylation with specificity for this change in tumor cells (cell culture and primary tumors) versus normal cell counterparts and incidence for the hypermethylation change in primary tumors.

3. Document the position of the hypermethylation change for tumor progression of given cancer types.

4. Document the potential significance for the gene silencing in tumorigenesis through gene reinsertion studies in cell culture and effects on soft agar cloning, growth of tumor cells in nude mouse explants, etc.

5. Establish function of the protein encoded by the silenced gene—either through known characteristics or testing for activity of recognized protein motifs in culture systems, etc.

6. Document tumor suppressor activity and functions of the gene for cell renewal, etc., especially for totally unknown genes, through mouse knock-out studies.

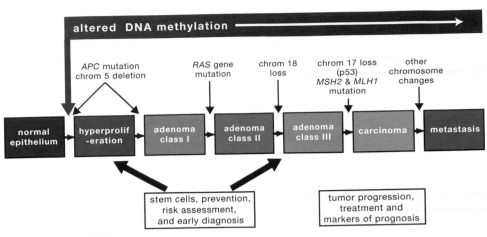

Figure 4. The Early Role for Abnormal DNA Methylation in Tumor Progression

This is depicted in the classic model (Kinzler and Vogelstein 1997) for genetic alterations during the evolution of colon cancer. The altered DNA methylation is shown to occur very early (*red arrow*), as discussed in the text, during conversion of normal to hyperplastic epithelium. This places it in a strategic position (*left bottom black arrow* and *left bottom box*) for channeling stem cells into abnormal clonal expansion (see Fig. 5) by cooperating with key genetic alterations. These epigenetic abnormalities also have marker connotations, as shown in the bottom left black box. The abnormal DNA methylation continues to accrue during progression from noninvasive to invasive and ultimately, metastatic tumors (*right bottom arrow* and *right box*). This has connotations for cancer treatment and for markers of prognosis.

genic knock-out approaches to establish the role of the gene as a tumor suppressor gene and to understand the functions of the encoded protein in development, adult cell renewal, etc. Mouse knock-out studies have proven extremely rewarding in documenting the function of HIC-1 as a tumor suppressor gene after it was identified by screening genomic regions that have undergone loss of heterozygosity (LOH) in cancerous cells (W.Y. Chen et al. 2003, 2004). These challenges, and especially step six, reveal the value of discovering genes epigenetically silenced in cancer, but create a major scope of work to be considered by investigators in the field.

5 Epigenetic Gene Silencing and Its Role in the Evolution of Cancer—Importance for Early Tumor Progression Stages

In the classic view of cancer evolution, as articulated by Vogelstein and colleagues (Kinzler and Vogelstein 1997), a series of genetic changes drives progression from early premalignant stages, through the appearance of invasive cancer, to onset of metastatic disease (Fig. 4). This progression does not necessarily occur in the same exact linear order from tumor to tumor. We know that throughout this course of events, epigenetic changes are occurring as well: There is early appearance of both the widespread loss of normal DNA methylation and more focal gains in gene promoters that we have been dis-

cussing. Thus, there is the potential for interaction of epigenetic and genetic events to drive progressive cellular abnormalities throughout the entire course of neoplastic progression. In this scenario, data for two epigenetic aspects—loss of imprinting (LOI) (as discussed in Chapter 23) and gene silencing—are proving to be extremely important for very early stages of cancer development.

LOI involves a process wherein the silenced allele of imprinted genes becomes activated during tumorigenesis such that biallelic expression of the gene, and excess gene product, are established (Rainier et al. 1993). The most studied example is for *IGF2* in tumors such as colon cancer (Kaneda and Feinberg 2005). In this case, the promoter hypermethylation event in the imprinted *H19* gene on chromosome 11p is the result of a complicated chromatin control process (see Chapter 19), to abnormally activate the silenced *IGF2* allele (Kaneda and Feinberg 2005). The resultant biallelic *IGF2* expression leads to excess production of the growth-promoting IGF2 protein. Experimental evidence suggests that this could play a role in very early progression steps of colon cancer (Kaneda and Feinberg 2005; Sakatani et al. 2005). In fact, recent mouse model studies suggest that LOI events alone may be sufficient to initiate the tumorigenesis process (Holm et al. 2005).

A second common neoplastic transition, the epigenetic silencing of genes, occurs in early phases of neoplastic development. This relates heavily to the questions posed about the roles of cellular stress and exposure in the development of disease states. The genes involved often appear to set the stage for stressed cells to survive DNA damage events and/or chronic injury settings, to clonally expand as stem/progenitor-type cells, and to then be predisposed to later genetic and epigenetic events to drive tumor progression (Fig. 5). The first evidence for such involvement comes from data for several classic tumor suppressor genes that can be either mutated or epigenetically silenced in human cancers. For example, the epigenetic silencing of $p16^{ink4a}$ occurs very early in populations of premalignant cells, during the early changes that precede tumors such as lung cancer (Belinsky et al. 1998) and in small populations of hyperplastic epithelial cells in otherwise normal breast in some women (Holst et al. 2003). In experimental settings where normal human mammary epithelial cells are grown in cell culture (on plastic), this type of $p16$ silencing is a prerequisite for very early steps toward cell trans-

formation (Kiyono et al. 1998; Romanov et al. 2001). This loss of gene function accompanies a failure of subsets of the mammary cells to reach a mortality checkpoint, and these cells then develop progressive chromosomal abnormalities and telomerase expression as they continue to proliferate.

A second example concerns the mismatch repair gene, *MLH1*. This gene is mutated in the germ line of families in which members are predisposed to a type of colon cancer with multiple genetic alterations and termed the "micro-satellite" instability phenotype (Liu et al. 1995). However, 10–15% of patients with nonfamilial colon cancers also have tumors with this phenotype, and the majority of these cancers harbor epigenetic silencing of a non-mutated *MLH1* gene (Herman et al. 1998; Veigl et al. 1998). In cell culture, reexpression of this silenced *MLH1* gene produces reappearance of a functional protein that restores a considerable portion of the damage mismatch repair (Herman et al. 1998).

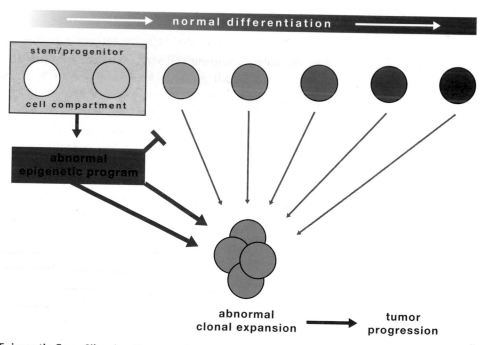

Figure 5. Epigenetic Gene Silencing Events and Tumorigenesis

The earliest steps in tumorigenesis are depicted as abnormal clonal expansion, which evolves during the stress of cell renewal. This is caused by factors such as aging and chronic injury, from, e.g., inflammation. These cell clones are those at risk of subsequent genetic and epigenetic events that would drive tumor progression. Abnormal epigenetic events, such as the aberrant gene silencing focused upon in this chapter, could be the earliest heritable causes, in many instances, for inducing the abnormal clonal expansion from within stem/progenitor cell compartments in a renewing adult cell system. The gene silencing is triggered by chromatin modifications that repress transcription, and the DNA hypermethylation of this chromatin serves as the tight lock, as discussed in the text, to stabilize the heritable silencing. The gene silencing, in turn, disrupts normal homeostasis, which prevents stem and progenitor cells from moving properly along the differentiation pathway for a given epithelial cell system (*top cells with deepening blue colors*) and channels them (*large red arrows*) into the abnormal clonal expansion.

Most recently, *Chfr*, a checkpoint-regulating gene that also controls another type of genomic integrity, chromosomal stability and ploidy, has been shown to be mutated in tumors but is more often silenced epigenetically in lung and other cancers and, importantly, early in progression stages of colon cancer (Mizuno et al. 2002). Mouse knock-out studies reveal a tumor suppressor role for this gene based on its function as an E3 ubiquitin ligase that regulates Aurora A, a control gene for mitosis. Embryonic cells from the mice have chromosomal instability and a predisposition to transformation.

As the list of hypermethylated genes in cancer has expanded, key silencing events in early tumor progression are now being defined for candidate tumor suppressor genes that only have a history of epigenetic change and not mutations. For example, the DNA repair gene, *O6-MGMT*, is silenced early in colon cancer progression (Esteller et al. 2001a), and this loss of function can predispose cells to persistence of alkylation damage at guanosines and, thus, G to A point mutations. Indeed, silencing of this gene occurs in premalignant colon polyps, prior to the appearance of a high rate of these mutations in both the *p53* and *RAS* genes in later colon tumor progression phases (Esteller et al. 2001a; Wolf et al. 2001). Similarly, the GST-Pi gene is silenced in virtually all premalignant lesions that are predisposing to prostate cancer, putting cells at risk of oxidative damage at adenines (Lee et al. 1994).

The third type of silenced genes—those discovered by approaches to randomly screening cancer genomes for epigenetically silenced genes—is also beginning to contribute significantly to our understanding of the early role of gene silencing in cancer. A particularly intriguing scenario has emerged in the progression of colon cancer: Epigenetic loss of function occurs in a family of genes, discovered through the microarray approach outlined earlier (Suzuki et al. 2002), which may allow early abnormal activation of a developmental pathway that is universally involved with the initiation and progression of this disease. Transcriptional silencing of the *secreted frizzled related protein* genes (*SFRPs*) (Suzuki et al. 2004) removes an antagonistic signal for interaction of Wnt ligands with their membrane receptors (Finch et al. 1997). This silencing correlates with Wnt-driven up-regulation of overall cellular levels of β-catenin, due especially to increased presence and activity of this transcription factor in the nucleus (Suzuki et al. 2004). Such transcription is the canonical readout for increased Wnt pathway activity (Morin et al. 1997; Gregorieff and Clevers 2005). Most important, *SFRP* silencing occurs in very early lesions

predisposing to colon cancer, before common mutations in downstream Wnt pathway proteins occur, which also result in activated β-catenin in the nucleus (Morin et al. 1997; Gregorieff and Clevers 2005). Thus, early activation of the Wnt pathway by epigenetic events appears poised to allow early expansion of cells, predisposed to activate the pathway further through mutational events. Persistence of both the epigenetic, through Wnt-driven increases in cellular β-catenin, and genetic alterations, through crippling of the protein complex that degrades β-catenin or activating Wnt mutations then seem to complement one another in driving progression of the disease (Suzuki et al. 2004).

Another example of this group of genes involves *HIC-1* (*hypermethylated-in-cancer 1*), which encodes a zinc finger transcriptional repressor. HIC-1 was discovered by random screening for hypermethylated CpG islands in a hot spot for chromosomal loss in cancer cells (Wales et al. 1995). This gene, which is silenced early in cancer progression but is not mutated, has proven to be a tumor suppressor in a mouse knock-out model (W.Y. Chen et al. 2003, 2004). It complements p53 mutations, partially through loss of function, which leads to up-regulation of SIRT1 (Chen et al. 2005), a key protein for sensing cell stress and contributing to stem/progenitor cell growth (Howitz et al. 2003; Nemoto et al. 2004; Kuzmichev et al. 2005).

Thus, the data discussed above contribute to the thematic hypotheses outlined in Figure 4. This suggests that some of the earliest heritable changes in the evolution of tumors may be epigenetic changes, which often involve the tight transcriptional silencing of genes, maintained by promoter DNA methylation. The challenges to understand these scenarios further are integrally linked to key challenges for the study of epigenetic changes in cancer, which are outlined in Table 5 and discussed more fully below. The meeting of such challenges, particularly for understanding the contribution of epigenetic changes in the very earliest steps in neoplastic progression, may strikingly enrich molecular strategies aimed at the prevention of, and early intervention for, cancer.

6 The Molecular Anatomy of Epigenetically Silenced Cancer Genes

Genes that are silenced in neoplastic cells are important for understanding the initiation and maintenance of cancer. They also serve as excellent models for understanding how gene silencing may be initiated and maintained, and how the mammalian genome is packaged to facilitate regions of

Table 5. Major research challenges for understanding the molecular events mediating epigenetic gene silencing in cancer

1. Elucidate links between simultaneous losses and gains of DNA methylation in the same cancer cells.

2. Determine the molecular nature of boundaries, and how they change during tumorigenesis, that separate areas of transcriptionally active zones encompassing gene promoters from the transcriptionally repressive areas that surround them and which may prevent the repressive chromatin from spreading through the active zone. Among the candidate mechanisms are roles that may be played by key histone modifications, by insulator proteins, by chromatin-remodeling proteins, etc.

3. What is the order of events for the evolution of gene silencing in cancer with respect to histone modifications, DNA methylation, etc.? Which comes first, and what are the key protein complexes that target the processes (DNA-methylating enzymes, histone-deacetylating and methylation enzymes, CpG methyl-binding proteins, Polycomb-silencing complexes, etc.) that determine the events?

4. Which specific DNA-methylating enzymes are required for initiating and/or maintaining the most stable gene silencing, and what protein complexes contain them, including their interaction with key components of the histone code?

5. Once established, what are all of the components of chromatin and DNA methylation machinery, and the hierarchy of their involvement, required to maintain the gene silencing, and how are they reversible?

transcription and repression of transcription. In turn, the understanding of chromatin function, which is a major emphasis of many of the chapters in this book, is facilitating our understanding of what may trigger aberrant gene silencing in cancer and how the components of this silencing maintain the attendant transcriptional repression.

Work of several laboratories has contributed to the current understanding of the chromatin configuration that surrounds hypermethylated CpG islands in promoters of multiple genes aberrantly silenced in cancer cells. These studies have also highlighted how this chromatin differs from those surrounding the same genes when they are basally expressed. In normal cells, or in cancer cells where the genes are not transcriptionally repressed, these genes are characterized by having a zone of open chromatin wherein the CpG islands are not DNA methylated, the nucleosomes are irregularly spaced such that hypersensitive sites can be detected, and key histone residues are marked by posttranslational modifications typical for active genes. Active covalent histone marks include acetylation of H3 at lysines 9 and 14 (H3K9ac and H3K14ac) and methylation of H3K4 (Nguyen et al. 2001; Fahrner et al. 2002).

At both the 5′ and 3′ borders of the above open chromatin region, there appears to be a stark transition in

chromatin structure, with characteristics of transcriptionally repressed genomic regions flanking the CpG island (Fig. 6). In these border regions, there is methylation of the less frequent CpG sites, and recruitment of methyl cytosine-binding proteins (MBDs) and their partners (e.g., histone deacetylases or HDACs) to the methylated CpGs (Chapter 18). The regions outside the CpG islands thus appear to be accessible to enzymes that catalyze histone methylation marks correlating with gene silencing. As a result of all of these factors, deacetylation of key histone residues, and presence of repressive histone methylation marks associated with transcriptional repression occur, most especially, H3K9me2 (Nguyen et al. 2001; Fahrner et al. 2002; Kondo et al. 2003).

These juxtaposed regions of active and repressive chromatin patterns (Fig. 6) suggest that the CpG island-containing promoters of active genes reside in a zone which is "protected," or alternatively, not targeted for repressive chromatin marks and DNA methylation (Nguyen et al. 2001; Fahrner et al. 2002; Kondo et al. 2003). Inherent to these concepts is the likelihood that molecular "boundaries" exist at the 5′ and 3′ borders of the promoter CpG islands in expressed genes. One major challenge is to define the precise nature of these boundaries. At present, candidates are the histone modifications themselves which mark the protected region of the promoter, the transcriptional activator and coactivator complexes which directly underpin active transcription, complexes of proteins which accomplish nucleosome placement and/or movement (i.e., nucleosome remodelers) that may mark genes for active transcription. These may promote access of transcriptional activating complexes, replacement of classic histones by variant histones such as H3.3 (Chapter 13), which appear to support active transcription, and action of insulator protein complexes and their recognition sequences. It is in the context of defining how one or more of these candidate processes maintain zones of transcriptionally permissive chromatin around the non-DNA-methylated CpG island containing promoters of active genes that genes silenced in cancers are superb research models for understanding modulation of gene expression in mammalian genomes.

The way in which the above transcriptionally active chromatin organization of CpG island containing promoters becomes converted during tumor progression remains largely unsolved, and dissection of this conversion remains one of the most important challenges in the cancer epigenetics field. As noted earlier, abnormally DNA-methylated gene promoter CpG islands, with attendant gene silencing, often appear in very early and prema-

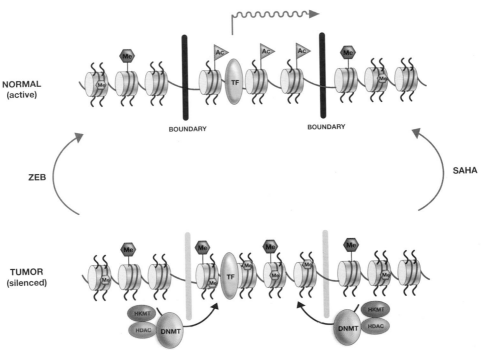

Figure 6. Model of the Relationships between DNA Methylation and Histone Modifications in the Promoter CpG Island Region of a Gene in Normal and Tumor Cells

In the expressed gene (*top*), a boundary is shown, the molecular nature of which is not yet characterized, which protects the CpG island surrounding the transcription start site (*green arrow*) from DNA methylation. CpG sites in CpG regions flanking this protective zone are, in contrast, DNA methylated (*pink hexagon marked M*) and associated with key silencing marks such as methylation of H3K9 (*red hexagon marked Me*). Key histone tail amino acids in the protected zone, such as H3K9, are in the acetylated state (*blue flags marked Ac*), and transcription factors (*yellow oval marked TF*) have access to the transcription start-site region. When the same gene is aberrantly silenced in a cancer cell (*bottom*), the CpG island is DNA-hypermethylated as the protective boundaries are now breached and not present. This methylation is maintained by DNA methyltransferase complexes (*pink ovals marked DNMT*), and methylcytosine-binding protein complexes that contain histone deacetylases (*blue ovals marked HDAC*), and histone methyltransferases (*red ovals marked HKMT*) that catalyze key silencing methylation marks on histone amino acid tails such as H3K9. TF complexes are no longer active (*lack of green arrow*). The major approaches currently underpinning ongoing cancer epigenetic clinical trials are depicted and consist of either DNA methyltransferase inhibitors to block DNA hypermethylation or HDAC inhibitors to restore the acetylation status of key histone amino acid residues. As discussed in the text, some of the most promising anticancer therapies include combinatorial use of DNMT1 and HDAC inhibitors.

lignant stages of tumor progression, making our understanding of the underlying factors potentially extremely important for our knowledge of cancer biology. It also opens new possibilities for assessing cancer risk factors, facilitating early cancer diagnosis, and considering new strategies for cancer prevention and early intervention.

Whatever the precise mechanisms involved, which are revisited later in this section, the end result appears to be during tumor progression, a "breakdown" in the mechanisms that maintain promoter CpG island regions from the incursion of the repressive type of chromatin located at the 5′ and 3′ borders. The result is conversion of the transcriptionally open configuration of the promoter to a closed one with more tightly compacted nucleosomes and loss of hypersensitive sites, and appearance of multi-

ple histone deacetylation and methylation states characteristic of transcriptionally repressive chromatin (Fig. 6). The hierarchy of all of these events, in terms of which are the most important in maintaining this repression, is still to be determined. However, one important feature that has emerged from analyses of multiple genes is that whatever the role of the DNA methylation component in initiating and/or maintaining the gene silencing, once it is in place it plays a dominant role in the heritability of the transcriptionally repressive state. Thus, maneuvers, such as inhibiting HDAC activity with specific drugs, fail to induce reexpression of cancer genes that harbor densely methylated CpG island containing promoters. However, if a low dose of a demethylating agent such as 5′-deoxy-aza-cytidine (DAC) is employed first, then HDAC inhibitors are additive and/or synergistic for causing reexpression of the silenced genes (Cameron et al. 1999; Suzuki et al. 2002). This paradigm, in association with microarray analyses searching for reexpression of basally silenced genes, has even proven valuable for identifying new epigenetically silenced genes in cancer (Suzuki et al. 2002). Likewise, gene reexpression following DAC administration can result in removal of histone methylation silencing marks from the promoter regions of epigenetically silenced cancer genes (Nguyen et al. 2001; Fahrner et al. 2002; Kondo et al. 2003). In this sense, as illustrated in Figure 6, DNA methylation may act as a "lock" to stabilize the epigenetic silencing of cancer genes and ensure its stable heritability for the loss of gene function.

Another key question in the field of epigenetic silencing of cancer genes concerns which of the DNMTs are responsible for establishing and maintaining the abnormal promoter DNA methylation that often accompanies this process. Current data suggest that these steps may be mediated somewhat differently than the classic role of *Dnmt*s, as established in studies of mouse development, would suggest. In mouse development, as outlined from knock-out studies of the three known biologically active DNA methyltransferases: Dnmt1, Dnmt3a, and Dnmt3b, the latter two enzymes are responsible for establishing methylation (de novo DNA methylation) whereas Dnmt1 is responsible for maintaining established patterns (maintenance DNA methylation) (Li et al. 1992; for more detail, see Chapter 18). However, in cultured colon cancer cells, genetic disruption studies of DNMTs indicate that maintenance of the majority of overall DNA methylation, including the promoter hypermethylation and its attendant gene silencing, require both DNMT1 and DNMT3b (Rhee et al. 2000, 2002). Studies of other cancer cell types have produced more variable results ranging from indication of some degree of partnership of these two enzymes for such maintenance to description of genes that are demethylated and transcriptionally reactivated with decrease in DNMT1 alone (Leu et al. 2003). Thus, the question of how DNMTs establish and maintain abnormal patterns of DNA methylation in cancer cells requires continued study. Most especially, we need clarification of the complexes through which these enzymes might act cooperatively and the mechanisms for their targeting to gene promoters. Several recent studies indicate that transcriptional repression complexes are key for such recruitment (Di Croce et al. 2002; Fuks et al. 2003; Brenner et al. 2005). Whatever the mechanisms involved, it must be remembered that, experimentally,

the mammalian DNMTs appear to have complex functions that include not only DNA methylation catalytic activity from the carboxy-terminal regions, but also direct transcriptional repression activities from amino-terminal domains (see Fig. 3 of Chapter 18) (Robertson et al. 2000; Rountree et al. 2000; Fuks et al. 2001). These enzymes can also bind to, and potentially recruit, key mediators of transcriptional repression including HDACs and methylcytosine-binding proteins (MBDs). Thus, a role for DNMTs potentially has many facets in transcriptional silencing, from initiation to maintenance, which may or may not include steps involving DNA methylation. Gene silencing in cancer will be an important scenario for sorting out these possibilities.

7 Summary of Major Research Issues for Understanding Epigenetic Gene Silencing in Cancer

It is clear from the above discussions that much has been learned about the molecular events which underlie the appearance of promoter DNA methylation and gene silencing during tumor progression, but even more remains to be elucidated. Table 5 summarizes at least some of the most important questions that must be resolved through future research. First, molecular events determining the simultaneous appearance of overall DNA hypomethylation and more localized promoter DNA hypermethylation must be elucidated. These juxtaposed states suggest a broad mistargeting of chromatin in cancer cells. The etiology for this should prove most illuminating for learning how mammalian cells package their genomes for proper patterns of gene expression and maintenance of chromosome integrity. Second, the determinants of chromatin boundaries around individual gene promoters and their correlation to both normal and abnormal states of transcription need to be outlined. Are insulators involved in protecting normally active genes from the incursion of transcriptionally repressive chromatin and/or do the histone modifications, themselves, set up boundaries? How would these boundary determinants be altered during tumor progression? Third, what is the order of events for the emergence of abnormal gene silencing in cancer cells? Is initial down-regulation of gene transcription required? Do key chromatin modifications, including deacetylation and methylation of histone amino acids and recruitment of the enzymes that mediate these processes (see Chapter 10), precede and target DNA methylation? If so, what are the molecular interactions underlying such targeting, and how do key silencing complexes important to normal and cancer stem cells, such as the Polycomb group proteins (see Chapter 11), play a role? Fourth, exactly which DNMTs are required for initiation and maintenance of abnormal promoter DNA methylation in cancer, and how do they interact? Which of the chromatin constituents discussed above may interact with and target these enzymes? Finally, once abnormal heritable gene silencing is established in cancer, what is the precise hierarchy of molecular steps that maintain it? This latter question is not only a key basic question, but is also central to the translational implication, discussed further below, for reversing abnormal gene silencing in cancer prevention and therapy.

8 Detection of Cancer by DNA Methylation

The fact that focal hypermethylation of CpG islands is so common in cancer cells, coupled with the ability to detect methylation with a high degree of sensitivity, has led to the development of several approaches for the detection of cancer in body fluids. Acquired changes in CpG island methylation can be detected in a background of normal cells following conversion of cytosines to uracil yet leaving 5-methylcytosine intact in DNA treated with sodium bisulfite. PCR approaches such as methylation-specific PCR (MSP), in which primers are designed to amplify only methylated regions, are very sensitive. Other methods include techniques based on real-time PCR such as "MethyLight," where a fluorescent probe that can only bind to methylated DNA is used to detect methylation patterns (as shown in the left title page image, with genes on the Y axis and types of tumors on the X). These techniques can detect one methylated allele in a background of about 1,000–10,000 alleles. Thus, the acquisition of an abnormal methylation pattern can be easily detected; these approaches are applicable to mixtures of cells or even various biological fluids such as plasma, urine, or sputum (Laird 2003). Cancer detection by identification of altered cytosine methylation is quite robust because of the inherent stability of DNA compared to RNA or proteins. Also, because altered methylation patterns are often cancer-specific, these approaches may be able to distinguish one type of cancer from another. There are now a host of studies, as noted above, providing "proof of principle" for the promising use of promoter DNA hypermethylated sequences as an extremely sensitive strategy for predicting cancer risk and/or detection. The final proof of the ultimate clinical value of this approach awaits larger studies in which the current hypotheses are fully validated. Such investigations will surely be conducted over the next several years.

9 Epigenetic Therapy

The heritable inactivation of cancer-related genes by altered DNA methylation and chromatin modification has led to the realization that silenced chromatin may represent a viable therapeutic target. Thus, a new therapeutic approach called "epigenetic therapy" has been developed in which drugs that can modify chromatin or DNA methylation patterns are used alone or in combination (Fig. 7) in order to affect therapeutic outcomes (Egger et al. 2004).

The nucleoside analogs, 5-azacytidine, 5-aza-2′-deoxycytidine, and zebularine, are powerful mechanism-based inhibitors of DNA cytosine methylation (Fig. 7a). These drugs are incorporated into the DNA of replicating cells after they have been metabolized to the appropriate deoxynucleoside triphosphate. Once incorporated into DNA, they interact with all three known DNA methyltransferases to form covalent intermediates, which ultimately inhibits DNA methylation in subsequent rounds of DNA synthesis. The mechanism of action of these compounds is quite well understood, and they have been used for some time to reactivate silenced genes in tissue culture or in xenograft models. More recently, however, they have found application in the treatment of certain hematological malignancies, particularly myeloid dysplastic syndrome (MDS), which is a pre-leukemic condition occurring mainly in elderly patients (Lubbert 2000; Wijermans et al. 2000; Silverman et al. 2002; Issa et al. 2004). Clinical responses for patients with this disorder, and with leukemias that may have progressed from the pre-leukemic stage, are becoming increasingly dramatic. Accordingly, drugs with the clinical names for 5-azacytidine and 5-aza-2′-deoxycytidine, Vidaza and Decitabine, respectively, have now been approved by the American Food and Drug Administration (FDA) for the treatment of patients with these disorders. Zebularine, which is also a mechanism-based inhibitor of DNA methyltransferases, is at an earlier stage of clinical development. To date, effective inhibitors that do not require incorporation into DNA have not been discovered, but these might be more desirable in the clinic because they might have fewer side effects. Numerous approaches to synthesize and/or discover such drugs are now ongoing.

Although Vidaza and Decitabine have been shown to be clinically efficacious, it has been more difficult to establish with clarity that the targets of drug action are methylated gene promoters. Preliminary experiments have suggested that the p15 tumor suppressor gene becomes demethylated following Decitabine treatment

a

Compound	Structure	Cancer Type	Clinical Trials
DNA METHYLATION INHIBITORS			
5-Azacytidine 5-Aza-CR Vidaza	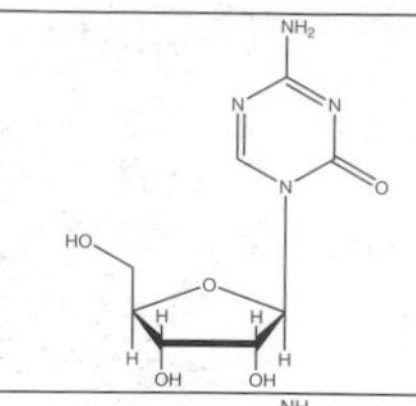	MDS; Hematologic malignancies	I, II, and III; FDA-approved for MDS
5-Aza-2'-deoxycytidine 5-Aza-CdR Dacogen	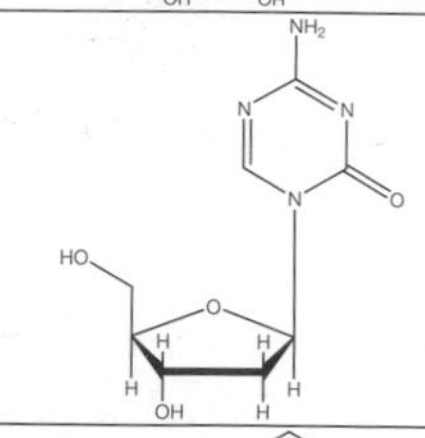	MDS; Hematologic malignancies	I, II, and III
Zebularine 1-β-D-ribofuranosyl-2(1H)- pyrimidinone	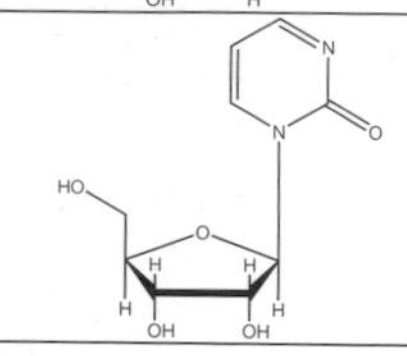	N/A	Preclinical

b

HISTONE DEACETYLASE INHIBITORS			
4-Phenylbutyrate (PBA)	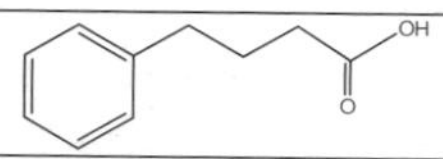	Refractory solid tumors	I
Suberoylanilide hydroxamic acid (SAHA)	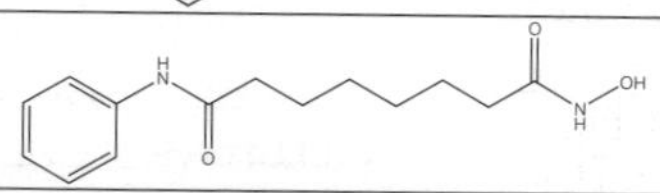	Solid tumors and hematologic malignancies	I, II
NVP-LAQ824	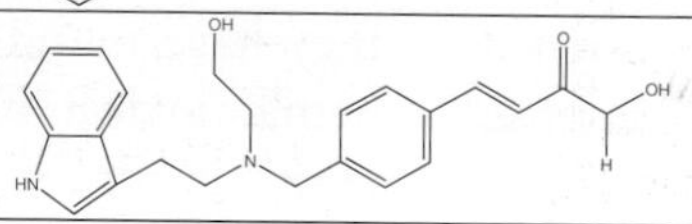	N/A	I
Depsipeptide FK-228 FR901228	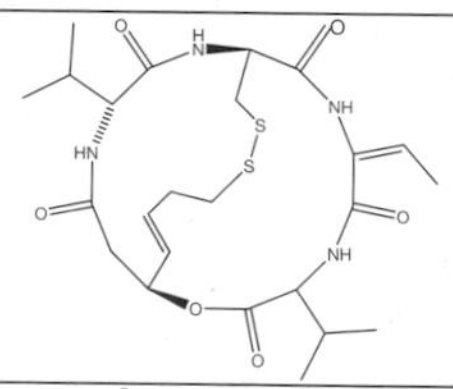	Advanced neoplasms, CLL, AML, and T-cell lymphoma	I, II
MS-275	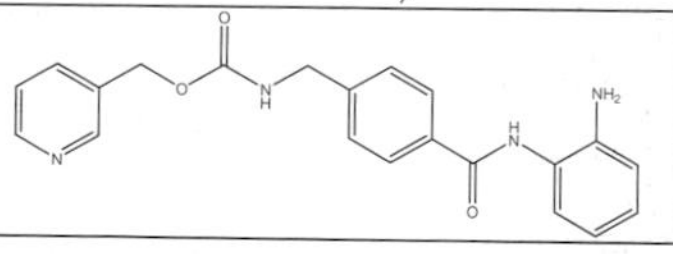	Solid tumors and lymphoma	I, II

Figure 7. Structures of Nucleoside Analog Inhibitors of DNA Methylation (a), and Inhibitors of Histone Deacetylation (b)

(a) Three nucleoside analogs are known that can inhibit DNA methylation after incorporation into DNA. 5-aza-CR (Vidaza) and 5-aza-2'-deoxycytidine (Decitabine) have been approved for the treatment of leukemia. Zebularine is at an earlier stage of development. (b) Some examples of the many histone deacetylase inhibitors, some of which are currently in clinical trials.

(Daskalakis et al. 2002); however, it remains to be shown whether the drugs act by inducing gene expression or by some other mechanism.

Clinical trials are also ongoing using inhibitors of HDACs (Fig. 7b). Several drugs are known to cause substantial inhibition of HDACs. Some of these, such as phenylbutyric acid or valproic acid, have been in clinical use to treat other conditions for some time (Marks et al. 2001; Richon and O'Brien 2002). These drugs inhibit all deacetylases, and it is not clear whether their efficacy in the treatment of certain kinds of malignancy is definitively due to inhibition of histone deacetylation rather than some

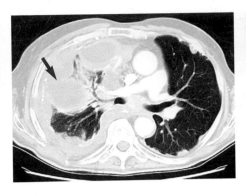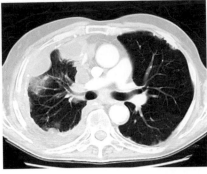

Figure 8. Histone Deacetylase Inhibitors Can Cause Tumor Regression

Computerized tomography (CT) scans of a patient with a mesothelioma. (*Left scan*) Before therapy there is a large lobulated white mass in the right lung compressing the aerated black lung (see *arrow*). (*Right scan*) 4 months after therapy with SAHA with shrinkage of the tumor and a much better expanded right lung. (Scans courtesy of Dr. Paul Marks, Memorial Sloan-Kettering Cancer Center, NY.)

other reaction. Newer compounds, such as suberoylanilide hydroxamic acid (SAHA) and depsipeptide, are more specific inhibitors of histone deacetylase and have shown good clinical outcomes (Fig. 8). Once again, however, it is difficult to be sure whether the drugs are performing in patients according to the theoretical mechanisms.

Newer clinical trials, based on the chromatin interactions with DNA methylation discussed in previous sections (Cameron et al. 1999; Suzuki et al. 2002), are focused on the combination of DNA methylation inhibitors and HDAC inhibitors, in an attempt to use lower doses of both drug classes, yet obtain strong synergistic effects. Several exciting clinical trials are currently under way to test these hypotheses; however, it is still not clear whether these approaches will work.

Whereas the approach of epigenetic therapy has a good basis in theory, the lack of specificity of some of the agents gives cause for concern. For example, the nucleoside analogs are nonspecific inhibitors of DNA methyltransferases and inhibit DNA methylation throughout the genome. Therefore, there is the possibility of the inadvertent reactivation of genes as the result of therapy, although this does not seem to be a major problem in those patients with serious diseases who have already been treated. Likewise, there are several additional chromatin modifications other than deacetylation, such as the methylation of key histone amino acids discussed in previous chapters, which can potentially be targets for therapy. There is, therefore, considerable interest in the discovery of new therapeutic targets with the goal of activating silenced genes, and exciting advances are anticipated in the coming years.

References

Amann J.M., Nip J., Strom D.K., Lutterbach B., Harada H., Lenny N., Downing J.R., Meyers S., and Hiebert S.W. 2001. ETO, a target of t(8;21) in acute leukemia, makes distinct contacts with multiple histone deacetylases and binds mSin3A through its oligomeriza-

tion domain. *Mol. Cell. Biol.* 21: 6470–6483.

Baylin S.B., Fearon E.R., Vogelstein B., de Bustros A., Sharkis S.J., Burke P.J., Staal S.P., and Nelkin B.D. 1987. Hypermethylation of the 5′ region of the calcitonin gene is a property of human lymphoid and acute myeloid malignancies. *Blood* 70: 412–417.

Belinsky S.A., Nikula K.J., Palmisano W.A., Michels R., Saccomanno G., Gabrielson E., Baylin S.B., and Herman J.G. 1998. Aberrant methylation of *p16^INK4a* is an early event in lung cancer and a potential biomarker for early diagnosis. *Proc. Natl. Acad. Sci.* 95: 11891–11896.

Brenner C., Deplus R., Didelot C., Loriot A., Vire E., De Smet C., Gutierrez A., Danovi D., Bernard D., Boon T., et al. 2005. Myc represses transcription through recruitment of DNA methyltransferase corepressor. *EMBO J.* 24: 336–346.

Burbee D.G., Forgacs E., Zochbauer-Muller S., Shivakumar L., Fong K., Gao B., Randle D., Kondo M., Virmani A., Bader S., et al. 2001. Epigenetic inactivation of RASSF1A in lung and breast cancers and malignant phenotype suppression. *J. Natl. Cancer Inst.* 93: 691–699.

Cameron E.E., Bachman K.E., Myohanen S., Herman J.G., and Baylin S.B. 1999. Synergy of demethylation and histone deacetylase inhibition in the re-expression of genes silenced in cancer. *Nat. Genet.* 21: 103–107.

Chen C.M., Chen H.L., Hsiau T.H., Hsiau A.H., Shi H., Brock G.J., Wei S.H., Caldwell C.W., Yan P.S., and Huang T.H. 2003. Methylation target array for rapid analysis of CpG island hypermethylation in multiple tissue genomes. *Am. J. Pathol.* 163: 37–45.

Chen W., Cooper T.K., Zahnow C.A., Overholtzer M., Zhao Z., Ladanyi M., Karp J.E., Gokgoz N., Wunder J.S., Andrulis I.L., et al. 2004. Epigenetic and genetic loss of *Hic1* function accentuates the role of *p53* in tumorigenesis. *Cancer Cell* 6: 387–398.

Chen W.Y., Wang D.H., Yen R.C., Luo J., Gu W., and Baylin S.B. 2005. Tumor suppressor HIC1 directly regulates SIRT1 to modulate p53-dependent DNA-damage responses. *Cell* 123: 437–448.

Chen W.Y., Zeng X., Carter M.G., Morrell C.N., Chiu Yen R.W., Esteller M., Watkins D.N., Herman J.G., Mankowski J.L., and Baylin S.B. 2003. Heterozygous disruption of *Hic1* predisposes mice to a gender-dependent spectrum of malignant tumors. *Nat. Genet.* 33: 197–202.

Dammann R., Li C., Yoon J.H., Chin P.L., Bates S., and Pfeifer G.P. 2000. Epigenetic inactivation of a RAS association domain family protein from the lung tumour suppressor locus 3p21.3. *Nat. Genet.* 25: 315–319.

Daskalakis M., Nguyen T.T., Nguyen C., Guldberg P., Kohler G., Wijermans P., Jones P.A., and Lubbert M. 2002. Demethylation of a hypermethylated P15/INK4B gene in patients with myelodysplastic syndrome by 5-Aza-2′-deoxycytidine (decitabine) treatment. *Blood* 100: 2957–2964.

Di Croce L., Raker V.A., Corsaro M., Fazi F., Fanelli M., Faretta M., Fuks F., Lo Coco F., Kouzarides T., Nervi C., et al. 2002. Methyltransferase recruitment and DNA hypermethylation of target promoters by an oncogenic transcription factor. *Science* **295:** 1079–1082.

Egger G., Liang G., Aparicio A., and Jones P.A. 2004. Epigenetics in human disease and prospects for epigenetic therapy. *Nature* **429:** 457–463.

Ehrlich M., Gama-Sosa M.A., Huang L.H., Midgett R.M., Kuo K.C., McCune R.A., and Gehrke C. 1982. Amount and distribution of 5-methylcytosine in human DNA from different types of tissues of cells. *Nucleic Acids Res.* **10:** 2709–2721.

Esteller M., Risques R.A., Toyota M., Capella G., Moreno V., Peinado M.A., Baylin S.B., and Herman J.G. 2001a. Promoter hypermethylation of the DNA repair gene O(6)-methylguanine-DNA methyltransferase is associated with the presence of G:C to A:T transition mutations in p53 in human colorectal tumorigenesis. *Cancer Res.* **61:** 4689–4692.

Esteller M., Fraga M.F., Guo M., Garcia-Foncillas J., Hedenfalk I., Godwin A.K., Trojan J., Vaurs-Barriere C., Bignon Y.J., Ramus S., et al. 2001b. DNA methylation patterns in hereditary human cancers mimic sporadic tumorigenesis. *Hum. Mol. Genet.* **10:** 3001–3007.

Fahrner J.A., Eguchi S., Herman J.G., and Baylin S.B. 2002. Dependence of histone modifications and gene expression on DNA hypermethylation in cancer. *Cancer Res.* **62:** 7213–7218.

Feinberg A.P. and Vogelstein B. 1983. Hypomethylation distinguishes genes of some human cancers from their normal counterparts. *Nature* **301:** 89–92.

Feinberg A.P., Gehrke C.W., Kuo K.C., and Ehrlich M. 1988. Reduced genomic 5-methylcytosine content in human colonic neoplasia. *Cancer Res.* **48:** 1159–1161.

Finch P.W., He X., Kelley M.J., Uren A., Schaudies R.P., Popescu N.C., Rudikoff S., Aaronson S.A., Varmus H.E., and Rubin J.S. 1997. Purification and molecular cloning of a secreted, Frizzled-related antagonist of Wnt action. *Proc. Natl. Acad. Sci.* **94:** 6770–6775.

Flatau E., Bogenmann E., and Jones P.A. 1983. Variable 5-methylcytosine levels in human tumor cell lines and fresh pediatric tumor explants. *Cancer Res.* **43:** 4901–4905.

Frost P., Liteplo R.G., Donaghue T.P., and Kerbel R.S. 1984. Selection of strongly immunogenic "tum" variants from tumors at high frequency using 5-azacytidine. *J. Exp. Med.* **159:** 1491–1501.

Fuks F., Hurd P.J., Deplus R., and Kouzarides T. 2003. The DNA methyltransferases associate with HP1 and the SUV39H1 histone methyltransferase. *Nucleic Acids Res.* **31:** 2305–2312.

Fuks F., Burgers W.A., Godin N., Kasai M., and Kouzarides T. 2001. Dnmt3a binds deacetylases and is recruited by a sequence-specific repressor to silence transcription. *EMBO J.* **20:** 2536–2544.

Gaudet F., Hodgson J.G., Eden A., Jackson-Grusby L., Dausman J., Gray J.W., Leonhardt H., and Jaenisch R. 2003. Induction of tumors in mice by genomic hypomethylation. *Science* **300:** 489–492.

Gazzoli I., Loda M., Garber J., Syngal S., and Kolodner R.D. 2002. A hereditary nonpolyposis colorectal carcinoma case associated with hypermethylation of the MLH1 gene in normal tissue and loss of heterozygosity of the unmethylated allele in the resulting microsatellite instability-high tumor. *Cancer Res.* **62:** 3925–3928.

Grady W.M., Willis J., Guilford P.J., Dunbier A.K., Toro T.T., Lynch H., Wiesner G., Ferguson K., Eng C., Park J.G., et al. 2000. Methylation of the CDH1 promoter as the second genetic hit in hereditary diffuse gastric cancer. *Nat. Genet.* **26:** 16–17.

Greenblatt M.S., Bennett W.P., Hollstein M., and Harris C.C. 1994. Mutations in the p53 tumor suppressor gene: Clues to cancer etiology and molecular pathogenesis. *Cancer Res.* **54:** 4855–4878.

Gregorieff A. and Clevers H. 2005. Wnt signaling in the intestinal epithelium: From endoderm to cancer. *Genes Dev.* **19:** 877–890.

Hanahan D. and Weinberg R.A. 2000. The hallmarks of cancer. *Cell* **100:** 57–70.

Herman J.G. and Baylin S.B. 2003. Gene silencing in cancer in association with promoter hypermethylation. *N. Engl. J. Med.* **349:** 2042–2054.

Herman J.G., Umar A., Polyak K., Graff J.R., Ahuja N., Issa J.P., Markowitz S., Willson J.K., Hamilton S.R., Kinzler K.W., et al. 1998. Incidence and functional consequences of *hMLH1* promoter hypermethylation in colorectal carcinoma. *Proc. Natl. Acad. Sci.* **95:** 6870–6875.

Holm T.M., Jackson-Grusby L., Brambrink T., Yamada Y., Rideout W.M., III, and Jaenisch R. 2005. Global loss of imprinting leads to widespread tumorigenesis in adult mice. *Cancer Cell* **8:** 275–285.

Holst C.R., Nuovo G.J., Esteller M., Chew K., Baylin S.B., Herman J.G., and Tlsty T.D. 2003. Methylation of *p16*^INK4a^ promoters occurs in vivo in histologically normal human mammary epithelia. *Cancer Res.* **63:** 1596–1601.

Howitz K.T., Bitterman K.J., Cohen H.Y., Lamming D.W., Lavu S., Wood J.G., Zipkin R.E., Chung P., Kisielewski A., Zhang L.L., et al. 2003. Small molecule activators of sirtuins extend *Saccharomyces cerevisiae* lifespan. *Nature* **425:** 191–196.

Issa J.P., Ottaviano Y.L., Celano P., Hamilton S.R., Davidson N.E., and Baylin S.B. 1994. Methylation of the oestrogen receptor CpG island links ageing and neoplasia in human colon. *Nat. Genet.* **7:** 536–540.

Issa J.P., Garcia-Manero G., Giles F.J., Mannari R., Thomas D., Faderl S., Bayar E., Lyons J., Rosenfeld C.S., Cortes J., and Kantarjian H.M. 2004. Phase 1 study of low-dose prolonged exposure schedules of the hypomethylating agent 5-aza-2′-deoxycytidine (decitabine) in hematopoietic malignancies. *Blood* **103:** 1635–1640.

Jones P.A. 1999. The DNA methylation paradox. *Trends Genet.* **15:** 34–37.

Jones P.A. and Baylin S.B. 2002. The fundamental role of epigenetic events in cancer. *Nat. Rev. Genet.* **3:** 415–428.

Jones P.A. and Laird P.W. 1999. Cancer epigenetics comes of age. *Nat. Genet.* **21:** 163–167.

Jones P.A. and Taylor S.M. 1980. Cellular differentiation, cytidine analogs and DNA methylation. *Cell* **20:** 85–93.

Kaminskas E., Farrell A., Abraham S., Baird A., Hsieh L.-S., Lee S.-L., Leighton J.K., Patel H., Rahman A., Sridhara R., et al. 2005. Approval summary: Azacitidine for treatment of myelodysplastic syndrome subtypes. *Clin. Cancer Res.* **11:** 3604–3608.

Kaneda A. and Feinberg A.P. 2005. Loss of imprinting of *IGF2*: A common epigenetic modifier of intestinal tumor risk. *Cancer Res.* **65:** 11236–11240.

Katzenellenbogen R.A., Baylin S.B., and Herman J.G. 1999. Hypermethylation of the DAP-kinase CpG island is a common alteration in B-cell malignancies. *Blood* **93:** 4347–4353.

Kinzler K.W. and Vogelstein B. 1997. Cancer-susceptibility genes. Gatekeepers and caretakers. *Nature* **386:** 761–763.

Kiyono T., Foster S.A., Koop J.I., McDougall J.K., Galloway D.A., and Klingelhutz A.J. 1998. Both Rb/p16^INK4a^ inactivation and telomerase activity are required to immortalize human epithelial cells. *Nature* **396:** 84–88.

Knudson A.G. 2001. Two genetic hits (more or less) to cancer. *Nat. Rev. Cancer* **1:** 157–162.

Kondo Y., Shen L., and Issa J.P. 2003. Critical role of histone methylation in tumor suppressor gene silencing in colorectal cancer. *Mol. Cell. Biol.* **23:** 206–215.

Kuzmichev A., Margueron R., Vaquero A., Preissner T.S., Scher M., Kirmizis A., Ouyang X., Brockdorff N., Abate-Shen C., Farnham P., and Reinberg D. 2005. Composition and histone substrates of polycomb repressive group complexes change during cellular differentiation. *Proc. Natl. Acad. Sci.* **102:** 1859–1864.

Laird P.W. 2003. The power and the promise of DNA methylation markers. *Nat. Rev. Cancer* **3:** 253–266.

Laird P.W., Jackson-Grusby L., Fazeli A., Dickinson S.L., Jung W.E., Li E., Weinberg R.A., and Jaenisch R. 1995. Suppression of intestinal neoplasia by DNA hypomethylation. *Cell* **81:** 197–205.

Lapeyre J.N. and Becker F.F. 1979. 5-Methylcytosine content of nuclear DNA during chemical hepatocarcinogenesis and in carcinomas which result. *Biochem. Biophys. Res. Commun.* **87:** 698–705.

Lee W.H., Morton R.A., Epstein J.I., Brooks J.D., Campbell P.A., Bova G.S., Hsieh W.S., Isaacs W.B., and Nelson W.G. 1994. Cytidine methylation of regulatory sequences near the π-class glutathione S-transferase gene accompanies human prostatic carcinogenesis. *Proc. Natl. Acad. Sci.* **91:** 11733–11737.

Leu Y.W., Rahmatpanah F., Shi H., Wei S.H., Liu J.C., Yan P.S., and Huang T.H. 2003. Double RNA interference of *DNMT3b* and *DNMT1* enhances DNA demethylation and gene reactivation. *Cancer Res.* **63:** 6110–6115.

Li E., Bestor T.H., and Jaenisch R. 1992. Targeted mutation of the DNA methyltransferase gene results in embryonic lethality. *Cell* **69:** 915–926.

Liu B., Nicolaides N.C., Markowitz S., Willson J.K., Parsons R.E., Jen J., Papadopolous N., Peltomaki P., de la Chapelle A., Hamilton S.R., et al. 1995. Mismatch repair gene defects in sporadic colorectal cancers with microsatellite instability. *Nat. Genet.* **9:** 48–55.

Lubbert M. 2000. DNA methylation inhibitors in the treatment of leukemias, myelodysplastic syndromes and hemoglobinopathies: Clinical results and possible mechanisms of action. *Curr. Top. Microbiol. Immunol.* **249:** 135–164.

Marks P., Rifkind R.A., Richon V.M., Breslow R., Miller T., and Kelly W.K. 2001. Histone deacetylases and cancer: Causes and therapies. *Nat. Rev. Cancer* **1:** 194–202.

Mizuno K., Osada H., Konishi H., Tatematsu Y., Yatabe Y., Mitsudomi T., Fujii Y., and Takahashi T. 2002. Aberrant hypermethylation of the CHFR prophase checkpoint gene in human lung cancers. *Oncogene* **21:** 2328–2333.

Morin P.J., Sparks A.B., Korinek V., Barker N., Clevers H., Vogelstein B., and Kinzler K.W. 1997. Activation of β-catenin-Tcf signaling in colon cancer by mutations in β-catenin or APC. *Science* **275:** 1787–1790.

Narayan A., Ji W., Zhang X.Y., Marrogi A., Graff J.R., Baylin S.B., and Ehrlich M. 1998. Hypomethylation of pericentromeric DNA in breast adenocarcinomas. *Int. J. Cancer* **77:** 833–838.

Nemoto S., Fergusson M.M., and Finkel T. 2004. Nutrient availability regulates SIRT1 through a forkhead-dependent pathway. *Science* **306:** 2105–2108.

Nguyen C.T., Gonzales F.A., and Jones P.A. 2001. Altered chromatin structure associated with methylation-induced gene silencing in cancer cells: Correlation of accessibility, methylation, MeCP2 binding and acetylation. *Nucleic Acids Res.* **29:** 4598–4606.

Pfeifer G.P., Tang M., and Denissenko M.F. 2000. Mutation hotspots and DNA methylation. *Curr. Top. Microbiol. Immunol.* **249:** 1–19.

Rainier S., Johnson L.A., Dobry C.J., Ping A.J., Grundy P.E., and Feinberg A.P. 1993. Relaxation of imprinted genes in human cancer. *Nature* **362:** 747–749.

Rhee I., Bachman K.E., Park B.H., Jair K.W., Yen R.W., Schuebel K.E., Cui H., Feinberg A.P., Lengauer C., Kinzler K.W., et al. 2002. DNMT1 and DNMT3b cooperate to silence genes in human cancer cells. *Nature* **416:** 552–556.

Rhee I., Jair K.W., Yen R.W., Lengauer C., Herman J.G., Kinzler K.W., Vogelstein B., Baylin S.B., and Schuebel K.E. 2000. CpG methylation is maintained in human cancer cells lacking *DNMT1*. *Nature* **404:** 1003–1007.

Richon V.M. and O'Brien J.P. 2002. Histone deacetylase inhibitors: A new class of potential therapeutic agents for cancer treatment. *Clin. Cancer Res.* **8:** 662–664.

Rideout W.M., III, Coetzee G.A., Olumi A.F., and Jones P.A. 1990. 5-Methylcytosine as an endogenous mutagen in the human LDL receptor and p53 genes. *Science* **249:** 1288–1290.

Robertson K.D., Ait-Si-Ali S., Yokochi T., Wade P.A., Jones P.L., and Wolffe A.P. 2000. DNMT1 forms a complex with Rb, E2F1 and HDAC1 and represses transcription from E2F-responsive promoters. *Nat. Genet.* **25:** 338–342.

Romanov S.R., Kozakiewicz B.K., Holst C.R., Stampfer M.R., Haupt L.M., and Tlsty T.D. 2001. Normal human mammary epithelial cells spontaneously escape senescence and acquire genomic changes. *Nature* **409:** 633–637.

Rountree M.R., Bachman K.E., and Baylin S.B. 2000. DNMT1 binds HDAC2 and a new co-repressor, DMAP1, to form a complex at replication foci. *Nat. Genet.* **25:** 269–277.

Sakai T., Toguchida J., Ohtani N., Yandell D.W., Rapaport J.M., and Dryja T.P. 1991. Allele-specific hypermethylation of the retinoblastoma tumor-suppressor gene. *Am. J. Hum. Genet.* **48:** 880–888.

Sakatani T., Kaneda A., Iacobuzio-Donahue C.A., Carter M.G., de Boom Witzel S., Okano H., Ko M.S., Ohlsson R., Longo D.L., and Feinberg A.P. 2005. Loss of imprinting of *Igf2* alters intestinal maturation and tumorigenesis in mice. *Science* **307:** 1976–1978.

Silverman L.R., Demakos E.P., Peterson B.L., Kornblith A.B., Holland J.C., Odchimar-Reissig R., Stone R.M., Nelson D., Powell B.L., DeCastro C.M., et al. 2002. Randomized controlled trial of azacitidine in patients with the myelodysplastic syndrome: A study of the cancer and leukemia group B. *J. Clin. Oncol.* **20:** 2429–2440.

Srinivasan P.R. and Borek E. 1964. Enzymatic alteration of nucleic acid structure. *Science* **145:** 548–553.

Suzuki H., Gabrielson E., Chen W., Anbazhagan R., van Engeland M., Weijenberg M.P., Herman J.G., and Baylin S.B. 2002. A genomic screen for genes upregulated by demethylation and histone deacetylase inhibition in human colorectal cancer. *Nat. Genet.* **31:** 141–149.

Suzuki H., Watkins D.N., Jair K.W., Schuebel K.E., Markowitz S.D., Chen W.D., Pretlow T.P., Yang B., Akiyama Y., Van Engeland M., et al. 2004. Epigenetic inactivation of *SFRP* genes allows constitutive WNT signaling in colorectal cancer. *Nat. Genet.* **36:** 417–422.

Toyota M., Ho C., Ahuja N., Jair K.W., Li Q., Ohe-Toyota M., Baylin S.B., and Issa J.P. 1999. Identification of differentially methylated sequences in colorectal cancer by methylated CpG island amplification. *Cancer Res.* **59:** 2307–2312.

Ushijima T. 2005. Detection and interpretation of altered methylation patterns in cancer cells. *Nat. Rev. Cancer* **5:** 223–231.

Veigl M.L., Kasturi L., Olechnowicz J., Ma A.H., Lutterbaugh J.D., Periyasamy S., Li G.M., Drummond J., Modrich P.L., Sedwick W.D., and Markowitz S.D. 1998. Biallelic inactivation of *hMLH1* by epigenetic gene silencing, a novel mechanism causing human

MSI cancers. *Proc. Natl. Acad. Sci.* **95:** 8698–8702.

Wales M.M., Biel M.A., el Deiry W., Nelkin B.D., Issa J.P., Cavenee W.K., Kuerbitz S.J., and Baylin S.B. 1995. p53 activates expression of HIC-1, a new candidate tumour suppressor gene on 17p13.3. *Nat. Med.* **1:** 570–577.

Wijermans P., Lubbert M., Verhoef G., Bosly A., Ravoet C., Andre M., and Ferrant A. 2000. Low-dose 5-aza-2′-deoxycytidine, a DNA hypomethylating agent, for the treatment of high-risk myelodysplastic syndrome: A multicenter phase II study in elderly patients. *J. Clin. Oncol.* **18:** 956–962.

Wolf P., Hu Y.C., Doffek K., Sidransky D., and Ahrendt S.A. 2001. O^6-methylguanine-DNA methyltransferase promoter hypermethylation shifts the *p53* mutational spectrum in non-small cell lung cancer. *Cancer Res.* **61:** 8113–8117.

Wolffe A.P. 2001. Chromatin remodeling: Why it is important in cancer. *Oncogene* **20:** 2988–2990.

Yamashita K., Upadhyay S., Osada M., Hoque M.O., Xiao Y., Mori M., Sato F., Meltzer S.J., and Sidransky D. 2002. Pharmacologic unmasking of epigenetically silenced tumor suppressor genes in esophageal squamous cell carcinoma. *Cancer Cell* **2:** 485–495.

Appendix 1

WWW Resources

Chromatin and epigenetic related Web sites

Chromatin Structure and Function	A chromatin resource page hosted by Jim Bone	http://www.chromatin.us
Epigenome Network of Excellence	Web site of the European interdisciplinary epigenetics research network	http://www.epigenome-noe.net
HEP	The human epigenome project research consortium	http://www.epigenome.org/
ENCODE	Encylopedia of DNA Elements: identifying functional elements in human	http://www.genome.gov/12513456
ChromDB	The plant chromatin database	http://www.chromdb.org/

Histone resources

The NHGRI histone database	Contains histone sequence information	http://research.nhgri.nih.gov/histones/
Abcam histone page	Contains downloads of histone modification maps	http://www.abcam.com/chromatin
Histone.com page (Upstate)	Histone modification map generated by Upstate	http://www.histone.com/modification_map.htm

Chromatin proteins

NPD-Nuclear Protein Database	Collection of known vertebrate proteins localized in the nucleus	http://npd.hgu.mrc.ac.uk/index.html
Chromatin proteins	Link to searches on the NPD site restricted to chromatin protein families	http://www.epigenome-noe.net/researchtools/structure.php
Flybase chromatin page	Covers Polycomb, trithorax, and other chromatin proteins in *Drosophila*	http://flybase.bio.indiana.edu/allied-data/lk/interactive-fly/aignfam/polycomb.htm

DNA methylation

MethDB	Human DNA methylation database	http://www.methdb.de/
DNA methylation society	Web site dedicated to aspects of biological methylation	http://www.dnamethsoc.com/
CpG island searcher	CpG island sequence search algorithm	http://www.uscnorris.com/cpgislands2/cpg.aspx

Imprinting

Mouse genomic imprinting	Chromosomal maps displaying known mouse imprinted regions	http://www.mgu.har.mrc.ac.uk/research/imprinting/
Gene Imprint	WWW information resource on genomic imprinting	http://www.geneimprint.com/index.html
Imprinted Gene Catalogue	Database of imprinted genes	http://igc.otago.ac.nz/home.html
Candidate imprinted transcripts	Human and mouse predicted imprinted transcriptome resource	http://fantom2.gsc.riken.go.jp/imprinting/

(continued on next page)

WWW Resources *(continued)*

RNA

NONCODE	Knowledge database dedicated to noncoding RNA	http://www.bioinfo.org.cn/NONCODE
RNAdb	Mammalian noncoding RNA database	http://research.imb.uq.edu.au/rnadb
MicroRNAdb	Database of eukaryotic microRNAs	http://166.111.30.65/micrornadb/
miRBase	microRNA data resource	http://microrna.sanger.ac.uk/
NARNA	Platform for research on natural antisense transcripts	http://www.narna.ncl.ac.uk/
ASRP	*Arabidopsis thaliana* small siRNA and miRNA data resource	http://asrp.cgrb.oregonstate.edu/

Commercial epigenetic resource sites

Abcam	Suppliers of histone antibodies	http://www.abcam.com/
Upstate	Suppliers of histone antibodies	http://www.upstatebiotech.com/
Epigenomics	DNA methylation diagnostic screening	http://www.epigenomics.com/

Organismal genome resources and databases

All eukaryotes

NCBI	Portal to multiple genome analysis and reference sites	http://www.ncbi.nlm.nih.gov
ENSEMBL	Eukaryotic genome browser	http://www.ensembl.org
UCSC Genome Bioinformatics	Genome sequence and resource portal	http://genome.ucsc.edu
EBI	Portal to various genomic and proteomic computational analysis resources	http://www.ebi.ac.uk/services
Sanger Institute	Portal to sequence, bioinformatic, and proteomic resources	http://www.sanger.ac.uk/
BCM Search Launcher	Portal for sequence searches and analysis tools	http://searchlauncher.bcm.tmc.edu/docs/sl_links.html
zPicture	Comparative sequence analysis tool	http://zpicture.dcode.org
iHOP	Information hyperlinked over proteins—protein resource	http://www.ihop-net.org/UniPub/iHOP/
Repeat Masker	Repeat sequence algorithm to identify repetitive DNA sequences	http://repeatmasker.org
Primer 3	Web-based primer design program	http://frodo.wi.mit.edu/cgi-bin/primer3/primer3.cgi

Individual organisms

S. cerevisiae	SGD—*Saccharomyces* genome database	http://www.yeastgenome.org
	ENSEMBL *S. cerevisiae* genome analysis portal	http://www.ensembl.org/Saccharomyces_cerevisiae/index.html
S. pombe	Portal to *S. pombe* related genomic sequence and analysis sites	http://www.sanger.ac.uk/Projects/S_pombe/
T. thermophila	TGD—*Tetrahymena* genome database	http://www.ciliate.org/
	Tetrahymena macronuclear genome sequencing project	http://www.genome.gov/12512294
A. thaliana	TAIR-the Arabidopsis information resource	http://www.arabidopsis.org
	GRAMENE—comparative grass genomics	http://www.gramene.org/
	Rice genome resource	http://www.tigr.org/tdb/e2k1/osa1/
C. elegans	Portal to *C. elegans* related resources	http://www.wormbase.org/
	ENSEMBL *C. elegans* genome analysis portal	http://www.ensembl.org/Caenorhabditis_elegans/index.html
	Portal to *C. elegans* related genomic sequence and analysis sites	http://www.sanger.ac.uk/Projects/C_elegans/
Drosophila	Database of the *Drosophila* genome	http://www.flybase.net
	ENSEMBL *D. melanogaster* genome analysis portal	http://www.ensembl.org/Drosophila_melanogaster/index.html
X. laevis	Information database	http://www.xenbase.org/
Mouse	Mouse genome informatics	http://www.informatics.jax.org
	Mouse strain resource	http://jaxmice.jax.org/index.html
	Baylor mouse genome project	http://www.mouse-genome.bcm.tmc.edu/
	Mouse ENSEMBL resource	http://www.ensembl.org/Mus_musculus/index.html
Human	Human genome resource portal	http://www.ncbi.nlm.nih.gov/genome/guide/human/
	ENSEMBL human genome analysis portal	http://www.ensembl.org/Homo_sapiens/index.html
	GDB human genome database	http://www.gdb.org

Appendix 2

Histone Modifications and References

HISTONE H2A

Site	Model	Enzyme	Function	Reference
K5ac	Hs,Sc	Tip60, p300/CBP,Hat1	Transcriptional activation	Yamamoto and Horikoshi 1997; Kimura and Horikoshi 1998; Verreault et al. 1998
K9bio		HCS Biotinidase	Acetylation and methylation dependent Involved in cell proliferation, gene silencing, and cellular response to DNA damage	Stanley et al. 2001; Kothapalli et al. 2005b; Chew et al. 2006
K7ac	Sc	Hat1, Esa1	Transcriptional activation	Suka et al. 2001
K13bio		HCS Biotinidase	Acetylation and methylation dependent Involved in cell proliferation, gene silencing, and cellular response to DNA damage	Stanley et al. 2001; Kothapalli et al. 2005b; Chew et al. 2006
K119ub1	Dm, Hs	dRing, RING1B	Polycomb silencing UV damage response	Wang et al. 2004; Kapetanaki et al. 2006
S121ph	Sc	Mec1 PIKK	DNA damage response Telomere silencing	Wyatt et al. 2003; Harvey et al. 2005
T125ph	Sc	Mec1 PIKK	DNA damage response Telomere silencing	Wyatt et al. 2003
K126bio	Hs	HCS Biotinidase	Acetylation and methylation dependent Involved in cell proliferation, gene silencing, and cellular response to DNA damage	Stanley et al. 2001; Kothapalli et al. 2005b; Chew et al. 2006
K126su	Sc		Transcriptional repression Blocks histone acetylation and histone ubiquitination	Nathan et al. 2006
K127bio	Hs	HCS Biotinidase	Acetylation and methylation dependent Involved in cell proliferation, gene silencing, and cellular response to DNA damage	Stanley et al. 2001; Kothapalli et al. 2005b; Chew et al. 2006
S128ph	Sc	Mec1 PIKK	DNA damage response Telomere silencing	Downs et al. 2000; Redon et al. 2003; Wyatt et al. 2003; Downs et al. 2004
K130bio	Hs	HCS Biotinidase	Acetylation and methylation dependent Involved in cell proliferation, gene silencing, and cellular response to DNA damage	Stanley et al. 2001; Kothapalli et al. 2005b; Chew et al. 2006

Additional H2A modifications:

S1ph, K4ac, K13me, K15ac, K21ac, K36ac, K74me, K75me, R77me, K95me, T120ph, H2A carbonylation (Pantazis and Bonner 1981; Song et al. 2003; Alhara et al. 2004).

HISTONE H2B

Site	Model	Enzyme	Function	Reference
K5ac	Hs		Transcriptional activation	Puerta et al. 1995; Galasinski et al. 2002
S10ph	Sc	Ste20	Apoptosis	Ahn et al. 2005
S14ph	Hs, Mm	Mst1/krs2 kinase	Apoptosis Somatic hypermutation and class switch recombination	Ajiro 2000; Cheung et al. 2003; Odegard et al. 2005
K16su	Sc		Gene repression	Nathan et al. 2006
K17su	Sc		Gene repression	Nathan et al. 2006
S33ph	Dm	CTK	Transcriptional activation	Maile et al. 2004
K120ub	Hs		Cell cycle progression in concert with SAGA for transcriptional activation through H3 methylation, meiosis	Robzyk et al. 2000; Sun and Allis 2002; Kao et al. 2004
K123ub	Sc	Rad6(E2) Bre1(E3)-ub1	Telomeric silencing by lowering histone methylation at H3K4 and H3K79	Emre et al. 2005

Additional H2B modifications:

E2arn, K5me, K6su, K7su, K11ac, K12ac, K15ac, K16ac, K20ac, K23me, K24ac, S32ph, K43me, K85ac, R99me, K108ac, K116ac, H2B carbonylation, H2B biotinylation (Rouleau et al. 2004).

HISTONE H3

Site	Model	Enzyme	Function	Reference
R2me	Hs Mm	CARM1 -me2a	Gene expression	Chen et al. 1999; Schurter et al. 2001
T3ph	Hs At	Haspin	Centromere mitotic spindle function	Polioudaki et al. 2004; Dai et al. 2005
K4me	Sc	Set1-me3	rDNA silencing, telomeric silencing Transcriptional activation	Briggs et al. 2001; Roguev et al. 2001; Nagy et al. 2002; Bryk et al. 2002; Bernstein et al. 2002; Santos-Rosa et al. 2002
	Tt		Transcriptional activation	Strahl et al.1999
	Hs	SET7/Set9-me1	Transcriptional activation	H. Wang et al. 2001a; Nishioka et al. 2002a; Wilson et al. 2002; Zegerman et al. 2002
	Ds Hs	MLL MLL2 -me3 MLL3	Trithorax activation	Milne et al. 2002; Nakamura et al. 2002
	Ds Hs	Ash1-me2	Trithorax activation	Beisel et al. 2002; Sanchez-Elsner et al. 2006
	Hs	SMYD3-me3	Transcriptional activation	Hamamoto et al. 2004
	Mm	Meisetz-me3	Meiotic prophase progression	Hayashi et al. 2005
K9ac	Sc	SAGA	Transcriptional activation	Grant et al. 1999
	Hs	SRC1	Nuclear receptor coactivator	Spencer et al. 1997; Schubeler et al. 2000; Vaquero et al. 2004
	Dm		Transcriptional activation	Nowak et al. 2000
K9me	Mm, Hs	G9a-me1,me2	Transcriptional repression Imprinting	Tachibana et al. 2001, 2002; Ogawa et al. 2002; Xin et al. 2003
	Dm	Su(var)3-9-me2	Dominant PEV modifier	Czermin et al. 2001; Schotta et al. 2002; Ebert et al. 2004
	Mm	Suv39h1-me3 Suv39h2-me3	Pericentric heterochromatin	O'Carroll et al. 2000; Rea et al. 2000; Lachner et al. 2001; Peters et al. 2001
	Hs	SUV39H1-me3	Rb-mediated silencing	Nielsen et al. 2001; Vandel et al. 2001
	Sp	Clr4-me1,me2	Centromeric and mating-type silencing	Bannister et al. 2001; Nakayama et al. 2001

(continued on next page)

HISTONE H3 (continued)

Site	Model	Enzyme	Function	Reference
	Nc	Dim5-me3	DNA methylation	Tamaru and Selker 2001
	At	KRYPTONITE-me2	DNA methylation	Jackson et al. 2002, 2004
	Hs	EuHMTase1-me1,me2	Transcriptional repression	Ogawa et al. 2002; Tachibana et al. 2005
	Hs,Mm	ESET-me2,me3	Transcriptional repression	Schultz et al. 2002; Yang et al. 2002; Dodge et al. 2004; Wang et al. 2004
	Dm, Hs	Ash1-me2	Trithorax activation	Beisel et al. 2002
	Hs	RIZ1-me2	Tumor suppression and response to female sex hormones	Kim et al. 2003; Carling et al. 2004
S10ph	Sc	Snf1	Transcriptional activation	Lo et al. 2001
	Dm	Jil-1	Transcriptional up-regulation of male X chromosome	Jin et al. 1999; Y. Wang et al. 2001
	Hs	Rsk2 Msk1 Msk2	Transcriptional activation of immediate early genes (in concert with H3K14 acetylation)	Sassone-Corsi et al. 1999; Thomson et al. 1999; Cheung et al. 2000; Clayton et al. 2000
	Hs	IKKα	Transcriptional up-regulation	Anest et al. 2003; Yamamoto et al. 2003
	Sc, Ce	Ip11/AuroraB	Mitotic chromosome condensation	Hendzel et al. 1997; Wei et al. 1999; Hsu et al. 2000
	An	NIMA	Mitotic chromosome condensation	De Souza et al. 2000
	Hs, Ce	Fyn kinase	UVB induced MAP kinase pathway	He et al. 2005
T11ph	Hs	Dlk / ZIP	Mitosis specific phosphorylation	Preuss et al. 2003
K14ac	Sc, Tt, Mm	Gcn5	Transcriptional activation	Brownell et al. 1996; Kuo et al. 1996
	Hs, Dm	TAF$_{II}$230 TAF$_{II}$250	Transcriptional activation	Mizzen et al. 1996
	Hs	p300	Transcriptional activation	Schiltz et al. 1999
	Hs	PCAF	Transcriptional activation	Schiltz et al. 1999
	Mm	SRC1	Nuclear receptor coactivator	Spencer et al. 1997
R17me	Hs, Mm	CARM1	Transcriptional activation (in concert with H3K18/23 acetylation)	Chen et al. 1999; Schurter et al. 2001; Bauer et al. 2002; Daujat et al. 2002
K18ac	Sc	SAGA Ada	Transcriptional activation	Grant et al. 1999
	Hs	p300	Transcriptional activation	Schiltz et al. 1999
	Hs	CBP	Transcriptional activation (in concert with H3R17 methylation)	Daujat et al. 2002
K23ac	Sc	SAGA	Transcriptional activation	Grant et al. 1999
	Hs	CBP	Transcriptional activation (in concert with H3R17 methylation)	Daujat et al. 2002
R26me	Hs	CARM1	In vitro methylation site	Chen et al. 1999; Schurter et al. 2001
K27me	Hs, Dm	E(z)/EZH2-me3	Polycomb repression Early B-cell development X-chromosome inactivation	Cao et al. 2002; Czermin et al. 2002; Kuzmichev et al. 2002; Muller et al. 2002; Su et al 2003
S28ph	Hs	Aurora-B	Mitotic chromosome condensation	Goto et al. 1999,2002
	Hs	MSK1	UVB induced phosphorylation	Zhong et al. 2001
K36me	Sc	Set2-me2	Gene repression	Strahl et al. 2002; Kizer et al. 2005
	Nc	Set2-me2	Transcriptional activation	Adhvaryu et al. 2005
	Sp	Set2-me2	Transcriptional elongation	Morris et al. 2005
K79me	Sc, Hs	Dot1/DOT1L-me2	Telomeric silencing, pachytene checkpoint	Feng et al. 2002; Lacoste et al. 2002; Ng et al. 2002; van Leeuwen et al. 2002

Additional H3 modifications:
K14me, K23me, K27ac, T32ph, K37me, K56me, K64me, K115ac, K118ac, K118me, K122ac, R128me (Hyland et al. 2005).

HISTONE H4

Site	Model	Enzyme	Function	Reference
S1ph	Hs, Sc	Casein kinase II	DNA damage response	Ruiz-Carrillo et al. 1975; Cheung et al. 2005; van Attikum and Gasser 2005
R3me	Hs, Sc	PRMT1	Transcriptional activation	H. Wang et al. 2001a,b
K5ac	*Tt, Dm, Hs*	Hat1	Histone deposition	Sobel et al. 1995; Parthun et al. 1996; Taplick et al. 1998; Turner 2000; Kruhlak et al. 2001
	Sc	Esa1/NuA4	Cell cycle progression	Smith et al. 1998; Allard et al. 1999; Clarke et al. 1999; Bird et al. 2002; Miranda et al. 2006
	Hs, Mm	ATF2	Sequence-specific TF	Kawasaki et al. 2000a
	Hs	p300	Transcriptional activation	Turner and Fellows 1989; Schiltz et al. 1999
K8ac	Hs,Mm	Y - ATF2	Excluded from Xi Sequence-specific transcription factor	Jeppesen et al. 1993; Choy et al. 2001; Kruhlak et al. 2001; Kawasaki et al. 2000b
	Hs	PCAF/ p300	Transcriptional activation	Turner and Fellows 1989; Schiltz et al. 1999
K12ac	Sc, Hs	Hat1	Excluded from Xi Histone deposition	Turner and Fellows 1989; Jeppesen et al. 1993; Kleff et al. 1995; Sobel et al. 1995; Parthun et al. 1996; Chang et al. 1997; Kruhlak et al. 2001
	Sc	NuA4	Mitotic and meiotic progression	Choy et al. 2001
K12bio	Hs	HCS Biotinidase	Decrease in response to DNA double-strand breaks Effects on cell proliferation	Stanley et al. 2001; Kothapalli et al. 2005a,b
K16ac	Mm		Excluded from Xi Cell cycle dependent acetylation	Jeppesen et al. 1993; Taplick et al. 1998
	Dm	MOF	Transcriptional up-regulation of male X chromosome	Akhtar and Becker 2000; Hsu et al. 2000
	Hs, Mm	ATF2	Sequence-specific transcription factor	Turner and Fellows 1989; Kawasaki et al. 2000a; Turner 2000; Kruhlak et al. 2001; Vacquero et al. 2004
K20me	Mm, Dm	Suv4-20h1-me2,me3 Suv4-20h2-me2,me3	Gene silencing	Schotta et al. 2004
	Hs, Dm	Pr-SET7/Set8-me1	Transcriptional silencing, mitotic condensation	Fang et al. 2002; Nishioka et al. 2002b; Rice et al. 2002
	Dm	Ash1-me2	Trithorax activation in concert with H3K4 and H3K9 methylation	Beisel et al. 2002
K59me	Sc		Silent chromatin formation	Zhang et al. 2003
Su	Hs	SUMO-1 SUMO-3	Transcriptional repression	Shiio and Eisenman 2003

Additional H4 modifications:

K12me, S47ph, K31ub1, K59me, K77ac, K79ac, K79me, K91ub1, R92me (Hyland et al. 2005).

HISTONE H1

Site	Model	Enzyme	Function	Reference
E2ar1	Rn	PARP-1	Involved in neurotrophic activity	Ogata et al. 1980; Visochek et al. 2005
T10ph	Hs		Mitosis specific Transcriptional activation H1b	Chadee et al. 1995; Garcia et al. 2004; Sarg et al. 2006
E14ar1	Rn	PARP-1	Involved in neurotrophic activity	Ogata et al. 1980; Visochek et al. 2005
S17ph	Hs		Interphase specific Transcriptional activation H1b	Garcia et al. 2004; Chadee et al. 1995; Sarg et al. 2006
K26me	Hs	EZH2-me2	Mediates HP1 binding	Kuzmichev et al. 2004; Daujat et al. 2005
S27ph	Hs		Blocks HP1 binding	Garcia et al. 2004; Daujat et al. 2005
T137ph	Hs		Mitosis specific Transcriptional activation H1b	Chadee et al. 1995; Garcia et al. 2004; Sarg et al. 2006
T154ph	Hs		Mitosis specific Transcriptional activation H1b	Chadee et al. 1995; Garcia et al. 2004; Sarg et al. 2006
S172ph	Hs		Interphase specific Transcriptional activation H1b	Chadee et al. 1995; Garcia et al. 2004; Sarg et al. 2006
S188ph	Hs		Interphase specific Transcriptional activation H1b	Chadee et al. 1995; Garcia et al. 2004; Sarg et al. 2006
K213ar1	Rn	PARP-1	Involved in neurotrophic activity	Ogata et al. 1980; Visochek et al. 2005

Additional H1 modifications:
K26ac, E114arn, H1 ubiquitination, H1 carbonylation (Pham and Sauer 2000; Rouleau et al. 2004).

HISTONE H2AX

Site	Model	Enzyme	Function	Reference
S139ph	Hs, Sc, Dm, Xl	ATM DNA-PK	DNA repair M-phase related Also known as γH2AX	Rogakou et al. 1998,1999; Burma et al. 2001; Stiff et al. 2004; Ichijima et al. 2005; Mukherjee et al. 2006

Additional H2AX modifications:
T136ph.

HISTONE macroH2A

Site	Model	Enzyme	Function	Reference
K17me	Hs			Chu et al. 2006
K115ub1	Hs			Ogawa et al. 2005; Chu et al. 2006
K122me	Hs			Chu et al. 2006
T128ph	Hs			Chu et al. 2006
K238me	Hs			Chu et al. 2006

HISTONE H3.3

Site	Model	Enzyme	Function	Reference
K4me	Dm	me1, me2, me3	Transcriptional activation	McKittrick et al. 2004
K9me	Dm	me1, me2	Transcriptional repression	McKittrick et al. 2004
K9ac	Dm, Hs		Transcriptional activation	McKittrick et al. 2004; Hake and Allis 2006
K14me	Dm	me1, me2		McKittrick et al. 2004
K14ac	Dm, Hs		Transcriptional activation	McKittrick et al. 2004; Hake and Allis 2006
K18ac	Hs		Transcriptional activation	Hake and Allis 2006
K23ac	Hs		Transcriptional activation	Hake and Allis 2006
K27me	Dm	me1, me2, me3	Transcriptional repression	McKittrick et al. 2004
K36me	Dm, Hs	me1, me2, me3	Transcriptional activation	McKittrick et al. 2004; Hake and Allis 2006
K37me	Dm	me1, me2		McKittrick et al. 2004
K79me	Dm, Hs	me1, me2	Transcriptional activation	McKittrick et al. 2004; Hake and Allis 2006
S31ph	Mammals		Mitosis specific phosphorylation	Hake et al. 2005

CEN-H3/ CENP-A

Site	Model	Enzyme	Function	Reference
S7ph	Hs		Mitosis	Zeitlin et al. 2001

Additional CENP-A modifications:
S17ph.

Abbreviations for model organisms:

(An) *Aspergillus nidulans, (At) Arabidopsis thaliana, (Ce) Caenorhabditis elegans, (Dm) Drosophila melanogaster,* (Hs) *Homo sapiens,* (Mm) *Mus musculus,* (Nc) *Neurospora crassa,* (Rn) *Rattus norvegicus,* (Sc) *Saccharomyces cerevisiae,* (Sp) *Schizosaccharomyces pombe,* (Tt) *Tetrahymena thermophila,* (Xl) *Xenopus laevis.*

The histone modifications follow the nomenclature as proposed by Turner (2005).

The tables list all known histone modifications and the known enzymes with their primary references (until May 2006). Additional modifications with currently unknown function are listed below the tables. These modifications were obtained from a combination of sources, which comprise review articles, information from Abcam Cambridge, UK and Upstate Charlottesville, USA and unpublished data by the Reinberg laboratory.

The table was compiled based on an original setup from Lachner et al. (2003) and significantly extended by Roopsha Sengupta and Mario Richter (Jenuwein laboratory IMP Vienna). Dr. Patrick Trojer (Reinberg laboratory HHMI, New Jersey) confirmed the table contents.

This table was validated and combined with additional information provided by Dr. Steven Gray (Dept. of Clinical Medicine, Institute of Molecular Medicine, St. James Hospital, Dublin).

References

Adhvaryu K.K., Morris S.A., Strahl B.D., and Selker E.U. 2005. Methylation of histone H3 lysine 36 is required for normal development in *Neurospora crassa*. *Eukaryot. Cell* **4:** 1455–1464.

Ahn S.H., Cheung W.L., Hsu J.Y., Diaz R.L., Smith M.M., and Allis C.D. 2005. Sterile 20 kinase phosphorylates histone H2B at serine 10 during hydrogen peroxide-induced apoptosis in *S. cerevisiae*. *Cell* **120:** 25–36.

Aihara H., Nakagawa T., Yasui K., Ohta T., Hirose S., Dhomae N., Takio K., Kaneko M., Takeshima Y., Muramatsu M., and Ito T. 2004. Nucleosomal histone kinase-1 phosphorylates H2A Thr 119 during mitosis in the early *Drosophila* embryo. *Genes Dev.* **18:** 877–888.

Ajiro K. 2000. Histone H2B phosphorylation in mammalian apoptotic cells. An association with DNA fragmentation. *J. Biol. Chem.* **275:** 439–443.

Akhtar A., and Becker P.B. 2000. Activation of transcription through histone H4 acetylation by MOF, an acetyltransferase essential for dosage compensation in *Drosophila*. *Mol. Cell* **5:** 367–375.

Allard S., Utley R.T., Savard J., Clarke A., Grant P., Brandl C.J., Pillus L., Workman J.L., and Cote J. 1999. NuA4, an essential transcription adaptor/histone H4 acetyltransferase complex containing Esa1p and the ATM-related cofactor Tra1p. *EMBO J.* **18:** 5108–5119.

Anest V., Hanson J.L., Cogswell P.C., Steinbrecher K.A., Strahl B.D., and Baldwin A.S. 2003. A nucleosomal function for IκB kinase-α in NF-κB-dependent gene expression. *Nature* **423:** 659–663.

Bannister A.J., Zegerman P., Partridge J.F., Miska E.A., Thomas J.O., Allshire R.C., and Kouzarides T. 2001. Selective recognition of methylated lysine 9 on histone H3 by the HP1 chromo domain. *Nature* **410:** 120–124.

Bauer U.M., Daujat S., Nielsen S.J., Nightingale K., and Kouzarides T. 2002. Methylation at arginine 17 of histone H3 is linked to gene activation. *EMBO Rep.* **3:** 39–44.

Beisel C., Imhof A., Greene J., Kremmer E., and Sauer F. 2002. Histone methylation by the *Drosophila* epigenetic transcriptional regulator Ash1. *Nature* **419:** 857–862.

Berger S.L. 2002. Histone modifications in transcriptional regulation. *Curr. Opin. Genet. Dev.* **12:** 142–148.

Bernstein B.E., Humphrey E.L., Erlich R.L., Schneider R., Bouman P., Liu J.S., Kouzarides T., and Schreiber S.L. 2002. Methylation of histone H3 Lys 4 in coding regions of active genes. *Proc. Natl. Acad. Sci.* **99:** 8695–8700.

Bird A.W., Yu D.Y., Pray-Grant M.G., Qiu Q., Harmon K.E., Megee P.C., Grant P.A., Smith M.M., and Christman M.F. 2002. Acetylation of histone H4 by Esa1 is required for DNA double-strand break repair. *Nature* **419:** 411–415.

Briggs S.D., Bryk M., Strahl B.D., Cheung W.L., Davie J.K., Dent S.Y., Winston F., and Allis C. D. 2001. Histone H3 lysine 4 methylation is mediated by Set1 and required for cell growth and rDNA silencing in *Saccharomyces cerevisiae*. *Genes Dev.* **15:** 3286–3295.

Bryk M., Briggs S.D., Strahl B.D., Curcio M.J., Allis C.D., and Winston F. 2002. Evidence that Set1, a factor required for methylation of histone H3, regulates rDNA silencing in *S. cerevisiae* by a Sir2-independent mechanism. *Curr. Biol.* **12:** 165–170.

Brownell J.E., Zhou J., Ranalli T., Kobayashi R., Edmondson D.G., Roth S.Y., and Allis C.D. 1996. *Tetrahymena* histone acetyltransferase A: A homolog to yeast Gcn5p linking histone acetylation to gene activation. *Cell* **84:** 843–851.

Burma S., Chen B.P., Murphy M., Kurimasa A., and Chen D.J. 2001. ATM phosphorylates histone H2AX in response to DNA double-strand breaks. *J. Biol. Chem.* **276:** 42462–42467.

Cao R., Wang L., Wang H., Xia L., Erdjument-Bromage H., Tempst P., Jones R.S., and Zhang Y. 2002. Role of histone H3 lysine 27 methylation in Polycomb-group silencing. *Science* **298:** 1039–1043.

Carling T., Kim K.C., Yang X.H., Gu J., Zhang X.K., and Huang S. 2004. A histone methyltransferase is required for maximal response to female sex hormones. *Mol. Cell. Biol.* **24:** 7032–7042.

Chadee D.N., Taylor W.R., Hurta R.A., Allis C.D., Wright J.A., and Davie J.R. 1995. Increased phosphorylation of histone H1 in mouse fibroblasts transformed with oncogenes or constitutively active mitogen-activated protein kinase kinase. *J. Biol. Chem.* **270:** 20098–20105.

Chang L., Loranger S.S., Mizzen C., Ernst S.G., Allis C.D., and Annunziato A.T. 1997. Histones in transit: Cytosolic histone complexes and diacetylation of H4 during nucleosome assembly in human cells. *Biochemistry* **36:** 469–480.

Chen D., Ma H., Hong H., Koh S.S., Huang S.M., Schurter B.T., Aswad D.W., and Stallcup M.R. 1999. Regulation of transcription by a protein methyltransferase. *Science* **284:** 2174–2177.

Cheung P., Tanner K.G., Cheung W.L., Sassone-Corsi P., Denu J.M., and Allis C.D. 2000. Synergistic coupling of histone H3 phosphorylation and acetylation in response to epidermal growth factor stimulation. *Mol. Cell* **5:** 905–915.

Cheung W.L., Turner F.B., Krishnamoorthy T., Wolner B., Ahn S.H., Foley M., Dorsey J.A., Peterson C.L., Berger S.L., and Allis C.D. 2005. Phosphorylation of histone H4 serine 1 during DNA damage requires casein kinase II in *S. cerevisiae*. *Curr. Biol.* **15:** 656–660.

Cheung W.L., Ajiro K., Samejima K., Kloc M., Cheung P., Mizzen C.A., Beeser A., Etkin L.D., Chernoff J., Earnshaw W.C., and Allis C.D. 2003. Apoptotic phosphorylation of histone H2B is mediated by mammalian sterile twenty kinase. *Cell* **113:** 507–517.

Chew Y.C., Camporeale G., Kothapalli N., Sarath G., and Zempleni J. 2006. Lysine residues in N-terminal and C-terminal regions of human histone H2A are targets for biotinylation by biotinidase. *J. Nutr. Biochem.* **17:** 225–233.

Choy J.S., Tobe B.T., Huh J.H., and Kron S.J. 2001. Yng2p-dependent NuA4 histone H4 acetylation activity is required for mitotic and meiotic progression. *J. Biol. Chem.* **276:** 43653–43662.

Chu F., Nusinow D.A., Chalkley R.J., Plath K., Panning B., and Burlingame A.L. 2006. Mapping post-translational modifications of the histone variant MacroH2A1 using tandem mass spectrometry. *Mol. Cell. Proteomics* **5:** 194–203.

Clarke A.S., Lowell J.E., Jacobson S.J., and Pillus L. 1999. Esa1p is an essential histone acetyltransferase required for cell cycle progression. *Mol. Cell. Biol.* **19:** 2515–2526.

Clayton A.L., Rose S., Barratt M.J., and Mahadevan L.C. 2000. Phosphoacetylation of histone H3 on *c-fos*- and *c jun*-associated nucleosomes upon gene activation. *EMBO J.* **19:** 3714–3726.

Coffee B., Zhang F., Warren S.T., and Reines D. 1999. Acetylated histones are associated with FMR1 in normal but not fragile X-syndrome cells. *Nat. Genet.* **22:** 98–101.

Czermin B., Melfi R., McCabe D., Seitz V., Imhof A., and Pirrotta V. 2002. *Drosophila* enhancer of Zeste/ESC complexes have a histone H3 methyltransferase activity that marks chromosomal Polycomb sites. *Cell* **111:** 185–196.

Czermin B., Schotta G., Hulsmann B.B., Brehm A., Becker P.B., Reuter G., and Imhof A. 2001. Physical and functional association of SU(VAR)3-9 and HDAC1 in *Drosophila*. *EMBO Rep.* **2:** 915–919.

Dai J., Sultan S., Taylor S.S., and Higgins J.M. 2005. The kinase haspin is required for mitotic histone H3 Thr 3 phosphorylation and normal metaphase chromosome alignment. *Genes Dev.* **19:** 472–488.

Daujat S., Zeissler U., Waldmann T., Happel N., and Schneider R. 2005. HP1 binds specifically to Lys26-methylated histone H1.4, whereas simultaneous Ser27 phosphorylation blocks HP1 binding. *J. Biol. Chem.* **280:** 38090–38095.

Daujat S., Bauer U.M., Shah V., Turner B., Berger S., and Kouzarides T. 2002. Crosstalk between CARM1 methylation and CBP acetylation on histone H3. *Curr. Biol.* **12:** 2090–2097.

De Souza C.P., Osmani A.H., Wu L.P., Spotts J.L., and Osmani S.A. 2000. Mitotic histone H3 phosphorylation by the NIMA kinase in *Aspergillus nidulans. Cell* **102:** 293–302.

Dodge J.E., Kang Y.K., Beppu H., Lei H., and Li E. 2004. Histone H3-K9 methyltransferase ESET is essential for early development. *Mol. Cell. Biol.* **24:** 2478–2486.

Downs J.A., Lowndes N.F., and Jackson S.P. 2000. A role for *Saccharomyces cerevisiae* histone H2A in DNA repair. *Nature* **408:** 1001–1004.

Downs J.A., Allard S., Jobin-Robitaille O., Javaheri A., Auger A., Bouchard N., Kron S.J., Jackson S.P., and Cote J. 2004. Binding of chromatin-modifying activities to phosphorylated histone H2A at DNA damage sites. *Mol. Cell* **16:** 979–990.

Ebert A., Schotta G., Lein S., Kubicek S., Krauss V., Jenuwein T., and Reuter G. 2004. Su(var) genes regulate the balance between euchromatin and heterochromatin in *Drosophila. Genes Dev.* **18:** 2973–2983.

Emre N.C., Ingvarsdottir K., Wyce A., Wood A., Krogan N.J., Henry K.W., Li K., Marmorstein R., Greenblatt J.F., Shilatifard A., and Berger S.L. 2005. Maintenance of low histone ubiquitylation by Ubp10 correlates with telomere-proximal Sir2 association and gene silencing. *Mol. Cell* **17:** 585–594.

Fang J., Feng Q., Ketel C.S., Wang H., Cao R., Xia L., Erdjument-Bromage H., Tempst P., Simon J.A., and Zhang Y. 2002. Purification and functional characterization of SET8, a nucleosomal histone H4-lysine 20-specific methyltransferase. *Curr. Biol.* **12:** 1086–1099.

Feng Q., Wang H., Ng H.H., Erdjument-Bromage H., Tempst P., Struhl K., and Zhang Y. 2002. Methylation of H3-lysine 79 is mediated by a new family of HMTases without a SET domain. *Curr. Biol.* **12:** 1052–1058.

Galasinski S.C., Louie D.F., Gloor K.K., Resing K.A., and Ahn N.G. 2002. Global regulation of post-translational modifications on core histones. *J. Biol. Chem.* **277:** 2579–2588.

Garcia B.A., Busby S.A., Barber C.M., Shabanowitz J., Allis C.D., and Hunt D.F. 2004. Characterization of phosphorylation sites on histone H1 isoforms by tandem mass spectrometry. *J. Proteome Res.* **3:** 1219–1227.

Goto H., Yasui Y., Nigg E.A., and Inagaki M. 2002. Aurora-B phosphorylates Histone H3 at serine28 with regard to the mitotic chromosome condensation. *Genes Cells* **7:** 11–17.

Goto H., Tomono Y., Ajiro K., Kosako H., Fujita M., Sakurai M., Okawa K., Iwamatsu A., Okigaki T., Takahashi T., and Inagaki M. 1999. Identification of a novel phosphorylation site on histone H3 coupled with mitotic chromosome condensation. *J. Biol. Chem.* **274:** 25543–25549.

Grant P.A., Eberharter A., John S., Cook R.G., Turner B.M., and Workman J.L. 1999. Expanded lysine acetylation specificity of Gcn5 in native complexes. *J. Biol. Chem.* **274:** 5895–5900.

Hake S.B. and Allis C.D. 2006. Histone H3 variants and their potential role in indexing mammalian genomes: The "H3 barcode hypothesis". *Proc. Natl. Acad. Sci.* **103:** 6428–6435.

Hake S.B., Garcia B.A., Kauer M., Baker S.P., Shabanowitz J., Hunt D.F., and Allis C.D. 2005. Serine 31 phosphorylation of histone variant H3.3 is specific to regions bordering centromeres in metaphase chromosomes. *Proc. Natl. Acad. Sci.* **102:** 6344–6349.

Hamamoto R., Furukawa Y., Morita M., Iimura Y., Silva F.P., Li M., Yagyu R., and Nakamura Y. 2004. SMYD3 encodes a histone methyltransferase involved in the proliferation of cancer cells. *Nat. Cell Biol.* **6:** 731–740.

Harvey A.C., Jackson S.P., and Downs J.A. 2005. *Saccharomyces cerevisiae* histone H2A Ser122 facilitates DNA repair. *Genetics* **170:** 543–553.

Hayashi K., Yoshida K., and Matsui Y. 2005. A histone H3 methyltransferase controls epigenetic events required for meiotic prophase. *Nature* **438:** 374–378.

He Z., Cho Y.Y., Ma W.Y., Choi H.S., Bode A.M., and Dong Z. 2005. Regulation of ultraviolet B-induced phosphorylation of histone H3 at serine 10 by Fyn kinase. *J. Biol. Chem.* **280:** 2446–2454.

Hendzel M.J., Wei Y., Mancini M.A., Van Hooser A., Ranalli T., Brinkley B.R., Bazett-Jones D.P., and Allis C.D. 1997. Mitosis-specific phosphorylation of histone H3 initiates primarily within pericentromeric heterochromatin during G2 and spreads in an ordered fashion coincident with mitotic chromosome condensation. *Chromosoma* **106:** 348–360.

Hsu J.Y., Sun Z.W., Li X., Reuben M., Tatchell K., Bishop D.K., Grushcow J.M., Brame C.J., Caldwell J.A., Hunt D.F., et al. 2000. Mitotic phosphorylation of histone H3 is governed by Ipl1/aurora kinase and Glc7/PP1 phosphatase in budding yeast and nematodes. *Cell* **102:** 279–291.

Hyland E.M., Cosgrove M.S., Molina H., Wang D., Pandey A., Cottee R.J., and Boeke J.D. 2005. Insights into the role of histone H3 and histone H4 core modifiable residues in *Saccharomyces cerevisiae. Mol. Cell. Biol.* **25:** 10060–10070.

Ichijima Y., Sakasai R., Okita N., Asahina K., Mizutani S., and Teraoka H. 2005. Phosphorylation of histone H2AX at M phase in human cells without DNA damage response. *Biochem. Biophys. Res. Commun.* **336:** 807–812.

Jackson J.P., Lindroth A.M., Cao X., and Jacobsen S.E. 2002. Control of CpNpG DNA methylation by the KRYPTONITE histone H3 methyltransferase. *Nature* **416:** 556–560.

Jackson J.P., Johnson L., Jasencakova Z., Zhang X., PerezBurgos L., Singh P.B., Cheng X., Schubert I., Jenuwein T., and Jacobsen S.E. 2004. Dimethylation of histone H3 lysine 9 is a critical mark for DNA methylation and gene silencing in *Arabidopsis thaliana. Chromosoma* **112:** 308–315.

Jeppesen P. and Turner B.M. 1993. The inactive X chromosome in female mammals is distinguished by a lack of histone H4 acetylation, a cytogenetic marker for gene expression. *Cell* **74:** 281–289.

Jin Y., Wang Y., Walker D.L., Dong H., Conley C., Johansen J., and Johansen K.M. 1999. JIL-1: A novel chromosomal tandem kinase implicated in transcriptional regulation in *Drosophila. Mol. Cell* **4:** 129–135.

Kao C.F., Hillyer C., Tsukuda T., Henry K., Berger S., and Osley M.A. 2004. Rad6 plays a role in transcriptional activation through ubiquitylation of histone H2B. *Genes Dev.* **18:** 184–195.

Kapetanaki M.G., Guerrero-Santoro J., Bisi D.C., Hsieh C.L., Rapic-Otrin V., and Levine A.S. 2006. The DDB1-CUL4ADDB2 ubiquitin ligase is deficient in xeroderma pigmentosum group E and targets histone H2A at UV-damaged DNA sites. *Proc. Natl. Acad. Sci.* **103:** 2588–2593.

Kawasaki H., Taira K., and Yokoyama K. 2000. Histone acetyltransferase (HAT) activity of ATF-2 is necessary for the CRE-dependent transcription. *Nucleic Acids Symp. Ser.* **2000:** 259–260.

Kawasaki H., Schiltz L., Chiu R., Itakura K., Taira K., Nakatani Y., and Yokoyama K.K. 2000. ATF-2 has intrinsic histone acetyltransferase activity which is modulated by phosphorylation. *Nature* **405:** 195–200.

Kim K.C., Geng L., and Huang S. 2003. Inactivation of a histone methyltransferase by mutations in human cancers. *Cancer Res.* **63:** 7619–7623.

Kimura A. and Horikoshi M. 1998. Tip60 acetylates six lysines of a specific class in core histones in vitro. *Genes Cells* **3:** 789–800.

Kizer K.O., Phatnani H.P., Shibata Y., Hall H., Greenleaf A.L., and Strahl B.D. 2005. A novel domain in Set2 mediates RNA polymerase II interaction and couples histone H3 K36 methylation with transcript elongation. *Mol. Cell. Biol.* **25:** 3305–3316.

Kleff S., Andrulis E.D., Anderson C.W., and Sternglanz R. 1995. Identification of a gene encoding a yeast histone H4 acetyltransferase. *J. Biol. Chem.* **270:** 24674–24677.

Kothapalli N., Sarath G., and Zempleni J. 2005a. Biotinylation of K12 in histone H4 decreases in response to DNA double-strand breaks in human JAr choriocarcinoma cells. *J. Nutr.* **135:** 2337–2342.

Kothapalli N., Camporeale G., Kueh A., Chew Y.C., Oommen A.M., Griffin J.B., and Zempleni J. 2005b. Biological functions of biotinylated histones. *J. Nutr. Biochem.* **16:** 446–448.

Kruhlak M.J., Hendzel M.J., Fischle W., Bertos N.R., Hameed S., Yang X.J., Verdin E., and Bazett-Jones D.P. 2001. Regulation of global acetylation in mitosis through loss of histone acetyltransferases and deacetylases from chromatin. *J. Biol. Chem.* **276:** 38307–38319.

Kuo M.H., Brownell J.E., Sobel R.E., Ranalli T.A., Cook R.G., Edmondson D.G., Roth S.Y., and Allis C.D. 1996. Transcription-linked acetylation by Gcn5p of histones H3 and H4 at specific lysines. *Nature* **383:** 269–272.

Kuzmichev A., Jenuwein T., Tempst P., and Reinberg D. 2004. Different EZH2-containing complexes target methylation of histone H1 or nucleosomal histone H3. *Mol. Cell* **14:** 183–193.

Kuzmichev A., Nishioka K., Erdjument-Bromage H., Tempst P., and Reinberg D. 2002. Histone methyltransferase activity associated with a human multiprotein complex containing the Enhancer of Zeste protein. *Genes Dev.* **16:** 2893–2905.

Lachner M., O'Carroll D., Rea S., Mechtler K., and Jenuwein T. 2001. Methylation of histone H3 lysine 9 creates a binding site for HP1 proteins. *Nature* **410:** 116–120.

Lacoste N., Utley R.T., Hunter J.M., Poirier G.G., and Cote J. 2002. Disruptor of telomeric silencing-1 is a chromatin-specific histone H3 methyltransferase. *J. Biol. Chem.* **277:** 30421–30424.

Lo W.S., Duggan L., Emre N.C., Belotserkovskya R., Lane W.S., Shiekhattar R., and Berger S.L. 2001. Snf1—A histone kinase that works in concert with the histone acetyltransferase Gcn5 to regulate transcription. *Science* **293:** 1142–1146.

Maile T., Kwoczynski S., Katzenberger R.J., Wassarman D.A., and Sauer F. 2004. TAF1 activates transcription by phosphorylation of serine 33 in histone H2B. *Science* **304:** 1010–1014.

McKittrick E., Gafken P.R., Ahmad K., and Henikoff S. 2004. Histone H3.3 is enriched in covalent modifications associated with active chromatin. *Proc. Natl. Acad. Sci.* **101:** 1525–1530.

Milne T.A., Briggs S.D., Brock H.W., Martin M.E., Gibbs D., Allis C.D., and Hess J.L. 2002. MLL targets SET domain methyltransferase activity to *Hox* gene promoters. *Mol. Cell* **10:** 1107–1117.

Miranda T.B., Sayegh J., Frankel A., Katz J.E., Miranda M., and Clarke S. 2006. Yeast Hsl7 (histone synthetic lethal 7) catalyses the in vitro formation of ω-N^G-monomethylarginine in calf thymus histone H2A. *Biochem. J.* **395:** 563–570.

Mizzen C.A., Yang X.J., Kokubo T., Brownell J.E., Bannister A.J., Owen-Hughes T., Workman J., Wang L., Berger S.L., Kouzarides T., et al. 1996. The TAF$_{II}$250 subunit of TFIID has histone acetyltransferase activity. *Cell* **87:** 1261–1270.

Morris S.A., Shibata Y., Noma K., Tsukamoto Y., Warren E., Temple B., Grewal S.I., and Strahl B.D. 2005. Histone H3 K36 methylation is associated with transcription elongation in *Schizosaccharomyces pombe*. *Eukaryot. Cell* **4:** 1446–1454.

Mukherjee B., Kessinger C., Kobayashi J., Chen B.P., Chen D.J., Chatterjee A., and Burma S. 2006. DNA-PK phosphorylates histone H2AX during apoptotic DNA fragmentation in mammalian cells. *DNA Repair* **5:** 575–590.

Muller J., Hart C.M., Francis N.J., Vargas M.L., Sengupta A., Wild B., Miller E.L., O'Connor M.B., Kingston R.E., and Simon J.A. 2002. Histone methyltransferase activity of a *Drosophila* Polycomb group repressor complex. *Cell* **111:** 197–208.

Nagy P.L., Griesenbeck J., Kornberg R.D., and Cleary M.L. 2002. A trithorax-group complex purified from *Saccharomyces cerevisiae* is required for methylation of histone H3. *Proc. Natl. Acad. Sci.* **99:** 90–94.

Nakamura T., Mori T., Tada S., Krajewski W., Rozovskaia T., Wassell R., Dubois G., Mazo A., Croce C.M., and Canaani E. 2002. ALL-1 is a histone methyltransferase that assembles a supercomplex of proteins involved in transcriptional regulation. *Mol. Cell* **10:** 1119–1128.

Nakayama J., Rice J.C., Strahl B.D., Allis C.D., and Grewal S.I. 2001. Role of histone H3 lysine 9 methylation in epigenetic control of heterochromatin assembly. *Science* **292:** 110–113.

Nathan D., Ingvarsdottir K., Sterner D.E., Bylebyl G.R., Dokmanovic M., Dorsey J.A., Whelan K.A., Krsmanovic M., Lane W.S., Meluh P.B., et al. 2006. Histone sumoylation is a negative regulator in *Saccharomyces cerevisiae* and shows dynamic interplay with positive-acting histone modifications. *Genes Dev.* **20:** 966–976.

Ng H.H., Xu R.M., Zhang Y., and Struhl K. 2002. Ubiquitination of histone H2B by Rad6 is required for efficient Dot1-mediated methylation of histone H3 lysine 79. *J. Biol. Chem.* **277:** 34655–34657.

Nielsen S.J., Schneider R., Bauer U.M., Bannister A.J., Morrison A., O'Carroll D., Firestein R., Cleary M., Jenuwein T., Herrera R.E., and Kouzarides T. 2001. Rb targets histone H3 methylation and HP1 to promoters. *Nature* **412:** 561–565.

Nishioka K., Chuikov S., Sarma K., Erdjument-Bromage H., Allis C.D., Tempst P., and Reinberg D. 2002a. Set9, a novel histone H3 methyltransferase that facilitates transcription by precluding histone tail modifications required for heterochromatin formation. *Genes Dev.* **16:** 479–489.

Nishioka K., Rice J.C., Sarma K., Erdjument-Bromage H., Werner J., Wang Y., Chuikov S., Valenzuela P., Tempst P., Steward R., et al. 2002. PR-Set7 is a nucleosome-specific methyltransferase that modifies lysine 20 of histone H4 and is associated with silent chromatin. *Mol. Cell* **9:** 1201–1213.

Nowak S.J. and Corces V.G. 2000. Phosphorylation of histone H3 correlates with transcriptionally active loci. *Genes Dev.* **14:** 3003–3013.

O'Carroll D., Scherthan H., Peters A.H., Opravil S., Haynes A.R., Laible G., Rea S., Schmid M., Lebersorger A., Jerratsch M., et al. 2000. Isolation and characterization of *Suv39h2*, a second histone H3 methyltransferase gene that displays testis-specific expression. *Mol. Cell. Biol.* **20:** 9423–9433.

Odegard V.H., Kim S.T., Anderson S.M., Shlomchik M.J., and Schatz D.G. 2005. Histone modifications associated with somatic hypermutation. *Immunity* **23:** 101–110.

Ogata N., Ueda K., Kagamiyama H., and Hayaishi O. 1980. ADP-ribosylation of histone H1. Identification of glutamic acid residues 2, 14, and the COOH-terminal lysine residue as modification sites. *J. Biol. Chem.* **255:** 7616–7620.

Ogawa H., Ishiguro K., Gaubatz S., Livingston D.M., and Nakatani Y. 2002. A complex with chromatin modifiers that occupies E2F- and Myc-responsive genes in G_0 cells. *Science* **296:** 1132–1136.

Ogawa Y., Ono T., Wakata Y., Okawa K., Tagami H., and Shibahara K.I. 2005. Histone variant macroH2A1.2 is mono-ubiquitinated at its histone domain. *Biochem. Biophys. Res. Commun.* **336:** 204–209.

Pantazis P. and Bonner W.M. 1981. Quantitative determination of histone modification. H2A acetylation and phosphorylation. *J. Biol. Chem.* **256:** 4669–4675.

Parthun M.R., Widom J., and Gottschling D.E. 1996. The major cytoplasmic histone acetyltransferase in yeast: Links to chromatin replication and histone metabolism. *Cell* 87: 85–94.

Peters A.H., O'Carroll D., Scherthan H., Mechtler K., Sauer S., Schofer C., Weipoltshammer K., Pagani M., Lachner M., Kohlmaier A., et al. 2001. Loss of the *Suv39h* histone methyltransferases impairs mammalian heterochromatin and genome stability. *Cell* 107: 323–337.

Pham A.D. and Sauer F. 2000. Ubiquitin-activating/conjugating activity of TAF$_{II}$250, a mediator of activation of gene expression in *Drosophila. Science* 289: 2357–2360.

Polioudaki H., Markaki Y., Kourmouli N., Dialynas G., Theodoropoulos P.A., Singh P.B., and Georgatos S.D. 2004. Mitotic phosphorylation of histone H3 at threonine 3. *FEBS Lett.* 560: 39–44.

Preuss U., Landsberg G., and Scheidtmann K.H. 2003. Novel mitosis-specific phosphorylation of histone H3 at Thr11 mediated by Dlk/ZIP kinase. *Nucleic Acids Res.* 31: 878–885.

Puerta C., Hernandez F., Lopez-Alarcon L., and Palacian E. 1995. Acetylation of histone H2A.H2B dimers facilitates transcription. *Biochem. Biophys. Res. Commun.* 210: 409–416.

Rea S., Eisenhaber F., O'Carroll D., Strahl B.D., Sun Z.W., Schmid M., Opravil S., Mechtler K., Ponting C.P., Allis C.D., and Jenuwein T. 2000. Regulation of chromatin structure by site-specific histone H3 methyltransferases. *Nature* 406: 593–599.

Redon C., Pilch D.R., Rogakou E.P., Orr A.H., Lowndes N.F., and Bonner W.M. 2003. Yeast histone 2A serine 129 is essential for the efficient repair of checkpoint-blind DNA damage. *EMBO Rep.* 4: 678–684.

Rice J.C., Nishioka K., Sarma K., Steward R., Reinberg D., and Allis C.D. 2002. Mitotic-specific methylation of histone H4 Lys 20 follows increased PR-Set7 expression and its localization to mitotic chromosomes. *Genes Dev.* 16: 2225–2230.

Robzyk K., Recht J., and Osley M.A. 2000. Rad6-dependent ubiquitination of histone H2B in yeast. *Science* 287: 501–504.

Rogakou E.P., Boon C., Redon C., and Bonner W.M. 1999. Megabase chromatin domains involved in DNA double-strand breaks in vivo. *J. Cell Biol.* 146: 905–916.

Rogakou E.P., Pilch D.R., Orr A.H., Ivanova V.S., and Bonner W.M. 1998. DNA double-stranded breaks induce histone H2AX phosphorylation on serine 139. *J. Biol. Chem.* 273: 5858–5868.

Roguev A., Schaft D., Shevchenko A., Pijnappel W.W., Wilm M., Aasland R., and Stewart A.F. 2001. The *Saccharomyces cerevisiae* Set1 complex includes an Ash2 homologue and methylates histone 3 lysine 4. *EMBO J.* 20: 7137–7148.

Rouleau M., Aubin R.A., and Poirier G.G. 2004. Poly(ADP-ribosyl)ated chromatin domains: Access granted. *J. Cell Sci.* 117: 815–825.

Ruiz-Carrillo A., Wangh L.J., and Allfrey V.G. 1975. Processing of newly synthesized histone molecules. *Science* 190: 117–128.

Sanchez-Elsner T., Gou D., Kremmer E., and Sauer F. 2006. Noncoding RNAs of trithorax response elements recruit *Drosophila* Ash1 to Ultrabithorax. *Science* 311: 1118–1123.

Santos-Rosa H., Schneider R., Bannister A.J., Sherriff J., Bernstein B.E., Emre N.C., Schreiber S.L., Mellor J., and Kouzarides T. 2002. Active genes are tri-methylated at K4 of histone H3. *Nature* 419: 407–411.

Sarg B., Helliger W., Talasz H., Forg B., and Lindner H.H. 2006. Histone H1 phosphorylation occurs site-specifically during interphase and mitosis: Identification of a novel phosphorylation site on histone H1. *J. Biol. Chem.* 281: 6573–6580.

Sassone-Corsi P., Mizzen C.A., Cheung P., Crosio C., Monaco L., Jacquot S., Hanauer A., and Allis C.D. 1999. Requirement of Rsk-2 for epidermal growth factor-activated phosphorylation of histone H3. *Science* 285: 886–891.

Schiltz R.L., Mizzen C.A., Vassilev A., Cook R.G., Allis C.D., and Nakatani Y. 1999. Overlapping but distinct patterns of histone acetylation by the human coactivators p300 and PCAF within nucleosomal substrates. *J. Biol. Chem.* 274: 1189–1192.

Schotta G., Ebert A., Krauss V., Fischer A., Hoffmann J., Rea S., Jenuwein T., Dorn R., and Reuter G. 2002. Central role of *Drosophila* SU(VAR)3-9 in histone H3-K9 methylation and heterochromatic gene silencing. *EMBO J.* 21: 1121–1131.

Schotta G., Lachner M., Sarma K., Ebert A., Sengupta R., Reuter G., Reinberg D., and Jenuwein T. 2004. A silencing pathway to induce H3-K9 and H4-K20 trimethylation at constitutive heterochromatin. *Genes Dev.* 18: 1251–1262.

Schubeler D., Francastel C., Cimbora D.M., Reik A., Martin D.I., and Groudine M. 2000. Nuclear localization and histone acetylation: A pathway for chromatin opening and transcriptional activation of the human β-globin locus. *Genes Dev.* 14: 940–950.

Schultz D.C., Ayyanathan K., Negorev D., Maul G.G., and Rauscher F.J., III. 2002. SETDB1: A novel KAP-1-associated histone H3, lysine 9-specific methyltransferase that contributes to HP1-mediated silencing of euchromatic genes by KRAB zinc-finger proteins. *Genes Dev.* 16: 919–932.

Schurter B.T., Koh S.S., Chen D., Bunick G.J., Harp J.M., Hanson B.L., Henschen-Edman A., Mackay D.R., Stallcup M.R., and Aswad D.W. 2001. Methylation of histone H3 by coactivator-associated arginine methyltransferase 1. *Biochemistry* 40: 5747–5756.

Shiio Y. and Eisenman R.N. 2003. Histone sumoylation is associated with transcriptional repression. *Proc. Natl. Acad. Sci.* 100: 13225–13230.

Smith E.R., Eisen A., Gu W., Sattah M., Pannuti A., Zhou J., Cook R.G., Lucchesi J.C., and Allis C.D. 1998. ESA1 is a histone acetyltransferase that is essential for growth in yeast. *Proc. Natl. Acad. Sci.* 95: 3561–3565.

Sobel R.E., Cook R.G., Perry C.A., Annunziato A.T., and Allis C.D. 1995. Conservation of deposition-related acetylation sites in newly synthesized histones H3 and H4. *Proc. Natl. Acad. Sci.* 92: 1237–1241.

Song O.K., Wang X., Waterborg J.H., and Sternglanz R. 2003. An N^{α}-acetyltransferase responsible for acetylation of the N-terminal residues of histones H4 and H2A. *J. Biol. Chem.* 278: 38109–38112.

Spencer T.E., Jenster G., Burcin M.M., Allis C.D., Zhou J., Mizzen C.A., McKenna N.J., Onate S.A., Tsai S.Y., Tsai M.J., and O'Malley B.W. 1997. Steroid receptor coactivator-1 is a histone acetyltransferase. *Nature* 389: 194–198.

Stanley J.S., Griffin J.B., and Zempleni J. 2001. Biotinylation of histones in human cells. Effects of cell proliferation. *Eur. J. Biochem.* 268: 5424–5429.

Stiff T., O'Driscoll M., Rief N., Iwabuchi K., Lobrich M., and Jeggo P.A. 2004. ATM and DNA-PK function redundantly to phosphorylate H2AX after exposure to ionizing radiation. *Cancer Res.* 64: 2390–2396.

Strahl B.D., Ohba R., Cook R.G., and Allis C.D. 1999. Methylation of histone H3 at lysine 4 is highly conserved and correlates with transcriptionally active nuclei in *Tetrahymena. Proc. Natl. Acad. Sci.* 96: 14967–14972.

Strahl B.D., Briggs S.D., Brame C.J., Caldwell J.A., Koh S.S., Ma H., Cook R.G., Shabanowitz J., Hunt D.F., Stallcup M.R., and Allis C.D. 2001. Methylation of histone H4 at arginine 3 occurs in vivo and is mediated by the nuclear receptor coactivator PRMT1. *Curr. Biol.* 11: 996–1000.

Strahl B.D., Grant P.A., Briggs S.D., Sun Z.W., Bone J.R., Caldwell J.A., Mollah S., Cook R.G., Shabanowitz J., Hunt D.F., and Allis C.D. 2002. Set2 is a nucleosomal histone H3-selective methyltransferase that mediates transcriptional repression. *Mol. Cell. Biol.* **22:** 1298–1306.

Su I.H., Basavaraj A., Krutchinsky A.N., Hobert O., Ullrich A., Chait B.T., and Tarakhovsky A. 2003. Ezh2 controls B cell development through histone H3 methylation and *Igh* rearrangement. *Nat. Immunol.* **4:** 124–131.

Suka N., Suka Y., Carmen A.A., Wu J., and Grunstein M. 2001. Highly specific antibodies determine histone acetylation site usage in yeast heterochromatin and euchromatin. *Mol. Cell* **8:** 473–479.

Sun X.J., Wei J., Wu X.Y., Hu M., Wang L., Wang H.H., Zhang Q.H., Chen S.J., Huang Q.H., and Chen Z. 2005. Identification and characterization of a novel human histone H3 lysine 36-specific methyltransferase. *J. Biol. Chem.* **280:** 35261–35271.

Sun Z.W. and Allis C.D. 2002. Ubiquitination of histone H2B regulates H3 methylation and gene silencing in yeast. *Nature* **418:** 104–108.

Sung Y.J. and Ambron R.T. 2004. PolyADP-ribose polymerase-1 (PARP-1) and the evolution of learning and memory. *Bioessays* **26:** 1268–1271.

Swerdlow P.S., Schuster T., and Finley D. 1990. A conserved sequence in histone H2A which is a ubiquitination site in higher eucaryotes is not required for growth in *Saccharomyces cerevisiae*. *Mol. Cell. Biol.* **10:** 4905–4911.

Tachibana M., Sugimoto K., Fukushima T., and Shinkai Y. 2001. Set domain-containing protein, G9a, is a novel lysine-preferring mammalian histone methyltransferase with hyperactivity and specific selectivity to lysines 9 and 27 of histone H3. *J. Biol. Chem.* **276:** 25309–25317.

Tachibana M., Ueda J., Fukuda M., Takeda N., Ohta T., Iwanari H., Sakihama T., Kodama T., Hamakubo T., and Shinkai Y. 2005. Histone methyltransferases G9a and GLP form heteromeric complexes and are both crucial for methylation of euchromatin at H3-K9. *Genes Dev.* **19:** 815–826.

Tachibana M., Sugimoto K., Nozaki M., Ueda J., Ohta T., Ohki M., Fukuda M., Takeda N., Niida H., Kato H., and Shinkai Y. 2002. G9a histone methyltransferase plays a dominant role in euchromatic histone H3 lysine 9 methylation and is essential for early embryogenesis. *Genes Dev.* **16:** 1779–1791.

Takechi S. and Nakayama T. 1999. Sas3 is a histone acetyltransferase and requires a zinc finger motif. *Biochem. Biophys. Res. Commun.* **266:** 405–410.

Tamaru H. and Selker E.U. 2001. A histone H3 methyltransferase controls DNA methylation in *Neurospora crassa*. *Nature* **414:** 277–283.

Taplick J., Kurtev V., Lagger G., and Seiser C. 1998. Histone H4 acetylation during interleukin-2 stimulation of mouse T cells. *FEBS Lett.* **436:** 349–352.

Thomson S., Clayton A.L., Hazzalin C.A., Rose S., Barratt M.J., and Mahadevan L.C. 1999. The nucleosomal response associated with immediate-early gene induction is mediated via alternative MAP kinase cascades: MSK1 as a potential histone H3/HMG-14 kinase. *EMBO J.* **18:** 4779–4793.

Turner B.M. 2000. Histone acetylation and an epigenetic code. *Bioessays* **22:** 836–845.

———. 2005. Reading signals on the nucleosome with a new nomenclature for modified histones. *Nat. Struct. Mol. Biol.* **12:** 110–112.

Turner B.M. and Fellows G. 1989. Specific antibodies reveal ordered and cell-cycle-related use of histone-H4 acetylation sites in mammalian cells. *Eur. J. Biochem.* **179:** 131–139.

van Attikum H. and Gasser S.M. 2005. The histone code at DNA breaks: A guide to repair? *Nat. Rev. Mol. Cell Biol.* **6:** 757–765.

van Leeuwen F., Gafken P.R., and Gottschling D.E. 2002. Dot1p modulates silencing in yeast by methylation of the nucleosome core. *Cell* **109:** 745–756.

Vandel L., Nicolas E., Vaute O., Ferreira R., Ait-Si-Ali S., and Trouche D. 2001. Transcriptional repression by the retinoblastoma protein through the recruitment of a histone methyltransferase. *Mol. Cell. Biol.* **21:** 6484–6494.

Vaquero A., Scher M., Lee D., Erdjument-Bromage H., Tempst P., and Reinberg D. 2004. Human SirT1 interacts with histone H1 and promotes formation of facultative heterochromatin. *Mol. Cell* **16:** 93–105.

Verreault A., Kaufman P.D., Kobayashi R., and Stillman B. 1998. Nucleosomal DNA regulates the core-histone-binding subunit of the human Hat1 acetyltransferase. *Curr. Biol.* **8:** 96–108.

Visochek L., Steingart R.A., Vulih-Shultzman I., Klein R., Priel E., Gozes I., and Cohen-Armon M. 2005. PolyADP-ribosylation is involved in neurotrophic activity. *J. Neurosci.* **25:** 7420–7428.

Wang H., Cao R., Xia L., Erdjument-Bromage H., Borchers C., Tempst P., and Zhang Y. 2001a. Purification and functional characterization of a histone H3-lysine 4-specific methyltransferase. *Mol. Cell* **8:** 1207–1217.

Wang H., Wang L., Erdjument-Bromage H., Vidal M., Tempst P., Jones R.S., and Zhang Y. 2004. Role of histone H2A ubiquitination in Polycomb silencing. *Nature* **431:** 873–878.

Wang H., An W., Cao R., Xia L., Erdjument-Bromage H., Chatton B., Tempst P., Roeder R.G., and Zhang Y. 2003. mAM facilitates conversion by ESET of dimethyl to trimethyl lysine 9 of histone H3 to cause transcriptional repression. *Mol. Cell* **12:** 475–487.

Wang H., Huang Z.Q., Xia L., Feng Q., Erdjument-Bromage H., Strahl B.D., Briggs S.D., Allis C.D., Wong J., Tempst P., and Zhang Y. 2001b. Methylation of histone H4 at arginine 3 facilitating transcriptional activation by nuclear hormone receptor. *Science* **293:** 853–857.

Wang Y., Zhang W., Jin Y., Johansen J., and Johansen K.M. 2001. The JIL-1 tandem kinase mediates histone H3 phosphorylation and is required for maintenance of chromatin structure in *Drosophila*. *Cell* **105:** 433–443.

Wei Y., Yu L., Bowen J., Gorovsky M.A., and Allis C.D. 1999. Phosphorylation of histone H3 is required for proper chromosome condensation and segregation. *Cell* **97:** 99–109.

Wilson J.R., Jing C., Walker P.A., Martin S.R., Howell S.A., Blackburn G.M., Gamblin S.J., and Xiao B. 2002. Crystal structure and functional analysis of the histone methyltransferase SET7/9. *Cell* **111:** 105–115.

Wyatt H.R., Liaw H., Green G.R., and Lustig A.J. 2003. Multiple roles for *Saccharomyces cerevisiae* histone H2A in telomere position effect, Spt phenotypes and double-strand-break repair. *Genetics* **164:** 47–64.

Xin Z., Tachibana M., Guggiari M., Heard E., Shinkai Y., and Wagstaff J. 2003. Role of histone methyltransferase G9a in CpG methylation of the Prader-Willi syndrome imprinting center. *J. Biol. Chem.* **278:** 14996–15000.

Yamamoto T. and Horikoshi M. 1997. Novel substrate specificity of the histone acetyltransferase activity of HIV-1-Tat interactive protein Tip60. *J. Biol. Chem.* **272:** 30595–30598.

Yamamoto Y., Verma U.N., Prajapati S., Kwak Y.T., and Gaynor R.B. 2003. Histone H3 phosphorylation by IKK-alpha is critical for cytokine-induced gene expression. *Nature* **423:** 655–659.

Yang L., Xia L., Wu D.Y., Wang H., Chansky H.A., Schubach W.H., Hickstein D.D., and Zhang Y. 2002. Molecular cloning of ESET, a novel histone H3-specific methyltransferase that interacts with ERG transcription factor. *Oncogene* **21:** 148–152.

Zegerman P., Canas B., Pappin D., and Kouzarides T. 2002. Histone H3 lysine 4 methylation disrupts binding of nucleosome remodeling and deacetylase (NuRD) repressor complex. *J. Biol. Chem.* **277:** 11621–11624.

Zeitlin S.G., Barber C.M., Allis C.D., and Sullivan K.F. 2001. Differential regulation of CENP-A and histone H3 phosphorylation in G$_2$/M. *J. Cell Sci.* **114:** 653–661.

Zhang L., Eugeni E.E., Parthun M.R., and Freitas M.A. 2003. Identification of novel histone post-translational modifications by peptide mass fingerprinting. *Chromosoma* **112:** 77–86.

Zhong S., Jansen C., She Q.B., Goto H., Inagaki M., Bode A.M., Ma W.Y., and Dong Z. 2001. Ultraviolet B-induced phosphorylation of histone H3 at serine 28 is mediated by MSK1. *J. Biol. Chem.* **276:** 33213–33219.

Reviews

Bannister A.J. and Kouzarides T. 2005. Reversing histone methylation. *Nature* **436:** 1103–1106.

Berger S.L. 2002. Histone modifications in transcriptional regulation. *Curr. Opin. Genet. Dev.* **12:** 42–148.

Davie J.R. and Spencer V.A. 2001. Signal transduction pathways and the modification of chromatin structure. *Prog. Nucleic Acid Res. Mol. Biol.* **65:** 299–340.

Cosgrove M.S., Boeke J.D., and Wolberger C. 2004. Regulated nucleosome mobility and the histone code. *Nat. Struct. Mol. Biol.* **11:** 1037–1043

Dobosy J.R. and Selker E.U. 2001. Emerging connections between DNA methylation and histone acetylation. *Cell. Mol. Life Sci.* **58:** 721–727.

Dunleavy E., Pidoux A., and Allshire R. 2005. Centromeric chromatin makes its mark. *Trends Biochem. Sci.* **30:** 172–175.

Elgin S.C. and Grewal S.I. 2003. Heterochromatin: Silence is golden. *Curr. Biol.* **13:** R895–R898.

Emre N.C. and Berger S.L. 2004. Histone H2B ubiquitylation and deubiquitylation in genomic regulation. *Cold Spring Harbor Symp. Quant. Biol.* **69:** 289–299.

Esteller M. 2006. Epigenetics provides a new generation of oncogenes and tumour-suppressor genes. *Br. J. Cancer* **94:** 179–183.

Feinberg A.P. and Tycko B. 2004. The history of cancer epigenetics. *Nat. Rev. Cancer* **4:** 143–153.

Fischer A., Hofmann I., Naumann K., and Reuter G. 2006. Heterochromatin proteins and the control of heterochromatic gene silencing in *Arabidopsis*. *J. Plant Physiol.* **163:** 358–368.

Fischle W., Wang Y., and Allis C.D. 2003. Binary switches and modification cassettes in histone biology and beyond. *Nature* **425:** 475–479.

———. 2003. Histone and chromatin cross-talk. *Curr. Opin. Cell. Biol.* **15:** 172–183.

Grewal S.I. and Elgin S.C. 2002. Heterochromatin: New possibilities for the inheritance of structure. *Curr. Opin. Genet. Dev.* **12:** 178–187.

Grunstein M. 1997. Histone acetylation in chromatin structure and transcription. *Nature* **389:** 349–352.

———. 1997. Molecular model for telomeric heterochromatin in yeast. *Curr. Opin. Cell Biol.* **9:** 383–387.

———. 1998. Yeast heterochromatin: Regulation of its assembly and inheritance by histones. *Cell* **93:** 325–328.

Henikoff S. and Ahmad K. 2005. Assembly of variant histones into chromatin. *Annu. Rev. Cell Dev. Biol.* **21:** 133–153.

Hild M. and Paro R. 2003. Anti-silencing from the core: A histone H2A variant protects euchromatin. *Nat. Cell Biol.* **5:** 278–280.

Jenuwein T. and Allis C.D. 2001. Translating the histone code. *Science* **293:** 1074–1080.

Kimmins S. and Sassone-Corsi P. 2005. Chromatin remodelling and epigenetic features of germ cells. *Nature* **434:** 583–589.

Kurdistani S.K. and Grunstein M. 2003. Histone acetylation and deacetylation in yeast. *Nat. Rev. Mol. Cell Biol.* **4:** 276–284.

Lachner M., O'Sullivan R.J., and Jenuwein T. 2003. An epigenetic road map for histone lysine methylation. *J. Cell Sci.* **116:** 2117–2124.

Luger K. and Richmond T.J. 1998. The histone tails of the nucleosome. *Curr. Opin. Genet. Dev.* **8:** 140–146.

Mellone B.G. and Allshire R.C. 2003. Stretching it: Putting the CEN (P-A) in centromere. *Curr. Opin. Genet. Dev.* **13:** 191–198.

Millar C.B., Kurdistani, S.K. and Grunstein M. 2004. Acetylation of yeast histone H4 lysine 16: A switch for protein interactions in heterochromatin and euchromatin. *Cold Spring Harbor Symp. Quant. Biol.* **69:** 193–200.

Nightingale K.P., O'Neill L.P., and Turner B.M. 2006. Histone modifications: Signalling receptors and potential elements of a heritable epigenetic code. *Curr. Opin. Genet. Dev.* **16:** 125–136.

Peterson C.L. and Laniel M.A. 2004. Histones and histone modifications. *Curr. Biol.* **14:** R546–551.

Reinberg D., Chuikov S., Farnham P., Karachentsev D., Kirmizis A., Kuzmichev A., Margueron R. Nishioka K., Preissner T.S., Sarma K., et al. 2004. Steps toward understanding the inheritance of repressive methyl-lysine marks in histones. *Cold Spring Harbor Symp. Quant. Biol.* **69:** 171–182.

Sarma K. and Reinberg D. 2005. Histone variants meet their match. *Nat. Rev. Mol. Cell Biol.* **6:** 139–149.

Spencer V.A. and Davie J.R. 2000. Signal transduction pathways and chromatin structure in cancer cells. *J. Cell. Biochem. Suppl.* **35:** 27–35.

Sternglanz R. 1996. Histone acetylation: A gateway to transcriptional activation. *Trends Biochem. Sci.* **21:** 357–358.

Turner B.M. 2000. Histone acetylation and an epigenetic code. *Bioessays* **22:** 836–845.

van Attikum H. and Gasser S.M. 2005. The histone code at DNA breaks: A guide to repair? *Nat. Rev. Mol. Cell Biol.* **6:** 757–765.

Vaughn M.W., Tanurdzic M., and Martienssen R. 2005. Replication, repair, and reactivation. *Dev. Cell* **9:** 724–725.

Wade P.A. and Wolffe A.P. 1997. Histone acetyltransferases in control. *Curr. Biol.* **7:** R82–R84.

Wade P.A., Pruss D., and Wolffe A.P. 1997. Histone acetylation: Chromatin in action. *Trends Biochem. Sci.* **22:** 128–132.

Zilberman D. and Henikoff S. 2005. Epigenetic inheritance in *Arabidopsis*: Selective silence. *Curr. Opin. Genet. Dev.* **15:** 557–562.

Index

A

A gene, 138–140
Abf1, 70, 74
Abnormal chromosome 10 (Ab10), 285
Abnormalities, cloning of animals and, 420–421
Absent Small or Homeotic (ASH) family, 177–178, 224, 233–234, 238, 242
Acetylation
    active chromatin and, 37
    DNA repair and, 270
    fragile X syndrome and, 448
    histone variants and, 133–134
    histones and, 17–18, 30–31
    PRC2 and, 219
    regulation of transcription by, 193–194
    SIR complex spread and, 73
    sperm reprogramming and, 387
    transcription and, 194–196
    trxG proteins and, 242–243
    V(D)J recombination and, 404
ACF, dosage compensation and, 316
Achiasmate segregation, 281–283
Actin-related protein 4 (Arp4), 261
Activation-induced deaminase (AID), 42, 398, 408
Active chromatin, 37, 132–134, 256–258
Acute myeloid leukemia, 460
Acute promyelocytic leukemia, 46, 460
Adaptive evolution, 256
Ade2, as yeast reporter gene, 65–66
Ade6, 106
Adrenoleukodystrophy, X-linked, 437
Adult stem cells, pluripotent, 378
Aging, 47, 76–77, 370, 449–451
Air gene, 118, 368–369, 372
Alcohol dehydrogenase (Adh) transgenes, 94
Allelic exclusion, 406–408
Altered Nuclear Transfer (ANT), 432
Alternative rearrangements, 138–139
am genes, RIP and, 116
Amphibians, 417, 419–420, 425–427
Amyloid fibers, prions and, 9–10
Aneuploidy, defined, 266
Angelman syndrome, 439–441
Angiogenesis, 459
ANT. See Altered Nuclear Transfer (ANT); Antennapedia complex (ANT-C)
Antennapedia complex (ANT-C), 233
Antibody diversity
    antigen receptor genes and, 399–400
    epigenetics and, 20
    lymphocyte development and, 403–408
    methylation and, 42
    recombination and, 399–400, 403–405
    somatic cells and, 17

Antigen receptor genes, 399–400, 403–405
Antigen specificity, 398
Antigen-induced receptor signaling, 408
Anti-repressors, 244–245
Antisense transcription, 348, 405
Anti-silencing, 224
Apoptosis, 269, 459
Arabidopsis
    cell proliferation and, 226–227
    components of epigenetic regulation in, 172–173
    genome organization of, 45, 155
    as model organism, 168
    PcG genes in, 216–217
    RNAi in, 160–162
    Web sites for information on, 478
Archaeal DNA, packaging of, 251–253
Arginines, methylation of, 37–38, 52, 193, 201–202
Argonaute proteins, 142–143, 146, 153–154, 180–185, 386
Argonaute-like proteins, 118–120, 160–163, 182
Arp. See Actin-related protein 4
Arthropods, genomic imprinting and, 358
Ascospores, 112
ASH family. See Absent Small or Homeotic
Asm-1, 119–120, 285–286
ATP-dependent remodeling complexes, 32, 39, 236, 238–242, 315–316
ATP-hook protein D1, 96
ATRX gene, 446
Aubergine, 95
Autism, 444
Autogamy, 129, 131, 138
Autonomous pathway, 217
Autonomously Replicating Sequences (ARS), 268
Autosomal genes, 292–293, 295–296, 312–313, 335
Autosomal signal elements (ASEs), X:A ratio assessment and, 295–296
5-Aza-2′-deoxycytidine, 471–472
5-Azacytidine, 471–472

B

B cells, 398–403, 408–411
Barr bodies
    as example of epigenetic phenotype, 26
    H2A and, 250, 262
    X inactivation and, 48, 322, 324
BCR, 403, 407
Beadle and Tatum, 112
Beckwith-Wiedemann syndrome, 440–442
Belts and braces mechanism, 337
Bias, 328–329

Binding factors, protection of CpG islands and, 347
Bisulfite modification, 346
Bithorax complex (BX-C), 233
Blastocysts, ES and TC cells from, 389
Blimp1, 382–383, 385, 408–411
Blocking factor model, X inactivation and, 329
Blood cells, PcG repression and, 224
Bmi1, 224, 226–227
Boundary elements, 5, 7, 72, 74, 468
Brahma, 39, 234, 238, 240–241
Brain-derived neurotrophic factor (Bdnf), 351
Breakage-fusion-bridge cycle, 271
BRG1, 237–238, 427
BRM. See Brahma
Bromodomains, 32, 195, 240
Brown dominant (bwD), 93
Bubbles, 268
Budding yeasts, 66. See also Saccharomyces cerevisiae

C

C. elegans
    anatomies of, 293
    centromere structure and function in, 273–274
    dosage compensation complex in, 293–297
    germ-cell specification in, 297–298, 383–384
    heterochromatin marks on single X of, 299–300
    imprinting of sperm X chromosome in, 302–304
    MES histone modifiers in, 301–302
    MSUD in, 299–301
    non-cell-autonomous silencing and transitivity in, 182
    PcG genes in, 216–217
    PRC2 and, 216
    RNAi and, 163
    sex chromosome imbalance in, 293
    sex determination in, 292
    silencing in, 286
    Web sites for information on, 478
    X chromosomal regulation in, 294, 297–299, 301–302
    X:A ratio assessment in, 295–296
Caedibacter toxins, 130
CAF. See Chromatin Assembly Factor complex
Caloric restriction, 77
Cancers
    causes of, 458–459
    chromatin and, 459–461
    CpG island methylation, 347
    demethylases and, 201

Cancers (continued)
    DNA methylation and, 342, 353, 461–462, 471
    epigenetic therapies for, 471–473
    H3K27 methylation and, 200
    historical symposia and, 7
    hypermethylation and, 51, 462–465
    methyltransferases and, 52, 398, 404, 462, 471
    MLH1 gene and, 449, 463, 466
    nuclear transplantation and, 424–425
    overview of, 50–52
    PRC complexes and, 46–47, 224, 226
    research issues for, 470–471
    silencing and, 465–471
    trxG genes and, 224, 237
Capping, 73–74
Carcinogens, cytosine methylation and, 462
CARM1, 201
Cats, cloned, 26
Cd79a cells, 402–403
CDKN1C, Beckwith-Wiedemann syndrome
        and, 440, 442
Cdx2, 388
Cell cycle, 40, 75–76, 222, 422
Cellular fates. See Somatic reprogramming
Cellular memory
    DNA methylation and, 343–344, 352
    Hox genes and, 212, 233
    nuclear transplantation and, 428–429
    overview of, 212–213, 232, 234
    PcG and trxG and, 45, 47
    repetitive elements and, 44–45
CenH region. See CENP-A
CENP-A
    centromere evolution and, 279–280
    centromeric chromatin and, 255–256,
        275–278
    fission yeast centromeres and, 104–107
    maintenance of, 112
    overview of, 250
    RNAi and, 156
    silencing and, 111
Centromeres
    achiasmate segregation and, 281–283
    chromatin composition in, 275–277
    chromosomal inheritance and, 268
    composition of in fission yeasts, 104–107
    eukaryotic structure and function of,
        273–274
    evolution of, 279–280
    function of, 34
    H3 variants and, 254–256
    regulation of identity of, 279
    structure, function, and propagation of, 278
Centromeric outer repeats, silencing and,
        107–109
Cerebellar Granule Neuron Progenitors
        (CGNPs), 226–227
Chaperones, 76, 387
CHD proteins, 198, 240–241
Chfr, 467
Chiasmatic segregation, 281–282
Chimeras, plants and, 169–170
Choice, random X inactivation and, 328–329
Chp1, 105–106, 108–109, 155–157
Chp2, 105–106
CHRAC, 316

Chromatin. See also Silent chromatin
    ATP-dependent remodeling complexes and,
        236
    cancers and, 459–461
    cellular memory and, 53–54
    in centromeres, 275–277
    changes in with reprogramming, 425
    defined, 192
    diseases and, 438
    DNA methylation and, 347–348
    DNA vs., 30
    effect of RNAi on structure of, 156–157,
        159–160, 162–163
    effect of structure on transcription,
        205–206
    epimutations and, 449
    euchromatin vs. heterochromatin and,
        34–36
    facioscapulohumeral dystrophy and, 449
    fragile X syndrome and, 438, 448–449
    function of, 24
    histone modifications and, 39–41, 192
    history of epigenetics and, 3, 7, 18–19
    ICF syndrome and, 446–447
    methylene tetrahydrofolate reductase
        deficiency and, 447
    molecular components of in plants,
        175–180
    organization of, 29, 31–34
    overview of, 24, 28, 192
    PcG and trxG and, 45–47
    position-effect variegation and, 28
    Rett syndrome and, 444–445
    Rubinstein-Taybi syndrome and, 443–444
    Schimke immuno-osseous dysplasia and, 447
    T-cell receptors (TCR) and, 398
    as template, 29–31
    thalassemias and, 445–448
    trxG proteins and, 237–243
    Web sites for information on, 477
    X inactivation and, 331–333
    Xist gene and, 330
Chromatin Assembly Factor (CAF) complex,
        76, 179–180, 253, 278, 387
Chromatin immunoprecipitation assays
        (ChIP), 38–39, 194
Chromatin remodeling
    antigen receptor diversity and, 403–405
    Arabidopsis epigenetic regulation and, 172
    ATP-dependent complexes, 32, 39, 236,
        238–242, 315–316
    B cells and, 408
    defined, 24
    diseases and, 443
    DNA methylation and, 347–348
    dosage compensation and, 309–312,
        315–317
    histone variants and, 39–41
    historical symposia and, 18
    neoplastic transformation and, 51
    PRC2 and, 219
    proteins for in plants, 179
    remodeling factors and, 38
    transcriptional control and, 205
    transposable elements and, 373
    trxG proteins and, 238–242

    Xist and, 48
Chromatin-remodeling complexes, 39–41
Chromatoid bodies, 386
Chromodomains
    dosage compensation and, 311
    H2AZ and, 259–260
    H3K36 methylation and, 199
    ORF targeting and, 196
    PcG and trxG and, 45–46, 220
Chromomethylases, 176
Chromoshadow domains, 311
Chromosomal imprinting. See Genomic
        imprinting
Chromosome breakage sequences (CBS),
        135–136
Chromosome fragmentation, 134–136
Chromosomes. See also Segregation
    abnormal, 266
    breakage of and diseases, 446
    changes in with reprogramming, 426
    ciliar fragmentation of, 135–136
    distribution of trxG proteins in, 241
    DNA methylation and, 352
    elements required for inheritance of,
        267–268
    evolution of Y, 323
    knob sequences and, 284–285
    mechanism of inheritance of, 267
    methylation and, 41
    organization of, 322
    overview of organization of, 24
    parental origin and, 4
    paternal loss of, 284
    polyploidy and, 168
    polytene, 26
    rearrangements in, 275
    stability of, 352
    targeting of dosage compensation to, 313–316
Cid12, 109, 158–159
Ciliates
    active vs. silent chromatin in, 132–134
    conjugation and, 129
    cortical patterning in, 131–132
    cytoplasmic inheritance in, 17, 130–131
    genome defense systems of, 144–146
    homology-dependent silencing in, 136–137
    life cycle of, 130
    maintenance of two genomes by, 128
    as model organism, 17, 27
    overview of, 129
    rearrangements in, 134–142
    RNAi in, 142–144
Circles, extrachromosomal rDNA repeat, 76–77
cis modification patterns
    diseases and, 438, 447–449
    genomic imprinting and, 361–363, 371–373
    H2AZ and, 260
    histone posttranslational modification and,
        204
    MSL complex and, 314
Citrulline, 202
CKcnp1ot1, 368–369
c-Kit, 400–401
Cleavage, 160, 427–428
Cleavage Polyadenylation Specificity Factor A
        (CPSF-A), 160

CLF. *See* Curly leaf
Clones, 26, 420, 426, 427
Cloning. *See* Nuclear transplantation; Reproductive cloning
Clr4
    H3K9 methylation and, 199–200
    heterochromatin formation and, 105–106, 108, 155–157
    recruitment of chromatin-modifying enzymes and, 159–160
    siRNA production and, 110
CMT3, 176
Cohesins, 76, 109–110, 276
Collochores, 280–281
Colon cancer, 463, 465–467
Common myeloid progenitors (CMPs), 400
Compaction, histone evolution and, 262
Compartmentalization in ciliates, 128
Compensation, 430. *See also* Dosage compensation
Condensation, 24, 69–70, 299, 303
Condensins, similarity of to DCC, 293–295
Conjugation, 129, 131
Constitutive heterochromatin, 31, 36, 47–48, 333
Contraction, 405–407
Controlling elements, 2–3
Coordination, chromatin remodeling and, 39
Corepressors, 351
Cortical patterning, 131–132
Cosuppression, 118, 175. *See also* Posttranscriptional gene silencing
Counting, 324, 327–328
Covalent modification, 24, 243
CpG dinucleotides
    cancers and, 458, 461
    DNA methylation and, 41, 333, 342–345, 369, 437
    MeCP2 and, 350–351, 445
    in plants, 168
CpG islands
    cancers and, 461, 463–464, 467–469
    DNA methylation and, 345–347, 349–350, 352, 469
    genomic imprinting and, 370
    histone modifications and, 469
    hypermethylation of, 462
    Web sites for information on, 477
CPSF-A. *See* Cleavage Polyadenylation Specificity Factor A
CREBBP, 444
Cross-fertilization, 129
Croziers, 113
CTCF, 349, 370, 372
Cul4, 160
Curly leaf (CLF), 178, 215, 217
Cytoplasmic bridges. *See* Plasmodesmata
Cytoplasmic inheritance
    ciliar rearrangements and, 139–142
    in ciliates, 130–131, 138–139
    implications of, 142
    Mendelian inheritance vs., 131
    scnRNA model and, 142–144
    of serotypes, 130
    transcriptome-scanning model and, 145–146
Cytosine
    cancers and, 462, 471

deamination of, 42, 398, 408
methylation and, 42, 175–177, 342, 343

D

D1, 96
d48 cell line, 138–140
Darwin's theory of evolution, 25
DDM1. *See* Decreased DNA methylation 1
DDP1, 96
De novo methylation, 175–176, 184–185, 344–345
Deacetylation, 68, 194–196, 472–473. *See also* Histone deacetylases
Deaminases, 42, 398, 408
Decitabine, 471–472
Decontraction, 407
Decreased DNA methylation 1 (DDM1), 161–162, 179, 348
Defective in RNA-directed methylation 1 (DRD1), 179
Defense mechanisms, 41, 140, 144–146
Definition of epigenetics, 2, 16, 24, 28–29
Deimination, 37–38, 193, 202, 388
Deletions. *See* Internal eliminated segments (IES)
Demeter (DME), 176, 388
Demethylation
    abnormal in cloned embryos, 427–428
    jumonji histone demethylase (JHDM1) and, 37–38
    of lysines, 37–38, 178, 200–201
    mammalian development and, 346
    oocytes and, 428
    in plants, 176
    sperm reprogramming and, 387–388
    of zygotic paternal genome, 346–347
Depsipeptide, 472
Deubiquitylation, overview of, 193, 203–204
Development
    of B cells, 400–403
    DNA methylation and, 346
    history of epigenetics and, 16–17
    of organ from single cell, 399
    PcG repression and, 218–220, 224–227
    regulation of X inactivation in mammals and, 325
    reprogramming and, 427–428
Dicer-like (DCL) proteins, 180, 182
Dicers
    heterochromatin modifications and, 160–161
    miRNAs and, 183
    MSUD and, 120
    quelling and, 119
    RNAi-mediated silencing in plants and, 180–185
    scnRNA model and, 142–144
    siRNAs and, 108, 157–159
    spermatogenesis and, 386
Diet, 449–451
Differentially DNA-Methylated Regions (DMR), 365–370, 387
Differentiation, 50–52, 334, 386–389, 408–410, 421–423
DIM-2, 115–116, 119

DIM-5, 117, 119
Dimerization, 252, 270
Dimorphism, 27, 129, 132–134
Diseases. *See also specific diseases*
    chromatin structure in *cis* and, 447–449
    chromatin structure in *trans* and, 443–447
    cloning and, 421
    environment and, 449–451
    genomic imprinting and, 365, 439–443
    human case studies and, 438–439
    methylation and, 41, 353
    overview of involvement of epigenetics in, 437
    PcG and trxG and, 47
Disomy, uniparental, 436, 438–439
Diversity. *See* Antibody diversity
Dlk1 cluster, 365–368
DMRs, 365–370, 387
DNA demethylases. *See* Demethylation
DNA methylation
    allelic exclusion and, 407
    cancers and, 458, 460–462, 465, 471
    cellular memory and, 343–344
    chromosomal stability and, 352
    CpG islands and, 469
    diseases and, 436–438, 443
    in embryonic stem cells, 390
    environment and, 353, 450–451
    genomic imprinting and, 360–361, 369–371
    in ICM cells, 388
    inhibitors of, 471–472
    in mammals, 342
    mapping of, 346
    mutations and, 352
    N. crassa and, 102
    origins of patterns of, 344–348
    in plants, 168
    primordial germ cell development and, 384–385
    regulation of gene expression by, 348–352
    reprogramming and, 388–389
    siRNAs and, 152
    Web sites for information on, 477
    X inactivation and, 333
DNA methyltransferases (DNMTs)
    cancers and, 52, 462, 471
    in *Drosophila*, 91
    exclusion of from nucleus, 388–389
    genomic imprinting and, 370, 386
    H3K9 methylation and, 200
    hemimethylated DNA and, 343
    homology in, 114
    ICF syndrome and, 446
    inhibition of, 471
    mammalian, 343–345
    methylation and, 41
    primordial germ cell development and, 384–385
    protection of CpG islands and, 347
    quelling and, 119
    RIP and, 115–116
DNA repair
    cancers and, 459
    chromatin structure and, 269–270
    H2AX and, 258–259
    MBD4 and, 352

DNA repair *(continued)*
  methylation and, 197, 199–200
DNA replication process
  control of, 268–269
  histone deposition and, 253–254
  histone variants and, 40
  MBD1 and, 350
  methylation and, 20
  in plants, 180
  retrotransposons and, 271
Dnmts. *See* DNA methyltransferases (DNMTs)
Docking domains, 259
Dogma, central, 25
Dolly, 49, 418
Domains-rearranged methyltransferases
        (DRM), 176, 185
Dosage compensation
  in *C. elegans*, 293–298
  cloning and, 428
  defined, 292
  in *Drosophila*, 308–316
  in mammals, 292, 322, 324–338
  mechanisms of, 48–49
Dosage compensation complex (DCC),
        293–296
DOT1, 199, 270, 302
Double-stranded RNA (dsRNA). *See also* RNAi
  breakage of, 270
  genomic imprinting and, 372
  RNAi-mediated silencing in plants and,
        180–185
  silencing in ciliates and, 137–138
  synthesis of, 157–159
DPY proteins, 294
dRing1 (dRing1/SCE) proteins, 220–221
DRM. *See* Domains-rearranged
        methyltransferases
Drosha, 183
Drosophila
  cellular memory and, 212
  centromeric chromatin and, 277
  chromosomal proteins and modifiers in,
        85–88, 91–93, 271
  dosage compensation in, 292, 308–309
  genome organization of, 45
  germ-cell specification in, 383–384
  heterochromatin in, 88, 95–96, 280–281
  histone modifications, silencing and, 88–91
  karyotype of, 309
  kismet (kis) and, 240–241
  methylation and, 41
  as model organism, 27–28
  MSUD absence in, 286
  neocentromeres in, 274–275
  paternal chromosome loss in, 284
  PcG genes in, 46, 216–217, 220
  polyhomeotic (ph) duplication in, 221
  PRC1 targeting in, 221
  RNAi and, 163
  segregation disorder and, 283–284
  segregation process in, 282
  sex chromosome dysfunction in, 286
  silencing and, 88–91, 96–97
  targeting of heterochromatin formation in,
        93–95
  telomere function regulation in, 282
  variegating phenotypes in, 83–85

Web sites for information on, 478
dsRNA. *See* Double-stranded RNA
dsx genes, 312–313
DTip60, 261
Duplication, 115–116, 221. *See also* Repeat-
        induced point mutation

E

E2A, 402–403
E3 ubiquitin ligase, 440–441, 467
EBF1, 401–403
EC2A, 401–402
Eggs, 386–387, 417–418, 427–428
Elongation, PcG silencing and, 49
Embryogenesis, 49, 329–330
Embryonic ectoderm expression (Eed),
        224–225
Embryonic Flower (EMF), 178–179, 217
Embryonic germ cells, pluripotent, 378
Embryonic stem cells
  de novo methylation and, 344
  generation of, 389–390
  pluripotent, 378, 385, 390
  somatic reprogramming and, 49–50
  therapeutic cloning and, 429–430
  therapeutic uses of, 431–432
  X inactivation and, 327–328, 334–335,
        337–338
Embryos, 363–364, 417–419, 428
EMF. *See* Embryonic Flower (EMF)
ENCODE, 477
Endonucleases, mating-type switching and,
        66–67
Enhancer of zeste (E(Z)) family, 90–91,
        177–178, 215, 226, 301–302
Enhancers
  allelic exclusion and, 407
  silencing in *C. elegans* and, 301–302
  of variegation (E(var)), 84–88
  of zeste (E(Z)) family, 90–91, 177–178, 215,
        226, 301–302
Environment, 55–56, 342, 353, 436, 449–451
Enx1, 224–225, 333–334
Enzymes, 7, 37, 39. *See also specific enzymes*
Epialleles, 27, 175
Epigenesis, 378, 381
Epigenetic code, 29
Epigenetic control, 52–54, 67–69
Epigenetic imprints, defined, 303
Epigenetic landscape, development of, 29
Epigenetic memory. *See* Cellular memory
Epigenetic states, defined, 28
Epigenetic therapies, for cancers, 471–473
Epigenetics, defined, 24
Epigenomes, defined, 436
Epimutations, human disease and, 449
Epiphenotypes, 26
Epistates, TSA-induced silencing and, 109
ERC, 76–77
Erythroid cells, CMPs and, 400
ESC. *See* Extra sex combs (ESC)
Essential (E) silencer, 70–71
Estrogen signaling, arginine methylation and, 202
Euchromatin
  defined, 24, 31, 64
  H3K4 methylation and, 197

  H3K9 methylation and, 200
  H3K27 methylation and, 200
  heterochromatin vs., 34–36, 92, 95, 174–175
  silent chromatin and, 48
E(var) (*Enhancer of variegation*), 84–88
Eversporting displacements, 2, 16. *See also*
        Position-effect variegation
Evolutionarily conserved chromosome
        segments (ECCS), 256. *See also*
        Adaptive evolution
Excisions. *See* Internal eliminated segments
Expression
  in B cells and plasma cells, 410–411
  cloning and, 428
  control of in lymphocytes, 402–403
  developmental programs and, 386–389
  genomic imprinting and, 358
  histone acetyltransferases (HATs) and, 18
  of imprinted genes, 225
  location and, 9
  PcG gene overexpression and, 226
  regulation of by DNA methylation, 348–351
  regulation of X chromosomes, 295
Extra sex combs (ESC), 75, 215–216, 301–302
Extracellular signaling, 400–401
Extrachromosomal rDNA repeat circles (ERC),
        76–77
E(Z) family. *See* Enhancers of zeste (E(Z))
        family
E(Z)-ESC complex, 301–302. *See also* PRC2
EZH2, 200, 226, 333–334, 398, 404
Ezh2/Enx1, 333–334

F

Facial abnormalities, methylation and, 41
Facilitate Chromatin Transcription (FACT), 40,
        205
Facioscapulohumeral dystrophy, 447, 449
FACT. *See* Facilitate Chromatin Transcription
Facultative heterochromatin, 31, 47–49,
        331–333
Fates. *See* Somatic reprogramming
Female-sterile homeotic (fsh) gene, 233–234
Fertilization, 129, 169, 387–388
Fertilization-independent endosperm (fie), 216
Fertilization-independent seed (fis), 178–179,
        215–217, 225–226
Fibroblast cells, 418–419
FK-228, 472
Flamenco, 162–163
Flowering, vernalization and, 217–219
Flt3, 400–401
FMRP protein, 448–449
Folate, 447, 450
Foreign sequences, elimination of, 140
Forks, DNA replication and, 268
FOX-1, 295–296
FR901228, 472
Fragile X syndrome, 438, 447–449
fragilis, 381
Fragmentation. *See* Chromosome fragmentation
FWA, 162

G

G9a, 410
GAGA factor, 244–245

Gametic imprints, 363, 365–368
Gametogenesis, 385–386, 427
GASC1. *See* Jumonji histone demethylase (JHDM1)
Gene silencing. *See* Silencing
Genetics, epigenetics vs., 16–17, 25
Genome organization
    comparison of, 44–45
    epigenetic control and, 52
    heterochromatin and, 35
    Web sites for information on, 478
Genome scanning model. *See* scnRNA model
Genome sizes, 44–45
Genomic imprinting
    Angelman syndrome and, 439–441
    Beckwith-Wiedemann syndrome and, 440–442
    as *cis*-acting regulatory system, 361–363, 371–372
    components of in plants, 179
    defined, 358, 378
    diseases and, 437
    DNA methylation and, 360–361, 369–371, 388
    embryonic and neonatal growth control by, 363–364
    function of in mammals, 364–365
    gene function and, 364
    historic overview of, 4–5, 359–361
    historical symposia and, 7
    imprint control elements and, 365–368
    loss of, 51, 465
    as model for mammalian epigenetic regulation, 372–373
    as monoallelic silencing, 48
    ncRNAs and, 368–369
    Prader-Willi syndrome and, 439–441
    PRC2 and, 225
    pseudohypoparathyroidism and, 440, 443
    Silver-Russell syndrome and, 440, 442–443
    of sperm X chromosome in *C. elegans*, 302–304
    Web sites for information on, 477
Germ cells. *See also* Primordial germ cells
    development of, 379–382
    inheritance of ciliar rearrangements and, 138–142
    mechanisms regulating specification of, 381–386
    pluripotency and, 391
    totipotency and, 44–45
    X reactivation in, 337
Gibberellins, 217
Glycosylases, 176, 352
Gnas family, 365–368, 440, 443
GNAT family, 195
Groucho family, 410
Growth, regulation of by genomic imprinting, 363–364
Growth factor receptors, 236
Growth signals, cancers and, 459
Gtl2, 368–369

H

H1, 263, 483
H19 gene, 361, 369, 371–372, 440, 442

H2.2, sperm reprogramming and, 387
H2A
    defined, 250
    deposition and replacement of, 262
H2AX, 258–259, 270, 483
H2AZ, 250, 259–261
    location of, 252
    macro, 250, 262, 334, 483
    table of, including references, 479
    variant formation and, 251
H2A Barr body deficient (H2ABbd), 250, 262
H2AX, 258–259, 270, 483
H2AZ, 250, 259–261
H2B, 262, 480
H3, 196, 252, 254–256, 276, 480–481
H3.3, 250, 256–258, 278
H3K27 methylation
    heterochromatic silencing and, 90–91
    MES proteins and, 301–302
    overview of, 200
    PRC targeting and, 223
    primordial germ cell development and, 384–385
    V(D)J recombination and, 404
    X inactivation and, 332, 335
H3K36, 198–199, 302
H3K4, 197–198, 333–334
H3K79, 199, 270
H3K9
    centromeric chromatin and, 276
    DNA methylation and, 348
    heterochromatin targeting and, 82
    in male single X chromosomes of *C. elegans*, 299–300
    overview of, 199–200
    in plants, 177–178
    primordial germ cell development and, 384–385
    repressed DNA partitioning and, 136
    silencing and, 91–92, 97, 104, 108, 142–144
    siRNAs and, 152
    spermatogenesis and, 386
    su(var) and, 89–91
    telomere elongation and, 271
H4
    KYP/SUVH4, 178
    MSL complex and, 309–312
    overview of, 200
    SIR complexes and, 72–73
    table of, including references, 482
    variant formation and, 251
Hairpin RNA, 157
Hairpin-tail mice, 359–360
HBRM, 238
HDAC. *See* Histone deacetylases (HDACs)
HDOT1L, 199
Helicases, 109, 158–159, 311
Hematopoietic stem cells (HSC), 226
Hematopoietic system. *See* Lymphocytes
Hemimethylated DNA, 41, 343
Heptoblastoma, 442
her-1 gene, 294, 297
Heritability, 343, 379
Hermaphrodites, 129, 294–299
Heterochromatin. *See also* Silent chromatin
    assembly of, 70–71, 156–157
    chromosome segregation and, 111

constitutive, 31, 36, 47–48, 333
defined, 24, 31, 64
euchromatin vs., 34–36, 92, 95, 174–175
facultative, 31, 47–49, 331–333
genome rearrangements and, 136
H2AZ and, 259
H3K9 methylation and, 82
historical symposia and, 5
in human neocentromeres, 254
meiotic drive and, 283–285
methylation and, 41
packaging of, 85
pairing and, 280–282
perinuclear attachment of, 74–75
proteins associated with, 88–89
repression of gene activity and, 69–70
RITS and RNAi and, 156–157
RNAi and, 42–43, 152, 155–156, 160–161
targeting of, 95
variability in, 95–96
X inactivation and, 326, 331–333
yeast centromeres and, 105–106
yeast mating types and, 67–69
Heterochromatin Protein 1 (HP1)
    function of, 90–94
    H2AZ and, 259–260
    H3K9 methylation and, 199–200
    PcG and, 45–46
    RNAi and, 117
    silencing and, 180, 223
    su(var) and, 28, 85–88, 96, 199–200, 316–317, 387
    telomere elongation and, 271
Heterogametic sex, defined, 323
Heterokaryotic phase, 112–113, 118
Heterothallic yeast strains, 67
Heterozygosity, loss of, 465
HIC-1, 467
High Mobility Group (HMG) proteins, 262
HIRA. *See* Histone regulator A
Histone acetyltransferases (HATs)
    in ciliates, 128, 133
    dosage compensation and, 309–311
    gene expression and, 18, 73, 194–195
    histone modifications and, 37
    purification of, 27
    Rubinstein-Taybi syndrome and, 444
Histone code hypothesis
    antigen receptor diversity and, 405
    heterochromatin-associated proteins and, 257
    overview of, 8, 36–39, 204, 437
    trxG proteins and, 243
    variation in, 97
Histone deacetylases (HDACs)
    cancers and, 52, 353, 469–470
    gene regulation and, 194–196, 401
    heterochromatin formation and, 105, 155–157
    histone modifications and, 37
    historical symposia and, 7
    inhibitors of, 472–473
    leukemias and, 460
    MeCP2 and, 445
    NuRD and, 351–352
    PRC2 and, 215–216
    Sir2 family of, 68–69, 72, 77

Histone deacetylases (HDACs) *(continued)*
  X inactivation and, 333–334
Histone lysine methyltransferases (HKMTs)
  B cells and, 408
  cancers and, 52, 398, 404
  ciliar histone variants and, 133–134
  CpG methylation and, 348
  DNA methylation in *N. crassa* and, 117
  DNA repair and, 270
  dosage compensation and, 316–317
  Drosophila and, 28
  Ezh2 and, 200, 226, 333–334, 398, 404
  heterochromatin formation and, 105–106,
    155–157
  histone modifications and, 37–38
  MLL as, 236–237
  in plants, 177–178
  primordial germ cell development and, 382
  RNA polymerase II and, 198
  SET domains and, 238
  silencing and, 88–91
  SUVH proteins and, 96
  trxG proteins and, 236
Histone modification
  Arabidopsis epigenetic regulation and, 172
  B cells and, 408
  centromeric chromatin and, 276–277
  constitutive vs. facultative heterochromatin
    and, 333
  CpG islands and, 469
  enzymes for in plants, 177–178
  fragile X syndrome and, 448
  genomic imprinting and, 363, 370
  locus contraction and, 406
  overview of, 26–27, 36–39, 192
  paternal chromosome loss and, 284
  PcG and trxG and, 46–47, 242–243
  RAG2 and, 405
  silencing in *C. elegans* and, 301–302
  silencing in *Drosophila* and, 88–92
  silencing in *S. pombe* and, 104
  sites for, 31
  sperm reprogramming and, 387–388
  spermatogenesis and, 303
  table of, including references, 479–484
  themes in, 203–205
  transcription factors and, 399
  X inactivation and, 301–302, 332–334
Histone regulator A (HIRA), 278, 387
Histone variants. *See also specific variants*
  CENP-A and, 278
  centromeres and, 254–256
  defined, 24
  enzymes for, 7, 27, 156
  genomic imprinting and, 363
  overview of, 39–41, 250
  roles for, 132–134
  sperm reprogramming and, 387–388
  transcriptional states, 205
  X inactivation and, 325, 332–334
Histones
  acetylation and, 17–18
  CENP-A and, 276
  chromosomal organization and, 24
  diversity in, 250
  evolution of, 262–263
  function of, 251

historical symposia and, 6
linker, 254, 263, 386
maintenance methylation and, 117
methylation of, 8–9
overview of, 192
overview of modifications of, 36–39
RAG2 and, 405
structure of, 29–30
variants of in ciliates, 132–134
Web sites for information on, 477
History of epigenetic research
  chromatin and, 18–19
  Cold Spring Harbor and, 2–8
  genetics and development and, 16–17
  genomic imprinting and, 359–361
  hypermethylation and, 463
  mechanistic interrelatedness and, 19–21
  methylation and, 17–18, 461, 463
  nuclear transplantation and, 417
  overview of, 16
  somatic cell DNA and, 17
HKMTs. *See* Histone lysine methyltransferases
    (HKMTs)
HML, 70–71
HMR, 70–71
HO endonuclease, 66–67
HOAP (HP1/ORC associated protein), 271
Holocentric chromosomes, 268, 280
Holokinetic chromosomes, 255
Homeless, 94–95
Homeologous pairing, 283
Homeotic genes, 236. *See also* Trithorax (trxG)
    proteins
Homeotic transformations, 213–214
Homogametic sex, defined, 323
Homologous recombination, 65, 363
Homology, 136–138, 140–142
Homology-dependent silencing, 136–137
Homothallic yeast strains, 67
HOTHEAD gene, 10
Hox genes
  cellular memory and, 212, 233
  PcG proteins and, 213–214
  PcG silencing and, 220
  PREs and, 222–223
  primordial germ cell development and, 382
  regulation of in *Drosophila*, 233–234
  Sex combs reduced (Scr) and, 235
HP1. *See* Heterochromatin Protein 1
HpaII tiny fragments, 345, 348
HPone, 117
Hrr1, 109, 158–159
hSET, 242
HU proteins, 251
Human artificial chromosomes (HACs), 274
Hyperhomocysteinemia, 447, 450
Hypermethylation, 51, 257, 462–465
Hypermutation, 42
Hypomethylation, 51, 446, 462
Hypoparathyroidism, 443

I

IAP retrotransposon, 385, 387
ICE, 368, 370
ICF syndrome. *See* Immunodeficiency
    Centromeric Instability

IFNB1, 410
IGF2
  Beckwith-Wiedemann syndrome and, 440,
    442
  colon cancer and, 465
  genomic imprinting and, 361, 365–368,
    371–372
IgH, 403
Ikaros association, 408
IL-7 receptor, 400–401
Imaginal discs, 213
Immunodeficiency Centromeric Instability
    (ICF), 41, 344–345, 444, 446–447
Immunoglobulins, 398, 405–406
Immunoprecipitation, 38–39, 194
Immunorejection, 431
Importance of epigenetics, 25
Important (I) silencer, 70–71
Imprecise deletions, 134–135
Imprint control elements, 365–368
Imprinted X inactivation, 325–327, 329–330
Imprinting. *See* Genomic imprinting
In(1)wm4, 85–86. *See also white* gene
Inheritance, 25, 75–76, 131–132, 186. *See also*
    Cytoplasmic inheritance
Inhibitors, 52, 471–473
INI1, cancers and, 237
Inner cell mass (ICM), 325, 380, 388
INO80-C, 261
Insulator model, 371–372
Intergenic repeats. *See* Imprecise deletions
Internal eliminated segments (IES)
  ciliar rearrangements and, 135
  elimination of foreign sequences and, 140
  heterochromatin partitioning and, 135
  inhibition of elimination of, 140–142
  maternal inheritance and, 141
  siRNAs and, 142–144
  transcription and, 43
Invasive elements, 44
ISWI complexes, 240, 315–316, 427. *See also*
    Brahma family; SNF2H family

J

JIL1 kinase, 92–93, 98, 311
Jumonji histone demethylase (JHDM1), 37–38,
    201

K

K79 methylation, overview of, 199
Karyogamy, 113, 129
Karyotype of *Drosophila*, 309
KCNQ, 365–368, 442
killer strains, 130
Kinases
  histone modifications and, 37–38
  JIL1 kinase, 92–93, 98, 311
  Phosphoinositol 3-kinase-like kinase family,
    258
Kinetochores
  CenH3 and, 255–256
  CENP-A and, 275–276
  centromere identity and, 278
  chromosomal inheritance and, 268
  fission yeast centromeres and, 105
  formation and function of, 274–275

function of, 273
silencing and, 104, 106–107, 110
Kismet (kis), 235, 240–241
Knobs, 243, 284–285
Knockouts, 120
Knudson two-hit theory, 50–51
KTO protein, 243
KYP/SUVH4, 178

L

Lambs, cloning of, 418
Leukemia, 237, 460
Life span. *See* Aging
Light, 93
Like heterochromatin protein (LHP1), 180
Linaria vulgaris, 168
Lineage commitment, 400–403
Lineage priming, 401
Linker histones, 254, 263, 386
Location, gene expression and, 9
Locus contraction, 405–407
Long interspersed repeats (LINES), 335
Long terminal repeat (LTR) retrotransposons, 353
Looping, telomere, 73–74
Loss of heterozygosity, 465
Loss of imprinting (LOI), 51, 465
LSD1. *See* Lysine-specific demethylase
LSD1 aminoxidase, 93
LSH2, 348
Lung cancer, 466
Lymphocytes
antigen receptor diversity and, 403–408
B-cell differentiation and, 408–411
cloning of mice from, 423–424
ICF syndrome and, 446
lineage commitment in, 400–403
pluripotency and, 399–400
Lymphomagenesis, 224
Lysines. *See also* Histone lysine
methyltransferases (HKMTs)
acetylation of, 195
demethylation of, 37–38, 178, 200–201
methylation of, 197–200
Lysine-specific demethylase (LSD1), 37–38, 178, 201

M

M deletion elements, 141–142
MacroH2A, 250, 262, 334, 483
Macronuclei
active vs. silent chromatin and, 132–134
chromosome fragmentation and, 135–136
cillial rearrangements and, 138–142
defined, 128
division of in ciliates, 129
induction of specific deletions in ciliates, 137–138
rescue of inherited deletions in, 140
Maintenance methylation, 114, 176, 343–344, 369
Maize, 283–285
Major groove, 348
Male specific lethal (MSL) gene complex, 309–315
Malignancy, nuclear transplantation and, 424–425

Mammals
changes associated with nuclear
reprogramming in, 427–428
derivation of from terminally differentiated
cells, 421–425
dosage compensation in, 292, 322, 324–338
genome organization of, 45
genomic imprinting in, 358, 361–373
neocentromere formation in, 274
nuclear transfer procedures for, 417–419
phenotypes of cloned, 420–421
reprogramming in, 380
trxG proteins in, 236–237
Web sites for information on, 478
Marsupials, 41, 325–326, 336, 364
Mating types (MAT). *See also* Conjugation
of ciliates, 130, 140–141
epigenetic control of repression of, 67–69
of *N. crassa*, 112
RNAi and, 156
S. pombe and, 103
switching of, 26, 66–67
mb-1 gene, 402–403
MBD proteins. *See* Methyl-CpG-Binding
proteins
McClintock, Barbara, 271
MeCP2, 350–351, 444–445
Medea (MEA), 176, 178–179, 225
Medea/Fertilization independent seed
formation (MEA/FIS1), 178
Medicine, nuclear transplantation and, 429–432
Medulloblastomas, 227
MEIDOS, 217
Meiosis
chromosomal inheritance and, 267
epigenetics and, 20
germ cell epigenetic mechanisms and, 385–386
heterochromatin and, 283–285
process of, 267, 280–281
Meiotic drive, 280, 283–285
Meiotic silencing by unpaired DNA (MSUD)
in *C. elegans*, 299–301
in *Drosophila*, 286
in mice, 285–286
in *Neurospora crassa*, 112, 119–121, 285–286
RNAi and, 154
Meisetz, 385–386
Memory, 10. *See also* Cellular memory
Mendelian inheritance, 25, 131, 186
MES proteins, 216, 301–302
Methylation. *See also* DNA methylation
abnormal in cloned embryos, 427–428
Arabidopsis epigenetic regulation and, 172
of arginines, 37–38, 52, 193, 201–202
B-cell development and, 402–403
control of in *Neurospora*, 116–117
defined, 24
diseases and, 446
DNA repair and, 270
DNA replication and, 20
function of, 24
heritability of, 343
heterochromatin modifications and, 162
heterochromatin silencing in *Drosophila*
and, 88–91
histone code hypothesis and, 8

histone modifications and, 30, 37, 133
historical symposia and, 4–5, 7
history of epigenetics and, 17–18
imprinted gene clusters and, 366–368
lack of in *S. pombe* and, 104
maintenance, 114, 176, 343–344, 369
MSUD and, 119
neoplastic transformation and, 50–51
in *Neurospora crassa*, 113–115
overview of, 41–42, 196–200
plant silencing and, 43
PRC2 and, 219
regulators of in plants, 43, 175–177
repressed DNA partitioning and, 136
RIP and, 115–117
silent chromatin and, 257
sites of in histone tails, 31
telomere elongation and, 271
transcriptional control and, 238
in trophectoderm cells, 389
trxG proteins and, 243
X inactivation and, 325, 334
Methylation induced premeiotically (MIP), 116, 154
Methyl-CpG binding proteins
DNA repair and, 352
DNA replication and, 350
function of, 342
methylation and, 41, 176–177, 349–350
Rett syndrome and, 350–351, 444–445
summary of functions of, 350
transcriptional control and, 351–352
5-methylcytosine, 461–462
Methylene tetrahydrofolate reductase
deficiency, 447
Methyltransferase (MET1) family, 176
Methyltransferases. *See also* DNA
methyltransferases (DNMTs);
Histone lysine methyltransferases
(HKMTs)
arginine methylation and, 37–38, 52, 201–202
de novo, 344–345
diseases and, 447
domains-rearranged, 176, 185
heterochromatin modifications and, 160–161
histone modifications and, 117, 236, 398
mammalian maintenance, 343
in plants, 175–178
RNA and, 41–42, 175, 184–185
X inactivation and, 333
Mice
Blimp1 and, 382–383
cancer research and, 465
early germ cell determination in, 381
generation of monoclonal from mature
cells, 423–424
genome organization of, 45
genomic imprinting and, 359–360, 364
as model organism, 28
MSUD in, 286
PcG genes in, 216–217
silencing in, 285–286
telomere function regulation in, 271, 282
Web sites for information on, 478
X inactivation in, 326, 338

Microarrays, 346, 464
Micrococcal nuclease, 84–86, 106–107
Microinjection, 419
Micronuclei, 128–129, 132–134
MicroRNAs, 386
MiniChromosome Maintenance (MCM)
    proteins, 269
Minus (–) mating type, 103
MIP. *See* Methylation induced premeiotically
miRNAs, 162, 182–184, 368
Mitosis, 266–267, 277
Mitotic checkpoint. *See* Spindle Assembly
    Checkpoint (SAC)
Miwi-like proteins, 386
MIX-1 proteins, 294
MLE, 310–311
MLH1 gene, 449, 463, 466
MLL, 236–237, 242
MNase. *See* Micrococcal nuclease
Model systems, 26–28, 170, 173
Modifications, 50–51, 325. *See also* Histone
    modification; *Specific processes*
MOF, 310–311, 312
Moira (mor), 235–236, 238
Morpheus' Molecule (MOM), 179, 186
Movable genetic elements. *See* Transposable
    elements
MS-275, 472
MSCI, 326–327
MSP, 471
MSUD. *See* Meiotic silencing by unpaired DNA
mtF genes, 140–141
Muller, H.J., 309–310
Muller's ratchet, 324
Mutagenesis, 4, 170–171, 174, 352
Myeloid cells, 400, 471
Myeloid dysplastic syndrome, 471
MYST family, 195

N

n–1 rule, 324, 327–328
Nanochromosomes, 135
Nanog, 384, 388, 390
Nasonia vitripennis, 284
ncRNA-mediated silencing model, 371–372
Ndj1, 271–272
Necdin, 441
Neocentromeres, 274–275, 280
Neoplastic transformation, 26, 50–51, 226, 370,
    465–466. *See also* Cancers
NESP55, 443
Neural stem cells (NSC), 226–227
Neural tube defects, 436
Neurons, 20, 424, 446
Neurospora crassa
    images of, 114, 120
    life cycle of, 113
    methylation in, 113–115
    as model organism, 26–27
    MSUD in, 112, 119–121, 285–286
    overview of, 112–113
    quelling in, 117–121
    RIP and, 115–117, 120–121
    silencing in, 285–286
    as tool, 102
Non-cell-autonomous silencing, 181–182
NONCODE Web site, 368, 478

Noncoding RNAs
    diseases and, 443
    dosage compensation and, 313–315
    function of, 25, 437
    genomic imprinting and, 366–369,
        371–372
    heterochromatinization and, 44
    siRNAs and, 108
    spermatogenesis and, 386
    X inactivation and, 325
Non-genetic differences, examples of, 26
Nonrandom X inactivation, 331. *See also*
    Imprinted X inactivation
Notch1, 401
Nowa1, 144
NT-ES, 429–432
NuA4/Tip60, 261
Nuclear dimorphism, 27, 129, 132–134
Nuclear envelopes, 74–75
Nuclear equivalence, 416
Nuclear organization, overview of, 9
Nuclear transfer embryonic stem cells (NT-ES),
    429–432
Nuclear transplantation
    cancers and, 424–425
    changes associated with, 425–428
    cloned animal phenotypes and, 419–425
    epigenetic memory and, 428–429
    genomic imprinting and, 360
    medical implications of, 429–432
    procedures for, 417–419
    somatic reprogramming and, 49–50
Nuclear transplantation therapy, 429–432
Nucleases, 65–67, 84–86, 106–107
Nucleolar dominance, 175
Nucleosomes
    acetylation and, 195
    archaeal vs. eukaryotic, 251–253
    CENP-A and, 278
    chromatin remodeling and, 39
    defined, 29
    H3K36 methylation and, 198–199
    historical symposia and, 3
    structure of, 30
    trxG proteins and, 242–243
NuRD, 351–352
NURF, 316, 441
NVP-LAQ824, 472

O

Oct4, 388, 390, 427–428
Octamer loss, 205
Oenothera blandina, 96
Olfactory neurons, 20, 424
Oncogenesis, 7, 50–51, 237. *See also* Cancers
Oocytes
    asymmetric distribution and, 386–387
    DNA demethylases and, 428
    IAP retrotransposon in, 387
    mammalian, 379
    nuclear transfer procedures and, 417–418,
        427–428
Organismal genome resources and databases,
    478
Origin firing, 75
Origin of recognition complexes (ORC), 68, 70,
    74, 268–269, 271

Origins, 268–269
OSA, 240

P

P elements, 84
P53BP1, 199
p55, 215
Packaging, 251
PADI4, 202, 388
PAF complex, 197–198
PAI2, 160–161
Paramecia, 27, 128–130. *See also* Ciliates
Paramutation, 3, 20, 27, 175, 186
Parathyroidism, 443
Parental conflict theory, 179, 365
Parental imprinting. *See* Imprinting
Parp, 390
Parthenogenesis, 363–364, 378
Pasha, 183
Paternal genomes, 346–347
Pax5, 401–402, 406, 409–411
PcG proteins. *See* Polycomb proteins
PCR, 346, 471
Pericentromeric foci, 46
Petunias, 118, 175
Ph1 locus, 283
4-Phenylbutyrate (PBA), 472
PHERES1, 217
Phosphatases (PPTases), 37, 93
Phosphoinositol 3-kinase-like kinase family,
    258
Phosphorylation, 31, 196, 258–259
Photoperiod pathway, 217
Photoperiod-independent early flowering
    (PIE), 179
Pickle (PKL), 179
Piwi, 94–95, 154, 162–163
Plant Chromatin Database, 174, 179
Plants. *See also* Arabidopsis
    benefits of use for research, 169–175
    centromere evolution in, 279
    epigenetic regulation in, 168
    genomic imprinting and, 358
    involvement of PRC2s in development of,
        218
    life cycle of, 171
    as model organisms, 27
    molecular components of chromatin in,
        175–180
    non-RNA epigenetic regulation of, 186
    RNAi-mediated silencing pathways of,
        180–186
Plasma, *Blimp1* and, 382–383
Plasma cells, 408–410
Plasmodesmata, 169–170
Plasticity, 44, 47, 49–50, 410–411
Pleiohomeotic (PHO) proteins, 223
Pleiohomeotic-like (PHOL) proteins, 223
Pluripotency
    defined, 378
    lymphocyte development and, 399–400
    lymphocytes and, 399–400
    primordial germ cell development and,
        381–382, 384–385
    reprogramming and, 49–50, 391
    spermatogenesis and, 386
    stem cell development and, 389–392

Pluripotent stem cells, 389–391
Plus (+) mating type, *S. pombe* and, 103
PML-leukemia, 46
Pol IV. *See* RNA polymerase IV
Polar granule component (pgc) gene, 384
Polycentric chromosomes, defined, 268
Polycomb (PcG) proteins
    Arabidopsis epigenetic regulation and, 173
    cellular memory and, 47, 212–213, 232
    chromatin modification in animals and, 163
    chromatin silencing marks and, 215–220
    CpG methylation and, 348
    function of, 45–47
    genetic identification of, 213–215
    genomic imprinting and, 370
    HP1 and, 90
    mammalian development and, 224–227
    overview of, 44, 220
    plant chromatin and, 178–179
    plant histone methyltransferases and, 177–178
    primordial germ cell development and, 384–385
    transcriptional silencing maintenance and, 220–224
    X inactivation and, 48, 333–334
Polycomb repressive complexes (PRCs)
    cellular memory and, 212
    chromatin modification by, 219
    components and evolutionary conservation of, 215–219
    components of, 220–222
    function of in development, 219–220
    H3K27 methylation and, 200
    in plants, 178–179
    prevention of heritable repression by, 224
    repressive functions and, 223–224
    targeting of to silenced genes, 222–223
    transcriptional repression and, 233
    X inactivation and, 48, 225
Polycomb response elements (PREs)
    anti-silencing and, 224
    cellular memory and, 212
    GAGA factor and, 245
    H3K27 methylation and, 200
    PcG and, 46
    PRC1 targeting and, 222–223
    transcriptional repression and, 233
Polycomb-like (PCL) protein, 216
Polyhomeotic (PH) proteins, 220
Polymerases, 185. *See also specific RNA polymerases*
Polyploidy, 168, 170
Polytene chromosomes, 26
Position-effect variegation (PEV)
    chromosomal proteins and, 85–88
    defined, 82
    Drosophila and, 27–28, 83–85, 91–93
    historical symposia and, 2–6
    overview of, 96–97
    suppression of, 93–95
    TPE and, 271
    X inactivation and, 335
Posterior Sex Combs (PSC) proteins, 220–221
Posttranscriptional gene silencing (PTGS). *See also* RNAi
    Drosophila and, 94–95

fission yeasts and, 109–110
    overview of, 152
    as plant equivalent of RNAi, 175
    RNA degradation and, 42
    RNAi-mediated silencing in plants and, 180–182
Posttranslational modifications
    acetylation, deacetylation and, 194–196
    deimination, 37–38, 193, 202, 388
    histones and, 193–194
    methylation and, 196–202
    phosphorylation and, 196
    themes in, 203–205
    ubiquitination, sumoylation and, 202–203
Prader-Willi syndrome, 439–441
PRC2. *See* Prereplication complex 2
Prdm1, 409–410
Pre-BCR, 407
Precise deletions, 134–135
Preformationism, 378, 381
Prereplication complex 2 (PRC2)
    chromatin remodeling and, 219
    DNA replication and, 269
    genomic imprinting and, 225
    histone deacetylases (HDACs) and, 215–216
    plant development and, 218
    retinoblastoma (Rb) proteins and, 226
    X inactivation and, 333–335
Primary nonrandom X inactivation, 331
Priming, lineage, 401
Primordial germ cells
    Blimp1 and, 382–383, 409–410
    determination of, 380
    development of, 381–382
    pluripotent, 378, 385
    pluripotent cell lines from, 390–391
    X reactivation in, 337
Prions, 9–10, 21, 25
PRMTs. *See* Protein arginine methyltransferases
Progeria, 47
Programmed DNA elimination, 27
Proliferation, 45, 222, 224, 226–227
Promoters, CpG islands and, 345–346
Protein arginine methyltransferases (PRMTs), 37–38, 52, 201–202
Protoperithecia, 112
Pseudoautosomal region (PAR), 325
Pseudohypoparathyroidism, 440, 443
Pseudokinetochores, 284–285
PSR chromosome, 284
Psychiatric illnesses, environment and, 436
PTGS. *See* Posttranscriptional gene silencing
Pws cluster, 365–368
Pyrimidine dimers, 270

Q

Quelling, 112, 116–121
Quiescence, reprogramming and, 49–50

R

R deletion elements, 141–142
RAG, 403–404, 431–432
Random X inactivation, 327–330
RanGAP gene, 283–284
Rap1. *See* Repressor activator protein 1 (Rap1)
Rastan's blocking factor model, 328–329

RC histones, 253–254
RdDM. *See* RNA-directed DNA methylation
rDNA, 76–77, 269, 280–281
rDNA rings, 76–77
RDRP. *See* RNA-dependent RNA polymerase
Reagents, Web sites for information on, 478
Rearrangements, 131, 134–142, 275
Recognition models, RNAi and, 159
Recombination, 65, 363, 398–400, 403–405
Redundancy, silencer elements and, 70
Relocation, of immunoglobulin genes, 405
Remodeling. *See* Chromatin remodeling
Repair. *See* DNA repair
Repeat elements, 74
Repeat-induced gene silencing (RIGS), 96
Repeat-induced point mutation (RIP)
    historical symposia and, 5
    methylation and, 41
    as model, 27
    MSUD and, 119
    in *Neurospora crassa*, 112, 115–117, 120–121
    RNAi and, 154
Repetitive elements, 44–45, 104, 107–108, 136
Replication. *See* DNA replication process
Replication processivity clamp (PCNA), 253
Replication Protein A complex, 180
Replication-independent (RI) nucleosome assembly, 253–254
Reporter genes, yeast and, 65–66
Repression. *See also* Polycomb repressive complexes
    B cell development and, 402–403, 409
    Bcl6 and Blimp1 and, 409
    conservation of, 383–384
    DNA methylation and, 347–352
    of heterochromatin, 69–70
    histone modifications and, 192
    LSD1 and, 201
    methylation and, 200
    primordial germ cell development and, 382
    SIR protein concentration and, 75
    sumoylation and, 203
    telomere looping and, 73–74
    trxG proteins and, 244–245
    yeast mating types and, 67–69
Repressor activator protein 1 (Rap1), 68–71
Repressor of silencing (ROS1), 176
Reproductive cloning, 429–432
Reprogramming. *See also* Nuclear transplantation
    after fertilization, 387
    in amphibians, 425–427
    in clones, 420, 426
    in mammals, 427–428
    overview of, 49–50
    pluripotency and, 49–50, 385, 391
    potential for totipotency and, 423
    state of donor cell differentiation and, 421–423
    of stem cells, 391
    totipotency and pluripotency and, 379
    X inactivation stability and, 336–337
    from zygote to blastocyst, 388–389
Research questions, 55–56
Restriction enzymes, 346
Restriction Landmark Genomic Sequencing, 463
Retinoblastoma (Rb) proteins, 226
Retrodifferentiation, 411

Retrotransposons, 44, 94, 271, 385, 387
Rett syndrome, 342, 350–351, 444–445
Reversibility, 176, 385
Reversion, 10
Rhabdomyosarcoma, 442
Rid gene, 115
Rik1, 160
Ring proteins, 214, 220–221, 244, 333–334
RIP. *See* Repeat-induced point mutation
RISC. *See* RNA-induced silencing complex
RITS. *See* RNA-induced transcriptional
    silencing effector
RNA knockdown. *See* Posttranscriptional
    silencing
RNA methyltransferases, 42
RNA polymerase II
    centromere repeats and, 109–110
    factors affecting elongation and, 205
    H3K4 methylation and, 197–198
    H3K36 methylation and, 198–199
    HKMTs and, 198
    PcG proteins and, 244
RNA polymerase IV, 43, 185
RNA polymerases, 160–161
RNA-dependent RNA polymerase (RDRP),
    118–119, 160–161, 182
RNA-dependent silencing, in *C. elegans*, 286
RNA-directed DNA methylation (RdDM),
    41–42, 175, 184–185
RNA-directed RNA polymerase complex
    (RDRC), 109, 154, 157–159
RNAi, 271. *See also* Meiotic silencing by
    unpaired DNA (MSUD);
    Posttranscriptional gene silencing;
    Quelling
    Arabidopsis and, 160–162
    Arabidopsis epigenetic regulation and,
      172–173
    Chp1 and, 106
    chromatin modification and, 109–110,
      162–163
    chromatin structure and, 156–157
    in ciliates, 128, 137, 142–146
    Drosophila and, 94–95
    dsRNA and siRNA and, 157–159
    evidence for role in silencing, 154
    fragile X syndrome and, 449
    genome rearrangements and, 136
    heterochromatin assembly and, 82, 97, 152,
      155–157
    HP1 localization and, 117
    mammalian DNA methylation and, 348
    metazoan development and, 28
    N. crassa and *S. pombe* and, 102
    ncRNAs and, 368
    overview of, 42–44, 153–154
    PTGS in plants vs., 175
    recognition models and, 159
    recruitment of chromatin-modifying
      enzymes by, 159–160
RITS and heterochromatin assembly and,
    156–157
    silencing and, 26
    silencing in ciliates and, 137, 142–144
    silencing in plants and, 180–185
    silencing in *S. pombe* and, 104, 111–112
    silent chromatin assembly and, 108–109

spermatogenesis and, 386
    Web sites for information on, 478
RNA-induced silencing complex (RISC),
    153–154, 182, 184
RNA-induced transcriptional silencing (RITS)
    effector, 106, 108–110, 153–157,
    199–200, 448
    Rolled, 93
    roX RNAs, 48–49, 308, 313–315
    Rsp, 283
RTT. *See* Rett syndrome
Rubinstein-Taybi syndrome, 443–444

S

S phase of mitosis, 75–76, 193–194, 269
Saccharomyces cerevisiae
    centromere identity and, 278
    centromere structure and function in, 273
    description of, 64
    genetic tools of, 65–66
    genome organization of, 45
    heterochromatin of, 67–71
    histone acetylation and, 73
    histone deacetylation and, 71–72
    inheritance of epigenetic states in, 75–76
    life cycle of, 66–67
    as model organism, 26–27
    overview of, 64
    rDNA repeat instability in, 76–77
    replication timing in, 269
    repression and, 74
    silencing in vs. in *S. pombe*, 103–104
    telomere looping and, 73–74
    trans-interaction in, 74–75
    trxG proteins and, 237–238
    Web sites for information on, 478
Sad-1 (Suppressor of ascus dominance) gene,
    120–121, 285
S-adenosyl-methionine (SAM), 177, 221
SAGA, 40, 259
Sallimus (sls) gene, 244
Sas2, 73, 309–310
Satellite DNA, ICF syndrome and, 446–447
Scc1, 76
Schimke immuno-osseous dysplasia, 444, 447
Schizosaccharomyces pombe
    centromere composition of, 104–107
    centromere identity and, 273
    centromere repeat transcription in, 109–110
    chromatin silencing in, 103–104
    epigenetic inheritance of centromere state
      in, 110–111
    gene silencing and, 42–43
    genome organization of, 45
    heterochromatic chromosomal organization
      in, 155
    heterochromatin and RNAi and, 155–156
    kinetochores and, 274
    as model organism, 26–27
    overview of, 102–103
    RNAi in, 108–110
    silent chromatin assembly and, 107–109
    Sir2 protein and, 68–69
    suppression of epigenetic variegation in, 53
    Web sites for information on, 478
Schmoos, 66

Sciarid flies, 284
SCID mice, 431–432
Scmx gene, 336
scnRNA model, 43, 142–146
SCNT. *See* Therapeutic cloning
SDC proteins, 293–295
sea-1, 295
Segregation
    achiasmate, 281–283
    CenH3 and, 255–256
    in *Drosophila*, 282–284
    H2AZ and, 259
    heterochromatin loss and, 111, 155
    heterochromatin pairing and, 281–282
    in meiotic process, 281–285
    normal, 267
    results of abnormal, 266
Selection, 25, 329
Senescence, 47, 76–77
SEP domains. *See* S-adenosyl-methionine (SAM)
Serotypes, 130
SET domains
    chromatin remodeling and, 219
    germ cell development and, 385–386
    histone modifications and, 177–178,
      198–199, 242
    plasma cell formation and, 410
    trxG proteins and, 238
Sex bodies, 259
Sex combs reduced (Scr), 221, 233, 235
Sex determination, 292–293, 295–296, 312–313,
    323–324
Sex reversal, 284
SEX-1, 295–296
Sexual reproduction, advantages of, 323
SFRPs, 467
Short interfering RNAs (siRNA), 42, 108–109,
    137
Signaling, lymphopoeisis and, 400–403
Signatures, histone modifications and, 37
Silencer elements, 69–71
Silencing. *See also* Dosage compensation;
    Posttranscriptional silencing; RNAi;
    Transcriptional gene silencing
    (TGS); Virus-induced gene
    silencing; X inactivation
    accumulation of in Xi, 334
    cancers and, 465–471
    centromeric outer repeats and, 107–109
    chromosomal proteins and, 85–88
    DNA methylation and, 349–350
    dosage compensation and, 292
    epigenetics and, 20
    at fission yeast centromeres, 104
    fragile X syndrome and, 448
    genomic imprinting and, 362, 360–372
    heterochromatin and, 31, 83–85
    histone modifications and, 88–91
    historical symposia and, 6–7
    homology-dependent in ciliates, 136–137
    inheritance of ciliar rearrangements and, 139
    kinetochore domain proteins and, 107
    mechanisms of, 111–112, 175
    meiotic by unpaired DNA (MSUD), 119–121
    methylation and, 41, 43
    movement of expressed genes and, 34
    non-cell-autonomous, 181–182

patterns of in telomeres, 74
PcG and trxG and, 46, 224
plants and, 43
PRC1 and, 224
redundancy and, 337
RNAi and, 26, 42–44, 136–137
in *S. pombe*, 103–104
subcompartment formation and, 76
subnuclear relocation and, 405
temperature and, 85
transcription of unpaired DNA and, 285
by unpaired DNA during meiosis, 285–286
of X chromosomes in *C. elegans*, 293,
298–299, 301–302
X inactivation as, 322
Silent cassettes, mating types and, 67–69
Silent chromatin
active chromatin vs., 132–134
H2AZ and, 259
historical symposia and, 6
hypermethylation and, 257
methylation and, 42
overview of, 8–9
PcG and, 45
RNA transcripts and, 48
in *S. pombe* nuclei, 107
spreading of, 335
Silent Information Regulator (SIR) proteins
acetylation, deacetylation and, 68–69, 73,
195–196
aging and, 76–77
as example, 26
family structure of, 68–69
heterochromatic repression and, 69–70
heterochromatin anchoring and, 75
histone acetylation and, 73
histone deacetylation and, 71–72
historical symposia and, 6, 7
nuclear localization of, 74–75
repression and, 68, 75
silent chromatin and, 71
telomere position effect (TPE) and, 72
Silent mating cassettes, historical symposia and, 4
Silver-Russell syndrome, 440, 442–443
Single-nucleotide polymorphisms (SNPs),
361–362
SIR proteins. *See* Silent Information Regulator
proteins
siRNAs
dicer-like (DCL) proteins and, 182
generation of, 157–159
heterochromatin assembly and, 157
quelling and, 119
RITS and, 156–157
scnRNAs vs., 144
silencing in ciliates and, 138, 140
sources of, 152
telomeres and, 43
trans-acting (ta-siRNAs), 182
SKD protein, 243
Skeletal system, PcG repression and, 224
SLC22A1L, 442
SMARCAL1, 447
SMYD3, 198
SNF2, 39, 205, 348
snoRNAs, 368, 437, 440–441
SNURF-SNRPN, 441

Solenoid models of chromatin organization, 33
Somaclonal variation, plants and, 169, 171
Somatic cells, 17, 138–142, 416, 418–419
Somatic DMRs, genomic imprinting and,
369–370
Somatic hypermutation, methylation and, 42
Somatic reprogramming. *See* Reprogramming
Sonic hedgehog (Shh), 226–227
Sonneborn, T.M., 129
Sox2 , 388, 390
Speciation, MSUD and, 120
Spermatogenesis
epigenetic modification during, 386
genomic imprinting and, 362
H2AX and, 259
heterochromatin pairing sites and, 281
histone replacement and, 262
hypercondensation during, 303
IAP retrotransposon in, 387
X inactivation and, 327
Spindle Assembly Checkpoint (SAC), 273
Spirotrichs, 135
Splayed (SPD), 179
SPM domains, 2–3. *See also* S-adenosyl-
methionine (SAM)
Sporulation, 66
Spread and retreat model, 335
Spreading of heterochromatin, 72–73, 83–85,
91–93
SRY gene, 323–324
SubH2Bv, 262
stella, 94, 381–382
Stem cells, 45, 226–227, 389–392. *See also*
Embryonic stem cells
Stem-loop-binding protein (SLBP), 253
Stichotrichs, 135
Structural inheritance, 131–132
Suberoylanilide hydroxamic acid (SAHA), 473
Subtraction techniques, 464
Sumoylation, 193, 202–203, 270
SUPERMAN, 160–161, 185
Suppressor genes, cancers and, 463
Su(var) genes
chromatin silencing marks and, 215
dosage compensation and, 316–317
Drosophila and, 84–88
H3K9 methylation and, 89–91
position-effect variegation and, 28
sperm reprogramming and, 387
spermatogenesis and, 386
SUVH proteins, 96, 178, 199–200
SWI2/SNF2, chromatin repression and,
238–241
Swi6, 105–106, 110, 155–157
SWI/ISWI enzymes. *See* Brahma family
Switch2/Sucrose Non-Fermentable2
(SWI2/SNF2)
chromatin remodeling and, 179
DNA methylation and, 348
H2AZ and, 259, 261
heterochromatin modifications and, 160–161
MeCP2 and, 445
methylation and, 41
nucleosome structure and, 205
SWR1-C, 261
SXL, 312–313
Synaptonemal complexes (SC), 282

T

T. thermophila, Web sites for information on, 478
"TA" internal eliminated segments (IES), 135
ta-siRNAs, 184
Tatum and Beadle, 112
T-cell development, 398
TCRβ, 403
Telomerases, 71, 270
Telomere position effect (TPE), 65, 88, 271
Telomere-associated sequences (TAS), 94, 108,
271
Telomeres
assembly of, 71
chromosomal inheritance and, 268
control of structure and function of,
270–273
Drosophila and, 94
function of, 34
functional regulation of, 282
historical symposia and, 6
looping of, 73–74
repression and, 75
siRNAs and, 43
trans-interaction of, 74–75
Temperature, silencing and, 85
Terminal flower 2 (TFL2), 180
Tetrahymena, 27, 43, 130. *See also* Ciliates
TGS. *See* Transcriptional gene silencing (TGS)
Thalassemias, 445–448
T-helper cells, 347, 352
Therapeutic cloning, 429–432
Therapies, epigenetic, 471–473
Tip60, 261, 311
Titin, 244
Toll-like receptors, 42
Tonalli (Tna) gene, 244
Totipotency, 44–45, 378, 423
Toxins, of *P. aurelia*, 130
tra genes, 312–313
Trans-acting siRNAs (ta-siRNAs), 184
Transcription
aberrant activation of, 225–226
acetylation and deacetylation and, 194–196
Bcl6 and Blimp1 and, 382–383, 409
changes in chromatin structure and, 204–205
changes in with reprogramming, 425–426
DNA methylation and, 348
equalization of gene products and, 309,
311–312
euchromatin and, 35
histone modifications and, 30
internal eliminated segment (IES) and, 43
mechanism of, 308
methylation-dependent repression of,
351–352
PRC1 and, 223
regulation of by histones and acetylation,
193–194
silencing and, 42–44
SWI/SNF complexes and, 239–240
trithorax (trxG) proteins and, 243–244
trxG proteins and, 237
Transcription factor (TF) occupancy, 35
Transcription factors
ciliar histone variants and, 133
DNA methylation interference with, 348–349

Transcription factors *(continued)*
  histone modifications and, 192
  lymphopoeisis and, 399, 401–402
  Rap1, 68–71
  role of in epigenetic control, 400–401
  SWI/SNF complexes as, 239–240
  trithorax (trxG) proteins as, 244–245
  trxG proteins as, 244–245
Transcriptional control, 259–261, 351–352
Transcriptional gene silencing (TGS). *See also*
    DNA and histone methylation
  Alcohol dehydrogenase (Adh) transgenes
    and, 94
  heterochromatin and, 42
  overview of, 152
  PcG complexes and, 220–224
  RNAi and, 108
  RNAi-mediated silencing in plants and,
    180–181
  X inactivation and, 332
Transcriptional interference, 372
Transcriptional memory. *See* Cellular memory
Transcriptome-scanning model, 145–146
Transdetermination, 213
Trans-effects, 32, 192, 260, 438, 443–447
Transforming DNA, 117–119
Transgene-induced silencing, 137
Transitivity, 181–182
Translocations, 335, 460
Transplantation. *See* Nuclear transplantation
Transposable elements. *See also* Retrotransposons
  chromatin remodeling and, 163, 373
  DNA methylation and, 41
  genome rearrangements and, 136
  heterochromatin modifications and, 162
  historical symposia and, 3, 17
  mammalian genomes and, 44
  N. crassa and, 113
  neoplastic transformation and, 51
  P elements as, 84
  plants and, 27, 175, 185
  RIP and, 120–121
  RNAi in plants and, 185
  silencing and, 44
  targeting of heterochromatin formation and,
    94
Trans-sensing, 286
Trichlorostatin A (TSA), 109, 111, 117
Trichogynes, 112
Trithorax (trxG) proteins
  as activators or anti-repressors, 244–245
  cellular memory and, 44, 47, 212, 232
  chromatin and, 237–243
  functional interactions between, 244
  histone methyltransferases and, 177–178
  histone modifications and, 242
  identification of, 233–236
  overview of, 45–47, 233, 236–237, 244–246
  PcG protein recruitment and, 223

SET domains in, 238
  transcription and, 237, 243–244
tRNA, 107
Trophectoderm (TE) layers, 380, 388
Trophoblast defense theory, 365
Tsix gene, 325, 327, 329
Tumor suppressor genes, 460
Tumors, 26, 50–51, 226, 370, 465–466. *See also*
    Cancers
Twi1 gene, 142–144, 154
Twins, variability in, 26

U

U7 small nuclear ribosomal complex, 253
Ube3aas, 368–369
Ubiquitin E3 ligase *(UBE3A)*, 440–441, 467
Ubiquitination
  DNA repair and, 270, 459
  overview of, 193, 202–203
  sites of in histone tails, 31
  X inactivation and, 332, 334
Uniparental disomy, 436, 438–439
Unpaired DNA, 285–286
Ura3, 65–66
Uracil DNA glycosylases, 352
Uremia, 450

V

Valproic acid, 472
Variable Surface antigen Genes (VSG), 4
Variant histones. *See* Histone variants
Variegated expression events. *See* Position-effect
    variegation (PEV)
V(D)J recombination, 399–400, 403–405
Vernalization (VRN), 178–179, 217–219
Vidaza, 471
Viroids, 184
Virus-induced gene silencing (VIGS), 175,
    181–182
VRN complex, 178–179, 217–219

W

Wasps, paternal chromosome loss in, 284
Way stations, X inactivation and, 335
Web sites, 368, 477–478
white gene, 65, 83–86, 93–94
Wild-type reversion. *See* Reversion
Wilms' tumors, 442. *See also* Prader-Willi
    syndrome
Window of opportunity, 337
Winged helix domains, 254
w^{m4}. See white gene
Wnt pathway, 467

X

X elements, 74
X inactivation
  in *C. elegans*, 298–299, 301–302

cloning and, 428
  defined, 292
  diseases and, 444–445
  facultative heterochromatin and, 47–49
  H2A variants and, 262
  historical symposia and, 3–5
  identification of in mammals, 324
  imprinted, 325–327, 329–330
  initiation of in mammals, 325–330
  maintenance of, 380
  overview of, 17–18, 47–49, 322, 325
  PcG repression and, 224–225
  propagation and maintenance of in
    mammals, 330–336
  random vs. imprinted, 325–330
  reactivation and reprogramming and,
    336–338, 346, 384, 389
  regulation of in mammals, 326–329
  switching modes of in mammals, 329–330
X inactivation center (XIC), 48, 325, 335
X reactivation, 336–337
X signal elements (XSEs), 295–296, 304
X:A ratio, 292–293, 295–296, 312–313
XBP1, 410
Xce, 329
Xenopus, Web sites for information on, 478. *See*
    also Amphibians
Xist gene
  embryonic stem cells and, 337–338
  genomic imprinting and, 372
  historical symposia and, 5
  pluripotent embryonic stem cells and, 390
  regulation of in mammals, 334–335
X inactivation and, 48, 225, 322, 325–328, 330
X reactivation and, 336
X-linked adrenoleukodystrophy, 437
X-linked genes, possible equalization methods
    for, 324. *See also* Dosage
    compensation
X-linked mental retardation, 446
XO embryos, 327–328
XOL-1, 293–295, 296
Xp imprints, 303–304, 325
XY bivalents, 259. *See also* Pseudoautosomal
    region (PAR)

Y

Y chromosomes, 85, 323
Y' elements, 74
Yeast, 236. *See also* Mating types (MAT); *specific*
    species
Yellow gene, 86
yKu, 69, 71, 75

Z

Zebularine, 471–472
ζ-η region, 115
Zigzag models of chromatin organization, 33
Zygotes, 379